Mercury Fate and Transport in the Global Atmosphere

Nicola Pirrone • Robert Mason
Editors

Mercury Fate and Transport in the Global Atmosphere

Emissions, Measurements and Models

Springer

Editors
Nicola Pirrone
CNR-Institute for Atmospheric Pollution
Rome
Italy

Robert Mason
Department of Marine Sciences
University of Connecticut
Groton, Connecticut, USA

ISBN: 978-0-387-93957-5 e-ISBN: 978-0-387-93958-2
DOI: 10.1007/978-0-387-93958-2
Springer Dordrecht Heidelberg London New York

Library of Congress Control Number: 2008942799

Printed on acid-free paper

Springer is part of Springer Science+Business Media (www.springer.com)

Preface

During the last decade the importance of environmental issues related to mercury released to the atmosphere by major anthropogenic sources, which include, but are not limited to, power plants for energy production and a variety of industrial plants, has gained growing attention for their effects on human health and ecosystems.

In this framework the UNEP Mercury Programme started, beginning in 2002, a process to assess to what extent contamination by mercury released from anthropogenic and natural sources may affect human health and ecosystems. A number of concerted initiatives have been undertaken on a global scale to assess the current state of our knowledge regarding atmospheric mercury emissions and transport, its deposition to and evasion from terrestrial and aquatic ecosystems, and also to evaluate the relative contributions of natural and anthropogenic sources to the global atmospheric mercury budget. At the beginning of 2005 the Governing Council of the United Nations Environment Programme (UNEP-GC) urged (Decision 23/9 IV), governments, inter-governmental and nongovernmental organizations and the private sector to develop and implement partnerships as one approach to reducing the risks to human health and the environment from the release of mercury and its compounds, by improving global understanding of international mercury emission sources, fate and transport. In this framework, the UNEP Global Partnership for Mercury Air Transport and Fate Research (UNEP-MFTP) was initiated in 2005 aiming to encourage collaborative research activities on different aspects of atmospheric mercury cycling on local to hemispheric and global scales.

Members of the UNEP-MFTP are, Italy (lead), Canada, Japan, South Africa, United States, Electric Power Research Institute (EPRI), Natural Resources Defence Council (NRDC) and UNEP. Since 2005, the UNEP-MFTP has met four times. The first meeting was held in Madison, Wisconsin in conjunction with the 8th International Conference of Mercury as a Global Pollutant, followed by the meeting in Gatineau, Quebec, Canada (9-10 January 2007) which aimed to discuss and define the elements included in Decision 23/9 IV. A 3rd meeting was held in Washington, D.C. on 10-11 October 2007 to review the Business Plan of the UNEP-MFTP, which was submitted (February 2008) to UNEP Chemicals, whereas the 4th meeting of the partnership was held in Rome (7-11 April 2008) in conjunction with the international workshop jointly organised by the UNEPMFTP and the Task Force on Hemispheric Transport of Air Pollution (TF HTAP) of the UNECE

Convention on Long-Range Transboundary Air Pollution, in which leading scientists from all over the world presented their contributions to the UNEP-MFTP Technical Report.

This book is the technical report the UNEP-MFTP provided to UNEP Chemicals, governments, inter-governmental and non-governmental organisations as well as the private sector, as a contribution to the ongoing work within the Open-Ended Working Group (OEWG) in charge of preparing the input to the next UNEP Governing Council. The book highlights major issues related to the interactions of mercury with terrestrial and aquatic ecosystems, and evaluates the relative contribution of anthropogenic and natural sources to the global atmospheric mercury budget. The preparation of this report has been made possible thanks to the contributions of all the members of the UNEP-MFTP, and of more than 70 scientists from leading universities and research institutions recognised as worldwide experts on different aspects related to the emissions, monitoring and modelling of mercury in the atmosphere and other environmental compartments. The draft of this report was delivered to UNEP Chemicals in February 2008 as a contribution to the Open-Ended Mercury Working Group of UNEP Chemicals (meeting held in Bangkok in November 2008) and to the UNEP Governing Council (meeting held in Nairobi, February 2009).

Dr. Nicola Pirrone
Chair of the UNEP-MFTP
CNR-Institute for Atmospheric Pollution
Rome, Italy

Acknowledgments

As Chair of the UNEP-MFTP I would like to express my gratitude to the members of the partnership for their continued support to the activity of the partnership during the last three years.

I would like to acknowledge the support received from the Italian Ministry for the Environment, Land and Sea; special thanks to Dr. Corrado Clini, Director General of the Department for Environmental Research and Development for his continued support.

The preparation of this technical report would have not been possible without the great contribution of all colleagues that have led the preparation of the chapters and to all co-authors that have provided valuable contributions. The contribution of the external reviewers is greatly acknowledged. Special thanks to the co-Editor, Dr. Robert Mason who has provided continued and full support during these months.

I would like to acknowledge the great contribution from both co-chairs of the TF HTAP, Dr. Andre Zuber and Dr. Terry Keating who have provided timely input to the activity of the UNEP-MFTP and to the preparation of this technical report.

I would like to express special thanks to my staff, Dr. Sergio Cinnirella, Dr. Ian M. Hedgecock, Dr. Gerlinde Jung and Dr. Francesca Sprovieri for their hard work and time dedicated to the preparation of this book, and to my secretary, Mrs Maria Orrico, for her hard work in formatting and preparing the final edited version of the book.

Dr. Nicola Pirrone
Chair of the UNEP-MFTP
CNR-Institute for Atmospheric Pollution
Rome, Italy

Contents

About the Editors

Nicola Pirrone is Director of the Institute for Atmospheric Pollution of the Italian National Research Council (CNR-IIA) and Adjunct Professor at the Department of Environmental and Health Sciences of the University of Michigan. He is Chair of the UNEP Global Partnership for Mercury Air Transport and Fate Research, Chair of the WG on Global Atmospheric Mercury Models Intercomparison within the Task Force on Hemispheric Transport of Air Pollutants (TF HTAP) of the UNECE-LRTAP convention and Chair of the CEN-TC264 WG that is preparing the European Standard Methods for monitoring mercury concentrations in ambient air and precipitations. He has been Chair of the European WG that prepared the "Air Quality Position Paper on Mercury" that is one of the scientific background documents of the Forth Air Quality Daughter Directive of the European Union. He has published over 100 peer-reviewed articles on different topics associated to atmospheric transport, chemistry and policy relevant issues related to major atmospheric pollutants.

Robert Mason is a Professor at the University of Connecticut with a joint appointment in the Departments of Marine Sciences and Chemistry. His research focus is the cycling of mercury in the biosphere with emphasis on the atmosphere and on air-sea exchange and redox reactions in the atmospheric boundary layer and the surface ocean. In addition, his research has a focus on mercury methylation in coastal ecosystems and other environments. Prof. Mason has an active research group with currently 4 PhD students and a post-doc. He is the author/coauthor of about 100 papers on mercury and is actively involved in a number of national and international mercury initiatives. Prof. Mason obtained his undergraduate and Master's degree in South Africa and his PhD at the University of Connecticut in 1991. He was a professor at the University of Maryland before returning to the University of Connecticut in 2005.

Contributing Authors

Amyot, Marc	Dép. de Sciences Biologiques, Université de Montréal, Montréal, Canada
Andersson, Maria	Department of Chemistry, Göteborg University, Göteborg, Sweden
Arellano, Avelino F.	National Center for Atmospheric Research, Boulder, CO, USA
Ariya, Parisa A.	Dep. of chemistry, and Dep. of Atmospheric and Oceanic Sciences, McGill University, Montreal, Canada
Ashton, Peter J.	Department of Marine Sciences, University of Connecticut, Connecticut, USA
Banic, Catharine	Environment Canada, Science and Technology Branch, Toronto, Canada
Beauchamp, Steve	Environment Canada, Meteorological Service of Canada, Dartmouth, Canada
Bhattacharya, Prosun	Royal Institute of Technology, Stockholm, Sweden
Brooks, Steve	National Oceanic and Atmospheric Administration, Oak Ridge, USA
Bullock, Russell	NOAA, Air Resources Laboratory,USA
Cinnirella, Sergio	CNR-Institute for Atmospheric Pollution, Division of Rende, Rende, Italy
Dastoor, Ashu P.	Air Quality Research Division, Science and Technology Branch, Canada
Dabrowski, James	CSIR – Natural Resources and the Environment, Pretoria, South Africa
Davignon, Didier	Environment Canada, 2121 Trans Canada Highway, Dorval, Quebec, Canada
Dommergue, Aurélien	Laboratoire de Glaciologie et Géophysique de l'Environnement, Saint Martin d'Hères, France
Donohoue, D. L.,	Division of Marine and Atmospheric Chemistry, University of Miami, Miami, USA
Ebinghaus, Ralf,	GKSS Research Centre Geesthacht, Institute for Coastal Research, Geesthacht, Germany

Euripides, Rico	GroundWork Non-governmental Organisation, Durban, South Africa
Feng, Xinbin	Chinese Academy of Sciences, Guiyang, China
Ferrari, Christophe P	Laboratoire de Glaciologie et Géophysique de l'Environnement, Saint Martin d'Hères, France
Finkelman, Robert B.	University of Texas, Dallas, USA
Friedli, Hans. R.	National Center for Atmospheric Research, Boulder, CO, USA
Gericke, Gerhard	ESKOM, ERID, Johannesburg, South Africa
Ginster, Martin	SASOL, Rosebank, Gauteng, South Africa
Goodsite, Michael E.	Department of Physics and Chemistry, University of Southern Denmark, Odense, Denmark
Hao, Jiming	Tsinghua University, Beijing, China
Hedgecock, Ian M.	CNR-Institute for Atmospheric Pollution, Division of Rende, Rende, Italy
Herbert, Kock Hans	GKSS Research Centre Geesthacht, Institute for Coastal Research, Geesthacht, Germany
Hynes, Anthony.J.	Division of Marine and Atmospheric Chemistry, University of Miami, Miami, USA
Ilyin, Ilia	Meteorological Synthesizing Centre – East of EMEP, Moscow, Russia
Jacob, Daniel	Harvard University - Department of Earth & Planetary Sciences, Cambridge, USA
Jaeglé, Lyatt	University of Washington, Dep. of Atmospheric Sciences, Seattle, USA
Jaffe, Dan	Interdisciplinary Arts and Sciences Dep. University of Washington, Bothell, USA
Jung, Gerlinde	CNR-Institute for Atmospheric Pollution, Division of Rende, Rende, Italy
Keeler, Gerald J.	Department of Environmental Health Sciences, University of Michigan, Ann Arbor, USA.
Leaner, Joy	CSIR – Natural Resources and the Environment, Stellenbosch, South Africa
Levin, Leonard	EPRI, Palo Alto, CA, USA.
Li, Guanghui,	Tsinghua University, Beijing, China
Lohman, Kristen	Atmospheric & Environmental Research, Inc., San Ramon, USA
Masekoameng, Elizabeth	CSIR – Natural Resources and the Environment, Pretoria, South Africa
Mason, Robert	Department of Marine Sciences, University of Connecticut, Groton, USA
Mukherjee, Arun B.	University of Helsinki, Helsinki, Finland
Kevin, Murray	Dep. of Marine Sciences, University of Connecticut, Connecticut, USA
Koyo, Ogasawara	National Institute for Environmental Studies, IDEA Consultant, Inc., Japan

Kirk, Peterson	Dep. of Biological Sciences, Universite de Montreal, 90 Vincent d'Indy, Montreal, Canada
Petersen, Chantel R.	CSIR - Natural Resources and the Environment, Stellenbosch, South Africa
Pirrone, Nicola	CNR-Institute for Atmospheric Pollution, Rome, Italy
Poissant, Laurier	Environment Canada, Science and Technology Branch, Montréal, Canada
Resane, Tabby	Department of Environmental Affairs and Tourism, Pretoria, South Africa
Richardson, Marguerite	Department of Environmental Affairs and Tourism, Pretoria, South Africa
Sarkar, Atanu	India Habitat Centre, New Delhi, India
Seigneur, Christian	Atmospheric & Environmental Research, Inc., San Ramon, CA, USA
Selin, Noelle E.	Massachusetts Institute of Technology, Dep. of Earth, Atmospheric, and Planetary Sciences, Cambridge, USA
Yasuyuki, Shibata	National Institute for Environmental Studies, Japan.
Sillman, Sanford	Dep. Atmospheric Sciences, University of Michigan, MI, USA
Snider, Graydon	Dep. of chemistry, and Dep. of Atmospheric and Oceanic Sciences, McGill University, Montreal, Canada
Sprovieri, Francesca	CNR-Institute for Atmospheric Pollution, Division of Rende, Rende, Italy
Steffen, Alexandra	Environment Canada, Air Quality Research Division, Toronto, Canada
Stracher, Glen	Department of Geology, East Georgia College, Swainsboro, Georgia, USA
Streets, David G.	Argonne National Laboratory, Argonne, USA
Strode, Sarah A.	University of Washington, Dep. of Atmospheric Sciences, Seattle, USA
Suzuki, Noriyuki	National Institute for Environmental Studies, IDEA Consultant, Inc., Japan
Telmer, Kevin H.	-School of Earth and Ocean Sciences, University of Victoria, Canada
Travnikov, Ralf Oleg	Meteorological Synthesizing Centre – East of EMEP, Moscow, Russia
Veiga, Marcello M.	Department of Mining Engineering, University of British Columbia, Canada
Vijayaraghavan, Krish	Atmospheric & Environmental Research, Inc., San Ramon, CA, USA
Wang, Shuxiao	Tsinghua University, Beijing, China
Weiss-Penzias, Peter	Department of Chemistry, University of Washington, Seattle, USA
Wu, Ye	Tsinghua University, Beijing, China
Zevenhoven Ron	University Turku/Åbo, Finland

External reviewers

Marianne Bailey	U.S. Environmental Protection Agency, OIA	USA
Robin Dennis	U.S. Environmental Protection Agency, NERL	USA
Bob Dyer	U.S. Environmental Protection Agency	USA
Arthur E. Dungan	Chlorine Institute, Inc.	USA
Stanley Durkee	U.S. Environmental Protection Agency, ORD	USA
Marilyn Engle	U.S. Environmental Protection Agency, OIA	USA
Luis E. Fernandez	Stanford University	USA
Mark Freeman	U.S. Department of Energy	USA
Charles French	U.S. Environmental Protection Agency	USA
Wendy Graham	U.S. Environmental Protection Agency	USA
Loren Habegger	Argonne National Laboratory	USA
Allen Kolker	U.S. Geological Survey	USA
Karissa Kovner	U.S. Environmental Protection Agency	USA
Bruce J. Lawrence	Bethlehem Apparatus Co. Inc	USA
Steve Lindberg	Environmental Sciences Division, ORNL	USA
Bian Liu	School of Public Health, Harvard University	USA
Carl Mazza	U.S. Environmental Protection Agency	USA
Peter Maxson	Concorde Cons.	Belgium
Elsie Sunderland	U.S. Environmental Protection Agency, ORD	USA
Ming H. Wong	Hong Kong Baptist University	China

Part I
Sources of Mercury Released to the Global Atmosphere

Chapter 1
Global Mercury Emissions to the Atmosphere from Natural and Anthropogenic Sources

Nicola Pirrone, Sergio Cinnirella, Xinbin Feng, Robert B. Finkelman, Hans R. Friedli, Joy Leaner, Rob Mason, Arun B. Mukherjee, Glenn Stracher, David G. Streets, and Kevin Telmer

Summary This chapter provides an up-to-date overview of global mercury emissions from natural and anthropogenic sources at country and regional/continental scale. The information reported in Chapters 2–8 is the basis of the assessment reported in this chapter, however, emissions data related to sources and regions not reported in chapters 2–8 have been derived, to the extent possible, from the most recent peer-reviewed literature and from official technical reports. Natural sources, which include the contribution from oceans and other surface waters, rocks, top soils and vegetation, volcanoes and other geothermal activities and biomass burning are estimated to release annually about 5207 Mg of mercury, part of which represent previously deposited anthropogenic and natural mercury from the atmosphere to ecosystem-receptors due to historic releases and part is a new contribution from natural reservoirs. Current anthropogenic sources, which include a large number of industrial point sources are estimated to release about 2909 Mg of mercury on an annual basis, the major contribution is from fossil fuel-fired power plants (1422 Mg yr^{-1}), artisanal small scale gold mining (400 Mg yr^{-1}), waste disposal (187 Mg yr^{-1}), non-ferrous metals manufacturing (310 Mg yr^{-1}) and cement production (236 Mg yr^{-1}). Our current estimate of global emissions suggest that summing up the contribution from natural and anthropogenic sources nearly 8116 Mg of mercury is released annually to the global atmosphere. The evaluation of global emissions presented in this report differs from previous published assessments because in the past, emissions from several sources, i.e., forest fires and coal-bed fires have not been accounted for, and also because of improved knowledge of some anthropogenic and natural sources (i.e., emissions from oceans, vegetation) as suggested by the most up-to-date literature.

1.1 Introduction

Mercury is ubiquitous in the atmosphere, it has ground level background concentrations which are almost constant over hemispheric scales; the southern hemisphere having a slightly lower concentration than the northern. Recent measurements of

N. Pirrone and R. Mason (eds.), *Mercury Fate and Transport in the Global Atmosphere*,
DOI: 10.1007/978-0-387-93958-2_1, © Springer Science+Business Media, LLC 2009

free tropospheric air, from high altitude sites and from measurements made on board aircraft indicate that its concentration changes little up to the tropopause. In the stratosphere mercury has been identified associated with the stratospheric aerosol. The transport of mercury, therefore, occurs in the boundary layer, in the free troposphere and stratosphere; the fate of mercury, therefore, is determined by the different chemical environments that these regions of the atmosphere represent, the different physical and meteorological processes which occur in them and also by exchange between them (Pirrone et al., 2000; Pirrone et al., 2005; Hedgecock et al., 2006; Lindberg et al., 2007).

The impact of energy resources exploitation, especially fossil fuel exploitation, on ecosystems in terms of mercury contamination is threefold. Firstly, because fossil fueled power plants are the highest emitting anthropogenic emission source of mercury to the atmosphere. Secondly, because the other pollutants emitted as a result of fossil fuel exploitation, such as NO_x and SO_2, have an impact on the atmospheric chemistry of mercury and influence its deposition patterns. While the previous two impacts are observable in the short term, the third is the medium to long term impact that exploitation of fossil fuels has on atmospheric mercury cycling, as a result of the release of greenhouse gases which contribute to climate change (Hedgecock and Pirrone, 2004; Eisenreich et al. 2005).

Improved information on emissions, particularly emissions in Europe and North America, have contributed to further progress in assessment of the regional impacts of mercury on terrestrial and aquatic environments (Pirrone et al., 2001a). Major international activities to assess source - receptor relationships for mercury in the environment are developed as part of international conventions (i.e., UNECE-LRTAP, OSPAR, HELCOM) and programmes (i.e., past EU funded projects, ACAPs, MERSA, UNEP). Policy makers in Europe have also taken the advantage of improved information on emissions to assess the effectiveness of measures aimed to reduce the impact of this highly toxic contaminant on human health and ecosystems. Following the preparation of the EU Position Paper on Ambient Air Pollution by mercury (Pirrone et al., 2001c), the EU adopted the European Mercury Strategy which is aimed to phase out the use of mercury in goods and industrial applications and reduce to the extent possible mercury emissions to the atmosphere from fossil fuels power plants and industrial facilities. In 2002 UNEP Chemicals released the first assessment (Global Mercury Assessment Report, GMA) on global mercury contamination (UNEP, 2002). Since then, a number of activities have been developed in order to support the achievement of objectives set by the UNEP Governing Council (decisions 23/9 IV in 2005 and 24/3 IV in 2007) to continue and elaborate possible strategies and mechanisms aimed to phase out the use of mercury in a wide range of products and reduce, to the extent possible, the emissions from industrial plants.

The aim of this chapter is to provide an overview of mercury emissions to the atmosphere from major natural and anthropogenic sources during the last decade in each region of the world.

1.2 Mercury Emissions from Natural Sources

Natural sources of mercury released to the atmosphere include volcanoes, soil and water surfaces, weathering processes of Earth crust and forest fires. Contributions vary in time and space depending on a number of factors including the presence of volcanic belts or geothermal activities, geological formations with high mercury concentration like cinnabar deposits, exchange processes between water and atmosphere, re-emission of previously deposited mercury from top soils and plants and forest fires (Pirrone et al., 2001b; Mason, 2008).

The ratio between the relative contributions of anthropogenic and natural source categories may vary within a region and time of the year. On a global scale the contribution from industrial sources have been found ranging between 1660 and 2200 Mg yr^{-1} (Pirrone et al., 1996; Pacyna et al., 2003, 2006b), whereas emissions from natural processes (which include emissions due to natural emissions as well as re-emissions of historic anthropogenic mercury) have been indicated as the major contribution (up to 60% of the total) to the global atmospheric mercury budget (Pirrone et al., 1996; 2001b). Among natural sources the emission from volcanoes, forest fires and surface waters represent a significant contribution and also emissions from contaminated soils in ancient mining industrial areas or particular geologic units rich in Hg (i.e. capgaronnite, cinnabar, cordierite) can also be significant (e.g., Ferrara et al., 1998; Ferrara et al., 2000a; Ferrara et al., 2000b; Gustin et al., 2002). Current estimate of mercury emissions from natural sources, without considering the contribution from biomass burning, is estimated to be 4532 Mg yr^{-1} (Mason, 2008).

1.2.1 Volcanoes and Geothermal Activities

The contribution of volcanoes, which may be an important source at the local scale, varies over time depending if they are in the degassing or eruption phase. Mercury is emitted from volcanoes primarily as gaseous Hg^0 and the Hg/SO_2 ratio is generally adopted to estimate mercury emission, although it is very controversial because of the paucity of relevant data and the orders of magnitude variation in all data types (Nriagu and Becker, 2003; Pyle and Mather, 2003; Mather and Pyle, 2004). In our estimate an Hg/SO_2 ratio of 1.18×10^{-5} for erupting volcanoes, 1.16×10^{-5} for continuously degassing volcanoes and 5.88×10^{-6} for ash rich plumes has been used (Ferrara et al., 2000b; Nriagu and Becker, 2003; Mather and Pyle, 2004).

Mercury emissions from calderas may represent an important natural source of mercury. The Phlegrean fields (Pozzuoli, Italy) have been monitored with a LIDAR system and steam fluxes of Hg associated as Hg-S complexes were in the range of 0.9 to 4.5 g day^{-1} (Ferrara et al., 1998). Concentrations in condensed steam fluxes reported by Ferrara et al. (1994) were in the range of 2 to 690 ng m^{-3}.

On average, volcanoes and geothermal activities release to the atmosphere ~90 Mg yr^{-1} of mercury (Mason, 2008), accounting for about< 2% of the contribution from natural sources.

1.2.2 *Water Surfaces*

Several studies suggest that the evasion of elemental mercury from surface waters is primarily driven by (1) the concentration gradient of mercury between top-water microlayer and air above the surface water, (2) solar irradiation, which is responsible for the photo-reduction of oxidised mercury in the top-water microlayer and (3) the temperature of the top-water microlayer and air above the surface water (air-water interface) (Pirrone et al., 2003; Pirrone et al., 2005; Hedgecock et al., 2006).

The evasion of mercury from lake surfaces is generally higher than that observed over the sea. Average emission rates in the North Sea were found in the range of 1.6 to 2.5 ng m^{-2} hr^{-1} (Cossa et al., 1996), whereas higher values (5.8 ng m^{-2} hr^{-1}) have been observed in the Scheldt outer estuary (Belgium) and over lakes in Sweden (up to 20.5 ng m^{-2} hr^{-1}). In open sea, mercury emission rates were much lower (1.16-2.5 ng m^{-2} hr^{-1}) and less variable between day and night, though dissolved mercury concentrations in the top water microlayer (6.0 ng L^{-1}) were very similar to that observed in unpolluted coastal areas. On average, coastal waters and Mediterranean Sea have the highest evasional flux with 1.83 and 1.96 ng m^{-2} hr^{-1}, respectively, while internal waters show a maximum net evasion of 2.39 ng m^{-2} hr^{-1} (Pirrone et al., 2003; Hedgecock et al., 2006).

Mason (2008) (Chapter 7 of this report) reports recent estimates of mercury evasion from ocean basins and lakes, which account for 2778 Mg yr^{-1} of net evasion to the atmosphere (Table 1.1). The estimated contribution of the Mediterranean Sea

Table 1.1 Summary of gaseous mercury fluxes for oceans and lakes (from Chapter 7)

Region	Net Evasion (average) (*Mg yr^{-1}*)	Ratio[1](%)
Atlantic Ocean	840	18.5
Pacific and Indian Ocean	1700	37.5
Antarctic Ocean	12	0.3
Mediterranean	70*	1.5
Coastal waters	60	1.3
Lakes	96	2.1
Total	2778	–

*as estimated by Hedgecock et al., (2006)
[1]calculated over the total evasion from natural sources which sum 4532 *Mg yr^{-1}*

was updated to 70 Mg yr^{-1} on the basis of recent measurements/modeling estimates done by Hedgecock et al. (2006). Hedgecock et al. (2006) compared measured mercury concentrations in air and modelled total deposition flux over the Mediterranean Sea on a monthly basis (Figure 1.1) to obtain monthly total emissions from and deposition fluxes to the Mediterranean Sea (Figure 1.2).

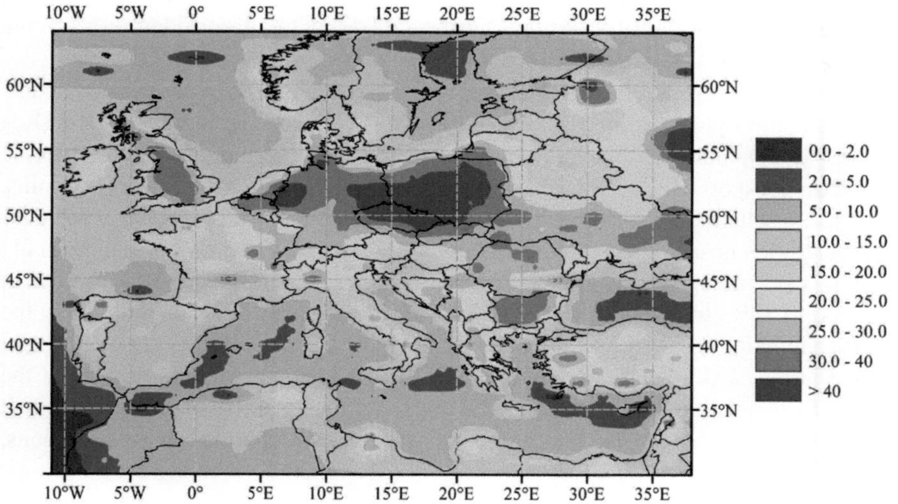

Figure 1.1 Modeled annual total deposition flux in *g km*$^{-2}$ (equivalent to *ng m*$^{-2}$) over the modeling domain (from Hedgecock et al., 2006)

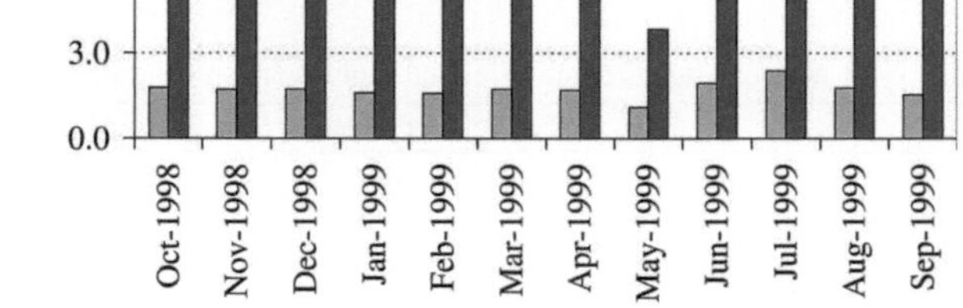

Figure 1.2 Modeled monthly total emission and deposition fluxes (*1 ton = 1 Mg*) to the Mediterranean Sea (from Hedgecock et al., 2006)

1.2.3 Rocks, Soils and Vegetation

Mercury emissions from top soils and vegetation are significantly influenced by meteorological conditions, historical atmospheric deposition and type of vegetation and top soil. Mercury fluxes from unaltered or background sites in North America have been observed in the range of -3.7 to 9.3 ng m^{-1} hr^{-1} (Nacht and Gustin, 2004) and are similar to other locations in which substrate contains background levels of Hg (Zhang et al., 2001). In altered geologic sites the mercury fluxes range from 15.5 ± 24.2 ng m^{-1} hr^{-1} (Nacht and Gustin, 2004), without considering a calcine waste that show a flux of 3334 ng m^{-1} hr^{-1}.

Mercury in vegetation originates from several mechanisms, including the uptake from the atmosphere, atmospheric deposition to foliage and uptake from roots (Rea et al., 2002); however, the proximity of vegetation to natural or anthropogenic sources (hot spots or contaminated sites) may increase its mercury content (Lodenius, 1998; Carballeira and Fernandez, 2002; Lodenius et al., 2003). The total mercury concentration observed in conifer sap flow is around 12.3 - 13.5 ng L^{-1} (Bishop et al., 1998), while the uptake of ground vegetation shows highest accumulation values in roots (82 - 88%) followed by rhizome (8 - 17%) and leaf (0.03 - 4%), highlighting the barrier function for the transport of inorganic mercury (Cavallini et al., 1999; Patra and Sharma, 2000; Schwesig and Krebs, 2003) and showing that almost all of the mercury in foliar tissue is originated from the atmosphere (Eriksen et al., 2003; Eriksen and Gustin, 2004). Summing up all the net evasional fluxes for all regions, the total net mercury evasion is 1664 Mg yr^{-1} (Mason 2008) (Table 1.2).

1.2.4 Biomass Burning

As a consequence of mercury content in vegetation, mercury emissions from biomass burning are significant (Veiga et al., 1994; Carvalho et al., 1998; Roulet et al., 1999; Friedli et al., 2001, 2003; Sigler et al., 2003; Pirrone et al., 2005). Their contribution is not often well considered in regional emissions estimates especially in very dry regions such as the south Mediterranean and several countries of Africa, which may represent an important contribution to the global atmospheric mercury budget

Table 1.2 Summary of mercury fluxes from terrestrial regions (from Chapter 7)

Region	Net Evasion (average) ($Mg\ yr^{-1}$)	Ratio[1](%)
Forest	342	7.5
Tundra/Grassland/Savannah/Prairie/Chaparral	448	9.9
Desert/Metalliferrous/ Non-vegetated Zones	546	12.0
Agricultural areas	128	2.8
Evasion after mercury Depletion Events	200	4.4
Total	1664	-

[1]calculated over the total evasion from natural sources which sum 4532 $Mg\ yr^{-1}$

(Veiga et al., 1994; Carvalho et al., 1998; Friedli et al., 2001; Friedli et al., 2003, Cinnirella and Pirrone, 2006; Wiedinmyer and Friedli, 2007; Cinnirella et al., 2008).

The mercury released to the atmosphere from vegetation is primarily related to mercury concentrations in foliage, the forest combustion efficiency and the efficiency of mercury released to the atmosphere. Mercury released to the atmosphere during a fire is strongly controlled by the mercury substrate concentration, light intensity and temperature (Ferrara et al., 1997; Engle et al., 2001; Zehner and Gustin, 2002). Field measurements in forests suggest that mercury release from the top soil during a fire is primarily dependent upon the increase of temperature caused by the activation of complex flux processes from the lower to the upper soil horizons (Iglesias et al., 1997), leading to an Hg flux of 1 to 5 mg m^{-2} (Woodruff et al., 2001).

Table 1.3 reports the estimate of mercury emissions form biomass burning at regional scale. A recent estimate suggests that on global scale nearly 675 Mg of mercury are released to the atmosphere every year (annual average for the period

Table 1.3 Mercury emissions from biomass burning (from Chapter 8) compared with that reported in literature

Forest	Total Hg release (*Mg* yr^{-1})	Avg. burnt area (10^6 *ha* yr^{-1})	Avg. fuel consumption (*Tg* yr^{-1})	Reference
Temperate/Boreal				
Canadian boreal	3.5	2.3	55	Sigler et al., 2003
Russian boreal	13.3	2.1	119	Cinnirella and Pirrone, 2006[a]
Russian boreal	16.1 (3.4-24.8)	3.9	260	Cinnirella and Pirrone, 2006[b]
Temperate/ Mediterranean (Europe)	2.4 (0.9-3.6)	0.5	46	Cinnirella and Pirrone, 2006[a]
Mediterranean (Europe/Africa)	2.3 (0.4-4.0)	0.5	45	Cinnirella and Pirrone, 2006[a]
Mediterranean (Europe/Africa)	4.3	0.4	366	Cinnirella et al., 2008
Temperate/boreal	59.5	-	530	Friedli et al., 2003
Boreal forests	22.5	5.0-15.0	240	Sigler et al., 2003
Tropical				
Tropical (Amazonian)	88.0	3.0-5.0	1404	Veiga et al., 1994
Tropical (Amazonian)	17.0	1.5-2.1	486	Lacerda, 1995
Tropical (Amazonian)	6.0-9.0	2.0-3.0	843	Roulet et al., 1999
World				
All forests	20.0	-	-	Nriagu, 1989
All forests	930 (510-1140)	-	621	Brunke et al., 2001[c]
All forests	590 (380-1330)	-	3460	Brunke et al., 2001[d]
Global biomass burning	675 ± 240	332		Friedli et al., 2008

Derived from:
[a]Ground-based data (1996–2002)
[b]Remote sensing data (1996–2002)
[c]Hg/CO emission ratio
[d]Hg/CO2 emission ratio

1997–2006), which accounts for about 8% of all current anthropogenic and natural emissions (Friedli et al., 2008).

1.3 Mercury Emissions from Anthropogenic Sources

Mercury is released to the atmosphere from a large number of man-made sources which include fossil fuels fired power plants, ferrous and non-ferrous metals manufacturing processes, chemicals production, processing of ores and waste disposal facilities and cement plants (Table 1.4).

Fossil fuels used in electric power generation facilities, especially those that use coal, is the largest source category of mercury released to the atmosphere, though other emission sources provide an important contribution in many regions of the world (Figure 1.3).

1.3.1 Anthropogenic Emissions by Source Category

The combustion of fossil fuels, roasting and smelting of ores, kilns operations in cement industry as well as uncontrolled incineration of urban and industrial wastes and production of certain chemicals release several volatile trace contaminants, such as mercury, into the atmosphere.

1.3.1.1 Coal and Oil Combustion, Wood and Wood Wastes

Coal represents the primary fuel in electrical power generation facilities, accounting for approximately 43% of total fuel used worldwide (EIA, 2008). Although it is very difficult to generalize on the mercury concentration in coal, the literature data

Table 1.4 Main source categories of mercury released annually in the environment

Releases from mobilisation of mercury impurities	Releases from intentional extraction and use of mercury	Releases from waste treatment, cremation etc.
Coal-fired power and heat production plants	Mercury extraction	Waste Incinerators
Energy production from other fuels	Artisanal gold mining	Landfills
Cement production	Caustic soda production	Cremation and cemeteries
Mining and other metallurgic activities	Use of fluorescent lamps,instruments and dental amalgam fillings	
Traffic activity (Gasoline, diesel, kerosene, biofuels)	Manufacturing of products containing mercury	

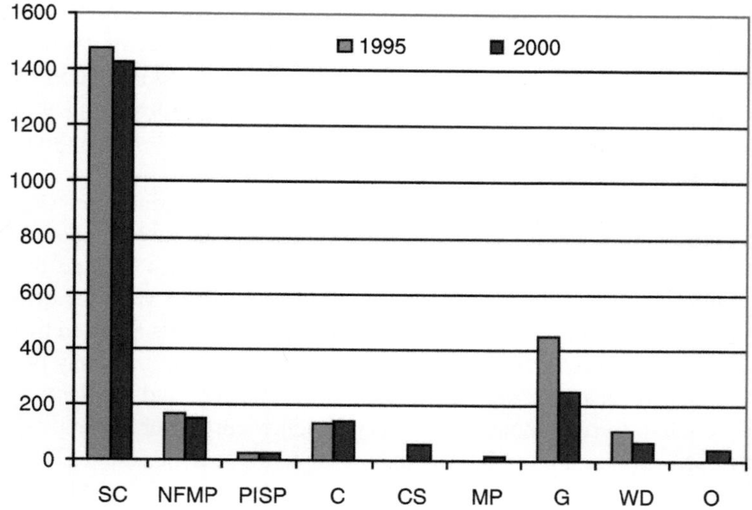

Figure 1.3 Global mercury emissions to the atmosphere by source category in 1995 and 2000 (based on Pacyna et al., 2003; Pacyna et al., 2006b). SC: stationary combustion; NFMP: non-ferrous metal production; PISP: pig-iron and steel production; C: cement production; CS: caustic soda production; MP: primary mercury production; G: gold production; WD: waste disposal; O: other sources

indicate that the mercury concentrations in coals vary between 0.01 and 1.5 g per Mg of fuel and that the concentration of mercury is somewhat lower in lignites than in bituminous and sub-bituminous coals (see Table 1.9 for details). But the lower heating values of lignite than bituminous and sub-bituminous coals may increase its consumption to generate an equivalent amount of energy and thus may release more mercury into the atmosphere (Tewalt et al., 2001). It should be noted, moreover, that concentrations of mercury within the same mining field may vary by one order of magnitude or more (Mukherjee et al., 2008a).

Canada, most European countries and Japan have regulations limiting emissions of various pollutants from coal fired power plants which help limit mercury emissions in these countries. In the United States currently there are no regulations for power plants that specifically target mercury emissions (because the Clean Air Mercury Rule (CAMR) was recently vacated by the courts), but there are regulations for other pollutants (such as SOx), which assume some mercury reductions as a co-benefit. Also, regulations in other countries often do not specifically target mercury emissions but rather may control SOx, particulate matter, etc., releases and as a co-benefit get some mercury emission reduction. Countries such as China, India, Russia and other countries of the Former Soviet Union (FSU), are taking some measures to reduce emissions from coal burning, but they have actually not been very effective. However, Russia, other countries of the FSU as well as other countries in Eastern Europe have decreased coal burning, in part because of depressed economies, but also because of shifts of energy production from coal to natural gas (Sznopek and Goonan, 2000). Wood and wood wastes are used as fuel in both industrial and residential sectors. In the industrial sector, wood waste is fired in industrial boilers to provide process heat, while wood is used in fireplaces and wood stoves in the residential sectors with no emission control technology. Insufficient data are available, however, to estimate the typical mercury content of wood and wood wastes.

Oil burning, as part of fossil fuels, is a contributor to mercury emissions to the global atmosphere, though it accounts for a much less amount compared to coal. The top five consumers of oil for power generation facilities include the United States, Japan, Russia, China and Germany. Relatively large volumes of distillate and residual oils are burned each year in the World. These fuels are used by utilities, commercial and industrial boilers (which, depending on their size, may be fired by either residual or distillate oils or a combination thereof) and residential boilers. Fuel oils contain trace amounts of mercury which occurs naturally in crude oils at levels that are very variable and which may relate to the specific source. These values range from 0.007 to 30 g Mg^{-1}, with a typical value being 3.5 g Mg^{-1} (USEPA, 1993; Pirrone et al., 2006; Mukherjee et al., 2008a). It is expected that mercury concentrations in residual oils are higher than those found in distillate oils, being the latter produced at an earlier stage in oil refineries. Heavier refinery fractions, including residual oils, contain higher quantities of mercury.

Natural gas may contain small amounts of mercury but the element is normally removed from the raw gas during the recovery of liquid constituents as well as during the removal of hydrogen sulfide. Therefore, it is assumed that mercury emissions during the natural gas combustion are insignificant (Pirrone et al., 1996; Pirrone et al.,1998; Pirrone et al., 2001c).

The most updated estimate (referred to 2000) of atmospheric releases of mercury from stationary combustion of fossil fuels is reported in (Table 1.5) (Pacyna et al., 2006b). Values should be intended with a ±25% uncertainty as recently suggested by Swain et al. (2007). A most recent assessment is being prepared as a part of the AMAP/UNEP report. New estimates for selected countries (see Table 1.29) show 640 Mg yr^{-1} of mercury released to the atmosphere

Table 1.5 Global atmospheric releases of mercury from stationary combustion of fossil fuels for the year 2000 (Pacyna et al., 2006b)

Continent	Stationary combustion (*Mg* yr^{-1})
Europe	88.8
Africa	205.2[a]
Asia + Russia	905.2
North America	79.6
South America	31
Australia and Oceania	112.6
Total	1422.4

[a]Despite the data is reported in the cited paper, it has been recognized as a mistake because the erroneous reference used (Finkelman R.B., personal communication).

1.3.1.2 Petrol, Diesel and Kerosene

Mercury emission from mobile sources has been reported as an important source for which there are few estimates (EPA, 1997). However, recent estimates in the United States indicate that the overall emissions quantities are relatively small compared to many other categories. For example, in the U.S. EPA National Emissions Inventory (NEI) for year 2002, it is reported that less than 1 Mg of mercury per year is emitted from mobile sources in the U.S.A., which is less than 1% of the total mercury emission of the country. Nevertheless, a significant effort has been made to assess mercury emissions from vehicular traffic (Liang et al., 1996; Wilhelm and Bloom, 2000; Wilhelm 2001; Hoyer et al., 2004; Conaway et al., 2005; Landis et al., 2007).

Mercury concentrations in petroleum and refined petroleum products are summarized in Table 1.6. In addition, emission factors for elemental, vapor-phase mercury plus particulate mercury for the light-duty gasoline vehicles have been found ranging

Table 1.6 Mercury concentration in crude oil and refined products of different geographic origin (from Wilhelm and Bigham, 2001)

Product	Range ($ng\ g^{-1}$)	Mean ($ng\ g^{-1}$)	St.Dev ($ng\ g^{-1}$)	Origin	Reference
crude oil	1–7	4			Liang et al, 2000
crude oil	0.1–12	<1		Asia	Tao et al., 1998
crude oil	NDi–1560	146		North America	Magaw et al., 1999
crude oil	1.0–3.2	1.7		Africa	Morris, R., 2000
crude oil	2.4–5.7	4.3		Middle East	Morris, R., 2000
crude oil	1.9	1.9		Canada	Morris, R., 2000
crude oil	2.5–9.3	5		North Sea	Morris, R., 2000
crude oil	0.1–2.7	1.4		Mexico	Morris, R., 2000
crude oil	0.8–12.3	5.2		South America	Morris, R., 2000
crude oil	3.1	3.1		n.a.	Morris, R., 2000
crude oil	<2–9	1.6		Canadian refineries	Duo et al., 2000
light distillates	0.1		2.8	n.a.	Wilhelm and Bigham, 2001
utility fuel oil	0.7		1.0	n.a.	Wilhelm and Bigham, 2001
Gasoline	0.72–3.2	1.5		n.a.	Liang et al., 1996
Gasoline	0.22–1.43	0.7		n.a.	Liang et al., 1996
Gasoline	0.08–1.4	0.50	0.40	n.a.	Conaway et al., 2005
Diesel	0.4			n.a.	Liang et al., 1996
Diesel	2.97			n.a.	Liang et al., 1996
Diesel	0.034		0.026	n.a.	Kelly et al., 2003
Diesel	0.05–0.34	0.15	0.06	n.a.	Conaway et al., 2005
Kerosene	0.04	0.04		U.S.	Liang et al., 1996
Heating oil	0.59	0.59		U.S.	Liang et al., 1996
Light distillates		1.32	2.81	U.S.	Bloom, 2000
Asphalt		0.27	0.32	U.S.	Bloom, 2000
Naphtha	3–40	15		Asian	Olson et al., 1997
Naphtha	8–60	40		U.S.	Olson et al., 1997
Petroleum coke	0–250	50	0.05	U.S.	USEPA, 2001

from 2.47 to 11.44 ng L^{-1} and from 70.92 to 123.84 ng L^{-1} for diesel vehicles (Hoyer et al., 2004). Experiments suggest that some of the factors influencing mercury emissions from mobile sources include oil consumption, driving conditions (including brake wear) and fuel consumption. In contrast, Landis et al. (2007) have measured 62 ng L^{-1} of mercury released from diesel powered vehicles and 284 ng L^{-1} from petrol powered vehicles (regular grade gasoline 89 octane). Gasoline vehicles were, therefore, found to be a significant source of Hg^0, RGM and Hg(p).

Field measurements in S. Francisco Bay (USA) indicate that petroleum products contribute with 0.7–13 kg yr^{-1} of mercury (3% of the total in the Bay) to the environment (Conaway et al., 2005). While on a national basis the mercury tailpipe emissions (based only on fuel mercury content) from on road motor vehicles was estimated in 148 kg yr^{-1} (Landis et al., 2007).

A very conservative global assessment on mercury emission from petroleum fuel consumption for 2000 has been done by considering emission factors reported by Landis et al. (2004) and the world consumption of petrol (1060.436 Gigalitres) and diesel (622.417 Gigalitres) (IEA, 2006).

Petrol combustion contributed with 238 kg yr^{-1} (121-281 kg yr^{-1}) of mercury emission while diesel contributed with 140 kg yr^{-1} (71-209). The total mercury emission was around 378 kg yr^{-1} (192-564 kg yr^{-1}) with a growing trend due to the increase of gasoline and diesel consumption. On a country basis, North America released 156 kg, Asia 94 kg and Europe 80 kg (Figure 1.4).

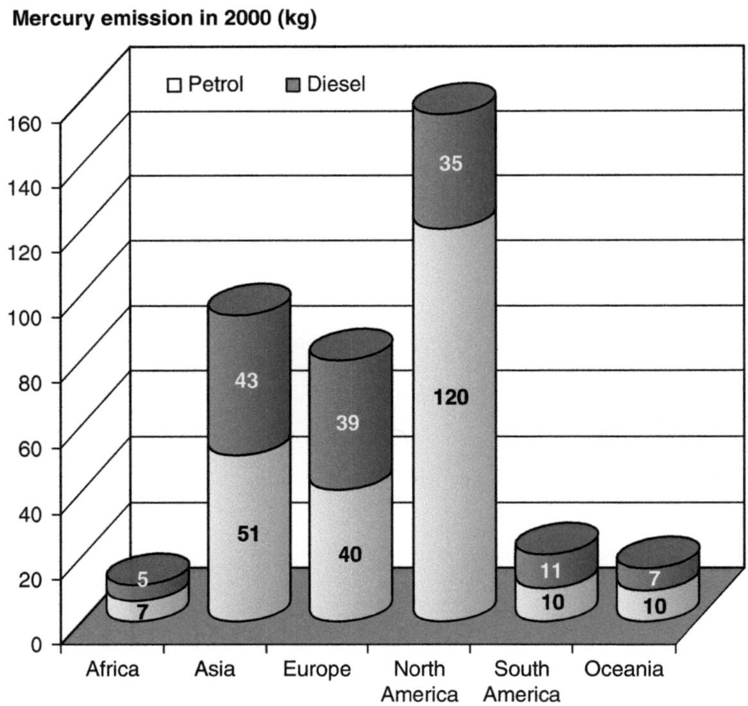

Figure 1.4 Global mercury emissions to the atmosphere by petrol and diesel consumption in 2000

The global contribution to atmospheric emission from petroleum fuels combustion represented 0.015% of the total anthropogenic emission. It should be kept in mind that these estimates are very conservative because not all countries are reported in the EIA database and because fuel consumption in the fishing industry and for military use is not included in the database. In addition, the contribution form biodiesel consumption is not considered in this assessment.

1.3.1.3 Iron-Steel Manufacturing

Emissions of mercury from primary and secondary pig iron and steel manufacturing are very much related to the overall production of these industrial goods and the efficiency of emission control measures. For selected countries, around 43 Mg of mercury per year are released to the environment and no major changes in the mercury emissions have been reported for this sector during the 1990 and beginning of 2000 (Pirrone et al., 2001b; Pacyna et al., 2006b). Asia (14.4 Mg yr^{-1}), Europe and North America (12.5 Mg yr^{-1} each) are the major emitting regions.

1.3.1.4 Primary & Secondary non-ferrous Metal Smelters

Mercury appears as an impurity of copper, zinc, lead and nickel ores as well as in gold ores. The production of these metals are known to be large sources of mercury released to the atmosphere, especially in developing countries (UNEP, 2002; Jones and Miller, 2005; Telmer and Vega, 2008). As shown in the GMA report, emissions of about 170 Mg per year from this sector can be considered as an underestimate.

Trends in non-ferrous metal production by different processes, with a special focus on new emerging economies, are leading to an increase of mercury releases to the atmosphere. Combustion temperature in boilers, furnaces and roasters are key parameters affecting the amounts of mercury released into the atmosphere as well as the chemical form and particle size distributions (Pirrone et al. 1996; Pirrone et al. 2001c; ZMWG, 2007).

It is very difficult to discuss the average content of mercury in the copper, zinc, lead, nickel and gold ores as very little information is available in the literature. However, some estimates are available, such as concentrations in gold ores (Jones and Miller, 2005). On the basis of the USGS mineral survey (USGS, 2004), the Zero Mercury Working Group (ZMWG) estimate mercury emissions from ore processing worldwide (Table 1.7) (ZMWG, 2007). Best estimates of mercury emitted from non-ferrous ore processing are about 310 Mg yr^{-1} which is different from the previous estimates (Pacyna et al., 2006) due to the contribution from China (Streets et al., 2008).

1.3.1.5 Caustic Soda Production (Chlor-alkali plants)

Approximately 150 chlor-alkali plants were in operation worldwide in 2004, though in Europe after recent legislation approval most plants have been closed or are going

Table 1.7 Estimates of mercury emissions from ore processing worldwide (ZMWG, 2007)

	Hg emissions *(Mg y⁻¹)*	
Metal	Lower-bound estimate	Upper-bound estimate
Copper	77.7	138.7
Gold	21.0	32.3
Lead	9.8	9.8
Zinc	68.5	330.1
Total	177.1	510.9

to be closed in a few years (Concorde East-West, 2006). Most of chlor-alkali plants use mercury cell technology with chlor-alkali processes that differs substantially from country to country (Eurochlor, 2007):

- Western Europe, predominance of mercury cell process (June 2000): 55%
- United States, predominance of diaphragm cell process: 75%
- Japan, predominance of membrane cell process: >90%
- India, predominance of membrane cell process: 86% (since 2000s) (Mukherjee et al., 2008b, chapter 4)

The remaining chlorine production capacity in western Europe consists of diaphragm cell process 22%, membrane cell process 20% and other processes 3% (Table 1.8) (Eurochlor, 2007). Global production capacity of chlorine in 2000 was about 12 Tg, the EU accounting for about 54% of the total. In 2000 the chlorineproduction capacity in western Europe was 6.6 Tg, 1.4 Tg in USA and 4.2 Tg in the rest of the world (EC, 2002; EC, 2004).

Due to the process characteristics, mercury can be emitted from the mercury cell process through air, water and wastes. The total mercury emission from chlor-alkali plants in western Europe was 9.5 Mg in 1998, ranging from 0.2-3.0 g of mercury per Mg of chlorine capacity at the individual plants (EC, 2001; EC, 2002).

In literature, significant discrepancies can be found between the amount of emissions reported and the amount of mercury purchased to replace mercury in cells. This missing mercury is reported to be in the range of 0.069 to 0.35 kg per Mg of NaOH produced with very different figures in new emerging countries (i.e. India) that loose up to 25 times more mercury than the global best figure. The estimate of 65.1 Mg yr^{-1} of mercury emission from chlor-alkali plants proposed by Pacyna et al. (2006b) differs from our estimate of 162.9 Mg yr^{-1} that considers recent updates for China and India (Mukherjee et al., 2008b; Streets et al., 2008).

1.3.1.6 Cement Production

Cement kilns, based on coal combustion and other mercury-containing materials, is a significant mercury emitting source category. Mercury out flowing the process derives from mercury existing in raw materials, limestone and coal fuel. Some alternative fuels also have a substantial mercury content. Mercury emissions monitoring data from cement kilns are very limited, therefore, the collection of new emissions data could be important for this source category. The evaluation of Hg

Table 1.8 Number of chlor-alkali plants, total chlorine production and percentage of processes that use mercury cells in some EU countries in 2005 (Eurochlor, 2007)

Country	Number of installations	Total chlorine Capacity *(Tg)*	Mercury cell process as of total capacity *(%)*
Belgium	3	752	74
Finland	1	115	35
France	7	1686	52
Germany	10	3972	37
Greece	1	37	100
Italy	9	982	83
Netherlands	1	624	11
Portugal	1	89	48
Spain	9	802	95
Sweden	2	310	71
United Kingdom	3	1091	78
Switzerland	3	104	100
Bulgaria	1	105	100
Czech Republic	2	183	100
Hungary	1	125	100
Poland	3	460	50
Romania	1	633	14
Slovak Republic	1	76	100
Total	59	12375	201

emissions on the basis of emission rates should be performed having in mind that large differences may occur in cement kilns technology which influence, ultimately, the emission rates. For example, in China, around 90% of the cement kilns are vertical shaft types, while in Western countries more energy efficient rotary kilns are used.

With respect to the mercury in the stack of cement kilns, the average mercury concentration is about 13 μg Nm^{-3} (Pirrone et al., 2001b). Pacyna et al. (2006b) reported an emission factor of 0.1 g per Mg of cement produced, which lead to 140.4 Mg yr^{-1} of mercury emitted to the environment (Pacyna et al., 2006b). Our best estimate based on cement production in 2005 (2315 Gg) and the reported emission coefficient is 232 Mg yr^{-1}.

1.3.1.7 Coal Bed Fires

Coal-bed fires have occurred since at least the Pliocene (Coates and Heffern, 2000). Clinker records evidence for extensive burning in the western U.S., most notably in Wyoming, Colorado, Utah and Montana. These ancient fires were initiated by natural causes including spontaneous combustion, lightning strikes and forest fires. Herring (1989, 1994) postulated that these ancient coal-bed fires could have been the source of a substantial portion of the nitrogen in earth's atmosphere.

Coal-bed fires and burning coal waste have proliferated worldwide since the Industrial Age, primarily as a consequence of anthropogenic activities. Spontaneous combustion due to exothermic reactions and the self-heating of coal exposed during mining is particularly problematic, making further mining dangerous while polluting surrounding communities. In other cases, coal fires are caused by mining activities including welding and electrical work, burning trash in abandoned mines, smoking, etc. (Stracher and Taylor, 2004). Today, tens of thousands of uncontrolled coal fires are burning around the world, emitting enormous amounts of the greenhouse gases, methane (during heating of the coal) and carbon dioxide, as well as CO, mercury, aerosols, sulphur compounds, and hydrocarbons including n-alkanes, iso-alkanes, cyclo-alkanes, alkyl aromatics, alkenes, ketones, ethers, halogenated hydrocarbons and additional volatile organics (Stracher et al., 2007; 2008). In China more coal fires are burning out of control than in any other country. Estimates for the amount of coal consumed annually by the fires in China range from 20 to 30 million Mg (Kuenzer, 2008) to 200 million Mg (Rosema et al., 1993; Discover, 1999) and may account for as much as 2-3% of the annual world emission of atmospheric CO_2 attributed to the burning of fossil fuels (ITC, 2008). There may be as many as 10,000 small coal and peat fires in Indonesia (A. Whitehouse, personal communication, 2004).

In the U.S. about 140 underground coal-mine fires and 58 burning gob piles have been reported from Pennsylvania (Stracher and Taylor, 2004). In India, South Africa, Russia, Eastern Europe and elsewhere, research reveals the mobilization of large volumes of potentially toxic elements including arsenic, selenium, fluorine and sulphur, in addition to smaller amounts of lead, copper, bismuth, tin, germanium and mercury. These data are based primarily on the analyses of solids precipitated around coal-fire gas vents. Analyses of coal-fire gas from South Africa show high concentrations of benzene, toluene, xylene, ethyl benzene and dozens of other organic compounds (Leaner, 2008 and herein references). In addition, an unpublished ICP-MS analysis from a coal-fire-gas vent in Colorado showed a compound that contains Hg. No attempt has been made to quantify the amount of mercury mobilized during these fires. Given the estimates for CO_2 emissions, it is possible that uncontrolled coal fires are responsible for a significant proportion of global mercury.

If we take the mean of the estimates for the amount of coal consumed annually by uncontrolled coal-bed fires in China it is estimated that about 112.5 million Mg of coal is burned. In addition to China, Walker (1999) reports coal fires from the United States, Canada, Australia, India, Indonesia, South Africa, England, Germany, Poland, Czech Republic, Russia, Ukraine, Turkey, Thailand, and other countries. If we assume an additional 87.5 million Mg of coal are consumed each year by these uncontrolled coal-bed fires plus coal-waste-pile, and coal-stock-pile fires we get a total of 200 million Mg. Multiplying this figure by the average mercury content in coal, 0.16 g Mg^{-1} (Table 1.9), the amount of mercury release annually to the atmosphere by uncontrolled coal fires would be about 32 Mg as an upper bound. If we accept the lower estimate for China of 20 million Mg of coal and assume that this represents half of the global uncontrolled fires (40 million Mg) we get a lower bound for mercury emissions of about 6.4 Mg of mercury released by uncontrolled coal fires. Because none of the emissions from uncontrolled coal

Table 1.9 Mercury content in coals from selected countries (USGS, unpublished data).

Country	Coal Type	Mean (*g Mg*$^{-1}$)	Range (*g Mg*$^{-1}$)	No. Samples	Comments
Argentina	Bituminous	0.1	0.03-0.18	2	
Botswana	Bituminous	0.09	0.04-0.15	11	
China	Anthracite - Bituminous	0.15	<0.02-0.69	329	Belkin et al. 2004
Colombia	Subbituminous	0.04	<0.02-0.17	16	
Egypt	Bituminous	0.12	0.04-0.36	14	
Peru	Anthracite - Bituminous	0.27	0.04-0.63	15	
Philippines	Subbituminous	0.04	<0.04-0.1	6	
Romania	Lignite - Subbituminous	0.21	0.07-0.46	11	
Slovak Rep.	Bituminous	0.08	0.03-0.13	7	
South Korea	Anthracite	0.30	<0.02-0.88	11	
Taiwan	Anthracite - Bituminous	0.67	0.07-2.3	4	Mean = 0.12 w/o the 2.3
Tanzania	Bituminous	0.12	0.04-0.22	24	
Turkey	Lignite	0.11	0.03-0.66	143	St.Dev = 0.093
US	All	0.17		7649	Finkelman, 1993
Vietnam	Anthracite	0.28	<0.02-0.67	6	
VV.	Bituminous	0.19	0.04-0.67	39	
Yugoslavia	Lignite	0.11	0.07-0.14	3	
Zambia	Bituminous	0.6	<0.03-3.6	12	
Zimbabwe	Bituminous	0.08	<0.03-0.5	3	

fires are attenuated by pollution-control systems, most of the mercury escapes into the atmosphere.

1.3.1.8 Waste

Hazardous or non-hazardous waste production containing mercury is governed by the yield and consumption of goods and their recycling process in society. Several products contain mercury and as calculated by Maxson (2003), the mercury supply from 1994 – 2000 for all products and processes production has averaged 3600 Mg per year, which one could take as a rough estimate of global mercury supply in 2000. A recent assessment for 2005 (UNEP, 2006b) estimate 3000 – 3800 Mg yr^{-1} of mercury supply. Small-scale artisanal gold mining, vinyl chloride monomer (VCM) production, chlor-alkali production and batteries production, plants and artisanal mining adsorb around 75% of the total supply (Figure 1.5).

Not all supplied mercury is finally converted in waste. The amount of mercury in waste depend upon the mercury content in products, products' lifetime and waste disposal mechanisms. The knowledge on mercury in different types of wastes is scarce and this implies that also the knowledge of mercury emissions from waste disposal practices (i.e., incinerators, landfills) is affected by a large uncertainty.

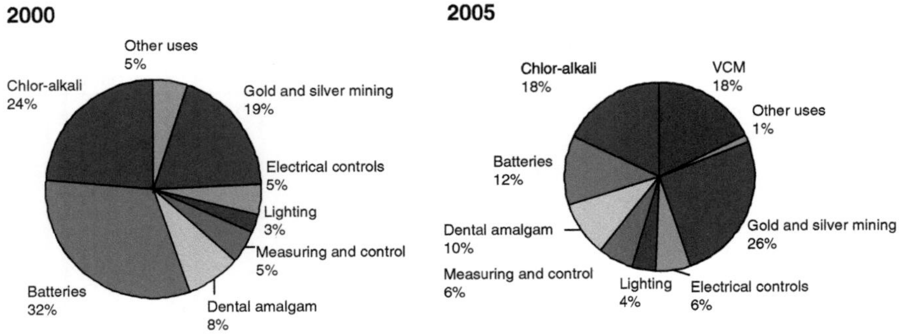

Figure 1.5 Percentages of global mercury consumption for different sector in 2000 and 2005 (from Maxson 2003 and UNEP, 2006b)

The causal factors of mercury in waste are categorised as follow (UNEP, 2007):

- Industrial equipments using mercury and consumer products;
- Typology of waste treatment processes used;
- Thermal process of natural mercury impurities in raw materials; and
- Extracting processes used in Artisanal and small scale gold mining.

1.3.1.9 Mercury in Waste Derived from Industrial Processes

Due to the phasing out of industrial mercury processes and mercury-containing products, a large amount of equipments for industrial mercury processes and mercury-containing products are expected to become mercury waste (UNEP, 2007). For example, the most qualified estimate of EU-15 mercury cell chlor-alkali plants amount to some 11,800 Mg. Waste from the Cl_2 industry has been estimated to contain 10 to 17 g of mercury per Mg of chlorine, showing variations within years (Garny, 2001; EC 2001b) (Figure 1.6). Based on a Cl_2 production capacity of 12.2 Gg (Eurochlor, 2008) it can be estimated that in 2000 mercury waste from chlor-alkali plants was 183 Mg.

1.3.1.10 Mercury in Waste Derived from Consumer Products

Most of mercury in waste derives from items that has been used and now are discontinued or are still used. Household categories of mercury sources are (EPA, 1992): batteries, electrical lighting, electrical equipment, instrument, pigments, paper coating, pharmaceuticals, dental amalgams and plastic catalysts. Some items pertain to both municipal and hospital solid wastes. The most critical products discharged in solid waste are lamps, batteries, diuretics, dental fillings, pigments, thermometers and plastics. Based on the lamp manufacturing industry's data

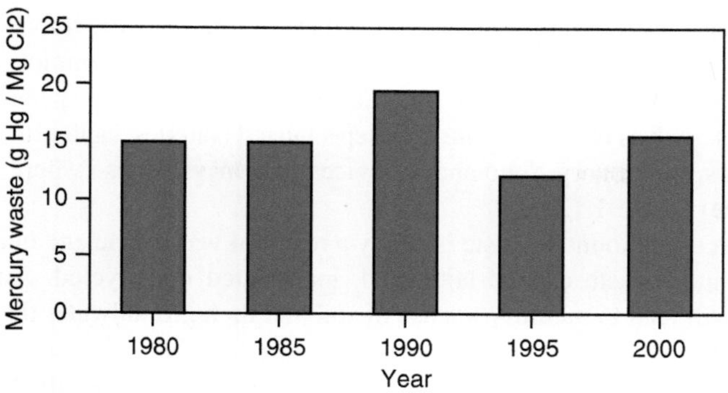

Figure 1.6 Mercury in solid waste from chlor-alkali plants in EU-15+Switzerland (from Garny, 2001)

(NEMA, 2001), the most common type of fluorescent lamp (4-foot T-8 model) was estimated to have an average mercury composition of about 10 mg per lamp (Maine Department of Environmental Protection, 2003; Culver 2007). A significant reduction of mercury content occurred from 1994 when the amount of mercury in lamps decreased from 44 to 20 mg. Mercury content in most common lamps is reported in Table 1.10.

It should be noted that because of the significant energy savings, using high efficiency fluorescent lamps containing mercury to replace incandescent lamps or older fluorescent lamps results in a net reduction in mercury emissions from fossil fuels fired power plants. EPA estimates that full implementation of the Green Lights program in USA would result in a reduction of close to 10 Mg of mercury per year due to reduced power generation. After being used, lamps can be disposed off in a municipal or hazardous waste landfill, recycled to recover mercury and other lamp materials or placed in a municipal waste incinerator. Landfilling has been a traditional mean of disposing off spent lamps. Due to their extremely low mercury content, lamps have historically accounted for only 3.8 percent of all the mercury deposited in municipal landfills (EPA, 1992). On the other side the incineration of mercury-containing lamps release up to 90% of the mercury to the air. A global estimate of mercury release from this category is not yet available.

Mercuric oxide (mercury zinc) batteries and button batteries are intended for use in medical devices. Despite their production being stopped in 1990, some of the medical devices may still require mercuric oxide batteries including cardiac monitors, pH meters, oxygen analysers and monitors and telemetry instruments. The alternative for mercuric oxide batteries, the zinc air batteries, may not be mercury-free. A zinc air button battery may contain up to 25 mg of mercury. Larger zinc air batteries are made up of stacked button batteries, each of which may contain up to 25 mg of mercury (Table 1.11). Any global estimate has been done on mercury release by use of batteries but the amount of batteries used worldwide (e.g. in the UK currently over 20,000 Mg of batteries are thrown away each year) should give a consistent contribution to mercury emissions. Thermometers, thermostats, fluorescent lamps, electrical switches, pharmaceuticals and dental fillings are common products in which mercury is still being used. Mercury is also used in many speciality medical applications, such as blood pressure cuffs, specialized batteries, cantor tubes, oesophageal dilators, pulmonary Scholander devices and in vaccines (where it acts as a preservative) (Table 1.12).

The mercury amount in waste is partly a result of waste type and quantity. Once generated solid waste can be landfilled, incinerated or recycled. Any adequate treatment will lead to mercury release in soil and then groundwater through leaching and in atmosphere through volatilisation. On the other side wastewater can be discharged directly into rivers, lakes and sea or treated in waste treatment plants. Also in this case the treatment process affects directly the amount of mercury released in waters.

1.3.1.11 Mercury Emissions from Industrial Incinerators

Industrial waste incinerators emit the highest amounts of mercury to the atmosphere (in some countries) due to the high mercury content of industrial hazardous wastes (Sung et al., 2004). Non-homogeneity of the industrial waste characteristics

Table 1.10 Mercury in fluorescent lamps (from Culver, 2007)

Lamp Type	Min (*mg*)	Max (*mg*)
4', Linear T8, (Low-Hg)	3.5	10
8' Linear T8	3.5	31-65
4' T12, TCLP	4.4	10
8' T12	6.8	31-65
U-bent T8	3.5	31-65
Preheat T8 (F15T8)	1.4	11-30
Compact fluorescent	1.4	11-30

Table 1.11 Mercury in button cell batteries

Button Cell Type	Mercury per Unit (*mg*)
Zinc-air	9.0
Silver-oxide	3.5
Alkaline	10.9

Table 1.12 Mercury content in common household item

Product	Mercury (*mg*)
Dental amalgams	500
Home thermometer	500–2000
Float switches in sump pumps	2000
Tilt thermostat	3000
Electrical tilt switches and relays	3500

result in the fluctuation in effluent mercury concentration. In Korea, mercury emissions are in the range of 619 to 1318 mg m^{-3} at the inlet of wet type particulate control device and mercury concentrations at the stack have been observed to be in the range of 40 to 325 mg m^{-3}.

Industrial waste incinerators with dry- and semidry-type APCDs (scrubber bag filter) may release mercury at a concentration of 14.3–59.3 mg m^{-3} at the inlet of APCDs and 17.8–58.8 mg m^{-3} at the stack.

1.3.1.12 Mercury Emissions from Municipal Waste Incinerators

Municipal waste incinerators (MWIs) can be a significant source of mercury emissions. For example, in the U.S.A. in year 1990, this sector was the largest emitting source category, with an estimated 52 Mg of mercury emissions for year 1990 (USEPA, 2006). However, emissions from municipal waste incineration were reduced by more than 90% in the U.S.A. from 1990 to 2002 due to regulations to control emissions as well as actions to reduce the use of mercury in products (e.g., batteries and paints), thereby reducing the mercury content of waste (USEPA, 2006). In addition, slag is produced from waste incineration processes. This product is mostly used for road construction, noise barriers, concrete production or landfilled. The slag unfortunately contains high concentrations of trace metals. Hg concentration varies from 0.02 to 7.75 mg kg^{-1} (IAWG, 1997). Assuming a Hg content of 4 mg kg^{-1}, total Hg in slag from waste incinerators in Europe (EU-15 countries+3 non-EU countries) varies between 24 and 54 Mg (Table 1.13).

1.3.1.13 Mercury Emissions from Medical Waste Incinerators

At present, any global estimate is available for operative medical waste incinerators. As consequence any global assessment has been done on mercury emission from this particular source because emissions are often included in the overall waste incinerators estimate (UNEP, 2002).

In the United States, about 5000 medical waste incinerators are in operation with most of them (90%) operating on-site. It is also assumed that at least larger MWIs are equipped with a simple or more efficient emission control devices. Most of the device systems employed in the medical waste incinerators are either wet or dry systems.

Wet systems typically comprise of a wet scrubber for controlling the emission of particles in combination with a packed-bed scrubber for removal of acidic gases and a high efficiency mist elimination. Dry systems include ESPs or FFs in combination with the sorbent injection. Concerning the removal of mercury, appearing mostly in a gaseous form, the efficiency of this process is rather limited. An improvement was obtained through adding activated carbon to the sorbent material in the sorbent injection/FF systems (Pirrone et al., 2001c). Medical waste incinerators are, therefore, a large source of mercury to the environment because

Table 1.13 Estimated mercury content ($Mg\ y^{-1}$) in waste of the EU member countries (from Mukherjee et al., 2004)

Source	Batteries	Lamps	Amalgam	Incen. waste	Sewage sludge [b]	Thermo-meters	Instru-ment	Sludge, Cl_2 prod.	Power plants	Lab waste	Metal industry	Total	Year	Ref.
Austria	0.9	0.06	5	30.2	0.34	–	–	0	–	–	–	36.5	1999/2000	1, 2
Belgium[a]	1.09	0.009	n.a.	–	0.18	0.01	–	–	–	–	–	1.29	1993	1, 2
Denmark	1.6	0.15	1	1.8	0.22	1.75	2.2	0	–	0.12	–	8.84	1996	2, 3
Finland	1	0.12	0.42	–	0.21	0.2	0.54	–	–	0.01	0.09	2.59	1996	2, 3
France	6.8	0.2	9	–	2.67	9	0.4	–	–	0.9	18	46.97	1993	2, 4
Germany	13	1.37[c]	29	–	2.66	–	5	–	–	3	–	54.03	1993	1, 2, 5
Ireland	11.45	0.03	0.5	–	0.03	–	–	–	–	–	–	12	1996	1, 2
Netherlands	15.7	0.083	11	15	0.57	–	–	2	–	–	4.75	49.1	1997/1999	1, 2
Spain	14.4	0.002	–	–	0.4	1	–	–	–	–	123 + 224[e]	362.48	1999	1, 6
Sweden	1	0.3	0.7	–	0.45	1.5	–	39	7	–	9.5	59.45	1996	1, 2
UK	4.4	4	10.5	14.4	2.86 44.5[d]	1.3	–	–	22.8	3.7	10.2	118.66	1996	1, 2
Total												752		

References: (1) National submission; (2) EEA (2002); (3) Huse et al. (1999); (4) Genie Urbain-Genie Rural (1999); (5) Maxson and Vonkeman (1996); (6) Anon. (2000).

[a]The Estimate is based on: batteries; 50% contain Hg-oxide (Hg content 30%) and the rest (0.7% Hg); weight of a lamp 250 g containing 15 mg Hg/unit; thermometer: weight 50 g

contain ing 1 g Hg (Huse et al., 1999).

[b]Sewage sludge values are based on 1996/1997 years.

[c]Estimated as follow - fluorescent discharge lamps: 15 mg Hg/lamp; compact lamps: 5mg Hg/lamp; high pressure Hg-vapour: 30mg Hg/lamp; metal halide: 30 mg Hg/lamp; high pressure sodium: 25mg Hg/lamp; 90M units (Genest, 1997). Holland—batteries: half of the total amount (2700 Mg) contain 0.2% Hg and the rest amount contain 1.0% Hg (Bjønstad and Linde, 1994). Electrofilter sludge contains 2900 mg Hg/kg sludge (dry matter); residue contains 20% Hg; jarosite waste from Zn-plant contains 50 mg Hg/kg; amalgam waste contains 50% Hg (Meijer, 2001); weight of lamp was estimated by multiplying 250 g/lamp; total Hg in lamps: 5.5×10^6 units x15 mg = 0.083 Mg.

[d]Hg in municipal waste

mercury in medical waste can be as much as 50 times higher than mercury in municipal solid waste (EPA, 2008).

1.3.1.14 Mercury Emissions from Landfill Process

Measurements of mercury concentrations have been performed upwind and downwind of the working face (where waste is trucked in from transfer stations, deposited, compacted and covered with inert filling material) at several Florida landfills. Downwind mercury concentrations (100 ng m^{-3}) were significantly higher (30 to 40 times) than that measured at upwind locations (Lindberg, 1999a; 1999b). Once buried, some of the inorganic mercury in the landfill is converted by bacteria into the more toxic methylmercury.

On the basis of direct quantification at landfill gas vent of giant landfill facilities in Seoul, Kim and Kim (2002) evaluated 420 ng m^{-3} (range 3.45 – 2952 ng m^{-3}) of mercury concentrations. On an annual basis, the computed fluxes of mercury from the whole site were on the order of 23 g.

Historical data from landfills, including a rough estimate of the total masses of deposited metals at the landfills since start-up, based on the metal-content data show 0.29-0.83 Mg per landfill in Finland.

High TGM concentrations (1100 – 1500 ng m^{-3}) have been measured at daily municipal solid waste generation in the Mexico City Metropolitan Area.

1.3.1.15 Mercury Emissions from Wastewater Treatment Process

Sometimes liquid mercury contained in products is intentionally or accidentally discharged into wastewater reaching at the end of the process the aquatic environment. Without proper treatment of wastewater, the dynamics of mercury includes the following steps:

1. During collection and transport of wastewater, $Hg^{(II)}$ is often in reducing conditions (caused by anoxia and various bacteria), leading to elemental mercury formation;
2. In the primary settling tank, mercury adsorbed to and incorporated into settleable solids is removed in the sludge;
3. In the mixed liquid aeration basin or other biological unit, bacteria, protozoa and other microorganisms proliferatively and effectively convert dissolved organic material and colloidal particles with associated mercury to a flocculent biological material which is eventually removed as waste sludge;
4. Bacterial action in anaerobic or aerobic digestion to stabilize sludge would produce additional transformations of elemental mercury. Elemental mercury formed may be stripped from solution by gas mixing systems (in the case of anaerobic digesters) or forced aeration. After stabilization, sewage sludge is often thickened or dewatered to reduce volume prior to ultimate disposal by land spreading, landfill or

incineration which are the anthropogenic sources of mercury emission (Huber 1997a; 1997b).

Sewage sludge is thus a residual product from industrial and urban wastewater. In recent years production of sewage sludge is increasing due to better effluent treatment methods and strict national regulation. For example in 1998, EU countries (except Italy) generated 7175 Gg (dry matter) of sewage sludge. The mercury concentration in sewage sludge, supplied only by seven countries for 1995/1998, varied between 0.6 and 3 mg kg^{-1}

1.3.1.16 Process of Natural Mercury Impurities in Raw Materials and Mercury Waste

Thermal processes (burn of raw material containing trace amount of mercury) include calcinations, combustion, crematoria, incineration, pyroprocessing, pyrometallurgy, retort, roasting, melting and smelting. Excluding the vapour phase of mercury in the flue gas released into atmosphere, mercury accumulates in solid incineration residues and flue gas cleaning residues, ash and slag which are finally landfilled, stabilised as concrete, or recycled as construction materials (Pirrone et al., 2001c).

1.3.1.17 Artisanal and Small Scale Gold and Mercury Mining

Mercury waste, called "tailings", released from artisanal and small scale gold mining activities has been becoming one of the hot issues, because almost all activities are in developing countries and countries with economies in transition and very often miners do not consider health effect due to their activities. Mercury in artisanal gold mining is used to form an amalgam that binds with gold (Chapter 7) and it is usually discharged with tailings and/or volatilized into the atmosphere. The magnitude of loss and means of mercury release from a specific site are defined by the Au-Hg separation procedures. A variety of amalgamation methods are used in artisanal mining operations (GMP, 2006).

1.3.1.18 Mercury Mining

The evaluation of the global primary mercury production is very uncertain because most countries do not report their mercury production in the official statistical yearbook. The USGS estimated 2795 Mg of mercury produced globally, whereas Gobi International assessment was higher with 3337 Mg of mercury annually produced. Maxson (2005) reported 3386 Mg of mercury used in different processes or products (Table 1.14). At present, producing primary mercury mines are located in

Table 1.14 Global mercury demand in 2000 by sector and by region (Maxson, 2005)

Mercury use category	EU-15 (*Mg*)	USA (*Mg*)	Rest-of-the-world (*Mg*)	Global (*Mg*)
Chlor-alkali industry	95	72	630	797
Small-scale gold/ silver mining	0	0	650	650
Batteries	15	16	1050	1081
Dental	70	44	158	272
Measuring & control	26	35	105	166
Lighting	21	17	53	91
Electrical control & switching	25	50	79	154
Other uses	50	50	75	175
Total	302	284	2800	3386

Algeria, People's Republic of China, Kyrgyzstan and Spain. Italy, Mexico, Slovakia, Slovenia and Turkey were all active minor producers before the world-wide collapse of mercury markets in the early 1990's and the banning process. Although none of these countries are presently producing mercury from primary mines, each retains significant reserves.

Table 1.15 gives information on recorded global primary production of mercury since 1981. There are also reports of small-scale artisanal mining of mercury in China, Russia (Siberia), Outer Mongolia, Peru and Mexico. It is likely that this production serves robust local demand for mercury, often for artisanal mining of gold – whether legal or illegal. World production of mercury is decreasing rapidly due to the banning policy adopted in several countries. Current mercury production on annual basis shows that nearly 1800 Mg of mercury was produced in 2000.

1.3.1.19 Artisanal Gold Mining

Mercury releases from Artisanal Small scale Gold Mining (ASGM) is due to the process used for gold extraction. Several countries have gold mining sites that release mercury in the atmosphere (Table 1.16). Actual estimates are based on available data on mercury and gold exports and imports by country and reported production and technology of extraction from all the countries known to have active ASGM communities.

The quality of the estimates ranges from good to poor across the countries. ASGM sector contributes with 640 to 1350 Mg of mercury per year (averaging 1000 Mg) to the environment from at least 70 countries. A significant fraction (350 Mg) is directly emitted to the atmosphere while 50 Mg are expected to be released latently for a total of 400 Mg. The rest is discharged to rivers and lakes (see Chapter 7)

Table 1.15 World production of mined mercury (*Mg*) as reported by the USGS (Jasinski, 1994; Reese, 1997–1999)

Country	1981–1985[1]	1986–1989[1]	1990	1991	1992	1993	1994	1995	1996	1997	1998	1999	2000
Algeria	386-877	587-764	637	431	476	459	414	292	368	447	224	200	240
China	800	850-1200	1000	760	580	520	470	780	510	830	230	200	200
Finland[2]	65-130	135-160	141	74	85	98	89	90	88	63	80	80	45
Kyrgyzstan	-	-	-	-	300	1000	379	380	584	610	620	620	600
Mexico	221-394	124-651	735	340	21	12	12	15	15	15	15	15	25
Russia	-	-	-	-	70	60	50	50	50	50	50	50	-
Slovakia / Czecho-slovakia	144-158	131-168	126	75	60	50	50	0	0	0	20	0	0
Slovenia	-	-	-	-	7	?	6	0	5	5	5	0	0
Spain	1416-1560	967-1471	-	-	-	643	393	1497	862	863	675	600	237[3]
Tajikistan	-	-	-	-	100	80	55	50	45	40	35	35	40
Ukraine	-	-	-	-	100	50	50	40	30	25	20	-	-
USA	570-962	140-520	562	58	64	w	w	w	65	w	-	-	15
USSR	1600-1700	1500-1650	800	750	-	-	-	-	-	-	-	-	-
Country	1981-1985[1]	1986-1989[1]	1990	1991	1992	1993	1994	1995	1996	1997	1998	1999	2000
Yugoslavia	0-88	51-75	37	9	-	-	-	-	-	-	-	-	-
Other countries	200-400	100-200	-	-	-	-	223	200	-	-	830	380	448
Total	5500-7100	4900-6700	4000	2500	1900	3000	2200	3400	2600	2900	2800	2200	
By Hylander & Meili (2003)	5600-6100	6100-6600	6100	3700	3100	3000	2000	3300	2800	2500	2000	2100	1800

[1]Reference: Metallgesellschaft (1992), as cited by OECD (1994). This reference's totals for 1990 and 1991 were 400-900 Mg higher than the presented totals from USGS.

[2]Numbers for Finland from 1990–1997 are from Finnish Environment Institute (1999).

[3]Spain has reported a production in 2000 of 237 Mg from the Spanish mercury mines.

w withheld in the references

Table 1.16 Mercury consumption from artisanal small scale gold mining by region

Gold mining site	Period *(since)*	Average annual *(Mg yr⁻¹)*	Total *(Mg)*	Reference
Amazon, Brasil	1979	180	3000	Telmer & Veiga, 2008
Mindanao, Phillipines	1985	26	260	Telmer & Veiga, 2008
Southern Brasil	1985	1	10	Telmer & Veiga, 2008
Puyango River, Peru	1987	3	24	Telmer & Veiga, 2008
North Sulavesi, Indonesia	1988	15	120	Telmer & Veiga, 2008
USA	1969	6	150	Telmer & Veiga, 2008
Canada	1976	1	14	Telmer & Veiga, 2008
Choco region, Colombia	1987	30	240	Telmer & Veiga, 2008
Narino, Colombia	1987	1	8	Telmer & Veiga, 2008
Victoria Fields, Tanzania	1991	6	24	Telmer & Veiga, 2008
Pando Dep., Bolivia	1979	20	300	Telmer & Veiga, 2008
Jia pi Valley, China	1938	2	116	Telmer & Veiga, 2008
Dixing region, China	1992	120	480	Telmer & Veiga, 2008
Guyana Shield, Venezuela	1989	45	360	Telmer & Veiga, 2008
Asmara, Eritrea	1907	0.1	10	Cinnirella and Pirrone*
Total		456.1	5106	

*estimates based on an historical paper on gold extraction in Eritrea (Zaccaria, 2005).

1.3.2 Anthropogenic Emissions by Region

In the last decade a considerable amount of research has been done to improve mercury emission inventories for developed and developing countries. In the following sections is reported a summary of mercury emissions by regions and source category.

1.3.2.1 Europe

In Europe, mercury emissions from anthropogenic sources in the year 2000 were near 240 Mg, with the highest contribution from the combustion of coal and other fossil fuels (48%) (Table 1.17). The second contributing category consisted of several industrial processes, including chlor-alkali production, non-ferrous and ferrous metal production and cement production (41%), while other sources, which include waste incineration and various uses of mercury, account for about 11% of the total.

Atmospheric mercury emissions in Europe decreased from the ‘80s to 2000 (Table 1.18) because of 1) the implementation of the FGD equipment in large power plants and other emission control devices in other industrial sectors, particularly in Western Europe and 2) the decline of economy in Eastern and Central Europe due to the switch of their economy from centrally planned to market oriented system (Pacyna et al. 2005; 2006b).

Table 1.17 Anthropogenic emissions of mercury in Europe in 2000 (*Mg yr*$^{-1}$) (from Pacyna et al., 2006a)

	Source category	Hg (*Mg yr*$^{-1}$)
Coal combustion	Power plant	63.5
	Residential heat	48.7
Oil combustion		1.7
Cement production		30.2
	Lead	7.6
	Zinc	7.8
	Pig & iron	12.5
Caustic soda		40.4
Waste disposal		11.6
Other		15.3
Total		239.3

Table 1.18 Trends in anthropogenic emissions of mercury in Europe since 1980 (*Mg yr*$^{-1}$)

Source category	1980 [a]	1985 [a]	1990 [a]	1995 [b]	2000 [c]
Combustion of fuels	350	296	195	186	114
Industrial processes	460	388	390	143	99
Other sources	50	42	42	59	26
Total	860	726	627	338	239

References:
[a] Pirrone et al., 1996
[b] Pacyna et al., 2001
[c] Pacyna et al., 2006a

1.3.2.2 North and Central America

Coal combustion and incineration of solid wastes account for most of mercury emissions in the USA National Emission Inventory (NEI), while nonferrous metal production accounts for most of mercury emissions in Canada and Mexico (Table 1.19).

In Canada, mercury releases can typically be attributed to waste incineration, coal combustion, smelting of ores and chlor-alkali industry (Environment Canada, 2008). Between 1990 and 1995, Canadian anthropogenic mercury emissions dropped from 36 to 11 Mg primarily as a result of technological improvements in the base metal mining and smelting industry. In 1995, this industry was the largest source of mercury into the atmosphere, contributing approximately 40% of total emissions. From 1995 to 2003, Canadian anthropogenic mercury emissions dropped to nearly 7 Mg. Three sectors, which include electricity generation, non-ferrous mining & smelting and incineration were responsible for 71% of mercury emissions into the atmosphere, accounting for 35%, 19% and 17% of Canadian emissions, respectively.

The mining industry, with both non-ferrous metal and gold sectors, is the highest source of mercury in Mexico, which represents 72% of the total (CEC, 2001; Gbor et al., 2007). Chlor-alkali production and combustion of heavy fuel oils are also potentially significant sources of mercury emissions. The total anthropogenic

Table 1.19 Mercury emission ($Mg\ yr^{-1}$) in USA, Canada and Mexico for major source categories

	USA (1990)[1]	USA (2002)[2]	Canada (1990)[3]	Canada (2003)[4]	Mexico (1990)[3]	Mexico (1999)[5]	TOTAL (USA 2002; Canada 2003; Mexico 1999)
Coal combustion[6]	56.0	55.0	1.3	2.4	1.6	2.2	59.6
Oil combustion[7]	3.0	5.0			0.15 - 22.9	0.7	5.7
Non-ferrous metal mining and refining (including large industrial scale gold mining)	7.0	10.7	23.9	1.3	12.5 - 31.1	22.8	34.7
Iron and steel industry		12.3[a]	0.6	0.4	0.4		12.8
Chlor-Alkali Plants	10.0	5.4				4.9	10.3
Manufacturing sources	14.0		1.7	0.8	3.7	0.7	1.4
Cement manufacturing		6.6		0.4			7.0
Waste incineration	113.0	11.6	3.4	1.4			13.0
Miscellaneous	17.0		5.3	0.3			0.3
Total	220.0	118.6	36.2	6.9	18.4 - 59.7	31.3	144.7

[1]Estimates for 1990 for U.S.A. were obtained from the US EPA Roadmap (US EPA, 2006) and/or the US EPA "1990 Emissions Inventory of Forty Potential Section 112(k) Pollutants". Final Report. May 21, 1999 (US EPA, 1999).

[2]National Emission Inventory (2002) - v3

[3]Pai et al., 2000

[4] National Pollutant Release Inventory of Canada (2003)

[5]Estimates were obtained from Commission for Environmental Cooperation report (CEC, 2001).

[6]For USA this category includes coal-fired power plants and industrial/commercial/institutional boilers that burn coal.

[7]For USA this category includes oil-fired power plants and industrial/commercial/institutional boilers that burn oil.

[a]This estimate includes primary and secondary steel production (e.g., Electric Arc Furnaces) and Iron and steel foundries.

mercury emission in North America is estimated to be 144.7 Mg yr^{-1}. The estimate is incomparable to previous regional and worldwide estimates of North American emissions that have ranged from 240 Mg yr^{-1} to 333 Mg yr^{-1} (Pirrone et al., 1996) because there was a strong reduction in emission mainly in the waste incineration sector. Inconsistencies have been found for waste incineration and oil combustion when comparing with that reported in the NEI.

1.3.2.3 Russia

The total intentional consumption of mercury in the Russian Federation in 2001/2002 was estimated to be in the range of 151 – 160 Mg yr^{-1} and since than the trend has declined. As the intentional consumption of mercury decreases, the mobilization of mercury impurities (trace element) increasingly account for a larger part of the total anthropogenic mercury flow. The total mobilization of mercury impurities in Russia in 2001 was estimated at 138 Mg (66 – 198 Mg) and the majority was mobilized in coal, oil and non-ferrous metals ore (Table 1.20). According to the official data the total emission of mercury from Russian enterprises that have the obligation to report their annual mercury emissions was 2.9 Mg in 2001. Besides these sources, significant amount of mercury is released from area sources and from processes in which mercury is present as a natural impurity in the raw materials.

Table 1.20 Mercury emissions (*Mg* yr^{-1}) to air and water in Russia (from ACAP, 2005)

Emission Source	Consumption/ mobilization	To air	Disposed in landfill/ waste dumps
Chlor-alkali production	103.0	1.2	39
Production of VCM	7.5	0.02	0
Gold mining using the amalgamation method, mining of sec. placers	5.5	3.1	1.1
Production of thermometers	26.0	0.009	0.1
Production of light sources	7.5	0.2	0.001
Other intentional uses	5.8	0.06	2.4
Total (intentional uses)	155.3	4.5	42.6
Coal - electricity producing sector	10.0	8.0	2.0
Coal - other uses (incl. waste from extraction)	12.0	6.3	3.6
Oil processing and use of petroleum products	33.0	3.4	
Gas, oil-shale and bio-fuels	8.0	1.0	
Zinc and lead production	31.0	1.9	8.5
Nickel and copper production	28.0	5.3	6.6
Production of other metals	7.8	2.6	4.2
Cement and lime	2.0	1.6	0.4
Total (mobilisation as impurity)	131.8	30.1	25.3
Waste incineration		3.5	
Landfilling			24.0
Sewage sludge		0.1	5.7
Total (waste treatment)		3.6	29.7

The present assessment uses the estimates from the 2005 Arctic Council study, prepared by a Russian expert group assisted by international experts, in which the total Russian emissions are estimated to be 39 Mg yr^{-1}, with 77% being the contribution from processes where mercury is mobilized as impurity (ACAP, 2005). (This effort was the first comprehensive assessment of mercury releases at the national level in the Russian Federation).

1.3.2.4 China

Mercury emissions in China are estimated to be 623.1 Mg in 2003 (Table 1.21). A large fraction of the emission (41%) is due to coal combustion, which in China includes three major subcategories: coal-fired power plants, industrial boilers and residential uses. Emissions from this categories increased from 202.4 Mg in 1995 to 334.0 Mg in 2005 (with the largest contribution from power plants and industries) (Streets et al., 2008, chapter 2). Approximately 40% of the mercury is released from non-ferrous metals smelters (Feng et al., 2008). Cement production (6%) and mercury mining (4%) represent minor contributions. In addition to indus-

Table 1.21 Mercury emission from different source categories in China in 2003 (Feng et al., 2008; Streets et al., 2008)

Source category	Emissions *(Mg yr^{-1})*
Coal combustion	256.7
Power plants	100.1
Industrial use	124.3
Residential use	21.7
Other uses	10.6
Nonferrous metals smelting	248.0
Zinc	115.0
Copper	17.6
Lead	70.7
Gold: large scale	16.2
Gold: artisanal	28.5
Fuel oil for stationary sources	0.6
Gasoline, diesel and kerosene	7.6
Biofuel combustion	10.7
Grassland/savanna burning	4.2
Agricultural residue burning	3.9
Household waste burning	10.4
Cement production	35.0
Iron and steel production	8.9
Caustic soda production	0
Mercury mining	27.5
Battery/fluorescent lamp production	3.7
Forest burning [a]	2.8
Coal mines spontaneous burning [a]	3.0
Total	623.0

[a] Forest burning and coal mines burning are obtained from Streets et al., 2005.

trial sources, this estimate accounts for annual emissions from biomass burning (2.8 Mg, without including natural sources which include re-emission of previously deposited mercury) and coal mines spontaneous burning (3 Mg) as reported by Street et al. (2005). Because the most updated emissions from industrial sources are for 2003, emissions from coal combustion reported here refer to for the same year despite they are available for 2005.

1.3.2.5 Australia

Mercury emissions from Australian coal fired power plants account for 2 to 8 Mg yr^{-1}. This is larger than the National Pollution Inventory estimate of 1.1 Mg yr^{-1}, because of the issues with brown coal fired power plants, but very significantly lower than the GMA estimate of 97 Mg yr^{-1} for stationary combustion sources in 2005 (Table 1.22). It is, however, in relatively good agreement with the earlier estimate of 6.3-8.6 Mg yr^{-1} quoted by Pirrone et al. (1996) for emissions between 1983 and 1992 (Nelson, 2007).

1.3.2.6 India

Mercury contamination is widespread in India and a recent study (Mukherjee et al., 2008b, chapter 4) has dealt with industrial emissions of mercury from coal combustion, iron and steel industry, non-ferrous metallurgical plants, chlor-alkali plants, cement industry, waste disposal and others minor sources (i.e. brick manufacturing, instruments, clinical thermometers). No information were found in the literature for the pulp and paper industry or for the oil and petrochemical industry in India as well as natural sources. Therefore, no estimates are provided in the regional budget. It should be stated that the lack of true emission data make very uncertain the estimate of anthropogenic emissions of mercury for this country.

Table 1.22 Emissions of mercury to the atmosphere from point sources in Australia (> 5 $kg\ yr^{-1}$) as reported in the Australian National Pollutant Inventory (Nelson, 2007)

Source category	Emissions $(Mg\ yr^{-1})$
Stationary power - Total	1.1
Stationary power - Black coal	0.98
Stationary power - Brown coal	0.1
Smelting/mineral processing	11.6
Iron and steel production	0.8
Petroleum refining	0.3
Cement	0.3
Waste	0.2
Coal mining	0.02
Other mining	0.3
Total (point sources)	15.7

The highest contributing source category is the coal combustion (48%) followed by waste disposal (31%). Industrial Hg emissions in India have decreased from 321 Mg in 2000 to 253 Mg in 2004. The Ministry of Environment and Forest in New Delhi has informed that 86% of Hg-cell chlorine plants have been converted to membrane process. This change implied that Hg emissions have decreased from 132 Mg in 2000 to 6.2 Mg in 2004 (see Chapter 4 for further information). Mercury emissions from biomass burning and brick industry have also been discussed by Mukherjee et al (2008) (Table 1.23). Based on industrial activities and socio-economic trends, mercury emissions vary from one region to another and they are projected to increase as coal combustion and waste generation increase (disposed off in landfills without proper treatment).

1.3.2.7 South Africa

Limited information is available for African countries. Also very few data and information on actual mercury emissions or levels of mercury in products and resources exists for South Africa, which is the most industrialized country of Africa. Nevertheless most of the mercury released in the environment comes from artisanal gold mining (Telmer and Veiga, 2008). The South African Mercury Assessment (SAMA) Programme (Leaner et al., 2006) has undertaken some limited mercury inventory development and monitoring studies in South Africa. The Country is a primary producer of many important and strategic metals (e.g. gold, platinum, lead, zinc) and is a major producer and consumer of coal (DME, 2003). Although these minerals and materials are known for their contribution to mercury pollution, detailed mercury emission inventories for these sources are unavailable. Pacyna

Table 1.23 Mercury emission from different source categories in India for 2000 and 2004 (*Mg yr*$^{-1}$)

Source	2000	2004
Coal fired power plants	100.44	120.85
Residential & Commercial boiler	3.65	3.70
Pig iron & steel production	3.84	4.56
Cu-production	3.84	11.78
Pb-production	2.49	1.83
Zn-production	1.41	1.90
Residual fuel oil consumption	0.52	0.47
Cement production	4.2	4.66
Municipal solid waste	50	70.00
Medical waste	6.6	6.60
E-waste	-	0.82
Biomass burning		
- Forest	7.74	7.74
- Crop	4.76	4.76
Chlor-alkali plants	132	6.2
Brick manufacturing	-	7.49
Total	321.49	253.36

et al. (2006b) suggested that in South Africa mercury release to the atmosphere accounts for 256.7 Mg, with most mercury emissions originating from industrial facilities, followed by stationary combustion. Leaner et al. (2008) critically revised previous estimates of mercury release from major anthropogenic sources, giving a global assessment of 40.2 Mg yr^{-1}. Most of emissions are associated with power generation that accounts for 77% of the total (Table 1.24).

The gasification process of coal accounts for a 4% of the total emission. Mercury is likely released during gasification of the coal, but there is a potential for its removal in conjunction with the removal of other pollutants. Coal for firing cement kilns and producing clinker are the major sources of mercury in cement production, contributing with 9% to the total emission.

1.3.2.8 South America

In Brazil, the amount of mercury entering the environment was estimated to be about 200 Mg yr^{-1} (Trade and Environment Database (TED) case 132). As described in TED case 132, gold recovery is performed by removing sediments from river bottoms and adjacent areas and feeding them through a number of mercury-coated sieves. Roughly 1.0 kg of mercury enters the environment for every kilogram of gold produced by artisans (Farid and others, 1991). Another estimate according to research by Veloso de Araujo (1995), in the Alta Floresta area, State of Mato Grosso, Brazil, was that a typical month's gold production of 230 kg emitted 240 kg of mercury to the atmosphere as elemental mercury vapour and 60 kg of mercury into rivers. Considering that coal consumption in South American Countries is near 28 Tg yr^{-1} (Mukherjee et al., 2008a) emissions of Hg from coalfired power plants is about 5.6 Mg yr^{-1} (emission factor 0.2 mg kg^{-1}).

Table 1.24 Mercury emissions (*Mg* yr^{-1}) from major anthropogenic sources in South Africa during 2004

Source Category	Hg Emissions
Power Plants	31.0
Coal Gasification	1.7
Consumer Products	0.1
Crude Oil Refining	0.5
Ferrous Metals: Iron and Steel	1.3
Coke Production / Alloy Steel	1.0
Pig Iron	0.3
Residential Heating	0.8
Cement Production	3.8
Non-Ferrous Metals: Primary smelters	0.6
Gold	0.3
Zinc, copper, lead	0.3
Incineration of wastes	0.6
Total	40.2

1.4 Global Assessment

Our knowledge of mercury emissions on a global and regional scale is still incomplete. Global emission estimates by Pirrone et al. (1996) for the year 1990–1992 (Table 1.25), by Pacyna et al. (2003) for the year 1995 (Table 1.26) and by Pacyna et al. (2006b) for the year 2000 (Table 1.27) indicate that Europe and North America

Table 1.25 Global emissions of total mercury from major anthropogenic sources in 1990 (*Mg yr*$^{-1}$) (Pirrone et al., 1996)

	Coal comb.	Oil comb.	Zn prod.	Pb prod.	Wood comb.	SWI	Misc.	Total
Africa	24.6	7.5	7.6	0.6	18.5	34.2	14.0	107.0
Asia	412.0	61.5	51.3	3.1	31.8	288.9	127.3	975.9
Western	95.4	36.2	67.7	5.0	1.6	97.1	45.5	348.5
Eastern	166.6	38.2	49.6	4.9	3.6	40.4	45.5	348.8
North	81.4	27.9	42.0	6.3	4.4	127.2	43.4	332.6
Oceania	8.1	1.4	10.1	1.6	0.4	7.0	4.3	32.9
Central	5.6	11.8	11.1	0.7	8.7	24.2	9.3	71.4
Total	793.7	184.5	239.4	22.2	69.0	619.0	289.3	2217.1

Table 1.26 Global emissions of total mercury from major anthropogenic sources in 1995 (*Mg yr*$^{-1}$) (Pacyna et al., 2003)

	SC	NF	PI	C	WD	TOT
Africa	197.0	7.9	0.5	5.2		210.6
Asia	860.4	87.4	12.1	81.8	32.6	1074.3
Australia	97.0	4.4	0.3	0.7	0.1	102.5
Europe	185.5	15.4	10.2	26.2	12.4	249.7
North America	104.8	25.1	4.6	12.9	66.1	213.5
Oceania	2.9			0.1		3.0
South America	26.9	25.4	1.4	5.5		59.2
Total	1474.5	165.6	29.1	132.4	111.2	1912.8

SC = Stationary combustion; NF = Non-ferrous metal production; PI = Pig iron and steel production; C = Cement production; WD = Waste disposal; TOT = Total.

Table 1.27 Global emissions of total mercury from major anthropogenic sources in 2000 (*Mg yr*$^{-1}$) (Pacyna et al. 2006b)

	SC	NF	PI	C	CS	M	G	WD	O	TOT
Africa	205.2	7.9	0.4	5.3	0.3	0.1	177.8		1.4	398.4
Asia[1]	878.7	87.6	11.6	89.9	30.7	0.1	47.2	32.6	0.9	1179.3
Australasia	112.6	4.4	0.3	0.8	0.7		7.7	0.1		126.6
Europe[1, 2]	113.9	30.2	15.4	12.5	40.4			11.6	15.3	239.3
North America	79.6	6.4	4.3	7.7	8.0	0.1	12.2	18.7	8.8	145.8
Russia	26.5	6.9	2.7	3.7	8.0		3.1	3.5	18.2	72.6
South America	31.0	25.4	1.4	6.5	5.0	22.8				92.1
Total	1447.5	168.8	36.1	126.4	93.1	23.1	248	66.5	44.6	2254.1

[1] Excluded Russia

[2] Data for Europe herein reported have been updated from Pacyna et al. 2006a.

CS= Caustic soda production; M= Mercury production; G = Gold production; O = Other.

seem to contribute less than 20% to the global anthropogenic emissions to the atmosphere, whereas Asia and Africa account for about 70% of global emissions and show a steady increase due to the fast economic development (Figure 1.7).

Our current estimate suggests that summing up the contribution from natural and anthropogenic sources nearly 8116 Mg of mercury is released annually to the global atmosphere (Table 1.28). Total mercury emission from anthropogenic sources account for 2909 Mg (36%). The present assessment shows that the majority of mercury emissions originate from combustion of fossil fuels (18%), particularly in the Asian countries including China and India (14% on total anthropogenic basis) (Table 1.29), where energy production from coal combustion is increasing at a rate of nearly 10% per year. Among industrialized countries Europe and North America account for 15% of the total emission that is mostly related to fossil fuel combustion. In addition to stationary combustion (18%), other sectors contribute to the global budget with an additional 18%. Combustion of coal is and will remain in the near future as the main source of energy in these countries.

Recent projections of energy production suggest that energy consumption will increase in all continents. Sharp increases are foreseen for Asian countries where most of future energy demand will still be supplied by fossil fuels, which will lead to an increase in annual emissions of mercury and other primary pollutants (i.e., NOx, SO2, aerosols, VOCs, CO2). Natural sources contribute to the global budget by 64% (5207 Mg) with oceans releasing most of the mercury (33%) followed by

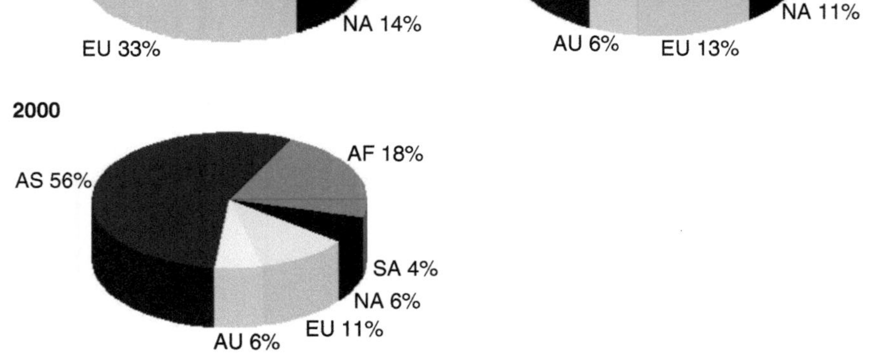

Figure 1.7 Trend of global anthropogenic emissions by region and year (from Pirrone et al., 1996 (1990); Pacyna et al., 2003 (1995); Pacyna et al., 2006b (2000)). AF, Africa; AS, Asia; AU, Australia; EU, Europa; NA, North America; SA, South America

Table 1.28 Total mercury emissions to the atmosphere by source category

Region	Hg emission *(Mg yr^{-1})*	Reference year	Reference
Natural			
Oceans	2682	2008	Mason, 2008
Lakes	96	2008	Mason, 2008
Forest	342	2008	Mason, 2008
Tundra/Grassland/ Savannah/Prairie/ Chaparral	448	2008	Mason, 2008
Desert/Metalliferrous/ Non-vegetated Zones	546	2008	Mason, 2008
Agricultural areas	128	2008	Mason, 2008
Evasion after mercury depletion events	200	2008	Mason, 2008
Biomass burning	675	2008	Friedli et al., 2008
Volcanoes and geothermal Areas	90	2008	Mason, 2008
Total (Natural)	5207		
Anthropogenic			
Coal combustion, oil Combustion	1422	2000	Pacyna et al., 2006b
Pig iron and steel Production	43	2000	This work
Non-ferrous metal Production	310	2007	USGS, 2004
Caustic soda Production	163	2000	This work
Cement production	236	2000	This work
Coal bed fires	32	2007	This work
Waste disposal	187	2007	This work
Mercury production	50	2007	This work
Artisanal gold mining production	400	2007	Telmer and Veiga, 2008
Other	65	2007	This work
Total (Anthropogenic)	2909		
Total (Natural + Anthropogenic)	8116		

biomass burning (forest 8% and agriculture 2%), deserts and metalliferous zones (7%), tundra and grassland (6%), forest (4%), evasion after mercury depletion events (3%) and volcanoes (1%).

1.5 Further Research

Additional research is required to reduce uncertainty in both anthropogenic and natural emissions estimates. The uncertainty of anthropogenic emission estimates is mostly related to the rapid economic development in emerging economies, particularly South and South-East Asia in which the impact of fossil fuel use in energy production is threefold. Firstly because fossil fuel power plants are the single most important anthropogenic emission source of mercury to the atmosphere, secondly because the other pollutants emitted as a result of fossil fuel exploitation such as NO_x and SO_2 have an impact on the atmospheric chemistry of mercury and influence its deposition patterns. A specific concern is for regions that are inadequately

Table 1.29 Global emissions ($Mg\ yr^{-1}$) of total mercury from major anthropogenic sources in major emitting countries/regions

	SC	NF	PI	C	CS	M	G	WD	O	Tot.	Year	Ref.
South Africa	32.6	0.3	1.3	3.8			0.3	0.6	1.3	40.2	2007	1
China	268.0	203.3	8.9	35.0		27.5	133.4	10.4	11.3	697.8	2003	2
India	120.9	12.1	2.9		106.0		0.5	77.4	8.0	327.8	2004	3
Australia	2.2	11.6	0.8	0.9			0.3	0.2	0.6	16.6	2005	4
Europe	113.9	15.4	12.5	27.3	40.4			11.6	15.3	236.4	2000	5
Russia	18.7	7.2	2.6	1.6	1.2		3.3	3.6	0.3	38.5	2001	6
North America	65.2	34.7	12.8	15.1	10.3			13.0	1.7	152.8	2003	7
South America	31.0	25.4	1.4	9.4	5.0	22.8	67.6			162.6	2000	8
Total	652.5	310.0	43.2	93.1	162.9	50.3	205.4	116.8	38.5	1672.7		
Rest of the World	769.9	0.0	0.0	142.6	0.0	0.0	194.6	70.2	26.5	1203.8		
Total	1422.4	310.0	43.2	235.7	162.9	50.3	400.0	187.0	65.0	2876.5		

References: (1) Leaner et al., 2008; (2) Street et al., 2008; (3) Mukherjee et al., 2008; (4) Nelson, 2007; (5) Pacyna et al., 2006a; (6) ACAP, 2005; (7) This work; 8 Pacyna et al., 2006b.

* Mercury emissions from artisanal mining here reported have been estimated by Telmer and Veiga (2008). Global mercury emission account for 300 $Mg\ yr^{-1}$ (Africa 24 $Mg\ yr^{-1}$, Asia 208 $Mg\ yr^{-1}$, South America 68 $Mg\ yr^{-1}$)

SC = Stationary combustion; NF = Non-ferrous metal production; PI = Pig iron and steel production; C = Cement production; CS= Caustic soda production; M= Mercury production; G = Gold production; WD = Waste disposal; O = Other; TOT = Total.

described in terms of point sources (Africa, South America) or exhibit unusually large uncertainties (Asia). These uncertainties affect model and policy development, and human welfare.

Atmospheric mercury models developed in recent years for assessing the relationship between emission source regions and receptor regions show a limited accuracy. The ability to determine the accuracy of current models is severely limited by the lack of a unified global emission inventory that accounts for a better emission source characterisation related to fossil fuel power plants in fast developing countries where energy demand is increasing at an annual rate of 10 per cent or even more.

The improvement of the mercury emission inventory on global scale, with special attention to fossil fuels fired power plants in countries characterised by a fast economic growth (i.e, China, India) will lead to a better assessment of the impact of different energy exploitation strategies foreseen in major environmental outlooks elaborated by leading institutions such as UNEP, World Bank, World Watch Institute and International Energy Agency (IEA).

Detailed mercury emission inventory may help nations to shape future energy management strategies that, among other things, will lead to a better assessment of countries' potential for renewable and non-renewable energy production; this is in agreement with recommendations and requirements of major international conventions and programmes aimed to reduce the impact of human activities on ecosystems quality and human health related to energy productions.

Socio-economic impacts associated with mercury pollution are very considerable with long-term impacts on human health, welfare and productivity. The impact of mercury pollution on human welfare can be immediate or in the years or decades to come. Delineation of mercury "hot spots" and knowledge of ecological processes that lead to their formation can reduce uncertainty and can help mitigation and outreach efforts of reducing health costs. This improved knowledge of the sources, transport and fate of environmental mercury would lead to realistic risk assessments, efficient mitigation efforts, and effective outreach to minimize adverse impacts on coastal ecosystems and human populations that consume seafood.

The emission to the atmosphere of mercury via natural processes constitutes an important part of the global Hg input and is a dominant part of the global mercury cycle. However, while there is an ongoing and continued effort to quantify these fluxes, the magnitude of their extent is still relatively poorly constrained. It must be emphasized that while the fluxes are due to natural processes, they constitute mercury that has originated from different sources, and because of the potential for deposited mercury to be re-emitted to the atmosphere from both terrestrial and aquatic surfaces, these fluxes include both primary sources and secondary (recycled) mercury.

These fluxes are comparable to the estimates of mercury inputs to the atmosphere from point anthropogenic source emissions. It appears that the mercury emission estimates from natural sources are within the range and confidence intervals of the results of a number of box and numerical modeling studies, and with empirical estimates.

Of great urgency is the development and validation of models for mercury cycling in forests, accounting for the biogeochemistry for each region because the average emissions from the land exceed the ocean on an areal basis. This would provide an understanding of the source/sink relationship and thus mercury accumulation or loss in ecosystems.

Such models could then be coupled with the fire carbon emission models. Fires that are certainly of great importance in terms of mercury emissions. Mercury in forests originates largely from deposition from the global atmospheric pool and thus is a global concern. The release of mercury from biomass burning is partially under direct human control. Knowing anthropogenic mercury emission would address restrictions to the global release and reduce the atmospheric and vegetation/soil pools.

References

ACAP, 2005. Arctic Mercury Releases Inventory. Reduction of Atmospheric Mercury Releases from Arctic States. Arctic Council Action Plan to Eliminate Pollution of the Arctic (ACAP), Danish Ministry of the Environment, Danish Environmental Protection Agency, Copenhagen, Denmark,: 116 pp.

ACAP, 2005. Assessment of Mercury Releases from the Russian Federation. Arctic Council Action Plan to Eliminate Pollution of the Arctic (ACAP). Danish Ministry of the Environment, Danish Environmental Protection Agency, Copenhagen, Denmark: 332 pp.

Andersson M.E., Gårdfeldt K., Wängberg I., Sprovieri F., Pirrone N., Lindqvist O., 2007. Seasonal and daily variation of mercury evasion at coastal and off shore sites from the Mediterranean Sea. Marine Chemistry, Vol.104, 214-226. Reprint on the Special Issue, Vol. 107, pp.104–116.

Belkin H.E., Finkelman R.B., Wang Q., Wang B., Zheng B., 2004. Mercury in China coals: abstracts of the 21st Annual Meeting of the Society for Organic Petrology, v. 21, p. 28.

Bishop K.H., Lee Y.H., Munthe J., Dambrine E., 1998. Xylem sap as a pathway for total mercury and methyl mercury transport from soil to tree canopy in a boreal forest. Biogeochemistry, 40, 101–113.

Bjønstad S.L., Linde M.R., 1994. Materialstrømanalyse av kvikksølv. Vurdering av alternativer, Utkast til SFT-rapport, Oslo, Norway.

Bloom N.S., 2000. Analysis and Stability of Mercury Speciation in Petroleum Hydrocarbons. Fresenius' J. Anal. Chem., 366, 438–443.

Brunke E.G., Labuschagne C., Slemr F., 2001. Gaseous Hg emissions from a fire in the Cape Peninsula, South Africa, during January 2000. Geophysical Research Letters, 28 (8): 1483–1486.

Carballeira A., Fernandez J.A., 2002. Bioconcentration of metals in the moss Scleropodium purum in the area surrounding a power plant. Chemosphere, 47: 1041–1048.

Carvalho J.A., Higuchi N., Araujo T., Santos J.C., 1998. Combustion completeness in a rain forest clearing experiment in Manaus, Brazil. Journal of Geophysical Research, 103 (D11): 13195–13200.

Cavallini A., Natali L., Durante M., Maserti B.E., 1999. Mercury uptake, distribution and DNA affinity in durum wheat (Triticum durum Desf.) plants, Sci. Tot. Env., 243/244, 119.

CEC, 2001. Preliminary Atmospheric Emissions Inventory of Mercury in Mexico. Final Report. Acosta y Asociados Project CEC-01 prepared for Commission for Environmental Cooperation No. 3.2.1.04.

Cinnirella S., Pirrone N., 2006. Spatial and temporal distribution of mercury emission from forest fires in Mediterranean region and Russian federation. Atmospheric Environment 40:7346–7361.

Cinnirella S., Pirrone N., Allegrini A., Guglietta D., 2008. Modeling mercury emissions from forest fires in the Mediterranean region Environmental Fluid Mechanics, 8: 129–145.

Coates D.A., Heffern E.L., 2000. Origin and geomorphology of clinker in the Powder River Basin, Wyoming and Montana. In: Coal bed methane and Tertiary geology of the Powder River Basin (Miller R., Ed.), Wyoming Geological Association 50th annual Field Conference Guidebook, Casper, WY, 211–229.

Conaway C.H., Mason R.P., Steding D.J., Flegal A.R., 2005. Estimate of mercury emission from gasoline and diesel fuel consumption, San Francisco Bay are, California. Atmospheric Environment, 39: 101–105.

Concorde East-West, 2006. Status Report: Mercury-cell Chlor-alkali Plants in Europe. Prepared for the European Environmental Bureau. October.

Cossa D., Coquery M., Gobeil C., Martin J.M., 1996. Mercury fluxes at the ocean margins. In Global and regional mercury cycles: sources, fluxes and mass balances, pp. 229–247. Ed. by W. Baeyens et al. Kluwer Academic Publishers.

Culver A., 2007. How to Specify Low-mercury and Lead-free Lighting Equipment. Available at: www.abag.ca.gov/abagenergywatch/pdfs/EnviroSpec-Mercury.pdf

Discover, 1999. China's on fire. R–D News, 20 (10): 20 pp.

DME, 2003. Integrated Energy Plan for the Republic of South Africa. Department of Minerals and Energy, South Africa, Pretoria, South Africa.

EC, 2001a. Integrated Pollution Prevention and Control (IPPC) Reference Document on Best Available Techniques in the Chlor-Alkali Manufacturing industry. Available at: eippcb.jrc.es

EC, 2001b. Pollutants in urban waste water and sewage sludge. European Commission, Luxemburg: 244 pp. Available at www.ec.europa.eu/environment/ waste /sludge /sludge_ pollutants.htm.

EC, 2002. Final report from the commission of the council concerning mercury from the Chlor-Alkali Industry. European Commission, Brussels.

EC, 2004. Mercury flows in Europe and the World: the impact of decommissioned chlor-alkali plants. Available at: ec.europa.eu/environment/chemicals/mercury/pdf/report.pdf

EIA, 2008. International Energy Outlook 2007. Available at: www.eia.doe.gov/oia f/ ieo /pdf/coal.pdf

Eisenreich S.J., Bernasconi C., Campostrini P., De Roo A., George G., Heiskanen A.S., Hjorth J., Hoepffner N., Jones K.C., Noges P., Pirrone N., Runnalls N., Somma F., Stilanakis N., Umlauf G., van de Bund W., Viaroli P., Vogt J., Zaldivar J.M., 2005. Climate Change and the European Water Dimension. A Report to the European Water Directors 2005. EU Report No. 21553, European Commission-Joint Research Centre, Ispra, Italy, pp.253.

Engle M.A., Gustin M.S., Zhang H., 2001. Quantifying naturalsource mercury emissions from the Ivanhoe Mining District, north-central Nevada, USA. Atmospheric Environment, 35: 3987-3997.

Environment Canada, 2008. Mercury and the environment. Available at: www.ec.gc.ca /MERCURY/ EN/ndex.cfm

EPA, 1992. Characterization of Products Containing Mercury in Municipal Solid Waste in the United States, 1970 to 2000. Environmental Protection Agency, Municipal Solid Waste Program, Office of Solid Waste, OSW-EPA-530-R-92-013.

EPA, 2008. Mercury in Medical Waste. Mercury Fact Sheet # 1. Environmental Protection Agency, Region 5 Air and Radiation Division. Available at: www.epa.gov/ARD-R5/glakes/fact1.htm

Ericksen J.A., Gustin M.S., Schorran D.E., Johnson D.W., Lindberg S.E., Coleman J.S., 2003. Accumulation of atmospheric mercury in forest foliage. Atmospheric Environment, 37 (12): 1613–1622.

Ericksen J.A., Gustin M.S., 2004. Foliar exchange of mercury as a function of soil and air mercury concentrations. Science of the Total Environment, 324: 271–279.

Eurochlor, 2008. Chlorine plants, January 2005. Chlorine online information resource. Available at: http://www.eurochlor.org/plants

Feng X., Streets D.G., Hao J., Wu Y., Li G., 2008. Mercury emissions from industrial sources in China. In: Mercury Fate and Transport in the Global Atmosphere: Measurements, models and policy implications (Pirrone N. and Mason R. Eds.), UNEP, 2008.

Ferrara R., Maserti B.E., De Liso A., Cioni R., Raco B., Taddeucci G., Edner H., Ragnarson P., Sverberg S., Wallinder E., 1994. Atmospheric mercury emission at Solfatara Volcano, Pozzuoli, Phlegraean Fields-Italy. Chemosphere, 29: 1421–1428.

Ferrara R., Maserti B.E., Andersson M., Edner H., Ragnarson P., Svanberg S., 1997. Mercury degassing rate from mineralized areas in the Mediterranean basin. Water, Air and Soil Pollution, 93: 59–66.

Ferrara R., Mazzolai B., Edner H., Svanberg S., Wallinder E., 1998. Atmospheric mercury sources in the Mt. Amiata area, Italy. The Science of the Total Environment, 213 (1–3): 13-23.

Ferrara R., Mazzolai B., Lanzillotta E., Nucaro E., Pirrone N., 2000a. Volcanoes as Emission Sources of Atmospheric Mercury in the Mediterranean Basin. The Science of Total Environment, 259, 115–121.

Ferrara R., Mazzolai B., Lanzillotta E., Nucaro E., Pirrone N., 2000b. Temporal trends in gaseous mercury evasion from the Mediterranean Seawaters. The Science of Total Environment, 259, 183–190.

Finkelman R.B., 1993. Trace and minor elements in coal. In: Organic Geochemistry (M.H. Engel and S.A. Macko, Eds.), Plenum Press, New York. p. 593–607.

Finnish Environment Institute, 1999. Atmospheric emissions of heavy metals in Finland in the 1990's. The Finnish Environment No. 329, Finnish Environment Institute, Helsinki (in Finnish).

Friedli H.R., Radke L.F., Lu J.Y., 2001. Mercury in Smoke from Biomass Fires. Geophysical Research Letters, 28 (17): 3223– 3226.

Friedli H.R., Radke L.F., Lu J.Y., Banic C.M., Leaitch W.R., MacPherson J.I., 2003. Mercury emissions from burning of biomass from temperate North American forests: laboratory and airborne measurements. Atmospheric Environment, 37 (2): 253–267.

Friedli H.R., Arellano A.F. Jr., Cinnirella S., Pirrone N., 2008. Mercury emissions from global biomass burning: spatial and temporal distribution. In: Mercury Fate and Transport in the Global Atmosphere: Measurements, models and policy implications (Pirrone N. and Mason R. Eds.), UNEP, 2008.

Genest W., 1997. Recycling of fluorescent tubes in Germany. Federal Environmental Agency, Berlin, Germany, p. 1–6.

Genie Urbain-Genie Rural, 1999. Les déchets mercuriels en France. Mercury waste in France, Parts 1 and 2, Nos. 7-8:20–48 and 17–53, France.

GMP, 2006. Manual for Training Artisanal and Small-Scale Gold Miners, UNIDO, Vienna, Austria.

Gustin M.S., Harald B., Christopher S., 2002. Investigation of the light-enhanced emission of mercury from naturally enriched substrates. Atmospheric Environment, 36 (20): 3241–3254.

Hedgecock I.M., Pirrone N., 2004. Chasing Quicksilver: Modeling the Atmospheric Lifetime of Hg0 (g) in the Marine Boundary Layer at Various Latitudes. Environmental Science and Technology, Vol.38, 69–76.

Hedgecock I.M. and Pirrone N., 2005. Modelling chemical and physical processes of Hg compounds in the marine boundary layer. In: Dynamics of Mercury Pollution on Regional and Global Scales, N. Pirrone and K. Mahaffey (Editors), Springer Verlag Publishers, Norwell, MA, USA. Chapter 13, pp. 295–317.

Hedgecock I.M., Pirrone N.,Trunfio G.A., Sprovieri F., 2006. Integrated mercury cycling, transport and air-water exchange (MECAWEx) model. Journal of Geophysical Research, 111 (D20302).

Herring J.R., 1989. Fires as cause, effect and feedback on the crustal cycles of carbon, phosphorus and nitrogen: Abstract of papers, 198th American Chemical Society National Meeting, American Chemical Society, Washington DC, GEOC 7.

Herring J.R., Ciener J.S., Been J.M., Szari S.L., Reime M.G., Rice D.D., Boyce B.C., 1994. Methane, carbon dioxide, oxygen and nitrogen in soil gas overlying coal beds of the Upper Cretaceous Fruitland Formation in the San Juan Basin, La Plata County, southwestern Colorado: Open-file report, U.S. Geological Survey, Reston, Virginia, 12 pp.

Hoyer M., Baldaufl R.V., Scarbro C., Barres J., Keeler G.J., 2004. Mercury Emissions from Motor Vehicles. 13th International Emission Inventory Conference. "Working for Clean Air in Clearwater". Clearwater, FL, June 8 – 10. Available at: www.epa.gov/ttn/chief/conference /ei13/

Huber K., 1997a. Mercury Use: Wastewater Treatment Plants, Great Lakes Binational Toxics Strategy - The Wisconsin Mercury Source Book. (Available at: www. epa.gov/glnpo/ nsdocs/ hgsbook/)

Huber K., 1997b. Wisconsin Mercury Sourcebook, U.S. EPA. Available at: www. p2pays.org/ef/04/03851.htm

Huse A, Lindmark GM, Sørensen PL, Weholt Ø, Mroueh U-M, Wahlström M., 1999. Investigation of categories and quantities of mercury waste and treatment capacity in the Nordic countries. Tema Nord No. 546. Nordic Council of Ministers, Copenhagen, 90 p.

Hylander L.D., Meili M., 2003. 500 years of m Nacht ercury production: global annual inventory by region until 2000 and associated emissions. Sci Total Environ, 304 (1–3): 137–144.

IAWG, 1997. Municipal Solid Waste Incinerator Residues. International Ash Working Group. Brussels.

IEA, 2006. Energy Balances of OECD Countries (2006 edition)--Extended Balances and Energy Balances of Non-OECD Countries (2006 edition)--Extended Balances. International Energy Agency (IEA) Statistics Division, Paris. Available at data.iea.org/ieastore/default.asp.

Iglesias T., Cala V., Gonzalez J., 1997. Mineralogical and chemical modifications in soils affected by a forest fire in the Mediterranean area. The Science of the Total Environment, 204(1): 89–96.

ITC, 2008. Overview about the coal fire problem. International Institute for Geoinformation Science and Earth Observation. Available at: www.itc.nl/~coalfire/problem/hazards.html

Jasinski S.M., 1994. The materials flow of mercury in the United States. The United States Department of the Interior, Bureau of Mines, Circular 9412. (Available at minerals.usgs.gov/minerals/pubs/commodity/mercury/).

Jones G., Miller G., 2005. Mercury and Modern Gold Mining in Nevada. Dept. of Natural Resources and Environmental Sciences. University of Nevada. Available at: www.unep.org

Kim K.-H., Kim M.-Y., 2002. Mercury emissions as landfill gas from a large-scale abandoned landfill site in Seoul. Atmospheric Environment Volume: 36, Issue: 31, October, 2002, pp. 4919–4928

Kuenzer C., 2008. (Lead Author), Galal H., Galal H., (Topic Editor), 2008. Coal fires. In: Encyclopedia of Earth. Eds. Cutler J. Cleveland (Eds.) Available at: www.eoearth.org/rticle/Coal_ fires

Lacerda L.D., 1995. Amazon mercury emissions. Nature, 374: 20–21.

Landis M.S., Lewis C.W., Stevens R.K., Keeler G.J., Dvonch J.T., Tremblay R.T., 2007. Ft. McHenry tunnel study: Source profiles and mercury emissions from diesel and gasoline powered vehicles. Atmospheric Environment, 41: 8711–8724.

Leaner J.J., Ashton P.J., Mason R.P., Kim E-H., Murray K., Dabrowski J.M., 2006. The Status of Mercury Research in South Africa. 8th International Conference on Mercury as a Global Pollutant, Madison, Wisconsin, August 2006.

Leaner J., Dabrowski J., Mason R., Resane T., Richardson M., Ginster M., Euripides R., Masekoameng E., 2008. Mercury emissions from point sources in South Africa. In: Mercury Fate and Transport in the Global Atmosphere: Measurements, models and policy implications (Pirrone N. and Mason R. Eds.), UNEP, 2008.

Liang L., Horvat M., Danilchik P., 1996. A Novel Analytical Method for Determination of Picogram Levels of Total Mercury in Gasoline and Other Petroleum Based Products. Sci Tot Environ., 187, 57–64.

Liang L., Lazoff S., Horvat M., Swain E., Gilkeson J., 2000. Determination of Mercury in Crude Oil by In Situ Thermal Decomposition Using a Simple Lab Built System. J. Anal. Chem., 367, 8–11.

Lindberg S.E., Price J.L., 1999a. Airborne emissions of mercury from municipal landfill operations: A short-term measurement study in Florida. J. Air & Waste Management Association, 49: 520–532.

Lindberg S.E., Roy K., Owens J., 1999b. Pathways of mercury in solid waste disposal. ORNL sampling operations summary and preliminary data report for PaMSWaD-I, Brevard County Landfill, February 6.

Lindberg S., Bullock R., Ebinghaus R., Engstrom D., Feng X., Fitzgerald W., Pirrone N., Prestbo E., Seigneur C., 2007. A Synthesis of Progress and Uncertainties in Attributing the Sources of Mercury in Deposition. Ambio, Vol. 36, No. 1, pp.19–32.

Lodenius M., 1998. Dry and wet deposition of mercury near a chlor-alkali plant. The Science of the Total Environment, 213 (1–3): 53-56.

Lodenius M., Tulisalo E., Soltanpour-Gargar A., 2003. Exchange of mercury between atmosphere and vegetation under contaminated conditions. The Science of the Total Environment 304 (1–3), 169–174.

Maine Department of Environmental Protection, 2003. A Strategy to Reduce the Mercury Content of Products: Report to the Joint Standing Committee on Natural Resources, January 2003, http://mainegov-images.informe.org/dep/mercury/productsweb.pdf

Mason R., 2008. Mercury Emissions from Natural Sources and their Importance in the Global Mercury Cycle. In: Mercury Fate and Transport in the Global Atmosphere: Measurements, models and policy implications (Pirrone N. and Mason R. Eds.), UNEP, 2008.

Mather T.A., Pyle D.M., 2004. Comment on Volcanic emissions of mercury to the atmosphere: global and regional inventories Sci. Tot. Env., 327, 323–329.

Maxson P., Vonkeman G., 1996. Heavy metals in products. Ministry of Housing, Spatial Planning and Environment, Directorate-General for Environmental Protection, The Hauge', Publikatiereeks Produktenbeleid nr. 1996/17.

Maxson P., 2003. Mercury flows in Europe and the world: The impact of decommissioned chlor-alkali plants. Draft final report 22 August 2003

Meijer P.J., 2001. Short survey of dangerous waste containing mercury in The Netherlands. Laboratory for Waste and Emissions, National Institute of Public Health and the Environment, Unpublished data.

Metallgesellschaft, (1939–1998). Metallstatistik 1929–1938; Metallstatistik 1957–1966; Metallstatistik 1981–1991; Metallstatistik 1985–1995; Metallstatistik 1987–1997. Annual volumes of metal statistics. Frankfurt-am-Main, 1939–1998 (in German).

Mukherjee A.B., Zevenhoven R., Brodersen J., Hylander L.D., Bhattacharya P., 2004. Mercury in waste in the European Union: sources, disposal methods and risks. Resources, Conservation and Recycling, 42 (2): 155–182.

Mukherjee A.B., Zevenhoven R., Bhattacharya P., Sajwan K.S., Kikuchi R., 2008a. Mercury flow via coal and coal utilization by-products: A global perspective. Resources, Conservation and Recycling, 52: 571–591.

Mukherjee A.B., Bhattacharya P., Sarkar A., Zevenhoven R., 2008b. Mercury emissions from industrial sources in India. In: Mercury Fate and Transport in the Global Atmosphere: Measurements, models and policy implications (Pirrone N. and Mason R. Eds.), UNEP, 2008.

Munthe J., Wangberg I., Pirrone N., Iverfeld A., Ferrara R., Ebinghaus R., Feng R., Gårdfeldt K., Keeler G.J., Lanzillotta E., Lindberg S.E., Lu J., Mamane Y., Prestbo E., Schmolke S., Schroder W.H., Sommar J., Sprovieri F., Stevens R.K., Stratton W., Tuncel G., Urba A., 2001. Intercomparison of Methods for Sampling and Analysis of Atmospheric Mercury Species. Atmospheric Environment.Vol. 35, 3007–3017.

Nacht D.M., Gustin M.S., 2004. Mercury emissions frombackground and altered geologic units throughout Nevada. Water, Air and Soil Pollution, 151: 179-193.

Nelson P.F., 2007. Atmospheric emissions of mercury from Australian point sources. Atmospheric Environment 41: 1717– 1724.

NEMA, 2001. Fluorescent Lamps and the Environment. National Electrical Manufacturers Association, NEMA01BR.

Nriagu J.O., 1989. A global assessment of natural sources of atmospheric trace metals. Nature 338, 47–49.

Nriagu J., Becker C., 2003. Volcanic emissions of mercury to the atmosphere: global and regional inventories. Sci. Tot. Env., 304, 3–12.

Pacyna E.G., Pacyna J.M., Pirrone N., 2001. European emissions of atmospheric mercury from anthropogenic sources in 1995. Atmospheric Environment, 35 (17): 2987–2996.

Pacyna J.M., Pacyna E.G., Steenhuisen F., Wilson S., 2003. Mapping 1995 global anthropogenic emissions of mercury. Atmospheric Environment, 37 (S1): S109–S117.

Pacyna J.M., Munthe J., Larjava K., Pacyna E.G., 2005. Mercury emissions from anthropogenic sources: estimates and measurements for Europe. In: Dynamics of Mercury Pollution on Regional and Global Scales. Atmospheric Processes, Human Health and Policy, (Pirrone and Mahaffey Eds.), Springer Verlag Publishers, Norwell, MA, USA, Chapter 3: 14 pp.

Pacyna E.G., Pacyna J.M., Fudala J., Strzelecka-Jastrzab E., Hlawiczka S., Panasiuk D., 2006a. Mercury emissions to the atmosphere from anthropogenic sources in Europe in 2000 and their scenarios until 2020. Science of the Total Environment, 370: 147–156.

Pacyna E.G., Pacyna J.M., Steenhuisen F., Wilson S., 2006b. Global anthropogenic mercury emission inventory for 2000. Atmospheric Environment, 40: 4048–4063.

Pai P., Niemi D., Powers B., 2000. A North American inventory of anthropogenic mercury emissions. Fuel Processing Technology, 65–66: 101–115.

Patra M., Sharma A., 2000. Mercury Toxicity In Plants. Botanical Review, 66: 379–422.

Pirrone N., Keeler G.J., Nriagu O., 1996. Regional differences in worldwide emissions of mercury to the atmosphere. Atmospheric Environment, 30 (17): 2981–2987.

Pirrone N., Allegrini I., Keeler G.J., Nriagu J.O., Rossmann R., Robbins J.A., 1998. Historical Atmospheric Mercury Emissions and Depositions in North America Compared to Mercury Accumulations in Sedimentary Records. Atmospheric Environment, 32, 929– 940.

Pirrone N., Hedgecock I., Forlano L., 2000. The Role of the Ambient Aerosol in the Atmospheric Processing of Semi-Volatile Contaminants: A Parameterised Numerical Model (GASPAR). Journal of Geophysical Research, 105, D8, 9773–9790.

Pirrone N., 2001. Mercury Research in Europe: Towards the preparation of the New EU Air Quality Directive. Atmospheric Environment, 35: 2979–2986.

Pirrone N., Pacyna J.M., Barth H., 2001a. Atmospheric Mercury Research in Europe. Atmospheric Environment, 35 (17): 2997–3006.

Pirrone N., Costa P., Pacyna J.M., Ferrara R., 2001b. Mercury Emissions to the Atmosphere from Natural and Anthropogenic Sources in the Mediterranean region. Atmospheric Environment. Vol. 35, 2997–3006.

Pirrone N., Munthe J., Barregård L., Ehrlich H.C., Petersen G., Fernandez R., Hansen J.C., Grandjean P., Horvat M., Steinnes E., Ahrens R., Pacyna J.M., Borowiak A., Boffetta P., Wichmann-Fiebig M., 2001c. Ambient Air Pollution by Mercury (Hg) - Position Paper. European Commision, Bruxelles. Available at: europa.eu.int/comm/environment/ ir/ackground.htm# mercury

Pirrone N., Wichmann-Fiebig M., 2003. Some Recommendations on Mercury Measurements and Research Activities in the European Union. Atmospheric Environment.Vol. 37, S-1, 3–8.

Pirrone N., Ferrara R., Hedgecock I.M., Kallos G., Mamane Y., Munthe J., Pacyna J.M., Pytharoulis I., Sprovieri F., Voudouri A., Wangberg I., 2003. Dynamic Processes of Mercury Over the Mediterranean Region: results from the Mediterranean Atmospheric Mercury Cycle System (MAMCS) project. Atmospheric Environment.Vol. 37-S1, 21–39.

Pirrone N., Hedgecock I.M., 2005. Climate Change and the Mercury Biogeochemical Cycle. In: Climate Change and the European Water Dimension: A Report to the European Water Directors 2005. EU Report No. 21553, Eisenreich (Eds.), European Commission- Joint Research Centre, Ispra, Italy, Chapter VI-C, 190–196.

Pirrone N., Mahaffey K., 2005. Where we Stand on Mercury Pollution and its health effects on Regional and Global Scales. In: Dynamics of Mercury Pollution on Regional and Global Scales (Pirrone N. and Mahaffey K. Eds.), Springer Verlag Publishers, Norwell, MA, USA. Chapter 1, pp.1– 21.

Pirrone N., Sprovieri S., Hedgecock I., Trunfio A., Cinnirella S., 2005. Dynamic Processes of Atmospheric Mercury and its Species in the Mediterranean Region. In: Dynamics of Mercury Pollution on Regional and Global Scales. Atmospheric Processes, Human Health and Policy, (Pirrone and Mahaffey Eds.), Springer Verlag Publishers, Norwell, MA, USA, Chapter 23: 41

Prakash A., 2007. Coal Fires in China. Available at: www. gi.alaska.edu/~prakash/coalfires/ coal-fires.html

Pyle D.M., Mather T.A., 2003. The importance of volcanic emissions for the global atmospheric mercury cycle. Atmos. Environ., 37: 5115–5124.

Rea A.W., Lindberg S.E., Scherbatskoy T., 2002. Mercury accumulation in foliage over time in two northern mixed-hardwood forests. Water, Air and Soil Pollution, 133 (1–4): 49–67.

Reese, 1981–2000. USA Geological Survey Minerals Yearbook: Mercury. All years 1981–2000.

Rosema A., van Genderen J.L., Schalke H.J.W.G., 1993. Environmental monitoring of coal fires in north China: Project identification Mission Report, BCRS 93-29, ISBN 90 5411 1054.

Roulet M., Lucotte M., Farella N., Serique G., Coelho H., Sousa Passos C.J., De Jesus da Silva E., Scavone de Andrade P., Mergler D., Guimarães J.R.D., Amorim M., 1999. Effects of recent human colonization on the presence of mercury in Amazonian ecosystems. Water, Air and soil pollution, 112: 297–313.
Schwesig D., Krebs O., 2003. The role of ground vegetation in the uptake of mercury and methylmercury in a forest ecosystem. Plant and Soil, 253: 445–455.
Sigler J.M., Lee X., Munger W., 2003. Emission and long-range transport of gaseous mercury from a large-scale Canadian boreal forest fire. Environmental Science and Technology, 37: 4343–4347.
Stracher G.B., Taylor T.P., 2004. Coal fires burning out of control around the world: thermodynamic recipe for environmental catastrophe. In: Coal Fires Burning around the World: a Global Catastrophe (Stracher G.B., Ed.), International Journal of Coal Geology, p. 7–17.
Stracher G.B., 2007. Coal fires burning around the world: Opportunity for innovative and interdisciplinary research, GSA Today, Geological Society of America, 17 (11): 36–37.
Stracher G.B., Lindsley-Griffin N., Griffin J.R., Renner S., Schroeder P., Viellenave J.H., Masalehdani M.N.-N., Kuenzer C., 2008. Revisiting the South Cañon Number 1 Coal Mine fire during a geologic excursion from Denver to Glenwood Springs, Colorado. In: [title] (Raynolds R.G., Ed.) Geological Society of America Field Guide 11, doi: 10.1130/2007.fld011(XX). (In press).
Streets D.G., Hao J.M., Wu Y., Jiang J.K., Chan M., Tian H.Z., Feng X.B., 2005. Anthropogenic mercury emissions in China. Atmospheric Environment, 39: 7789–7806.
Streets D.G., Hao J., Wang S., Wu Y., 2008. Mercury emissions from coal combustion in China. In: Mercury Fate and Transport in the Global Atmosphere: Measurements, models and policy implications (Pirrone N. and Mason R. Eds.), UNEP, 2008.
Swain E.B., Jakus P., Lupi F., Maxson P., Pacyna J., Penn A., Rice G., Spiegel S., Veiga M, 2007. Socioeconomic Consequences of Mercury Use and Pollution. Ambio: A Journal of the Human Environment, XXXVI (1).
Sznopek J.L., Goonan T.G., 2000. The material flow of mercury in the economies of the United States and theWorld'. US Geological Survey Circular, 1197: 28 p. Available at: greenwood.cr.usgs.gov/circulars/c1197/.
Telmer K., Vega M., 2008. Knowledge Gaps in Mercury Pollution from Gold Mining. In: Mercury Fate and Transport in the Global Atmosphere: Measurements, models and policy implications (Pirrone N. and Mason R. Eds.), UNEP, 2008.
Tewalt S.J., Bragg L.J., Finkelman R.B., 2001. Mercury in U.S. Coal; abundance, distribution and modes of occurrence: U.S. Geological Survey Fact Sheet FS-095-091. Available at: pubs.usgs.gov/factsheet/fs095-01.
UNEP, 2002. Global Mercury Assessment, UNEP, Geneva, Switzerland
UNEP, 2005. Toolkit for Identification and Quantification of Mercury Releases, UNEP, Geneva, Switzerland
UNEP, 2006a. Guide for Reducing Major Uses and Releases of Mercury. Available at: www.chem.unep.ch/ mercury/Sector%20Guide%202006.pdf.
UNEP, 2006b. Summary of Supply, Trade and Demand Information on Mercury, UNEP Chemicals, Geneva, Swizerland. Available at: www.chem.unep.ch/mercury/Trade-information.htm
UNEP, 2007. Draft technical guidelines on the environmentally sound management of mercury wastes. UNEP/CHW/OEWG/6/INF/16. Updated after OEWG6 (3rd Draft). Available at: www.basel.int/techmatters/mercury/guidelines/301007.doc
USEPA, 1993. Locating and estimating air emissions from sources of mercury and mercury compounds. September 1993. As cited by Scoullos et al, 2000.
USEPA, 1997. Mercury Study Report to Congress. Office of Air Quality Planning and Standards and Office of Research and Development. U.S. Environmental Protection Agency
USEPA, 2001. Mercury in Petroleum and Natural Gas: Estimation of Emissions from Production, Processing and Combustion; Technical Report Prepared by National Risk Management Research Laboratory, EPA-600/R-01/066.
USEPA, 2002. National Emission Inventory (NEI). Available at: www.epa.gov
USEPA, 2006. Mercury Roadmap. Available at: www.epa.gov/mercury.

USGS, 2004. Minerals Yearbook. U.S. Geological Survey. Available at: minerals.usgs.gov/ minerals/pubs/myb.html.

Veiga M.M., Meech J.A., Onante N., 1994. Mercury pollution from deforestation. Nature, 368: 816–817.

Walker S., 1999. Uncontrolled fires in coal and coal wastes. International Energy Agency, London. Report CCC/16, 72 p.

Weidinmyer C., Friedli H., 2007. Mercury emission estimates from fires: An initial inventory for the United States, Environ. Sci. Technol., 41, 8092–8098.

Wilhelm S.M., 2001. Estimate of Mercury Emissions to the Atmosphere from Petroleum. Environmental Science & Technology, 35 (24): 4704-4710

Wilhelm S.M., Bigham G.N., 2001. Concentration of Total Mercury in Crude Oil Refined in the United States. Paper Presented at the 6th International Conference on Mercury as a Global Pollutant, Minamata, Japan, Oct 15–19, 2001.

Wilhelm S.M., Bloom N., 2000. Mercury in petroleum. Fuel Processing Technology, 63–1: 1–27.

Won J.H., Park J.Y., Lee T.G., 2007. Mercury emissions from automobiles using gasoline, diesel and LPG. Atmospheric Environment, 41: 7547–7552.

Woodruff L.G., Harden J.W., Cannon W.F., Gough L.P., 2001. Mercury loss from the forest floor during wildland fire. Eos Transactions AGU, 82(47), Fall Meeting Suppl., Abstract B32B–0117.

World Coal Institute, 2007, http://www.worldcoa1.org/pages/content/index.asp?PageID=104

Zaccaria M., 2005. L'oro dell'Eritrea, 1897–1914. Africa, LX, 1, pp. 65–110 (in italian).

Zehner R.E., Gustin M.S., 2002. Estimation of Mercury Vapor Flux from Natural Substrate in Nevada. Environmental Science & Technology, 36: 4039–4045.

ZMWG, 2007. Comments on the planned content of the atmospheric emissions report. Zero Mercury Working Group (ZMWG). Available at: www.chem.unep.ch/mercury

Chapter 2
Mercury Emissions from Coal Combustion in China

David G. Streets, Jiming Hao, Shuxiao Wang, and Ye Wu

Summary This chapter reviews the magnitude and spatial distribution of mercury emissions from coal combustion in China. Due to the large quantities of coal burned and the relatively low level of technology, particularly in industry, emissions are high. Emissions were stable at about 200-210 Mg during the period 1995-2000, but because of rapid economic growth starting in 2001, mercury emissions grew quickly to a value of 334 Mg in 2005. The annual average growth rate for the period 1995-2005 was 5.1%. The uncertainty in emission estimates is about ±35% (95% confidence intervals). Emissions are concentrated in those provinces with high concentrations of mercury in coal (like Guizhou Province) and provinces in which a lot of coal is burned (like Shanxi Province). Because significant amounts of coal are burned in homes and small industrial facilities, without any kind of emission control at all, emissions of particulate mercury are higher in China than in the developed world; the speciation profile nationwide is: 64% $Hg^{(II)}$, 19% Hg(p), and 17% Hg^0. In the future, growth in mercury emissions is expected to be limited by the application of FGD for SO_2 control and other advanced technologies. Estimates of emissions are hampered by the lack of comprehensive and reliable emissions testing programs in China.

2.1 Introduction

Mercury pollution has been recognized by Chinese researchers and government officials for some time. However, it is only relatively recently that researchers have begun to quantify the releases of mercury and measure the concentrations of mercury in the air, water and land. The serious nature of the pollution levels in China has now begun to raise issues that could lead to regulation of mercury emissions in the future. Feng (2005), Jiang et al. (2006), and Zhang and Wong (2007) have summarized the state of knowledge about mercury pollution in China. In addition, concern has been raised about transport of mercury away from the Asian continent and its contribution to regional and hemispheric background levels (see, e.g., Friedli et al., 2004; Jaffe et al., 2005; Pan et al., 2006). Coal combustion and nonferrous metals smelting are

N. Pirrone and R. Mason (eds.), *Mercury Fate and Transport in the Global Atmosphere*, DOI: 10.1007/978-0-387-93958-2_2,

roughly equally responsible for mercury releases in China, supplemented by other industrial operations (Streets et al., 2005; Wu et al., 2006a). This chapter only addresses mercury releases from coal combustion. We discuss the contextual background for estimating emissions of mercury from coal combustion in China and present estimates for the period 1995 to 2005 with a forward glance to 2020.

2.2 Results and Discussion

2.2.1 Coal use Trends, 1995-2005

The major determinant of mercury emissions from coal combustion is the amount of coal burned. Coal consumption data for China are available by sector, coal type, and province from the China Energy Statistical Yearbooks (NBS, 1998-2005). It is important to distinguish between coal that is combusted directly and coal that is diverted to other uses, because this has major implications for mercury release rates. Trends in total raw coal consumption are shown in Figure 2.1. In 1995, total raw coal consumption was 1460 Tg, of which the industrial sector consumed 482 Tg (33%) for direct combustion, slightly more than the power sector, 446 Tg (31%). The residential sector consumed 138 Tg (9%). The remaining 27% of coal was used for coal washing, coking, industrial feedstocks, briquettes, and miscellaneous types of combustion. Coal use declined during the period 1996-1999 due to a variety of economic and other reasons (see, e.g., Sinton and Fridley, 2000), but subsequently began to increase quickly, as the economy of China underwent rapid expansion. By 2005, total raw coal consumption had risen to 2650 Tg, of which the major contributing sectors were: power plants 1050 Tg (40% of total), industrial combustion 718 Tg (27%), and residential use 138 Tg (5%).

Among the major coal-consuming sectors, the power sector was the leading sector in total coal growth, increasing by an average of 8.9% annually during the period 1995-2005. The industrial coal-combustion sector showed a moderate increase in coal use, 4.1% annually. Coal use in the residential sector was the same in 2005 as in 1995 in absolute terms, meaning that its share had been slowly decreasing (-0.1% per yr) mainly due to fuel transitions to cleaner gaseous and liquid fuels. Other uses of coal have grown as well, notably a tremendous annual-average growth rate of 17.8% in the use of coal as industrial feedstock, mostly achieved during the past five years. Figure 2.2 shows how the various uses of raw coal in the industrial sector have changed during the period 1995-2005.

2.2.2 Mercury in Coal

A reliable determination of the average or typical concentration of mercury in Chinese coals by province or nationwide is hampered by the innate heterogeneity of mercury in coal, as well as the relative paucity and unrepresentativeness of

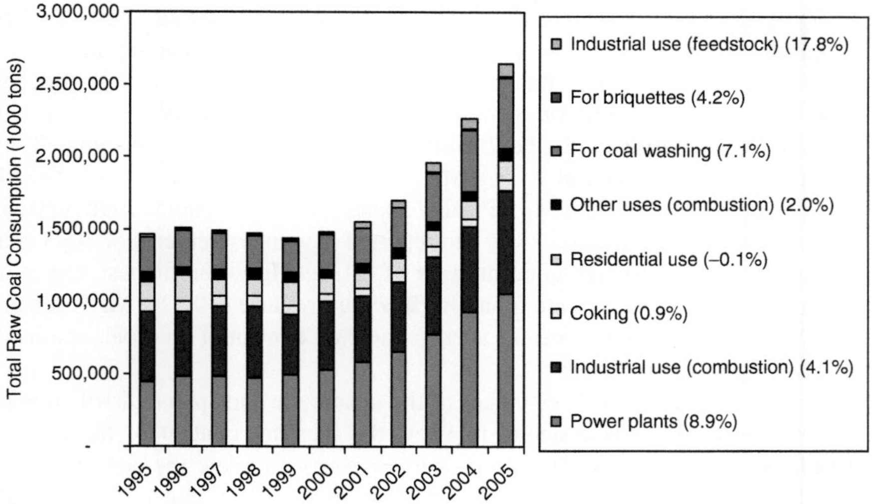

Figure 2.1 Trends in total raw coal consumption in China, 1995-2005; annual-average growth rates for the entire period are shown in the caption

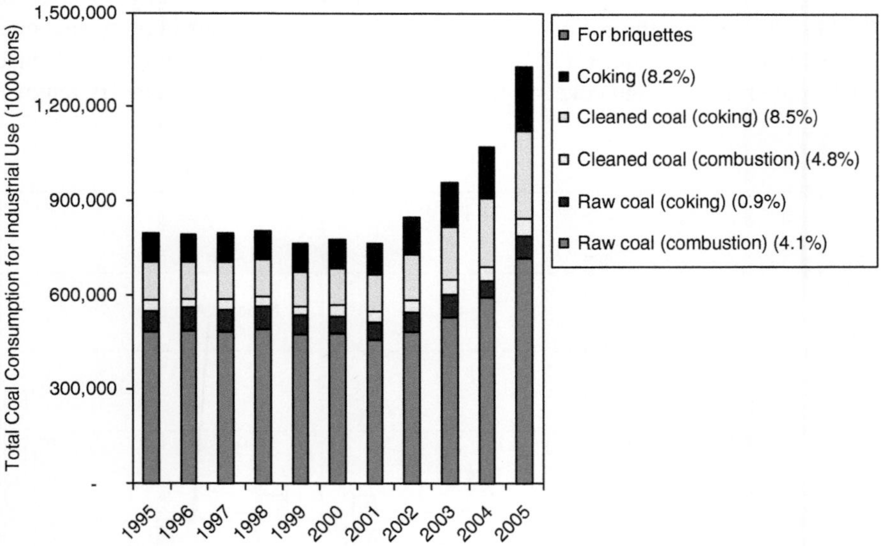

Figure 2.2 Trends in industrial raw coal consumption in China, 1995-2005; annual-average growth rates for the entire period are shown in the caption

measurements. Early Chinese studies on mercury emissions relied on limited sampling data. Wang et al. (1999, 2000) and Zhang et al. (2002) used an average value for the mercury concentration of Chinese coals of 0.22 g Mg^{-1}, with a wide range of 0.02-1.92 g Mg^{-1}, based on samples from fourteen provinces. Other estimates from the Chinese literature are 0.15 g Mg^{-1} (Huang and Yang, 2002) and 0.16 g Mg^{-1}

(Zhang et al., 1999). All these estimates were based on sampling of raw coal in coal-producing areas. An advancement in our understanding of mercury in Chinese coals occurred through an initiative by the U.S. Geological Survey, as part of the World Coal Quality Inventory, to measure about 300 coal samples from around China in collaboration with the Institute of Geochemistry in Guiyang. They obtained an average value of 0.15 g Mg^{-1} with a 1Σ standard deviation of 0.14, within a range of <0.2–0.69 g Mg^{-1}. Finally, Zheng et al. (2007) summarized previous studies of mercury in Chinese coals and reported new measurements of 1699 coal samples, having an average concentration of 0.19 g Mg^{-1}. The highest values of mercury content in raw coal are found in Guizhou Province (~0.52 g Mg^{-1}) (Zheng et al., 2007). Figure 2.3 presents the average mercury content of raw coal, as mined, for coal-producing provinces.

In order to obtain reliable estimates of the magnitude and spatial distribution of mercury emissions, it is essential to know the mercury content of the coal as burned, not just as mined. Therefore, it is necessary to relate the coal produced (mined) in particular provinces to its consumption in each province. Streets et al. (2005) and Wu et al. (2006a) developed a coal transportation matrix to link coal production to coal consumption. Using a merged data set from the USGS data and the Chinese literature data, they determined that the average mercury content of coal as burned was 0.18-0.19 g Mg^{-1}, varying very slightly in the range of 0.180 to 0.189 g Mg^{-1} during the period 1995 to 2003, due to fluctuations in provincial coal production. Streets et al. (2005) also calculated the mercury content of cleaned coal, coal briquettes, and coke, as produced. They assumed an average Hg removal

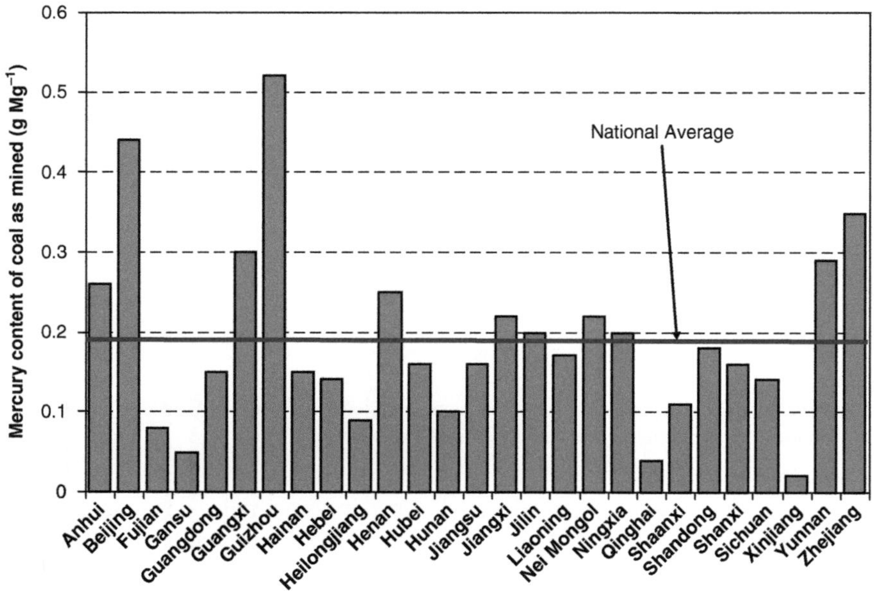

Figure 2.3 Mercury content of raw coal, as mined

efficiency for coal cleaning of 30% that is independent of the mercury content; during this period about 16% of total coal was cleaned in China. It was further assumed that 10% of the mercury contained in a given coal remains in coke after the coking process. Because there is no evidence of mercury removal during the briquette production process, it was assumed that 100% of the mercury in the raw coal or cleaned coal is transferred to the briquettes. Further tests on these and other coal-derived products are clearly called for.

2.3 Mercury Released to the Atmosphere

Because mercury release rates and the speciation profiles depend greatly on combustion technology, combustion conditions, and emission control technology, it is necessary to define Chinese coal utilization practices rather carefully. Streets et al. (2005) and Wu et al. (2006a) developed a model containing 65 individual source types for coal combustion, 22 of which are for coal-fired power plants, 30 for industrial use, nine for residential use, and four for other uses. The partitioning of each combustion technology/control device/fuel type by province and sector over time is built into their model based on a wide literature review.

In the past decade, the installation of particulate matter (PM) control devices in boilers has increased significantly in China, especially in the power sector. Since the mid-1980s, electrostatic precipitators (ESP) increased their share by 4-5% annually, to replace wet particle scrubbers and cyclones in power plants. Now the share of ESP installation in the total coal-fired power capacity is about 95% nationwide. However, in the industrial sector, the penetration of PM control installation lags behind. Although installation of wet particle scrubbers increased during the past decade, the fraction of industrial coal use without any PM control device is still large at present, close to 30%. The reasons are: (a) a large number of small boilers are scattered throughout China, especially in the poorer and more remote provinces such as Guizhou and Yunnan, without PM control; and (b) coke ovens, consuming a large amount of raw coal and clean coal, are generally without PM control. Since the mid 1990's, flue-gas desulfurization (FGD) also began to be installed in power plants to reduce SO_2 emissions. In 1995, FGD installed capacity was only 0.7 GW, rising to 5 GW in 2000; however, by the end of 2005, the FGD capacity had reached 53 GW, mostly in Sichuan (including Chongqing), Beijing, Shandong, Guangdong, Heilongjiang, Jiangsu, and Zhejiang Provinces. It is essential to reflect the rapidly changing mix of technologies in the coal-consuming sectors of China in calculations of mercury emissions over time. Figure 2.4 illustrates how the mix of PM controls changed during the period 1995-2003 in the power and industrial sectors. Residential use is also an important coal-consuming sector in China, representing 7% of raw coal, 5% of cleaned coal, and 90% of briquettes in 1999. Traditional cookstoves and improved cookstoves are the major combustion types for residential cooking and heating, both of which are without any PM control device. In the big cities, however, many residents obtain heat from centralized heating systems that

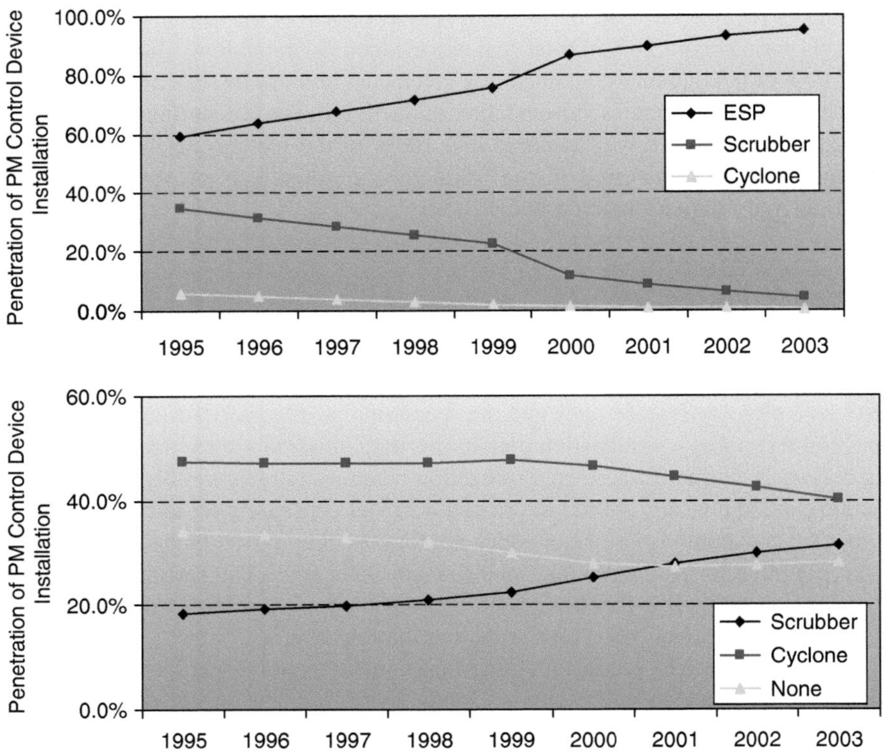

Figure 2.4 Time development of the penetration of PM control devices in China in the power sector (upper) and the industrial sector (lower), 1995-2003

use mid- or large-sized boilers. This part of the coal consumption for residential heating use assumes the use of stoker boilers with cyclone controls. For farming, construction, transportation, and commerce, the coal consumption is combined and assumes the use of small stokers without any PM control.

The typical scheme for calculating mercury emissions from coal combustion is illustrated in Figure 2.5. A fraction of the mercury contained in the fuel is not emitted to the air but is retained in the bottom ash and disposed of as solid waste. The share of Hg remaining in the bottom ash is different for different boiler types. Studies in China (Huang et al., 2003; Zhu et al., 2002) indicate that only 1-2% of Hg remains in the bottom ash for pulverized coal (PC) boilers in power plants; however, the ratio may increase to 7-9% for industrial PC or fluidized-bed boilers and 17-18% for industrial stoker-fired boilers (Wang et al., 2000; Wang and Ma, 1997).

The control technologies used to reduce traditional air pollutant emissions (e.g., PM and SO_2) from coal-combustion boilers also remove some of the mercury from the flue gas; however, the removal efficiencies vary widely. Until very recently, there were few measurements of mercury removal efficiencies for Chinese boilers. Wang et al. (1999, 2000) reported from measurements on two power plants in

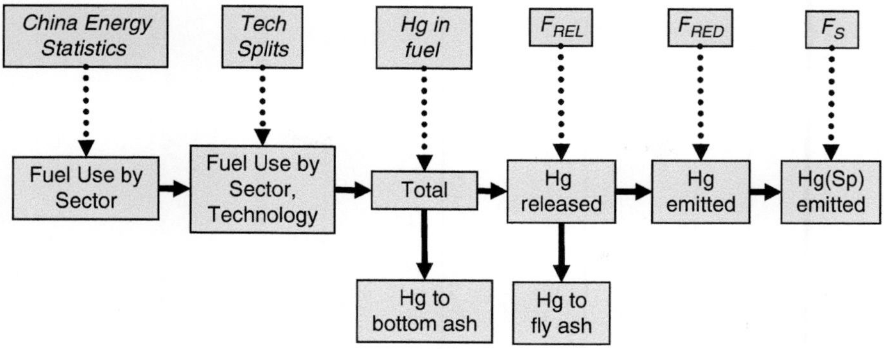

Figure 2.5 Calculation procedure for mercury emissions; F_{REL} = fraction released to the air during combustion; F_{RED} = fraction reduced by emission control devices; F_S = fraction emitted by species type

Changchun that mercury collection efficiency averaged 26% within a range of 7-47%. Zhu et al. (2002) suggested the following mercury removal efficiencies of the three predominant types of PM control devices installed in boilers in China: (a) ESP has a moderate removal efficiency of ~30%; (b) wet PM scrubbers show very little benefit, with mercury removal efficiency of ~4-8%; and (c) cyclones remove essentially no mercury (<0.1%). However, there is very little information about mercury removal efficiencies on devices other than ESPs on PC power plants in China.

Recently, programs of testing mercury emissions from Chinese sources have begun at Zhejiang University, Tsinghua University, and the Institute of Geochemistry in Guiyang, and studies are beginning to be published in the peer-reviewed literature. Tang et al. (2007), Chen et al. (2007), Yi et al. (2008), and Zhou et al. (2007) have all reported test data, including investigation of the roles of chlorine and ash in mercury release, but further work is needed to digest these results and generalize them to the population of source types in China.

2.4 Emission Trends in China

Wang et al. (1999, 2000) and Zhang et al. (2002) were the first to report mercury emissions from coal combustion in China, citing a value of 213.8 Mg for the year 1995. They further reported an annual average growth rate of ~4.8% a year for emissions in the 17 years prior to 1995, rising from ~95 Mg in 1980 to ~160 Mg in 1990.

Predicted emissions for 2000 were 273 Mg. Streets et al. (2005) conducted a detailed examination of mercury emissions from all sources and reported a value of 202.4 Mg for mercury from coal combustion in 1999, 38% of total emissions of mercury in China (535.8 Mg). Wu et al. (2006a), using the same methodology and data as in Streets et al. (2005), developed an emission trend from 1995 to 2003.

This trend incorporates the coal consumption and technology trends presented earlier. The results of that study are presented in Table 2.1 and Figure 2.6.

Table 2.1 Mercury emissions from coal combustion (*Mg* yr^{-1})

	1995	1996	1997	1998	1999	2000	2001	2002	2003	2004	2005	AAGR (%)
Power plants	63.4	68.7	67.2	66.2	67.8	70.1	76.3	84.2	100.1	114.3	124.8	7.0
Industrial use	104.7	106.3	107.8	108.3	103.2	104.2	101.9	109.9	124.3	136.4	169.4	4.9
Residential use	23.1	23.5	22.7	21.5	19.7	19.6	19.9	19.7	21.7	23.7	26.2	1.3
Other uses	11.2	10.8	10.5	11.6	11.5	10.5	10.7	11.8	10.6	11.7	13.6	2.0
Total	202.4	209.3	208.2	207.6	202.2	204.4	208.8	225.6	256.7	286.1	334.0	5.1

Data for 1995-2003 are from Wu et al. (2006a). This trend has been extended to 2004 and 2005 using the same methodology and data sources.
AAGR = Average Annual Growth Rate.

Additionally, for this present report, the trend has been extended from 2003 to 2005.

Wu et al. (2006a) find that mercury emissions from coal combustion increased from 202.4 Mg in 1995 to 334.0 Mg in 2005, an annual-average growth rate of 5.1%. The largest growth in mercury emissions (7.0% per year) has been in the power sector, consistent with the growth in coal combustion in the power sector, from 63.4 Mg in 1995 to 124.8 Mg in 2005.

Emissions from industrial coal combustion have grown by 4.9% per year, from 104.7 Mg in 1995 to 169.4 Mg in 2005. A formal uncertainty analysis has been conducted on these estimates, following the method described in Streets et al. (2003), and is shown in Figure 2.7. The 95% confidence intervals are approximately ± 35%, changing little over the time period.

The mercury emissions have also been speciated as described in Streets et al. (2005) and Wu et al. (2006a) across all emitting source types. The net result for coal combustion – varying over all sectors, source types and technologies – is shown in Figure 2.8 by province. This figure shows that the fraction of mercury emitted in particulate form is particularly high in Guizhou Province and to a lesser extent in Qinghai and Xinjiang Provinces.

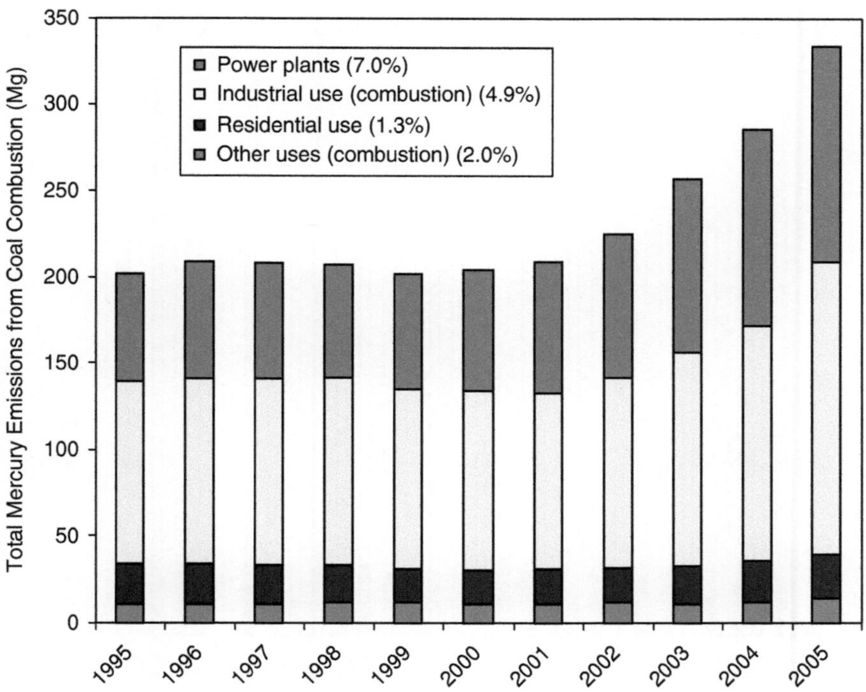

Figure 2.6 Trends in mercury emissions in China, 1995-2005; annual-average growth rates for the entire period are shown in the caption

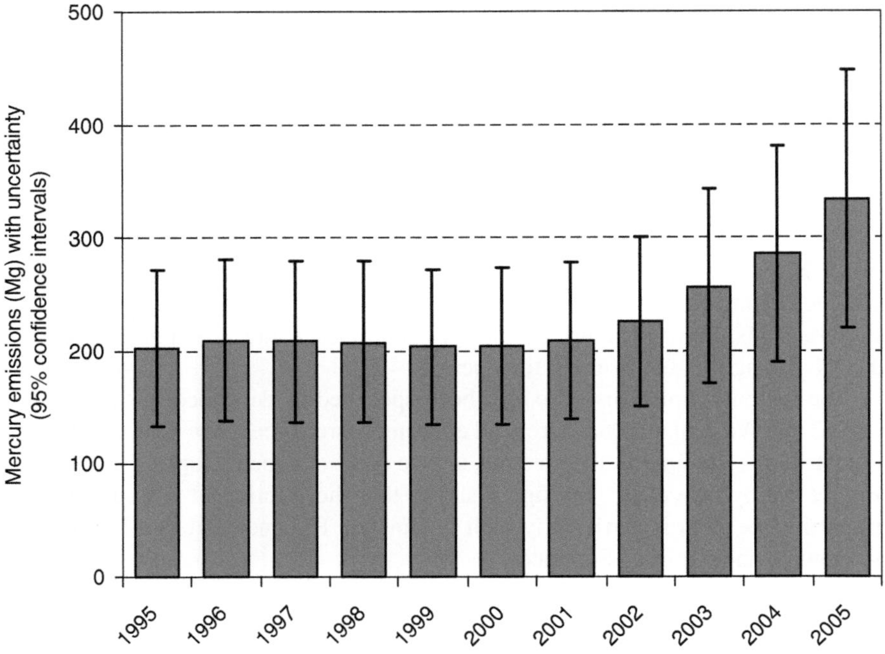

Figure 2.7 Uncertainty in mercury emission estimates for coal combustion, as 95% confidence intervals

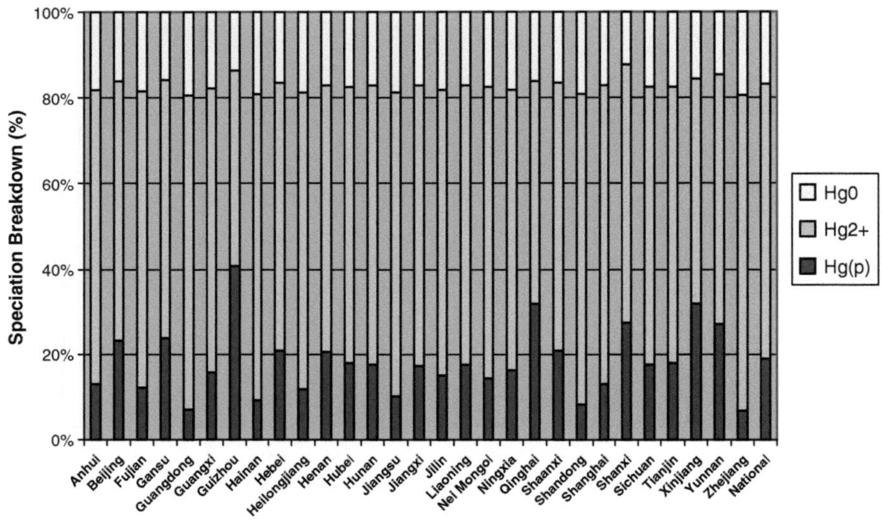

Figure 2.8 Speciation of mercury emitted from coal combustion in 1999, by province

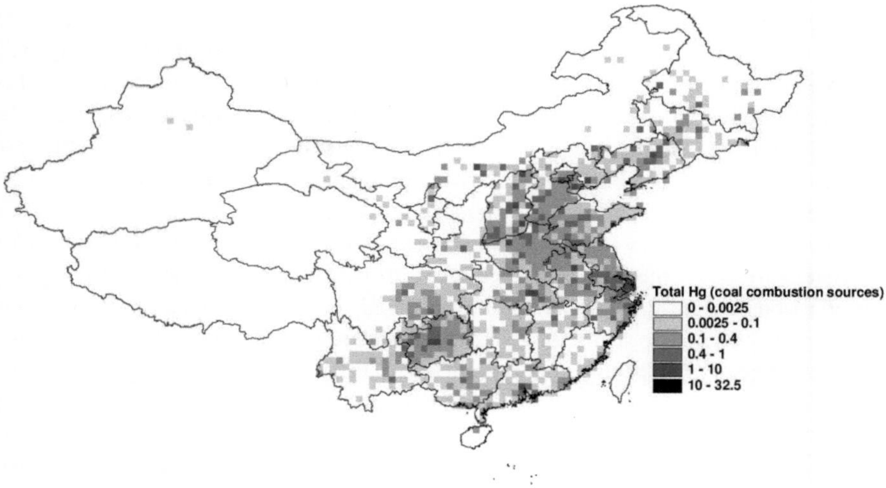

Figure 2.9 Gridded mercury emissions from coal combustion for the year 1999 at 30 min × 30 min spatial resolution (units are *Mg yr*$^{-1}$ per grid cell)

This is a combination of high Hg content of coals and extensive use of coal in small, uncontrolled facilities. In the developed provinces – Guangdong, Shanghai, and Zhejiang, for example – particulate mercury releases are low. For the nation as a whole, the average speciation profile for mercury from coal combustion is: 64% $Hg^{(II)}$, 19% Hg(p), and 17% Hg^0. It should be noted that the mercury speciation profiles used thus far have relied on western data sources and are subject to change when new Chinese test data become available.

Streets et al. (2005) showed that the three provinces emitting the largest amounts of mercury from coal combustion in 1999 were Guizhou (18 Mg), Shanxi (15 Mg), and Henan (14 Mg) areas with heavy coal use and relatively low levels of technology. Figure 2.9 presents the spatial distribution of mercury emissions from coal combustion in 1999 at 30 min × 30 min resolution.

2.5 Future Mercury Emissions from Coal Combustion

Though mercury emissions from coal combustion have grown dramatically since 2001, there is hope for a change in the trend through the expected implementation of FGD on power plants. China announced in its 11th Five-Year Plan a renewed and concerted effort to control SO_2 emissions, intending to achieve a 10% reduction in 2010 emission levels relative to 2005. This goal will mostly be achieved by the installation of FGD units on a large number of power plants.

Reduction targets have been agreed upon with provincial governments and power companies, favorable electricity rate pricing and loans have been granted, and SEPA has been given ministerial status (Ministry of Environmental Protection) with greater enforcement powers. Even by 2006, the installed FGD capacity had doubled relative to 2005, from 53 GW to 104 GW. Figure 2.10 shows that implementation of FGD is expected to reach 58% nationwide by 2010 and 67% nationwide by 2020 starting with the developed coastal provinces and then spreading to the rest of the country. Because FGD also removes some mercury along with the SO_2, there will be a significant co-benefit for mercury reduction.

The key question is whether the implementation of FGD on power plants will be sufficient to offset the expected continued growth in power generation and coal combustion. Figure 2.11(a) shows the expected growth in power generation and coal use in the power sector out to 2020. Fast growth continues, with electricity generation growth outpacing coal growth due to improvements in energy efficiency. Coal use is expected to reach 1290 Tg in 2010 and 1770 Tg in 2020; electricity use rises to 2.62 billion MWh in 2010 and 3.80 billion MWh in 2020. Although more testing is needed to determine typical mercury removal efficiencies in Chinese power plants, we assume 74% reduction from an ESP + FGD configuration.

With this assumption, annual mercury emissions from coal-fired power plants are effectively held to about 2003 levels by 2010 (approximately 105 Mg) (Wu et al., 2006b). About 61 Mg of mercury emissions are avoided through FGD in this scenario, as shown in Figure 2.11(b). Emissions thereafter would begin to rise again, but with a modest additional investment in selective catalytic reduction (SCR) and activated carbon injection (ACI) technologies, 2020 emissions could also be held to the level of 100 Mg or thereabouts (Wu et al., 2006b). So the prospect of stabilizing mercury emissions from power plants is at hand.

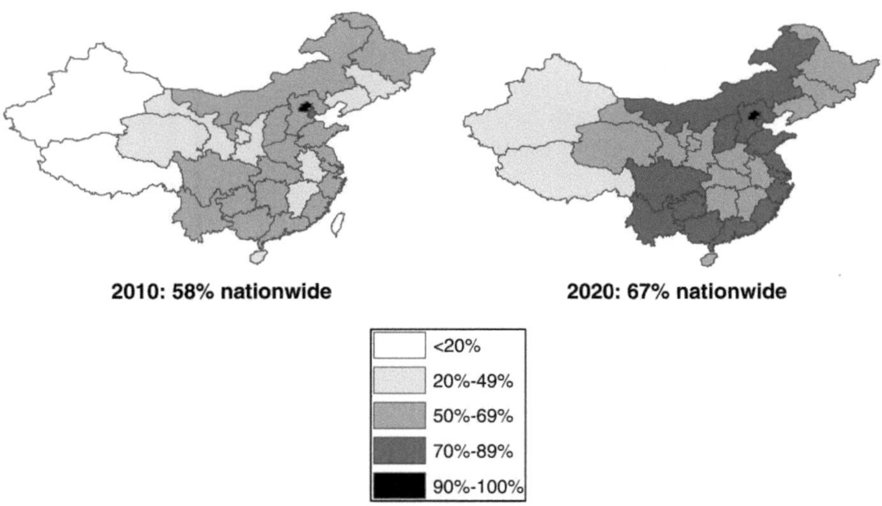

Figure 2.10 Expected extent of FGD implementation on coal-fired power plants in China in 2010 and 2020, showing percentage implementation rates in each province

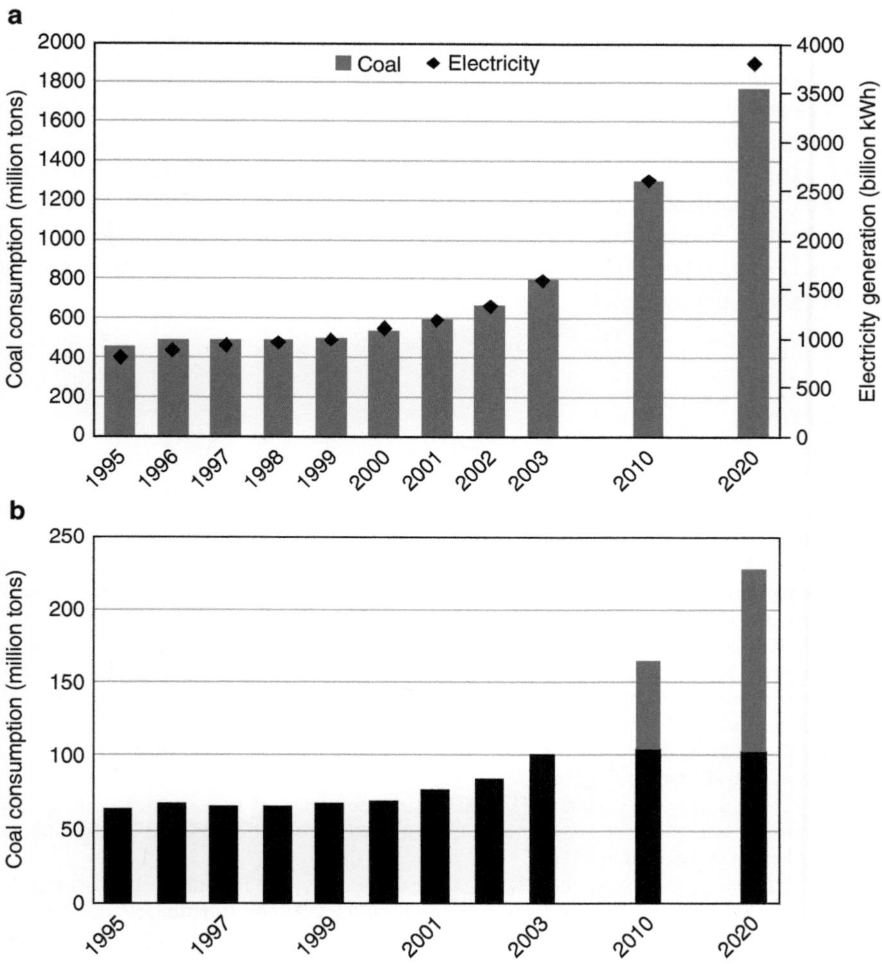

Figure 2.11 (a) Anticipated growth in power generation and coal use in power plants out to 2020; and (b) the effect of FGD and other controls on future mercury emission levels (grey), showing avoided emissions through application of emission control technology (black)

2.6 Future Research and Policy Implications

The existence of a mercury pollution problem has been known in China for several decades, but it is only recently that quantification has been attempted. Estimates of mercury emissions from coal combustion were first made about ten years ago, and they have improved over time. Considerable progress has been made in understanding the mercury content of Chinese coals. However, to a large extent, emissions quantification has had to rely on technology performance data as measured in the West. This is a considerable drawback, as we are not at all sure that facilities in

China achieve the same level of performance as they do in the West-and many of the special Chinese technologies have never been sampled at all. It was only in 2007 that a number of papers were published by Chinese researchers reporting on field testing of mercury emissions and collection in Chinese plants. Over the next few years, these results need to be collated, compared with western data, extrapolated to the whole of China, and supplemented with additional test data. In particular, the mercury collection efficiencies of PM control devices and FGD need to be refined when burning coals with the special chlorine and ash specifications of Chinese coals. When this has been accomplished, we will be able to have greater confidence in the estimates.

References

Chen, L., Duan, Y., Zhuo, Y., Yang, L., Zhang, L., Yang, X., Yao, Q., Jiang, Y., Xu, X., 2007. Mercury transformation across particulate control devices in six power plants of China: The co-effect of chlorine and ash composition. Fuel, 86, 603-610.

Feng, X., 2005. Mercury pollution in China-an overview. In: Dynamics of Mercury Pollution on Regional and Global Scales: Atmospheric Processes, Human Exposure Around the World, N. Pirrone and K. Mahaffey (Editors), Springer Publishers, Norwell, MA, USA., 657-678.

Friedli, H.R., Radke, L.F., Prescott, R., Li, P., Woo, J.-H., Carmichael, G.R., 2004. Mercury in the atmosphere around Japan, Korea, and China as observed during the 2001 ACE-Asia field campaign: measurements, distributions, sources, and implications. J. Geophys. Res., 109, D19S25, doi:10.1029/2003JD004244.

Huang, W., Yang, Y., 2002. Mercury in coal in China. Coal Geology of China, 14, 37-40 (in Chinese with abstract in English).

Huang, Y., Jin, B., Zhong, Z., Zhou, H., 2003. Study on groups of trace elements during coal combustion. Journal of Southeast University, Natural Science Edition, 33, 148-152 (in Chinese with abstract in English).

Jaffe, D., Prestbo, E., Swartzendruber, P., Weiss-Penzias, P., Kato, S., Takami, A., Hatakeyama, S., Kajii, Y., 2005. Export of atmospheric mercury from Asia. Atmos. Environ., 39, 3029-3038.

Jiang, G.-B., Shi, J.-B., Feng, X.-B., 2006. Mercury pollution in China. Environ. Sci. Technol., 40, 3672-3678.

NBS, 1998-2007. China Energy Statistical Yearbooks (1991-2005). National Bureau of Statistics of China, China Statistics Press, Beijing, P.R. China.

Pan, L., Woo, J.-H., Carmichael, G.R., Tang, Y., Friedli, H.R., Radke, L.F., 2006. Regional distribution and emissions of mercury in east Asia: A modeling analysis of Asian Pacific Regional Aerosol Characterization Experiment (ACE-Asia) observations. J. Geophys. Res., 111, D07109, doi:10.1029/2005JD006381.

Sinton, J.E., Fridley, D.G., 2000. What goes up: Recent trends in China's energy consumption. Energy Policy, 28, 671-687.

Streets D.G., Bond T.C., Carrnichael G.R., Fernandes S.D., Fu Q., He D., Klimont Z., Nelson S.M., Tsai N.Y., Wang M.Q., Woo J.-H., Yarber K.F., 2000. An inventory of gaseous and primary aerosol emissions in Asia in the year 2000. J. Geophys. Res., 108, D21, 8809, doi:10.1029/2002JD003093.

Streets, D.G., Hao, J.M., Wu, Y., Jiang, J.K., Chan, M., Tian, H.Z., Feng, X.B., 2005. Anthropogenic mercury emissions in China. Atmos. Environ., 39, 7789-7806.

Tang, S., Feng, X., Qiu, J., Yin, G., Yang, Z., 2007. Mercury speciation and emissions from coal combustion in Guiyang, southwest China. Environmental Research, 105, 175-182.

Wang, Q., Ma, R., 1997. The mercury in coal and its cinder. China Environ. Sci., 17, 76-79 (in Chinese with abstract in English).

Wang, Q., Shen, W., Ma, Z., 1999. The estimation of mercury emission from coal combustion in China, China Environ. Sci., 19, 318-321 (in Chinese with abstract in English).

Wang, Q., Shen, W., Ma, Z., 2000. Estimation of mercury emission from coal combustion in China, Environ. Sci. Technol., 34, 2711-2713.

Wu, Y., Wang, S., Streets, D.G., Hao, J., Chan, M., Jiang, J., 2006a. Trends in anthropogenic mercury emissions in China from 1995 to 2003. Environ. Sci. Technol., 40, 5312-5318.

Wu, Y., Wang, S., Streets, D.G., Hao, J., Chan, M., 2006b. Mercury emissions from China: Current status and future trend. In: Proceedings of the 2006 Air & Waste Management Association 99th Annual Conference, New Orleans, LA.

Yi, H., Hao, J., Duan, L., Tang, X., Ning, P., Li, X., 2008. Fine particle and trace element emissions from an anthracite coal-fired power plant equipped with a bag-house in China. Fuel, 87, 2050-2057.

Zhang, J., Ren, D., Xu, D., Zhao, F., 1999. Mercury in coal and its effect on environment. Advances in Environmental Science, 7, 100-104 (in Chinese with abstract in English).

Zhang, M.Q., Zhu, Y.C., Deng, R.W., 2002. Evaluation of mercury emissions to the atmosphere from coal combustion, China. Ambio, 31, 482-484.

Zhang, L., Wong, M.H., 2007. Environmental mercury contamination in China: sources and impacts. Environ. International, 33, 108-121.

Zheng, L., Liu, G., Chou, C.-L., 2007. The distribution, occurrence and environmental effect of mercury in Chinese coals. Sci. Total Environ., 384, 374-383.

Zhou, J., Luo, Z., Hu, C., Cen, K., 2007. Factors impacting gaseous mercury speciation in post-combustion. Energy & Fuels, 21, 491-495.

Zhu, Z., Xu, L., Tan, Y., 2002. Research on characteristics of mercury distribution in combustion products for a 300MW pulverized coal fired boiler. Power Engineering, 22, 1594-1597 (in Chinese with abstract in English).

Chapter 3
Mercury Emissions from Industrial Sources in China

Xinbin Feng, David Streets, Jiming Hao, Ye Wu, and Guanghui Li

Summary In this chapter, we reviewed mercury emissions from industrial sources in China. The industrial sources included fuel oil for stationary sources; gasoline; diesel and kerosene; biofuel combustion; Grassland/ savannah burning; waste and residue burning; cement production; iron and steel production; caustic soda production; non-ferrous metal smelting (Zn, Pb, Cu, and Au); mercury mining; and battery and fluorescent lamp production. Mercury emission factors from most source categories were obtained according to measurement data from Europe and North America. The mercury emission factor for zinc smelting, which was believed to be the largest industrial source, was adopted from the data of recent studies in China. We used the information published in the literature to estimate the emission of different mercury species. The total mercury emission from industrial sources in China was 253.07 Mg in 1999. Non-ferrous metal smelting (including zinc, lead, copper and gold smelting) is the largest industrial mercury emission source in China and the total mercury emissions reached 167.8 Mg. The total mercury emissions from industrial sources in China in 1995 was 296.4 Mg, increasing to 360.5 Mg in 2003, at an average annual growth rate of 2.90%. Due to lack of field measurement data to quantify mercury emission factors for most of industrial sources, a large uncertainty is associated with the current emission inventory. A number of studies need to be undertaken to reduce the uncertainties. Surveys are needed to evaluate mercury contents in raw materials of different industrial categories. Mercury balance studies are necessary for representative plants of different industrial sources. The speciation of mercury emissions from different industrial sources are also urgently needed in order to better understand the atmospheric fate of mercury emitted from these sources.

3.1 Introduction

Global emissions of anthropogenic mercury to the atmosphere have been estimated to be 1900 Mg in 1995 (Pacyna and Pacyna, 2002), of which 77% was from coal combustion, with the remainder divided among non-ferrous metals production, cement production, and waster disposal. However, the scenario of mercury emissions from China may be quite different. Mercury emissions from industrial sources

N. Pirrone and R. Mason (eds.), *Mercury Fate and Transport in the Global Atmosphere*,
DOI: 10.1007/978-0-387-93958-2_3,

other than coal combustion contribute a significant portion of the total emissions (Streets et al., 2005; Wu et al., 2006).

China has been regarded as one of the largest atmospheric mercury emission sources from a global perspective (Pirrone et al., 1996; Pacyna and Pacyna, 2002) due to rapid economic development. The increasing mercury emissions in China have resulted in the elevation of mercury concentrations in ambient air both in urban and rural areas in China. Average concentrations of total gaseous mercury (TGM) in Guiyang, Guizhou Province, have been measured in the range of 5–15 ng m^{-3} (Feng et al., 2002; 2003; 2004a; 2004b), attributed to uncontrolled coal-burning in the residential and industrial sectors. In Beijing, Liu et al. (2002) measured TGM concentrations in the range of 6–10 ng m^{-3} during winter. Fang et al. (2001) measured average particulate mercury concentrations of about 0.5 ng m^{-3} in North eastern Changchun City, Jilin Province, rising to as high as 2 ng m^{-3} during the heating season. Xiu et al. (2005) measured somewhat lower levels of Hg in Total suspended particulate (TSP) in Shanghai, in the range of 0.2–0.5 ng m^{-3}. The average TGM concentrations in rural areas such as the Gongga Mountain area in South western China (Fu et al., 2008) and the Changba Mountain area in North eastern China (Wan et al., 2008) reached 3.98 and 3.22 ng m^{-3}, respectively, which were significantly elevated compared to the values of 1.5 to 2.0 ng m^{-3} measured in rural areas in Europe and North America (Ebinghaus et al., 2002; Schroeder et al., 2001; Lindberg et al., 2007). These measurement data demonstrated that mercury emissions from anthropogenic activities have resulted in elevated TGM concentrations in ambient air in China. There are, however, tremendous uncertainties in mercury emission inventory estimates for China simply because of the lack of direct measurement data to establish reliable emission factors for different anthropogenic sources. Currently, the mercury emission factors from different anthropogenic sources are generally adopted from the data obtained from the studies conducted in Europe and North America. However, mercury measurements of major emission sources in China are known to have been taken or are underway, even if the data are, as yet, unpublished and unavailable, e.g., the cement plant measurements made by US-EPA in cooperation with Chinese entities and others, and power plants measurements made by US-DOE and Chinese universities and authorities. In Chapter 2 of this report, Streets et al. (2008) reviewed mercury emission from the coal combustion sector in China, while mercury emissions from industrial sources will be evaluated in this chapter.

The major industrial mercury emission sources in China include fuel oil for stationary sources; gasoline; diesel and kerosene; biofuel combustion; Grassland/savanna burning; waste and residue burning; cement production; iron and steel production; caustic soda production; non-ferrous metal smelting (Zn, Pb, Cu, and Au); mercury mining; and battery and fluorescent lamp production.

A model has been developed to calculate mercury emissions from different industrial sources in China (Streets et al., 2005; Wu et al., 2006). The basic concept of the mercury emissions calculation is described by the following equation:

$$E_t = \sum \sum \left[ef_{i,j,t} A_{iij,j,t} F_{RELj,t} \left(1 - F_{REMj,t}\right)\right] \quad (1)$$

where E_t is the mercury emission; $ef_{i,j,t}$ is emission factor for other fuels or non-combustion processes; $A_{i,j,t}$ is the amount of fuel consumption or production yield of non-combustion processes; $F_{RELj,t}$ is the fraction of mercury emitted to the atmosphere; $F_{REMj,t}$ is the fraction of mercury removed by emission control devices; j is the type of combustion with/without control devices; i is the province; and t is the year.

3.2 Emission Factors from Different Industrial Sources in China

Table 3.1 lists mercury emission factors from various industrial activities in China and the majority of the data are adopted from Streets et al. (2005). Mercury emission factors from the zinc smelting sector which is currently believed to the one of the largest

Table 3.1 Emission factors for total Hg from industrial sources in China

Source category	Unit	Emission factor
1. Fuel oil for stationary sources (e.g., power plants, industrial use)	$g\ Mg^{-1}$ oil	0.014[a]
2. Gasoline, diesel, and kerosene	$g\ Mg^{-1}$ oil	0.058[a]
3. Biofuel combustion	$g\ Mg^{-1}$ biofuel	0.020[b]
4. Grassland/savanna burning	$g\ Mg^{-1}$ grass burning	0.080[c]
5. Waste and Residue Burning		0.037[d]
Agricultural residue	g Mg^{-1} residue 0.037[d]	2.80[e]
Household waste	g Mg^{-1} waste 2.80[e]	
6. Cement production	$g\ Mg^{-1}$ cement	0.040[f]
7. Iron and steel production	$g\ Mg^{-1}$ steel	0.04[g]
8. Caustic soda production	$g\ Mg^{-1}$ caustic soda	20.4[h]
9. Non-ferrous metal smelting		
Zinc (Zn)	g Mg^{-1} Zn	5.7-155[i]
Copper (Cu)	g Mg^{-1} Cu	9.6[j]
Lead (Pb)	g Mg^{-1} Pb	43.6[j]
Gold (Au): large-scale production	$Mg\ Mg^{-1}$ Au	0.79[j]
Gold (Au): artisanal production	$Mg\ Mg^{-1}$ Au	15.0[k]
10. Mercury mining	$kg\ Mg^{-1}$ Hg	45.0[k]
11. Battery and fluorescent lamp production	$Mg\ Mg^{-1}$ Hg used	0.05[k]

[a]From US EPA (1995).
[b]From Friedli et al. (2003a).
[c]Average emission factor for forests is 0.113 g Mg^{-1} (Friedli et al., 2003b). We assume that grasslands are generally like forests in terms of long-term exposure to Hg, but with typically rather shorter lifetimes for Hg uptake. This value is therefore lowered to 0.080 g Mg^{-1} for grassland burning.
[d]From Friedli et al. (2003b).
[e]From UNECE/EMEP (2004).
[f]Coal related Hg emissions for cement production are excluded from this category. Energy intensity of 0.196 Mg of coal $Mg^{-1}$of cement produced (Zhou et al., 2003) is used here to adjust emission factor of 0.065 g Mg^{-1} of cement (US EPA, 1997) to 0.040 g Mg^{-1} of cement produced.
[g]From Pacyna and Pacyna (2002).
[h]From Qi et al. (2000).
[i]From Li (2007) and Feng et al. (2004).
[j]From Jiang (2004).
[k]From Qi (1997).

mercury emission sources in China is adopted from recent studies by Li (2007) and Feng et al. (2004). It is obvious that no field measurement data in China regarding mercury emission studies is currently available in the open literature for most source categories. Therefore, mercury emission factors from most source categories were obtained according to measurement data obtained in Europe and North America.

Zinc smelting processes in China can be divided into two major types, namely the pyro-metallurgic process (PMP) and the electrolytic process (EP). The pyro-metallurgic process can then be divided into four sub-types, such as the imperial smelting process (ISP), retort zinc smelting process (RZSP), electric zinc furnace (EZF), and artisanal zinc smelting process (AZSP). Total zinc production in China reached 2.71 million Mg in 2004. It is estimated (Jiang, 2006) that 71.8% of total zinc production is based on EP technology, 7.7%, 5.9%, 13%, and 1.6% of total zinc production used ISP, RZSP, EZF and AZSP techniques, respectively.

Using the mass balance method, Feng et al. (2004) calculated mercury emission factors from artisanal zinc smelting using both oxide and sulphide ores and the data are shown in Table 3.2. Applying the same method, Li (2007) investigated mercury emission factors from four large scale zinc smelters using EP with flue gas mercury removal devices and without flue gas mercury removal devices, RZSP and ISP techniques, respectively and one artisanal zinc smelter using oxide ores and the emission factors are listed in Table 3.2. We can see that mercury emission factors varied significantly with different smelting processes and decreased dramatically if mercury removal devices were applied for the smelters. Streets et al. (2005) and Wu et al. (2006) used an averaged emission factor of 86.6 g Mg^{-1} Zn to estimate mercury emission from zinc smelting sources. From Table 3.2, we can clearly see that in most cases mercury emission factors were much lower than the value that is currently used. Therefore, we have applied these mercury emission factors in Table 3.2 to estimate mercury emissions from different zinc smelting factories using different zinc smelting processes in China in this paper.

Table 3.2 Mercury emission factors from zinc smelting using different smelting processes in China

Methods	EF_{Hg} ($\mu g\ Mg^{-1}$)	Information Source
Artisanal Zn Smelting using oxide ore	75	Li, 2007
Artisanal Zn Smelting using oxide ore	79	Feng et al., 2004
Artisanal Zn Smelting using sulfide ore	155	Feng et al., 2004
Electrostatic Process (EP) with mercury removal device	5.7	Li, 2007
Electrostatic Process (EP) without mercury removal device	54	Li, 2007
Retort Zn Smelting Process RZSP)	34	Li, 2007
Imperial Smelting Process (ISP)	122	Li, 2007

3.3 Speciation of Mercury Compounds from Different Industrial Sources in China

Primary emissions are classified according to gaseous elemental mercury (Hg^0), divalent gaseous mercury ($Hg^{(II)}$), and particulate mercury (Hg(p)). Generally no field measurement data on the speciation of mercury from the industrial sources is available from open literature. Streets et al. (2005) and Wu et al. (2006) used measurements from Pacyna and Pacyna (2002), and Friedli et al. (2001, 2003a, b) for different industrial sources as shown in Table 3.3.

3.4 Emissions from Different Industrial Sources in China in 1999

Streets et al.(2005) estimated anthropogenic mercury emissions in China for the year 1999 and the total emission from industrial sources other than coal combustion reached 327.95 Mg. We recalculated the total mercury emissions from industrial sources according to the emission factors listed in Table 3.2 and the total mercury emission was 253.07 Mg as shown in Table 3.4. The difference between these data sets is from the estimate of mercury emission from zinc smelting. We used the new mercury emission factors to estimate mercury emissions from zinc smelting factories using different processing techniques according the studies by Li (2007) and Feng et al. (2004).

It can be seen that non-ferrous metal smelting (including zinc, lead, copper and gold smelting) is the largest industrial mercury emission source in China and the total mercury emissions reached 167.8 Mg, which constituted 66% of the total

Table 3.3 Speciation of total mercury for each major source type (as fraction of the total)

Source category	Hg^0	$Hg^{(II)}$	Hg(p)
1. Fuel oil for stationary sources (e.g., power plants, industrial use)[a]	0.50	0.40	0.10
2. Gasoline, diesel, and kerosene combustion[a,b]	0.50	0.40	0.10
3. Biofuel combustion[b]	0.96	0.00	0.04
4. Grassland/savanna burning[b]	0.96	0.00	0.04
5. Waste and residue burning[b]	0.96	0.00	0.04
6. Cement production[a]	0.80	0.15	0.05
7. Iron and steel production[a]	0.80	0.15	0.05
8. Caustic soda production[a]	0.70	0.30	0.00
9. Non-ferrous metal smelting[a]	0.80	0.15	0.05
10. Mercury mining[c]	0.80	0.15	0.05
11. Battery and fluorescent lamp production[c]	0.80	0.15	0.05

[a] From Pacyna and Pacyna (2002).
[b] From Friedli et al. (2001, 2003a, 2003b).
[c] Assumed to be the same profile as other non-combustion sources.

Table 3.4 Summary of Hg emission estimates (*Mg*) for industrial sources associated with fuel consumption and materials production and use in 1999

Source category	Fuel consumption or material yield (*Mg*)	Hg	Hg^0	$Hg^{(II)}$	Hg(p)
Fuel oil for stationary sources	33.8×10^6 [a]	0.47	0.24	0.19	0.05
Gasoline, diesel, and kerosene	96.8×10^6 [a]	5.61	2.81	2.25	0.56
Biofuel combustion	413.0×10^6 [b]	8.26	7.93	0.00	0.33
Grassland/savanna burning	52.1×10^6 [c]	4.17	4.00	0.00	0.17
Waste and residue burning		5.94	5.71	0.00	0.25
Agricultural residue	105.3×10^6 [c]	3.90	3.74	0.00	0.17
Household waste	0.7×10^6 [d]	2.05	1.96	0.00	0.08
Cement production	566.9×10^6 [e]	22.68	18.14	3.40	1.13
Iron and steel production	123.0×10^6 [e]	4.92	3.94	0.74	0.25
Caustic soda production	9.3×10^3 [d]	0.19	0.13	0.06	0.00
Non-ferrous metal smelting		167.8	134.23	25.17	8.39
Zinc (Zn)	1.7×10^6 [f]	73	58.4	10.95	3.65
Copper (Cu)	1.1×10^6 [f]	10.12	8.09	1.52	0.51
Lead (Pb)	0.9×10^6 [f]	40.08	32.06	6.01	2.00
Gold (Au): large scale	20.4 [d]	6.10	12.88	2.41	0.80
Gold (Au): artisanal	1.9 [g]	28.50	22.80	4.28	1.43
Mercury mining	195.0 [f]	8.78	7.02	1.32	0.44
Battery/fluorescent lamp production	485.0 [h]	24.25	19.40	3.64	1.21
Total		253.07	203.55	36.77	12.78

[a] From NBS (2001).
[b] From ECCCEY (2000).
[c] From Streets et al. (2003b).
[d] From Jiang (2004).
[e] From NBS (2000).
[f] From ECCNMI (2000).
[g] Artisanal gold smelting activities were officially banned in September 1996, but some mines continue to operate surreptitiously. In our study, we assume artisanal gold production in 1999 is 1.9 Mg, one-third of 1995 artisanal gold production (Feng, 2005).
[h] This is the amount of Hg used in battery and fluorescent lamp production (Jiang, 2004; Yang et al., 2003).

mercury emissions from all industrial sources excluding coal combustion. Zinc production in China is increasing significantly, reaching 1.7 million Mg in 1999. We estimated that total mercury emissions in 1999 were 73 Mg which is less than the value of 147.6 Mg estimated by Streets et al. (2005). Copper production in China was about 1.1 million Mg in 1999. Total mercury emissions from copper smelting were 10.12 Mg. Hg emissions from copper smelting are much lower than those of zinc smelting due to the use of a lower emission factor for copper smelting (9.6 g Mg^{-1} of copper produced) in the estimation, which mainly results from much lower mercury contents in copper concentrate ore than those in zinc concentrate ore. Lead production in China was about 0.9 million Mg in 1999. It is estimated that total mercury emissions from lead smelting were 40.08 Mg.

In 1999, the 15 largest lead smelting plants contributed 57% of total lead yield. Most of these large plants are located in Hunan, Yunnan, Henan, and Guangdong Provinces (Streets et al., 2005). Because mercury emissions from gold smelting using amalgamation technology are strongly affected by the size of the smelting plant, the gold smelting process is separated into two parts: large-scale gold smelting in industrial plants and small-scale artisanal gold smelting.

Amalgamation technology is gradually being phased out in the large-scale gold smelting plants. In 1999, only about 20 Mg of gold were produced from large-scale plants using amalgamation technology, which resulted in 16.10 Mg of total mercury emissions. Although artisanal gold production was small in 1999, mercury emissions were still large due to the high emission factor for this process (Feng, 2005). It is estimated that total mercury emissions from small artisanal gold smelting were 28.50 Mg in 1999. Artisanal gold smelting was officially banned in September 1996, though it persists in remote areas. It is difficult to get precise gold production estimates from these small activities, and the mercury emission estimates from this activity are subject to large uncertainties. Battery/fluorescent lamp production and cement production emitted 24.3 and 22.7 Mg of mercury, which constituted 10% and 9% of the total mercury emissions from all industrial sources except coal combustion, respectively. Other sources contributed about 15% of the of the total mercury emissions from all industrial sources excluding coal combustion.

As shown in Table 3.4, among the total emission of 253.1 Mg from various industrial sources, 203.55 Mg, 36.77 Mg and 12.8 Mg were emitted as Hg^0, $Hg^{(II)}$ and Hg(p), respectively. Gaseous Hg^0 is the major form of mercury emitted from industrial sources other than coal combustion, and it constituted 80% of total mercury emissions. Divalent gaseous mercury ($Hg^{(II)}$) and particulate mercury (Hg(p)) constituted 15% and 5% of total mercury emissions, respectively.

3.5 Mercury Emission Trends from 1995 to 2003

Wu and co-workers (2006) developed multiple-year inventories of anthropogenic mercury emissions in China from 1995 through 2003. After updating mercury emissions estimates from zinc smelting using the new emission factors (Li, 2007 and Feng et al. 2004), we also analysed mercury emission trends for the industrial sources from 1995 to 2003 as shown in Table 3.5.

The total mercury emissions from industrial sources in China in 1995 was 296.4 Mg, increasing to 360.5 Mg in 2003, at an average annual growth rate of 2.90%. Non-ferrous metals smelting operations are known to be one of the largest sources of mercury in China (Streets et al., 2005; Wu et al., 2006). Due to fast economic development in China, the demand for non-ferrous metals is increasing significantly. As a result, total mercury emissions from the non-ferrous metals smelting category (zinc, lead, copper and gold) increased rapidly at an annual average rate of 4.30%, from 182.5 Mg in 1995 to 248.0 Mg in 2003.

Table 3.5 Summary of total mercury emission estimates (*Mg*) from industrial sources from 1995 to 2003

Source category	1995	1996	1997	1998	1999	2000	2001	2002	2003	AAGRa (%)
Non-ferrous metals smelting	182.5	158.9	152.8	155	167.8	186	213.8	221.1	248	4.3
1) Zinc (Zn)	50	49	66	69	73	85	105	105	115	11.6
2) Copper (Cu)	10.4	10.7	11.3	8.4	10.1	12.7	13.7	14.8	17.6	6.9
3) Lead (Pb)	26.5	30.8	30.9	33	40.1	48	54.3	57.8	70.7	13.0
4) Gold (Au): large scale	10.1	11.4	16.1	16.1	16.1	11.8	12.3	15	16.2	6.0
5) Gold (Au): artisanal	85.5	57	28.5	28.5	28.5	28.5	28.5	28.5	28.5	-12.8
Fuel oil for stationary sources	0.5	0.5	0.5	0.5	0.5	0.5	0.5	0.5	0.6	2.3
Gasoline, diesel, and kerosene	4.3	4.6	4.6	5	5.6	6.1	6.4	6.8	7.6	7.2
Biofuel combustion	10.1	9.1	8.7	8.7	8.3	8.6	9.5	10.6	10.7	0.7
Grassland/ savanna burning[b]	4.2	4.2	4.2	4.2	4.2	4.2	4.2	4.2	4.2	0.0
Agricultural residue burning[b]	3.9	3.9	3.9	3.9	3.9	3.9	3.9	3.9	3.9	0.0
Household waste burning	0.6	0.6	0.6	2	2	2.8	3.2	7.7	10.4	42.5
Cement production	19.9	20.5	21.3	21.4	22.7	23.9	27	29.4	35	7.4
Iron and steel production	3.8	4.1	4.4	4.6	4.9	5.1	6.1	7.3	8.9	11.2
Caustic soda production	2.4	2.4	2.5	1.3	0.2	0.2	0.2	0.2	0	N/A
Mercury mining	35.1	22.9	37.6	10.1	8.8	9.1	8.7	22.3	27.5	-3.0
Battery/fluorescent lamp production	29.1	34.1	49.7	37.6	24.5	16.2	8.7	6.2	3.7	-22.7
Total	296.4	265.8	290.8	254.3	253.4	266.6	292.2	320.2	360.5	2.9
a) Hg^0	238.4	213.7	233.6	204.7	203.8	214.4	235.0	258.2	290.6	2.9
b) $Hg^{(II)}$	43.2	38.8	42.7	36.9	36.8	38.7	42.5	45.9	51.7	2.7
c) Hg(p)	14.8	13.2	14.5	12.7	12.8	13.5	14.7	16.1	18.1	3.0

[a] Annual average growth rate.
[b] Assumed no change over time due to lack of data

In this category, zinc smelting is the largest single sector in total mercury emissions, reaching 115.0 Mg in 2003 at an average annual rate of 11.6%. However, lead smelting was the leading sector in mercury emissions growth: from 26.5 Mg in 1995 to 70.7 Mg in 2003, increasing by 13% annually. Total mercury emissions from copper smelting increased to 17.6 Mg in 2003, increasing at an annual rate of 6.9%. Gold smelting is the only sector with decreasing mercury emissions in this category, attributed to an official ban of artisanal gold production in China since 1996. In 2003, total mercury emissions from gold smelting were 44.7 Mg, compared to 95.6 Mg in 1995. It should be noted that the estimate of mercury emissions from metals smelting is subject to a high uncertainty due to limited test samples, lack of detailed information on metal smelting processes in typical Chinese plants, and lack of precise production estimates from small activities (Wu et al., 2006).

Besides non-ferrous metals smelting, cement production, mercury mining, battery and fluorescent lamp production, household waste burning, and biofuel burning are also major contributors of mercury emissions during the whole period or part of the period (1995-2003). Total mercury emissions from cement production (coal-related emissions are excluded to avoid double-counting with industrial coal use) increased steadily from 19.9 Mg in 1995 to 35.0 Mg in 2003, at an annual rate of 7.4%. In China, domestic mercury mining shrank dramatically in the late 1990s, but has rebounded since 2002. As a result, the total Hg emissions from mercury mining fluctuated significantly, decreasing from 35.1 Mg in 1995 to 8.7 Mg in 2001, then back up to 27.5 Mg in 2003. However, mercury emissions from artisanal mercury mining activities in Guizhou during that period of time were not included because it is very difficult to obtain the precise mercury production from this small scale mercury mining activity. It was estimated that the annual mercury emissions from artisanal mercury mining activity in the Wuchuan mercury mining area in Guizhou reached 3.9 to 9.8 Mg (Li et al., 2006). Mercury containing batteries are being phased out in China due to the release of a stringent standard in December 1997. Therefore, total mercury emissions from this sector increased initially, from 29.1 Mg in 1995 to 49.7 Mg in 1997, then decreased significantly from 1998, to as low as 3.7 Mg in 2003. Biofuels dominate rural energy supply in China. Total mercury emissions from biofuel burning have remained nearly constant, at around 10 Mg annually. Although household waste burning contributed only 0.6 Mg of mercury emissions in 1995, it is the leading sector among all of the mercury source sectors in emission growth, reaching 10.4 Mg in 2003 at an annual growth rate of over 40%. This is simply because the living standards of Chinese people has increased latterly and consequently the amount of waste produced by each family increased. Among other miscellaneous small sources, liquid fuels (gasoline, diesel, and kerosene) and iron and steel production are two sectors with high mercury emission growth, at annual average growth rates of 7.2 and 11.2%, respectively.

It is estimated that 81% of total mercury from industrial sources in 2003 is released as Hg^0, 14% as $Hg^{(II)}$, and 5% as Hg(p), compared to 80% as Hg^0, 15% as $Hg^{(II)}$, and 5% as Hg(p) in 1995.

3.6 Uncertainties

Quantifying Hg emissions is more difficult than quantifying, say, SO_2 or NO_x, because the emissions come from so many source types, not primarily combustion sources. In this respect Hg emissions are similar to VOC emissions. It is acknowledged that for some types of sources very little is known about actual activity levels and emission factors, and the choices in such cases rely heavily on inferences of activity levels from quite limited and uncertain statistical information. On the other hand, at least for combustion sources and releases from mercury containing ores, total emissions are constrained by the Hg content of the raw material, in a similar way to the sulfur content of fossil fuels, and this acts to reduce the uncertainty. Several factors influence the estimation of emissions, including emission factor and activity level. We estimate the uncertainty for each emitting sector by combining the coefficients of variation (CV, or the standard deviation divided by the mean) of the contributing factors. We then combine these uncertainties to estimate the total uncertainty of Hg emission estimates by quadrature average when the source estimates are uncorrelated. We follow the same detailed methodology for uncertainty analysis that was described in the TRACE-P inventory paper of Streets et al. (2003a). Figure 3.1 shows the results of uncertainty estimation in Hg emissions by source type.

The general findings are that Hg emissions are known least well in the artisanal gold smelting sector (±450%), followed by the mercury mining sector (±340%).

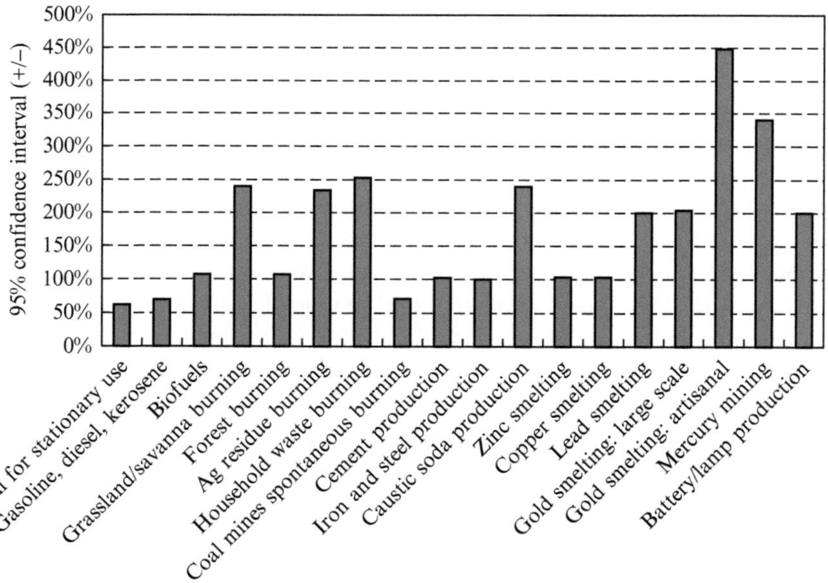

Figure 3.1 Uncertainty (%) in Hg emission estimates (95% confidence intervals, ±) by sector (modified from Streets et al., 2005)

As the confidence intervals are frequently greater than the mean, the presentation of relative confidence intervals >±100% might suggest that the lower confidence interval is negative. However, the true confidence interval is not symmetric about the mean because some of the underlying variables are log-normally distributed. A better interpretation of "±400%", for example, might be "within a factor of five" so that the confidence interval would be 20-500% of the mean given.

3.7 Future Research and Policy Implications

Due to the fact that high uncertainties are associated with the current mercury emission inventory from industrial sources other than the coal combustion sector, a great number of studies need to be undertaken to reduce the uncertainties. First of all, surveys are needed to evaluate the mercury content of the raw materials of different industrial categories. Knowing mercury contents in raw materials, we can easily constrain the upper limit of mercury emissions from industrial sources. However, information regarding the mercury content of raw materials in industrial sectors is extremely scarce in China. Secondly, mercury balance studies are necessary for representative plants of different industrial sources. Until we know the mercury balance in an industrial process, it is impossible for us to determine what percentage of mercury in raw materials is emitted to the atmosphere. This kind of study is rarely reported in open literature in China. Thirdly, the speciation of mercury emissions from different industrial sources are also urgently needed in order to better understand the atmospheric fate of mercury emitted from these sources. Once we have all the above mentioned information, we will have accurate mercury emission inventory from industrial sources in China.

References

Ebinghaus, R., H. H. Kock, A. M. Coggins, T. G. Spain, S. G. Jennings, and C. Temme, 2002, Long-term measurements of atmospheric mercury at Mace Head, Irish west coast, between 1995 and 2001, Atmospheric Environment, 36, 5267–5276.

Editorial Committee of China Mechanical Industry Yearbook (ECCMIY), 2000. China Mechanical Industry Yearbook, 1986–1999. China Mechanical Industry Press, Beijing (in Chinese).

Fang, F., Wang, Q., Liu, R., Ma, Z., Hao, Q., 2001. Atmospheric particulate mercury in Changchun City, China. Atmospheric Environment 35, 4265–4272.

Feng, X., 2005. Mercury pollution in China-an overview. In: Pirrone, N., Mahaffey, K. (Eds.), Dynamics of Mercury Pollution on Regional and Global Scales: Atmospheric Processes, Human Exposure Around the World. Springer Publishers, Norwell, MA, USA, pp. 657–678.

Feng, X., Li, G., Qiu, G., 2004. A preliminary study on mercury contaminations to the environment from artisanal zinc smelting using indigenous method in Hezhang county, Guizhou, China. Part I mercury emissions from zinc smelting and its influences on the surface water. Atmospheric Environment 38, 6223–6230.

Feng, X., Shang, L., Wang, S., Tang, S., Zheng, W., 2004a. Temporal variation of total gaseous mercury in the air of Guiyang, China. Journal of Geophysical Research 109, D03303.

Feng, X., Sommar, J., Lindqvist, O., Hong, Y., 2002. Occurrence, emissions and deposition of mercury during coal combustion in the province Guizhou. Water, Air, and Soil Pollution 139, 311–324.

Feng, X., Tang, S., Shang, L., Yan, H., Sommar, J., Lindqvist, O., 2003. Total gaseous mercury in the atmosphere of Guiyang, PR China. The Science of the Total Environment 304, 61–72.

Feng, X., Yan, H., Wang, S., Qiu, G., Tang, S., Shang, L., Dai, Q., Hou, Y., 2004b. Seasonal variation of gaseous mercury exchange rate between air and water surface over Baihua reservoir, Guizhou, China. Atmospheric Environment 38, 4721–4732.

Friedli, H.R., Radke, L.F., Lu, J.Y., 2001. Mercury in smoke from biomass fires. Geophysical Research Letters 28 (17), 3223–3226.

Friedli, H.R., Radke, L.F., Lu, J.Y., Banic, C.M., Leaitch, W.R., MacPherson, J.I., 2003a. Mercury emissions from burning of biomass from temperate North American forests: laboratory and airborne measurements. Atmospheric Environment 37, 253–267.

Friedli, H.R., Radke, L.F., Prescott, R., Hobbs, P.V., Sinha, P., 2003b. Mercury emissions from the August 2001 wildfires in Washington State and an agricultural waste fire in Oregon and atmospheric mercury budget estimates. Global Biogeochemical Cycles 17 (2), 1039.

Fu X., Feng X., Zhu W., Wang S., Lu J., Total gaseous mercury concentrations in ambient air in the eastern slope of Mt. Gongga, South-Eastern fringe of the Tibetan plateau, China. Atmospheric Environment, 2008 (in press)

Jiang J., 2006, The technological progress and the situation of Zinc smelting in China, 5: 19-23 (in Chinese with English abstract).

Li G., 2007, Mercury emission from Zinc Smelting in China and environmental impacts, PhD thesis, Institute of Geochemistry, Chinese Academy of Sciences, pp 1-110.

Li P., Feng X., Qiu G., Wang S., 2006, Estimate of mercury emissions from artisanal mercury mining in Wuchuan, Guizhou, China. Environmental Science, 27(5): 837-840 (in Chinese with English abstract).

Lindberg S., Bullock R., Ebinghaus R., Engstrom D., Feng X., Fitzgerald W., Pirrone N.,, Prestbo E., and Seigneur C., 2007, A synthesis of progress and uncertainties in attributing the sources of mercury in deposition. Ambio, 36(1): 19-32.

Liu, S., Nadim, F., Perkins, C., Carley, R.J., Hoag, G.E., Lin, Y., Chen, L., 2002. Atmospheric mercury monitoring survey in Beijing, China. Chemosphere 48, 97–107.

National Bureau of Statistics of China (NBS), P.R. China, 2000. China Statistical Yearbook (1999). China Statistics Press, Beijing, China.

Pacyna, E.G., Pacyna, J.M., 2002. Global emission of mercury from anthropogenic sources in 1995. Water, Air, and Soil Pollution 137, 149–165.

Pirrone N., Keeler G.J., Nriagu J.O. 1996. Regional differences in worldwide emissions of mercury to the atmosphere. Atmospheric Environment 30(17), 2981-2987.

Qi, X., 1997. Development and application of an information administration system on mercury. Masters Dissertation, Research Center for Eco-Environmental Sciences, Beijing, China (in Chinese with abstract in English).

Schroeder, W. H., A. Steffen, G. Lawson, and W. Strachan, 2001, Mercury measurements at Alert, in Synopsis of Research Conducted Under the 2000/2001 Northern Contaminants Program, edited by S. Kalhok, pp. 130– 135, Indian and North. Affairs Can., Ottawa, Ontario.

Streets, D. G., Hao J., Wang S., Wu Y., 2008, Mercury emission from coal combustion in China. Edited by Nicola Pirrone, UNEP MFTP Report, pp ??

Streets, D.G., Bond, T.C., Carmichael, G.R., Fernandes, S.D., Fu, Q., He, D., Klimont, Z., Nelson, S.M., Tsai, N.Y., Wang, M.Q., Woo, J.-H., Yarber, K.F., 2003a. An inventory of gaseous and primary aerosol emissions in Asia in the year 2000. Journal of Geophysical Research 108 (D21), 8809.

Streets, D.G., Yarber, K.F., Woo, J.-H., Carmichael, G.R., 2003b. Biomass burning in Asia: annual and seasonal estimates and atmospheric emissions. Global Biogeochemical Cycles 17 (4), 1099.

Streets, D.G., Hao, J.M., Wu, Y., Jiang, J.K., Chan, M., Tian, H.Z., Feng, X.B. Anthropogenic mercury emissions in China. Atmos. Environ. 2005, 39, 7789-7806.

UNECE/EMEP, 2004. Atmospheric Emission Inventory Guidebook, third ed http://reports.eea.eu.int/EMEPCORINAIR4/en/page002.html.

US Environmental Protection Agency (US EPA), 1995. Compilation of Air Pollutant Emission Factors, AP-42, fifth ed., vol.I: Stationary Point and Area Sources.

US Environmental Protection Agency (US EPA), 1997. Mercury Study Report to Congress, vol. II: An Inventory of Anthropogenic Mercury Emissions in the United States. EPA-452/R-97-004.

Wan Q., Feng X., Lu J., Zheng W., Han S., Xu H., Total gaseous mercury in ambient air at a remote site in Changbai Mountain area, Northeast China. Journal of Geophysical Research, 2008 (in review)

Wu, Y., Wang, S., Streets, D.G., Hao, J., Chan, M., Jiang, J. Trends in anthropogenic mercury emissions in China from 1995 to 2003. Environ. Sci. Technol. 2006, 40, 5312-5318.

Xiu, G.L., Jin, Q., Zhang, D., Shi, S., Huang, X., Zhang, W., Bao, L., Gao, P., Chen, B., 2005. Characterization of sizefractionated particulate mercury in Shanghai ambient air. Atmospheric Environment 39, 419–427.

Zhou, D., Dai, Y., Yu, C., 2003. China's Sustainable Energy Scenarios 2002. China Environmental Science Press, Beijing (in Chinese).

Jiang, J., 2004. Preliminary studies on emission and control of atmosphere mercury in China. Masters Dissertation, Tsinghua University, Beijing, China (in Chinese with abstract in English).

National Bureau of Statistics of China (NBS), P.R. China, 2000. China Statistical Yearbook (1999). China Statistics Press, Beijing, China.

Editorial Committee of China Nonferrous Metals Industry (ECCNMI), 2000. China Nonferrous Metals Industry Statistical Yearbook, 1999. China Nonferrous Metals Industry Press, Beijing (in Chinese).

Yang, F., Liu, J., Wang, R., 2003. Estimation on the amount of mercury used in domestic batteries and potential amount of mercury emissions in China. Shanghai Environmental Sciences 22 (5), 322–328 (in Chinese with abstract in English

Chapter 4
Mercury Emissions from Industrial Sources in India and its Effects in the Environment

Arun B. Mukherjee, Prosun Bhattacharya, Atanu Sarkar, and Ron Zevenhoven

Summary This study describes the atmospheric mercury (Hg) emissions from industrial sources in India for the years 2000 to 2004. In India emission inventories of Hg and other trace elements from anthropogenic sources have been largely neglected, although the GDP (Gross Domestic Products growth) has touched 9.6% at the beginning of the 21st century. In coal production India is the third largest in the world, whereas Indian cement and brick production have reached second place in the world. With increased industrial development, acute pollution problems have been identified in the subcontinent. There is no consistent earlier information for Hg emissions to the environment for any sectors of industry. This paper may be the first road map in which we have tried to find out the total emission of Hg from a wide range of sources, e.g. from coal combustion to clinical thermometers broken during production or packing. There is a lack of basic data and in an attempt to correct this, emission factors suitable for Asian countries have been selected to complete this study. Before this document, there were some efforts in Europe to develop emission inventories for Hg from coal combustion or chlor-alkali plants for India. In this study it was found that total atmospheric emission from industrial sources has decreased from 321 Mg in 2000 to 253 Mg in 2004 due to a switch for the membrane cell process in the chlor-alkali industry. In 2004 the largest part of the Hg emissions stemmed from coal combustion in thermal power plants. Hg-cell technology had been used earlier in chlorine and sodium hydroxide production, as a result of which Hg concentration in terrestrial and aquatic species are nowadays quite high in coastal areas. India can thus be referred to as a mercury "hot spot". We have received limited information on emissions of Hg from industrial sources in India. Estimates are based on emission factors and the values taken from the literature. Against a background of limited data and information, this paper gives an overview of Hg emissions in India and of the recent steps undertaken by authorities to curb the emissions of Hg and its subsequent trans-boundary movement in the global environment.

N. Pirrone and R. Mason (eds.), *Mercury Fate and Transport in the Global Atmosphere*,
DOI: 10.1007/978-0-387-93958-2_4,

4.1 Introduction

Economic advancement of any country whether developed or developing depends on capital to provide the rapid growth of manufacturing industries, infrastructure, and for the modernization of economies and societies. For the past three decades, India has achieved increased production of metals, cement, fertilizers, chlorine, pulp and paper as well as heat and electricity, through burning of coal, natural gases and oil (Table 4.1). Hence the country became one of the most rapidly growing economies with an average annual growth of 9.6 percent and it has later crossed the ten percent level (Choi, 2003). During the course of development, industrial management and the government authorities did not pay adequate attention at the regional or central levels, to pollution problems due to mining operations, metal smelting, electroplating, energy and fuel consumption, sludge dumping and many others operations causing pollution problems in the terrestrial and aquatic environments. Examples of soil pollution as well as other pollution problems in the aquatic and terrestrial environment are well documented in India (Kumar et al., 1995; Choi, 2003). In the industrial area of Chhattisgarh state, water discharged from different industries such as thermal power plants, the steel industry, the cement industry, sulfuric acid plants, rice mills, coal washing etc, was cited to contain total Hg between 6.7 – 678 ng mL^{-1} with mean and median values of 118 and 49.3 ng mL^{-1}, respectively. High concentrations of Hg in human hair have also been reported in Chhattisgarh state. Human activities redistribute Hg in a manner that causes elevated concentrations of pollutants in the human food chain (Nriagu and Pacyna, 1988).

Table 4.1 Production of metals, coal, residue fuel oil and cement in India, 2000-2004 (*Tg*)

Element	2000	2001	2002	2003	2004
Copper, Cu[1]	0.256	0.293	0.385	0.391	0.401
Sec. Copper[4]					0.007
Lead, Pb[1]	0.057	0.074	0.064	0.078	0.042
Sec. Pb[4]					0.040-0.050
Zinc, Zn[1]	0.176	0.207	0.232	0.254	0.238
Sec. Zn[4]					*0.065*
Pig Iron, Fe[1]	21	22	24	24	25
Raw Steel[1]	27	27	29	32	32
Hard coal[2]	310	312.5	333.7	340e	373
Residue fuel oil production[3]	7.965	8.308	7.855	6.905	7.267
Cement[1]	100	100	100	110	111e

In italic = estimated value

[1]http://minerals.usgs.gov/minerals/pubs/commudity

[2]USGS, 2005; www.worldcoal.org accessed on 11.29.2007

[3]IEA (2007);

[4]Secondary metal production for 2004.

Information received from Indian Copper Development Center, Kolkata (2007) and Indian Lead and Zinc dev Association, New Delhi (2007). Note: The authority mentioned that production of secondary lead varies between 40000 to 50000 Mg in a year

Among the several trace elements in the periodic table, Hg is considered as a toxic trace element to humans, animals, and the ecosystem because of its unique geochemical characteristics. It is a natural element that cannot be created or destroyed and the same amount has existed on Earth since the planet was formed. It exists in several states: Hg-metallic liquid, Hg vapour, inorganic $Hg^{(I)}$ (mercurous salts) inorganic $Hg^{(II)}$ (mercuric salts) and gaseous Me-Hg (Drasch et al., 2004). There is global concern regarding the recycling of this element, especially in the Indian subcontinent and China where less attention has been paid to the environmental consequences of increased production of chlorine, metals, waste incineration and coal combustion as well as brick manufacturing in India. It has been suggested that the overall amount of Hg mobilized and released in the Indian atmosphere has decreased slightly in recent years, although there is no emission inventory of Hg from these facilities and the data availability is scarce. This is further complicated by the fact that the decision makers still pay minimal attention to the issues concerning Hg emission and related environmental hazards. Even in the international journals little attention has been paid to the atmospheric Hg emission into the techno sphere in India. In general, the state of Hg and other trace metal research is not well established as compared to the western world.

In India, there are no cinnabar ore resources for the production of Hg and neither is there information that indicates whether Hg is recovered as a by-product from certain processes. There is controversy regarding the amount of Hg imported from the European Union (EU) and other countries, but the total import amount has decreased from 253.7 Mg in 1996 to 123.4 Mg in 2004 (Pandey, 2006). This imported Hg is generally used in chlor-alkali plants and the leading Hg users are shown in Table 4.2.

Table 4.2 Leading mercury users in India (1998 – 2001)

Sector	Hg content per unit	Units produced	Total Hg (*Mg*)
Chlor-alkali	~ 200 g Hg used per Mg of Cl_2 produced	450,000 [a]	70
Thermometers	0.6 – 1.0 g	8957,000 [b]	7.2
Batteries	Alkaline not more than 25 mg	NA	
Hg-Zinc	Total 33 to 50% by wt of the battery	1,650 million [c]	25
Zn-Carbon	Total 1% Hg by wt of the battery	NA	
Fluorescent lamps	0.0252 – 0.080 g per lamp [f]	150 million [c]	7.89
Thermostat switches	3 – 6 g	4051,000 [d]	18.23
Alarm clocks	Average 0.6 – 0.7 g $unit^{-1}$	1481,000P [b]	0.96
Hearing aids	0.4 g $unit^{-1}$	95,500 [c]	0.04
Sum			129.32

[a]Environmental rating of Indian Caustic-Chlorine Sector, Green Rating Project (2002), Center for Science and Environment.
[b]Industrial Handbook, Centre for Industrial & Economic Research /Delhi), 1998
[c]Industrial Handbook, Centre for Industrial & Economic Research /Delhi), 2000–2001
[d]http://www.Indianfoline.com/auto/db01.html
[e]Telephone conversion with Battery Industrial Official
[f]Draft Wisconsin Mercury Sourcebook, Wisconsin Department of Natural Resources (USEPA, May 1997)

There is no information of cross-boundary flow of Hg vapour from India to other parts of the world except for the Himalayas (covering an area of roughly 6×10^5 km^2). The study by Banic et al. (2003) suggests that Hg has the capacity to move to high altitudes. During snow deposition, $Hg^{(II)}$ can be photo-reduced to elemental Hg and remitted back into the atmosphere. It is still unknown how Hg species at the Himalayan region precipitates in the terrestrial and aquatic environment in mountain areas (Loewen et al., 2005).

A recent emission inventory of Hg by Jaffe et al. (2005) indicated that Asian Hg accounts for more than 50% of the global anthropogenic release of Hg. These authors confirmed that the ratio of Hg/CO is a good indicator of Asian industrial flow, including India. These authors suggested that it is possible to calculate Hg emission based on the Hg/CO ratio and the inventory of CO emissions. Lindberg et al. (2007) pointed out that biogeochemical cycling of Hg is similar to that of carbon (C), sulphur (S) and nitrogen (N). However, levels of Hg emissions in the form of aqueous and atmospheric Hg on the Indian subcontinent are alarming. Recently, Srivastava (2003) outlined the sources of Hg and its risks to the Indian environment. Studies by Weiss-Penzias et al. (2003) indicated that industrial emissions of Hg from Asia can be transported across the Pacific within five days.

Mercury in the air consists of two main chemical forms, being elemental Hg (Hg^0) and divalent Hg compounds which are in gaseous forms or are bound to particles in the atmosphere. Mercury can also exist in the environment in the form of organo-metallic compounds e.g. methyl mercury (MeHg). But, the speciation of Hg in the aquatic environment determines its chemical reactivity, mobility and biological activity. Mercury is deposited over land, water, and forest regions either by wet (Figure 4.1) or dry deposition. But the enhanced wet deposition rates close to major Hg sources through cloud-droplet activation and precipitation scavenging

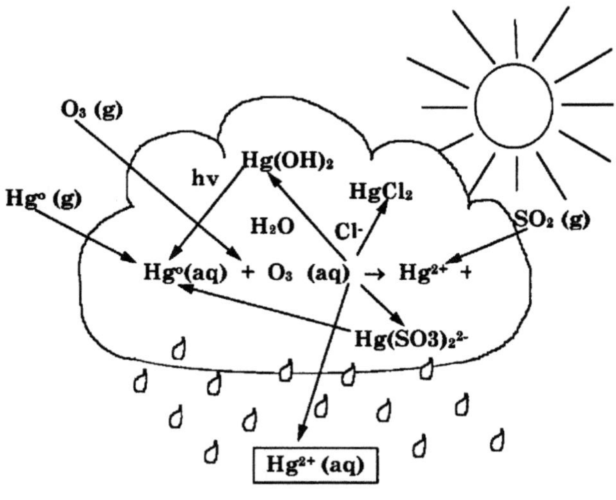

Figure 4.1 Chemistry of wet deposition of mercury (reproduced from Lindqvist et al., 1991)

have been confirmed (Dvonch et al., 1998; Munthe et al., 2001). However, the deposition rate is highest in regions where elevated rain fall or snow fall occurs. The humidity in India is quite high especially in the south, middle and eastern parts, where deposition of Hg is expected to be highest. In northern Europe in the 1990s, an ~ 40% decrease of wet deposition in southern Sweden was a result of a decrease in air concentrations, and thus in the wet deposition rate of Hg (Munthe et al., 2001).

International transport of Hg in the Indian subcontinent has not been studied sufficiently, as a result of which it is impossible to predict its effects in terrestrial, aquatic and freshwater ecosystems. Atmospheric Hg can be deposited to aquatic systems in the Himalayas region, which may be the source of Hg in river water, sediments and head water systems. Subraminium et al. (2003) measured Hg in fish species (0.069 – 3.920 mg kg^{-1} wet wt), sediment (0.16 – 5.71 mg kg^{-1}) and water (0.17 – 2.351 mg L^{-1}) from the Indus River. These authors did not mention Me-Hg in Indian fish. It should be remembered, however, that many variables must be considered, such as atmospheric processes mixed with aquatic variables, which will dictate the overall levels of Me-Hg in fish tissue (Downs et al., 1998).

After the first international conference on "Hg as a Global Pollutant" in 1990 at Gävle, Sweden, understanding of Hg chemistry in the ecosystem has increased among the scientific community. Still our scientific understanding of Hg in ecosystems is not absolute and is rarely complete (Lindberg et al., 2007). Weiss-Penzias et al. (2003) measured gas-phase elemental Hg (Hg^0), inorganic reactive gaseous mercury (RGM) and particulate Hg (Hg(p)) in the marine boundary layer of Washington state, USA in 2001 – 2002. It has also been observed that in the Polar regions, Hg^0 can be converted to RGM by chemical reaction with halogen species (Ebinghaus et al., 2002).

One rather difficult question is whether emission inventories are consistent with observations. Emission factors can be used to estimate the emission of an element, but this approach is far from perfect (Pacyna et al., 2006a). In the view of those authors, the accuracy of emission estimations is based on the accuracy of emission factors available in the Emission Inventory Guidebook (UN ECE, 2000, (http://www.epa.gov/ttn/chief/ap42/ndex.html). However, emission factors will give a useful guideline to the emission of an element or species from industrial or natural sources.

Many authors have identified natural sources of Hg emissions (not in India), for example, from forest fires, sands, oceanic mist, volcanic activities, photo reduction of divalent Hg in natural waters. The weathering of bed rocks may contribute to high concentrations of Hg (100 ng g^{-1}). Wide ranges of Hg in rocks (Table 4.3) determine equally wide ranges of natural background levels in soils and sediments, impacting on Hg bioaccumulation in aquatic and terrestrial species. In addition, laboratory studies have quantified the Hg emissions from soils where solar radiation has enhanced Hg emissions (Gustin et al., 2002). We must understand the natural background levels of Hg species before any conclusions on anthropogenic Hg input may be drawn. The Hg in the freshwater food chain may be due to the global increase in the background level of total Hg (Rohde, 1996).

This study (November 2007 – May 2008) is based on a literature survey with a final goal to estimate Hg emission from industrial sources in India. Stack emission measurements are not mandatory in Indian industry. Hence no emissions data are

Table 4.3 Mercury concentration in different rock samples (from different sources)

Source	Hg (*ng g*$^{-1}$)	Range (*ng g*$^{-1}$)	Location
Ocean ridge basalt	10*		
Granite porphyry	117	5 – 468	Maine
Gabbro, Granite	10	-	Minnesota
Granite, (Granodiorit)	30*		-
Granite,(Rhyolite)	3.5	1.4 – 281	Sweden
Limestone	6	0.8 – 31.2	Sweden
Limestone	-	40 – 50*	-
Sandstone	110	-	-
Sandstone	-	40 – 100*	
Black shale	234	31.9 – 340	Sweden
Shale	5.9	0.9 – 33.5	Sweden
Shale	-	180 – 400*	
Mafic (Basalt)	3.9	0.2 – 17.7	-
Mafic	-	4 – 10*	-

*Source: Kabata-Pendias and Pendias (2001)

available in the literature for trace metals, including Hg. In addition information on atmospheric Hg emissions is very limited or non-existent in sources presented by the government. Several Indian organizations and companies were contacted for information during this study but the response was very poor or nil.

In the present study, the estimated atmospheric Hg emissions are based on emission factors for the European Union (EU), for the U.S., from the literature and from the limited information received from India. An emission factor is a representative value that attempts to relate the quantity of a pollutant released to the atmosphere with an activity associated with the release of that pollutant. The general equation for emissions estimation is:

$$E = A \times EF \times (1 - ER/100) \quad (1)$$

where:
E = emission;
A = activity rate;
EF = emission factor, and
ER =overall emission reduction efficiency, %

Here we did not deal with emissions from secondary metal production as there are no emission data available. This document is by no means complete, but gives a first road map on how to deal with emission patterns of Hg in India.

4.2 Results

There is no doubt that industrial development has contributed to momentous economic growth in India over the last few decades, which has not, however, spread uniformly in society. Industrialization, population growth, urbanization, and unbalanced uses of raw materials have created enormous air pollution, causing acute environmental problems (Garg et al., 2006). Never before in the history of mankind have such vast environmental risk factors from Hg, or natural danger to humans, terrestrial and aquatic species been reported. Hence, education and awareness programs must be launched across the Indian subcontinent to educate the population on the risks from Hg and other trace element exposure, addressing especially the most vulnerable sectors of the population e.g. pregnant women and children (Srivastava, 2003). It is well recognized that Hg is widely spread in India and in this study we have dealt with industrial emissions of Hg from the following sources:

- Coal combustion
- Iron & Steel Industry
- Non-ferrous metallurgical plants
- Chlor-alkali plants
- Cement industry
- Wastes
- Biomass burning
- Others (e.g. brick manufacturing, instruments, clinical thermometers)

In this study, no information was available from the pulp & paper industry or from the oil and petrochemical industry in India.

4.2.1 Coal Combustion

Coal reserves are distributed widely across the planet, but recoverable reserves are reported for only seventy countries. It has been estimated that world coal reserves may be sufficient for at least another 2-3 centuries whereas the figures for oil and gas are 41 and 65 years, respectively, at current production levels. In India, the coal mining area covers some 855 km^2 and the total number of coal mines is 572 (March 2004), of which 170 are opencast, 359 underground and 33 mixed (Mine Closure, 2005), Figure 4.2. India is the third largest hard coal producer in the world after the PR China and the USA. Coal production has increased from 310 Tg in 2000 to 373 Tg in 2004. About 70% of the heat and electricity production in India depends on indigenous coal. From time to time, steam coal (11 Tg in 2001) and coking coal (9.8 Tg in 2001) have been imported, which in 2005 had increased to 41 Tg of steam and coking (19 Tg) coals (GOI, 2006). Coking coals are primarily consumed in the iron and steel industry. There are 81 thermal power plants in India, three of which are not operating currently.

The occurrence and distribution of Hg in different compartments of ecosystems has been studied by many authors. Mercury is a chalcophile element, having great affinity for sulphur-containing compounds. This element (which in pure form and at ambient conditions is a liquid) is generally incorporated in pyrite (iron sulphide) and the concentrations of Hg vary with mineral paragenesis (Kolker et al., 2006). Due to its high vapour pressure and physicochemical properties, the element

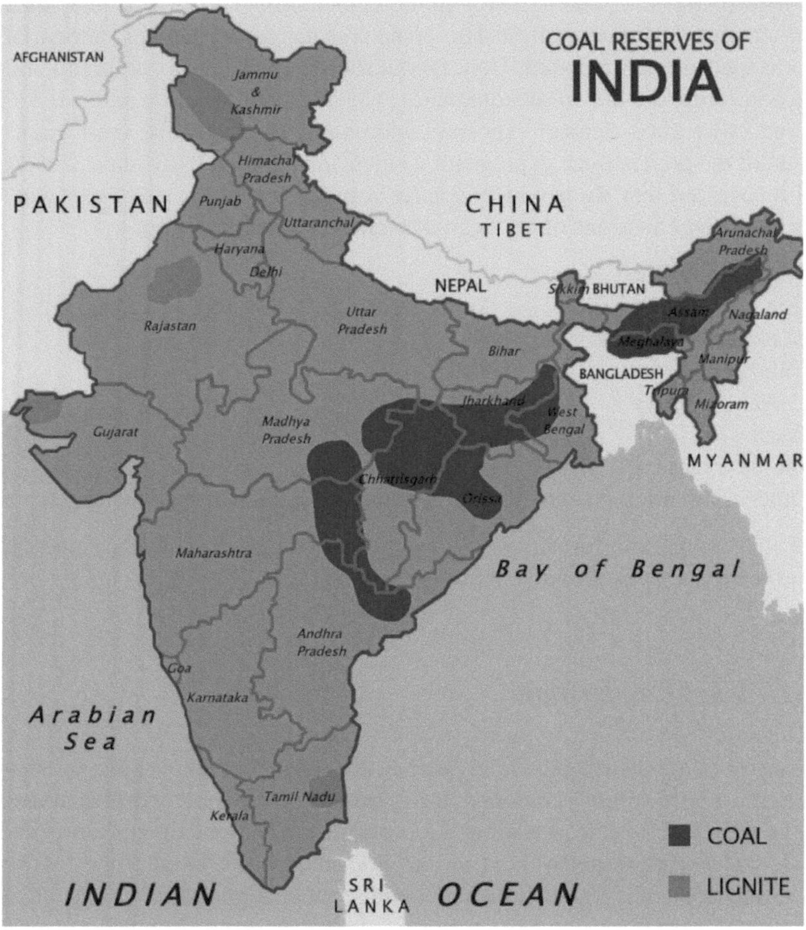

Figure 4.2 Coal reserves in India, 2004

vaporizes easily during processing and thereby is released into the atmosphere. In India, gas cleaning equipment is not modern (involving almost exclusively only fly ash (FA) emissions control) and there are no flue gas desulphurisation (FGD) plants.

For estimation of Hg emission from coal and other products, emission factors have been selected and occasionally estimated in the current study (Table 4.4). Mercury in coal has been measured by the Pollution Control Research Institute of Bharat Heavy Electricals Ltd, India (Table 4.5) whereas the Centre for Science and Environment (CSE, 2005) pointed out that the concentration of Hg in Indian coal varies between 0.01 and 1.1 mg kg^{-1} (= ppm-wt). For estimation of Hg emission

Table 4.4 Emission factors of mercury from industrial sources used for India

Source Category	Unit	EF	Reference
Coal fired power plants	$g\ Mg^{-1}$Coal	0.3, 0.324*	1, 2
Residential and commercial boiler	$g\ Mg^{-1}$ Coal	0.5	1
Crude steel production	$g\ Mg^{-1}$ Steel	0.08*	2
Residual fuel oil combustion	$g\ Mg^{-1}$oil	0.065	10, 11
Non-ferrous metal production			
- Copper	$g\ Mg^{-1}$ Cu production	15	3, 4
- Zinc	$g\ Mg^{-1}$ Zn production	8	1
- Lead	$g\ Mg^{-1}$ Pb production	43.6	5, 6
Caustic soda production	$g\ Mg^{-1}$ NaOH production	20.4	7
Cement production	$g\ Mg^{-1}$ Cement	0.042*	2
Wastes			
- Municipal solid waste (MSW)	$g\ Mg^{-1}$ MSW	1.0	1, 3
- Medical waste	$g\ Mg^{-1}$ Medical waste	20	3
- Electronic waste**			8
Miscellaneous			
- Bricks	$g\ Mg^{-1}$ brick	0.0214	9
Chlor-alkali plants	$g\ Mg^{-1}$ of NaOH	20.4	7
Forest burning	$g\ Mg^{-1}$ of fuel	0.242	12
Non-forest burning	$g\ Mg^{-1}$ of fuel	0.041	13

References: (1) Pacyna et al. (2006b); (2) This study; (3) Pirrone et al. (1996); (4) Nriagu and Pacyna (1988); (5) Li (2007); (6) Feng et al. (2004); (7) Qi et al. (2000); (8) Sarkar (2007); (9) USEPA (1997a); (10) Mukherjee et al. (2000); (11) Sunderland and Chmura (2000); (12) Veiga and Meech (1994); (13 Friedli et al. (2008).

*Estimated in this study

**Sarkar (2007): Hg in each computer is 0.0022%, weight of Hg in each computer is 0.00059 kg (assuming average weight of a computer is 27 kg), recycling efficiency is 0%, estimated obsolete computer is 1.38 million. Total annual production of Hg releasing to the environment is (0.00059 kg x 1380000)/1000 = 0.82 Mg.

Table 4.5 Samples collected from eight coal based power plants in India (BHEL, 2004vide Pandey, 2006)

Names of power plants	Hg in coal (*mg kg*$^{-1}$)
GHTTP, Lehra, Mohabatt	0.26
Anpara, UP (BTPS)	0.26
North Chennai	0.33
NLC-TPS II	0.18
Chandrapura STPS	0.325
Kolaghat TPS (West Bengal)	0.61
Talchar TPS	0.33
Gandhinagar TPS	042
Mean (Range)	0.376 (0.18 – 0.61)

Note: Number of samples analyzed are unknown.

from coal it is assumed that Indian coal contains on average 0.376 mg Hg/kg. The emission of Hg to the atmosphere will be more if we use the emission factor 0.5 g Hg Mg^{-1} calculated by Pacyna and Pacyna (2000), and also the efficiency of

Table 4.6 Atmospheric mercury emissions from industrial sources in India for 2000 and 2004, respectively (this study)

Source category	Consumption/ Prod. 2000 (*Tg*)	Hg emissions 2000 (*Mg*)	Consumption/ Prod. 2004 (*Tg*)	Hg emissions 2004 (*Mg*)
Coal fired power plants	310	100.44	373	120.85
Residential & Commercial boiler	7.3	3.65	7.4	3.7
Pig iron & steel production	48	3.84	57	4.56
Cu-production	0.256	3.84	0.401	11.78
Pb-production	0.057	2.49	0.042	1.83
Zn-production	0.176	1.41	0.238	1.90
Residual fuel oil Consumption	7.96	0.52	7.27	0.47
Cement production	100	4.2	111	4.66
Municipal solid waste	50	50	70	70
Medical waste	0.33	6.6	0.33	6.6
E-waste	-	-	0.146	0.82
Biomass burning	32 (16 – 61)	7.74	32 (16 – 61)	7.74
- Forest	116 (58 – 289)	4.76	116 (58 – 289)	4.76
- Crop				
Chlor-alkali plants	0.476	132[a]	0.304[b]	6.2
Brick manufacturing	-	-	350 (140 × 10^9 pieces) [c]	7.49
Sum		321.49		253.36

[a]We have assumed that in 2000, average Hg emissions from Chlor-alkali plants were (185 + 79 = 264/2 = 132 Mg (see section 3.4);

[b]In 2004, the data for 2006 has been used to estimate Hg emissions form Hg-cell plants. The abnormal reduction of Hg emission in 2004 was due to conversion of Hg-cell process to Membrane cell process where no Hg is used.

[c]We have assumed that weight of a Indian brick is 2.5 kg. Based on this information, the total weight of bricks in this study: 140 × 10^9 pieces × 2.5 kg wt of a brick i.e. 350 Tg of bricks.Weight of a brick is obtained by personal communication with TERI, New Delhi on 11.01.2008.

Regarding biomass burning, Venkataraman et al. (2006) mentioned unit: Tg yr^{-1}. For this reason, we have assumed that the same amount of biomass was burnt in 2000 and 2004, respectively.

gas cleaning equipment in power plants is in question. The emissions of Hg from coal fired power plants in Indian in the years 2000 and 2004 are shown in Table 4.6. The emission of Hg from power plants has increased by 17% since 2000 and the amount of Hg discharged into the environment may still be increasing as many plants have no gas cleaning equipment, similar to the situation in China. It is necessary to add, however, that at lower temperatures (350 – 400°C) in the flue gas duct, in the presence of chlorine, sulphur and calcium part of the Hg^0 vapour is oxidized to $Hg^{(II)}$, and/or reacts with carbonaceous ash particles and is deposited as Hg(p). Indian coal typically contains over 35%-wt fly ash (FA), and this particle surface offers an important site for Hg absorption (Mukherjee et al., 2008). In a recent study by USGS on 102 selected coal samples from different basins of India concentrations of Hg were found to be in the range between 0.02.to 0.16 mg kg^{-1}. Further details will be available at the project website of the USGS (http//energy. er.usgs.gov/coal quality/wocqi/collaborators.html) (A. Kolker, USGS, and S. Dunkee USEPA, personal communication, 5, May, 2008).

Currently, the energy sector contributes over 120 Tg of fly ash (FA) across the country. Several recent studies have identified heat, humidity, solar radiation and the presence of water as important factors in the release of Hg from FA, FGD solids or a mixture of these. The current estimation of Hg in FA is about 41 Mg (Mukherjee et al., 2008), although there is a lack of information on Hg in Indian coals and coal FA (Mukherjee and Zevenhoven, 2006).

4.3 Iron and Steel industry

In the 21st century, iron and crude steel production in India has increased from 21.3 Tg in 2000 to 25.0 Tg in 2004, and from 26.9 Tg in 2000 to 32.0 Tg in 2004, respectively. Steel is manufactured mainly by integrated steel manufacturing processes using the chemical reduction of iron ore, and conversion of iron from the blast furnace in a basic oxygen furnace (BOF). Steel can also be produced by melting steel scrap (e.g. from shredded cars) in an electric arc furnace (EAF). Coke, necessary in the iron and steel industry, is obtained by coking in ovens at 1000 °C or more. Here, Hg from coal is passed into the gas and other products of solid, liquid and gaseous by-product phases of the coking process. Coal consumption for the production of iron and steel in India accounts for about 13% of the total consumption i.e. 48.5 Tg in 2004. The emission factor calculated for Hg emission is 0.08 g Mg^{-1} crude steel which is quite realistic (see Table 4.4). It should be stated here that coke still contains a small amount of Hg. Hence some Hg will pass into the atmosphere also from the sintering plant, blast furnace and steel production. Our emission factors are higher than the emission factors calculated by other authors. The simple reason for this is that the quality of coal in India is quite poor due to a high ash content (30-40%-wt). For this reason, more coal is needed per Mg of steel production than in the USA or in Europe. Residual fuel oil was also used in this sector, although the consumption of this decreased from 672 Gg in 2001 to 620 Gg in 2004.

In the electric arc furnace process metal from shredded cars is generally used as a raw material for conversion into steel. It has been reported that Hg-lamps are often removed when scrap cars are processed into crude steel (Personal communication with Dr. Pandey, TERI, New Delhi, April 2008). Otherwise, more Hg will be emitted from the iron and steel industry.

The atmospheric deposition of Hg in the vicinity of iron and steel works was cited as being in the range of 60 to 836 g/km^2/month whereas Hg concentration measured in dust was 56 mg kg^{-1}. In surface soil it varied between 40 and 72 mg kg^{-1} (Srivastava, 2003).

4.3.1 *Non-ferrous Metallurgical Iindustry in India*

4.3.1.1 Production of Metals by Different Processes and Emissions of Mercury

The primary non-ferrous metal industries are based on copper (Cu), lead (Pb) and zinc (Zn). There are four copper smelters in which the Flash Smelting Process, the Ausmelt process and the Imperial Smelting are practiced. In addition, there is one zinc production plant where Zn is produced by the hydrometallurgical process. The Cu-smelters are situated at Khetri, (Rajasthan), Ghatsila, (Jharkhand), Dahej (Gujarat), Tuticoran, (Tamil Nadu) where the Copper Flash Smelting Process and the Ausmelt Process are used, respectively, besides the later process at Tuticoran, where Australian Cu-concentrate containing 5.0 mg Hg kg^{-1} was reported to be used (Personal communication with Manger of Tuticoran Cu-smelter, December 13, 2007). The Outokumpu Flash Smelting Process was originally developed for Cu-concentrate in Finland. In this, dried concentrate is smelted in the Flash Smelting Furnace in the presence of pre-heated air and oxygen to produce high grade Cu-matte which is then converted into blister copper in the converter.

On the other hand, in the Ausmelt process, feed materials are fed through a port located in the roof of the furnace and fall into the molten bath, which favors material transfer and handling systems. Air and oxygen are necessary for combustion, and molten metal and slag are removed and off-gases from the Ausmelt furnace are cooled and cleaned in gas clean-up systems before discharge.

In Udaipur (Rajasthan), zinc is produced by the hydrometallurgical process which comprises the following steps: roasting, leaching, solution purification, zinc electro-winning, melting, casting, and alloying.

In Tundoo (Jharkhand) lead is produced by the Blast Furnace Process whereas at Chhattisgarh the Imperial Smelting Process has been erected for the co-production of zinc and lead. The total production amounts of copper, zinc and lead for 2000 – 2004 are shown in Table 4.1. For Cu production, the major part of the concentrates is imported from Australia. Emissions of Hg from the Cu, Zn and Pb - industries are shown also in Table 4.6.

Except for the hydrometallurgical process, Hg is evaporated at the high process temperatures that occur during the production of Cu, Zn, and Pb. When Hg is released from ores, concentrates or from fossil fuels and enters into the biosphere, it can be highly mobile, cycling between the Earth's surface and the atmosphere. Speciation of Hg is very important. Mercury as $HgCl_2$ may be captured by some gas clean-up devices (e.g. wet scrubbers), but elemental Hg (Hg^0) is not captured effectively. Once Hg is released from a process, it cycles between soils, the aqueous environment and the atmosphere. It has been confirmed that the common forms of Hg in the environment are: metallic Hg^0, $HgCl_2$ and MeHg (UNEP, 2002).

In India, secondary Cu is produced by Boliden's KALDO process (a Swedish process) where roasting, smelting and converting occur in the same converter, charged with Cu-scrap. Production data is shown in Table 4.1. Beside primary production, Pb and Zn metals are also produced through secondary routes from scrap, dross and residues. Secondary Pb production occurs in less organized sectors (Personal communication with the Director of Indian Lead and Zinc Development Association, New Delhi, December 6, 2007). The emissions of Hg from such secondary metals production are not known.

4.4 Chlor-alkali Industry in India

4.4.1 Chlorine and Caustic Soda Production

The basic raw materials for chlorine chemistry are sodium chloride, water and energy. There are three main electrolytic production technologies utilized in the chlor-alkali industry, being the diaphragm process, the mercury process and the membrane cell processes. Generally, chlorine and caustic soda are co-produced in a fixed ratio (1:1) by chlor-alkali plants and hydrogen is also produced. The primary product is chlorine. In each process, the electrolysed salt solution is directly converted from chloride ions to elemental chlorine by direct application of electric current, and the overall chemical reaction is as follows:

$$2NaCl + 2H_2O \rightarrow Cl_2 + H_2 + 2NaOH \quad (2)$$

According to the Alkali Manufacturers Association (AMA) in India, forty-two chlor-alkali plants in India have the capacity to produce a total of about 2.2 Tg of chlorine per year, whereas the world production was cited at 55 Tg per year. On the global market, the Middle East played an important role in the production and exporting of caustic soda in 2007. The western region of India is the largest manufacturer of chlorine (1.04 Tg) in twelve chlor-alkali plants, followed by the Southern region (0.44 Tg), the Northern region (0.28 Tg) and the Eastern region (0.24 Tg). Pandey (2006) indicated thirty-five plants in India of which twenty-five have been converted to the Membrane process. These plants are vital for the chemical industry and this industry sector has been in operation in India since 1941. The number of world chlor-alkali industry plants (Hg electrolysis units) has been reduced from eighty-six in 2002 to seventy-four in 2006 (Figure 4.3). In India 86% of the plants have been recently converted to membrane cell technology and the rest, 8 or 10 units, are in the process of conversion to membrane cell technology, which does not use Hg in the process (Pandey 2006, AMA, 2007). The present list of chlor-alkali plants operating in India is given in Appendix 1 and Figures 4.4 & 4.5 indicate Hg-Cell and converted Membrane Cell plants in India, respectively.

A non-governmental organization (NGO) in New Delhi has estimated that in 1999 – 2000, the loss of Hg from the Hg-cell process to be 394 g Hg per Mg of Cl_2 production. during the same period, chlorine production was 0.48 Tg. This correspond to an annual Hg emission of about 185 Mg. (http://www. toxicslink.org/docs/06035_publications-1-33-2.pdf). In another study, Srivastava (2003) estimated that the Hg loss from Hg-cell plants in India is about 142 g Mg^{-1} NaOH produced. From his estimate, Hg loss to the atmosphere between the years 1997-2000 was about 79 Mg yr^{-1}.

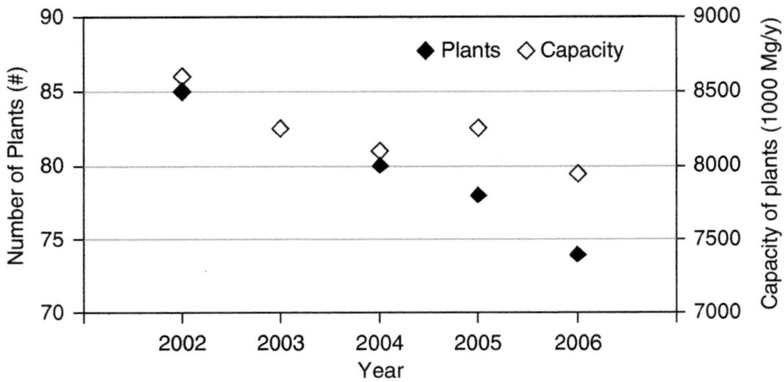

Figure 4.3 The scenario of world's chlorine plants and production capacity, 2006 (WCC, 2007; Reproduced with permission , Veronique Garny, 2007)

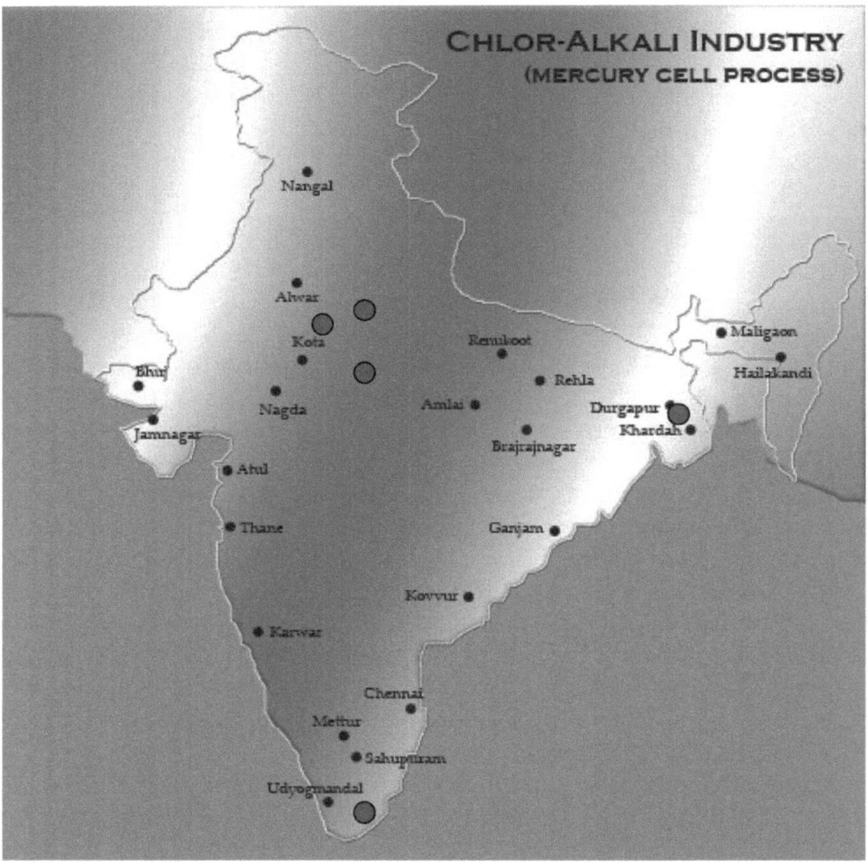

Figure 4.4 Mercury cell chlor-alkali industry in India. Big circles indicate Hg-based thermometer industry. Adapted from Toxicslink, New Delhi. See Appendix I which indicates conversion of major number of chlor-alkali plants from Hg-cell to Membrane cell process where no mercury is used (from http://www.toxicslink.org/docs/06035_publications-1-33-2.pdf)

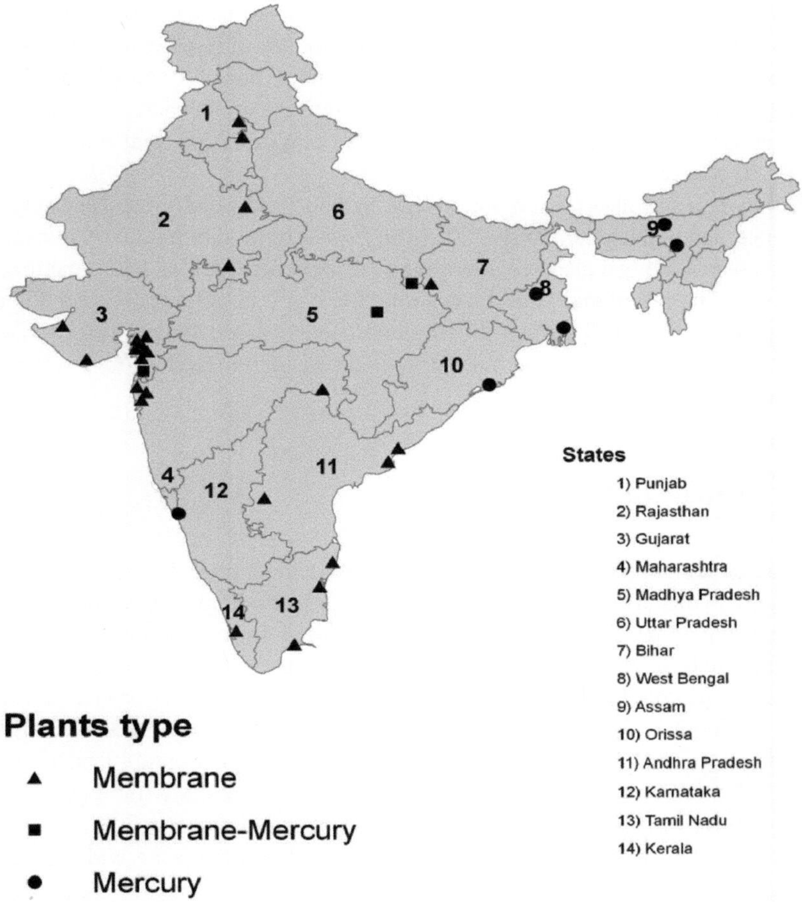

Figure 4.5 Locations of chlorine industries in India, 2008 (Information received from S. Sinha, Toxicslink, New Delhi, 2008: plotted in the map of India); See Appendix 1 for details of plants

Clearly, the information on Hg emissions from this particular chemical sector in the past is highly unreliable. However, more recent information from AMA (2007) indicates that NaOH production by the Hg-cell process was at 0.3 Tg for 2006-07 whereas for the same period the production by the Membrane Process it was 1.7 Tg NaOH. Considering the emission factor for Hg emissions from Chlor-alkali plants to be 20.4 g Hg Mg^{-1}, total Hg passed into the atmosphere from this chemical industry was 6.2 Mg yr^{1}. (Table 4.6). In the 1990s, chlor-alkali plants were the single largest Hg consuming industry in India, consuming about 70 – 80 Mg of Hg each year (CSE, 2005). Mercury has been detected in groundwater and surface water in the vicinity of Hg-cell chlor-alkali plants. In addition, Hg occurred also in the vicinity of dyes, paints and pigment manufacturing units which use Hg-based catalysts in the manufacturing processes (CSE, 2005). According to the Ministry of Environment and Forest (New Delhi), Indian chlor-alkali plants will be Hg-free in 2012 . In the near future, chlor-alkali plants may be free from Hg-cells, but this does not mean that at the same time Hg will disappear from the vicinity of chlor-alkali plants. The metal will not disappear from the environment, but will convert into MeHg, Hg^0 or mercuric chloride which will then pass into the atmosphere, to again fall on forest soils or the water shed. Speciation of Hg plays an important role in toxicity and the exposure to living organisms.

However, it has also been reported that about 170 Mg of Hg has been imported and consumed in the years 2004–2005 (http://www.dgft.delhi.nic.in/). Chlorine and sodium hydroxide use also provide a range of benefits, such as PVC manufacture which is an important material. It has a long life e.g. it lasts for more than 35 years. It emits about 50% less carbon dioxide and needs less oil for production. Beside these, in many energy saving buildings, foam insulation and PVC windows are based on chlorine chemistry.

4.5 Cement Industry

The Indian cement industry is the second largest cement producer in the world with an installed capacity of 144 Tg annually. Due to technological development some Portland cement production plants are well advanced. In cement production, energy consumption is quite high. In the Indian cement industry, the capacity of kilns varies between 10 Mg day^{-1} and 7,500 Mg day^{-1}. Most cement, 94%, is produced in large (capacity 600 Mg day^{-1}) plants. At present there are 124 large rotary kiln plants. In India, in general Ordinary Portland Cement (OPC) (56%) and blended cement (43%) are manufactured. The dry process (93%) route besides some (much more energy intensive) wet and semi-dry processes (7%) are practiced.

In the cement industry, Hg was found to be emitted from the wide range raw materials and other resources used. There are more than thirty raw different materials used in the manufacture of Portland cements. These materials can be classified as: (a) calcareous, (b) siliceous, (c) argillaceous, and (4) ferriferous. A variety of calcareous raw materials are used in Portland cement including: limestone, chalk, marl, sea shells etc. The thermal treatment of raw materials for the manufacturing of Portland cement is carried out in kilns. It is not known if any plants in India use waste as an alternate fuel in a cement kiln.

However, there are four steps in production:

- Evaporation of uncombined water;
- Dehydration e.g. at temperature 430 ºC, formation of oxides of Si, Al and Fe occurs;
- Formation of calcium oxide at 980 ºC;
- In the burning zone of the rotary kiln, the cement clinker is formed at 1510 ºC

In determining the emission factor, we followed the material balance used by Smith (1999), the input data being as follows:

- 0.40 mg kg^{-1} (Hg / coal)
- 0.1 Mg coal / cement

- 1.2 Mg limestone / cement
- 0.03 mg kg^{-1} (Hg /limestone) (UNEP, 2005)
- Emission factor of coal fired kiln 1.5×10^{-6} kg /Mg cement (UNEP, 2005)

Based on the above information, the emission factor for Hg per Mg of cement production is 45.6 mg Hg Mg^{-1} cement (the EPA's value is 65 mg Hg Mg^{-1} cement, USEPA (1997)). In a dry process, Hg will leave the kiln in gaseous form, but in the pre-heater tower, it may be adsorbed both on the kiln feed and cement kiln dust (CKD). The speciation of Hg is again very important as retention of Hg in gas cleaning equipment depends upon: a) the gas cleaning equipment; b) the form of Hg; c) the temperature and retention time (Senior et al., 2003; Mukherjee and Zevenhoven, 2006).

4.6 Wastes Disposal

4.6.1 Municipal Solid Waste (MSW)

People in India live on 3.28 million km^2 of land. Due to population growth, there is also an increase in municipal solid waste (MSW) in India. The growth rate of MSW is reported to be 1.33% per capita per annum (EPTRI, 1995). The collection efficiency of MSW is about 72.5%, but still waste transport capacity is lacking in 70% of the cities (TERI, 1998 vide Singhal and Pandey, 2001). MSW consists mainly of household garbage, and other commercial, institutional and industrial solid wastes. In household wastes, broken thermometers, instruments, Hg-vapor lamps, toys, electric switches, fluorescent tube lights, Hg-batteries etc. are the expected to be the Hg containing products. In addition, most MSW contains large amounts of organic species.

MSW Rules 2000 indicate that the municipal solid waste should be disposed of in a environment-friendly manner such as: pelletisation, combustion/incineration, land filling, bio-methanation and composting from which power could be produced for local industry (timesofIndia_Indiatimes.Com/articleshow/ 134243.cms). However, MSW generally disposed of by the following ways:

1. Landfill practice
2. Open dumping
3. Open burning

The generation figures for MSW in India are based on the recent study conducted by Singhal and Pandey (2001) (see Figure 4.6). Between the years 2001 and 2004, 50 to 70 Tg per year MSW was generated in India, with most waste generated in Uttar Pradesh where 166 million people live. It has been estimated by those authors that in the year 2047 the amount of MSW might reach 260 Tg. An estimated emission factor for Hg in MSW is 1.0 g Mg^{-1}, without emission control. In India garbage and mixed waste are often burnt near the road side. 94% of MSW is dumped in landfills

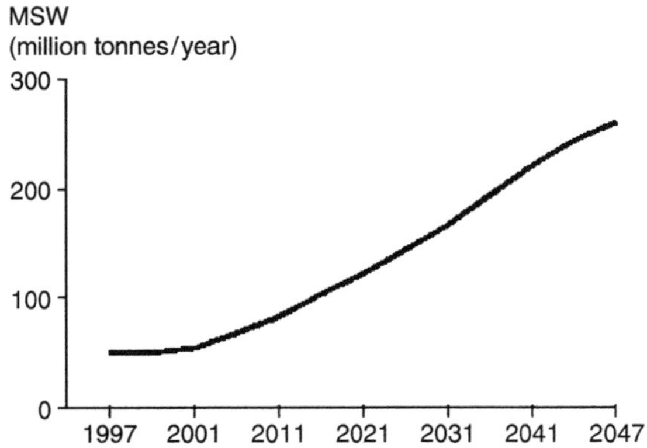

Figure 4.6 Solid waste generation in India (Reproduced from Singhal and Pandey, 2001)

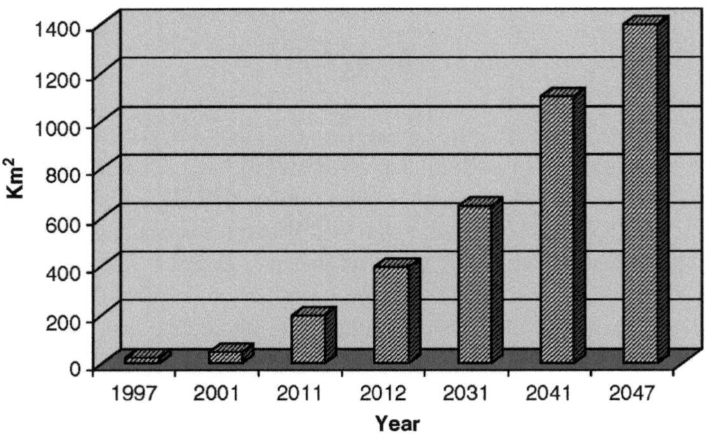

Figure 4.7 Land requirement (km^2) for disposal of municipal solid waste (reproduced from Singhal and Pandey 2001)

without any proper systems and 5 % of wastes are used for composting (CPCB, 2000). Due to the humid climate and rainy seasons, leaching of Hg can be expected.

The huge amount of MSW produced is a serious problem in India. Unfortunately, little attention is being paid to proper management of MSW. Problems with wastes are not particular to India but occur in most Asian countries. Asian countries face serious problems in the solving of disposal problems for wastes. It is not only a technical problem, but many political, legal and environmental factors are also involved. In addition, land requirements for the disposal of MSW will increase and it has been estimated that by the middle of the 21st century 1400 km^2 of land many be needed for MSW (Figure 4.7). Sharholy et al. (2007) determined the constituents of MSW for Allahabad where 1.1 million people live (2006) and these are shown in Table 4.7. This may represent a rough picture on MSW for whole India.

Table 4.7 Estimation of the essential parts of MSW in India based on the study for Allahabadcity (After Sharholy et al. 2007)

Elements in MSW	Weight %	% weight based on 21 class 1st cities*
Paper	3.6	5.7
Cardboard	1.09	-
Metal, tin cans	2.54	2.1
Glass	0.73	2.1
Food wastes	45.3	41.80
Textile rags	2.22	3.5
Plastic (Poly bag)	2.86	3.9
Miscellaneous (bricks, ash, fine dust, rubber, wood, leather, wastewater etc.	41.66	41.1
Total	100	100
Moisture	25 %	

*(CPCB, 1999)

4.6.2 *Medical Wastes*

There is limited information available regarding medical wastes in India. It is generally expected that medical wastes contain more Hg than MSW does. In this sector, there are many instruments in hospitals which contain large amounts of Hg. There are basic regulations related to health care in India, but unfortunately these regulations are not followed properly. Visvanathan (2006) estimated medical waste generation in Asia, the figure for India being 0.33 Tg yr^{-1}. Based on 20 g Hg emission per Mg of medical waste (USEPA 1997), the total Hg emission to the atmosphere from this source is estimated to be 6.6 Mg per year. The average health care waste generation per bed per day in India has been estimated at 1 – 2 kg (Table 4.8). According to law experts in India there is no lack of legislation, but the problem lies with implementation. Often medical waste is disposed of together with MSW, due to which the waste stream becomes hazardous. Technologies available for handling medical waste include: a) incineration; b) autoclave; c) microwave; d) chemical disinfection and e) plasma pyrolysis. In India there are incineration plants at some hospitals to handle medical wastes. The capacity of these varies from 50 to 175 kg hr^{-1} of infectious and non-infectious wastes generated. In Delhi, there are 61 medical waste incinerators, but there is no information regarding the handling of hospital wastes in other parts of the country.

4.6.3 *Electronic Waste (E-waste)*

In recent years, discarded electronic waste known as "E-waste" often enters into the waste streams in India as in many other (Asian) countries. Automation, increased demand for electronic equipment including computers and increased consumer

Table 4.8 Estimated medical waste generation in selected Asian countries (Visvanathan, 2006)

Country	Waste *(kg bed^{-1}day^{-1})*	Total wastes *(Mg yr^{-1})*
Bangladesh	0.8 – 1.67	93,075 (only in Dhaka)
Bhutan	0.27	73
China	-	730,000
India	1 – 2	330,000
Malaysia	1.9	-
Nepal	0.5	365
Pakistan	1.06	250,000
Sri Lanka	0.36	6,600 (Only in Colombo)
Thailand	0.68	-
Metro Manila (Philippines)	-	17,155
Vietnam	2.27 (Hanoi)	60,000

choice are believed to be the major reasons for growing quantities of E-waste. Recently its production in India has increased to 380 Gg in 2007 and it has been forecast by the authorities that its production will increase to 470 Gg in 2011. There is also a huge amount of E-waste imported from the West, possibly as much as 50,000 Mg annually; mainly discarded computers and accessories (http://www.physorg.com/news116912274.html accessed January 21, 2008). Before or while burning E-wastes, scrap dealers can recover valuable metals, which is however not without impact on health and the environment. In India, there is a general lack of recycling technology. Computer scrap is often reused and many different types of E-waste end up in landfills, creating health and environmental problems since this method of disposal is not well developed. In addition, India has dumped E-waste in other countries.

Sarkar (2007) studied E-waste generation (excluding imported) in India and observed that 146 Gg electronic wastes were generated from PCs, refrigerators, TVs and washing machines in 2006, and the amount is expected to rise to 1.6 Tg in 2012. In addition, India receives a large volume of E-wastes. As there is no national level policy for management of E-waste, most of the recycling facilities are unorganised and do not use suitable, state-of-the-art technologies for the recovery of toxic metals. Selected trace elements from E-wastes in India have been addressed by Sarkar (2007), indicating that 0.82 Mg Hg yr^{-1} escapes into the environment (Table 4.9), and the values will increase due to increased annual E-waste and the lower lifetimes of newer computers. The total Hg released from the above mentioned three types of wastes is given in Table 4.6.

4.7 Biomass Burning

We consider Hg emissions from biomass burning to be anthropogenic and it is therefore necessary to understand its effects on the atmosphere on a regional (i.e. in Asia) and global scale (Reddy and Venkataraman, 2002; Streets et al. 2003). Venkataraman et al. (2006) focused forest and crop waste burning in India between

Table 4.9 Mercury in electronic wastes in India (from Sarkar, 2007)

Waste type	2004 *(Tg)*	Total Hg *(Mg yr^{-1})*
E-waste	0.146	0.82*

E-waste in India – annual and total production and entry into environment (Referring to the estimated number of annual production of obsolete computers (1.38 million), total market size (15.5 million), MCC data on average weight of computer (27 kg), proportion of presence of toxic chemicals in each computer and recycling efficiency - annual and total production of various toxic substances from e-waste generated from discarded computers and related materials have been estimated.

	TW (%)	AW *(mg)*	RE	YP *(Mg)*	AR *(Mg)*	EE *(Mg)*	TP *(Mg)*	TR *(Mg)*	TE *(Mg)*
				A	B	A-B	C	D	C-D
Hg	0.0022	0.00059	0	0.82□	0.00	0.82	9.19	0.00	9.2

Source (adapted): MCC (Microelectronics and Computer Technology Corporation), 1996, Electronics Industry Environmental Roadmap, Austin, Texas, details available at www.svtc.org/cleancc/pubs/sayno.htm (last accessed on 20th March 2006) and Boralkar D.B. (2006), Perspective of electronic waste management, Green Business Opportunities, Vol 12 (1), pp 7 – 10.
TW = Total weight of Hg in each computer; AW = Average weight of Hg in each computer; RE = Recycling efficiency; YP = Yearly production of toxin material; AR = Max recycled annually; EE = Entry into environment; TP = Tot production (based on market size); TR = Total max recycled possible; TE = Total entry into environment.

1995 – 2000 using forest burnt areas and biomass density for Indian ecosystems. It has been estimated that Indian forest is burned at a rate of 32 (16 – 61) Tg yr^{-1} and in open and dense forest with low density biomass cover (Streets et al. 2003). Crop waste burning, including cereal, sugarcane waste, oilseeds, fiber crops and pulses were also estimated at 116 (58 – 289) Tg yr^{-1} (Venkataraman et al. (2006). Mercury emissions during biomass burning have been shown in Table 4.6. However, biomass is the main source of energy for villagers and about half of all energy in India used for cooking food. It is interesting to note that the firewood consumption (kg/capita/yr) increases with increased altitude in the region of Garwal Himalaya. The summer time average consumption at 500 altitudinal range (m a.s.l.) is 392.28 kg/person/year whereas at 2000 altitudinal range it is 1019 kg/person/year (Bhatt and Sachan, 2004). It means Hg emissions will be higher as well when firewood is burnt at higher altitude in the mountain region.

4.8 Miscellaneous

4.8.1 Brick Industry

The Indian brick industry is the second largest in the world after China. The Energy and Resources Institute (TERI) has estimated over 100,000 units producing 140 billion bricks per year. There are three types of brick works, based on production capacity i.e. small (<1 million bricks per year); medium (1 – 2.5 million bricks per years) and large facilities (>2.5 million bricks per year). Smaller brick facilities are generally situated in the village areas whereas the medium and larger types are located near urban areas. It is an energy intensive process and coal is the major fuel used. It has been estimated by TERI that about 24 Tg of coal, containing 0.376 g Hg Mg^{-1}, are fed to the brick kilns for the production of 140 billion bricks (e.g. 350 Tg of bricks considering the weight of a brick in India to be 2.5 kg, based on TERI, 2008). In addition, Indian brick kilns consume a considerable amount of biomass and fuel oil, but the amounts of these are not known (Garg et al., 2006). Based on the use of low grade coal, the uncontrolled emissions of Hg from these processes to the atmosphere are 7.5 t Hg yr^{-1} (Table 4.6). In addition there are PIC (product of incomplete combustion) emissions. In 1996, the Indian government enforced regulations which have caused some technological improvements e.g. reduced dust emissions, and improvement of firing technology, especially in large brick works (Maithel and Uma, 2000). It has been reported by TERI that new technologies such as vertical shaft brick kilns (VSBK) has been introduced in several brick production facilities. The process claims to both lower investment and to meet emission standards.

4.8.2 Instruments, Batteries and Thermometers

Substantial amounts of Hg are used in the production of instruments, batteries and clinical thermometers. Table 4.2 indicates the amount of Hg used in the manufacture of instruments between 1998 and 2001. Clinical thermometers and barometers may often be broken during manufacture. Broken products are put aside and often cause fugitive emission of Hg. However, there are no measurements by which the total amount of Hg thus lost to the environment can be detected. In addition, bookkeeping is quite poor. One of the largest thermometer companies, at Kodaikanal in the Tamil Nadu state, was forced to close due to illegally dumping Hg-bearing waste into the surroundings. Before closing the estimated Hg emission from broken pieces at the plant, was 3.5 to 4.2 t yr^{-1} (Each thermometer contains 1.0 g Hg; total production 10-12 million pieces per year; breakage during production and handling 35%.) The Hg consumption in different instruments is shown in Table 4.10.

Table 4.10 Total mercury consumption in instrument manufacturing Industry (from different sources)

Instruments	Unit *(kg)*	Used/unit *(g)*
Clinical thermometers	3,100	0.61
Lab thermometer	900	3.0
Blood pressure monitors	12,000	60.0
Barometers (in kg)	125	5
Total	16,125	

4.9 Mercury in the Indian Environment and the Cycling in the Bio-geosphere

The soil, water and air are not only a part of the ecosystem but also play an important role for humans, animals and aquatic species because the survival of human, aquatic species and plants is tied up with uncontaminated soil, water, and air. Releases of the non-essential element Hg from industrial sources are well documented in developed countries. This study addresses emission of this element in the Indian ecosystem and indicates that its presence in soils, plants, air and water and aquatic species and sediments are at alarming levels, which is also supported by the work of Srivastave (2003) and many other authors in the West (Hylander and Meili, 2003; Pacyna and Pacyna, 2001, 2002; Pirrone et al. 1998). Coastal areas are often contaminated by the discharge of Hg from all Hg-cell chlor-alkali plants. At one well known clinical thermometer works in Tamil Nadu, an ambient air concentration of 1.32 μg Hg m^{-3} was reported. Outside the factory, lichen (Parmelia sulcata) and moss (Funaria hygrometrica) samples contained 7.9 μg kg^{-1} and 8.3 μg kg^{-1}, respectively. Fish in lake waters contained 120 to 290 mg kg^{-1}, whereas total Hg and MeHg in waters were measured to be 356 – 465 ng L^{-1}, and 50 ng L^{-1} respectively (Karunasagar et al., 2006). Concentration of Hg_T in sediments in the same lake near the factory situated 2130 m above mean sea level, varied from 279 to 350 mg kg^{-1}.

High concentrations of Hg are reported in fish that grow in saline or fresh water in coastal areas (1.1 – 700 mg kg^{-1}) Table 4.11) for many states of India (WHO's permissible value is 0.5 mg Hg kg^{-1}). Sinha et al. (2007) studied the Hg concentration in different samples from the river Ganges (Table 4.12) and a high concentration was reported for various species although not in the water samples themselves.

Seasonal variation was also reported by these authors. These authors collected about 61 fish samples for the river Ganges near Varanasi, where they found that the Hg concentration in the fish (*Macrognathus pancalus*) varied up to 91.7 mg kg^{-1}.

Table 4.11 Average and maximum mercury concentration in fish and other species

Place	Fish/species	Hg *(mg kg^{-1})*	Hg max. *(mg kg^{-1})*	Reference
North Koel river, Jharkhand	Fish		600 - 700	1
Mumbai, East Coast,	Fish	0.03 – 0.082	1.6	2
Maharashtra	Bivalves	0.13 – 10.82	21.6	2
Sagar Island,	Gastropods	1.05 – 3.60	7.2	2
East coast West Bengal	Crabs	1.42 – 4.94	9.9	2
	Bivalves	0.06 – 2.24	4.5	2
Binage, Karwar, Karnataka	Oysters	0.18 – 0.54	1.1	3

1. "Mercury concentration of fishes in north koel river, Rehela, Bihar, India". Indian Biologist 23(2) 1992; 58 – 60.
2. Chemosphere, vol 33 147 – 158 (1996), cited in Global Mercury Assessment, UNEP Chemicals, 2002.
3. "Heavy metal distribution in the biotic and abiotic matrices along Karnataka coast, West coast of India". Indian Journal of Marine Sciences, 27, June 1998, 201–205.

Table 4.12 Mercury concentration of different samples of the Ganges River collected at Varanasi, India (Sinha et al., 2007)

Season	Water ($\mu g\ kg^{-1}$)	Sediment ($\mu g\ kg^{-1}$)	Benthos ($\mu g\ kg^{-1}$)	Fish ($mg\ kg^{-1}$)	Soil ($\mu g\ kg^{-1}$)	Vegetation ($\mu g\ kg^{-1}$)
Winter	0	106 ± 113	144 ± 252	4.048 ± 18.676	95± 114	254 ± 397
Summer	0.37 ± 0.29	80 ± 87	108 ± 167	0.205 ± 0.531	126 ± 111	98 ± 81
Post monsoon	0.32 ± 0.48	0	92 ± 129	4.369 ± 16.067	0	245 ± 127

The observed Hg concentration in their study was more than that in the fish samples collected from the western coast, Mumbai (0.03 – 0.82 mg kg^{-1}). Mercury in fish resides as MeHg (which affects humans) bound to the proteins and muscle tissues of the fish. Besides MeHg, concentration of elemental Hg is also observed in fish and shellfish species. It was reported also that rice fields contaminated by Hg from coal combustion power plants contain MeHg. In many parts of the world, including India, high concentrations of MeHg and Hg have been demonstrated in oceans, rivers, lakes and reservoirs. Pirrone and Mahaffey (2005) reviewed Hg in global fish populations. In many regions of India, fish is an important daily food item and the authorities should make people understand the effects of high concentrations of Hg and MeHg in fish. In the 1960s in Japan, Minamata disease was caused by consuming Hg contaminated fish and rice. It has been reported in the panel discussion of the 8th International Conference on Mercury as a Global Pollutant at Madison, WI, USA that MeHg in human hair is 250 to 300 times more than Hg in the blood for those who eat fish regularly or frequently. Patel (2003) studied the health impacts on humans at contaminated sites of the central region, Chhattisgarh state of India where 50 Tg of coal and minerals are exploited annually by various industries and thermal power plants. In addition over 600 rice mills are in operation producing > 2 Tg of rice contaminated by Hg, As and Pb. Twenty-two human hair samples were analysed from the contaminated areas and were found to contain Hg from 2.6 – 37.8 mg kg^{-1} with mean and median values of 12.3 and 10.4 mg kg^{-1}, respectively. This author believes that in this region, the toxic effects of Hg due to contaminated of food and water are expected to show up shortly. However, scientific knowledge on MeHg exposure and its effects on humans are still not complete.

Das et al. (1998) indicated that 90% of MSW in India is directly dumped on the land in an improper manner and the problem with disposal facilities in towns and cities is that they cannot keep pace with the quantity of waste generated. It has also been forecasted that by the middle of the 21st century, MSW will be generated at a rate of 250 – 300 Tg yr^{1} (Sharholy et al., 2007; Das et al. 1998). Sewage sludge, pesticides, composts and fertilizers are often used on agricultural lands and these are also sources of Hg in soils and groundwater. Often, untreated effluent is pumped into rivers, lakes and groundwater, and through bore wells, (such as in Andhra Pradesh and Gujarat states). The Hg leached from landfills contaminates groundwater and streams, from which it moves into soils, and agricultural lands.

Mercury emission scenarios for coal combustion, MSW, and the non-ferrous metallurgical industry are alarming. The highest Hg emission from coal combustion has been estimated in this document for the year 2004.

4.10 Discussion

This study indicates that the atmospheric Hg emissions in India from industrial sources ranged between 321 to 253 Mg annually during the past years. Occasionally, Hg emissions have been given for a single year, e.g. in brick manufacturing, medical wastes and E-wastes. The highest Hg emissions occurred from the combustion of fossil fuels in power plants, followed by three types of waste. The Hg emissions given for Chlor-alkali plants should be used with caution. In addition, the Hg emissions from residual fuel oil use over the period 2001 to 2004 were negligible e.g. 0.54 – 0.47 Mg (based on the uncontrolled emission factor 0.065 g Mg^{-1}). No information was available from the pulp & paper industry or the oil and petrochemical industry in India. It has been reported by the Ministry of Environment and Forests, New Delhi, that major Hg-cell plants have been converted to the Membrane Cell process (Pandey, 2006) which means that Hg use and emission in chlor-alkali plants are expected to drop. At the end of the 1990s, Indian chlor-alkali plants discharged Hg-contaminated wastewaters containing Hg in the range of 0.08 to 2 mg L^{-1} (the Indian standard for Hg in industrial waters is 0.001 mg L^{-1}). Mercury is used often in electrical and electronic devices, fluorescent lamps, laboratory and medical equipment, clinical thermometers and computer components. However, strict enforcement of the regulations inside the country is needed. If necessary, a trade ban should be implemented to curb the use of Hg in India. Mercury emissions vary widely from one region to another in India, based on industrial activities. Often, activities such as coal mining and the burning of coal in power plants, metallurgical industry, chlor-alkali facilities, and waste disposal will produce trace element problems, including Hg, in the region. We have also identified Hg emissions from the cement industry and brick manufacturing, where 24 Tg of coal are used. In addition, Hg containing E-waste, medical waste and the MSW add Hg to the ecosystem of India.

Two scientists from the Department of Botany, Sri Krishnadevaraya University, Anantapur, India studied Hg in plant species (*Tephrosia purpurea, Cassia auriculata* and *Arachis hypogaea*) near a cement works at Bethamacharla, Andhra Pradesh. They found more Hg, 0.76±0.04 (SD) ng/mg in the leaves than roots (0.750±0.02) or stem (0.541±0.01) in *Cassia auriculata.* It is also confirmed that the mobility of Hg is greater when it enters the plants through the stem or leaf. The accumulation of Hg indicates that the cement industry emits Hg to the vicinity of the plants, but the accumulation levels vary with species, wind direction, soil pH, aeration as well as soil moisture.

Villagers and ordinary people in cities or towns are often not aware of the toxicity of Hg and how this element enters into human food chain. Hence proper education is necessary so that people in India can understand the effects of Hg concentration in fish or in drinking water. Mercury is notorious for its toxicity to biological organisms. Excessive releases of Hg and its compounds may lead to severe environmental and health consequences (Wong et al., 2006).

In India Hg emissions from most industrial sources are still increasing. The coastal areas as well as the inside of the country are highly polluted due to chlor-alkali plants and the process by which Cl_2/NaOH have been produced since the 1940s

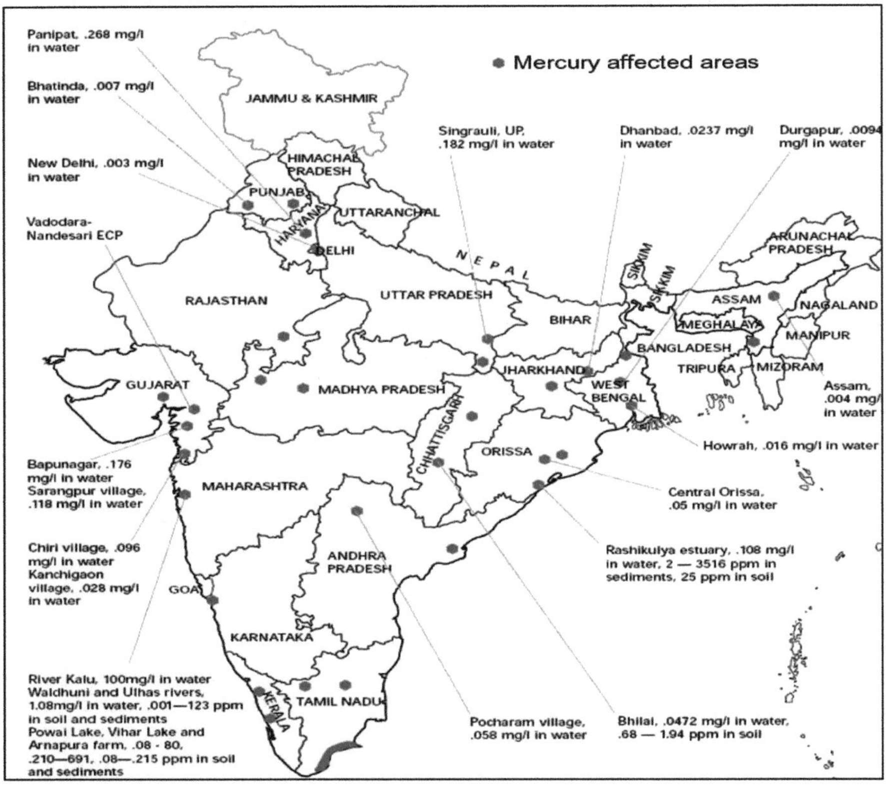

Figure 4.8 Mercury pollution due to use of Mercury-Cell Chlor-alkali plants in India in the 20th entury (Reproduce with permission, Center for Science and Environment, New Delhi 2008)

(Figure 4.8). The six Asian countries, India, China, Pakistan, Bangladesh, Japan, and Indonesia are all densely populated. Mercury emissions from industrial sources may cause future epidemics among the population. Over time, excessive Hg emissions will cause bioaccumulation and bio-magnification in the food chain, which may create serious problems with human health in India and surrounding countries. It has already been mentioned regarding the high concentration of Hg^0 and MeHg in fish, shellfish and seafood, that there are signs of Hg in the human hair when Hg - contaminated fish is eaten often. Excessive Hg emissions in India can cause Hg deposition problems on a global scale.

Industrial progress is necessary if the economic growth of a country is to be maintained. If India follows this pattern, the country must adopt the best available technology to control Hg emissions from industry. In coal fired power plants, electrostatic precipitators are not enough. This gas cleaning equipment does not capture vapour - phase Hg^0 from the process. Management in India should understand that there is a lack of scientific information concerning the emission, distribution, and biogeochemical behaviour of Hg in India. The imbalance between India and the

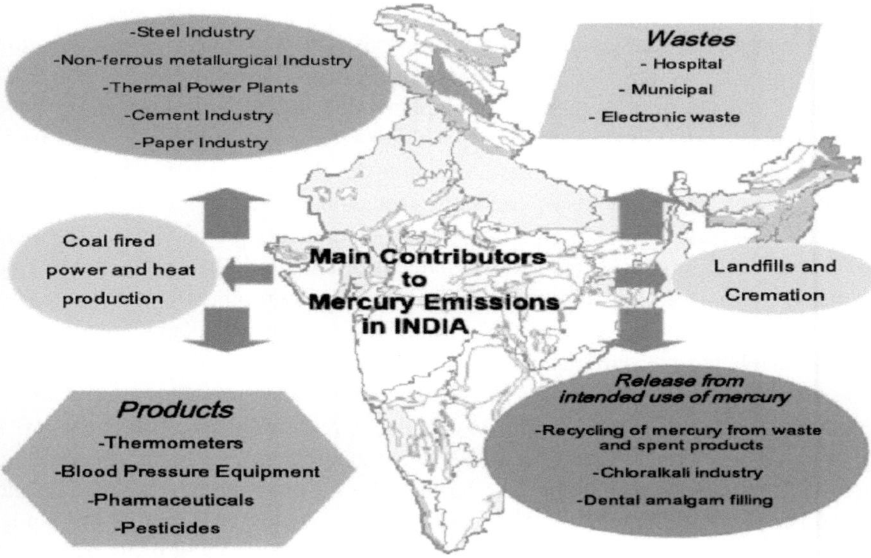

Figure 4.9 Sources of mercury in India (Modified from Srivastava, 2003)

developed countries regarding Hg emissions from industrial sources is due to technical motivation, environmental awareness and socio-economic conditions. India is, however, trying to curb environmental trace elements including Hg, by the formulation of discussions, mass education, strict regulations and effective control technologies.

On this subcontinent, detailed studies on e.g. Hg emission rates from specific industries, improved analytical techniques especially for Hg in coal, data bases, studies of biogeochemical properties of Hg, should all be encouraged for environmental and health reasons. In addition, there should be intense and continuous exchange programs among researchers and scientists between the developed countries, India and other countries in the region.

It should also be stated here that due to the lack of true emission data there will be uncertainties in the estimation of anthropogenic emissions of Hg not only in India but also in other parts of the globe. In India the Hg emissions from certain industrial sources, such as the chlor-alkali industry, has decreased very recently. However, the summary of the estimated Hg emissions is depicted in Table 4.6 and sources are shown in Figure 4.9.

4.11 Future Directions

Our knowledge of Hg and its compounds has improved quite a lot since the 1st International Conference on Mercury as a Global Pollutant, held in Gävle, Sweden in 1990. The European Union (especially EU-15), the U.S., Canada and Japan have formulated rules and regulations to curb Hg emissions from industry and the results can be seen now in many parts of these countries. The technology for gas cleaning equipment such as the flue gas desulphurisation process and others have been improved and the improvements also implemented. Local emissions of Hg have been reduced, but cross-boundary deposition of Hg by dry and wet methods is increasing. This means that increased economic development in India or in Asia and the burning of large amounts of coal and other industrial developments associated with this will result in long range transport of elemental Hg and MeHg from Asia to America and Europe. At present we understand the speciation of Hg and its role in the ecosystem, but the behaviour of MeHg in the environment is less clear.

In India, it is vital that scientists determine the sources of Hg and its emissions. Based on reliable measurements, it is possible to build up emission factors from which emissions of an element can be calculated. Stack measurements are expensive, but material balances can be applied to calculate emissions of Hg. This is not 100% correct but shows the path of how emissions of an element like Hg are developing. In the 20th century the Hg-cell process was used for the production of $NaOH/Cl_2$, and Hg emission was reported to be 150 – 200 Mg yr^{-1}. During the course of this study, we have understood that the majority (86%) of Indian chlor-alkali pants have been converted to the membrane-cell process with the remaining 8-10 plants still using the Hg-cell process. The Ministry of Environment and Forests, New Delhi, has stated that the total Hg release to the environment should be <2.0 g Mg^{-1} of product by December 2005 (Pandey, 2006). This information is encouraging. It is necessary for India to generate more reliable data, which can be used by scientists during modeling and formation of emission inventories. Computer modeling is often used in simulating global Hg scenarios. Often information on emissions from India and other Asian countries report emissions that are too low, as a result of which large differences occur between expectation and reality with respect to Hg cycles. Hence lack of scientific data in Asia has caused inaccurate assessments on Hg amounts and its association with environmental and health effects.

Asia is a region in which great diversity in climate is found. One area may be hot and humid while another region has very high precipitation such as in Bangladesh (1400 mm rain per year). In addition there is acid rain (pH < 4.5) SO_4^{2-} loading on soils and the aquatic environment. These diversities will have their effect on the bio-accumulation, bio-magnification and the uptake of Hg. There have been few studies, or none at all, carried out in Asian countries and these now have more focus on the Hg question than the Western countries. We have known for quite some time that high concentrations of Hg occur in pike in lakes in Sweden, Finland and Norway and that many lakes have been blacklisted. Scientists are working to understand the cause of this and produce remedies. In India, however people eat fish caught in lakes, rivers and the sea, unaware of the Hg problem in the fish and of how high concentrations may affect, particularly the high risk population i.e. pregnant women, and children. Remedial measures can be promoted by introducing non-Hg medical equipments such as digital thermometers, blood pressure equipment, avoiding amalgam fillings and for the large-scale use of alternative energy sources (solar, hydro etc.) to reduce coal use in the thermal power industry.

The scientific community has largely overlooked Hg emissions from the brick industry, which has been in operation now for decades. In our present study, we have observed that India, after China, is the second highest producer of bricks. In this industry 24 Tg of poor quality coal and bio-fuels (accurate amounts unknown) are consumed. Detailed study is necessary to improve the processes and to determine emissions of Hg and unburned particulates and their effects on the environment.

Appendix 1 Current lists of chlor-alkali plants in India (Personal communication with Toxicslink, New Delhi on 01.17. 2008)

No.	Name of the Unit	Location	Technology Adopted
A	**Eastern Region**		
1	Bihar Caustic & Chem.	Jharkhand	Membrane
2	Durgapur Chemicals	Durgapur, WB	Mercury
3	Hindustan Heavy Chem.	Kolkata	Mercury
4	Hindustan Paper (Nagaon & Cachar)	Assam	Mercury
5	HJI-Prop: GMMCO Ltd	Amlai, MP	Membrane + Mercury
6	Jayshree Chemicals Ltd.	Ganjim, Orissa	Mercury
7	Kanoria Chemicals Ltd.	Renukoot, UP	Membrane + Mercury
B	**Western Region**		
8	Atul Ltd.	Valsad, Gujarat	Membrane + Mercury
9	Ballarpur Industries	Ballarshah	Membrane
10	Century Rayons	Thane, Maharashtra	Membrane
11	Grasim Industries	Nagada, MP	Membrane
12	Gujarat Alkalies & Chem.	Dahej & Baroda, Gujarat	Membrane
13	Indian Rayon	Veveral, Gujarat	Membrane
14	NRC Ltd.	Thane, Maharashtra	Membrane
15	Reliance (IPCL)	Dahej, Gujarat	Membrane
16	Standard Industries	Mumbai, Maharashtra	Membrane
17	Shriram Alkalies	Jhagadia, Gujarat	Membrane
18	Tata Chemicals	Jamnagar, Gujarat	Membrane
19	United Phosphorus	Bharuch, Gujarat	Membrane
C	**Northern Region**		
20	Lords Chloro Alkarli Ltd.	Alwar, UP	Membrane
21	Punjab Alkarlies & Chem.	NayaNangal, Punjab	Membrane
22	Shriram Vinyl Chemicals	Kota, Rajasthan	Membrane
23	Siel Chemicals Complex	Rajpura, Punjab	Membrane
D	**Southern Region**		
24	Chemplast Sanmar (Mettur)	Mettur Dam, Tamil Nadu	Mercury
25	Chemplast Sanmar (Karaikal)	Karaikal	Membrane
26	Chemfab Alkalis	Pondicherry	Membrane
27	DCW Ltd.	Sahupuram, Tamil Nadu	Membrane
28	Solaris Chemtech.	Karwar, Karnataka	Mercury
29	Sreee Rayalaseema	Kurnool, AP	Membrane
30	Tamilnadu Petroproducts	Chennai, Tamil Nadu	Membrane
31	The Andhra Sugars	Kovvur & Saggonda, AP	Membrane
32	The Travancore Cochin Chem.	Kochi, Kerala	Membrane

References

AMA (Alkali Manufactures Association) 2007. http://www.ama-india.org/

Banic, C.M., Beauchamp, S.T., Tordon, R.J., Schroeder, W.H., Steffen, A., Anlauf, K.A., Wong, H.K.T. 2003. Vertical dstribution of gaseous elemental mercury in Canada. J Geophys. Res. 108, ACH 61 – 613.

Bhatt, B.P., Sachan, M.S. 2004. Firewood consumption along an altitudinal gradiant in mountain village of India. Biomass and Bioenergy 27, 69 – 75.

BHEL. 2004. Report No. PCI/001/2004 "Assessment and development of environmental standards of heavy metals and trace elements emissions from coal based thermal power plant. Pollution control Research Institute, Bharat Heavy Electricals., Haridwar, India.

Choi, Y.J. 2003. Recent development in India's IT industry and its implications. KIEP Publications, Seoul, Korea, 73 pp.

CPCB, 2000. Status of municipical solid waste generation collection and disposal in Class 1-cities. Central Pollution Control Board, Ministry of Environment & Forests, new Delhi.India

CSE (Center for Science and Environment), 2005. The mercurial menace. CSE, Delhi 3 (3), 1-11.

Das, D., Srinivasu, M. and Bandyopadhyay, M. 1998. Solid state acidification of vegetable waste. Indian Journal of Environmental Health 40(4): 333 – 342.

Downs, S.G., Macleod, C.L. and Lester, J.N. 1998. Mercury in precipitation and its relation to bioaccumulation in fish: A literature review. Water, Air, & Soil Pollut. J. 108 (1-2), 149 – 187.

Drasch, G., Horvart, M., Stoeppler, M. 2004. Mercury. In: Merian E., Anke, M., Ihnat, M., Stoeppler, M. (eds.) Elements and their compounds in the environment. 2nd ed., Wiley-VCH, Weinheim, 931 – 1005.

Dvonch, J.T., Graney, J.R., Marsik, F.J., Keeler, G.J., Stevens, R.K. 1998. An investigation of source-receptor relationships for mercury in South Florida using event precipitation data. Sci. Total Environ. 213, 95 – 108.

Ebinghaus, R., Kock, H.H., Temme, C., Einax, J.W. et al. 2002. Antarctic Springtime Depletion of Atmospheric Mercury. Environ. Sci. Technol. 36, 1238 – 1244.

EPTRI 1995. Status of solid waste disposal in metropolis Hyderabad. Environment Protection, Training & Research Institute.

Feng, X., Li, G., Qiu, G. 2004. A preliminary study of mercury contaminations to the environment from artisanal zinc smelting using indigenous method in Hezhang county, Guizhou, China. Part 1 mercury emissions from zinc smelting and its influences on the surface water. Atmospheric Environ. 38, 6223 – 6230.

Friedli, H.R., Arellano, A. F., Jr., Cinnirella, S., Pirrone, N. 2008. Mercury emissions from global biomass burning: Spatial and temporal distribution. In: Mercury fate and transport in the global atmosphere: Measurements, models and policy implications (Pirrone N. and Mason, R. Eds.) UNEP, Chapter 8.

Garg, A., Shukla, P.R. and Kapshe, M. 2006. The sectoral trends of multi gas emissions inventory of India. Atmos. Environ. 40, 4608 – 4620.

GOI, 2006. Coal directory of India, Part 1b, Coal Statistics, Ministry of Coal Controllers Organization, Kolkata, p. VII 5, IV6.

Gustin, M.S., Biester, H. and Kim, C.S. 2002. Investigation of the light-enhanced emission of mercury from naturally enriched substances. Atmos. Environ. 36, 3241 – 3254.

Hylander, L.D., Meili, M. 2003. 500 years of mercury production: global annual inventory by region until 2000 and associated emissions. Sci. Total Environ. 304 (1 – 3), 13 – 27.

Jaffe, D., Prestbo, E., Swartzendruber, P., Weiss-Penzias, P., et al. 2005. Export of atmospheric mercury from Asia. Atmos. Environ. 39, 3029 – 3038.

Kabata-Pendias, A., Pendias, H. 2001. Trace elements in soils and plants. 3rd Ed., CRC Press, Boka Raton, FL.

Karunasagar, D., Balaram Krishna, M.V., Anjaneyulu, Y. and Arunachalam, J. 2006. Studies of mercury pollution in a lake due to thermometer factory situated in a tourist resort: Kodaikkanal, India. Environ. Pollut. 143: 153 – 158.

Kolker, A., Senior, C.L. and Quick, J.C. 2006. mercury in coal and the impact of coal quality on mercury emissions from combustion systems. Appl. geochem. 21, 1821 – 1836.

Kumar, P.B.A.N., Dushenkov, V., Motto, H. and Raskin, I. 1995. Phytoextraction: The use of plants to remove heavy metals from soils. Environ. Sci. Technol. 29, 1778 – 1783.

Li, G., 2007. Mercury emission from Zink Smelting in China and environmental impacts. PhD theses, Institute of Geochemistry, Chinese Academy of Sciences, 1 – 110.

Lindberg, S., Bullock, R., Ebinghaus, R. et. al. 2007. A synthesis of progress and uncertainties attributing the sources of mercury in deposition. Ambio 36 (1), 19 – 32.

Lindqvist, O., Johansson, K., Aastrup, M., Andersson, A., Bringmark, L., Hovesenius, G., Hakonson, L., Iverfeldt, A. et al. 1991. Mercury in the Swedish environment: Recent research on causes, consequences and corrective methods. Water, Air & Soil Pollution, 55, 261 pp.

Loewen, M., Subodh, S., Gregg, T., Feiyue, B., Frank, W. (2005) Persistent organic pollutants and mercury in the Himalaya. Aquatic Ecosystem Health & Management, 8 (3), 223 - 233.

Maithel, S., Uma, R. 2000. Environmental regulations and the Indian Brick Industry. *Environ. Practice* 2 (3), 230 – 231

Mine Closure 2005. 2nd Indo-US Coal Working Group Meeting, Washington, November 2005.

Mukherjee, A.B., Melanen, M., Ekqvist, M., Verta, M. 2000. Assessment of atmospheric mercury emissions in Finland. Sci Total Environ 259, 73 – 83.

Mukherjee, A.B., Zevenhoven, R. 2006. Mercury in coal ash and its fate in the Indian subcontinent: A synoptic review. *Sci. Tot. Environ.* 368, 323 – 334.

Mukherjee, A.B., Zevenhoven, R., Bhattacharya, P., Sajwan, K.S. and Kikuchi, R. 2008. Mercury flow via coal and coal utilization by-products: A global perspective. *Res., Conserv. Recycl.* 52, 571 – 591.

Munthe, J., Kindbom, K., Kruger, O., Peterson, G., Pacyna, J. and Iverfeldt, Å. 2001. Examining source-receptor relationships for mercury in Scandinavia. *Water, Air, Soil Pollut. Focus* 1, 99 – 110.

Nriagu, J.O. and Pacyna, J.M. 1988. Quantitative assessment of worldwide contamination of air, water and soils by trace metals. *Nature* 333, 134 – 139.

Pacyna, E.G., Pacyna, J.M. 2000. Global emission of mercury from anthropogenic sources in 1995. *Water, Air, Soil, Pollut.* 137 (1-4), 149 – 165.

Pacyna, J.M., Pacyna, E.G. 2001. Assessment of global and regional emissions of trace metals to the atmosphere from anthropogenic sources worldwide. *Environ. Rev.*, 9(4), 269 - 298.

Pacyna, J.M., Pacyna, E.G. 2002. Global emission of mercury from anthropogenic sources in 1995. *Water, Air, Soil Pollut.* 137, 149 – 165.

Pacyna, J.M., Pacyna, E.G., Steenhuisen, F., Wilson, S. 2003. Mapping 1995 global anthropogenic emissions of mercury. Atmos Environ 37-S, 109 – 117.

Pacyna, E.G., Pacyna, J.M., Fudala, J. et al. 2006a. Mercury emissions to the atmosphere from anthropogenic sources in Europe in 2000 and their scenarios until 2020. *Sci. Total Environ.*, 370, 147 – 156.

Pacyna, E.G., Pacyna, J.M., Steenhuisen, F., Wilson, S. 2006b. Global anthropogenic mercury emission inventory for 2000. Atmos. Environ. 40, 4048 – 4063.

Pandey, G.K. 2006. Initiatives taken for estimation and control of mercury from various sources in India. Ministry of Environment & Forests, Government of India, New Delhi. Paper presented in BAQ (*Better Air Quality*) 2006, Indonesia.

Patel, K.S. 2003. Health impacts of mercury cycling in contaminated environment of central India.

Pirrone, N., Keeler, G.J. and Nriagu, J.O. 1996. Regional differences in worldwide emissions of mercury to the atmosphere. *Atmos. Environ*. 30(17), 2981 – 2987.

Pirrone, N., Allegrini, I., Keeler, G.J., Nriagu, J.O., Rossmann, R., Robbins, J.A. 1998. Historical atmospheric mercury emissions and depositions in North America compared to mercury accumulations in secondary records. Atmos. Environ. 32 , 929 – 940.

Pirrone, N. and Mahaffey, K.R. (eds.) 2005. Dynamics of mercury pollution in regional and global scale: Atmospheric process and Human Exposures around the World. Springer-Verlag, New York, 748 pp.

Qi, X.F., Lin, Y.H., Chen, J.H. 2000. An evaluation of mercury emissions from the chlo-alkali industry in China. J Environ Sci-China, 12, 24 – 30 (Sup.)

Reddy, M.S., Venkataraman, C. 2002. A 0.25° X 0.25° inventory aerosol and sulfur dioxide emissions from India: II. Biomass combustion. Atmos Environ 36, 699 – 712.

Rohde, H. 1996. *Plenary Season Speech.*, 4th International Conference on Mercury as a Global Pollutant, Hamburg, September 1996.

Rohde, H. 1996. Plenary Season Speech: The global mercury cycle – how large is the human impact ? 4th International Conference on Mercury as a Global Pollutant. Congress Center Hamburg, Germany, Aug 4 – 8, 1996, p. 2.

Sarkar, A. 2007. E-waste in India: Environmental health perspective and possible actions. In: Rajeshwari, K.V., Basu, S. & Johri, R. (eds.) Tackling E-waste: towards efficient management techniques. TERI Press, New Delhi, India, 19 – 34.

Senior, C., Saroflim, A., Eddings, E. 2003. Behavior and measurement of mercury in cement kilns. Presented in: IEEE-IAS/PCA 45th Cement Industry Technical Conference, May 2003, Dallas, Texas 233 – 248.

Sharholy, M., Ahmed, K., Vaishya, R.C., Gupta, R.D. 2007. Municipal solid waste characteristics and management in Allahabad, India. *Waste Management* 27: 490 – 496.

Singhal, S. and Pandey, S. 2001 Solid waste management in India: Status and future directions. *TERI Information Monitor on Environmental Science*. 6(1), 1 – 4.

Sinha, R.K., Sinha, S.K., Kedia, D.K., Kumar, A., Rani, N., Sharma, G., Prasad, K. 2007. A holistic study on mercury pollution in the Ganga River system at Varanasi, India. *Current Science* 92 (9), 1223 – 1228

Srivastava, R.C. 2003. Guidance and awareness raising materials under the UNEP Mercury Program (Indian Scenario), UN Chemicals.

Streets, D.G., Bond, T.C., Carmichael, G.R., Fernandes, S.D., Fu, Q., He, D. et al. 2003. An inventory of gaseous and primary aerosol emissions in Asia in the year 2000. J. Geophys. Res., 108(D21), 8809.

Subramianium, V., Madhavan, N., Saxena, R. and Lundin, L.C. 2003. Nature of distribution of mercury in the sediments of the River Yamuna (tributary of the Ganges), India. *J. Environ. Monit.*, 5, 427 – 434.

Sunderland, E.M., Chmura, G.L. 2000. The history of mercury emission from fuel combustion in Maritine Canada. Environ Pollut 110, 297 – 306.

UN ECE 2000. Joint EMEP/CORINAIR atmospheric emission inventory guidebook. Geneva, Switzerland. The United Nations Economic Commissions for Europe, 2000.

UNEP, 2002. Global mercury assessment, (United Nations Environmental Programme – Chemicals), Geneva, Switzerland.

UNEP, 2005. Toolkit for identification and qualification of Hg release.

USEPA 1997. Mercury study report, Vol II, (U.S. Environmental Protection Agency) EPA-452/ R – 97 – 004, Washington, DC.

Veiga, M.M., Meech, J.A. 1994. Mercury pollution from deforestation. Nature 368, 816 – 817.

Venkataraman, C., Habib, G., Kadamba, D., Shrivastava, M., Leon, J.-F., Crouzile, B., Boucher, O., Streets, D.G. 2006. Emissions from open biomass burning in India: Integrating the inventory approach with high-resolution Moderate Resolution Imaging Spectroradiometer (MODIS) active-fire and land cover data. Global Biogeochemical Cycles, 20, GB2013, 1 – 12.

Visvanathan, C. 2006. Medical waste management issues in Asia. Presented in the International Conf "Asia 3R Conference", 30 October – 1 November, 2006, Tokyo, Japan.

WCC 2007. Chloralkali partnership data – 2006, World Chlorine Council Bruxelles

Weiss-Penzias, P., Prestbo, E.M. and Landis, M.S. 2003. Gaseous elemental mercury in marine boundary layer: Evidence for rapid removal in anthropogenic pollution. Environ. Sci. Techno. 37, 3755 – 3763.

Wong, Coby, S.C., Duzgoren-Aydin, N.S., Aydin, A. and Wong, M.H. 2006. Sources and trends of environmental mercury emissions in Asia. *Sci. Total Environ.* 368, 649 – 662.

Chapter 5
Mercury Emissions from Point Sources in South Africa

Joy J. Leaner, James M. Dabrowski, Robert P. Mason, Tabby Resane, Marguerite Richardson, Martin Ginster, Gerhard Gericke, Chantel R. Petersen, Elizabeth Masekoameng, Peter J. Ashton, and Kevin Murray

Summary As a first step towards assessing Hg levels in a systematic approach in South Africa, representatives from the South African government, academia, research councils and key industries recently initiated a South African Mercury Assessment (SAMA) Programme (Leaner et al., 2006). The SAMA Programme has undertaken some limited Hg inventory development and monitoring studies in South Africa. The preliminary results of those studies and that of Hg monitoring undertaken at Cape Point's Global Atmospheric Watch Station (Baker et al., 2002), are discussed in this paper.

5.1 Introduction

Mercury (Hg) emissions to the environment are increasing globally, particularly in developing countries (Pacyna et al., 2003; 2006). Important sources of Hg include amongst others, coal combustion, waste incineration, cement production and ferrous metals production (UNEP, 2002). Although these Hg sources have been identified for southern Africa (UNEP, 2002), information on specific Hg emissions and concentrations in the region are poorly understood. Mercury estimates are also often approximate, since most southern African countries do not have formal Hg emission inventories.

Coal combustion provides the largest source of energy to South Africa, yet the information on Hg emissions from this source is sparse. Many households, particularly in rural areas, burn some coal for heating and cooking purposes, and the emissions of various pollutants, including Hg pose a risk to human health. Due to South Africa's reliance on coal as a primary energy source, it is inevitable that emissions of compounds such as sulphur dioxide (SO_2), nitrogen oxides (NO_x) and Hg will increase unless improved control measures are applied or more renewable energy sources are used. Already, increased Hg emissions in Asia during 1990 and 2000 have been related to an increase in demand for energy in that region (Pacyna et al., 2003).

For South Africa, Pacyna et al. (2003; 2006) estimated that anthropogenic sources released about 256.7 Mg of Hg to the atmosphere during 2000, with most Hg emissions originating from industry production, followed by stationary combustion. Coal combustion and gold mining were regarded as the most important Hg sources. Furthermore, South Africa was estimated to account for about 16% of the total

N. Pirrone and R. Mason (eds.), *Mercury Fate and Transport in the Global Atmosphere*, DOI: 10.1007/978-0-387-93958-2_5, © Springer Science + Business Media, LLC 2009 global Hg emissions (1590.7 Mg), making it the second highest emitter of Hg after China (Pacyna et al., 2006). The assessment of Pacyna et al. (2006), combined with the limited available information on actual Hg emission measurements or Hg concentrations in products and resources for South Africa, signalled the need for a critical evaluation of the major Hg sources in the country.

Mercury monitoring programmes aimed at establishing the extent to which Hg emissions from point sources pose a problem to the South African environment are required. In addition, an evaluation of Hg emissions from point sources in South Africa is needed at a regional, national and global level. A number of published studies have focused on Hg in the South African environment (*e.g.* Van den Heever and Frey, 1996; McNab et al., 1997; Baker et al., 2002; Barratt and Combrink, 2002; Oberthür and Saager, 1986; Steenkamp et al., 2000; Oosthuizen and Ehrlich, 2001; Fatoki and Awofolu, 2003; Wagner and Hlatshwayo, 2005; Dalvie and Ehrlich, 2006), however none of these address the development of a Hg emissions inventory for South Africa.

In a co-ordinated attempt to assess the extent of Hg pollution in South Africa, representatives from the South African government, academia, research councils and key industries recently launched the South African Mercury Assessment (SAMA) Programme (Leaner et al., 2007). Stakeholders participating in the SAMA Programme have already undertaken limited Hg inventory development, while monitoring of total gaseous Hg has recently been undertaken in Pretoria in the Gauteng Province. In addition, total gaseous Hg monitoring at the Cape Point Global Atmospheric Watch Station in the Western Cape Province (Baker et al., 2002) has been continuously monitored for several years. Recent efforts, however, have also focused on measuring total Hg concentrations in wet deposition at Cape Point and in Pretoria.

Aside from the Hg emission estimates from Pacyna et al. (2003; 2006), this report provides the first comprehensive assessment of Hg emissions for South Africa. This chapter attempts to refine published atmospheric Hg emission estimates for South Africa through the development of a national inventory and the use of data derived from past (Baker et al., 2002) and current Hg monitoring being undertaken.

5.2 Current Understanding of Mercury Emissions and Levels in South Africa

5.2.1 Priority areas Identified for Monitoring Air Pollution in South Africa

A number of air pollution "hot spots" exist in South Africa, where severe air quality problems have been experienced. Recently, the Department of Environmental Affairs and Tourism (DEAT) made special interventions to improve the air quality in two of these identified "hot spots" regions. In terms of the National Environmental Manage ment: Air Quality Act 2004 (Act No. 39 of 2004) (DEAT, 2004), DEAT declared two National Priority Areas in the country, namely the Vaal Triangle Air-shed Priority Area and the Highveld Priority Area (Figures 5.1 and 5.2) (DEAT, 2006a and 2007).

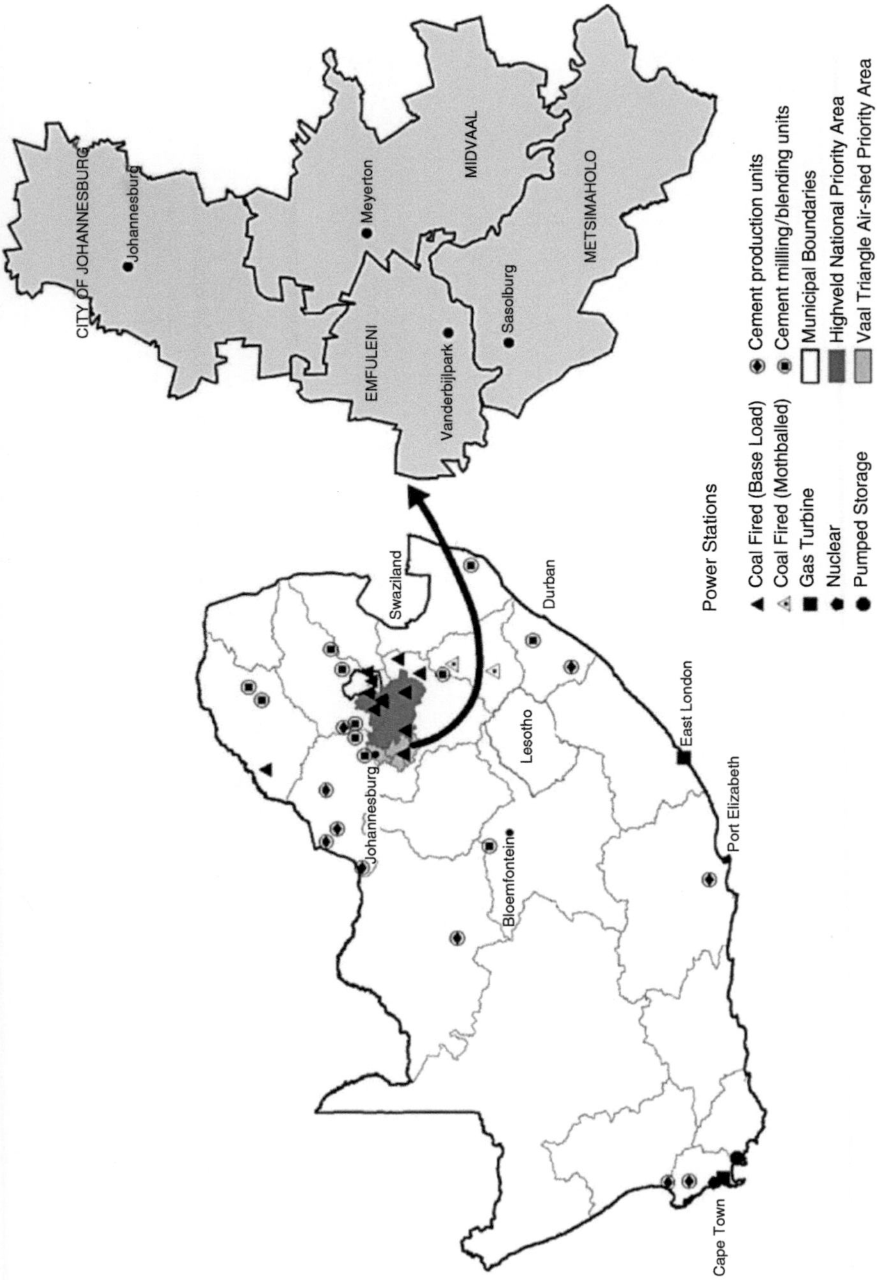

Figure 5.1 Location of the Vaal Air-shed Priority Area in South Africa

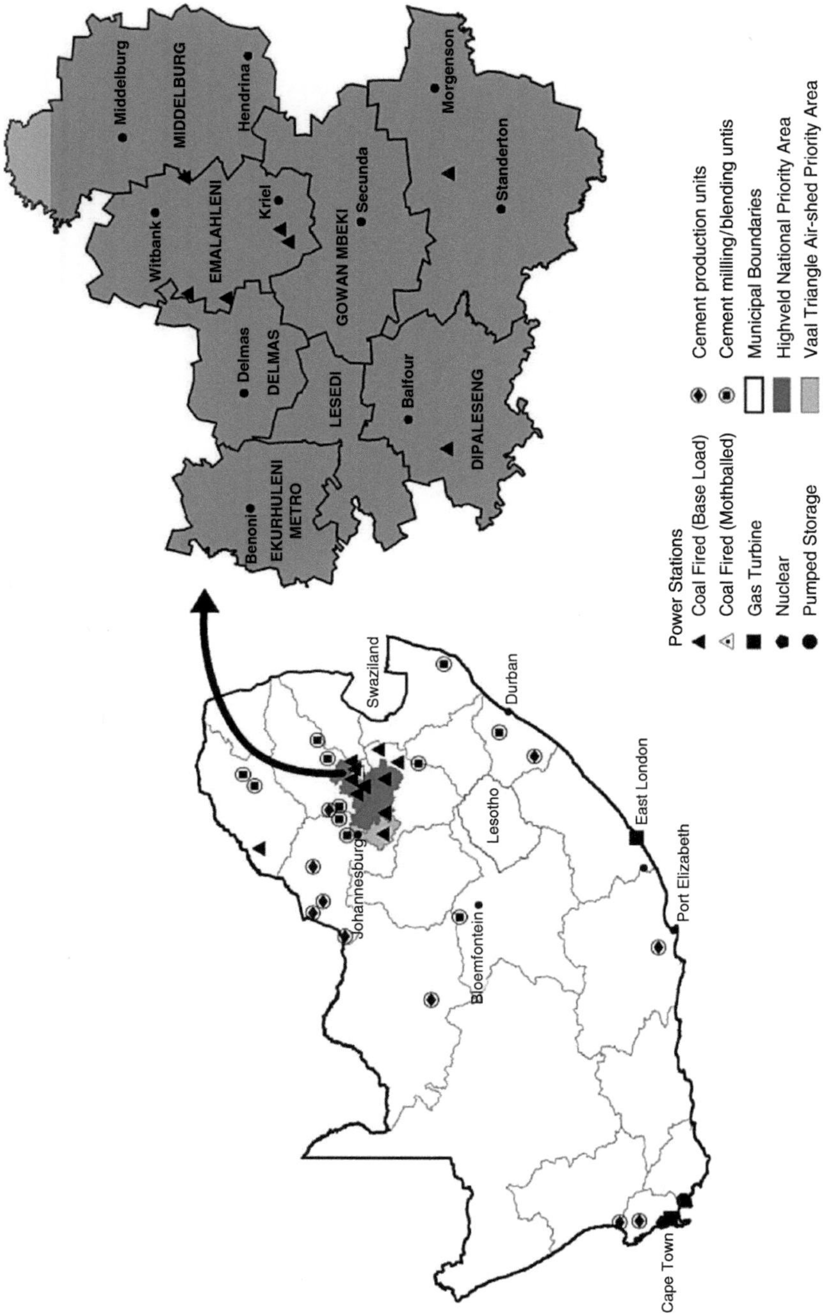

Figure 5.2 The Highveld Priority Area in South Africa

Table 5.1 Current air quality parameters monitored in the Vaal Air-shed Priority Area of South Africa

Responsible Authority	Location	Air Quality Parameters Monitored
City of Johannesburg	Jabavu and Orange Farm	PM_{10} and SO_2
Sedibeng District Municipality	Midvaal Local Municipality (Meyerton)	NO_2, NO, NOx, SO_2, O_3, CO and PM_{10}
Sedibeng District Municipality	Emfuleni Local Municipality (Vanderbijlpark)	NO_2, O_3, SO_2, C_6H_6, Xylene and Toluene

The National Priority Areas have been declared because the "proposed ambient air quality standards are being or may be exceeded in the area, or any other situation exists which is causing, or may cause, a significant negative impact on air quality in the area; and because the areas require specific air quality management action to rectify the situation" (DEAT, 2006a).

The Vaal Triangle Air-shed Priority Area comprises activities such as heavy industries; one coal-fired power station; several commercial operations and transportation; small-scale boiler operations; landfill and waste incineration; and domestic fuel burning (DEAT, 2006a). The Highveld Priority Area also has a range of industrial, mining and agricultural activities including: coal-fired power stations; timber and related industries; metal smelters; petrochemical plants; and heavy and small industrial operations (DEAT, 2007).

Air pollution monitoring activities in the identified National Priority Areas of South Africa focus primarily on particulate matter (PM_{10} and $PM_{2.5}$), SO_2 and other pollutants (Table 5.1), and generally do not include Hg emissions. However, given the similarity in the sources of these pollutants, the extent to which they are emitted is likely to be indicative of any atmospheric Hg emissions. Since most of South Africa's coal fired power plants are located in the Mpumalanga Province (Figure 5.2), where approximately 83% of South Africa's coal production takes place (EIA, 2006), any Hg emissions monitoring in the National Priority Areas will provide an improved Hg assessment for the country, as suggested by others (Leaner et al., 2007; Dabrowski et al., 2008). Moreover, other coal burning industries (*e.g.* cement production, coal gasification facilities) are also located in the identified National Priority Areas (Figures 5.1 and 5.2).

The DEAT is currently implementing an air quality monitoring network, with five monitoring stations, to measure the following pollutants: SO_2, nitrogen oxides (NOx), ozone (O_3), carbon monoxide (CO), particulate matter smaller than 10 μm (PM_{10}), particulate matter smaller than 2.5 μm ($PM_{2.5}$), lead (Pb), Hg, benzene, toluene, ethyl benzene and xylene (BTEX), at the two National Priority Areas. There are also plans to deploy atmospheric Hg samplers at other locations in South Africa, e.g. Eskom plans to install a Hg analyser at one of its monitoring sites.

5.2.2 Mercury Emissions Inventory for South Africa

South Africa is a primary producer of many important and strategic metals (e.g. gold, platinum, lead and zinc) and is a major producer and consumer of coal (DME, 2003;

Murkherjee et al., 2008). Although production of these minerals and materials is known to contribute to Hg pollution, detailed Hg emission inventories for these sources are unavailable for South Africa.

Atmospheric Hg emission estimates for different source categories were calculated as follows:

$$Hg_{(atomsphere)} = C_{(mass)} \times F_{[Hg]} \times 10^{-6} \times (1\text{-}ERF) \quad (1)$$

where: $C_{(mass)}$ is either the amount of coal combusted, commodity produced, or waste deposited or incinerated (Mg yr^{-1}); $F_{[Hg]}$ is the fraction of Hg emitted by $C_{(mass)}$ in mg / kg; and ERF is the emission reduction factor (UNEP, 2005), which is based on the type of emission control device, and whether present or absent.

Based on the above calculation and an analysis of different Hg source categories, coal consumption by coal-fired power plants was identified as the largest potential source of Hg emissions, given that approximately 112.20 Tg of coal was combusted during 2004 (CoMSA, 2004) (Table 5.2). The country is reliant on the combustion

Table 5.2 Total amount of coal consumed or commodity produced by major industries in South Africa during 2004

Source Category	Coal Consumed / Commodity Produced *(Tg yr^{-1})*	Total Amount per Source Category *(Tg yr^{-1})*	Reference
Coal-Fired Power Plants	-	112.200	CoMSA (2004)
Coal Gasification	-	41.444	DME (2008)
Consumer Products (Hg waste)[a]	-	9.145×10^{-9}	DTI (2004)
Fuel Production	-	20.225	-
- Minerals	2.129	-	DME (2008)
- Crude Oil Refining	18.096	-	DME (2008)
Cement Production	-	14.922	-
- Cement Production (clinker)[b]	1.946	-	DME (2005)
- Cement Production (new)	12.975	-	CNCI (2008)
Ferrous Metals: Iron & Steel	-	13.620	-
- Coke Production	2.717	-	DME (2008)
- Iron & Steel (scrap smelting)	4.904	-	DME (2008)
- Pig Iron & Steel (new)	6.000	-	SAISI (2007)
Residential Heating	-	4.996	DME (2008)
Non-Ferrous: Primary Metals	-	0.629	-
- Gold	2.55×10^{-4}	-	CoMSA (2006)
- Zinc	0.240	-	DME (2006)
- Copper[c]	0.346	-	DME (2006)
- Lead	0.042	-	DME (2006)
Waste Incineration (medical)	-	0.028	DEAT (2006b)
Total		208.064	

[a]Based on the importation of 1 829 066 double-ended fluorescent light tubes, and assuming that each tube contains 10 *mg* Hg.

[b]Based on annual cement production, and that 15 Mg of coal are required to produce 100 *Mg* of clinker.

[c]Based on 103 900 *Mg* copper produced, and the average copper content of copper concentrate being 30%.

of coal, with coal-fired power plants accounting for about 90% of the country's primary energy needs (Spalding-Fetcher and Matibe, 2003). Much of the coal used for energy production in South Africa is supplied by mines located on the Highveld coal-field (Wagner and Hlatswayo, 2005), which is also the second largest productive coal-field in the country.

Cement production (DME, 2005; CNCI, 2008), coal gasification (DME, 2008), fuel production (DME, 2008) and ferrous metals production (DME, 2008; SAISI, 2007) are other significant Hg source categories identified for South Africa (Table 5.2). Coal combustion in residential heating (DME, 2008), non-ferrous metals production (CoMSA, 2006; DME, 2006) and medical waste incineration (DEAT, 2006b) totalled an estimated 5.65 Tg during 2004 (Table 5.2). Consumer products such as fluorescent light tubes (DTI, 2004) that contain Hg, ranked the lowest in terms of Hg waste likely deposited to landfills (9.14×10^{-9} Tg) (Table 5.2). Mercury emissions from these potential sources were evaluated to the extent possible, given the limited information available for South Africa.

As discussed below, total atmospheric Hg emissions from all potential sources in South Africa were estimated to be about 40 Mg during 2004 (Figure 5.3). It should

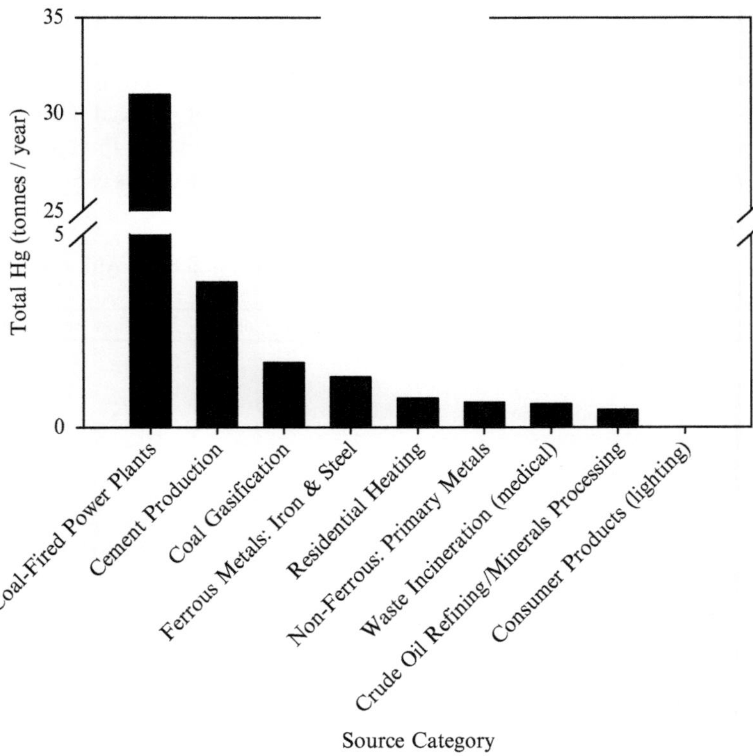

Figure 5.3 Average atmospheric Hg emissions (1 metric ton = 1 Mg) estimated for different source categories in South Africa during 2004

be noted that Hg emissions estimated from coal combustion (i.e. coal-fired power plants) and particularly non-ferrous metals: primary metals, of which gold contributes only a small fraction, are significantly lower (Figure 5.3) than values that have been reported previously for South Africa (see Pacyna et al., 2006).

5.2.2.1 Coal Combustion: Power Plants

Mercury emissions from coal-fired power plants and the degree of Hg speciation (Hg^0 or $Hg^{(II)}$) depend on the amount of coal combusted and the emission control devices used to remove SO_2 or NO_x, and other particulate and gaseous pollutants (Pacyna et al., 2006; Dabrowski et al., 2008). For South Africa, emission control devices in coal-fired power plants include electrostatic precipitators and fabric filters (Table 5.3). The atmospheric Hg emissions from coal-fired power stations were estimated to be 30.96 Mg during 2004 (Figure 5.3).

The estimate was based on Hg concentrations measured in coal used at South Africa's coal-fired power stations during 2001 (Table 5.3; Gericke et al., 2007), the amount of coal consumed at these power stations during 2004 (CoMSA, 2004; Table 5.2) and the reduction factors associated with the emission control devices used (Table 5.3). While the data on which these estimates are based is relatively small, and is somewhat higher than previously reported (average 0.15 ppm; Wagner and Hlatshwayo, 2005), the Hg concentrations in coal (Table 5.3) are within a much smaller range than that which would have been derived using the default values in the UNEP assessment tool (range 0.1 – 1 ppm Hg in coal; UNEP, 2005). Our previous Hg emission estimates of 9.75 Mg of Hg from coal-fired power plants were based on 0.15 ppm Hg in coal (Dabrowski et al., 2008). This estimate is likely

Table 5.3 Emission control devices used at coal-fired power plants of South Africa (adapted from Dabrowski et al., 2008)

Power Plant	Emission Control Device	Hg in coal *($g\ Mg^{-1}$)*	Emission Reduction Factor
Arnot	Fabric filters	0.17[a]	0.50
Duvha	Electrostatic precipitators and Fabric filters	0.23[a]	0.50
Hendrina	Fabric filters	0.21[a]	0.50
Kendal	Electrostatic precipitators	0.44[a]	0.50
Kriel	Electrostatic precipitators	0.34[a]	0.50
Lethabo	Electrostatic precipitators	0.36[a]	0.50
Majuba	Fabric filters	0.29[a]	0.50
Matimba	Cold sided - Electrostatic precipitators	0.45[a]	0.10
Matla	Electrostatic precipitators	0.29[a]	0.50
Tutuka	Cold sided – Electrostatic precipitators	0.29[a]	0.10
Sasol (Secunda)	Cold sided - Electrostatic precipitators	0.15[b]	0.10

[a] Gericke et al. (2007)
[b] Wagner and Hlatshwayo (2005)

conservative since the average value reported by Wagner and Hlatswayo (2005) excludes one of the coal samples (average 0.23 ± 0.03 ppm Hg) as unrepresentative and also the USGS average value of 0.2 ppm Hg. Clearly more analysis of South African coal samples and/or Hg emission measurements is required to improve the Hg emission estimates.

The Hg emissions estimated in this study and Dabrowksi et al. (2008) are significantly lower than previously published Hg emission estimates for South Africa (see Pacyna et al., 2006). Nevertheless, Hg emissions estimated from coal-fired power plants are substantially higher than all other sectors (Figure 5.3), particularly those that use coal combustion processes (e.g. cement production, coal gasification, residential heating, iron and steel processing). With South Africa's increasing demand for energy, the commissioning of two new coal-fired power plants and the de-mothballing of three existing coal-fired power plants (MRA, 2003), Hg emissions to the environment will inevitably increase over the next decade, unless strict emission control technologies and Hg reduction policies are implemented.

5.2.2.2 Coal Combustion: Coal Gasification Process

South Africa uses a significant amount of coal as a feedstock for the production of viable alternative fuels and chemicals. South Africa's largest synthetic fuels producer converts low-grade coal into petroleum products. Synthesis gas, rich in hydrogen and carbon monoxide derived from the gasification of coal, is converted to hydrocarbon products via the Fischer Tropsch process (Van Dyk et al., 2006). A large facility located near Secunda in the Mpumalanga Province produces about 30% of South Africa's liquid fuels requirements (180 000 barrels of fuels and chemicals per day). During 2004, total coal consumption for the production of alternative fuels and chemicals in South Africa was approximately 41 Tg (DME, 2008). Generally, about 70% of this coal used is gasified, while the remaining 30% is combusted to produce process steam and electricity (Wagner et al., 2008).

During coal gasification, most Hg is associated with the crude gas stream, although a small proportion is found in the ash fraction (2.5 – 20% of Hg), and less in the liquid hydrocarbon co-products fraction (Wagner et al., 2008). Bunt and Waanders (2008) reported that gaseous Hg^0 was the most volatile of all trace metals during fixed-bed gasification. The gas cleaning process, known as the rectisol process, is a functional sink that removes Hg^0 and other impurities from the crude gas, as a solid. The Hg that is removed during maintenance cleaning is disposed of as per requirements for Hg^0 containing waste. Of the 30% coal combusted for producing utilities (steam and electricity), Wagner et al. (2008) estimated that about 80% of the Hg associated with this coal is emitted. This estimate was based on both isokinetic sampling and mass balance calculations.

A total of 1.68 Mg Hg was estimated to be emitted from coal combustion in the coal based petrochemical process during 2004 (Figure 5.3). This estimate was based on the following: 30% of the coal consumed during 2004 in the coal-fired steam plants (Table 5.2); an average coal Hg concentration of 0.15 ppm Hg (Wagner and Hlatswayo,

2005); and taking into account the emission control devices used (Table 5.3). Overall, the Hg emissions estimated (1.68 Mg in 2004) is comparable to the estimate of Hg emissions (1.25 Mg per year) reported by Wagner et al. (2008). Additional atmospheric Hg measurements are, however, required to verify the above estimates.

5.2.2.3 Crude Oil Refining and Minerals Processing

South Africa is not a major oil producer but has the second largest oil refining capacity in Africa, following Egypt (EIA, 2007). Total Hg concentrations in crude oil used in the refining process range between 0.01 to 0.5 ppm (Pacyna et al., 2006). Mercury concentrations in crude oil vary substantially depending on its origin, and the affinity of Hg for mineral matter (Pacyna et al., 2006). Since South Africa imports more than 80% of its crude oil requirements from the Middle East (Iran and Saudi Arabia; EIA, 2007), Hg emissions from this source is considered to not vary significantly.

Overall, a total of 0.45 Mg Hg is estimated to have been released during crude oil refining and minerals processing in South Africa in 2004. Of this, crude oil refining is estimated to emit about 0.16 Mg Hg, while coal combustion during minerals processing accounts for the remaining fraction of 0.287 Mg Hg (Figure 5.3). These emissions were based on approximately 18 and 2 Tg of crude oil refining and minerals processing during 2004 (Table 5.2; DME, 2008), respectively; and their associated emission reduction factors (Table 5.4). As South Africa's oil and fuel consumption has increased steadily during 1986 – 2006 (EIA, 2007), with no signs of slowing in the near future, Hg emissions from crude oil refining and minerals processing are likely to continue to increase in the future.

Overall, a total of 0.45 Mg Hg is estimated to have been released during fuel production in South Africa in 2004. Of this, the coal combusted during minerals processing is estimated to emit about 0.287 Mg Hg, while the remaining fraction (0.16 Mg Hg) is estimated to be emitted during crude oil refining (Figure 5.3). These emissions were based on approximately 2 and 18 Tg of minerals processing and crude oil refining during 2004 (Table 5.2; DME, 2008), respectively; and their associated emission reduction factors (Table 5.4). As South Africa's oil and fuel consumption has increased steadily during 1986 – 2006 (EIA, 2007), with no signs of slowing in the near future, Hg emissions from crude oil refining and minerals processing are likely to continue to increase in the future.

5.2.2.4 Cement Production

Coal for firing cement kilns and producing clinker are the major sources of Hg in cement production. Using annual cement production data, the annual coal consumption was estimated, considering that approximately 15 Mg of coal is burned in order to produce 100 Mg of cement clinker (DME, 2005). Therefore, production of about 12.98 Tg

Table 5.4 Emission reduction factors used for estimating atmospheric total Hg emissions in different source categories in South Africa

Source Category	Emission Reduction Factor	Reference
Consumer Products (Hg waste)	0.95	UNEP (2005)
Fuel Production	0.10	UNEP (2005)
Cement Production	0.10	UNEP (2005)
Ferrous Metals: Iron & Steel (scrap smelting)	0.10	UNEP (2005)
Ferrous Metals: Pig Iron & Steel	0.05	UNEP (2005)
Non-Ferrous: Primary Metals (zinc, copper, lead)	0.90	UNEP (2005)
Waste Incineration (medical)	0.10	UNEP (2005)

cement in 2005 (CNCI, 2008) is equivalent to utilising approximately 1.95 Tg of coal for firing the kilns (Table 5.2). Using appropriate emission reduction factors (Table 5.4), approximately 3.77 Mg of Hg was released to the South African environment during this period (Figure 5.3), including emissions resulting from combustion of coal for firing of kilns.

Pacyna et al. (2006) reported that Hg emissions for cement production in Africa were 5.3 Mg in 2000. Thus, the Hg emissions reported in this assessment are over 50% of the total estimate for Africa. Cement production will likely increase as new infrastructure is required to support the growing South African economy. Thus, Hg emissions for South Africa and the continent will increase. Monitoring of Hg emissions from this source, however, will provide more robust Hg emission assessments for South Africa.

5.2.2.5 Ferrous Metal Production - Iron and Steel

Relative to other sectors, a small amount of coal is used in the South African iron and steel industry. The major source of Hg emissions from this activity is from coke production (Pacyna et al., 2006). Using appropriate emission reduction factors (Table 5.4), the combustion of approximately 7.62 Tg of coal for coke production and scrap smelting in the iron and steel industry during 2004 (Table 5.2; DME, 2008) is estimated to have released about 1 Mg of Hg to the environment (Figure 5.3). About one-third of the Hg emitted is from coke production, with the remaining two-thirds attributed to scrap smelting. In addition, about 0.29 Mg Hg (Figure 5.3) is estimated to be released from the production of about 6 Tg of pig iron and steel during 2004 (Table 5.2; SAISI, 2007).

The estimated Hg emissions from these sources reported in this study are higher than Hg emissions reported for the African continent in 2000 (0.4 Mg; Pacyna et al., 2006). As with all developing countries, South Africa is experiencing rapid industrial growth, particularly in this sector of the economy, and Hg emissions from this source are expected to increase in future. More detailed information on the type and efficiency of emission control devices used would improve our understanding of Hg emissions from this source.

5.2.2.6 Coal Combustion: Residential Heating

Although informal settlements and rural area households in South Africa use coal for heating and cooking (Spalding-Fecher and Matibe, 2003), this source of Hg is not well defined. Approximately 5 Tg of coal was used for heating and cooking during 2004 (Table 5.2; DME, 2008). In the absence of emission control devices, and assuming a Hg concentration of 0.15 ppm in Highveld coal (Wagner and Hlatswayo, 2005), about 0.75 Mg Hg is estimated to have been emitted to the atmosphere during this time (Figure 5.3). Since about 90% of Hg emitted from this source is gaseous Hg^0 and $Hg^{(II)}$ (Pacyna et al., 2003), individuals would be directly exposed to about 0.66 Mg of Hg during heating or cooking in their homes. Increased Hg emissions, concomitant with an increase in coal burning will likely occur during winter, similar to what has been reported in China (Wang et al., 2006).

5.2.2.7 Non-ferrous Metal Production: Primary Metals

Mercury emissions from different primary (virgin) metal ores vary according to the technology used to process the ore, the content of Hg in the ore and the type of emission control devices employed during processing. Overall, an estimated average of 0.64 Mg Hg was emitted during non-ferrous metal production (Figure 5.3), as determined as below. Most of the Hg emissions was estimated to be from gold (Au) production, followed by the production of copper (Cu), zinc (Zn) and lead (Pb).

In South Africa, the Witwatersrand ores reportedly have Au and Hg concentrations ranging between 80.9 – 92.9 wt.%, and 0.6 – 5.8 wt.%, respectively (Frimmel and Gartz, 1997). Gold from these ores was traditionally extracted using the Hg:Au amalgam method, until it was replaced with a cyanide-based process in 1890 when mining operations reached greater depths (Naickera et al., 2003). Previously, Schröder et al. (1982) estimated that Hg emissions from the entire gold-mining industry in South Africa is less than 0.2 Mg per year, with approximately 4% lost to the atmosphere. Others have suggested a 6% Hg loss to the atmosphere during gold recovery (Jones and Miller, 2005).

Taking into account Hg:Au ratios ranging between 0.01 – 0.06 in ores (Frimmel and Gartz, 1997), and assuming that between 4 – 6% Hg is lost to the atmosphere, approximately 0.10 – 0.93 Mg Hg was estimated to have been emitted when about 255 Mg of Au was produced in South Africa during 2004 (Table 5.2; CoMSA, 2006). Based on this data, the estimated average of 0.32 Mg Hg produced by the Au mining industry during 2004 (Figure 5.3) is significantly lower than values recently reported for South Africa (Pacyna et al., 2006). The Hg emission estimates from this industry, which uses cyanidation and not Hg amalgamation to extract gold from ore, are based on limited information and require Hg emission measurements at the source(s). The production processes for other primary (virgin) metals were estimated to emit a total of approximately 0.32 Mg Hg per year (Figure 5.3) from the production of 0.240 Tg Zn, 0.346 Tg Cu and 0.042 Tg Pb (Table 5.2; DME, 2006). Emission reduction factors for these metals are listed in Table 5.4. Overall, the Hg emissions from primary metal production in South Africa are much lower than emissions released during ferrous metal production.

5.2.2.8 Consumer Products, Waste Deposition (landfills) and Incineration

Waste deposition, land filling and incineration are important sources of Hg emissions to the environment, particularly since Hg-containing products (e.g. batteries, lamps, and electric switches) are often discarded as general waste to landfills or incinerated. Although the Hg content of municipal waste streams is thought to be decreasing in the developed world (Van Veizen et al., 2002), little is known about the situation for developing countries. In South Africa, about 95% of waste was disposed of in landfills prior to 2000 (DWAF, 1998). Landfills generally release landfill gas that contains varying quantities of heavy metals, including Hg (Lindberg et al., 2005; de la Rosa, 2006; Nguyen et al., 2007; Ilgen et al., 2007). Indeed, total Hg levels in landfill gas at a Florida landfill measured up to ~12 $\mu g/m^3$ (Lindberg et al., 2005). Total Hg levels in landfill gas have not yet been measured in South Africa. Currently, consumer products such as Hg-containing fluorescent light sources (double end tubes) are not separated from general waste and are assumed to be land filled in South Africa. A total of 1,829,066 fluorescent light tubes (double end) were imported into South Africa during 2004 (DTI, 2004); assuming an average Hg content of 10 mg per item (NEWMOA, 2006), this yields an estimated 0.018 Mg Hg produced by this source (Table 5.2). Assuming that about 0.009 Mg Hg (50% of the fluorescent light tubes) were land filled, and using appropriate emission reduction factors (Table 5.4), approximately 0.46 kg Hg (0.0005 Mg Hg) was estimated to be released from this source during 2004 (Figure 5.3). The estimate is conservative, and Hg emissions from fluorescent tubes (and compact fluorescent lights) are likely to increase, concomitant with the drive towards more energy efficient lighting in South Africa.

Medical waste is reported to be the fourth largest contributor of Hg to the global environment (Zimmer and McKinley, 2008). South Africa's National Waste Management Strategy requires medical waste to be sorted prior to disposal or incineration; and that the disposal of potentially hazardous medical waste to landfills should be avoided (DEAT, 1999). The authorised medical waste treatment capacity (commercial service providers; public and private hospitals) in South Africa was approximately 0.028 Tg of medical waste in 2005 (Table 5.2) (DEAT, 2006b). Assuming that this was the amount of medical waste incinerated during 2004, and using the relevant emission factors (Table 5.4), approximately 0.60 Mg of Hg is estimated to have been released to the South African environment (Figure 5.3). Poor on-site incinerators in public hospitals or clinics, if present; and the burning or illegal dumping of waste in residential areas will likely increase Hg emissions to the South African environment.

5.2.2.9 Artisanal and Small-scale Gold Mining Activities

Although artisanal gold miners operate in the Mpumalanga and Limpopo Provinces (CoMSA, 2006), the extent to which artisanal and small-scale gold mining activities contribute to Hg emissions in South Africa is unknown. This activity is illegal in South Africa, yet between 8 000 to 20 000 small-scale gold miners are estimated to be operating in the country. Mercury emissions from artisanal gold mining in other African countries are low (e.g. 3 to 5 Mg per year in Zimbabwe; Veiga, 2004). For South Africa, the Hg emissions from these activities are likely to be lower, considering the estimated number of gold miners in the country is far lower than that found in Zimbabwe.

5.3 Monitoring Hg Emissions in South Africa

Atmospheric monitoring of Hg concentrations in South Africa to date has mostly been made at Cape Point's Global Atmospheric Watch (GAW) Station in the Western Cape. There have been measurements of total gaseous Hg since 1995 at the Cape Point GAW Station (Baker et al., 2002; Slemr et al., 2006) and these have recently been supplemented with additional atmospheric Hg sampling. The average yearly concentrations of total gaseous Hg in the atmosphere between 1995 and 2004 ranged between 1 and 1.5 ng m^{-3}, similar to those measured on board ship in the South Atlantic, and only slightly elevated compared to those measured at Neumayer on the Antarctic Peninsula (Baker et al., 2002; Slemr et al., 2006). These concentrations therefore represent the regional background signal and do not show enhanced concentrations that are predicted by modelling studies that use the Pacyna et al. (2006) emission scenarios (e.g. Selin et al., 2007; Strode et al., 2007).

Initial atmospheric Hg studies are currently underway at the CSIR in Pretoria in the Gauteng Province. While these studies are just beginning, there is evidence that the concentrations of total gaseous Hg are occasionally elevated (~ 2 ng m^{-3}), especially during the day, and there is some indication of a diel variation (lower concentrations at night). Such concentrations and variations may reflect local and regional sources in the vicinity, and are not surprising given the high level of urban and industrial activity in the Gauteng Province.

More recently, rainfall collections have been made (weekly bulk phase) at Cape Point's GAW station. Because the collector is continuously open, the device collects both wet deposition and some fraction of the dry deposition. The average Hg concentration in rainfall for seven weeks of weekly rain sampling (July and August 2007) was 6.3 ± 3.0 ng L^{-1}. While this preliminary data should not be over-extrapolated, they are not substantially elevated for what may be expected from a coastal location on the South Atlantic Ocean, and are consistent with the air measurements. Scaling this data to a yearly flux suggests that wet deposition could amount to around 3 μg m^{-2} yr^{-1}, a value that is consistent with the estimates of Mason et al. (1994) for the remote southern Hemisphere, and lower than that of similar locations in the North

Atlantic (Bermuda, for example, is 8 μg m^{-2} yr^{-1}, or most locations on the east coast of the USA (e.g. Mason et al., 2000). Overall, the Hg concentrations in precipitation and atmospheric air samples measured at Cape Point are inconsistent with the predictions of modelling studies that use the Pacyna et al. (2006) emission scenarios (e.g. Selin et al., 2007; Strode et al., 2007). Weekly rain collections in Pretoria in Gauteng Province are also now underway and the preliminary results from these collections (August 2007 to February 2008) suggest higher concentrations (21 ±18 ng L^{-1}; volume weighted mean concentration 16.5 ng L^{-1}). The associated annual flux is 8.8 μg m^{-2} yr^{1} (for an annual rainfall of approximately 0.5 m yr^{1}). Thus, there is a contrast in Hg concentrations and fluxes between Pretoria and Cape Point, which reinforces the notion of more extensive anthropogenic emissions in the northern-most provinces of South Africa. The current data support the re-evaluation that while there are substantial Hg emissions for South Africa, their extent is lower than previously thought.

5.4 Gaps in Our Current Understanding

The industries mentioned above are all important in South Africa and further monitoring and research is required to verify the reported Hg emission estimates from these sources. Other important sources of Hg emissions, such as chlor-alkali production, are also part of South Africa's industries. In addition to this, there is a need to evaluate the contribution of biomass burning as a potential source of Hg to the South African environment. Measurements of total gaseous Hg at Cape Point, downwind of a fire on the Cape Peninsula, suggested that biomass burning could be a significant source of Hg in the southern Hemisphere (Brunke et al., 2001). The impacts of Hg from these sources have not been characterised in South Africa. In addition, a detailed examination of Hg levels in artisanal gold mining areas in South Africa is needed to evaluate the potential impact that such activities may have on human health and the surrounding terrestrial and aquatic ecosystems. Although these are potential sources of Hg, no information is available on the Hg content of emissions from these sources. In addition, the fate and transport processes of gaseous Hg through the entire electricity generation process require further investigation.

5.5 Research Needs

All estimates included in this study are based on the best available information. While data on the Hg content of coal exist, these data are generally only available for the Highveld coal field. The Hg content of coal and Hg emissions resulting from its use are likely to vary across the country. Major gaps in our understanding of point source Hg emissions include the Hg content of raw materials used in industry (i.e. in iron and base metal ores, limestone in the cement industry, etc.) and the type and efficiency of control devices used in various industrial sectors. Further research on the Hg point sources, and the actual Hg emission measurements from these sources, are required.

References

Baker, P.G.L., Brunke, E-G., Slemr, F. and Crouch, A.M. (2002). Atmospheric mercury measurement at Cape Point, South Africa. *Atmospheric Environment*, 36: 2459–2465.

Barratt, G.J. and Combrink, J. (2002). An assessment of the degree of mercury (Hg) bio-transformation in two river systems following discharges from a mercury recovery plant. *Water SA - Special Edition: WISA Proceedings*, 2005: 1-5.

Brunke, E-G., Labuschagne, C. and Scheel, H.E. (2001). Trace gas variations at Cape Point, South Africa, during May 1997 following a regional biomass burning episode. *Atmospheric Environment*, 35: 777-786.

Bunt, J.R. and Waanders, F.B. (2008). Trace element behaviour in the Sasol-Lurgi MKIV FBDB gasifier - Part 1 (The volatile elements: Hg, As, Se, Cd and Pb). *Fuel,* doi:10.1016/ j.fuel. 2008.01.017. (article in press).

CNCI (Cement and Concrete Institute) 2008. Statistics. Available [online] at: http://cnci. org.za. (Last accessed: April 2008).

CoMSA (Chamber of Mines of South Africa). (2004). Facts and Figures, 2004. Available [online] at: www.bullion.org.za. (Last accessed: January 2008).

CoMSA (Chamber of Mines of South Africa). (2006). Mining Education: Minerals & Metals. Available at: www.bullion.org.za/MiningEducation/Gold. (Last accessed: July 2006).

Dabrowski, J.A., Ashton, P.J., Murray, K., Leaner, J.J. and Mason, R.P. (2008). Anthropogenic mercury emissions in South Africa: coal combustion in power plants. Atmospheric Environment. (accepted for publication).

Dalvie, M.A. and Ehrlich, R. (2006). Community mercury levels in the vicinity of peri-urban waste disposal sites and fossil fuel burning operations. Environment International, 32: 493-499.

De la Rosa, D.A., Velasco, A., Rosas, A. and Volke-Sepulveda, T. (2006). Total gaseous mercury and volatile organic compounds measurements at five municipal solid waste disposal sites surrounding Mexico City Metropolitan Area. Atmospheric Environment, 40: 2097-2088.

DEAT (Department of Environmental Affairs and Tourism). (2004). National Environmental Management: Air Quality Act 2004 (Act No. 39 of 2004). Pretoria, South Africa.

DEAT (Department of Environmental Affairs and Tourism). (2006a). Vaal Triangle Airshed Priority Area – Macro Siting Report. Pretoria, South Africa. Report Number: DEAT_RN_060001.

DEAT (Department of Environmental Affairs and Tourism). (2006b). National Waste Management Strategy Implementation South Africa: Projections for Health Care Risk Waste Treatment. Pretoria, South Africa. Report Number: 12/9/6.

DEAT (Department of Environmental Affairs and Tourism). (2007). Highveld Priority Area Ambient Air Quality Monitoring Network – Micro-Scale Siting Report. Pretoria, South Africa. Report Number: SIA_RN_070002.

DME (Department of Mineral and Energy, South Africa). (2005). Dolomite and Limestone in South Africa: Supply and Demand – 2005. Directorate: Mineral Economics. Pretoria, South Africa.

DME (Department of Mineral and Energy, South Africa). (2006). South Africa's Mineral Industry 2005/2006. Directorate: Mineral Economics. DME: Pretoria, South Africa. Report Number:

DME (Department of Mineral and Energy, South Africa). (2008). Aggregate energy balances. Available at: www.dme.gov.za/energy/documents.stm#6. (Last accessed: April 2008).

DME (Department of Minerals and Energy, South Africa). (2003). Integrated Energy Plan for the Republic of South Africa. Pretoria, South Africa. Available [online] at: http://www.dme.gov. za/pdfs/energy/efficiency/ee_strategy_05.pdf.(Last accessed: March 2008).

DTI (Department of Trade and Industry). (2004). Import/Export data. Available at: www.thedti.gov.za/econdb/raportt/tradata.html. (Last accessed: April 2008).

DWAF (Department of Water Affairs and Forestry). (1998). Waste generation in South Africa, baseline studies, Waste Management series. Pretoria, South Africa. Report Number: WMB 306K/1508/4/5.

EIA (Energy Information Administration). (2006). Country Analysis Briefs: Southern Africa (SADC). Available at: www.eia.doe.gov/emeu/cabs/SADC/Coal.html. (Last accessed: December 2007).

EIA (Energy Information Administration). (2007). Country Analysis Briefs: South Africa. Available at: www.eia.doe.gov/emeu/cabs/South_Africa/Oil.html. (Last accessed: January 2008).

Fatoki, O.S. and Awofolu, R. (2003). Levels of Cd, Hg and Zn in some surface waters from the Eastern Cape Province, South Africa. *Water SA*, 29: 375-380.

Frimmel, H.E. and Gartz, V.H. (1997). Witwatersrand gold particle chemistry matches model of metamorphosed, hydrothermally altered placer deposits. *Mineralium Deposita*, 32: 523-530.

Gericke, G., Surender, D. and Delport, W. (2007). Executive summary of mercury research and trace element behaviour. Eskom Report Number C096501. ESKOM, South Africa.

Ilgen, G., Glidemann, D., Herrmann, R., Hertel, F. and Huang, J.H. (2007). Organometals of tin lead and mercury compounds in landfill gases and leachate from Bavaria, Germany. Waste Management, doi:10.1016/j.wasman.2007.06.020 (*in press*).

Leaner, J.J., Dabrowski, J.M., Murray, K., Ashton, P.J., Mason, R.P., MacMillan, P., Zunckel, M. and Oosthuizen, R. (2007). Mercury Research for Policy Development in South Africa. In: Biogeochemistry of Trace Elements: Environmental Protection, Remediation and Human Health. Eds: Y.G. Zhu, N. Lepp and R. Naidu. Tsinghua University Press, China. ISBN: 978-7-302-15627-7, pp. 1036.

Lindberg, S.E., Southworth, G., Prestbo, E.M., Wallschlager, A., Bogle, M.A. and Price, J. (2005). Gaseous methyl-and inorganic mercury in landfill gas from Florida, Minnesota, Delaware and California. *Atmospheric Environment*, 39: 249-258.

Mason, R.P., Lawson, N.M. and Sheu, G.R. (2000). Annual and seasonal trends in mercury deposition in Maryland. *Atmospheric Environment*, 34: 1691-1701.

McNab, N.J., Hughes, J.C. and Howard, J.R. (1997). Pollution effects of wastewater sludge application to sandy soils with particular reference to the behaviour of mercury. *Applied Geochemistry*, 12: 321-325.

MRA (Mining Review Africa). (2003). Eskom projects 2% annual growth in its coal uptake. Available at: www.miningreview.com/archive/033/16_1.htm. (Last accessed: December 2007).

Murkherjee, A.B., Zevenhoven, R., Bhattacharya, P., Sajwan, K.S. and Kikuchi, R. (2008). Mercury flow via coal and coal utilization by-products: A global perspective. *Resources, Conservation and Recycling*, 52: 571-591.

Naickera, K., Cukrows ka, E. and McCarthy, T.S. (2003). Acid mine drainage arising from gold mining activity in Johannesburg, South Africa and environs. *Environmental Pollution*, 122: 29-40.

NEWMOA (Northeast Waste Management Officials' Association). (2006). Mercury in Lighting, Boston. Available [online] at: http://www.newmoa.org/ prevention/mercury/imerc/ FactSheets/ lighting.cfm. (Last accessed: April 2008).

Nguyen, H.T., Ki-Hyun, K., Min-Yooung, K. and Zang-Ho, S. (2008). Exchange pattern of gaseous elemental mercury in an active urban landfill facility. *Chemosphere*, 70: 821–832.

Oberthür, T. and Saager, R. (1986). Silver and Mercury in Gold Particles from the Proterozoic Witwatersrand Placer Gold Deposits of South Africa: Metallogenic and Geochemical Implications. *Economic Geology*, 81: 20-31.

Oosthuizen, J. and Ehrlich, R. (2001). The impact of pollution from a mercury processing plant in KwaZulu-Natal, South Africa, on the health of fish-eating communities in the area: an environmental health risk assessment. *International Journal of Environmental Health Research*, 11: 41-50.

Pacyna, E.G., Pacyna, J.M., Steenhuisen, F. and Wilson, S. (2006). Global anthropogenic mercury emission inventory for 2000. *Atmospheric Environment*, 40: 4048-4063.

Pacyna, J.M., Pacyna, E.G., Steenhuisen, F. and Wilson, S. (2003). Mapping 1995 global anthropogenic emissions of mercury. *Atmospheric Environment*, 37: 109-117.

SAISI (South African Iron and Steel Institute) (2007) Statistics: Iron and Crude Steel Production. Available [online] at: http://www.saisi.co.za/statistics.php. (Last accessed: November 2007).
Schröder, H.H.E., Van Der Linde, A. and Strydom, N.B. (1982). The emission of mercury from gold reduction works in South Africa. *Journal of the South African Institute of Mining and Metallurgy*, July: 193-199.
Selin, N.E., Jacob, D.J., Park, R.J., Yantosca, R.M., Strode, S., Jaeglé, L., and Jaffe, D. (2007). Chemical cycling and deposition of atmospheric mercury: Global constraints from observations. *Journal of Geophysical Research*, 112: D02308, doi:10.1029/2006JD007450.
Slemr, F., Brunke, E.-G., Ebinghaus, R., Temme, C., Munthe, J., Wängberg, I., Schroeder, W., Steffen, A. and Berg, T. (2003). Worldwide trend of atmospheric mercury since 1977. *Geophysical Research Letters*, 30(10): 1516, doi:10.1029/2003GL016954, 2003.
Spalding-Fecher, R. and Matibe, D.K. (2003). Electricity and externalities in South Africa. *Energy Policy*, 31: 721–734.
Steenkamp, V., von Arb, M. and Stewart, M.J. (2000). Metal concentrations in plants and urine from patients treated with traditional remedies. *Forensic Science International*, 114: 89-95.
Strode, S.A., Jaeglé, L., Selin, N.E., Jacob, D.J., Park, R.J., Yantosca, R.M., Mason, R.P. and Slemr, F. (2007). Air-sea exchange in the global mercury cycle. *Global Biogeochemical Cycles*, 21: GB1017, doi:10.1029/2006GB002766.
UNEP (United Nations Environment Programme). (2002). Global Mercury Assessment Report. UNEP: Geneva, Switzerland. Available [online] at: http://www.chem.unep.ch/ MERCURY/ Report/ GMA-report-TOC.htm. (Last accessed: March 2008).
UNEP (United Nations Environment Programme). (2005). Toolkit for identification and quantification of mercury releases. UNEP: Geneva, Switzerland. Available at: www.chem.unep.ch/ mercury/ Toolkit/default.htm. (Last accessed: March 2008).
Van den Heever, D.J. and Frey, B.J. (1996). Human health aspects of certain metals in tissue of the African sharptooth catfish, Clarius gariepinus, kept in treated sewage effluent and the Krugersdrift Dam: Chromium and mercury. *Water SA*, 22: 73-78.
van Dyk, J.C., Keyser, M.J., Coertzen, M. (2006). Syngas production from South African coal sources using Sasol-Lurgi gasifiers. *International Journal of Coal Geology*, 65: 243-253.
Van Veizen, D., Langenkamp, H. and Herb, G. (2002). Review: Mercury in Waste Incineration. *Waste Management and Research*, 20: 556-568.
Veiga, M. (2004). Project Report: Equipment Specification for the Demonstration Units in Zimbabwe. *Project EG/GLO/01/*G34: Removal of Barriers to Introduction of Cleaner Artisanal Gold Mining and Extraction Technologies. Global Mercury Project. Vienna, Austria. Available [online] at: www.unido.org/file-storage/download?file_id=44425. (Last accessed: March 2008).
Wagner, N.J. and Hlatshwayo, B. (2005). The occurrence of potentially hazardous trace elements in five Highveld coals, South Africa. International Journal of Coal Geology, 63: 228-246.
Wagner, N.J., Hlatshwayo, B. and Ginster, M. (2008). A source apportioned mercury mass balance across a coal-based petrochemical complex. Fuels Processing, FU PROC-D-08-00064 (submitted).
Wang, Q., Shen, W. and Zhuangwei, M. (2000). Estimation of mercury emission from coal combustion in China. *Environmental Science and Technology*, 34: 2711-2713.
Zimmer, C. and McKinley, D. (2008). New Approaches to Pollution Prevention in the Healthcare Industry. *Journal of Cleaner Production*, 16: 734-742.

Chapter 6
World Emissions of Mercury from Artisanal and Small Scale Gold Mining

Kevin H. Telmer and Marcello M. Veiga

Summary We estimate mercury releases from artisanal and small scale gold mining (ASGM) based on available data about mercury and gold exports and imports by country and from field reports from the countries known to have active ASGM communities. The quality of the estimates ranges from reasonable to poor across the countries. This paper aims to give a first order estimate of the amount and location of mercury being released into the environment globally by ASGM, to motivate stakeholders to improve the quality of these estimates, to illustrate the linkages between global mercury trade and its use in ASGM, and the fourth objective is to provide a practical outline of the options available for reducing mercury use in ASGM. We estimate that artisanal and small scale gold mining releases between 640 to 1350 Mg of mercury per annum into the environment, averaging 1000 Mg yr^{-1}, from at least 70 countries. 350 Mg yr^{-1} of this are directly emitted to the atmosphere while the remainder (650 Mg yr^{-1}) are released into the hydrosphere (rivers, lakes, soils, tailings). However, a significant but unknown portion of the amount released into the hydrosphere is later emitted to the atmosphere when it volatilizes (latent emissions). Considering that ASGM is growing, latent emissions conservatively amount to at least 50 Mg yr^{-1} bringing the total emission of mercury to the atmosphere from ASGM to 400 Mg yr^{-1}. This estimate of emission to the atmosphere differs from the previous one provided in the 2002 UNEP Global Mercury Assessment both in terms of its magnitude (400 Mg yr^{-1}, versus 300 Mg yr^{-1}) and in the way the estimate has been made. The current estimate is based on a better understanding of ASGM and on a wider variety of information sources, more field evidence, better extrapolation methods, and independent testing by analysis of official trade data.

6.1 Introduction

We begin with a presentation of the intricacies of why mercury is used in ASGM and how it is released to the environment. A good understanding of the use of mercury in ASGM is needed in order to evaluate both the emission estimate and the options available for reducing mercury use.

N. Pirrone and R. Mason (eds.), *Mercury Fate and Transport in the Global Atmosphere*,
DOI: 10.1007/978-0-387-93958-2_6, © Springer Science + Business Media, LLC 2009

We then begin to build the database on mercury in ASGM by identifying the known localities of ASGM – documented to occur in 70 countries – by citing reports from governments, international bodies, NGOs, the peer reviewed literature, and from mining companies. This is followed by a section that uses case studies and field data collected from various intervention efforts, as well as arguments from later sections, to make an estimate of the consumption of mercury in ASGM by country. This is further broken down into an estimate of how much mercury is directly released to the atmosphere.

The next section examines the global trade in mercury and gold for the purposes of placing the magnitude of mercury consumption by the ASGM community into perspective. Because reporting is voluntary, this approach is imperfect but does provide some useful information on mercury in ASGM. It also re-enforces the notion that mproved reporting of mercury trade would greatly improve our ability to track flows of mercury around the world. For example, despite having active dental services that undoubtedly use mercury, there are 70 countries that do not report any trade in mercury. Analysing the trade data, allows some crude but independent constraints on the magnitude of mercury consumption in ASGM to be made.

We then explain the current knowledge gaps surrounding mercury use in ASGM. This is to point out that despite being one of the largest sources of mercury to the environment, research on mercury in ASGM has been relatively poorly funded and grossly unsophisticated relative to that carried out in the northern hemisphere, and that small scale mining communities are a good place to build knowledge about mercury. Aside from answering important questions about mercury's behaviour, working in these communities would additionally bring needed resources, raise awareness, and undoubtedly produce some innovative ideas. The current lack of understanding about mercury in ASGM puts a limitation on the development of innovative solutions towards prevention and remediation.

The final section examines the options available to reduce mercury use in ASGM and the estimates the magnitude of reductions for each of the options discussed.

6.2 Why Mercury is Used

Mercury is used in ASGM for the following reasons:

1. Mercury use is very easy – the easiest and quickest method to extract gold from many alluvial ores under the existing field conditions. This is sometimes debated by those who have not spent much time in the field, but it is a verity. A simple way to look at this is as follows. In the case study by Telmer and Stapper (2007), the effective ore grade (what is recoverable by the miners) was about 0.1 g Mg^{-1}; the miners processed about 100 Mg of ore per day to produce a gravity concentrate of 10 kg of ore. That represents a concentration factor of 10,000 times. The 10 kg of concentrate contains 10 g gold and so they need to further concentrate

by 1000 times. This can be done by manual gravity methods (like panning) but will require significant time and will risk the loss of some gold (particularly the finer fraction). For example, recreational small scale miners in Canada often spend 2 or more hours panning up their concentrate. Capturing the gold by amalgamating the concentrate takes about 10 minutes and produces more certain results. So in ASGM sites, the 2 hours is instead used to continue mining and produce another 2 g of gold.

2. Mercury is very independent – the whole mining process can be accomplished by just one person thereby eliminating the necessity of participating in undesirable and unfair labour practices (there is no need to be indentured). Often in more mature ASGM sites the bottom of the labour pool are still indentured to middle men or "a syndicate", but even so, their salaries are inevitably higher than those from their former occupation, and they always have the choice to strike out on their own – an important and desirable psychological condition for most people around the world.
3. Mercury is highly effective at capturing gold under the conditions found in ASGM sites. Again, the verity of this statement is occasionally debated by academics but under the circumstances found in ASGM sites, it is indisputably true. That is not to say it is technically always the "most" effective method to capture gold, but it can often be the "optimal" method under the socio-economic and political conditions found in ASGM sites. For example, in the first point (#1) above, a centrifuge or other technology may be more effective than mercury, but at what cost? and what infrastructure is needed to operate it? Often costs and infrastructure are prohibitive. This is particularly true when operations are illegal, which is most of the cases. Who is going to risk significant investment into an illegal operation?
4. Mercury is typically very accessible – it is as portable and easy to transport as gold and so moves across borders and into camps as easily as or more easily than many other contraband materials. As far as we know, eliminating mercury through local enforcement has never been successful. In fact it often has a detrimental effect on the miners. For example, in Indonesia, mercury was made illegal in 2006. This drove mercury trade underground and doubled the price paid in the ASGM sites but did nothing to stem the flow of mercury – in fact it made selling it more lucrative for merchants. However, it is also true that increased prices may have been an incentive to increase recycling efforts – keeping in mind that the affordable recycling technology was only made available through an intervention program, the GMP.
5. Mercury is relatively very cheap, as explained through the following perspective:

    - As of Jan 22, 2008, prices were: mercury (US$600/76 lb flask; US$17.40/kg); gold (US$874.00/ozt)
    - This is close to historical highs for both mercury and gold.
    - Therefore 1g mercury = US$0.017; and 1g gold = US$28.10
    - The mercury: gold price ratio is therefore 1:1,650
    - If 2 units of mercury were used to produce 1 unit of gold, the cost of the mercury would represent 0.1% of revenue. An invisible amount.

- In the mine fields, the price paid for gold is less than the international price, typically 8 to 10% less (~US$25/ozt) and the price paid for mercury is higher, particularly where it is illegal making gouging by suppliers easier. Some miners have reported paying as high as US$200/kg (US$0.20/g) (Creporizão, Brazil). Under these prices the cost of using 2 units of mercury to produce 1 unit of gold represents a mercury: gold price ratio of 1:125 or 0.8% of revenue – still remarkably cheap.
- However, once expenses are paid (fuel, equipment, food, shelter), and profits are divided – usually very inequitably with the lion's share going towards the top of the labour pyramid – the cost of mercury may become significant for labourers at the bottom, and so despite its apparent cheapness, an economic incentive to conserve mercury does exist for the lowest paid labourers and for those who deal in large quantities of mercury – often gold dealers.

6. Miners are not always aware of the health risks that mercury poses. Images of people carelessly exposing themselves to mercury in Figure 6.1 tragically show the truth of this.
7. Miners have no choice – in many cases miners are not aware of alternatives if they do exist, or do not have the capacity to practice them.
8. Mercury is most commonly used when simple gravity methods cannot produce concentrates greater than 10-20% gold. This is true of many simple hydraulic sluicing operations and many shallow colluvial or hard rock operations. If a concentrate of 20% can be produced, then direct gold smelting is possible.
9. Mercury is used when capital (cash) is needed quickly for subsistence or to purchase materials and supplies required for more sophisticated techniques like leaching with cyanide. This point is often a difficult one for citizens of developed nations to fully grasp. The miners – even the middle men – do not have bank accounts or credit cards or much, if any, access to social assistance like health care, and therefore often cannot wait to get paid. For example, miners who have made the transition to cyanide leaching and whom know that the maximum gold can be obtained through cyanide leaching alone, often return to using mercury when an emergency such as a family illness or wedding comes up, simply because they cannot wait until completion of the more time consuming, albeit more efficient, cyanide processing method (often a 1 month cycle).

In summary, using mercury is cheap, simple, fast, independent, and reliable. And so in many settings, it is hard to beat. That is why, as a first line of intervention, it may be more appropriate to try to reduce mercury consumption through conservation practices like retorting, fume hoods, and mercury re-activation or cleaning (making dirty mercury usable again and thereby preventing it from being discarded into the environment), rather than immediately aiming for the total elimination of mercury use. The introduction of conservation practises can easily reduce mercury consumption by 50 to 90% and it is an easily accepted change in practice – one that can even have the powerful incentive of being profitable (Agrawal, 2007).

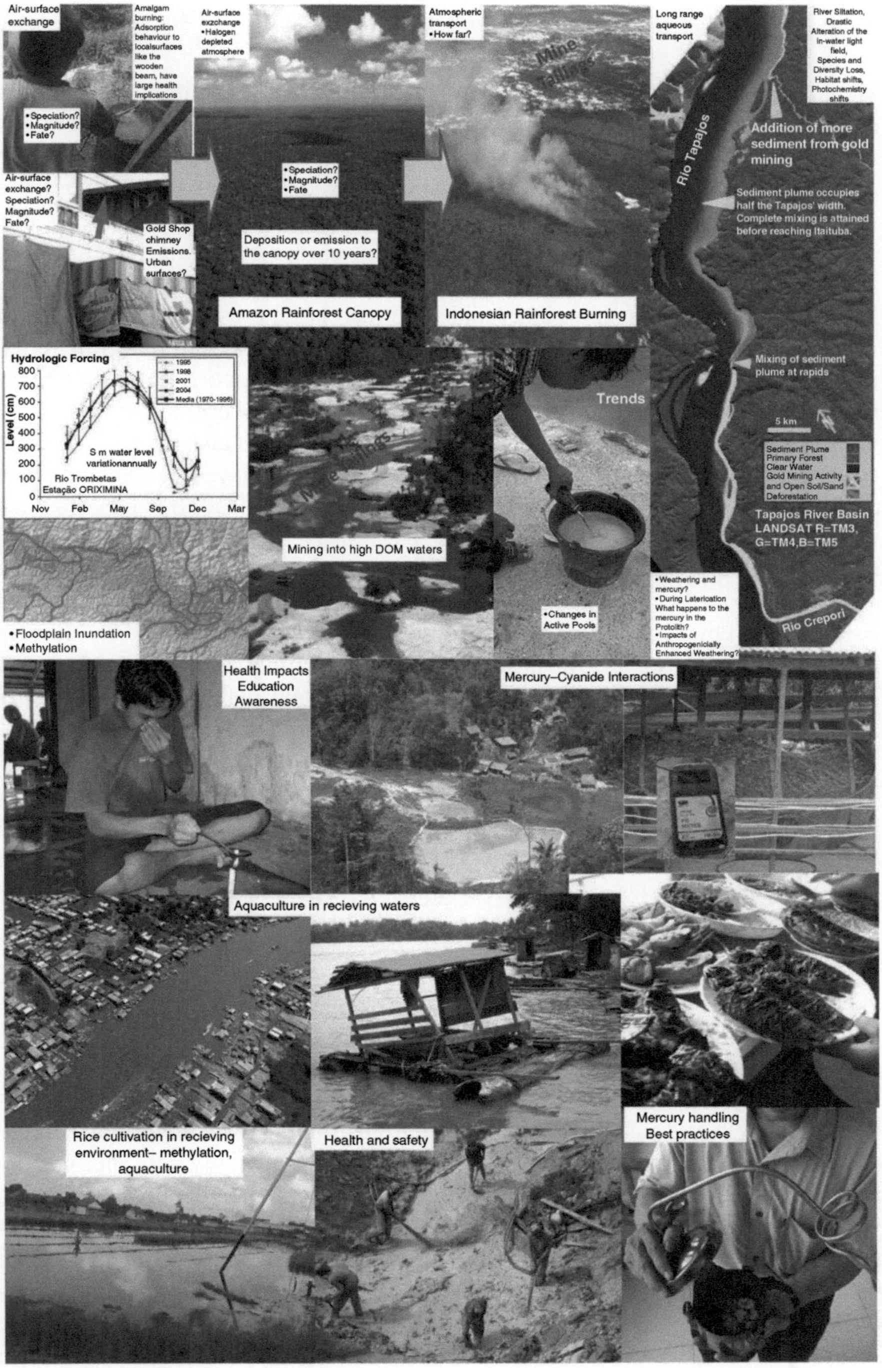

Figure 6.1 Illustration of some of the many knowledge gaps remaining about mercury in ASGM

6.2.1 *How Mercury is Released to the Environment*

Mercury is released to the environment during artisanal gold mining in a variety of ways. When it is used to amalgamate gold, some escapes directly into water bodies as elemental mercury droplets or as coatings of mercury adsorbed onto sediment grains. The mercury that forms the amalgam with gold is emitted to the atmosphere when the amalgam is heated – if a fume hood or retort is not used. As well naturally occurring mercury in soils and sediments that are eroded by sluicing and dredging becomes remobilised and bio available in receiving waters (Telmer et al. 2006). Finally, where a combination of cyanide and mercury are used, the formation of water soluble cyano-mercuric complexes enhances transport and bio-availability. Albeit the fate of mercury in any of these processes is poorly understood, the interactions of cyanide and mercury are the least understood at this time.

When miners use cyanide, this dissolves not only gold but also mercury, forming cyano-mercury complexes. These complexes are easily mobilized by rain and often, due to poor containment practices, quickly reach stream waters. It is expected that water-soluble mercury cyanide is either more bio available or easier to be biomethylated than elemental mercury. This possibility deserves more investigation, but indirect evidence collected by the Global Mercury Project sites in Indonesia, Zimbabwe and Brazil suggest this is the case. Dangerously high levels of mercury in fish (average 2.53 ± 3.91 mg Hg kg^{-1}; carnivorous fish: 4.16 ± 5.42 mg Hg kg^{-1}) were found in Brazil when mercury and cyanide were used together compared to when only mercury amalgamation was performed (UNIDO, 2006). Other similar investigations were carried out in Indonesia (Castilhos et al., 2006; Baker and Telmer, 2007).

Overall, therefore, the pathway that mercury from ASGM takes into the environment, whether it is emitted to the atmosphere, first released into surface water and soils and later emitted (latent emissions), or exported in products (see later section); as well as the amount of mercury consumed per unit of gold produced, varies greatly across ASGM operations and communities.

6.2.1.1 Whole Ore Amalgamation

Whole ore amalgamation is the process of bringing mercury into contact with 100% of the material being mined. Typically, mercury is either added when the ore is being ground in mills or the slurry produced from grinding is passed over a mercury coated copper plate. Amalgamating the whole ore uses mercury very inefficiently and so between 3 and 50 units of mercury are consumed to produce 1 unit of gold, with an average of around 5. Most of the mercury loss during whole ore amalgamation initially occurs into the solid tailings which are often discharged directly into receiving waters and soils. Importantly, however, it is well documented that this mercury continues to evade into the environment for centuries (Alpers and Hunerlach, 2000; Al et al., 2006; Shaw et al. 2006; Winch, 2006). Further, although little studied, it is certain that mercury in tailings that are subsequently leached with cyanide to recover more gold (a growing trend already observed in 10 countries) undergoes enhanced aqueous transport and emission to the atmosphere. This is because of the complexation of mercury by cyanide. It is well known that mercury and cyanide, like gold and cyanide, readily form soluble complexes, and that when cyano-mercury complexes degrade, mercury readily volatilises.

Immediate emissions to the atmosphere during whole ore amalgamation occur when the recovered amalgam is heated to produce the gold. In the simplest case, such as the use of mercury coated copper plates, immediate losses to the atmosphere are therefore roughly equal to the amount of gold produced. However, there can be significant additional emissions to the atmosphere on a time scale of weeks to months from tailings and in particular from operations that employ cyanide. For example, in a whole ore amalgamation operation like those in Indonesia documented in Sulaiman et al. (2007), if 20 g of mercury are consumed to produce 1 g of gold, then 19 g of mercury are lost to the tailings and 1 g of mercury is immediately emitted to the atmosphere. However, additional mercury is released to the atmosphere shortly thereafter from: (i) volatilisation from cyanide rich tailings; (ii) during cyanidation gold is adsorbed from the solution by activated carbon. Mercury is also unavoidably adsorbed. To recover the gold, the carbon is burnt and so any adsorbed mercury is emitted at that time; (iii) the "ash" produced by burning the activated carbon is often re-amalgamated with mercury and this amalgam is also thermally decomposed to produce the gold, releasing an additional amount of mercury to the atmosphere equal to the total gold produced. In such cases, immediate emissions to the atmosphere are minimally greater than the total gold produced and this includes the amount of gold produced via cyanide leaching.

6.2.1.2 Amalgamation of a Concentrate

In cases where only a gravity concentrate is amalgamated, losses are normally about 1 to 2 units of mercury for each unit of gold produced, but can be significantly lower if a mercury capturing system is used when the amalgam is burnt – retorts or fume hoods. For example, in Central Kalimantan, commonly 1.3 g of mercury is consumed to amalgamate 1 g of gold from a gravity concentrate produced by sluicing alluvial ore (Telmer and Stapper, 2007). In this case 0.3 g of mercury is discharged to water with the tailings and 1 g of mercury is emitted to the atmosphere when the amalgam is burnt. Consumption of mercury in Brazil as recorded by Sousa and Veiga (2007) is similar.

Sometimes the tailings are rich in minerals such as zircon which are valuable to the ceramics and abrasives industries and so the tailings are not discarded but rather are further processed and then export (often to China or Korea). During reprocessing the tailings are often amalgamated a second time to recover any residual gold, and then further processed to produce (i) a high grade heavy mineral concentrate which is contaminated in mercury and export, and (ii) a waste which is discarded. The mercury that is export with the zircon is certain to be emitted to the atmosphere during later industrial use. The fate of the mercury in the residual waste is unknown but may end up in aggregate products such as bricks or be discarded into local waterways. An additional cause of mercury pollution that is frequently overlooked is the discarding of "dirty mercury". When ore is amalgamated with mercury the products are (i) solid amalgam; (ii) tailings; and (iii) residual liquid mercury. For example, a miner may add 100 g of mercury to 10 kg of concentrate and then recover 20 g of amalgam (50% gold, 50% mercury), and 87 g of residual liquid mercury with 3 g lost to the tailings. They would then re-use the residual liquid mercury to amalgamate the next day's concentrate. However, the effectiveness of the liquid mercury is reduced as it becomes oxidized and contaminated with impurities – this is referred to as "dirty mercury". Typically, after 3 or 4 uses, mercury becomes much less effective at amalgamation and so it is discarded. In the case of dredge operations in Kalimantan, dredge operators just throw it into the river. This causes mercury consumption to be higher than the 1.3 units of mercury for every 1 unit of gold described above. When mercury is not recycled through re-activation (described in the final section), consumption is likely to be at least twice the ratio established by recording only the immediate losses that occur during amalgamation.

6.3 Where ASGM is Occurring

There is reasonably good information about where ASGM is occurring. The Information sources are: reports from the MMSD (2002); 16 years of archives from the Northern Miner (1992–2008); reports and conference materials from the World Bank's Secretariat on Communities and Small Scale Mining (CASM, 2007) up to 2007 (7 meetings); 5 years of reports and conference materials form the UNDP/GEF/UNIDO Global Mercury Project (GMP) up to 2007; reports from other intervention programs such as the Swiss Development Agency (SDA), the Canadian International Development Agency (CIDA), the World Wildlife Fund (WWF); reports and abstracts from the International Congresses on Mercury as a Global Pollutant (ICMGP) up to 2006 (8 congresses); numerous articles published in the peer reviewed literature; and personal communications with field operatives of intervention programs and people employed in the ASGM economy – miners and gold and mercury merchants. Table 6.1 (see Appendix 1) lists the countries and column 3 of Table 6.1 lists the sources of information that identify the presence of ASGM by country (note that these information sources are in some cases different from those used later to estimate current mercury consumption – column 7). Accordingly, ASGM has been documented to occur in 70 countries. Figure 6.2 illustrates the global distribution of ASGM based on data from Table 6.1. There are at least 6 more countries that are likely to have ASGM occurring bring the likely total to 76 countries but with no firm documentation for those countries we will use the more conservative number of 70.

6.4 Amount of mercury used in ASGM

Amounts of mercury consumed in ASGM can be determined primarily in 5 ways.

1. Direct measurements – using a balance to directly weigh amounts of mercury used.
2. Applying a mercury/gold (Hg:Au) ratio based on the style of operation (gravity concentrate or whole ore amalgamation) to estimates of gold production.

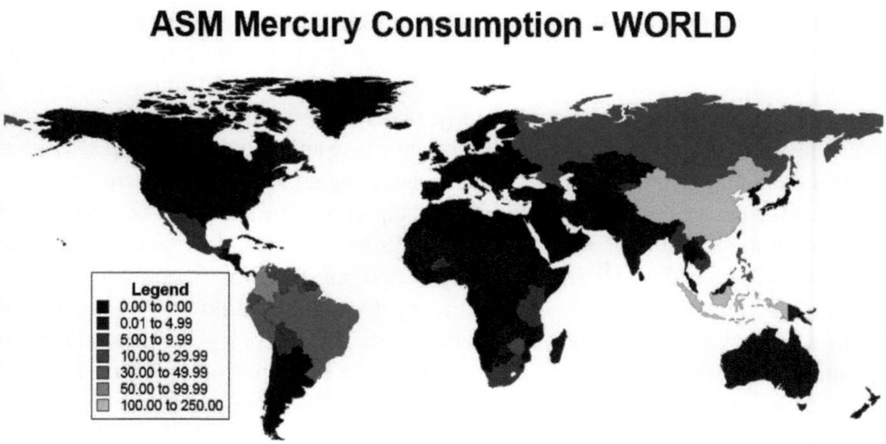

Figure 6.2 Map of mercury consumption by artisanal small scale gold mining globally

3. To get to number 2, estimate the number of miners actively mining and their average gold production.
4. Interviewing miners and gold merchants who buy or sell mercury.
5. Official trade data.

The first four approaches involve directly working with miners and gold merchants and gaining their trust.

Unfortunately, there is very little high quality information on amounts of mercury, size of operations, and what styles of operation are in use around the world in ASGM sites. Much of what exists is anecdotal. In part, this is because of ASGM's highly decentralized and remote nature and because it often exists outside the law. Specifically: (i) there is a lack of interest from governments about ASGM because miners are marginal citizens – they do not pay tax, do not vote, do not have permanent homes, etc.; (ii) miners are subjected to gold price cycles and gold rushes and unfair labour practices and so are very migratory and dispersed; (iii) many ASGM sites are in remote areas where there is no infrastructure and therefore no information; (iv) many clandestine (illegal) activities are involved in ASGM such as money laundering, tax evasion, weapon acquisition, etc., making it sometimes difficult to access miners and making the quality of information they provide sometimes questionable; (v) miners and mining and the use of mercury are often prohibited – perhaps more than 90% of all miners are operating in illegal ways.

But we have found that, in fact, many of these obstacles can be overcome and the lack of information is not only due to these reasons. It is also due to the differing cultures of various intervention efforts. Telmer and Stapper (2007) explain this as follows: "A good knowledge base is the required backbone to formulate solutions to the problems associated with mercury and ASGM. Indeed, many well meaning attempts to improve the livelihoods and living conditions of miners or to reduce the environmental impacts of ASGM have failed because of lack of appropriate knowledge about the ASGM community. There have been attempts to create alternative livelihoods or to introduce mercury-free technologies to miners based simply on the *idea* or *wish* that they should behave differently, rather than starting by understanding the financial burden that such interventions might cause and then building up a solution from there." They go on to explain that "In assessing an ASGM site, there are many useful bits and pieces of information that help constrain the socio-economic and environmental realities of small scale gold mining. Of these, perhaps some of the most useful quantities are: (i) how many people are mining? (ii) how much gold are they producing?; (iii) how much mercury do they use to do so?; and (iv) what is the scale of the impacts they are having on the landscape? – How much habitat (land and water) has been impacted? This basic information can then be used to constrain many other important aspects of ASGM, and then to educate the stakeholders and interest groups involved – including the miners themselves. This in turn helps immensely in guiding the formulation of appropriate intervention strategies, focusing resources, and avoiding costly and frustrating failures." And so unfortunately, despite years of efforts, most interventions in ASGM have either not attempted to, or have not been able to effectively measure the quantity of mercury consumed by miners in ASGM sites. There are however some cases where the amounts of mercury consumed have been well documented.

6.4.1 Indonesia

1. Telmer and Stapper (2007) together with Agrawal (2007) used a scale to directly weigh amounts of mercury used to amalgamate ore, and then extrapolated these statistics to Central Kalimantan by using aerial photography and satellite imagery. The estimate of mercury consumption since 1990 to 2006 for Central Kalimantan not including river dredging was 70 Mg of mercury with the lion's share (10 Mg yr^{-1}) being consumed in more recent times. River dredging consumes more mercury than land based work because the miners throw away the mercury once it becomes oxidized (refered to as "dirty mercury") and is no longer a strong amalgamator of gold – a habit that can be changed by teaching how to clean or re-activate mercury (Pantoja and Alvaarez, 2000; Wuerker, 2008) Sousa and Veiga (2007) estimate how much mercury this prevents from entering the environment for a case study in Brazil. Mercury consumption is estimated to at least double for the region when river dredging is included. Central Kalimantan is about 1/3 of Kalimantan but contains about 1/2 of Kalimantan's ASGM sites, and so by further extrapolation using satellite imagery, it is estimated that 40-60 Mg yr^{1} of mercury are consumed in Kalimantan. This is a minimum estimate because it does not include any high-grade underground workings which are known to occur in Kalimantan (Mansur Geiger, Kalimantan Gold Corporation, pers. comm., 2008) but difficult to see with publically

available satellite imagery. Further, there are many small operations up the many tributary river channels that cannot be easily seen by satellite imagery. Many of these were seen by low flying aerial survey performed while ground truthing the larger areas with aerial photography – small scale mining was ubiquitous, often appearing in the wake of illegal logging.

2. Sulaiman et al. (2007) examined a whole ore amalgamation operation in North Sulawesi, Indonesia, and also used a balance to directly weigh amounts of mercury used to amalgamate ore per mining operation. Mercury losses per unit of gold amalgamated were extremely high averaging 37.5 g mercury lost per 1 g gold produced. The consumption of mercury in just one small area that contained roughly 100 individual operators was 3 Mg yr^{-1}. [An important additional and worrisome consideration here is that once the ore has been subjected to amalgamation by mercury, it is subsequently leached with cyanide and then the final tailings are crudely disposed of into unlined ponds that leak into rivers and groundwater. It is known that cyanide complexes mercury as well as gold and so it is certain that the cyanide leaching is enhancing the transport and distribution of mercury in the environment. It is also known from large scale mining operations that cyanide leaching enhances mercury evasion to the atmosphere and so that too is certainly occurring.] Two more mining areas in North Sulawesi of equal magnitude were visited making a total of 9 Mg yr^{-1} mercury consumption only for the limited study area. However, it is known that there are more operations in Sulawesi making this a minimum for that island.

The mercury consumption for these two areas is 40-60 Mg yr^{-1} for Kalimantan plus 9 Mg yr^{-1} for a part of Sulawesi with a total between 50 and 70 Mg yr^{-1}. The MMSD report on indonesia by Clive Aspinal (2002) claims much higher losses of mercury in north Sulawesi – a total of 270 kg Hg per day which would make annual losses, based on 260 working days per annum, equal to 70 Mg of mercury – just for one area in North Sulawesi. Further, the report uses that estimate from North Sulawesi to extrapolate and make a hypothetical loss of mercury per annum for all Indonesia of 1400 Mg Hg yr^{-1}. Clearly, this is an overestimate. Nonetheless, the report does help give some useful information on the extent of ASGM in Indonesia claiming that in 2002 small scale gold miners were operating in Kalimantan, Sulawesi, Java, Sumatra, and Irian Jaya (now called Papua) – essentially all of the major islands. Through talking to miners, we learned that it occurs on several other islands as well. Considering the broad distribution of ASGM in Indonesia and the fact that ASGM has grown since the MMSD was completed in 2002 (the price of gold has tripled during that time increasing the incentive to mine), we feel it is reasonable to double the estimates from Kalimantan and Sulawesi for a total mercury consumption for Indonesia equal to 100 to 140 Mg yr^{-1}. To make the quality of this estimate clear, and to illustrate how poor the database on mercury in ASGM is, it is important to understand that despite the obviously loose nature of this estimate, it is perhaps our most certain figure. Scaling up from one operation to the country level inevitably involves significant assumptions; nonetheless, we have begun with quantitative data and used the tools that are available to scale up.

6.4.2 Brazil

Sousa and Veiga, (2007) have estimated that there are 40,000 miners in the Crepori area of the Tapajos basin (Reserva Garimpeira) and that they consume 40 g mercury/month for a total of 19.2 Mg yr^{-1}. Telmer and Stapper (2007) independently looked at a subset of this region representing about 1/2 of the area for the period 1979 to 2006, and only considered land based operations (i.e. no river dredging included) and estimated an annual mercury use of 4 Mg yr^{-1} for 2006 – the closest year to the work of Sousa and Veiga. By extrapolation to the whole area and including dredges an amount of 15 to 25 Mg yr^{-1} is possible, roughly corroborating the results of Sousa and Veiga. Brazil is a vast territory and has several other known ASGM sites including several new areas in the western state of Acre (Blore, 2007) and so we feel that doubling this estimate to 40 Mg yr^{-1} is reasonable.

6.4.3 Other Countries with Documented Estimates

Quantities of mercury have also been relatively well documented in Cambodia 7.5 Mg yr^{-1} (Murphy, 2006); Guyana 15 Mg yr^{-1}; Suriname 7.5 Mg yr^{-1}, French Guyana 7.5 Mg yr^{-1} (Vieira, 2008); and Monglia 11.5 Mg yr^{-1} (Grayson, 2007). As well, quantities of mercury have been estimated in four more countries that participated in the Global Mercury Project: Sudan 0.8 Mg yr^{-1} (Ibrahm, 2003); Zimbabwe 25 Mg yr^{-1}; Laos 1.3 Mg yr^{-1}, and Tanzania 6 Mg yr^{-1}. Gunson and Yue (2002) reported a minimum of 50 Mg yr^{-1} mercury released through ASGM in China, however this estimate was since revised to a min and max of 237 to 652 Mg yr^{-1} through more thorough research (Gunson, 2004) and seems reasonable based on the fact that China became the world's largest gold producer in 2007, much of its production is known to come from small mines, and that much of China's ASGM employs inefficient whole ore amalgamation where the consumption of mercury can be very high. Unfortunately at this time China officially admits no ASGM operations occur in its territory.

6.4.4 Other Countries - Direct Anecdotal Information

We have direct anecdotal information on ASGM operations in another 15 countries (Ghana, Mozambique, Guinea, Uganda, Peru, Colombia, Ecuador, Nicaragua, Bolivia, Venezuela, Suriname, Chile, Costa Rica, Guatemala, and Madagascar). These involved either visits or telephone conversations with various stakeholders and miners and gold merchants who, through personal communications provided estimates of mercury consumption by the ASGM community – listed in Table 6.1. As such these estimates are based entirely on anecdotal information gained through

Appendix 1

Table 6.1 Mercury consumption by country for 2008 in artisanal small scale gold mining (ASGM) estimated by the authors; official imports and exports of mercury and gold as recorded in the UN's COMTRADE database per annum for the five year period 2002-2006; the number of chlor-alkali plants per county in 2004 that use mercury (data from the Chlorine Institute). Note: Iraq and Libya do not report any trade in Hg or Gold. However Iraq has 3 mercury based Chlor-Alkali plants and Libya has 1

			ASGM Mercury ($Mg\ yr^{-1}$)				COMTRADE Data				Chlorine Inst.
	Country	ASGM Presence	min	max	mean	Basis for Estimate	Hg Export ($Mg\ yr^{-1}$)	Hg Import ($Mg\ yr^{-1}$)	Au Export ($Mg\ yr^{-1}$)	Au Import ($Mg\ yr^{-1}$)	Chlor-alk. Plants (#)
Amount			641.9	1352.5	997.2		3227.9	3202.7	5954.9	4884.3	229
Count	163	70	70	70	70	70	65	117	108	124	44
1	Albania							0.05			
2	Algeria						218.38	0.31	0.187	0.567	3
3	Andorra								0.015	0.254	
4	Anguilla										
5	Argentina						0.31	15.18	32.310	0.026	4
6	Armenia							0.00	5.422	4.812	
7	Australia	APLA (2004)	0	2	1.0	Guess	49.96	37.62	258.902	106.918	
8	Austria						1.70	3.84	13.071	20.078	
9	Azerbaijan	?	0.05	0.5	0.3	Min		21.22			
10	Bahrain							0.05	0.109	0.014	
11	Bangladesh							7.93		0.009	1
12	Barbados							0.04		0.005	
13	Belarus							1.06	11.375	2.806	
14	Belgium						17.12	37.23	20.974	12.314	4
15	Benin	Yager et al. (2002)	0.05	0.5	0.3	Min		0.00		0.002	
16	Belize								6.372		
17	Bolivia	Graham (2002); Hentschel et al. (2002)	5	10	7.5	MMSD, GMP, CASM		1.29	6.701	2.378	
18	Bosnia Herzegovina						38.20	0.04		0.055	4

(continued)

Table 6.1 (continued)

			ASGM Mercury ($Mg\ yr^{-1}$)				COMTRADE Data				Chlorine Inst.
	Country	ASGM Presence	min	max	mean	Basis for Estimate	Hg Export ($Mg\ yr^{-1}$)	Hg Import ($Mg\ yr^{-1}$)	Au Export ($Mg\ yr^{-1}$)	Au Import ($Mg\ yr^{-1}$)	Chlor-alk. Plants (#)
19	Botswana	Madawo (2007)	0.5	1	0.8	GMP					
20	Brazil	Blore (2007); Veiga (1997)	30	60	45.0	Sousa and Veiga (2007); Telmer and Stapper (2007)	0.09	54.60	32.009	0.712	11
										0.012	
21	Bulgaria						0.12	0.44			
22	Burkina Faso	Hiyate (2008); Saywell (2008); ILO (1999)	3	7	5.0	MMSD, Northern Miner		0.001	0.192		
23	Burundi	Cumming (1997) ; Priester and Hentschel (1992)	0.05	0.5	0.3	Min					
24	Cambodia	Sotham (2001)	5	10	7.5	Sotham (2004); Murphy (2006)				1.996	
25	Cameroon	?	0.05	0.5	0.3	Min		0.003			
26	Canada	Basque (1991)	0	2	1.0		8.16	9.03	203.737	150.442	
27	Central Africa Republic	Yager et al (2002); Northern Miner (2001, v87, no.2)	0.05	0.5	0.3	Min					
28	Chad	Mobbs (1996)	0.05	0.5	0.3	Min					
29	Chile	Castro and Sanchez (2003)	3	5	4.0		42.49	1.59	15.230	0.068	
30	China	Saywell (2007); Gunson and Veiga (2004)	237	652	444.5	Gunson (2004)	0.18	188.49	22.642	18.558	6
31	China, Hong Kong AR						11.93	41.79	278.843	132.759	

32	China, Macao SAR							0.00	0.002	0.032	
33	Colombia	Harris (2006); Lacerda (2003)	50	100	75.0	Gov. of Antioquia	0.03	63.18	35.744	0.247	1
34	Cook Isds									0.000	
35	Costa Rica	Northern Miner (1998, v 84, no. 17); Veiga (1997)	0.05	0.5	0.3	Min		0.31	194.969	135.707	
36	Côte d'Ivoire	Mobbs (1998)						0.04	1.518		
37	Croatia						0.46	0.05	0.259	1.907	3
38	Cuba							2.72	0.215	0.081	1
39	Cyprus							0.02	0.037	1.306	
40	Czech Rep.						74.50	3.36	6.006	5.102	2
41	Denmark						9.00	0.42	1.119	2.331	2
42	Dominica							0.002		0.006	
43	Dominican Republic	Veiga (1997)	0.05	0.5	0.3	Min					
44	Democratic Republic of Congo	Vaccaro (2007); ILO (1999)	1	2	1.5	Guess					
45	Ecuador	Robertson (2006); Betancourt et al (2005)	10	20	15.0	GMP	0.03	9.47	285.982	0.922	
46	El Salvador							0.70	0.057	1.488	
47	Estonia						2.79	0.003	0.047	1.589	
48	Ethiopia	Labonne (2002)	0.05	0.5	0.3	Min		0.05	5.278	0.256	
49	Faeroe Isds							0.004		0.005	
50	Fiji							0.19	4.829	0.684	
51	Finland						32.16	0.20	1.221	1.170	5
52	France						56.76	137.93	33.397	41.417	10
53	French Guiana	Fréry et al (2001)	5	10	7.5	Viera (2008)				0.081	
54	French Polynesia										
55	Gambia	Dolley (1996)	0.05	0.5	0.3	Min		0.005			
56	Georgia										

(continued)

Table 6.1 (continued)

			ASGM Mercury ($Mg\ yr^{-1}$)				COMTRADE Data				Chlorine Inst.
	Country	ASGM Presence	min	max	mean	Basis for Estimate	Hg Export ($Mg\ yr^{-1}$)	Hg Import ($Mg\ yr^{-1}$)	Au Export ($Mg\ yr^{-1}$)	Au Import ($Mg\ yr^{-1}$)	Chlor-alk. Plants (#)
57	Gabon	Northern Miner (2003, v. 89, no. 40); Priester and Hentschel (1992)	0.05	0.5	0.3	Min			6.694	0.063	
58	Germany						50.35	52.82	59.861	50.487	21
59	Ghana	Northern Miner (2007, v. 93, no. 39); Babut et al (2003)	3	6	4.5	GMP		5.41	80.243	1.717	
60	Greece							2.79	1.069	28.300	1
61	Guatemala	UNEP (2005)	1	2	1.5	GMP		27.81	0.620	2.327	
62	Guinea	Labonne (2002)	0.05	0.5	0.3	Min			79.313		
63	Guinea-Bissau	Dolley (1996); Bermudez-Lugo (2002)	0.05	0.5	0.3	Min		34.36	7.417	0.028	
64	Guyana	Couture and Lambert, (2003)	10	20	15.0	Viera (2008)					
65	Honduras	Attenborough (1999); Veiga (1997)	0.05	0.5	0.3	Min	0.65	0.51	5.955	0.106	
66	Hungary							4.11	0.764	0.736	2
67	Iceland							0.003	0.008	0.493	
68	India	Duval (2004); Siddaiah (2001)	1	2	1.5	Guess	11.10	186.81	0.594	743.774	48
69	Indonesia	Castilhos et al (2006)	130	160	145.0	Telmer and Stapper (2007)					4
70	Iran	?	0.05	0.5	0.3	Min	0.13	126.19		2.666	7
71	Ireland						0.40	4.91	0.118	2.444	
72	Israel						13.92	24.01	0.444	3.987	2
73	Italy						66.94	39.67	46.578	275.543	11

74	Ivory coast	Yager et al (2002)	0.05	0.5	0.3	Min					
75	Jamaica							0.04		0.037	
76	Japan						108.29	4.58	124.808	67.000	
77	Jordan							0.03	12.989	7.125	
78	Kazakhstan	?	0.05	0.5	0.3	Min	1.24		26.678	0.535	
79	Kenya	Yager et al (2002)	5	10	7.5	GMP	0.93	10.57	558.196	0.765	
80	Kyrgyzstan	Appel et al (2003)	5	10	7.5	Appel	460.48		12.521	0.079	
81	Laos	Boungnaphalom (2003)	0.5	2	1.3	GMP					
82	Latvia						0.00	0.01	0.001	0.147	
83	Lebanon							3.42	17.625	11.290	
84	Lesotho	Coakley (2002)	0.05	0.5	0.3	Min					
85	Liberia	DLI (2003)	0.05	0.5	0.3	Min					
86	Lithuania							0.003		0.012	
87	Luxembourg						0.06	0.11	6.625	6.435	
88	Madagascar	Rajaobelina (2003)	1	2	1.5	Guess	0.00	0.02	0.001	0.002	
89	Malawi	Dreschler (2001)	0.05	0.5	0.3	Min		0.02		0.004	
90	Malaysia	Priester and Hentschel (1992)	2	5	3.5	Google Earth	173.88	42.04	732.191	651.339	
91	Maldives							0.01		0.137	
92	Mali	Northern Miner (2008, v93., no. 47; no. 50); MMSD (2002)	1	2	1.5	MMSD			41.061	2.412	
93	Malta							0.003	0.406	9.461	
94	Mauritius										
95	Mayotte							0.06	0.114	0.778	
96	Mauritania	Mbendi (2004)	0.05	0.5	0.3	Min				0.006	
97	Mexico	Graham (2003); Veiga (1997)	5	10	7.5	Guess	7.64	221.27	138.191	646.113	
98	Mongolia	Grayson (2007)	8	15	11.5	Grayson (2007)			11.503		
99	Montserrat									0.003	
100	Morocco						16.94	2.18	0.915	2.472	3

(continued)

Table 6.1 (continued)

			ASGM Mercury $(Mg\ yr^{-1})$				COMTRADE Data				Chlorine Inst.
	Country	ASGM Presence	min	max	mean	Basis for Estimate	Hg Export *(Mg yr-1)*	Hg Import $(Mg\ yr^{-1})$	Au Export $(Mg\ yr^{-1})$	Au Import $(Mg\ yr^{-1})$	Chlor-alk. Plants (#)
101	Mozambique	Spiegel et al (2006)	3	5	4.0	GMP		0.004	1.201	0.006	
102	Myanmar	UNESCAP (2003)	5	8	6.5	UNESCAP (2003)			2.956	.454	1
103	Namibia							0.05			
104	Nepal							0.09			
105	Netherlands						198.36	592.26	16.818	20.764	1
106	New Caledonia							0.01		0.077	
107	New Zealand	?	0.05	0.5	0.3	Min	0.04	0.01	18.221	1.067	
108	Nicaragua	Attenborough (1999); Rosario and Ault (1997)	1	2	1.5	Silva (2008)		0.29	4.989	0.021	1
109	Niger	Alfa (2000)	0.05	0.5	0.3	Min			3.218		
110	Nigeria	Vaccaro (2006); Priester and Hentschel (1992)	0.05	0.5	0.3	Min					
111	Norway						0.02	0.08	3.732	2.509	
112	Oceania	?	0	1	0.5	Guess					
113	Oman							0.09	0.367	5.306	
114	Pakistan							34.36		27.650	3
115	Panama	Attenborough (1999)	1	2	1.5	Guess		0.14	0.232	0.572	
116	Papua New Guinea	Crispin (2003)	2	4	3.0	MMSD		3.74	28.230	0.571	
117	Paraguay							0.12	0.051	0.046	
118	Peru	Brooks et al (2006)	20	40	30.0	Brooks et al (2006)	75.46	82.67	415.032	0.026	8
119	Philippines	Israel and Asirot (2000)	20	30	25.0	Israel and Asirot (2000)					1
120	Poland						11.39	9.22	0.482	2.373	3
121	Portugal						15.76	0.28	0.184	2.344	
122	Qatar							0.05	0.016	1.762	

123	Rep. of Korea						8.57	131.15	117.824	156.265	5
124	Rep. of Moldova									0.018	
125	Romania						7.08	18.85	1.223	0.453	1
126	Russian Federation	Stepanov and Yusupov (2001)	7	15	11.0	Stepanov and Yusupov (2001)	138.59	38.45	0.020	0.002	4
127	Rwanda	Priester and Hentschel (1992)	0.05	0.5	0.3	Min					
128	Saudi Arabia						0.00	1.94	234.719	44.816	1
129	Senegal	Savornin et al (2007)	1	2	1.5	Guess		0.04	0.508	0.079	
130	Serbia						0.00	1.18	0.312	0.013	4
131	Sierra Leone	Thonae (2004)	0.05	0.5	0.3	Min					
132	Singapore						84.95	138.74	61.018	116.456	
133	Slovakia						0.90	3.07	0.283	2.993	
134	Slovenia						5.18	0.07	0.456	0.856	2
135	South Africa	Beales (2005); Mahlatsi and Guest (2003)	5	10	7.5	GMP	9.94	11.68	5.999	1.155	
136	Spain						668.75	454.43	7.831	24.311	9
137	Sudan	Ibrahim (2003)	0.5	1	0.8	GMP	0.03	6.01	0.208	7.167	
138	Sri Lanka										
139	Suriname	Heemskerk (2003)	5	10	7.5	Viera (2008)				0.001	
140	Sweden						0.33	4.03	14.461	1.441	5
141	Switzerland						124.07	9.83	42.253	9.038	2
142	Syria							19.96		0.005	
143	Tajikistan	Dawson (1996)	3	5	4.0	Dawson (1996)					
144	TFYR of Macedonia						18.02	0.01	0.016	0.054	
145	Thailand	Umbangtalad et al. (2007)	1	2	1.5	Guess	4.40	12.20	35.851	108.214	
146	Timor-Leste										
147	Togo	Yager et al (2002)	3	5	4.0	Yager et al (2002)	0.97	8.36			
148	Trinidad and Tobago						0.04	0.13	0.127	0.370	

(continued)

Table 6.1 (continued)

			ASGM Mercury $(Mg\ yr^{-1})$				COMTRADE Data				Chlorine Inst.
	Country	ASGM Presence	min	max	mean	Basis for Estimate	Hg Export $(Mg\ yr^{-1})$	Hg Import $(Mg\ yr^{-1})$	Au Export $(Mg\ yr^{-1})$	Au Import $(Mg\ yr^{-1})$	Chlor-alk. Plants (#)
149	Tunisia						0.00	0.05	0.409	2.207	
150	Turkey						0.03	5.25	13.413	222.339	2
151	Turks and Caicos Is.										
152	Uganda	World Bank (2003)	0.5	1	0.8	Hinton (2008)		0.05	10.462		
153	United Arab Emirates						1.19	9.70	52.822	117.468	1
154	United Kingdom						56.66	23.08	2.168	347.705	9
155	United Rep. of Tanzania	Kinabo (2003)	4	8	6.0	Kinabo, Ikingura		2.77	50.736	26.600	
156	Uruguay							1.96	3.792	0.100	2
157	USA	Weekend Prospector (2004)	1	2	1.5		319.83	130.75	1330.776	360.750	8
158	Uzbekistan	Trung (2001); Northern Miner (v. 83 no. 41, 1997)	0.05	0.5	0.3	Min				1.002	
159	Venezuela	Veiga et al (2005)	10	20	15.0	Veiga et al (2005)			1.002	0.705	
160	Viet Nam	Trung (2001); Northern Miner (v. 83 no. 41, 1997)	5	10	7.5				0.353	56.863	
161	Yemen								0.824	0.390	
162	Zambia	Kambani (2003)	0.5	1	0.8	Kambani (2003), Nyambe, (2008)		0.12	0.750	0.01	
163	Zimbabwe	Maponga. and Ngorima (2003)	20	30	25.0	GMP		11.12	16.266	41.005	

discussion. We therefore have relatively good information on ASGM mercury consumption from 2 countries, reasonable information from 7 more countries, and some but poor information from 14 more countries amounting to some knowledge by the authors on ASGM sites in 23 countries. These 23 countries represent in excess of 80% of ASGM mercury consumption.

6.4.5 Remaining Countries - Indirect Anecdotal Information

There is further information from the MMSD reports published in 2002 but as mercury was not necessarily a primary focus, the estimates, like the one from Indonesia discussed earlier, are of variable quality. Some seem to exaggerate the amounts of mercury consumed by the ASGM community. For example our estimate of 7.5 Mg yr^{-1} for Bolivia is far lower than the numbers given by the MMSD report (no total is given but 25 Mg yr^{-1} are ascribed to just one area). Some seem to understate the problem - the report on India by Chakravorty (2002), for example, claims that there is no "gold rush" in India and essentially no use of mercury in ASGM in India. However, other anecdotal reports from Indian colleagues claim this is not the case and clearly India is heavily involved in the gold industry.

According to goldnews (http://goldnews.bullionvault.com) India consumes nearly 800 Mg of Gold Bullion/a, accounting for about 20% of world gold consumption. Nearly 600 Mg of it goes into making jewellery representing $13.5 billion in fiscal 2006-07, and accounting for 8.3% of world jewellery sales by value. For the sake of being conservative, we have assigned India an almost impossibly small amount of mercury consumption through ASGM of 0.3 Mg yr^{-1}, however, we imagine that this could be substantially larger. Many of the other MMSD reports mention the use of mercury, even the intensive use of mercury, but do not estimate quantities used.

In total there is some form of information for 48 countries ranging from relatively good to reasonable to poor as described above. Those 48 countries have been assigned value for consumption of mercury by ASGM operations. For the remaining 25 countries, there is only information indicating the presence of ASGM. These countries have been assigned a minimum amount of mercury of 0.3 Mg yr^{-1} equalling a total of 7.5 Mg for those 25 countries.

6.5 Reported Trade in Mercury and Gold

In ASGM, mercury consumption (mercury purchased) is equal to the amount of mercury released to the environment as none that was purchased is ever returned to the commodity market. Notably, although mercury is traded freely as a commodity around the world, it is never officially purchased for gold amalgamation despite the fact that a large amount of what is traded ends up being used for that purpose. For example,

Brazil, French Guiana, and Indonesia, despite their large ASGM communities, are countries in which mercury is not allowed by law to be used in gold mining. Here, in order to understand the limits of our estimations on mercury use in ASGM, we analyse the existing global trade data on mercury and gold to make some crude observations and also to show how invisible the trade in mercury and gold from ASGM is.

The following analysis is based on the data in Table 6.1 which lists global trade in mercury and gold per annum using data from the United Nations Commodity Trade Statistics Database (COMTRADE*)* covering the five year period 2002-2006. The table also contains the number of chlor-alkali plants that use mercury per country as reported by the Chlorine Institute (2006), and the estimate of mercury consumption by ASGM made by the authors that is discussed later.

Figure 6.3a shows total reported global mercury trade by country for the years 2002 through 2006 (a 5 year period) as recorded by COMTRADE. The database relies on voluntary reporting and so is incomplete. For example, the number of years reported varied between 5 and 1 for the countries listed. Figure 6.3b shows trade for the same period but per annum by normalising to the number of years reported (mass of mercury/years reported). For some determinations, it is better to use Figure 6.3a, for other determinations, it may be more appropriate to use Figure 6.3b. For example, to determine the average price paid for mercury, it is better to use only the reported data, but the incomplete reporting in Figure 6.3a would underestimate the total trade in mercury. The data in Figure 6.3b, on the other hand, may over-estimate total mercury trade if some countries imported less mercury during the years for which no reporting was done. The trade data from COMTRADE is therefore, to some degree, complicated to interpret. Nonetheless, the following conclusions can be reached:

1. The number of countries that actively trade in mercury is 119 but there are 190 countries listed in the UN's COMTRADE database. Therefore there are 71 countries that either do not consume any mercury or do not report consumption. Due to dental practices alone (300 to 400 Mg mercury consumed per annum for amalgam fillings, Maxson, 2008a; P. Maxson, Concorde Cons., Belgium, 2008, pers. Comm.), it is unlikely that these 71 countries do not trade in mercury at least for dental use. This suggests that the database represents a minimum amount of trade.
2. The total reported mercury trade for the 5 year period of reporting from 2002 to 2006 (Figure 6.3a) is 12,750 Mg exported (2550 Mg yr^{-1}), and 14,870 Mg imported (2970 Mg yr^{-1}). Amounts of "re-export" (57 Mg or 11.5 Mg yr^{-1}) and "re-import" (0.04 Mg or 0.008 Mg yr^{-1}) are insignificant. This implies a surplus amount of import of 2120 Mg over the 5 year period. Under any trade or consumption scenario, this is unlikely. The reason for the discrepancy is unknown - perhaps tax avoidance, perhaps incomplete reporting, or both. As it is unlikely that importing is over-reported, this, as above in point 1, again suggests that mercury trade is underestimated by the UN COMTRADE database. Therefore, it is reasonable to assume that, a minimum of 2970 Mg of mercury per annum on average was imported during the years 2002-2006. Normalising by years reported (Figure 6.3b) the exports of mercury become 3,230 Mg yr^{-1} and imports 3,200 Mg yr^{-1}. The discrepancy is lessened and direction of imbalance reversed to surplus exports

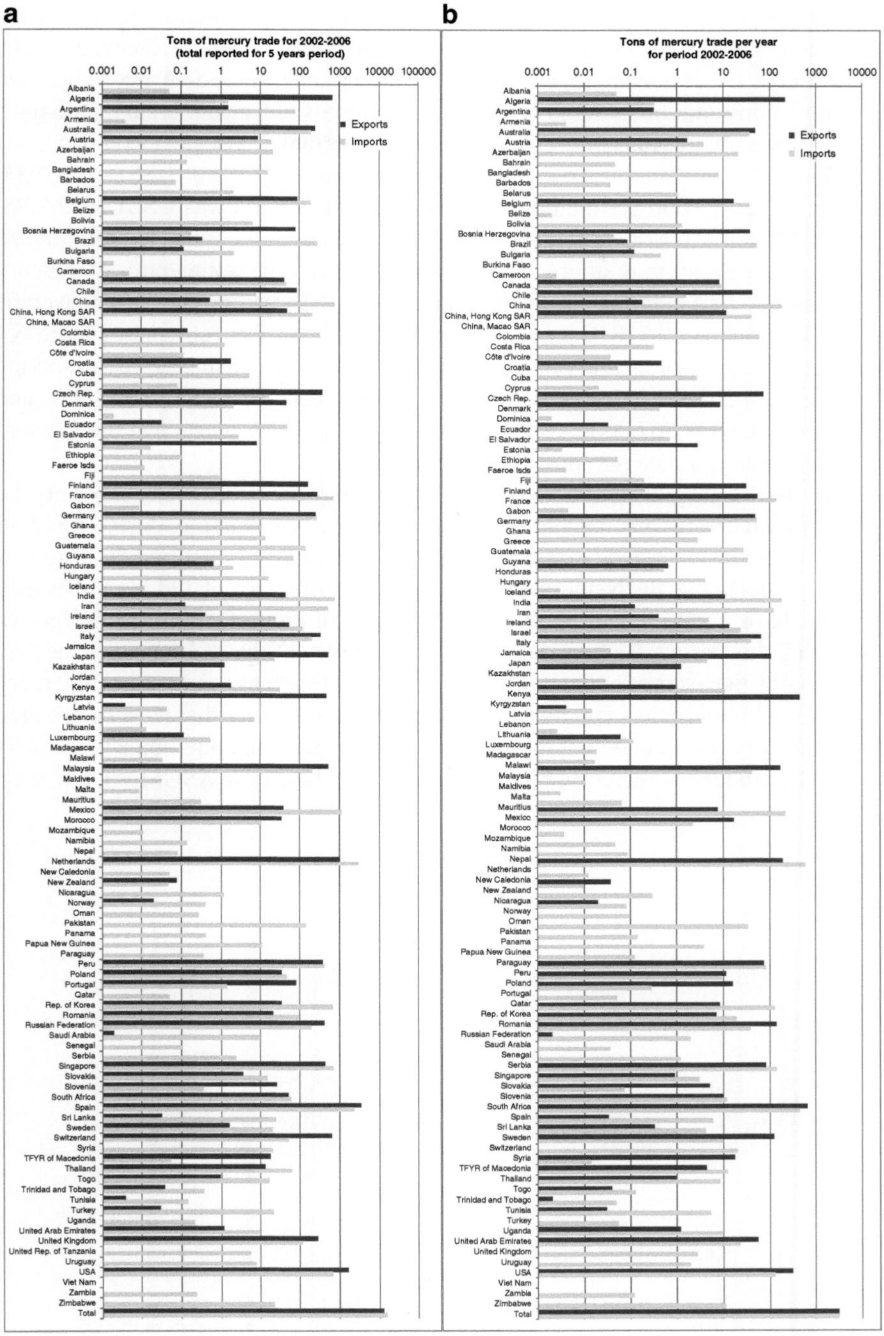

Figure 6.3 (**a**) International exports and imports of mercury by country in Mg for the 5 year period 2002-2006 (UN Comtrade, 2008). All reporting countries are listed. (**b**) Same in (**a**) but per annum by normalizing to the number of years reported – many countries did not report for all years

(30 Mg per annum greater export). The improved balance and more reasonable direction of surplus (more export than import) suggests that this is a better estimate, although by no means robust. It is also important to note that trade does not imply consumption as some of the trade represents recycled mercury, perhaps 5 to 10% globally (Maxson, Concorde Cons., Belgium, 2008, pers. comm.).

3. The total value of reported exports (based on Figure 6.3a) was US$113,587,000 or US$22,717,000 per annum; the total value of imports was US$132,593,000 or US$26,519,000/a. Although, for the above reasons, the totals are minimums, the average price may still be representative. The average selling price of mercury was US$8.91/kg, the average buying price was US$8.92/kg. The current average dealer price is US$18.33/kg (Northern Miner, 2008). For reference, Figure 6.4 shows the price of mercury over the last 108 years. Normalising by years reported (Figure 6.3b) the value of exports per annum becomes US$26,690,000, and imports US$28,567,000. Again the improved balance suggests that this is more reasonable estimate but must be a minimum.
4. The minimum consumption of mercury by human endeavours can only be approximated by this database - but not robustly. For example, during this 5 yr period, the Netherlands exported 198 Mg per annum and imported 592 Mg yr^{-1}. It is unlikely that the Netherlands consumed the difference of 394 Mg yr^{-1} as they only have 1 chlor-alkali plant (Table 6.1) and so their stock must have grown during this period. Rather the data gives some idea of the main mercury dealers globally, the average amount of trade per annum (the mass and value of mercury moving around), the main mercury importers, the countries that must be engaged in mercury trade but do not report it (e.g. Philippines), and puts some constraints on the amount of mercury that could possibly be used for ASGM – i.e. it must

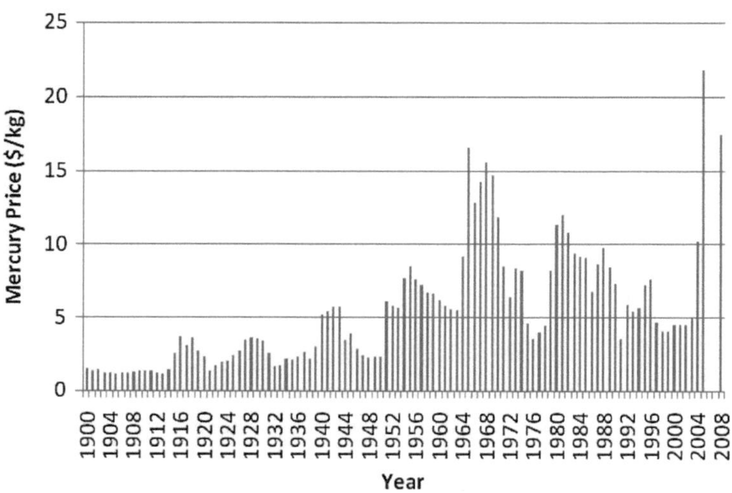

Figure 6.4 Price of mercury over the last 108 years (sources: Northern Miner, 2008; Reece, 2006)

be significantly lower than global trade. Figure 6.5a and 5b illustrate exporters and importers by country. There are 54 countries that only import mercury – these are clearly visible in Figure 6.5a – and there are 2 countries that only export mercury, Kyrgyzstan and Kazakhstan – visible in Figure 6.5b.

5. The minimum consumption of mercury by human endeavours can only be approximated by this database - but not robustly. For example, during this 5 yr period, the Netherlands exported 198 Mg per annum and imported 592 Mg yr^{-1}. It is unlikely that the Netherlands consumed the difference of 394 Mg yr^{-1} as they only have 1 chlor-alkali plant (Table 6.1) and so their stock must have grown during this period. Rather the data gives some idea of the main mercury dealers globally, the average amount of trade per annum (the mass and value of mercury moving around), the main mercury importers, the countries that must be engaged in mercury trade but do not report it (e.g. Philippines), and puts some constraints on the amount of mercury that could possibly be used for ASGM – i.e. it must be significantly lower than global trade. Figure 6.5a and 5b illustrate exporters and importers by country. There are 54 countries that only import mercury – these are clearly visible in Figure 6.5a – and there are 2 countries that only export mercury, Kyrgyzstan and Kazakhstan – visible in Figure 6.5b.
6. There is clearly a transfer of mercury from the developed countries and northern hemisphere to less developed countries and southern hemisphere. Veiga et al., (2006) also discuss this trend. Because overall, industrial consumption of mercury is dropping (Maxson, 2008a), consumption of mercury by ASGM is the most logical explanation for the direction of this transfer.
7. The non-ASGM consumption of mercury for 2005 is estimated to be a minimum of 2385 Mg yr^{-1}, and a maximum of 3365 Mg yr^{-1} (Maxson, 2008a; P. Maxson, Concorde Cons., Belgium, 2008, pers. Comm.). If the global trade data from COMTRADE represents a minimum of 2970 Mg yr^{-1}, and we assume that the 2005 data is a good average for the period 2002-2006 (reasonable), then that suggests that the minimum amount of mercury available for ASGM ranges between 585 Mg yr^{-1} and negative 395 Mg yr^{-1} – hardly a satisfying result. Obviously ASGM consumption is not zero and so perhaps the maximum non-ASGM mercury consumption estimate of 3365 Mg yr^{-1} is too high. That leaves us with a minimum ASGM consumption supported by official data of somewhere between 100 (arbitrary non-zero value) and 585 Mg yr^{-1} with an average minimum of 345 Mg yr^{-1} mercury consumed by ASGM. This is a hypothetical minimum based on available but clearly incomplete official trade data – a starting place. The real amount of mercury consumed by ASGM must be higher as becomes clear when field data is considered – discussed below.
8. The price of mercury has risen sharply since 2003 from ~US$5.00/kg to the current price in 2008 of US$18.33/kg, slightly down from a 2006 peak of around ~US$23.00/kg. That is, on average, a quadrupling of price. From 1975 until 2005, the price of mercury and gold correlated well, but this relationship broke in 2007 when the price of gold sharply increased and the price of mercury actually decreased. The increase in price of mercury to a peak in 2006 was likely a response to the announcement of mercury mine closures and an expectation that supplies would

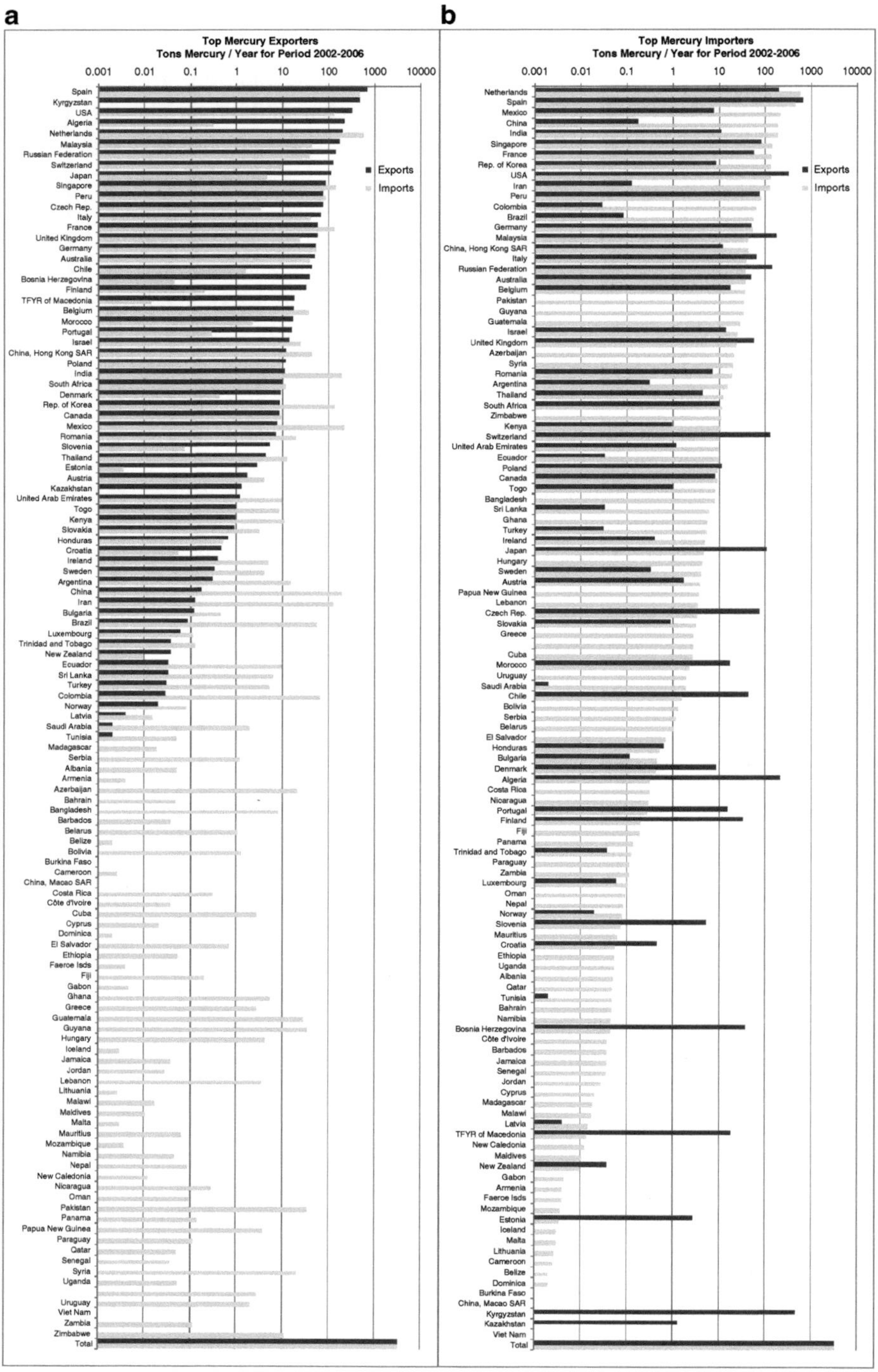

Figure 6.5 (**a**) International exports and imports of mercury per annum sorted by top exporters and (**b**) importers for the 5 year period 2002-2006 (UN Comtrade, 2008). All reporting countries are listed.

be short (Maxson, pers. Comm.). Demand for mercury for ASGM and VCM production in China (vinyl chloride monomer – a feedstock for polyvinyl chloride plastics (PVC)) also likely contributed to supporting higher prices. VCM use of mercury is large and growing (Maxson, 2008b).

9. There is a set of further observations that can be drawn from Table 6.1 as follows:

   - There are 28 countries with known ASGM sites that do not officially export any gold: Azerbaijan, Benin, Botswana, Burundi, Cambodia, Cameroon, Central Africa Republic, Chad, Dominican Republic, DRC, French Guiana, Gambia, Indonesia, Iran, Ivory coast, Laos, Liberia, Malawi, Mauritania, Nigeria, Oceania, Philippines, Rwanda, Sierra Leone, Suriname, Tajikistan, Togo, Uzbekistan
   - There are 16 countries with known ASGM sites that do not officially record any mercury or gold transactions whatsoever: Botswana, Burundi, Central Africa Republic, Chad, Dominican Republic, DRC, Indonesia, Ivory coast, Laos, Liberia, Nigeria, Oceania, Philippines, Rwanda, Sierra Leone, Tajikistan. Two of these countries, Indonesia and the Philippines, are known to have very large ASGM activities (see refs in Table 6.1).
   - There are 4 countries that only export gold: Belize, Guinea, Mongolia, Niger
   - There are 16 main mercury exporting countries (those who export more than 50 Mg yr^{-1}): Algeria, Czech Rep., France, Germany, Italy, Japan, Kyrgyzstan, Malaysia, Netherlands, Peru, Russian Federation, Singapore, Spain, Switzerland, United Kingdom, USA
   - Officially there are 54 countries that only import mercury (a total of 190 Mg yr^{-1}): Albania, Armenia, Azerbaijan, Bahrain, Bangladesh, Barbados, Belarus, Benin, Bolivia, Burkina Faso, Cameroon, China, Macao SAR, Costa Rica, Côte d'Ivoire, Cuba, Cyprus, Dominica, El Salvador, Ethiopia, Faeroe Isds, Fiji, Gambia, Ghana, Greece, Guatemala, Guyana, Hungary, Iceland, Jamaica, Jordan, Lebanon, Lithuania, Malawi, Maldives, Malta, Mayotte, Mozambique, Namibia, Nepal, New Caledonia, Nicaragua, Oman, Pakistan, Panama, Papua New Guinea, Paraguay, Qatar, Senegal, Syria, Uganda, United Rep. of Tanzania, Uruguay, Zambia, Zimbabwe.
   - In terms of countries that are potentially significant distributors of mercury for use in ASGM, there are 13 countries with no or few mercury using chloralkali plants that import significant amounts of mercury: Australia, Azerbaijan, China, Hong Kong SAR, Guatemala, Guyana, Kenya, Malaysia, Mexico, Singapore, South Africa, Syria, Thailand, Zimbabwe. Of these, Mexico and Singapore are by far the largest, importing 221 and 138 Mg yr^{-1}, respectively.
   - There are 14 countries that import more than 50 Mg yr^{-1} of mercury: Brazil, China, Colombia, France, Germany, India, Iran, Mexico, Netherlands, Peru, Rep. of Korea, Singapore, Spain, USA
   - Estimates of ASGM mercury consumption is greater than official mercury imports in 53 countries of the 70 known to have ASGM sites. The opposite is true for the other 21 (imports>ASGM consumption) indicating that the official imports of mercury to these countries is sufficient to meet ASGM demand.

6.5.1 *Using Gold Production to Estimate Mercury Consumption in ASGM*

Regarding the use of gold production to estimate ASGM consumption of mercury, this also is controversial. There have been unofficial estimates by gold dealers that ASGM gold production is around 150 Mg per annum. However, this is very difficult to know because the gold market is unlike most other commodity markets in that 80% of gold produced over the past 6000 years is still in existence today and available to be traded – a total amount of approximately 135,000 Mg (Schofield, 2007). Official gold production per annum is around 2500 Mg yr^{-1} (Schofield, 2007). So gold traded and gold produced officially by large scale mines do not have to match up and this makes it very difficult to constrain the amount of ASGM gold entering the market each year. This is particularly true because the vast majority of gold trade is done on an unallocated basis meaning that it is held in a vault in common with other gold and the customer has a general entitlement only. Essentially this means that gold from many sources is mixed into one pool by holding companies.

However, even if ASGM production were only 150 Mg per annum, considering that a significant portion of that is estimated to be produced using whole ore amalgamation (around 50%) which uses large amounts of mercury, then globally with a Hg: Au ratio of at least 3:1, a production of 150 Mg of gold per annum would imply 450 Mg yr^{-1} of mercury consumed in ASGM. This calculation is made as yet another hypothetical minimum and to further support later arguments that mercury consumption by the ASGM community must be significantly higher.

Others lines of evidence discussed in the MMSD (2002) and GMP (2007) documents, as well as some earlier work (Veiga, 1997) that used patterns of production per miner per region, suggest that global gold production by ASGM is much higher – 400 to 600 Mg Au per annum. At this level, the consumption of mercury must be near to 1000 Mg yr^{-1} as reflected in Table 6.1. In a reverse argument, if the estimate of mercury consumption in Table 6.1 is reasonable, then global gold production by ASGM must be about 1/3 of 1000 Mg = 350 Mg of gold/year, a value in-between that of gold traders and MMSD estimates.

6.6 Knowledge Gaps about Mercury in ASGM

In order to evaluate the significance of mercury emitted from ASGM, and to enable discussion about how best to reduce emissions, it is useful to elaborate the current gaps in our understanding about it.

The fate of mercury in the environment released from ASGM remains poorly understood. For example, of the portion emitted to the atmosphere, how much falls out locally, and how much travels long distances and over what time scale has never been adequately investigated and so remains poorly known. This is despite the fact that the long range transport of these emissions and subsequent deposition in other countries is a key interest of the UNEP Mercury Program and other parties concerned about global mercury pollution.

Further, what happens to the mercury emitted from ASGM following deposition is also not well known as most of the high calibre research that has been done on atmospheric mercury and its fate has been done in temperate or polar environments whereas most ASGM occurs in the tropics where hydrology, soils and vegetation, productivity, and rates of biogeochemical cycling are vastly different. The fate of the mercury from ASGM that is directly discharged into water is equally poorly known. How it is transported, how far it travels, how and where it becomes methylated, and ultimately how much of it enters the local versus global fisheries is poorly known.

In fact many of the general knowledge gaps about mercury that were highlighted by the plenary panellists at the 8th ICMGP (International Conference on Mercury as a Global Pollutant, "Mercury 2006") apply directly to mercury and gold mining. Some of the relevant gaps identified at that congress are:

- Air-surface exchange
- Role of Halogens
- Trends in active pools
- Hydrology
- How to scale up
- The role of dissolved organic matter (DOM)
- Modelling challenges
- Inorganic mercury vs. Methyl mercury contamination in fish
- Mercury in aquaculture

For a variety of reasons, small scale mining is a good place to build this knowledge. Perhaps even the best place as it would additionally bring needed resources, raise awareness, and undoubtedly produce some innovative ideas. The current lack of understanding about mercury in ASGM puts a limitation on the development of innovative solutions towards prevention and remediation. Table 6.2 lists the knowledge gaps highlighted by the plenary panellists at the 8th International Congress on Mercury as a Global Pollutant (Mercury 2006 in Madison, Wisconsin) and how they relate to ASGM as well as some additional important knowledge gaps that were not highlighted. A large part of mercury emitted to the atmosphere from ASGM has been thought to be deposited locally around gold shops and mining sites where amalgam is burnt. Part of the argument used to support this idea, are halos or "bulls eyes" around amalgamation burning centres. Using data from CETEM (1992), Telmer et al. (2006), made some mass balances for soils around gold shops in Alta Floresta, and found that the amount of mercury in the observed bulls eye may be as low as 1% of that emitted, suggesting that in fact, mercury emitted to the atmosphere is travelling long distances. This interpretation is supported by measurements of mercury in the atmosphere made by airplane over the Amazon Basin (Artaxo et al., 2000). They concluded that gold mining areas contribute 63% of the total atmospheric Hg over the Amazon. Telmer et al. also speculate that

Table 6.2 Knowledge gaps about mercury use in ASGM

	General relationship to ASGM	Specific example
Knowledge gap highlighted		
1. Air surface exchange	How far is mercury vapour from ASGM transported? How much mercury enters the global mercury cycle through the atmosphere? How much mercury from ASGM is emitted to the atmosphere? What are the latent emissions of mercury from ASGM to the atmosphere (other than directly during amalgam burning)? What is the fate of mercury emitted to the atmosphere from amalgam burning?	How does the tropical rainforest canopy and or thick clay rich tropical soils interact with atmospheric mercury released from ASGM? How do atmospheric conditions and processes in the tropics cause mercury transport to differ from conditions in the temperate and polar environments? Are the tropics a net source or net sink for atmospheric mercury?
2. The role of halogens	What role do halogens play in the fate of mercury released from ASGM?	The atmosphere is highly depleted in halogens in the middle of the Amazon basin and in other equatorial sites where much ASGM occurs. Does this effect long range transport?
3. Trends in active pools	Which active pools are growing due to ASGM?	Commodity storage? Elemental Mercury in soils? Adsorbed mercury in soils? Methyl Mercury in Floodplains?
4. The role of hydrology	How do annual inundation events in the world's tropical floodplains interact with mercury released from ASGM?	How much does inundation of floodplains downstream of mining operations cause methylation? Does annual inundation enhance or suppress mercury bioavailability?
5. Scaling up	How can the fate of emissions from one ASGM site be extrapolated to other sites and other regions in the world	Do emissions in rainforest ASGM behave similarly to semi-arid ASGM? How can global ASGM emissions be modelled?
6. The role of dissolved organic matter (DOM)	How does DOM interact with mercury released by ASGM?	The draining and burning of tropical peat lands which releases vast amounts of DOM directly interacts with mercury released from ASGM. How does this effect transport and fate?
7. Modelling	How to include mercury releases from ASGM into models of the global mercury cycle?	
8. Inorganic versus organic mercury contamination in fish	Is inorganic mercury a significant vector of mercury into fish exposed to ASGM discharges?	
9. mercury in aquaculture	Many aquaculture operations are in the receiving waters of ASGM. Are these fish impacted?	

Knowledge gaps that are important but not highlighted		
10. Human Health	Long term vapour mercury exposure	How persistent are mercury vapours in ASGM communities? How can locally made chelating agents detoxify contaminated people?
11. Mercury and Cyanide interactions	Many ASGM operations have begun to use cyanide to extract gold and this often occurs in conjunction with mercury use.	What are the Cyanide-Mercury complexes? How long do they persist? How can they most efficiently be destroyed? What are the fate of the decay products? Can they exacerbate methylation?
12. Best practices regarding the handling of mercury	Use of retorts and fume hoods, storage and disposal of mercury and mercury wastes	Retorts reduce emission of mercury to the atmosphere but do not always significantly reduce human exposure due to operation and handling practices. How can design, operation, and education be improved?

Note: It must be mentioned that there are many equally important knowledge gaps that are unrelated to mercury regarding best practices in ASGM, most importantly those involving health and safety.

factors such as atmospheric conditions at the site and time of amalgam burning play an important role in controlling entry of mercury into the regional or global atmospheric cycle. For example, if amalgam is burnt on a hot tropical afternoon when the atmosphere is turbulent well mixed, the likelihood of mercury entering higher levels of the atmosphere and being transported long distances will be greater than if the amalgam is burnt in the evening or early morning, when the atmosphere is less well mixed. It is also possible that mercury deposited locally at one time is quickly desorbed and transported at another as the tropical atmosphere is very energetic. As no firm scientific evidence has yet been provided to prove the distance mercury emitted from amalgam burning travels (Veiga and Baker, 2004), clearly more research on this topic is needed.

To illustrate this knowledge gap and others, Figure 6.1 shows examples of mercury being emitted into the environment from ASGM (some tragic) and how these relate to the identified knowledge gaps. Filling these gaps is required if we are to understand the impacts and costs of mercury emissions from ASGM at local, regional, and global scales.

6.6.1 River Siltation in ASGM

Another significant environmental impact caused by ASGM is river siltation. It is mentioned here because it does have a direct and large impact on mercury transport. Dredging and sluicing sediment and soils for gold extraction causes the discharge of huge amounts of sediment into rivers, lakes and oceans. For example, small scale mining is now the main source of sediment to Brazil's Tapajos River which is one of the Amazon's largest tributaries and one of the world's largest rivers (Telmer et al., 2006b). The Tapajos is about twice the size of Europe's largest river, the Danube. In the tropics sediments are very fine because they are rich in clays and amorphous oxides (mostly iron oxy-hydroxides). This is due to the nature of soil formation in the tropics. When discharged into rivers, a significant portion of these clay rich sediments remain in suspension indefinitely. Sediment discharged from ASGM is consequently transported hundreds to thousands of kilometers downstream and into the ocean.

These sediment discharges have severe environmental impacts. The increases in suspended sediment reduce the penetration of light into waters and change the nutrient supply. This drastically alters the natural habitat (Costa, et al., 2008):

- Biological productivity and diversity is reduced
- Shifts in species composition are extreme

However this also directly relates to mercury. The process of soil formation naturally concentrates and sequesters mercury. Soils around gold mining areas are both naturally rich in mercury (Jonasson and Boyle, 1972) and receive mercury released from amalgam burning.

The erosion of soils by mining releases mercury accumulated during soil formation into water bodies at hugely accelerated rates (Telmer et al., 2006) where it likely becomes available to be methylated and bioaccumulated in downstream floodplains. Forest clearing in the Amazon is also thought to contribute to this process (Roulet et al., 1998). Therefore mercury released into water bodies by soil erosion represents a large anthropogenic source of mercury into waters. The amount of mercury released by this process includes that added by miners but also the mercury that was naturally accumulated in the soils. In some cases, the latter can be the larger number.

6.7 Reducing Mercury use in ASGM

The amount of mercury consumed by artisanal small scale gold mining (ASGM) depends on three main factors: (i) the type of ore being mined; (ii) the technique used to process the ore; and (iii) the technique used to process amalgam to produce gold. To varying degrees, these factors are interdependent.

6.7.1 Reducing Emissions

In a few cases, mercury consumption has been significantly reduced through the use of fume hoods, retorts, and by re-activating dirty mercury. In Brazil and Indonesia, simple fume hoods have been adapted by some gold shops that trap about 90% of former atmospheric emissions (Sousa and Veiga, 2007; Agrawal, 2007; Chouinard, 2007, Argonne National Laboratory, 2008). The fume hoods in Indonesia are very cheap ($US35) and allow gold shop owners to recover and re-sell mercury, thereby recycling it and greatly reducing overall mercury consumption (Agrawal, 2007). They need to recover only 1kg of mercury in order to recover the cost of buying a fume hood. Brazilian efforts in collaboration with the USEPA, are producing similar results (Argonne National Laboratory, 2008). The USEPA led efforts have produced a detailed accounting of the functioning and efficiency of fume hoods constructed in some of the Brazilian Amazon gold mining communities.

As well, importantly, additional reductions in mercury consumption are occurring by teaching simple mercury re-activation and cleaning methods (Pantoja and Alvarez, 2000; R. Wuerker pers. comm., 2007). Using these methods, so called "dirty mercury" is never discarded and this reduces overall consumption and contamination. Pantoja and Alvarez (2000) use a simple electrochemical cell to operated with a 12 volt battery to reduce oxidized mercury to its elemental form. Ralph Wuerker is an astronomer with experience running liquid mercury telescopes which suffer surface oxidation that occasionally needs to be removed. The astronomers (as well as many chemists studying electrochemistry with mercury drop electrodes) simply pass the liquid mercury through a coffee filter to clean it. It is also worth mentioning that retorting mercury (evaporating and then condensing it) produces relatively clean mercury that is able to effectively amalgamate gold. For example, the gold shop owners who operate fume hoods in Kalimantan, sell their recovered mercury with no further cleaning procedure for direct use in mining and this is accepted by the miners.

Retorts also significantly reduce mercury consumption by facilitating mercury recycling. Rickford Vieira, a key person involved with the World Wildlife Fund's efforts to combat environmental degradation due to small scale mining in the Guyanas and Suriname has stated that overall mercury consumption has been reduced to 1:1 by use of retorts. UNIDO's Global Mercury Project, as well as other intervention efforts, have also introduced retorts in an effort to reduce mercury releases to the atmosphere. Although, even with a reduction of 90%, the levels of mercury released by ASGM are still quite unacceptable by modern environmental laws, such a reduction represents a vast improvement from the status quo.

Capturing direct emissions to the atmosphere is a positive development, but in order to have a more significant impact on mercury consumption in ASGM, the practice of whole ore amalgamation must be eliminated or reduced. That is because whole ore amalgamation is (i) the least efficient way to use mercury and so causes the greatest losses; and (ii) is likely to grow as the exploitation of colluvial and bedrock ores becomes more common – these types of ores are the ones that are wholly amalgamated. Eliminating whole ore amalgamation is a much more complicated endeavour than capturing direct mercury emissions to the atmosphere with fume hood and retorts. Most concepts about how to eliminate it involve: (i) introducing efficient processing which involves increasing the sophistication of the processing technology; (ii) increasing initial capital investment; and (iii) increasing the organisation of the labour pool – all big challenges for poor and transient communities that reside at the margins of legal society. However, if these steps can be accomplished, it is possible that more gold can be captured, or less mercury would be consumed, both of which would have monetary value to the miners and so there potentially are underlying economic incentives for such change. It is also important to mention that an increased mercury price, perhaps driven by legally binding export bans from the big exporters such as the European Union and the United States, would likely induce miners to use less mercury in order to reduce costs. Simply put, as the price of mercury rises, the economic feasibility of whole ore amalgamation is reduced.

In order to conceptualise possible reductions in mercury use in ASGM, it is useful to break down the possible approaches as follows:

1. We estimate that if fume hoods and retorts are adopted by any singular ASGM site, immediate emissions to the atmosphere can be reasonably reduced by 90% – less for operations that use cyanide. So where 1 g of mercury was emitted to the atmosphere for every g of gold produced, then only 0.1 g of mercury would be emitted.
2. If mercury re-activation or cleaning methods were adopted for any singular operation, then mercury consumption would be reduced by 50%. So where 2 g of mercury were used to capture 1 g of gold, only 1 g of mercury would be used.
3. If an operation is able to stop amalgamating the whole ore, then mercury consumption can be reduced by 90%. So where 10 g of mercury were used to capture 1 g of gold, only 1 g of mercury would be used.

Overall, if 50% of ASGM mercury use (50% of 1000 = 500 Mg yr^{-1}) is consumed through the amalgamation of concentrate (2:1 Hg: Au ratio – this includes losses incurred when dirty mercury is disposed); and 50% (50% of 1000 = 500 Mg yr^{-1}) is consumed through whole ore amalgamation (5:1 Hg: Au ratio – this is an average based on a mix processing with copper plates and milling with mercury in the grinding circuit), and for every unit of gold produced a unit of mercury is directly emitted to the atmosphere then, (i) 350 Mg of gold are produced by ASGM each year; (ii) 350 Mg of mercury consumed by ASGM are directly emitted to the atmosphere (35% of total mercury consumption), (iii) 250 Mg yr^{-1} of mercury (25%) are discarded because the mercury is dirty, and (iv) 400 Mg yr^{-1} (40%) are lost directly to tailings during whole ore amalgamation. Of these latter two (25% + 40% = 65%), some portion would be latently emitted to the atmosphere from tailings and waters, and some portion would remain in the hydrosphere. The rate of latent emission is unknown but is particularly high where mercury is used in combination with cyanide processing. Considering the growth in ASGM, the growth in the use of cyanide in ASGM, and the growth in the production of mercury contaminated waste from ASGM (multi-year accumulation of tailings), latent emissions conservatively amount to at least 50 Mg yr^{-1} bringing the total emission of mercury to the atmosphere from ASGM to 400 Mg yr^{-1}. Under such a scenario, then adoption of #1 (emission control that captures 90% of emissions) could reduce mercury consumption globally by a maximum of 0.9*35% = 31.5% or more with better emission capturing technology; adoption of #2 (mercury re-activation or cleaning) could reduce mercury consumption by a maximum 25%; and adoption of #3 (elimination of whole ore amalgamation) could reduce mercury consumption by 0.9*40% = 36%. The latter assumes that 10% of mercury used to amalgamate gravity concentrates (rather than whole ore) will still be lost to tailings. Also, note that the estimated reduction for emission control includes capturing 90% of the emissions caused by the burning of amalgam produced at whole ore amalgamation operations – i.e. 100% of ASGM sites. If all three of these approaches were adopted universally, mercury consumption by ASGM globally could be reduced by 96% (from 1000 to 40 Mg yr^{-1}), emissions to the atmosphere could be reduced by 90% (from 350 to 35 Mg yr^{-1}), and losses to tailings, rivers, lakes and soils, could be reduced by 99.2% (from 650 to 5 Mg yr^{-1}).

To put this further into perspective, if the top 10 countries using mercury in ASGM minus China, which are: Indonesia, Colombia, Brazil, Peru, Philippines, Zimbabwe, Ecuador, Guyana, Venezuela, and Mongolia, that together emit around 400 Mg mercury per annum, were to adopt emissions control measures (fume hoods and retorts, #1), and learn how to clean mercury (#2) then roughly 240 Mg less mercury per annum would be consumed. If China is included, the reduction in mercury consumption would increase to 500 Mg of mercury per annum. China is separated to highlight its importance. Considering that these two approaches are vastly simpler than #3 - elimination of whole ore amalgamation - and have been effectively demonstrated to be profitable for miners, their adoption by mining communities should be relatively simple and successful – perhaps even quick if governments cooperate. The elimination of whole ore amalgamation must also remain a focus, as the current trend is that this practice is increasing. We suggest that by working towards the three approaches above, it is reasonable to expect a 50% reduction in mercury use in ASGM globally on a time scale of perhaps 10 years.

6.8 Conclusions

In summary, the impacts of mercury use in artisanal and small scale gold mining (ASGM) are as follows:

- 400 Mg yr^{-1} of mercury per annum are volatilized to the atmosphere (350 directly; 50 through latent emissions).
- 650 Mg yr^{-1} are discharged into rivers, lakes, floodplains, soils, and tailings (50 Mg yr^{-1} of these are volatilized to the atmosphere).
- Global food chain contamination is likely to be occurring through long range atmospheric transport, deposition, and accumulation in global fisheries - global ecosystem damage is likely to be occurring
- Severe occupational hazards occur – mercury vapour
- Intense local food chain contamination is occurring
- Intense local ecosystem damage is occurring
- Neurological damage to people and animals is occurring
- Tens of thousands of polluted sites have been created that are long-term (centuries) health hazards to populations and ecosystems
- Overall the emissions of mercury from ASGM are leading to decreased capacity for innovation and prosperity for people at local regional and global levels – societal regression

The most significant environmental issues are: (i) mercury emissions to the atmosphere, transport of these emissions locally, regionally and globally, and ultimately leading to aquatic food chain contamination and human health impacts through fish consumption; (ii) health impacts through direct mercury vapour exposure, (iii) release of mercury into aquatic systems and the consequential development of mercury hotspots that last for centuries, and (iv) land-degradation and river siltation and the associated deforestation, loss of organic soil, modification of hydrologic regimes and loss of aquatic habitat.

Finding a resolution to mercury use in ASGM is complicated by the characteristics of the informal gold mining sector including that ASGM remains illegal in many of the areas where it operates; ASGM communities are remote and have a transient nature; miners move quickly when better gold areas are found; different mine types and gold purifying methods are used in different regions; and the general lack of communication within and between artisanal miners and society and government authorities. An approach that links field knowledge, a field presence, and community economic considerations with international stakeholders may have a chance at success where other efforts have failed. A key to this approach is building a reliable knowledgebase about ASGM communities, particularly how and why they operate as they do, and the economic drivers behind these operations. A good knowledge base is the required backbone to formulate solutions to the problems associated with mercury and ASGM.

One important function of this knowledgebase is to enable the determination of the financial implications that proposed changes in practice will cause. These are an important primary criterion for finding sustainable solutions. At the global level, the database on ASGM remains poor. How many people are mining, how much gold they are producing, how much mercury they use, what happens to the mercury, and how much habitat (land and water) has been impacted remains poorly known? Here, in recognition of the importance of good information in bringing the issue into focus and finding solutions, we have used the available data to make a first estimate of these quantities, and to point out the knowledge gaps surrounding mercury use in ASGM. We welcome any inputs that will improve the database or innovations that can contribute to solutions.

References

Agrawal S., 2007. Final Report, UNIDO Project No. EG/GLO/01/G34, Contract No. 16001054/ML, Community Awareness on Hazards of Exposure to Mercury and Supply of Equipment for Mercury-cleaner Gold Processing Technologies in Galangan, Central Kalimantan, Indonesia.

Al T.A., Leybourne M.I., Maprani A.C., MacQuarrie K.T., Dalziel J.A., Fox D. andYeats P.A., 2006.Effects of acid-sulfate weathering and cyanide-containing gold tailings on the transport and fate of mercury and other metals in Gossan Creek: Murray Brook mine, New Brunswick, Canada.Applied Geochemistry, 21:1969-1985.

Alfa, S., 2000. Child labour in Small-scale Mines in Niger. International Labour Organization, Sector Publications: Available at: www.ilo.org/public/english/dialogue/sector/papers/childmin/137e1.htm#1

Alpers, C.N. and Hunerlach, M.P., 2000. Mercury Contamination from Historic Gold Mining in California. USGS FS062-00, May 2000, 6 p.

APLA - Amalgamated Prospectors and Leaseholders Association of Western Australia (Inc), 2004.

Appel, P.W.U.; Dyikanova, C.; Esengulova, N.; Tagaeva, A., 2003. Baseline Survey of Artisanal and Small-scale Mining and Teaching Seminars for Small scale miners in Kyrgyz Republic. Report from Geological Survey of Denmark and Greenland, n.2004/11. Copenhagen, 27 p.

Appleton, J.D.; Taylor, H.; Lister, T.R.; Smith, B., 2004. Final Report for the Assessment of Environment in the Rwamagasa area, Tanzania. UNIDO Project EG/GLO/01/34. British Geological Survey, Commissioned Report CR/04/014. Nottingham, UK.159 p.

Argonne National Laboratory draft report 2008. "Technology Demonstration for Reducing Mercury Emissions From Small-Scale Gold Refining Facilities" prepared for U.S. Environmental Protection Agency

Artaxo, P.; Campos, R.C.; Fernandes, E.T.; Martins, J.V.; Xiao, Z., Lindqvist, O.; Fernandez-Jimenez, M.T.; Maenhaut, W., 2000.Large Scale Mercury and Trace Element Measurements in the Amazon Basin.Atmospheric Environment, 34:4085-4096.

Attenborough M., 1999. Social problems in developing countries pose challenge - Canadian companies learning that it pays to be a friendly neighbour The Northern Miner, Volume 85, Number 4.

Babut, M.; Sekyi, R.; Rambaud, A.; Potin-Gautier, M.; Tellier, S.; Bannerman, W.; Beinhoff, C., 2003.Improving the Environmental Management of Small-scale Gold Mining in Ghana: a Case of Dumasi.Journal of Cleaner Production, v.11, p.215-221.

Baker R. and Telmer K., 2007. Summary of Fish Mercury Data from Tanoyan Mining Area, Bolaang Mongodow North Sulawesi. UNIDO, Project EG/GLO/01/G34.

Basque, G., 1991.Gold Panner's Manual.Sunfire Pub. Ltd., Langley, BC, 108 p.

Beales P., 2005. Busy Indaba showcases African mining industry, Northern Miner, Volume 91 Number 1.

Bermudez-Lugo, O., 2002. The Mineral Industries of Gambia, Guine-Bissau, and Senegal. U.S. Geological Survey Minerals I

Chouinard R., 2007. Results of the Awareness Campaign and Technology Demonstration for Artisanal Gold Miners, Summary Report, Brazil – Indonesia – Laos – Sudan – Tanzania – Zimbabwe. UNIDO, Project EG/GLO/01/G34.

Coakley, G.J., 2002. The Mineral Industry of Lesotho and Swaziland. U.S. Geological Survey-Minerals Information. Available at: minerals.er.usgs.gov/minerals/pubs/country/africa. html#ir

Costa, M.P.F., Telmer, K., and Novo, E.M.L.M., 2008. Spectral light attenuation in Amazonian waters. Submitted to Limnology and Oceanography.

Couture, R. and Lambert, J.D., 2003. Source of Mercury in Small scale Mining Communities of Guyana. Presented at the GGMC Mining Week conference. August 2003. Georgetown, Guyana.

Crispin, G., 2003.Environmental Management in Small-scale Mining in PNG.J. Cleaner Production, v.11, n.2, p.175-183

Cumming J., 1997. Banro Resource on hold in eastern Zaire after purchase of Sominki, Northern Miner, Volume 83, Number 2.

Danielson V., 1992. Tanzania opens its doors to foreign investment. Northern Miner, Volume 78, Number 7.

Dawson S., 1996. Nelson mining gold along historic Silk Road. Northern Miner. Volume 82, Number 6.

DLI – Defense Language Institute, 2003. Liberia in Perspective. Available at: www.lingnet.org/areaStudies/perspectives/liberia/liberia.pdf.

Dolley, T.P., 1996. The Mineral Industry of Senegal, the Gambia, and Guinea Bissau. U.S. Geological Survey—Minerals Information. minerals. usgs.gov/ minerals/pubs / country/1994/9233094.pdf

Drasch, G.; Boese-O'Reilly, S.; Beinhoff, C.; Roider, G.; Maydl, S., 2001.The Mt. Diwata Study on the Philippines 1999 – Assessing Mercury Intoxication of the Population by Small-scale Gold Mining.The Science of the Total Environment, v.267, p.151-168.

Dreschler, B., 2001. Small-scale Mining and Sustainable Development within the SADC Region. MMSD Report 84. Available at: www.iied.org/mmsd/mmsd_pdfs/asm_southern_africa.pdf

Ecosystems Contamination Assessment. In: Mercury in the Tapajos Basin. p.75-94. Villas Boas, Beinhoff, Silva(eds.). GEF/UNIDO/CYTED/CETEM/IMAAC publication. Rio de Janeiro.

Fréry, N; Maury-Brachet, N.; Maillot, E.; Deheeger, M.; de Mérona, B.; Boudou, A., 2001.Gold-Mining Activities and Mercury Contamination of Native Amerindian Communities in French Guiana: Key Role of Fish in Dietary Uptake.Environmental Health Perspectives, v.109, p.449-456.

GMP, 2007. Partnership on Mercury Reductions in Artisanal and Small Scale Gold Mining (ASGM). Available at : www.chem.unep.ch/Mercury/partnerships/

Graham R., 2002. Eaglecrest advances San Simon, Northern Miner, Volume 88, Number 40.

Graham R.. 2003. Juniors eye Mexican gold prospects. The Northern Miner, Volume 89 Number 42.

Grayson, R. 2007.Anatomy of the Gold Rush in Modern Mongolia.World Placer Journal, v.7, p.1-65.

Gunson and Yue, 2002. Artisanal Mining in the People's Republic of China. Mining, Minerals and Sustainable Development (MMSD) report 74.

Gunson, A.J., 2004.Mercury and Artisanal and Small-scale Gold Miners in China. MASc Thesis. Dept. Mining Engineering, University of British Columbia, Vancouver, Canada.154 p.

Gunson, A.J. andVeiga, M.M., 2004.Mercury and Artisanal Gold Mining in China.Environmental Practice, v.6, n.2, p.109-120.

Heemskerk, M. 2003.Risks, Attitudes and Mitigation among Gold Miners and Others in the Suriname Rainforest.Natural Resources Forum, v.27, p.267-278.

Hentschel T., Hruschka,F., Priester M., 2002. Global Report on Artisanal & Small-Scale Mining. MMSD Summary Report.

Hinton, J.J.; Veiga, M.M.; Beinhoff, C., 2003.Women and Artisanal Mining: Gender Roles and the Road Ahead. In:The Socio-Economic Impacts of Artisanal and Small-Scale Mining in Developing Countries. Chapter 11.G. Hilson (ed.),Pub. by A.A. Balkema, Swets Publishers, Netherlands.

Hinton, J.J.; Veiga, M.M.; Veiga A.T., 2003.Clean Artisanal Mining, a Utopian Approach? Journal of Cleaner Production, v.11, p.99-115.

Hiyate A., 2008. Semafo banks on Mana Gold Mine to Take Company from Red to Black. Northern Miner, Volume 93 Number 51.

Hunerlach, M.P.; Rytuba, J.J; Alpers, C.N., 1999. Mercury Contamination from Hydraulic Placer-Gold Mining in the Dutch Flat Mining District, California. U.S. Geological Survey Water-Resources Investigations Report 99-4018B, p. 179-189.

Ibrahm, M.S., 2003. Information about the Project Sites in Sudan. Report to GEF/UNDP/UNIDO Global Mercury Project. October, 2003. 10 p.

Ikingura, J. R. andAkagi, H., 1996.Monitoring of Fish and Human Exposure to Mercury Due to Gold Mining in the Lake Victoria Goldfields, Tanzania.The Science of the Total Environment, v.191, n.1-2, p.59-68.

Ikingura, J.R.; Mutakyahwa, M.K.D.; Kahatano, J.M.J., 1997.Mercury and Mining in Africa with Special Reference to Tanzania.Water, Air & Soil Pollution, v.97, p.223-232.

Ikingura, J.R., 1998.Mercury Pollution Due to Small-scale Gold Mining in Tanzania Goldfields. In:Small-scale Mining in African Countries: Prospects, Policy and Environmental Impacts. p.143-158.Ed L. Landner, Dept. Of Geology, Univ.Dar es Salamm, Tanzania.

International Labour Office (ILO), 1999. Social and Labour Issues in Small-scale Mines, Report for the Tripartite Meeting on Social and Labour Issues in Small-scale Mines, Geneva 17-22 May, 1999.

Israel, D.C. and J.P. Asirot. 2000. Mercury Pollution in Small-Scale Gold Mining in the Philippines. Philippine Institute for Development Studies, Discussion Paper Series No. 2000-06. Economy and Environment Program for Southeast Asia, 2000.

Israel D.C., 2000. Mercury Pollution Due to Small-Scale Gold Mining: A Serious Menace. Philippine Institute for Development Studies, Policy Notes, No. 2000-03

Jonasson, I.R. and Boyle, R.W., 1972. Geochemistry of mercury and origins of natural contamination of the environment. Canadian Mining and Metallurgical Bulletin 65.

Kambani S.M., 2003.Small-scale mining and cleaner production issues in Zambia.Journal of Cleaner Production, 11; 141–146

Kinabo, C.P., 2002.Comparative Analysis of Mercury Content in Cosmetics and Soaps Used in the City of Dar Es Salaam. In:Proc. International Workshop on Health and Environmental Effects of Mercury: Impacts of Mercury from Artisanal Gold Mining in Africa. p.173-186. Tanzania, Nov. 19-20, 2002.Ed. National Institute of Minamata Disease, Japan.

Kinabo, C., 2003.Women Engagement and Child Labour in Small-scale Mining – Tanzanian Case Study.Urban Heath and Development Bull., v.6, n.4, p.46-56.South Africa.

Labonne, B., 2002. Seminar on Artisanal and Small-scale Mining in Africa: Identifying Best Practices and Bulding the Sustainable Livelihoods of Communities. Synthesis Report: Available at: www.naturalresources.org/minerals/smscalemining/docs.htm

Lacerda, L.D., 2003.Updating global Hg Emissions from Small-scale Gold Mining and Assessing its Environmental Impacts.Environmental Geology v.43, p.308–314

Madawo I., 2007. Canadians lead new Scramble for Africa. Northern Miner, Volume 93, Number 42.

Mahlatsi, S. andGuest, R., 2003.The iGoli Mercury-free Gold Extraction Process.Urban Heath and Development Bull., v.6, n.4, p.62-63.South Africa.

Maponga, O. andNgorima, C.F., 2003.Overcoming Environmental Problems in the Gold Panning Sector through Legislation and Education: the Zimbabwean Experience.Journal of Cleaner Production v.11, p.147–157

Maxson, P., 2004. Mercury flows in Europe and the world: The Impact of Decommissioned Chlor-alkali Plants.Prepared by Concorde EasMg/West Sprl for the European Commission (Environment Directorate), final report, Feb 2004, Brussels, Belgium. Available at: europa.eu.inMg/comm/environmenMg/chemicals/mercury/index.htm

Maxson, P., 2008a. AMAP study of air emissions & transport. Preliminary draft report for UNEP.

Maxson, P., 2008b. Global atmospheric Hg emissions from incineration of mercury products in the waste stream. Preliminary draft report for the Mercury Policy Project, February 2008.

Mbendi Information for Africa, 2004. Mauritania Mining Overview. Available at: www.mbendi.co.za/indy/ming/ af/mu/p0005.htm

MMSD – Mining, Minerals and Sustainable Development, 2002. Breaking New Ground. International Institute for Environment and Development and World Business Council for Sustainable Development. London, UK. 441 p.

Mobbs, P.M., 1996. The Mineral Industry of Chad. U.S. Geological Survey-Minerals Information. minerals.usgs.gov/minerals/ pubs/ country/1996/9206096.pdf -

Mobbs, P.M., 1998. The Mineral Industry of Côte D'Ivoire. U.S. Geological Survey-Minerals Information. Available at: minerals.er.usgs.gov/minerals/ pubs/country/1997/9208097.pdf

Murphy T., 2006. Mercury Contamination along the Mekong River, Cambodia. In: Book of Abstracts of the 8th International Conference on Mercury as a Global Pollutant. Madison, Wisconsin, USA. Aug 6-11, 2006.

Northern Miner, 1997. Gold output from East to rise in '98, FACTS 'N' FIGURES, Volume 83, Number 41.

Northern Miner, 2001. Vaaldiam raises Boungou funds, Volume 87, Number 2.

Northern Miner ,2003. SouthernEra advances Messina, raises CUS$77m Volume 89 Number 40.

Northern Miner, 2007. Exploration Annual Review, Artisanals point the way for African Gold Group, Volume 93 Number 39.

Northern Miner, 2008. Delta drills high-grade gold in Mali , Volume 93 Number 50.

Northern Miner, 2008. Delta drills high-grade gold in Mali. Daily News, Volume 93, Number 47.

Nyambe I., 2008. pers. Comm.

Pantoja F. And Alvaarez R., 2000. Decrease of pollution by mercury in gold mining in Latin Amercia. In: Mine Closure in Iberoamerica., eds.) Villas Boas R.C., Berreto M.L., CYTED/ IMAAC/UNIDO: Rio de Janeiro, pp. 178-190.

Priester, M. and Hentschel, T., 1992. Small-scale Gold Mining: Processing Techniques in Developing Countries. GTZ/GATE, Vieweg, Eschborn, Germany, 96 p.

Rajaobelina, S., 2003. Biodiversity and Small Scale Gold Mining in Madagascar. Paper presented at 3rd CASM Annual General Meeting and Learning Event, Elmina, Ghana, September 7-10, 2003. Available at: www.casmsite.org/events_Elmina.html

Robertson R., 2006. Aurelian Gold Discovery Takes Centre Stage Drilling Continues to Show Expansion Potential to the South. Northern Miner, Volume 92 Number 39.

Rosario J. And Ault S.K. 1997. Environmental and Health Impacts of Mercury and Cyanide in Gold-Mining in Nicaragua. Environmental Health Project, Contract No. HRN-5994-C-00-3036-00, Project No. 936-5994, sponsored by the Bureau for Global Programs, Field Support and Research, Office of Health and Nutrition, U.S. Agency for International Development, Washington, DC, ACTIVITY REPORT, No. 33.

Roulet, M., Lucotte, M., Canuel, R., Rheault, I., Tran, S., De Freitos Gog, Y.G., Farella, N., Souza do Vale, R., Sousa Passos, C.J., DeJesus da Silva, E., Mergler, D., Amorim, M., 1998.Distribution and partition of total mercury in waters of the Tapajos River Basin, Brazilian Amazon.The Science of the Total Environment 213, 203–211.

Savornin, O.; Niang, K.; Diouf, A., 2007. Artisanal Gold Mining in the Tambacounda Region of Senegal. Report to Blacksmith Institute. 16p.

Schofield N.C., 2007.Comodity Derivatives, Markets and Applications.John Wiley and Sons Canada, Mississauga, pp.315.

Shaw S.A., Al T.A., and MacQuarrie K.T.B., 2006. Mercury mobility in unsaturated gold mine tailings, Murray Brook Mine, New Brunswick, Canada. Applied Geochemistry, 21: 1986-1998.

Siddaiah, N.S., 2001. Mercury and Arsenic Pollution Dues to Gold Mining, Milling and Smelting in India: A Need to Assess Its Impact on Human Health and Ecosystem. . In: Book of Abstracts of the 6th International Conference on Mercury as a Global Pollutant. Minamata, Japan, Oct.15-19, 2001.

Silva A.C., 2008. pers. Comm.. based on Ph.D. work on community mining in Nicaragua.

Sotham, S., 2001.Artisanal Gold Mining in Cambodia. In:Small-scale Mining in Asia. p.31-40.S. Murao, V.B. Maglambayan, N. Cruz (eds.).Mining Journal Books Ltd., London, UK.

Sotham, S., 2004. Small-scale gold mining in Cambodia, a Situation Assessment. Ed. Dr. Carl Middleton, Additional research contributed by the NGO Forum on Cambodia's Environmental Forum Core Team. Oxfam, America.

Sousa and Veiga, 2007. Brazil Country Report, UNDP/GEF/UNIDO Project EG/GLO/01/G34. Final Report.

Spiegel, S.J.; Savornin, O.; Shoko, D.; Veiga, M.M., 2006.Mercury Reduction in Munhena, Mozambique: Homemade Solutions and the Social Context for Change.International Journal of Occupational and Environmental Health, v.12, n.13, p.215-221.

Stepanov, V.A. and Yusupov, D.V., 2001. Mercury Contamination of Soil-vegetative Cover in Zone of Influence of a Gold-concentration Enterprise. In: Book of Abstracts of the 6th International Conference on Mercury as a Global Pollutant. Minamata, Japan, Oct.15-19, 2001.

Sulaiman R., Baker R., Susulorini B., Telmer K., and Spiegel S., 2007. Indonesia Country Report, UNDP/GEF/UNIDO Project EG/GLO/01/G34. Final Report.

Telmer K., Stapper D., Costa M.P.F., Ribeiro C., Veiga M.M., 2006. Knowledge Gaps in Mercury Pollution from Gold Mining. In: Book of Abstracts of the 8th International Conference on Mercury as a Global Pollutant. Madison, Wisconsin, USA. Aug 6-11, 2006.

Telmer K., Costa M.P.F., Angélica R.S., Araujo E.S., andMaurice Y., 2006b.The source and fate of sediment and mercury in the Tapajós River, Pará, Brazilian Amazon: ground and space based evidence. Journal of Environmental Management, 81: 101-113.(invited, special issue)

Telmer K. and Stapper D., 2007. Evaluating and Monitoring Small Scale Gold Mining and Mercury Use: Building a Knowledge-base with Satellite Imagery and Field Work UNDP/GEF/ UNIDO Project EG/GLO/01/G34 Final Report.

Thomae B., 2004. Mano River's Pampana yields more gold. Northern Miner, Volume 90, Number 25.

Trung, N.X., 2001.Small-scale Gold Mining in Vietnam. In:Small-scale Mining in Asia. p.41-45. S. Murao, V.B. Maglambayan, N. Cruz (eds.).Mining Journal Books Ltd., London, UK.

Umbangtalad S., Parkpian P., Visvanathan C., Delaune R.D., andJugsujinda A. 2007.Assessment of Hg contamination and exposure to miners and schoolchildren at a small-scale gold mining and recovery operation in Thailand.Journal of Environmental Science and Health Part A:42:2071-2079.

UNEP – United Nations Environment Programme, 2002. Global Mercury Assessment

UNEP – United Nations Environment Programme, 2005. Strategic Approach to International Chemicals Management (SAICM). Status Report on Partnerships as One Approach to Reducing the Risks to Human Health and the Environment from the Release of Mercury and Its Compounds to the Environment. Vienna, 19–24 September 2005. Available at: www.chem.unep.ch/SAICM/ meeting/prepcom3/ en/INF18%20mercury.pdf

UNESCAP – United Nations Economic and Social Commission for Asia and the Pacific, 2003. UNESCAP Assists Myanmar in Formulating Environmental Policy and Guidelines for the Mining Industry. Available at: www.unescap.org/esd/water/mineral/2003/26_6February.asp

UNIDO, 2006. Summary of the Environmental and Health Assessment Reports. The Global Mercury Project. Compiled by A.J.Gunson et al. Available at: www.globalmercury project.org.

Vaccaro A., 2007. After surviving DRC's dog days, Banro nears its payday – Site Visit. Northern Miner, Volume 93, Number 42.

Veiga, M.M., 1997, Introducing New Technologies for Abatement of Global Mercury Pollution in Latin America. Ed. UNIDO/UBC/CETEM, Rio de Janeiro, 94 p.

Veiga M.M. and Baker R.F., 2004. Protocols for Environmental and Health Assessment of Mercury Released by Artisanal and Small-Scale Gold Miners. Vienna, Austria: GEF/UNDP/ UNIDO, ISBN 92-1-106429-5, 294p.

Veiga, M.M.; Bermudez, D.; Pacheco-Ferreira, H.; Pedroso, L.R.M.; Gunson, A.J.; Berrios, G.; Vos, L.; Huidobro, P.; Roeser, M. 2005.Mercury Pollution from Artisanal Gold Mining in Block B, El Callao, Bolívar State, Venezuela. In:Dynamics of Mercury Pollution on Regional and Global Scales, Atmospheric Processes and Human Exposures Around the World. p.421-450.Pirrone, N.; Mahaffey, K.R. (Eds.)ISBN: 0-387-24493-X, July 2005, Springer, Norwell, MA, USA

Veiga M.M., Maxson P., andHylander L., 2006.Origin of Mercury in Artisanal Gold Mining.Journal of Cleaner Production, 14:436-447.

Vieira R., 2008. WWF Guinas, Goldmining Pollution Abatement. World Wildlife Federation.

Weekend Proespector, 2004. Available at: www.geocities.com/Yosemite/Trails/1849/

Wickre J.B., Karagas M.R., Folt C.L., andSturup S., 2004.Environmental exposure and fingernail analysis of arsenic and mercury in children and adults in a Nicaraguan gold mining community Archives of Environmental Health, 59:400-409.

Winch S., Parsons M., Mills H., Fortin D., Lean D., Kostka J., 2006. Mercury Speciation and Sulfate-Reducing Bacteria in Mine Tailings. In: Book of Abstracts of the 8th International Conference on Mercury as a Global Pollutant. Madison, Wisconsin, USA. Aug 6-11, 2006.

Wuerker, 2008. The cleaning of mercury by filtration during the use of liquid mercury telescopes, (Personal communication).

World Bank, 2003. Project Information Document: Sustainable Management of Mineral Resources Project, Uganda, Report 90 p.

Yager, T.R.; Coakley, G.J.; Mobbs, P.M., 2002. The Mineral Industries of Benin, Cape Verde, the Central African Republic, Côte D'Ivoire, and Togo. U.S. Geological Survey-Minerals Information. Available at: minerals.er.usgs.gov/minerals/pubs/country/africa.html

Chapter 7
Mercury Emissions from Natural Processes and their Importance in the Global Mercury Cycle

Robert P. Mason

Summary The emission to the atmosphere of mercury (Hg) via natural processes constitutes an important part of the global Hg input and is a dominant part of the global mercury cycle. However, while there is an ongoing and continued effort to quantify these fluxes, the magnitude of their extent is still relatively poorly constrained. It must be emphasized that while the fluxes discussed in this chapter are due to natural processes, they constitute Hg that has originated from different sources, and because of the potential for deposited Hg to be re-emitted to the atmosphere from both the terrestrial and aquatic landscape, these fluxes include both primary sources and secondary (recycled) Hg. Thus the emissions are due to: 1) primary geogenic natural emissions (e.g. volcanoes); 2) recycled Hg from natural sources; and 3) recycled Hg from point source anthropogenic emissions. Overall, it is estimated that terrestrial inputs are 1850 Mg yr^{-1} while emission from the ocean is 2680 Mg yr^{-1}. On an area basis, emissions from land are higher than from the ocean. Forests constitute about 20% of these emissions, with total emissions from vegetated regions being about 60% of the total terrestrial inputs. Geogenic sources account for about 5% of current terrestrial inputs.

7.1 Introduction

The atmosphere is the most important pathway for the worldwide dispersion and transport of Hg (Fitzgerald et al., 1998; Mason et al., 1994; Mason and Sheu, 2002). Therefore, an understanding the global sources, speciation, fate and transport, and atmospheric transformations of Hg is necessary to understand the importance of emissions of Hg from natural and anthropogenic sources in contributing to methylmercury (MeHg) production in aquatic systems, and to its bioaccumulation through all levels of the food chain. Mercury is added to the atmosphere from both natural and anthropogenic emissions and human activity and associated inputs to the biosphere has substantially altered the global Hg cycle so that anthropogenic sources now dominate the atmospheric inputs (Fitzgerald et al., 1998; Lamborg et al., 2002; Mason and Sheu, 2002). While emissions of Hg are in the form of elemental Hg (Hg^0), and gaseous and particulate ionic Hg ($Hg^{(II)}$), most of the Hg in the atmosphere

N. Pirrone and R. Mason (eds.), *Mercury Fate and Transport in the Global Atmosphere*, DOI: 10.1007/978-0-387-93958-2_7,

is Hg^0 (typically >95% of the total), and most of the "natural" sources of Hg to the atmosphere are Hg^0, although there is a small signal due to inputs of particulate-associated Hg (Hg(p)) (e.g., volcanoes, dust). In the atmosphere, Hg^0 is relatively uncreative with an average atmospheric residence time of 0.5-1 year. In addition to Hg^0, two other atmospheric Hg fractions have been operationally defined based on physicochemical properties and current methods of measurement - the gaseous ionic $Hg^{(II)}$ fraction, termed reactive gaseous Hg (RGM), and Hg(p) (Landis et al., 2002; Lindberg and Stratton, 1998; Mason and Sheu, 2002; Sheu and Mason, 2001). The speciation of RGM is not known in detail but based on laboratory studies and the methods of its collection, it is assumed to consist of gaseous neutral $Hg^{(II)}$ complexes, such as $HgCl_2$, $HgBr_2$, and HgOBr (Ariya et al., 2002; Balabanov and Peterson, 2003).

Concentrations of Hg^0 in the remote atmosphere range from around 1-2.5 ng m^{-3}, with higher concentrations present in impacted regions (see Chapter 9). Also, Hg^0 is relatively insoluble (49.4×10^{-6} g L^{-1} at 20 °C) and its major loss pathway from the atmosphere is through oxidation and subsequent removal by wet and dry deposition. As most water bodies are typically saturated with Hg^0 relative to the atmosphere, and the Henry's Law coefficient for Hg^0 (1.37×10^{-3} mol m^{-3} Pa^{-1} at 20 °C) is relatively high, net dry deposition of Hg^0 to water is not an important removal mechanism. However, there is evidence for the uptake of Hg^0 by vegetation (Gustin and Lindberg, 2005; Rea et al., 2001). While global Hg models have identified wet and dry deposition, and evasion of dissolved gaseous Hg (DGM; principally Hg^0) from the ocean, as critical pathways for global Hg exchange at the Earth's surface (Hudson et al., 1994; Lamborg et al., 2002; Mason et al., 1994; Sunderland and Mason, 2007), emission of Hg^0 from terrestrial sources has not been as closely examined. A number of recent studies have examined the flux of Hg^0 from plants, and the uptake of Hg^0 by plants, and have quantified the net exchange of Hg^0 within some terrestrial ecosystems. These results suggest that these fluxes are substantial and therefore potential inputs from such "natural" processes to the atmosphere cannot be ignored, and need to be examined in more detail.

In the context of this chapter, the word "natural" will be used to refer to the input of Hg to the atmosphere from natural processes (e.g. evasion from the ocean, soils, and plants due to natural phenomena), even if these exchanges represent a mixed source contribution – for example, emissions are due both to the presence of background (pre-industrial) Hg and from Hg added to the environment in the last centuries due to human-related emissions. In addition, even in the pre-industrial world, these natural emissions had two components – inputs of primary (geogenic) Hg from sources such as volcanoes, and recycling of this deposited Hg from the oceans and terrestrial environment. Thus, there was substantial recycling of Hg at the Earth's surface even in pre-industrial times. Finally, given that Hg is not homogeneously distributed in the surface soils and rocks, there will be regional and local differences in emissions that it related to homogeneity in the terrestrial material (Mason and Sheu, 2002; Rasmussen et al., 2005). Thus, as recorded in some historical archives (Schuster et al., 2002), there has not been a constant global input of Hg from these natural processes over time.

This chapter, while focused on emissions from natural processes, will not specifically focus on the emissions that have been directly exacerbated by human activity, such as emission from biomass burning, from mine waste sites and other anthropogenic area sources (Mason and Sheu, 2002). Therefore, for example, given that much of the biomass burning that currently occurs globally is human-induced, this flux will not be specifically dealt with in this chapter, and is covered in another section of the report. Finally, in terms of the evasion of Hg^0 from plants, there is increasing evidence that this is a bi-directional exchange and that there is both uptake of Hg^0 via stomatal exchange and other processes, and also evasion of Hg^0 from plants, and that the extent and direction of net exchange is a function of temperature, season, solar radiation, surface wetness and other factors (Gustin and Lindberg, 2005). Furthermore, for a system such as a forest, there is exchange due to the plants as well as evasion of Hg^0 from the forest floor. All these different sources need to be considered and will be discussed in this chapter. However, overall it is the net evasion that is of importance and the major goal of this chapter is therefore to provide estimates for the net exchange of Hg^0 for each biomass type and for aquatic systems. Finally, the chapter is focused on natural processes that lead to the net evasion of Hg^0 and does not discuss in any detail the processes that may lead to deposition of Hg in forms other than Hg^0 (e.g. dry deposition of RGM, for example). Thus, the estimates will be specifically characterized as described in this section.

While most of the emissions discussed in this chapter are areal fluxes rather than point source inputs, there is one source, inputs from volcanic and related activity, that is natural and that can be considered a primary point source input. A number of early studies proposed a wide range of estimates for this flux (Fitzgerald et al., 1998; Mason et al., 1994; Nriagu, 1989) and since then there have been few detailed studies of these potential sources. However, recent papers have re-examined the information available on geogenic emissions and a more constrained range in values has been quoted in the recent literature (Ferrara, 2000b; Fitzgerald, 2003; Mather, 2004; Nriagu, 2003). In terms of emissions from other terrestrial sources, there is still much debate and these fluxes are not well constrained, as detailed more below (Gustin and Lindberg, 2005; Gustin et al., 2000; Rasmussen et al., 2005).

The processes controlling the evasion of Hg^0 from the terrestrial environment are an area of on-going research and the dynamic nature of this exchange between the terrestrial environment and the atmosphere has only recently been appreciated. The most important processes are depicted in Figure 7.1 for a forested environment, and the exchange processes will be similar for other types of vegetation (Bash et al., 2007; St Louis et al., 2001). Deposition of Hg to the Earth's surface can occur in a number of forms and these vary both spatially and temporally, making overall estimation of the magnitude of the fluxes difficult (Mason and Sheu, 2002). Overall, there is little data on which to base such estimates, and the data is also concentrated in Northern Hemispheric temperate and polar regions, and thus estimates in the literature that are made for some environments and locations have required extrapolation based on data from other ecosystems. Wet deposition of Hg contains both dissolved Hg, derived from scavenging of RGM from the atmosphere

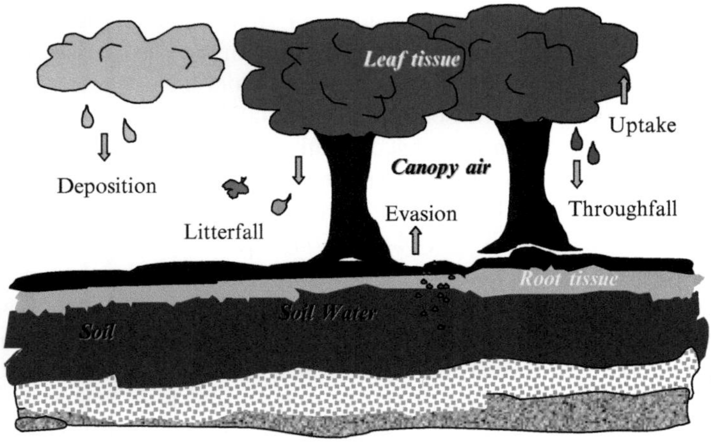

Figure 7.1 Schematic representation of the major processes involved in the exchange of mercury between the terrestrial environment and the atmosphere. Modified from (St Louis et al., 2001)

by rain droplets, and from formation of $Hg^{(II)}$ species in cloud droplets via atmospheric reactions. Particulate Hg is also scavenged into precipitation and some of the Hg associated with the Hg(p) fraction will be solubilized into the liquid phase. Dry deposition of both RGM and Hg^0 can also occur. The dry deposition of RGM is very efficient as the compounds that constitute this fraction are very soluble and reactive. Elemental Hg can also be deposited to the terrestrial environment via uptake into the substrate and into living biomass. It has been shown in a number of recent studies that Hg^0 can be taken up via vegetation, most likely through gas exchange across the stomata of the plants (Bash et al., 2007; Gustin and Lindberg, 2005). A number of studies suggest that the Hg found in litterfall is primarily associated with direct uptake from the atmosphere as apposed to translocation of Hg through the plant from the soil (Ericksen and Gustin, 2004; Ericksen et al., 2003). It appears that Hg can be taken up into roots but that, because of its strong binding capacity to organic ligands, and particularly to those with thiol groups, it is not transported to a large extent with the movement of water through the plant (evapotranspiration).

There is the potential for both RGM and Hg(p) to be deposited on surfaces such as the forest canopy, and this Hg can be subsequently washed off by wet deposition, and therefore deposited to the terrestrial surface. Such processing is the probable source of the enhanced Hg concentration in throughfall relative to wet deposition. A number of studies have examined the relative concentrations and found that throughfall concentrations are 1-2 times those of wet deposition (Lawson and Mason, 2001; Rea et al., 1996). This difference therefore suggests that dry deposition of Hg, likely as RGM and as Hg(p) can be an important process and can rival the input of Hg from wet deposition in some locations.

One important aspect of the cycling of mercury at the Earth's surface is the process of Hg^0 oxidation in the atmosphere and mercury depletion events in the polar regions that coincide with ozone depletion during "polar sunrise" (so-called MDE's)

(Ebinghaus et al., 2002; Lindberg et al., 2002; Schroeder et al., 1998). While these occurrences lead to an enhanced deposition of Hg to the icepack, it has been shown that this deposited Hg is rapidly reduced and re-emitted to the atmosphere (Ariya et al., 2004; Lalonde et al., 2002; Mason and Sheu, 2002; Poulain et al., 2004). This evasion is estimated to be on the order of 200 Mg yr^{-1} (Ariya et al., 2004; Mason and Sheu, 2002) and is therefore a significant evasion flux

Evasion of Hg^0 from the ocean is thought to be an important part of the global Hg cycle, and the magnitude of the oceanic evasion is of similar magnitude to the input from anthropogenic point sources (Mason and Sheu, 2002). Measurements of volatile Hg are often reported as DGM and there are two important volatile species – Hg^0 and dimethylmercury (Me_2Hg). While Me_2Hg has been quantified in ocean waters (Cossa et al., 1997; Kim and Fitzgerald, 1988; Lamborg, 1999; Mason et al., 1998; Mason and Sullivan, 1999), in most freshwaters examined and in the surface ocean, DGM is essentially equivalent to Hg^0. For the ocean, studies have focused on air-sea exchange in the Atlantic waters (Cossa et al., 1997; Gardfeldt, 2003; Kim and Fitzgerald, 1988; Lamborg, 1999; Mason et al., 1998; Mason and Sullivan, 1999), Pacific (Kim and Fitzgerald, 1988; Mason and Fitzgerald, 1993), the coastal regions (Baeyens and Leermakers, 1998; Mason et al., 1999) and the Mediterranean (Andersson, 2007; Ferrara, 2000a; Ferrara, 2003; Gardfeldt, 2003). Some studies have suggested that the estimated evasion rates for Hg from the ocean substantially exceed the current wet plus particulate dry deposition estimates and riverine inputs, suggesting another potential source for upper ocean Hg. While one potential reason for this lack of balance is the relative paucity (both temporally and spatially) of DGM data for the ocean and thus its potential unrepresentative nature, it has recently been hypothesized that there is a substantial input of Hg to the ocean via dry deposition of RGM (Mason and Sheu, 2002) and through its scavenging by salt particles and subsequent deposition (Selin, 2007; Strode, 2007).

Mercury in surface waters also exists in two oxidation states, Hg^0, as a dissolved gas, and $Hg^{(II)}$. The ionic Hg exists both in the dissolved phase and attached to particulate matter, both living (phytoplankton, zooplankton) and detritus. In addition, in most surface waters, a small fraction of the total Hg is $MeHg^{(II)}$, which also can be dissolved or particulate-associated (Mason and Benoit, 2003). The reactions leading to the transformation of Hg between the two oxidation states in surface waters can be both abiotically and biotically mediated, with the abiotic processes being mostly photo chemically driven. The reduction of $Hg^{(II)}$ to Hg^0 occurs in surface waters under a variety of conditions and recent studies, in both freshwater and saline waters, have provided evidence to suggest that, in most situations, while there is net Hg^0 formation, there is also Hg^0 oxidation occurring (Amyot et al., 1997; Lalonde et al., 2001; Mason et al., 2001; Whalin et al., 2007). A number of studies have also shown that $Hg^{(II)}$ reduction occurs in the presence of algae and bacteria (e.g. (Lanzillotta et al., 2004; Mason et al., 1995). The reduction of $Hg^{(II)}$ in the presence of organic matter has been shown by others (Costa and Liss, 2000). It has been suggested that the reduction of $Hg^{(II)}$ may occur through charge transfer reactions involving Hg-DOC complexes but organic matter complexation can also lead to a decrease in $Hg^{(II)}$ reduction (Rolfhus and Fitzgerald, 2001). Overall, however, higher DOC concentration is one factor that likely controls the overall lower concentrations and rates of Hg^0 evasion from coastal and estuarine environments compared to the open ocean. The potential for photochemical processes to dominate the reduction of $Hg^{(II)}$ is demonstrated by studies that have examined the short-term changes in Hg^0 concentration and have shown a diurnal cycle linked to photo synthetically active radiation (e.g. Amyot et al., 1997; (Amyot et al., 1994; Lanzillotta et al., 2002).

7.2 Estimates of Oceanic Evasion

A number of modeling papers have recently focused on the estimation of the evasion of Hg^0 from the ocean to the atmosphere (Mason and Sheu, 2002; Selin et al., 2008; Strode, 2007; Sunderland and Mason, 2007) and as these papers have reviewed and examined the available literature, they provide a reasonable summary and consensus for this report. Such data have been used to extrapolate fluxes to the global scale. While the process of Hg^0 oxidation in surface waters (Lalonde et al., 2001; Mason et al., 2001; Whalin and Mason, 2006) is a potential removal mechanism for surface water Hg^0, the flux estimates are based on the actual measurements of Hg^0, and as recent sample collections have been in the top meters of the ocean, these represent the steady state Hg^0 concentration that results from both oxidation and reduction processes. The flux estimates from a number of sources, as well as the measured concentrations of dissolved Hg and DGM for a number of ocean basins, was recently summarized by (Sunderland and Mason, 2007) and presented here in Table 7.1. It is likely that evasion rates are lower in winter, as suggested by the model developed for the far North Atlantic ((Mason et al., 1998), and the global model of Strode et al. (2007), although there is little open ocean data to support this notion. For coastal environments, there is evidence for a lower flux in winter compared to summer, which again suggests that there is seasonality in the fluxes (Baeyens and Leemakers, 1998; Rolfhus and Fitzgerald, 2001).

The data in Table 7.1 show that there is variability across ocean basins in both the concentration of Hg and Hg^0, with the Atlantic Ocean and Mediterranean Sea having overall higher concentrations than the Pacific and Indian Oceans. Such differences make sense in terms of the historical inputs of Hg to the atmosphere, which were previously highest for North America and Europe, and which therefore impacted the North Atlantic and Mediterranean more than the other ocean basins (Pacyna, 2006; Pirrone et al., 1996). There is also a clear difference in the concentration of Hg in the North versus the South Atlantic (Laurier et al., 2004) which is also likely a reflection of historic Hg inputs to the atmosphere. More recently, enhanced inputs to the North Pacific, due to increased industrialization in Asia and associated Hg emissions, have been occurring (Selin et al., 2008) but given the basin size, and the timing of these inputs (Pacyna, 2006), the ocean water concentrations have not yet changed substantially to reflect these increased inputs. Modeling of (Sunderland and Mason, 2007) suggests that the surface North Pacific

Table 7.1 Observed seawater Hg data (mean ± stdev) and estimated evasion fluxes as reported in the literature. Adapted from (Sunderland 2007) and (Mason 2005)

Ocean Basin	Total Hg (*pM*)	Hg^0 (*pM*)	Evasion (*nmol/m²/day*)	Notes
North Atlantic	2.4 ± 1.6	0.4 ± 0.3	1.9 ± 1.3	1,
	1.6 ± 0.4		0.15	2 , 4,
	2.1 ± 0.6			3
South and	1.7 ± 0.7	1.2 ± 0.8	0.32	5, 8,
Equatorial	2.9 ±1.7	0.08 – 0.16 0.11	9.6	6, 7, 9,
Atlantic				8
Mediterranean	2.2 ± 0.4	0.2 ± 0.1	0.14	10, 12, 13
	2.5 ± 1.3	0.08 – 0.2	0.23 – 0.48	8, 11, 13,
	1.5 ± 0.4		0.28 – 0.94	8, 12
North Pacific	0.6 ± 0.3	0.06 ± 0.03	0.9 ± 0.1	14, 15
South & Equatorial	1–2	0.1 ± 0.07	0.3 ± 0.3	4, 14, 15, 17,
Pacific		0.04 – 0.3	0.40	4, 17
		0.06 – 0.1		
Antarctic	0.7–1.1	no data	no data	16
Coastal Watersl	5–25	0.02–0.65	0.13–0.4	18, 19, 20

References

1. Profile averages for samples taken between 50–70°N from (Mason, Rolfhus et al. 1998)
2. North Atlantic Surface Water flowing into the Mediterranean Sea measured by (Cossa, Martin et al. 1997)
3. European continental shelf margin from (Cossa 2004)
4. (Mason and Fitzgerald 1991; Mason and Fitzgerald 1993)
5. Subsurface water from the South and Equatorial Atlantic from (Mason and Sullivan 1999)
6. Surface water samples from the South and Equatorial Atlantic from (Mason and Sullivan 1999)
7. (Mason, Lawson et al. 2001)
8. (Gardfeldt 2003)
9. (Lamborg 1999)
10. Seawater exiting the Strait of Gibraltar from (Cossa, Martin et al. 1997)
11. Western Mediterranean, (Cossa, Martin et al. 1997)
12. (Horvat 2003)
13. (Ferrara 2006)
14. (Laurier 2003)
15. (Laurier, Mason et al. 2004)
16. (Dalziel 1995); (Laurier, Mason et al. 2004) and (Mason and Sullivan 1999)
17. (Kim and Fitzgerald 1986)
18. (Mason, Lawson et al. 1999; Mason 2005; Whalin 2007)
19. (Rolfhus and Fitzgerald 2001)
20. (Baeyens and Leermakers 1998)

Ocean waters will take decades rather than years to respond to these documented changes in anthropogenic atmospheric inputs and will achieve a steady state given current levels of Hg input. modeled responses at present levels of atmospheric deposition indicate concentrations will rise substantially in the immediate future until a new steady state concentration is achieved.

Overall, these modeling studies support the notion that spatial differences in the inputs of Hg to the atmosphere are reflected in ocean water concentrations and provide further justification to support the data of the limited field studies (Sunderland

and Mason, 2007). Finally, while total Hg concentrations are higher in estuarine and coastal waters, fluxes of Hg^0 are lower as the Hg in these waters is more strongly complexed to organic matter and the water quality allows for less light penetration, which both decrease the overall rate of net Hg reduction in these waters, as discussed above (Whalin et al., 2007). Mason and Sheu (2002) have previously estimated the flux of Hg from the ocean as 2600 Mg yr^{-1}, by constraining the values from the field studies, through accounting for possible seasonal and latitudinal differences. Recent data (Laurier and Mason, 2007; Laurier, 2003) and modeling (Strode, 2007; Sunderland and Mason, 2007) does not contradict this estimate. The model evaluation of (Sunderland and Mason, 2007) supports the magnitude of this estimate, although there is the potential that the value may be either higher or lower. In generating the range in possible concentrations and fluxes, (Sunderland and Mason, 2007) used Monte Carlo simulations of their 16 box ocean model, and different estimates of atmospheric inputs to constrain and examine the potential variability in the fluxes. The estimated average values, and the ranges for each ocean basin, are shown in Table 7.2. As can be seen, there is variability in all of the estimates of almost an order of magnitude in some cases and the fluxes for the different ocean basins reflect both the differences in their concentration and the differences in their overall area.

The minimum estimate for the evasion of about 800 Mg yr^{-1} is lower than most estimates that are in the literature, but equivalent to that of (Lamborg et al., 2002). The average value is consistent with most other estimates of 2000-2800 Mg yr^{-1} (Bergan et al., 1999; Mason and Sheu, 2002; Seigneur, 2004; Selin, 2007; Shia et al., 1999). The upper bound is comparable with the recent modeling estimates of (Selin et al., 2008) (5000 Mg yr^{-1}). Therefore, it appears that the estimates are within the range and confidence intervals of the results of a number of box and numerical modeling studies, and with empirical estimates. As noted above, these fluxes are comparable to the estimates of Hg inputs to the atmosphere from point source anthropogenic emissions.

Table 7.2 Ranges (90% confidence intervals) in the estimated fluxes from the ocean to the atmosphere for the various ocean basins. Adapted from (Hedgecock et al., 2006; Mason and Sheu, 2002; Sunderland and Mason, 2007). Note that the ranges are not simply additive as they are simulated from individual probability distributions

Basin	Latitude Range	Area ($\times 10^{14}$ m^2)	Flux (*Mmol* yr^{-1})
North Atlantic	>55°N	0.20	120(40 – 220)
Surface Atlantic	35°S – 55°N	0.62	640(160 – 1300)
Intermediate Atlantic	65°S – 35°S	0.20	80 (20 – 160)
Surface Mediterranean	30 – 35°N	0.025	70 (8 –80)
North Pacific	>30°N	0.27	200 (80 – 360)
Surface Pacific & Indian	40°S – 30°N	1.48	1280 (380 – 2600)
Intermediate Pacific & Indian	65°S – 40°S	0.50	220 (60 – 440)
Surface Antarctic	>65°S	0.12	12 (4 – 22)
Coastal Waters	10% of total area	0.035	60 (30 – 100)
Total		3.45	2682 (782 – 5282)

7.2 Estimates of Net Terrestrial Evasion

As noted above, the discussions and estimates that will be included in this section refer to the exchange of Hg^0 and therefore include uptake of Hg^0 by vegetation as well as loss of Hg^0 to the atmosphere from vegetation. It appears that the exchange of Hg^0 with vegetation can be bi-directional and the notion that the direction of exchange is related to the atmospheric Hg^0 concentration has been discussed in terms of a "compensation point" (Gustin, 2003; Gustin and Lindberg, 2005; Lindberg, 1996; Lindberg et al., 1998a), which is defined as the minimum concentration in the atmosphere which will result in a release of Hg from a particular vegetation type. The value is not the same for all plants. When reviewing the literature it became apparent that the approach and experimental design in most cases was not sufficient to truly examine both uptake and emission simultaneously, but there is evidence that this must occur, and may depend on the time of day, presence of solar radiation, temperature and atmospheric concentrations. The evidence for uptake of Hg^0 by plant leaves is mostly derived from measurements of the concentration of Hg in leaves, and demonstration that the concentration increases over a growing season, and that transport of Hg from soils through the roots and stems cannot account for this accumulation (e.g. (Frescholtz et al., 2003; Gustin and Lindberg, 2005; Miller et al., 2005; Rea et al., 2002). Similarly, air concentrations of Hg^0, and other parameters, also impact the evasion of Hg^0 from soils and other substrates.

Additionally, in most studies, the estimates that are used to constrain the fluxes or are used to provide estimates of fluxes are either derived from short-term or small scale (mesocosm) experiments, or from the use of sampling devices, such as flux chambers, that can have some impact on the measurement being made e.g. flux chambers interfere with air flow at the surface. Also, there is always the potential for artifacts in such studies and while such problems have mostly been examined and are accounted for in the calculations, there is always a need for caution, as demonstrated by the recent finding of problems associated with the use of eco-chambers for Hg flux studies at low concentrations (Stamenkovic and Gustin, 2007). Alternatively, the flux estimates are derived indirectly, such as from the use of litterfall as a measure of the uptake of Hg^0 by plant leaves. Again, there is the potential that Hg may enter the leaf from other processes and it is likely that such a measure provides an over-estimate of the actual uptake. Again, these results need to be scaled to annual fluxes when, for example, in temperate and polar regions, many trees are deciduous and lose their leaves over winter. Also, the presence or absence of a canopy can alter the soil flux in such situations. Thus, there is the need for some caution in extrapolating from these results to regional or global scales. A number of papers have reviewed the older literature (Gustin and Lindberg, 2005; Gustin and Lindberg, 2000; Zhang and Lindberg, 1999) and these summaries will be built on for the current assessment. These papers, and a number of other studies (see list associated with Table 7.3), provide the information on fluxes that is gathered in Table 7.3, and the estimates of Hg^0 uptake by plant material that are gathered in Table 7.4. As noted, there is apparently contradictory information in the literature as some studies show evasion of Hg^0 from vegetation while other studies suggest uptake. It is not possible to determine

Table 7.3 Average fluxes, or in some cases the range of fluxes, for various ecosystems measured by a number of investigators. Values have all been converted to a common flux unit of *nmol m[2] month[1]*. In the table, negative values indicate uptake of mercury rather than release. Results from the older literature are combined in estimates given in various review papers

Location	Flux *(nmol m² mth¹)*	Ref.	Location	Flux *(nmol m² mth¹)*	Ref.
Vegetation Forest Hardwood	29 –238	1	Ground Level Forest Floor Sweden	1.4 – 7.2	1
Pine	3.6 – 126	1	Oak Ridge, TN	7.2 – 25	8
Maple	20	2	Michigan	5.0	9
Spruce	6.1	2	Brazil	<1	7
Poplar	9.7	2	"Deforested" site	50	7
Oak	16.4	2	Desert Soils	-3.6 to 10.8	6
Other Sagebush	-5.0	3	Mineralized Areas (Av. for Nevada)	13.3	10
Prairie grass	12.5	5	High Hg regions	max 1500	11
Cattail	60	6			
Model Estimates Hard wood forest	max 16	4	Model Estimates Forest Soil	5.4	4
Agricultural crops	max 11	4	Agricultural soil	8.3	4
"Plant-related emissions"	3.2	12	"Average global soil"	1.5	12

References: (1) Hanson, Lindberg et al. (1995); Lindberg, Hanson et al. (1998b); (2) Hanson, Lindberg et al. (1995); (3) Fay and Gustin (2007); (4) Bash et al., (2004); (5) Obrist et al., (2005); (6) Gustin et al., (2006) and references therein; (7) Magarelli and Fostier (2005); (8) Carpi and Lindberg (1997); (9): Zhang, et al. (2003); (10) Gustin (2003) and references therein; (11) Gustin and Lindeberg (2005); (12) Selin et al., (2008).

Table 7.4 Estimates of the uptake of mercury by vegetation, primarily trees as estimated from the concentration of Hg in litterfall collected at the end of the growing season, or from measurements of leaf concentration over time

Location	Flux *(nmol m² mth¹)*	Reference
Various studies	6.6 – 16.5	1, 2
Temperate forest	5.4	3
Temperate forest	5.0	4
Model: NE USA	0.8 – 3	5

References: (1) Lindberg et al. (2004); (2) Gustin and Lindberg, (2005); (3): Rea et al. 1996; Rea et al. (2002); (4) St Louis et al. (2001); (5) Miller, et al. (2007).

exclusively for all vegetation types the overall net uptake, nor to ascertain the impact of seasonal differences, or the impact of changing air Hg^0 concentrations, and the extent these may exceed a particular compensation point. Thus, the approach taken here is to scale both the uptake and release fluxes from the various studies and calculate a net overall flux (Table 7.5) that is a composite of these two processes, and reflects the net uptake or emission for a particular terrestrial habitat type, and for a particular region (polar, temperate or tropical).

As seen in Table 7.3, there is a wide range in the values in the literature for a vegetation type, or for soils. For example, the fluxes for forests and various tree types vary from low values around 6 to very high values of 240 nmol m^{-2} month^{-1}. In compiling the information from the literature it was necessary to convert the fluxes to comparable units for ease of comparison. Clearly, in making such calculations it is necessary to assume that the measurements made are valid for the region on an average basis. Clearly, in some cases this may not be truly valid, and in the calculations for Table 7.5 any seasonal effect is further considered, as detailed below. Furthermore, some of the variability in Table 7.3 is due to seasonal and other effects. A number of parameters are now known to influence the net uptake or emission of Hg^0 from both vegetation and soils, and this accounts to some degree for the variability as many of the studies have been done over short periods of time, or within a particular season. Variability between species is also a potential problem in extrapolating the results. For example, the study of (Fay and Gustin, 2007) using three different plant species in mesocosm experiments showed that under the same conditions there could be Hg^0 deposition (plant uptake), bi-directional exchange and no measurable flux for the different plant species. Again, as found for ocean environments, the individual fluxes in Table 7.3 appear much higher than those that can be sustained at steady state, and therefore modeling and other approaches are needed to further constrain these values.

Available estimates from modeling studies are also included in Table 7.3 where these are based on detailed formulations rather than simple mass balance approaches. The modeling of (Bash et al., 2007; Bash et al., 2004), for example,

Table 7.5 Estimates of net evasion of mercury from terrestrial ecosystems which were calculated from the information provided in Tables 7.3 and 7.4, as discussed in the text

Region	Estimated Area ($\times 10^{14}$ m^2)	Net Evasion (average) ($Mg\ yr^{-1}$)	Total evasion*
Polar/Boreal (>700)			
Boreal Forests	0.12	40 (12–60)	2.4
Tundra	0.08	48 (16–96)	2.9
Boreal Lakes	0.005	4.5 (2.7–9)	0.3
Ice Covered Polar Regions*	0.08	40 (24–80)	2.4
Temperate (30–700)			
Temperate Forest	0.12	102 (30–120)	6.2
Grassland/Prairie	0.08	89 (57–178)	5.4
Chaparral/Scrub	0.08	88 (56–176)	5.3
Temperate Agriculture	0.09	68 (36–145)	4.1
Desert/Metalliferrous Zones	0.18	269 (143–716)	16.3
Temperate Lakes	0.03	58 (35–117)	3.5
Tropical/Subtropical (0–300)			
Tropical Forests	0.25	200 (61–245)	12.1
Tropical Prairie/Grassland	0.009	12.6 (6.8–25)	0.8
Tropical Agriculture	0.063	52 (32–113)	3.2
Desert/Metalliferrous Zones	0.16	239 (159–637)	14.5
Savannah	0.15	210 (113–420)	12.7
Tropical Lakes	0.011	33 (20–68)	2.0
Volcanoes/Geothermal	NA	90 (60–600)	5.5
Total	1.5	1650 (863–3806)	

*These fluxes do not include re-emission of Hg to the atmosphere during polar sunrise in response to MDE's.

results in a diurnal variation in plant fluxes and thus the maximum values are given in Table 7.3 – the minima at night is around zero. The average soil evasion flux (18 nmol m^{-2} yr^{-1}; 3.6 μg m^{-2} yr^{-1}) of (Selin et al., 2008) was derived using a model incoporating solar radiation, temperature and canopy attenuation of surface temperature. These modelers also estimated fluxes related to plant evapotranspiration and also due to the re-emission of recently deposited atmospheric Hg (respectively, 18 and 20 nmol m^{-2} yr^{-1}; 3.6 and 4 μg m^{-2} yr^{-1}), which are combined together as a single value in Table 7.3.

Modeling results and correlations based on field studies suggest that fluxes are related to solar radiation, temperature and the impacts of moisture as well as the concentration of Hg^0, CO_2 and of oxidants in the air (e.g. (Bash et al., 2004; Gustin et al., 2006; Millhollen et al., 2006; Obrist, 2005; Selin et al., 2008). Fluxes are also strongly dependent on air concentration, and in many cases, fluxes for a particular plant species are related to the "compensation point" air concentration. At lower concentrations than the compensation point, plants release Hg while at higher air concentrations, Hg is taken up. Most studies to date suggest that the compensation point for many tree and plant species is within the typical background range (1-3 ng m^{-3}) (Gustin and Lindberg, 2005; Lindberg, 1996; Lindberg et al., 1998b) and so in many cases, the uptake is bi-directional, and likely overall there is less net uptake into the vegetation. Similarly, uptake into soil or release of Hg^0 from soil depends on the soil Hg content. In some studies (Xin and Gustin, 2007), soils emitted Hg during the day but there was deposition to soil at night, and the extent of uptake or release was a function of both air and soil Hg content. At typical air concentration, very little net release or uptake occurred. Again, many studies that suggest Hg^0 release from contaminated soils should be extrapolated with caution as there is the possibility for little or no release, or even uptake into low Hg soils. Emissions from soils are dependent on canopy cover as this can reduce the rate of evasion (Gustin and Lindberg, 2005).

The results from studies on the uptake of Hg by plant leaves are shown in Table 7.4. These studies are somewhat limited but they show in all cases that there is uptake of Hg into the leaves and that this Hg cannot be removed by simple washing or leaf surface extraction, which would indicate that this Hg was due to dry deposition of particulate material or from RGM deposition (Gustin and Lindberg, 2005; Miller et al., 2007; Miller et al., 2005; Rea et al., 2002). Overall the various studies are relatively consistent in that the estimated fluxes are of the same order, especially for the more recent studies in temperate forests. Miller et al. (2005) used these results to develop a regional model for the uptake of Hg^0 by vegetation in the Northeast USA and the modeled range in values for the different states in the region were lower than the specific studies, but these can be used to extrapolate the limited studies to a more global scale.

The information from Tables 7.3 and 7.4, and the modeling approaches discussed above, provide the basis for the estimations contained in Table 7.5. In scaling up the fluxes the following are taken into account: 1) seasonal impacts, such as snow cover on land and ice cover on lakes; 2) seasonal differences in solar irradiation and its impact on fluxes; 3) temperature effects; and 4) the variability in fluxes for different vegetation types. In determining the various regions in Table 7.5, global maps of ecosystem biomes were used to generate the estimated areas for the different vegetation types, and the various regions were divided by latitude as noted in Table 7.5.

Some interpolation was required to cover the various regions, and to take into account vegetation types, and to account for agricultural zones, and for the emissions from lakes. Also, given the literature on the uptake and release of Hg^0 from plants, the following assumptions were made: 1) evasion of Hg^0 from vegetation was: forest>grasslands/other terrain>agriculture and tropical>temperate>polar; 2) evasion from metalliferrous zones was higher than from vegetated areas; 3) evasion from urban areas was not specifically included; 4) uptake of Hg^0 by plant leaves is four times greater for forest compared to other vegetation (grasses, agriculture); 5) uptake was greater for tropical versus polar regions because of seasonal differences (e.g. temperature, solar radiation); and 6) there was only evasion, and no uptake to lakes and non-vegetated terrain. Using these assumptions, and the range of values in Tables 7.3 and 7.4, the following estimates were generated (Table 7.5).

While the evasion has been partitioned according to landscape type and latitude, it must be re-stated that these estimates are based on a relatively small dataset and are therefore likely to be highly uncertain. This is indicated by the range in values given for each flux estimate, which vary by a similar amount to the estimates for ocean evasion, up to an order of magnitude difference. Thus, these results and estimates form a basis for future work and require much more analysis and confirmation of their validity, but they do provide a first order estimation.

The recent global mercury model of Selin et al. (2008) estimates that non-point source evasion from land (excluding biomass burning) at around 2200 Mg yr^{-1}; somewhat higher than the estimates here but not substantially different and within the range of values given in Table 7.5. Other recent modeling efforts have used a value of 2000 Mg yr^{-1}, and have mostly assumed that 50% of this is from re-emission of anthropogenically-derived Hg deposition (Bergan et al., 1999; Lamborg et al., 2002; Mason and Sheu, 2002; Seigneur, 2004; Selin, 2007; Shia et al., 1999).

The value estimated here is most similar to that of Mason and Sheu (2002) (1600 Mg yr^{-1}). However, the range in the values in Table 7.5 encompasses all these model estimates. Overall, on average, forests account for 22% of the emissions, based on these estimates, agricultural locations about 8% and other vegetated regions 27%. Thus, more than 50% of the emissions are from these vegetated regions. Deserts and metal-rich locations are also important sources to the atmosphere, accounting for 30% of the emissions.

Volcanoes account for about 5% of the total emissions. Lakes are overall a minor source (~6%) to the atmosphere, compared to the terrestrial sources. Overall, most of the emissions are from the tropical regions (53%), compared to the temperate regions (39%) with the polar regions being a minor source (8%). Again, these variations fit with the limited data, and the latitudinal variation is consistent with the Selin et al. (2008) model, and makes sense given that the primary factors influencing the degree of emission are temperature and solar radiation, which are highest in the tropical zone.

The information in Tables 7.2 and 7.5 are further summarized in Table 7.6. In addition, the potential evasion flux from emissions of Hg^0 in response to enhanced deposition of RGM after MDE's is included in the table. This estimate is based on a number of recent estimates of the importance of this source (Ariya et al., 2004; Mason and Sheu, 2002), which are derived form an analysis of the data compiled in a number of recent studies (Lalonde et al., 2003; Lalonde et al., 2002; Poulain et al., 2007;

Table 7.6 Overall summary of fluxes by vegetation type and for aquatic systems. Compiled from the data in Tables 7.2 and 7.5

Region	Estimated Area ($\times 10^{14}$ *m*^{2})	Net Evasion (average) (*Mg yr*$^{-1}$)
Oceans	3.6	2682
Atlantic Ocean	1.02	840
Pacific and Indian Ocean	2.25	1700
Antarctic Ocean	0.12	12
Mediterranean	0.025	70
Coastal waters	0.035	60
Terrestrial	1.5	1850
Forest	0.49	342
Tundra/Grassland/Savannah/Prairie/Chaparral	0.40	448
Desert/Metalliferrous/ Non-vegetated Zones	0.42	546
Agricultural areas	0.15	128
Lakes	0.046	96
Evasion after Mercury Depletion Events	NA	200
Volcanoes/Geothermal	NA	90
Total		4532

Poulain et al., 2004). These estimates are uncertain but suggest that the flux to the atmosphere from these processes is 200 Mg yr^{-1}. The compilation in Table 7.6 emphasizes the importance of oceanic evasion in contributing Hg to the atmosphere as this accounts for about 60% of the total emissions from all the sources documented in this chapter. This again is consistent with the literature and all the modeling studies discussed above.

Overall, the evasion from the ocean, which constitutes 70% of the surface of the Earth is 2680 Mg yr^{-1}, or, on average ~36 nmol m^{-2} yr^{-1} (7.2 μg m^{-2} month^{-1}). In contrast, evasion from the terrestrial environment (30% of the surface) is 1850 Mg yr^{-1}, or, on average, ~63 nmol m^{-2} yr^{-1} (12.6 μg m^{-2} month^{-1}). Therefore the average emissions from the land exceed the ocean on an areal basis. This appears reasonable given the number of sources and mechanisms for the formation and evasion of Hg^0 from these environments. Furthermore, one needs to consider the importance of the impact of enhanced deposition to the land related to anthropogenic inputs, which are concentrated on the terrestrial landscape. The mass of Hg in soils (>50 nmol kg^{-1}) is much larger than that of the ocean (a few pmol kg^{-1}), and even though much of the terrestrial Hg is bound and uncreative in terms of reduction, there is clearly a substantial reservoir of Hg available for evasion. Indeed, using typical concentrations of Hg in soils (50-500 nmol kg^{-1}), the average yearly flux is equivalent to the removal of 1-10% of the Hg in the surface cm of such soils. Thus, there is indeed a large reservoir of Hg in the terrestrial environment that is supporting the evasion of Hg^0, and this is clearly enhanced by the presence of vegetation. Finally, the total evasion estimated here is about 4400 Mg yr^{-1}, which is still less than the estimated wet and dry deposition of Hg from the atmosphere, which ranges from 6000 to 6800 Mg yr^{-1} in different model simulations. The difference is consistent with the estimates of anthropogenic inputs, which is the other major source of Hg to the global atmosphere besides evasion of Hg^0 from the ocean and from terrestrial environments.

References

Amyot, M., Mierle, G., Lean, D.R.S. and McQueen, D.J., 1994. Sunlight-induced formation of dissolved gaseous mercury in lake water. Environ. Sci. Technol., 28: 2366–2371.

Amyot, M., Mierle, G. and McQueen, D.J., 1997. Effects of solar radiation on the formation of dissolved gaseous mercury in temperate lakes. Geochem. Cosmochem. *Acta.*, 61: 975.

Andersson, M.E., K. Gardfeldt, I. Wangberg, F. Sprovieri, N. Pirrone, and O. Lindqvist, 2007. Seasonal and daily variation of mercury evasion at coastal and off shore sites from the Mediterranean Sea. Mar. Chem., 104: 214–226.

Ariya, P.A., Khalizov, A. and Gidas, A., 2002. Reactions of gaseous mercury with atomic and molecular halogens: Kinetics, product studies, and atmospheric implications. Journal of Physical Chemistry A, 106(32): 7310–7320.

Ariya, P.A. et al., 2004. The Arctic: a sink for mercury. Tellus Series B-Chemical and Physical Meteorology, 56(5): 397–403.

Baeyens, W. and Leermakers, M., 1998. Elemental mercury concentrations and formation rates in the Scheldt estuary and the North Sea. Marine Chemistry, 60: 257–266.

Balabanov, N.B. and Peterson, K.A., 2003. Mercury and reactive halogens: The thermochemistry of Hg+{Cl-2, Br-2, BrCl, ClO, and BrO}. Journal of Physical Chemistry A, 107(38): 7465–7470.

Bash, J.O., Miller, D.R., Meyer, T.H. and Bresnahan, P.A., 2004. Northeast United States and Southeast Canada natural mercury emissions estimated with a surface emission model. Atmospheric Environment, 38(33): 5683–5692.

Bash, J.O., Bresnahan, P. and Miller, D.R., 2007. Dynamic surface interface exchanges of mercury: A review and compartmentalized modeling framework. *Journal of Applied Meteorology* and Climatology, 46(10): 1606–1618.

Bergan, T., Gallardo, L. and Rodhe, H., 1999. Mercury in the global troposphere: a three-dimensional model study. *Atmospheric Environment*, 33(10): 1575–1585.

Carpi, A. and Lindberg, S., 1997. Sunlight-mediated emission of elemental mercury from soil amended with municipal sewage sludge. *Environ. Sci. Technol.*, 31: 2085–2091.

Cossa, D., Martin, J.-M., Takayanagi, K. and Sanjuan, J., 1997. The distribution and cycling of mercury species in the western Mediterranean. *Deep Sea Research*, 44: 721–740.

Cossa, D., M.H. Cotte-Krief, R.P. Mason, and J. Bretaudeau-Sanjuan, 2004. Total mercury in the water column near the shelf edge of the European continental margin. *Mar. Chem.*, 90: 21–29.

Costa, M. and Liss, P., 2000. Photoreduction and evolution of mercury from seawater. *Science of the Total Environment*, 261(1-3): 125–135.

Dalziel, J., 1995. Reactive mercury in the eastern North Atlantic and southeast Atlantic. *Marine Chemistry*, 49: 307–314.

Ebinghaus, R. et al., 2002. Antarctic springtime depletion of atmospheric mercury. *Environmental Science & Technology*, 36(6): 1238–1244.

Ericksen, J.A. et al., 2003. Accumulation of atmospheric mercury in forest foliage. *Atmospheric Environment*, 37(12): 1613–1622.

Ericksen, J.A. and Gustin, M.S., 2004. Foliar exchange of mercury as a function of soil and air mercury concentrations. *Science of the Total Environment*, 324(1-3): 271–279.

Fay, L. and Gustin, M., 2007. Assessing the influence of different atmospheric and soil mercury concentrations on foliar mercury concentrations in a controlled environment. *Water Air and Soil Pollution*, 181(1-4): 373–384.

Ferrara, R., B. Mazzolai, E. Lanzillotta, E. Nucaro, and N. Pirrone , 2000a. Temporal trends in gaseous mercury evasion in the Mediterranean seawaters. *Sci. Total Environ.*, 259: 183–190.

Ferrara, R., Mazzolai, B, Lanzillotta, E, Nucaro, E, Pirrone, N, 2000b. Volcanoes as emission sources of atmospheric mercury in the Mediterranean Basin. *Science of the Total Environment*, 259: 115–121.

Ferrara, R., C. Ceccarini, E. Lanzillotta, K. Gardfeldt, J. Sommar, M. Horvat, M. Logar, V. Fajon, and J. Kotnik, 2003. Profiles of dissolved gaseous mercury concentration in the Mediterranean seawater. *Atmos. Environ.*, 37: Suppl. 1, S85–S92.

Fitzgerald, W.F., Engstrom, D.R., Mason, R.P. and Nater, E.A., 1998. The case for atmospheric mercury contamination in remote areas. *Environmental Science & Technology*, 32(1): 1–7.

Fitzgerald, W., Lamborg, CH, 2003. Geochemistry of Mercury in the Environment. In: Holland H., Turekian, KK (Editor), Treatise of Geochemistry. Elsevier, Amsterdam, pp. 107–148.

Frescholtz, T.F., Gustin, M.S., Schorran, D.E. and Fernandez, G.C.J., 2003. Assessing the source of mercury in foliar tissue of quaking aspen. *Environmental Toxicology and Chemistry*, 22(9): 2114–2119.

Gardfeldt, K., J. Sommar, R. Ferrara, C. Ceccarini, E. Lanzillotta, J. Munthe, I. Wangberg, O. Lindqvist, N. Pirrone, F. Sprovieri, E. Pesenti, and D. Stromberg (2003), , . 2003. Evasion of mercury from coastal and open waters of the Atlantic Ocean and the Mediterranean Sea. *Atmos. Environ.* , 37: Suppl. 1, S73–S84.

Gustin, M.S. and Lindberg, S.E., 2000. Assessing the contribution of natural sources to the global mercury cycle: The importance of intercomparing dynamic flux measurements. *Fresenius Journal of Analytical Chemistry*, 366(5): 417–422.

Gustin, M.S. et al., 2000. Assessing the contribution of natural sources to regional atmospheric mercury budgets. *Science of the Total Environment*, 259(1-3): 61–71.

Gustin, M.S., 2003. Are mercury emissions from geologic sources significant? A status report. *Science of the Total Environment*, 304(1-3): 153–167.

Gustin, M.S. and Lindberg, S., 2005. Terrestrial mercury fluxes: Is the net exchange up, down or neither?. In: Pirrone, N. Mahaffey, KR(Editor), Dunamics of Mercury Pollution on Regional and Global Scales. Springer, New York, pp. 241–259.

Gustin, M.S. et al., 2006. Mercury exchange between the atmosphere and low mercury containing substrates. *Applied Geochemistry*, 21(11): 1913–1923.

Hanson, P.J., Lindberg, S.E., Tabberer, T.A., Owens, J.G. and Kim, K.H., 1995. Foliar Exchange Of Mercury-Vapor - Evidence For A Compensation Point. *Water Air and Soil Pollution*, 80(1-4): 373–382.

Hedgecock, I.M.N., Pirrone, G. A., Trunfio., F., Sprovieri, 2006. Integrated mercury cycling, transport, and air-water exchange (MECAWEx) model. *J. Geophys. Res. – Atmos*, 111: D20302, doi:10.1029/2006JD007117.

Horvat, M., J.Kotnik J, M. Logar, V. Fajon, T. Zvonaric, and N. Pirrone 2003. Speciation of mercury in surface and deep-sea waters in the Mediterranean Sea. Atmos. *Environ.*, 37: S93–S108 Suppl. 1.

Hudson, R.J.M., Gherini, S., Watras, C. and Porcella, D., 1994. Modeling the biogeochemical cycling of mercury in lakes. In: Watras C.J. and Huckabee J.W. (Editors), Mercury as a Global Pollutant: Towards Integration and Synthesis. Lewis, Boca Raton,pp. 473–526.

Kim, J.P. and Fitzgerald, W.F., 1986. Sea-air partitioning of mercury over the equatorial Pacific Ocean. *Science*: 1131–1133.

Kim, J.P. and Fitzgerald, W.F., 1988. Gaseous mercury profiles in the tropical Pacific Ocean. *Geophysical Research Letter*, 15: 40–43.

Lalonde, J.D., Amyot, M., Doyon, M.R. and Auclair, J.C., 2003. Photo-induced $Hg^{(II)}$ reduction in snow from the remote and temperate Experimental Lakes Area (Ontario, Canada). Journal of Geophysical Research-Atmospheres, 108(D6): art. no.-4200.

Lalonde, J.D., Amyot, M., Kraepiel, A.M.L. and Morel, F.M.M., 2001. Photooxidation of Hg^0 in artificial and natural waters. *Environmental Science & Technology*, 35(7): 1367–1372.

Lalonde, J.D., Poulain, A.J. and Amyot, M., 2002. The role of mercury redox reactions in snow on snow-to-air mercury transfer. *Environmental Science & Technology*, 36(2): 174–178.

Lamborg, C.H., Rolfhus, and W.F. Fitzgeeiald 1999. The atmospheric cycling and air-sea exchange of mercury species in the south and equatorial Atlantic Ocean. *Deep-Sea Res.* II, 46: 957–977.

Lamborg, C.H., Fitzgerald, W.F., O'Donnell, J. and Torgersen, T., 2002. A non-steady-state compartmental model of global-scale mercury biogeochemistry with interhemispheric atmospheric gradients. *Geochimica Et Cosmochimica Acta*, 66(7): 1105–1118.

Landis, M.S., Stevens, R.K., Schaedlich, F. and Prestbo, E.M., 2002. Development and characterization of an annular denuder methodology for the measurement of divalent inorganic reactive gaseous mercury in ambient air. *Environmental Science & Technology*, 36(13): 3000–3009.

Lanzillotta, E., Ceccarini, C. and Ferrara, R., 2002. Photo-induced formation of dissolved gaseous mercury in coastal and offshore seawater of the Mediterranean basin. *Science of the Total Environment*, 300(1-3): 179–187.

Lanzillotta, E. et al., 2004. Importance of the biogenic organic matter in photo-formation of dissolved gaseous mercury in a culture of the marine diatom Chaetoceros sp. *Science of the Total Environment*, 318(1-3): 211–221.

Laurier, F.J.G., R.P. Mason, L. Whalin, and S. Kato, 2003. Reactive gaseous mercury formation in the North Pacific Ocean's marine boundary layer: A potential role of halogen chemistry. JGR-Atmos., 108 (D17): Art. # 4529.

Laurier, F.J.G., Mason, R.P., Gill, G.A. and Whalin, L., 2004. Mercury distributions in the North Pacific Ocean - 20 years of observations. *Marine Chemistry*, 90(1-4): 3–19.

Laurier, F. and Mason, R., 2007. Mercury concentration and speciation in the coastal and open ocean boundary layer. Journal of Geophysical Research-Atmospheres, 112(D6).

Lawson, N.M. and Mason, R.P., 2001. Concentration of mercury, methylmercury, cadmium, lead, arsenic,and selenium in the rain and stream water of two contrasting watersheds in Western Maryland. *Water Research*, 35(17): 4039–4052.

Lindberg, S., 1996. Forests and the global biogeochemical cycle of mercury. In: W.Baeyens, R. Ebinghaus and O. Vasiliev (Editors), Global and regional mercury cycles: sources, fluxes and mass balances. Kluwer Academic Publishers, Dordrecht, pp. 359–380.

Lindberg, S.E. and Stratton, W.J., 1998. Atmospheric mercury speciation: concentrations and behavior of reactive gaseous mercury in ambient air. *Environ. Sci. Technol.*, 32: 49–57.

Lindberg, S.E., Hanson, P.J., Meyers, T.P. and Kim, K.H., 1998b. Air/surface exchange of mercury vapor over forests - The need for a reassessment of continental biogenic emissions. *Atmospheric Environment*, 32(5): 895–908.

Lindberg, S.E. et al., 2002. Dynamic oxidation of gaseous mercury in the Arctic troposphere at polar sunrise. *Environmental Science & Technology*, 36(6): 1245–1256.

Magarelli, G. and Fostier, A.H., 2005. Influence of deforestation on the mercury air/soil exchange in the Negro River Basin, Amazon. *Atmospheric Environment*, 39(39): 7518–7528.

Mason, R.P. and Fitzgerald, W.F., 1991. Mercury speciation in open ocean waters. *Water, Air, and Soil Pollution*, 56: 779–789.

Mason, R.P. and Fitzgerald, W.F., 1993. The distribution and biogeochemical cycling of mercury in the equatorial Pacific Ocean. *Deep-Sea Research*, 40(9): 1897–1924.

Mason, R.P., Fitzgerald, W.F. and Morel, F.M.M., 1994. The Biogeochemical Cycling of Elemental Mercury - Anthropogenic Influences. *Geochimica Et Cosmochimica Acta*, 58(15): 3191–3198.

Mason, R.P., Rolfhus, K.R. and Fitzgerald, W.F., 1998. Mercury in the North Atlantic. *Marine Chemistry*, 61: 37–53.

Mason, R.P., Morel, F.M.M. and Hemond, H.F., 1995. The role of microorganisms in elemental mercury formation in natural waters. *Water, Air, and Soil Pollution*, 80: 775–787.

Mason, R.P. and Sullivan, K.A., 1999. The distribution and speciation of mercury in the South and equatorial Atlantic. *Deep-Sea Research Part Ii-Topical Studies in Oceanography*, 46(5): 937–956.

Mason, R.P. et al., 1999 Mercury in the Chesapeake Bay. *Marine Chemistry*, 65(1-2): 77–96.

Mason, R.P., Lawson, N.M. and Sheu, G.R., 2001. Mercury in the Atlantic Ocean: factors controlling air-sea exchange of mercury and its distribution in the upper waters. *Deep-Sea Research II*, 48: 2829–2853.

Mason, R.P. and Sheu, G.R., 2002. Role of the ocean in the global mercury cycle. Global Biogeochemical Cycles, 16(4): art. no.-1093.

Mason, R.P. and Benoit, J.M., 2003. Organomercury compounds in the environment. In: P.J. Craig (Editor), Organometallic Compounds in the Environment. John Wiley and Sons, Chichester, pp. 57–99.

Mason, R.P., 2005. Air-sea exchange and marine boundary layer atmospheric transformations of mercury and their importance in the global mercury cycle. In: N. Pirrone, KR. Mahaffey, (Editor), Dynamics of Mercury Pollution on Regional and Global Scales. Springer, New York, pp. 213-239.

Mather, T., Pyle, DM, 2004. Comment on Volcanic emissions of mercury to the atmosphere: Global and regional inventories. *Science of the Total Environment*, 327: 323–329.

Miller, E.K. et al., 2005. Estimation and mapping of wet and dry mercury deposition across northeastern North America. *Ecotoxicology*, 14(1-2): 53–70.

Miller, C.L., Mason, R.P., Gilmour, C.C. and Heyes, A., 2007. Influence of dissolved organic matter on the complexation of mercury under sulfidic conditions. *Environmental Toxicology and Chemistry*, 26(4): 624–633.

Millhollen, A.G., Gustin, M.S. and Obrist, D., 2006. Foliar mercury accumulation and exchange for three tree species. *Environmental Science & Technology*, 40(19): 6001–6006.

Nriagu, J.O., 1989. A global assessment of natural sources of atmospheric trace metals. *Nature*, 338: 47–49.

Nriagu, J.O., Becker, C, 2003. Volcanic emissions of mercury to the atmosphere: Global and regional inventories. *Science of the Total Environment*, 304: 3–12.

Obrist, D., Gustin, MS, Arnone, JA, Johnson, DW, Schorran, DE, Verburg, PJ, 2005. Measurements of gaseous elemental mercury fluxes over intact tallgrass prairie monoliths during one full year. *Atmospheric Environment*, 39: 957–965.

Pacyna, E.G., J.M. Pacyna, F. Steenhuisen, and S. Wilson, 2006. Global anthropogenic mercury emission inventory for 2000. *Atmos. Environ*, 40: 4048–4063.

Pirrone, N., Keeler, G.J. and Nriagu, J.O., 1996. Regional differences in worldwide emissions of mercury to the atmosphere. *Atmospheric Environment*, 30(19): 3379.

Poulain, A.J. et al., 2004. Redox transformations of mercury in an Arctic snowpack at springtime. *Atmospheric Environment*, 38(39): 6763–6774.

Poulain, A.J. et al., 2007. Biological and chemical redox transformations of mercury in fresh and salt waters of the high arctic during spring and summer. *Environmental Science & Technology*, 41(6): 1883–1888.

Rasmussen, P.E. et al., 2005. Measurement of gaseous mercury fluxes in the terrestrial environment. In: P.M.E.a.P. J.B. (Editor), Mercury: Sources, Measurements, Cycles and Effects. Mineralogical Society of America, Nova Scotia, CA.

Rea, A.W., Keeler, G.J. and Scherbatskoy, T., 1996. The deposition of mercury in throughfall and litterfall in the Lake Champlain watershed: a short-term study. *Atmospheric Environment*, 30(19): 3257–3263.

Rea, A.W., Lindberg, S.E. and Keeler, G.J., 2001. Dry deposition and foliar leaching of mercury and selected trace elements in deciduous forest throughfall. *Atmospheric Environment*, 35(20): 3453–3462.

Rea, A.W., Lindberg, S.E., Scherbatskoy, T. and Keeler, G.J., 2002. Mercury accumulation in foliage over time in two northern mixed-hardwood forests. *Water Air and Soil Pollution*, 133(1-4): 49–67.

Rolfhus, K.R. and Fitzgerald, W.F., 2001. The evasion and spatial/temporal distribution of mercury species in Long Island Sound, CT-NY. *Geochimica Et Cosmochimica Acta*, 65(3): 407–418.

Schroeder, W. et al., 1998. Arctic springtime depletion of mercury. *Nature*, 394: 331–332.

Schuster P.F., Krabbenhoft D.P., Naftz D.L., Cecil L.D., Olson M.L., Dewild J.F., Susong D.D., Green J.R., Abbott M.L., 2002. Atmospheric mercury deposition during the last 270 years: A glacial ice core record of natural and anthropogenic sources. *Environ. Sci. Technol.* 36: 2303–2310.

Seigneur, C., K. Vijayaraghavan, K. Lohman, P. Karamchandani, and C. Scott 2004. Global source attribution for mercury deposition in the United States. *Environ. Sci. Technol.*, 38: 555–569.

Selin, N.E., D.J. Jacob, R.J. Park, R.M. Yantosca, S. Strode, L. Jaegle, and D. Jaffe, 2007. Chemical cycling and deposition of atmospheric mercury: Global constraints from observations. J. Geophys. Res. – Atmos, 112: D02308, doi: 10.1029/2006JD007450.

Selin, N.E. et al., 2008. Global 3-D land-ocean-atmosphere model for mercury: present-day vs. preindustrial cycles and anthropogenic enhancement factors for deposition. Global Biogeochem. Cycles, Accepted.

Sheu, G.R. and Mason, R.P., 2001. An examination of methods for the measurements of reactive gaseous mercury in the atmosphere. *Environmental Science & Technology*, 35(6): 1209–1216.

Shia, R.L., Seigneur, C., Pai, P., Ko, M. and Sze, N.D., 1999. Global simulation of atmospheric mercury concentrations and deposition fluxes. *Journal of Geophysical Research-Atmospheres*, 104(D19): 23747–23760.

St Louis, V.L. et al., 2001. Importance of the forest canopy to fluxes of methyl mercury and total mercury to boreal ecosystems. *Environmental Science & Technology*, 35(15): 3089–3098.

Strode, S.A., L. Jaegle, N.E. Selin, D.J. Jacob, R.J. Park, R.M. Yantosca, R.P. Mason, and F. Slemr, 2007. Air-sea exchange in the global mercury cycle. Global Biogeochem. Cycles, 21: doi:10.1029/2006GB002766.

Sunderland, E.M. and Mason, R.P., 2007. Human impacts on open ocean mercury concentrations. Global Biogeochem. Cycles, 21: GB4022, doi:10.1029/2006GB002876.

Whalin, L.M. and Mason, R.P., 2006. A new method for the investigation of mercury redox chemistry in natural waters utilizing deflatable Teflon (R) bags and additions of isotopically labeled mercury. *Analytica Chimica Acta*, 558(1-2): 211–221.

Whalin, L., Kim, E.-H. and Mason, R., 2007. Factors influencing the oxidation, reduction, methylation and demethylation of mercury in coastal waters. *Mar. Chem.*, 107: 278–294.

Xin, M. and Gustin, M.S., 2007. Gaseous elemental mercury exchange with low mercury containing soils: Investigation of controlling factors. *Applied Geochemistry*, 22(7): 1451–1466.

Zhang, H. and Lindberg, S.E., 1999. Processes influencing the emission of mercury from soils: A conceptual model. *Journal of Geophysical Research-Atmospheres*, 104(D17): 21889–21896.

Zhang, H., Lindberg, S., Gustin, M. and Xu, X.H., 2003. Toward a better understanding of mercury emissions from soils, Biogeochemistry of Environmentally Important Trace Elements. Acs Symposium Series, pp. 246–261.

Chapter 8
Mercury Emissions from Global Biomass Burning: Spatial and Temporal Distribution

Hans. R. Friedli, Avelino F. Arellano, Jr., Sergio Cinnirella, and Nicola Pirrone

Summary This chapter represents a new addition to the UNEP global mercury budget: the mercury emissions from biomass burning, here defined as emissions from wildfires and prescribed burns, and excluding contributions from bio-fuel consumption and charcoal production and use. The results cover the 1997-2006 timeframe. The average annual global mercury emission estimate from biomass burning for 1997-2006 is 675 ± 240 Mg yr^{-1}. This accounts for 8% of all current anthropogenic and natural emissions. The largest Hg emissions are from tropical and boreal Asia, followed by Africa and South America. They do not coincide with the largest carbon biomass burning emissions, which originate from Africa. Our methodology for budget estimation is based on a satellite-constrained bottom-up global carbon fire emission database (GFED version 2), which divides the globe into regions with similar ecosystems and burn behaviour. To estimate mercury emissions, the carbon model output is paired with regional emission factors for Hg, EF(Hg). There are large uncertainties in the budget estimation associated with burned area, fuel mass, and combustion completeness. The discrepancy between the model and traditional ground based assessments (e.g. FRA, 2000) is unacceptably large at this time. Of great urgency is the development and validation of a model for mercury cycling in forests, accounting for the biogeochemistry for each region. This would provide an understanding of the source/sink relationship and thus mercury accumulation or loss in ecosystems. Limiting the burning of tropical and boreal forests would have two beneficial effects: reducing the source of mercury releases to the atmosphere from burning, and maintaining a sink for atmospheric mercury. Restricting the global release mercury would reduce the vegetation/soil pools, and the potential Hg release in case of fire.

8.1 Introduction

The importance of mercury emissions to the atmosphere from biomass burning was first recognized in South America by Veiga et al. (1994), probably as the result of the confluence of mercury pollution from artisan gold mining and ongoing clearing of tropical forests by burning for agricultural uses. After 2000, research describing

N. Pirrone and R. Mason (eds.), *Mercury Fate and Transport in the Global Atmosphere*,

DOI: 10.1007/978-0-387-93958-2_8,

laboratory and field experiments extended to other geographic regions with extensive wildfire activity. The concern for this newly recognized pathway of mercury is about its participation in the biogeochemical cycle, which includes conversion into methyl mercury, a toxic and bioaccumulating compound hazardous to humans, other mammals, and birds.

This same time period coincides with rapid advances in satellite remote sensing and retrieval algorithms, providing information on biomass burning on a global scale. We are now in a transition phase where remote sensing is adding to ground based reporting by providing critical data on fuel characterization, fire detection and burn area growth, fire intensity and smoke plume composition and transport.

Since this is the first inclusion of biomass burning into global mercury budgets, a description of the salient facts about the burn process and the biogeochemistry are briefly described. For more detailed insight, the reader is referred to the pertinent references in the text. We briefly discuss the following aspects:

- What is the global distribution and speciation of biomass?
- How does mercury enter the biomass?
- How is mercury distributed in biomass and organic soils?
- How does fire release mercury?
- How can we estimate emissions of mercury from fires?

8.1.1 Global Distribution of Vegetation

Figure 8.1 depicts the global vegetation mass distribution. Vegetation mass is concentrated in the tropical and subtropical zones (tropical forests and savanna and grass lands) and in the northern tiers of the globe (temperate and boreal forest). A related website (http://www.unep-wcmc.org/forest/global_map.htm) also describes all vegetation types.

8.1.2 Biogeochemistry of Mercury in Forests

Mercury in vegetation and organic soil is the result of bi-directional processes connecting atmosphere, plants, organic soils, and hydrology (Gustin et al., 2008; Lindberg, 1996; St. Louis et al., 2000; Driscoll et al, 2007). Mercury enters ecosystems mostly by wet and dry deposition of particulate Hg(p), ionic (RGM), and gaseous elementary mercury (GEM) onto live vegetation and soil surfaces, and by stomatic assimilation of GEM (Erickson et al., 2003; Frescholtz et al., 2003, Fay et al., 2007). Mercury on vegetation surfaces can be incorporated into plant tissue or photo-chemically reduced to GEM and released, or it can be washed off as throughfall. Xylem sap contribution to plant mercury is minor (Bishop et al., 1998) except for plants in soils with high mercury content, contaminated or naturally enriched. Upon deposition to the ground in throughfall or in senesced leaves,

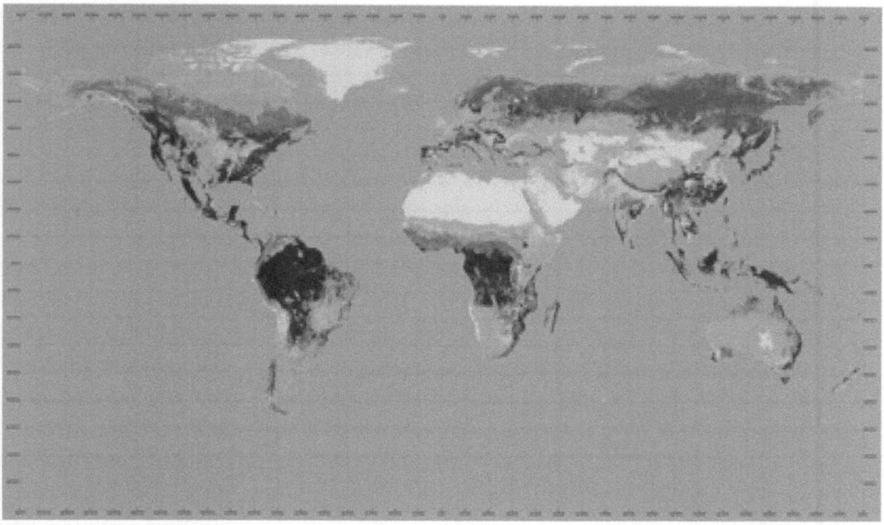

Figure 8.1 Olson's major world ecosystem complexes ranked by carbon in live vegetation: An updated database using the GLC2000 land cover product (cdiac.ornl.gov)

needles, bark and dead wood (litterfall), mercury is sequestered by reduced sulfur groups contained in the carbon pool (Skyllberg et al., 2003). This general behaviour was confirmed by the METAALICUS field experiments (Harris et al., 2007), where one mercury isotope sprayed onto live vegetation remained mostly immobilized in the organic soil, while a different isotope sprayed into water in the same region was rapidly converted into methyl mercury.

8.1.3 Mercury Distribution in Vegetation and Organic Soil by Region

Knowing the partitioning of mercury in the fuel pool is essential because only not all components of the fuel pool are combusted in a fire. The mercury content in plants is species- and plant-part specific. Mercury increases through the growing season with insignificant levels at leaf-out. Mature leaves and needles from deciduous, hardwood, coniferous trees in North American forests contain about 30-70 ng g^{-1} mercury determined by dry mass (dm). Similar ranges have been measured for Central- and South America, Australia and Africa. Some European oak leaves can reach 150-200 ng g^{-1} (dm). Bark contains higher mercury concentrations than leaves and the content differs between live and dead fractions. Bole wood is much lower in mercury, e.g. ~2 ng g^{-1} (dm) for aspen. Mercury in African savanna grasses ranges from 6-9 ng g^{-1} (dm), the same range as in agricultural residues (rice, corn).

Of critical importance is the partition of mercury above and below the forest surface. In temperate and boreal forests the mercury pool in the live plants amounts to <10% of the total pool (Grigal, 2002, 2003). As an example, in a boreal forest plot in Saskatchewan, Canada, 93-95% of the mercury pool was located in the organic soil above the mineral layer (Friedli et al., 2007). The mercury concentration in the organic soil reached 300 ng g^{-1} (dm), values typical for upland boreal forests. The mercury pool in a forest depends on stand age. In the above example the mercury pool increases from ~1.0 to 2.9 to 7.8 mg m^2 for stand ages 39, 133 and 180 years, respectively. Similar data are available for temperate North American forests (Harden et al., 2004; Engle et al., 2006; Biswas et al., 2007, 2008).

Because Hg is primarily located in the organic soil layers of a forest, it is important to understand the extent of the organic soil. Carbon accumulation (i.e. photosynthesis vs. decomposition) of an ecosystem depends on climate, vegetation type and stand age, burn frequency and hydrology. Upland boreal forests and temperate forests accumulate 5-20 cm of organic soils and burn frequently, whereas lowland peat, bogs, and permafrost may accumulate carbon over 100s of years between fires and reach several meters in thickness. As a consequence these ecosystems contain much larger mercury pools, commensurate with a large carbon pool (Turetsky et al., 2006). Of particular interest are the tropical peat fields in south East Asia because of their large size and the lack of mercury data.

African and Australian savannas accumulate only 0-2.5 cm of organic soil (Shea et. al., 1996), because of frequent (annual or biannual) burning and climatic conditions that favour rapid decomposition of organic matter. Vegetation and soil mercury in the African region is poorly defined: while the carbon emissions are the largest by far for all regions, few EF(Hg) are known. Since burning is frequent, modeled deposition rates for mercury may serve as an upper limits on what level annual mercury emissions could reach. Seigneur et al. (2003) estimated wet and dry deposition in northern and southern hemisphere Africa (NHAF, SHAF) to be 8-15 μg Hg m^{-2}. Assuming that evasion, stomatic uptake and bacterial and herbivore losses are small, and fuel mass burned ranges from 1-4 kg m^{-2} yr^{-1}, EF(Hg) of 2-3.75 μg Hg kg^{-1} could be expected. This compares with 6±3 μg Hg kg^{-1} determined from mercury content in grass (Friedli et al., 2008).

Undisturbed wet tropical forests also have shallow layers of organic soil because of rapid decomposition of the vegetation by insects and bacterial activity. At a site in the Amazon, 40% of the mercury resides in live vegetation, 60% in organic soil and litter (Michelazzo et al., 2008). This is very different from boreal forests, where >90% of the mercury is contained in the organic soil. Here the mercury pool is also smaller, 0.65 mg m^{-2} compared to 1-3 mg m^{-2} in upland boreal forests.

8.1.4 Mercury Release from Burning Biomass and Organic Soil in Different Landscapes

This section starts with a general description of the mercury release process as it was investigated in controlled laboratory experiments. It then discusses the emission behaviour in wildfires burning in different landscapes and under different fire weather conditions.

The mercury release process was studied in lab-scale experiments by Obrist et al. (2007) and Friedli et al. (2001; 2003a). These studies showed that essentially all mercury is released during the flaming phase of the combustion of dry leaves, needles and small twigs; little is known about the emission from smouldering fires, which are important for organic soil combustion. The speciation of the emitted mercury is dependent on the moisture content of the fuel: dry fuels emit GEM almost exclusively; green or wet fuels generate more smoke which can include up the 40% of the released mercury in particulate form Hg(p). The available data on mercury speciation are very limited but are important because speciation influences the location of the downwind deposition of the released mercury.

Wildfires are very complex phenomena and are dependent on fuel source (composition, mass, structure, moisture content), physical setting (slope, elevation, relief, soil structure), weather (temperature, wind, humidity, insolation) and climate (e.g. drought). Depending on conditions, different heat release rates are observed. Wildfires (and prescribed fires) behave spatially and temporarily differently because of regional climatic and weather differences. After ignition by lightning, spotting or an anthropogenic source, surface fires consume surface material (grasses, mosses and lichen, litter, downed wood) and shrubs, and can transition via ladder fuel into crown fires which consume needles, leaves, small branches and bark. Mercury contained in the burned matter is essentially fully released. If the ground is ignited, the resulting ground fire (organic soil horizons, logs, wood piles, stumps) can last for days to years and release partially or fully the mercury pool, creating a mosaic of burned and unburned areas with variable amounts of residual mercury.

Boreal wildfires, particularly under drought conditions, are often severe. They have regional and global environmental impacts because they frequently inject combustion products, including mercury, into the stratosphere (Fromm and Servranckx, 2003). The plumes undergo long-range transport and chemical transformations: mercury reactions include the conversion of GEM into Hg(p) and RGM, which have shorter lifetimes. Boreal fire plumes transported over long distances have been observed for North America (e.g. Sigler et al., 2003) and Western Russia (Witham and Manning, 2007). Intense boreal fires can generate pyro-cumulus clouds and intensive local precipitation, which could result in mercury deposition and hotspots on the ground.

Fires in dry tropical and extra-tropical temperate forests exhibit fire dynamics commensurate with fuel density, vegetation speciation, surface geography, and weather and climatic conditions.

In tropical savanna fires the heat release is smaller and stratospheric injection is less likely. The fuel mass available for burning in these ecosystems varies substantially depending on past rainfall or drought, e.g. on El Niño cycles for African savannas.

Undisturbed wet tropical forests have a 100 to 1000 year fire frequency. Where deposition mercury ends up in these ecosystems is unclear: there is little organic soil to sequester mercury as would happen in northern forests. Does it evade from the litterfall back into the atmosphere, is it absorbed into the mineral layer, or is it hydrologically removed? Commercial deforestation in tropical forests resulting in complete burning of all fuel mass removes all mercury, including from bole wood and stumps.

8.1.5 Estimation of Mercury Emissions from Biomass Burning

Mercury emission estimates from biomass burning are based on carbon budgets in combination with emission factors for mercury, EF (Hg), related to carbon released. To achieve a uniform, globally consistent treatment of carbon release for all observed fires, we selected a sophisticated carbon emission model, the Carnegie-Ames-Stanford-Approach (CASA) biogeochemical model, specifically modified to account for biomass burning (hereinafter termed as the Global Fire Emission Database version 2 (GFEDv2) as described by van der Werf et al., 2006), which partitions observed global fires into regions with similar fuel types and fire behaviour. Emission factors for mercury for different ecosystem types originate from two methods: (1) ground-based measurements of the difference in mercury pools before and after a fire, and (2) enhancement ratios (ER) of Hg and CO in plumes measured on the ground or by aircraft. Mercury emissions are then estimated as the product of carbon emissions and mercury emission factors.

8.1.6 Carbon Emission Model

The release of carbon from large-scale biomass burning is traditionally estimated from the amount of biomass (or fuel) burnt, which is calculated as a product of 1) the areal extent of the burn (burned area, BA), 2) the amount of fuel available for burning (fuel load, FL), and 3) the fraction of fuel load that has been combusted (combustion completeness, CC), integrated across space and time of interest (Seiler and Crutzen, 1980). Due to the inherent complexity of wildfires, and biomass burning in general, the estimates of these three components exhibit large spatial and temporal variability that limits our assessment of the accuracy of carbon emissions. Early studies estimated the carbon release by developing inventories (e.g. from fire management agencies) for each of these components and for different biome types, and extrapolate these inventories using global vegetation maps to continental or global-scale estimates (e.g. Hao and Liu, 1994). However, over the past two decades, these methods are being supplanted, primarily due to the increased availability of biogeochemical and fire-related observations from space (remotely-sensed), in conjunction with ground-based studies on fuel loads (and consumption), airborne measurements of smoke plumes, as well as the advancements in biogeochemical and transport models. We briefly describe below the current methods of estimating each of these components, with a particular focus on integrated approaches in estimating the carbon release on a regional to global spatio-temporal scale (e.g. GFEDv2, van der Werf et al. 2006). This type of approach provides a consistent and traceable representation of global carbon emissions, which can be very useful in evaluating and interpreting estimates with ground truth and other independent datasets. Please refer to van der Werf et al. (2006) for more details on the carbon emission model used in this report.

Burned Area

Global burned areas can be derived from detection of active fires from multi-sensor satellite imageries (ATSR, AVHRR, MODIS, VIRS) and the relationship of fire counts to burned area (e.g. Giglio et al. 2006). Estimates of burned area are now available for specific regions, typically from forest services and fire management agencies (e.g. Canada Forest Service) and from country reports (e.g. Forest Resource Assessment, or FRA). These regional estimates serve as basis for validating burned area estimates developed at a global scale. Most recently, burned area products derived from satellite observations of burn scars became available for the year 2000 at a monthly time scale and 1km x 1km spatial resolution. Of these products, the GBA 2000 (Gregoire et al., 2002) is based on the SPOT-VEGETATION instrument, while GLOBSCAR (Simon et al., 2004) is generated from data collected by the ATSR instrument. Detailed comparisons between these two products against country reports reveal key differences. Most notably, GBA 2000 has substantially higher burned area results in Africa and Australia, which Simon et al. (2004) attributed to inability of GLOBSCAR algorithms to detect large areas of burning in woodlands and shrub lands. Other than these two regions, the two products appear to be generally similar at continental scales (Kasischke and Penner, 2004).

From the evaluation of Giglio et al. (2006), the burned area reported in GFEDv2 appears to be in general agreement with Canadian Interagency Forest Fire Centre (CIFFC) compilations (slope about 70%) and US National Interagency Fire Center (NIFC) statistics (slope about 83%) with some degree of underestimation for very large burned areas. For Russia, the burned area estimate in 2001 is about 26% larger than reported by Sukhinin et al. (2004). Globally, the estimates are larger than GBA 2000 (by about 32%) and GLOBSCAR (by about 144%). The average annual burned area in GFEDv2 for the years 1997-2006 is about 332 ± 26 Mha yr^{-1}. The majority is due to burning in Africa (66%), followed Australia (13%), Asia (10%) and South America (5%).

On average, uncertainties in GFEDv2 global burned area estimates appear to be smaller (20-30%) than earlier estimates which had uncertainties over a factor of 2 (Kasischke and Penner, 2004). However, the differences in GFEDv2 burned area with other estimates can be large during some years and on local to regional scales. One of the shortcomings of these satellite-based estimates is their difficulty to detect small burned areas, particularly in deforestation regions with persistent cloud cover and mechanized clumping of fuels. In addition, the ability to globally validate these satellite-based estimates remains problematic due to the lack of validation data from ground-based measurements.

Fuel Loads

The available fuel load is defined here as organic matter available for combustion and includes all above-ground herbaceous biomass, above-ground woody biomass, coarse woody debris, and litter. Regional information on fuel loads are either derived from inventory-based compilations of fuel load maps (e.g. ECE-FAO, IFFN reports) or from satellite observations of vegetation indices (e.g. Barbosa et al., 1999). More recently, estimates on available fuel loads employ biogeochemical models that simulate the carbon fluxes across the different pools of the terrestrial biosphere (e.g. foliage, woody materials such as branches, stems, boles and roots, litter, active and passive soil), including carbon losses from fires, herbivores, and fuel wood collection. In most cases, these biogeochemical models employ added constraints from satellite-derived estimates of net primary production (e.g. from NDVI, normalized differential vegetation index) and leaf-area index (LAI), together with maps of soil moisture, temperature, precipitation, soil types, and vegetation types to realistically represent the spatio-temporal patterns and amount of biomass across different ecosystems. This approach integrates process-level information of large-scale vegetation dynamics with important drivers of fuel loads that will likely respond to global change processes. In GFEDv2, the biogeochemical model employed is the Carnegie-Ames-Stanford-Approach, CASA, (Potter et al. 1993) with a spatial resolution of $1^{\circ}x1^{\circ}$ and a temporal resolution of one month. The model uses the combination of satellite-derived estimates of fraction of incident photo synthetically active radiation (PAR) absorbed by the green plant canopy (fAPAR) from SeaWIFS (Behrenfield et al. 2001) and satellite-based estimates of solar insolation for PAR (Bishop and Rossow, 1991) to simulate net primary production (NPP). Please refer to van der Werf et al. (2006) for specific details on the allocation of NPP to different carbon pools. In GFEDv2, CASA was specifically modified to account for fire by the addition of satellite-based estimates of burned area (Giglio et al. 2006) to adjust for the potential loss of above-ground biomass and litter due to fire. In cases of fire, a direct loss of carbon is initiated from above-ground carbon pools depending on fire mortality and combustion completeness (van der Werf et al. 2006).

Other recent global studies using the biogeochemical approach, include Hoelzemann et al. (2004) who used the Lund-Postdam-Jena (LPJ) Dynamic Global Vegetation Model (DGVM), and Jain et al. (2006) who used the terrestrial component of the Integrated Science Assessment Model (ISAM). Regional-scale emission models have also been employed to refine estimates for specific regions of the globe that are highly susceptible to fires (e.g. Kasischke et al., 2005 for the boreal region, Hély et al., 2003 for Africa, Wiedinmyer et al., 2006 for North America).

The average fuel load (in kg dry matter m^{-2}), calculated from GFEDv2 for the year 2000 across regions of similar vegetation types, ranges from 1.3 ± 1 kg m^{-2} in non-forests, 18 ± 21 kg m^{-2} in tropical forests (mostly equatorial Asia), to 10 ± 8 kg m^{-2} in extra tropical forests (including boreal). These estimates are comparable on average to field measurements (Guild et al., 1998; Hobbs et al., 1996; Kasischke and Bruhwiler, 2002; Shea et al., 1996) and appear to have uncertainties on a regional scale, of about a factor of 2. A large part of the uncertainty is attributed to poor representation of soil organic carbon and below-ground biomass, especially in boreal regions and peat lands. In GFEDv2, highest fuel loads are predicted over equatorial Asia and boreal regions, where soil organic carbon represented a large fraction of the fuel load. The fuel loads in boreal North America and boreal Asia were approximately 8 kg m^{-2}, while in savanna regions, the fuel loads were highest in southern hemisphere Africa

(3 kg m^{-2}) where significant burning occurred in woodland areas and lowest in Australia (0.8 kg m^{-2}), where much of the burning occurred in grasslands.

Combustion Completeness

Combustion completeness (CC) is defined here as the ratio of fuel consumed from fires to total available fuels. Measurements of combustion completeness vary across a wide range of vegetation types (Shea et al., 1996; Hoffa et al., 1999, Carvalho et al., 2001). In general, they are associated with the types of fuel, fuel loads, and fuel configurations (including water content) but may also vary significantly depending on fire practices, fire severity and dynamics. The values range from 1 (complete combustion) for well-aired and dry litter (fine fuels) to about 0.2 for coarser fuels like stems and woody debris, which burn incompletely. Foliage and twigs in boreal regions, for example, have high CC while its living stems and boles have low CC. Studies have also shown that CC varies during the burning season, with a tendency to increase when fuels have more time to dry out (Shea et al. 1996).

In GFEDv2, CC is allowed to vary in the biogeochemical model across the different carbon pools and from month to month. CC values range from 0.8-1.0 for leaves, 0.2-0.3 for stems, 0.9-1.0 for fine leaf litter, 0.5-0.6 for coarse woody debris and 0.9-1.0 for soil organic carbon. In addition, CC is increased in stems and coarse litter in areas with high levels of fire persistence.

Uncertainties in CC can be attributed (yet difficult to quantify) to the lack of direct observations on fire behaviour across different ecosystems and throughout the burning season. Also, differences in fire practices across difference regions may increase the uncertainty. For example, in tropical forests undergoing deforestation where mechanized clumping is prevalent, it is observed that CC tends to approach to 1.0 over the course of the burning season as fuels are piled and ignited multiple times (van der Werf et al., 2006).

Fuel Consumption

The amount of biomass burned (fuel consumed) is a product of burned area, fuel load and combustion completeness. The release of carbon from biomass burning is then calculated by assuming a carbon content of 45% of the dry biomass (Andreae and Merlet, 2001). Thus, estimating the uncertainty in fuel consumption has a tendency to be multiplicative.

8.1.7 Mercury Emission Factors

Emission factor, EF(Hg), estimation for this work is based on two different methodologies: one is centered on plume composition, measured at ground level or by aircraft; the second is based on the change in the mercury pool in soils before and after a fire. Soil-based mercury is reported as g Hg released per area burned (ha or m^2); plume-based EF(Hg) have the units μg Hg per kg fuel burned. The conversion between EFs requires fire-specific values for fuel burned per unit area, which in most cases are estimates.

Mercury in plumes is measured in ng Hg/ (std) m^3 and the corresponding carbon equivalent in the same volume is calculated from measured CO_2 + CO (or estimated as 9:1 (molar) from CO measurements), corrected by a factor of 1.05 to account for carbon in trace components CH_4, particulates, non-methane organic compounds (NMOC). For the conversion of carbon to dry mass fuel, an average factor of 0.45 for carbon in fuel is assumed (consistently with the carbon model described above). All emitted mercury species (pHg, GEM, and RGM) must be included in the analysis. Mercury emission factors determined from plume observations are generally underestimates because some pHg is lost by deposition before it can be sampled. In nascent plumes, pHg is a significant fraction of the mercury emitted, up to 40% (Friedli et al., 2003a, Friedli et al., 2003b, Obrist et al., 2007). Measurements made far from a fire event may additionally miss the mercury lost by conversion to and deposition as pHg along the plume track. For these two reasons, most plume studies are underestimates.

A convenient method to assess mercury in fire plumes is to measure the enhancement ratio (ER) which is defined as Δ[Hg]/Δ[CO], Δ where Δ[Hg] is the sum of all species in excess of background concentration and Δ[CO] is the difference between CO concentration in the plume and the background. For this work we have assumed that the ratio between EF (μg/kg fuel) and ER (molar) is constant for all fires and related by a factor of 1425, derived from the most robust aircraft data available (Friedli et al. 2003b: 113 ± 59 μg/kg fuel burned and its corresponding ER of 0.793 × 10^{-7} (Δ [Hg]/Δ[CO]). To convert ER into emission estimates, a corresponding CO source term is required. Table 8.1 lists ER from ground and aircraft-based measurements for different regions.

The average of all measurements in Table 8.1 is 1.54 × 10^{-7} (mol/mol) corresponding to EF(Hg) of 220 μg Hg/kg fuel based on Friedli et al. (2003b), from which both ER (Δ[Hg]/Δ[CO] and EF(Hg) are available.

The advantage of the plume method is its integrative nature, averaging variation in fire dynamics and fuel composition, and giving a relatively broad spatial coverage. The disadvantage is the requirement for an instrumented aircraft on standby.

Table 8.1 Published molar enhancement ratios (ER) observed from fire plumes worldwide

Location	ER (Δ[Hg]/Δ[CO])	Reference
Washington State (ac)	(0.79 ± 0.04) × 10^{-7}	Friedli et al. (2003b)
Pacific NW (ground)	(1.46 ± 0.9) × 10^{-7}	Weiss-Penzias et al. (2007)
Alaska (ground)	(1.57 ± 0.67) × 10^{-7}	Weiss-Penzias et al. (2007)
Quebec (ac)	2.04 × 10^{-7}	Friedli et al. (2003a)
Quebec (ground)	0.86 × 10^{-7}	Sigler et al. (2003)
South Africa (ground)	(2.1 ± 0.21) × 10^{-7}	Brunke et al. (2001)
South America (ac)	(1.17 ± 0.15) × 10^{-7}	Ebinghaus et al. (2007)
South America (ac)	(2.39 ± 0.99) × 10^{-7}	Ebinghaus et al. (2007)

where ac = aircraft measurements and ground = ground measurements

The soil-based method applies to biomes where mercury resides predominantly in the organic soil, i.e. boreal and temperate forests. It is based on the difference in the mercury pools in organic soil in adjacent plots before and after a fire

Table 8.2 Emission factors (EF in *μg/kg* fuel) used in the emission calculations

Reference	Method	EF Range	EF Mean	Notes[2]	Fuel Burned[3]
Boreal Forest					
Harden et al. (2004)	S	0 - 138	69	Alaska, Conif. Forest, (PB)	2.5
Friedli et al. (2003a)	P	112 - 112	112	Quebec, Pine Forest, (W)	2.5
Sigler et al. (2003)	P	60 - 60	60	Quebec, Pine Forest, (W)	2.4
Cinnirella & Pirrone (2006)	P	62 - 112	87	Siberia, (W)	5.6
Weiss-Penzias et al. (2007)	P	136 - 278	207	Alaska, (W)	2.5
Turetsky et al. (2006)	S	90 - 297	193	Alaska, Upland, (W)	2.5
Turetsky et al. (2006)	S	535 - 2417	1476	Alaska, Lowland, (W)	2.9
Mean		142 - 488	315		
Temperate Forest					
Friedli et al. (2003b)	P	54 - 172	113	Washington, Mixed Forest, (W)	2.5
Brunke et al. (2001)	P	78.7 - 163.4	121	South Africa, Fynbos, (W)	2.5
Engle et al. (2006)	S	80 - 204	142	California, Conif. Forest, (PB)	2.5
Engle et al. (2006)	S	88 - 196	142	Nevada, Coniferous Forest, (W)	2.5
Biswas et al. (2006, 2008)	S	168 - 348	256	Washington, Mixed Forest, (R)	2.5
Biswas et al. (2007)	S	296 - 1012	654	Wyoming, Coniferous Forest, (W)	2.5
Biswas et al. (2007)	S	144 - 516	402	Wyoming, Aspen Forest, (W)	2.5
Woodruff et al. (2001)	S	80 - 80	80	Minnesota, (PB)	2.5
Mean		124 - 336	239		
Sage-Chaparral (Shrublands)					
Engle et al. (2006)	S	18.7 - 39.9	29.3	Nevada, Sage, W	1.84
Cinnirella & Pirrone (2006)	P	52.8 - 52.8	52.8	Mediterranean, W	1.25
Mean		35.8 - 46.4	41.1		
Grasslands and Ag. Waste					
Obrist et al. (2007)	L	7.8-9.8	8.8	Rice straw	
Friedli et al. (2008)	L	3-9	6	South African savanna	
Friedli et al. (2003)	S	38	38	Oregon, Wheat	3.06
Mean		3-38	18		

[1]Type of method used, soil (S), plume (P), or laboratory (L)

[2]Type of fire, wildfire (W), prescribed burn (PB), Rex fire (R)

[3]in kg fuel m^{-2} burned; used for EF(Hg) calculation. When nothing else was available, 2.5 *kg* fuel m^{-2} burned was assumed (based on Amiro et al. 2001). All others are from respective references.

(Harden et al., 2004; Biswas et al., 2007, 2008; Engle et al., 2006; Turetsky et al., 2006). In temperate or boreal forests, mercury resides >90% in the organic soil. Since the above ground mercury fraction is only <10% of the total Hg pool, it was neglected in the references given for boreal or temperate northern forests. Release of Hg from the underlying mineral layer is also negligible (Engle et al., 2006). The advantage of the soil-based method is that it requires only total mercury measurements in the organic soil. The disadvantage is poor statistics because of large short- and long-range spatial variability and inconsistent response to fire dynamics.

One direct comparison of the two EF(Hg) estimation techniques has been accomplished. Measurements in the fire plume from the Rex fire in Washington State were made from an aircraft during the burn (Friedli et al., 2003b) and soil measurements were made post-fire in adjacent burned and unburned sites (Biswas et al., 2006, 2008). As expected, the soil-based release estimate was higher, 6.4 ± 1.1 g Hg ha^{-1} as compared to 2.9 ± 2.2 g Hg ha^{-1} from the aircraft measurements. The available references for both methods are combined in Table 8.2 and arranged by ranges and means for different landscapes. The means are unweighted averages of all measurements in each vegetation type. The uncertainties in the reported values results from many contributions, including the influence of burn severity, inclusion of measurements of all mercury species, paucity of measurement for all fuel types (especially for grasses, shrubs and agricultural waste products) and uncertainties in the emission factor estimation technique.

The amount of fuel burned during a fire is highly variable among different ecosystems, and even within the same vegetation types (Amiro et al., 2001, French et al., 2004). Yet, a value for fuel burned is needed to determine emission factors. The means from Table 8.2 were applied to the three land classifications used in the carbon emission model, i.e. to tropical forests, extra tropical forests and non-forested land, and applied as indicated in Table 8.3 to the regions selected for the carbon emission model.

8.2 Results and Discussion

8.2.1 Global Distribution of Carbon Emissions

The average monthly spatial distribution of carbon emissions from GFEDv2 are shown in Figure 8.2 These represent the model-predicted mean seasonality of carbon released from large-scale biomass burning during the most recent decade (1997-2006). As has been noted in previous studies, the biomass burning activity across the globe varies, depending on vegetation types and climate conditions, as well as fire cultural patterns, which correlate well with agricultural practices and land use, particularly in Africa, South America and Asia (e.g. Duncan et al., 2003). Globally, biomass burning occurs at all times of the year, with a distinct peak in July-September and a lower peak in December-February. A large fraction of this peak is due to fires in Africa during the dry season. In Northern Hemisphere Africa, burning occurs in winter dry season within the savanna ecosystems (e.g. Sahara desert and central

Table 8.3 Emission factors used in this report (in μg Hg/kg fuel)

Region	Extratropical	Non-Forest	Tropical
BONA	315	41	
TENA	242	41	
CEAM	242	41	198
NHSA	242	41	198
SHSA	242	41	198
EURO	242		
MIDE	41		
NHAF	242	41	198
SHAF	242	41	198
BOAS	315	41	
CEAS	242	41	198
SEAS	242	41	
EQAS		41	315
AUST	242	41	198

Please refer to Figure 8.4 for definition of regions.

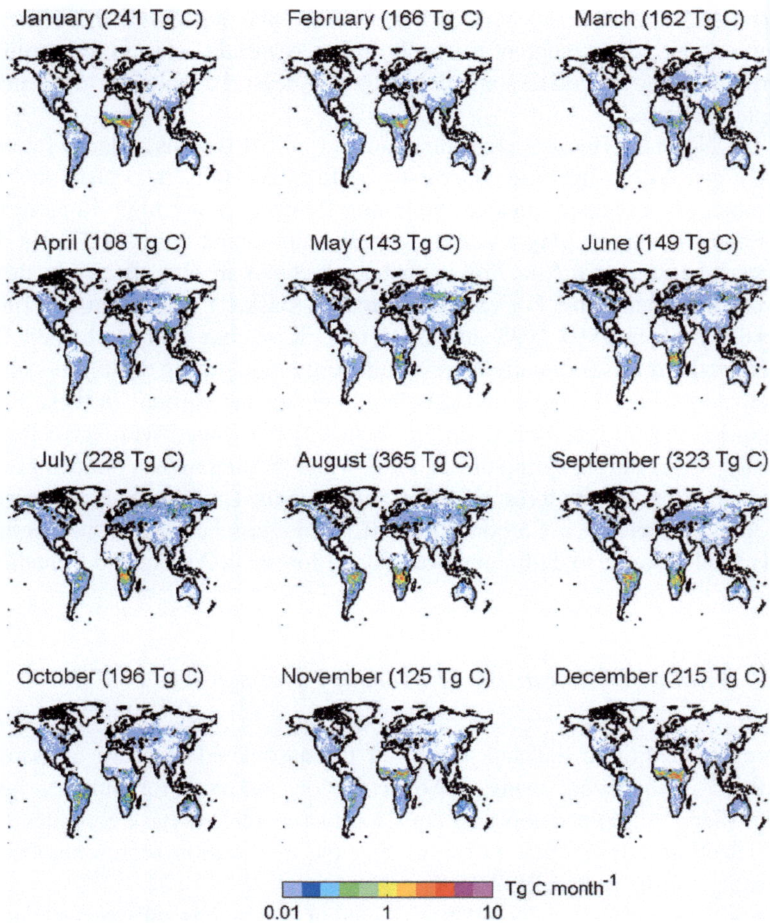

Figure 8.2 Average monthly carbon emissions for the period 1997-2006

African rain forests). It begins in the Sahel in October and spreads south through November with highest burning activity in December and January.

On the other hand, fires in the Southern Hemisphere Africa typically start in the woodlands of Zaire and Congo by early June and continue to peak across the southeast in the grasslands and shrub lands of Angola, Zambia and Tanzania during August through September and Mozambique in October. A similar fire pattern occurs in South America, mostly in the cerrados of Brazil and along the arc of fire (or deforestation) in the Amazon rainforest, generally in August through September.

Fires in tropical Asia (Indonesia, Malaysia and the rest of Southeast Asia) are highly variable and are usually associated with very dry conditions, like El Niño events, enabling land-owners to use fire more efficiently as a tool for land-clearing in the region. Because of the large peat deposits in Indonesia that are exposed during land-clearing, the carbon emissions in equatorial Asia is significantly high during El Niño years (e.g. 1997-1998). Burning typically occurs in March-April over Southeast Asia and in August-October in equatorial Asia.

Fires in Central America peak during April to June mainly due to deforestation and agriculture, while much of the bush fires in Australia occur in the shrub lands in northern Australia and to some degree in southeast Australia during September through December.

In the Northern Hemisphere temperate and boreal regions, a large fraction of fires are caused by lightning. Boreal forest fires generally take place in May to September when temperatures and lightning frequencies are high. Fires in boreal Asia typically start in May around Mongolia and spread north in Siberia during summer. Like equatorial Asia, fires in the boreal regions are also affected by droughts and very dry conditions. For example, fires in Alaska, Canada and Siberia were especially high in 1997-1998 and 2004 (e.g. Kasischke and Bruhwiler, 2002). Consequently, releases of carbon are significantly high during these years also due to characteristically higher fuel loads (e.g. soil organic carbon) in these regions. Burning in western Russia and Europe is mostly associated with agriculture and generally occurs during spring through fall, while burning in the continental U.S. is associated primarily with forest wild fires and prescribed burning generally occurring from June to September. Carbon release from biomass burning in the continental U.S. is small relative to emissions from boreal forests in Alaska and Canada.

8.2.2 Global Distribution of Mercury Emissions

We present in Figure 8.3 and Table 8.4 the associated mercury emissions as calculated in this work using carbon emissions as basis for biomass burning activity (Section 2.6) and applying emission factors that we have compiled for the globe (Section 2.7). Similar to Figure 8.2, the distribution represents the mean seasonality for the period 1997-2006.

Overall, we find a strong correlation of mercury emissions with carbon emissions from biomass burning shown in Figure 8.2, which as can be expected, highlights

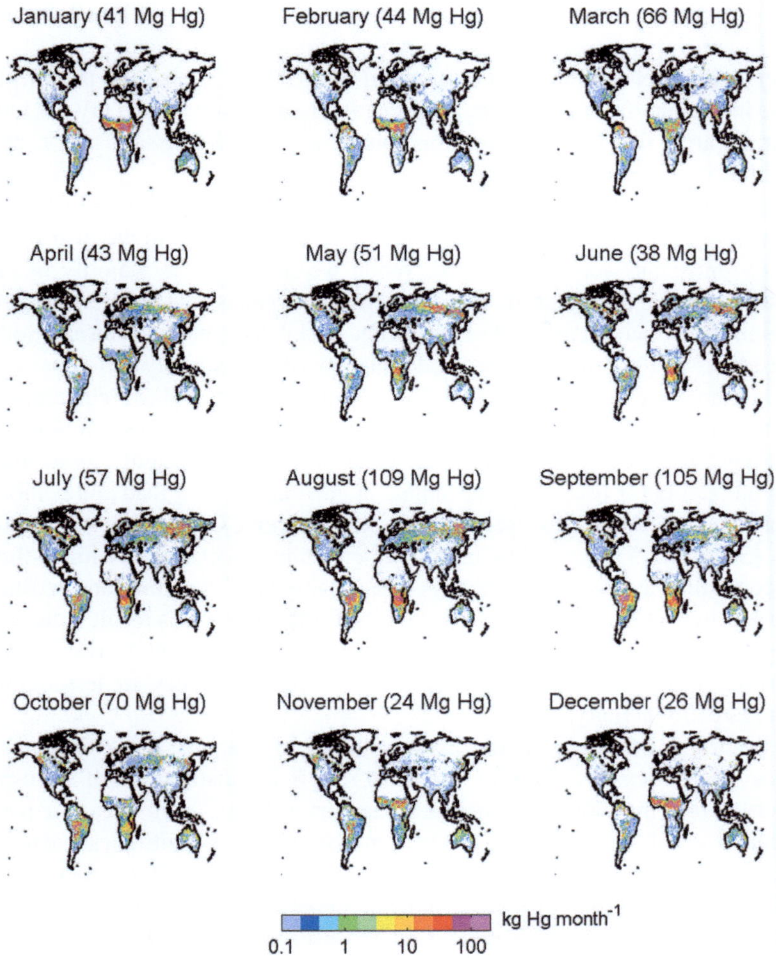

Figure 8.3 Average monthly mercury emissions for the period 1997-2006

Table 8.4 Mean seasonality of global mercury and carbon emissions (1997-2006)

	Hg (*Mg*)	Carbon (*Tg*)	Burned Area (*Mha*)	Effective EF(*μg Hg / kg fuel*)
Jan	41	241	46	77
Feb	44	166	23	120
Mar	66	162	15	184
Apr	43	108	11	177
May	51	143	15	160
Jun	38	149	16	116
Jul	57	228	24	112
Aug	109	365	34	134
Sep	105	323	35	147
Oct	70	196	26	161
Nov	24	125	31	86
Dec	26	215	55	56

the general influence of the spatio-temporal variability of global fire patterns to the release in mercury from vegetation. In this work, the emission of mercury is a function of burned area, fuel load, combustion completeness and emission factor, each of which has its associated uncertainties.

A key finding of this exercise is the significant impact of fires in the boreal region and in the tropical forests of Asia and South America, to the distribution of mercury emissions across the globe. While we find that the majority of carbon emissions can be attributed to fires in African savannas, our results for mercury emissions show major contributions from equatorial Asia, boreal Asia and Southern Hemisphere South America. The model also predicts a slightly different seasonal peak in mercury emissions, which are strongest in August and March compared to carbon emissions that are strongest in August and December. This is mainly due to the strong influence of high emission factors for these forested regions and low emission factors for fires in the savannas. In general, the fuel load and combustion completeness (CC) are inversely related, with lower CC in areas with high fuel loads (Section 2.5); hence having a compensating effect to emission estimates. As a consequence, this effect highlights the sensitivity of the emissions to estimates of burned area and assumed emission factors. Furthermore, this result indicates that while uncertainties on estimates of carbon emissions from fires in savanna (as well as burning from agricultural waste) play an important role, they are less significant with regards to mercury.

In addition, we note that the major sources of mercury from fires occur in regions where transport plays an important role in the distribution of atmospheric mercury across the globe. In particular, mercury released from fires in equatorial Asia and tropical South America can be transported to higher altitudes due to strong convection in these regions and can then be dispersed efficiently over a larger area. In a similar manner, mercury released in boreal regions can be injected at relatively higher altitudes, thereby largely influencing the distribution of mercury in the Northern Hemisphere middle to high latitudes.

8.2.3 Regional Estimates of Carbon and Mercury Emissions

We report in this section our estimates of carbon and mercury emissions from biomass burning for different regions. Here, we used the regional divisions from the carbon emission model (please refer to Figure 8.4) in generating a regional budget for annual sources of carbon and mercury as shown in Figure 8.5 with corresponding values in Table 8.5. This annual budget is an average for 1997-2006.

We estimate a global source of mercury from biomass burning of about 675 ± 240 Mg yr^{-1}. As noted earlier, there is a distinct shift in regional contributions of mercury emissions to the global budget relative to regional contributions of carbon emissions. On a process level, we find a clear correlation of burned area with carbon emissions (r=0.55) but not with mercury emissions, again indicating the dependency of mercury emissions on variability in EF(Hg)'s.

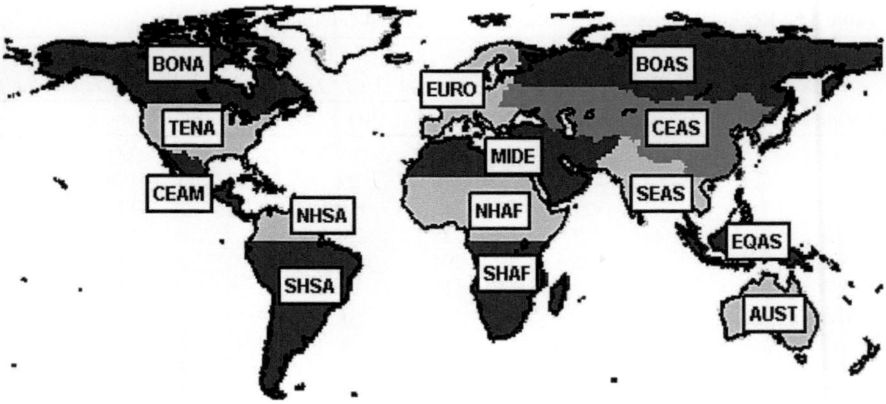

Figure 8.4 Map of regions used in GFEDv2 (from van der Werf et al. 2006)

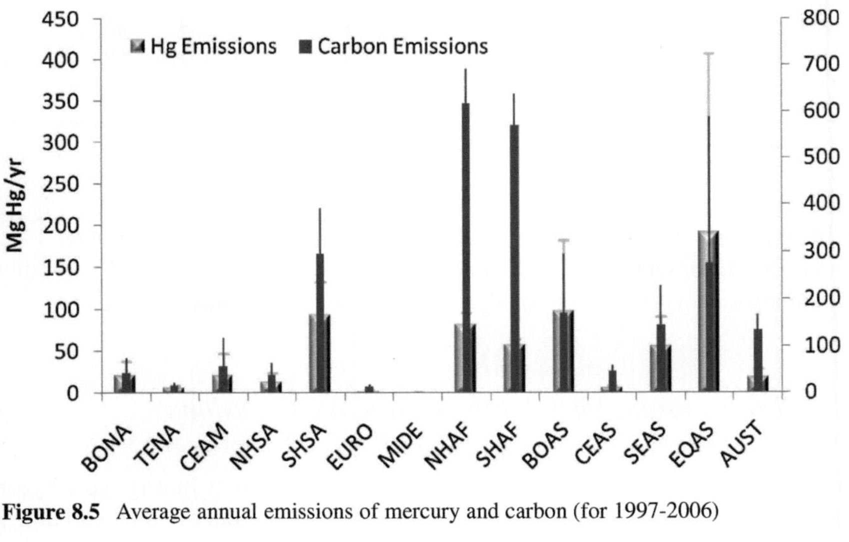

Figure 8.5 Average annual emissions of mercury and carbon (for 1997-2006)

The major sources of mercury come from equatorial Asia or EQAS (28%), boreal Asia or BOAS (15%) and southern hemisphere South America or SHSA (14%), and only in part from northern hemisphere Africa or NHAF (12%), southern hemisphere Africa or SHAF (9%), southeast Asia or SEAS (8%), central America or CEAM (4%) and Australia or AUST (3%). Temperate North America or TENA (1%), boreal North America or BONA (3%), and central Asia or CEAS, northern hemisphere South America or NHSA, Europe or EURO and Middle East (MIDE) combined (2%) contribute little to the global budget.

This result has important policy implications with regards to assessing and regulating the impact of mercury to ecosystems and human health. We note however

Table 8.5 Regional emission estimates for mercury and carbon (1997-2006)

	Hg *(Mg yr⁻¹)*		Carbon (*Tg yr⁻¹*)		Burned Area *(Mha yr⁻¹)*		Effective EF *(µg Hg/kg fuel)*
Regions	Mean	SD	Mean	SD	Mean	SD	Mean
BONA	22	16	42	30	2	1	233
TENA	6	3	16	6	2	0	178
CEAM	22	25	56	61	3	2	175
NHSA	13	10	38	25	4	1	157
SHSA	95	39	294	97	12	2	145
EURO	2	1	14	5	2	1	72
MIDE	0	0	1	0	0	0	17
NHAF	83	13	618	74	141	13	60
SHAF	58	7	571	68	78	9	46
BOAS	99	83	170	124	9	5	263
CEAS	7	2	47	12	17	5	67
SEAS	57	35	144	84	12	5	177
EQAS	192	216	276	312	4	4	312
AUST	19	9	133	35	46	18	65
Global	675	240	2420	382	332	26	279
Boreal (BONA+BOAS)	121	85	212	128	11	5	248
Temperate (TENA+EURO+MIDE)	9	3	30	7	4	1	89
Rest of the World	545	224	2178	360	316	26	134

\# SD for standard deviation

that there is a significant variability (as shown from the error bars) throughout the 10-year period, particularly for EQAS. We further discuss the inter-annual variability and comparisons with other regional estimates in the succeeding sections.

8.2.4 Inter-annual Variability of Mercury Emissions

Shown in Figure 8.6 are annual emissions for different regions during the 10-year period chosen in this report. As can be seen, there is a large inter-annual variability of mercury emissions across different regions, particularly in EQAS, SHSA, BOAS, BONA and CEAM. There were large amounts of mercury released during the strong El Niño year of 1997-1998 and during drought conditions in 2003-2004. The bulk of this inter-annual variability occurred in Indonesia where peat deposits were available as fuel loads, as well as in boreal region and in deforestation regions of tropical America. On the other hand, there is a more or less uniform contribution of mercury emissions to the global budget from Africa for the 10-year period. The difference in inter-annual variability is also reflected in the mean estimates discussed in the previous section and is consistent with previous studies on biomass burning (e.g. Duncan et al., 2003; van der Werf et al., 2006). Indeed, there is a clear indication of a strong relationship between biomass burning, precipitation and temperature in conjunction with fire practices in different regions.

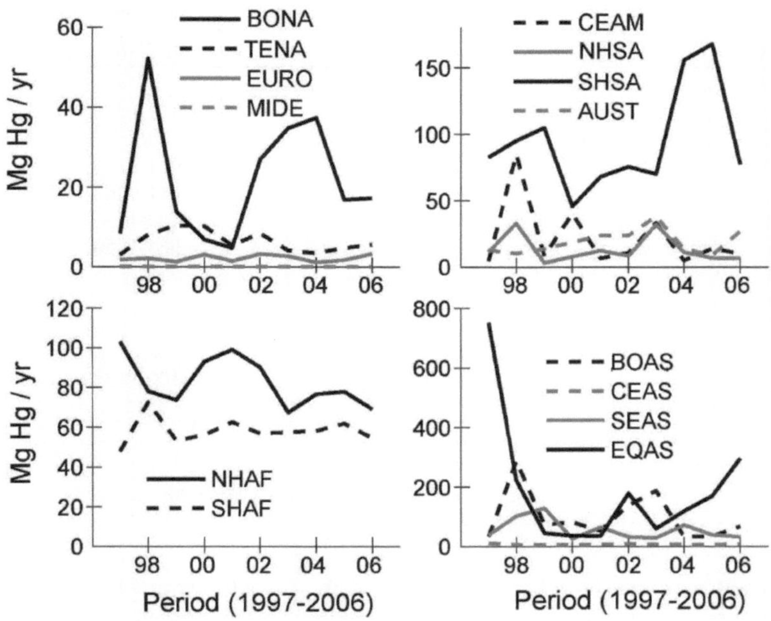

Figure 8.6 Annual mercury emissions for 1997-2006 (see map for region description)

On a regional scale, the interannual variability is significantly larger. This impact can be seen for example from fires in boreal North America, where large amounts of mercury were released during the extreme fire seasons of 1997-1998 and 2003-2004, relative to non-major fire years such as during 2000 (Turetsky et al., 2006). This is consistent with observations of other trace gases like CO, where it was observed that fires in Alaska contributed to poorer air quality in continental U.S (e.g. Pfister et al., 2004).

8.2.5 *Global Hg Emissions using Global CO Emission Estimates*

An alternative approach to estimating mercury emissions from biomass burning is to use data-constrained biomass burning inventories for trace gases. CO is highly correlated with mercury in smoke plumes from biomass burning (e.g. Friedli et al., 2003b). Inventories of CO from biomass burning have been compiled based on carbon emission models and observed emission factors (bottom-up approach). More recently, the availability of CO ground-based, airborne, and remotely-sensed measurements provided stronger constraints for CO sources (top-down approach), particularly CO from biomass burning which exhibits a large spatio-temporal variability.

Using a global average emission factor for mercury, as described in section 2.7, we show in Table 8.6 our global estimates of mercury emissions based on various

Table 8.6 Global Hg emissions based on global CO emission estimates

	CO^{a}*(Tg yr^{-1})*	Hg^{b}*(Mg yr^{-1})*
Bottom-up Approach		
Andreae and Merlet, 2001	465	1023
Brunke et al. 2001	612	1346
Duncan et al. 2003	437	961
Ito and Penner, 2004	264-421	581-926
Hoelzemann et al. 2004	202-571	444-1256
Jain et al. 2006	438-568	964-1250
van der Werf et al. 2006	433	953
Top-down Approach		
Bergamaschi et al. 2000	722	1588
Petron et al. 2004[c]	322	708
Arellano et al. 2006[c]	501-563	1002-1239
Muller and Stavrakou, 2005[c]	359	790

[a]climatological estimate
[b]using an average EF of 220 μg Hg/kg fuel (see Table 8.1)
[c]uses CO coentration (or burned area) data for the year 2000

CO inventories. We note here that some of the inventories are based on carbon emission models, but are constrained and updated by other independent datasets to better match atmospheric CO concentrations. Even so, these inventories have associated uncertainties, such as representation of CO transport in global chemical transport models, which need to be accounted for. Estimates for global mercury emissions range from (708 to 1346 Mg Hg yr^{-1}), which are higher than the 675±240 Mg yr^{-1} estimates derived from the GFEDv2 model and the selected EF(Hg), confirming uncertainty and variability in the emissions. While this approach considers a global emission factor for mercury along with a global CO emission estimate (and oversimplifies the spatio-temporal heterogeneity), this serves as independent general assessment of global mercury emission estimates and its uncertainties.

8.2.6 Comparison with Other Regional Emission Estimates

The essence of this work is the application of a globally consistent model to build up fuel pools and to estimate the carbon release resulting from combustion, to follow the process by remote sensing techniques and combine the carbon emissions with mercury emission factors, EF(Hg), to estimate global mercury emissions. Our model approach has limitation in fire detection and carbon emission uncertainties and does not always reflect regional fire practices, e.g. differences between wildfire and slash and burn fires. At this stage of sophistication, it would be unreasonable to expect full agreement with regionally collected estimates. However, global estimates appear to be reasonable as indicated by the fact that carbon and CO based approaches give similar results. In the following paragraph we compare results from the GFEDv2 model and literature values for the same regions. Shown on Table 8.7 are the key parameters used in our estimates and those from the literature.

Table 8.7 Comparison of estimates of carbon (C) and mercury (Hg) emissions with literature

Regions	Hg (*Mg yr*$^{-1}$)	C (*Tg yr*$^{-1}$)	Burned Area (*Mha yr*$^{-1}$)	Fuel Burned (*kg m-2) dm (normal)*	Effective EF *µg Hg kg*$^{-1}$ *dm (normal)*
Mediterranean[1]					
This work	2.30	9.9	1.87	1.2	104
Cinnirella et al. (2008)	4.3	17.3	0.3	12.4	112
Russia[2]					
This work	100 ± 83	177 ± 124	11 ± 4.9	3.54	254
Conard & Davidenko (1996)	(63)	82.1	7.3	2.5	345
Lavoué et al. (2000)	13.4-31.0	40.5	3.6	2.5	124-345
Cinnirella & Pirrone (2006)*	13.3 ± 10.5	52.92	2.1 ± 1.7	5.6	112
Cinnirella & Pirrone (2006)#	16.1 ± 7.3	115.83	3.9 ± 1.8	6.6	112
China[3]					
This work	2.22 ± 0.58	7.8 ± 1.5	1.9 ± 0.5	0.93	127
Streets et al. (2005)	10.9	82.1			60
USA[4]					
This work	7.2 ± 2.2	26 ± 11	2.33 ± 0.7	2.5	123
Wiedinmyer & Friedli (2007)	20.3 (43)	40.1	3.7	2.4	228
Amazon[5]					
This work	108			5.4	145-157
Michelazzo et al. (2008)	8.7-90			10.6	50/61

[1]for year 2006, with regions as described by Cinnirella et al. (2008)
[2]average for 1997-2006, with regions as described by Cinnirella and Pirrone, (2006)
[3]average across 1997-2006, with regions as described by Streets et al. (2003)
[4]average across 2002-2006, with regions as described by Wiedinmyer and Friedli (2007)
[5]average across 1997-2006, with SHSA region as defined in Figure 8.4
#from ground-based data for 1996-2001
*from remote-sensing data for 1996-2002

Mediterranean: The estimate from Cinnirella et al. (2008) for mercury release from the Mediterranean region is 4.3 Mg for the year 2006. This is about 80% higher than our estimate of 2.3 Mg for the same period. The difference is proportional to the ratio of in carbon emissions from the two estimates. The carbon emission value of Cinnirella et al. (2008) is the result of very low burn areas and high values for fuel burned m^{-2} which yield compensated values for carbon emissions. Here is a case where burn areas and fuel burned vary by factors 6-10, while the mercury emissions are within a factor of <2 because the EF(Hg) used are closely similar, 112 versus 104.

Russian Federation: Here the comparison is complicated because burned area (Mha yr^{-1}), fuel burned (kg m^{-2}) vary significantly, and different EF(Hg) have been used. The value given in parenthesis for Conard and Davidenko (1996), 63 Mg, is a calculated value based on the boreal forest EF(Hg) of 345 µg Hg/kg (dm). Recently, Sukhinin et al. (2004) reported a burned area estimate, similar to Conard and Davidenko (1996) for the period 1995-1997. They indicated that their estimate is still conservative given the shortcomings of a satellite-based approach to observe burned scars. Yet, their estimate is a factor of 2-5 higher than the reports from the Russia Federal Forest Service (RFFS), which Sukhinin et al. (2004) suggested to be an underestimate due to limited regions monitored by RFFS (i.e. there are uncontrolled and undocumented fires in unprotected zones). These regions have been decreasing in recent years due to fire suppression. As mentioned in the previous section, the GFEDv2 burned area is higher than Sukhinin et al. (2004). It is however, significantly lower relative to ground-based and satellite-based estimates by Cinnirella and Pirrone (2006) and climatological estimate by Lavoué et al. (2000). We note that the period of the data comparison is slightly different, and may in part, account for the discrepancies. The rest of the difference can be attributed to differences in fuel burned and EF(Hg)'s, which in some cases, have compensating effects to the estimates in Hg emissions.

China: There is a discrepancy between our and the Streets and al. (2003, 2005) assessments because of a 10 fold difference in burn areas for forest and grassland fires as well as burning of crop residues and agricultural wastes. Here is a case where methodology limitations may play the decisive role: the GFEDv2 model may underestimate significantly the extensive and frequent crop residue fires. However, the burn areas for forest and grassland fires by themselves are also in doubt. Yan et al. (2006) claimed that Streets et al. overestimated burn areas taken from 1950-1992 by a factor of 10. They reported that the burned area in recent decades had decreased dramatically relative to the averaging period (1950-1992) used by Streets et al (2003) (as much as a factor 10) due in part to fire suppression. We note that the bulk of the estimate in carbon as well as in mercury in China is attributed to crop residue burning rather than forest and grassland fires. This highlights the importance of applying appropriate EFs and points out the limitation of GFEDv2 (and other satellite-derived estimates) to detect small-scale biomass burning.

USA: In this case, the inputs for the two studies, GFEDv2 and that used by Wiedinmyer and Friedli (2007), i.e. burn area, fuel burned, are in reasonable agreement and lead to correspondingly reasonable agreements in carbon emissions. Discounting the difference in EF(Hg) used in the two calculations, there is still a factor of two difference in mercury emission, which must be attributed to the model assumption differences, not yet understood. The average mercury emission calculated by the two models are 20.3 and 43 (range 20-65) Mg Hg yr^{-1}, much larger than the 7.2 ± 2.2 Mg Hg yr^{-1} calculated with GFEDv2 model.

Amazon: This is an example where the application of GFEDv2 may be of limited use because of poor fire detection and burn area assessment for small-scale fires, smoke obscuration and incorrect assumptions about fuel consumption. In the slash and burn operations as practiced in the Amazon the amount of fuel burned is much higher, e.g. 10.6 kg m^{-2} as reported by Michelazzo et al. (2008), compared to 5.4 kg m^{-2} mean fuel consumption assumed in the GFEDv2 model for SHSA. The experimental EF(Hg)'s are 61.0 and 50.4 μg Hg kg^{-1} for 2005 and 2004, compared to the 1997-2006 averages of 145-157 μg Hg kg^{-1} estimated for the GFEDv2 model. Earlier estimates for SA slash and burn mercury emissions range from 8.7 to 90 Mg Hg yr^{-1}, compared to our estimate of 108 Mg Hg yr^{-1} for all of NHSA and SHSA.

8.3 Future Work

The highest priority for additional research is the reduction of uncertainties, both in carbon emissions and EF(Hg). In the carbon emission model all uncertainty sources, i.e. fire detection and burn area measurement, inclusion of all affected fuels in all biomes, and combustion completion for all fuels are to be addressed. A specific concern is for regions that are inadequately described in terms of vegetation speciation and fire behaviour, or exhibit unusually large uncertainties. A new opportunity is the inclusion of the fire dynamics data becoming available from satellite measurements (Roberts and Wooster, 2007). Validation with ground based statistics needs to continue, and the advances in remote sensing promise further progress and insights. Emission models are indispensable for unbiased global comparisons, although local areas may not always be correctly represented in a global model.

The EF(Hg) are more poorly defined than carbon emission uncertainties. Refinement in measuring techniques, conceptual understanding, and above all, more regional measurements are needed. Most available data is from Europe and North America, both minor contributors to global mercury emissions from biomass burning. EF(Hg) from the large carbon emitters, NHAF, SHAF, EQAS, SHAM and BOAS, are very sparse or non-existing. New data should include ground and/or plume measurements using ER or EF determinations. One area poorly understood is the effect of fire dynamics on mercury release in different biomes: e.g. mercury speciation in plumes burning in different biomes under flaming and smouldering conditions. A comprehensive biogeochemical model for mercury in forested areas would provide an understanding of the source/sink balance and thus mercury accumulation or loss in an ecosystem. Such models could then be coupled with carbon emission models and become components of an earth system model. It would also be useful to project climate change impacts. The interaction of fire with fuels and the consequences for mercury release must be better understood. This is particularly true because the mercury and carbon distribution is dramatically different among ecosystems. One of major unknowns is the emission expectations for forests with and without large carbon and mercury reservoirs. For Mediterranean vegetation, savannas, grass lands and agricultural waste, the assessment of above ground fuels by remote or ground-based measurements likely is sufficient to includes all fuel involved in mercury emission. By contrast, in temperate and boreal forests, most mercury is contained in organic soils, which dominate the mercury emissions to variable degrees, depending on fire severity. The role of the top few cm of soil in different landscapes is of paramount importance to mercury emission behaviour. Other possible options to obtain EF(Hg) result from the strong research interest and available data on particulate emissions from biomass burning: i.e. PM2.5, total carbon, organic carbon, total particulate matter, which, combined with mercury assays of the particulates, can yield independent EF(Hg). The fate and transport of emitted mercury is difficult to define because of the regionally different injection heights which lead to unique plume trajectories and associated chemistries and deposition. Some case studies for effluents from the major burn regions, e.g. Africa, Southeast Asia or Siberia, would be useful.

8.4 Policy Implications

Mercury in vegetation and organic soils originates largely from the deposition from the global atmospheric pool and thus must be of global concern. The release of mercury from biomass burning is partially under direct human control. Limiting the burning of tropical and boreal forests (EQAS, SHSA, BOAS) would have two beneficial effects: reducing the source of mercury releases to the atmosphere from burning, and maintaining a sink for atmospheric mercury in the vegetation and organic soil. Restricting the global release of anthropogenic mercury over time would reduce the atmospheric and vegetation/soil pools and thus the release potential in case of fires. Warming as a result of climate change will be felt particularly in boreal forests (Randerson et al., 2006), which harbour huge carbon and mercury pools, and may experience more frequent, larger and more severe wildfires.

Acknowledgments We would like to thank James T. Randerson of University of California, Irvine, Guido R. van der Werf of Vrije Universiteit Amsterdam, Louis Giglio of Science Systems and Applications, Inc., Maryland, G. James Collatz of NASA Goddard Space Flight Center, Maryland and Prasad S. Kasibhatla of Duke University for GFEDv2 emission data and Christine Wiedinmyer and Gabriele Pfister for valuable reviews of the manuscript. H. Friedli and A. Arellano are funded by the National Center for Atmospheric Research, which is sponsored by the National Science Foundation. Nicola Pirrone and Sergio Cinnirella would like to acknowledge the contribution of the Ministry of Environment for its support.

References

Amiro, B. D., Todd, J. B., Wotton, B. M., Logan, K. A., Fannigan, M. D., Stocks, B. J., Mason, J. A., Martell, D. L. and Hirsch, K. G., 2001. Direct carbon emissions from Canadian forest fires, 1959 – 1999. *Can. J. For. Res.*, 31, 512–525.

Andreae, M.O. and Merlet, P., 2001. Emission of trace gases and aerosols from biomass burning. *Global Biogeochem. Cyc.*, 15 (4): 955–96.

Arellano, A.F., Kasibhatla, P.S., Giglio, L., van der Werf, G.R., Randerson, J.T. and Collatz, G.J., 2005. Time-dependent inversion estimates of global biomass-burning CO emissions using Measurement of Pollution in the Troposphere (MOPITT) measurements, *J. Geophys. Res.*, 111, D09303, doi:10.1029/2005JD006613.

Barbosa, P.M., Stroppiana, D., Grégoire, J.-M. and Pereira, J.M.C., 1999. An assessment of vegetation fire in Africa (1981-1991): Burned areas, burned biomass, and atmospheric emissions, *Global Biogeochem. Cycles.*, 13(4), 933–950.

Berenfield, M.J., Randerson, J.T., McClain, C.R., Feldman, G.C., Los, S.O., Tucker, C.J., Falkowski, P.G., Field, C.B. and Frouin, R., 2001. Biosphere primary production during an ENSO transition, *Science*, 291 (5513), 2594–2597.

Bergamaschi, P., Hein, R., Heimann, M. and Crutzen, P.J., 2000. Inverse modeling of the global CO cycle: 1. Inversion of CO mixing ratios, *J. Geophys. Res.*, 105, 1909–1927.

Bishop, J.K.B. and Rosswo, W.B., 1991. Spatial and temporal variability of global surface solar irradiance, *J. Geophys. Res.*, 96 (C9), 16839–16858.

Biswas, A., Blum J., Keeler, J., 2006. A comparison of methods to estimate mercury emissions during wildfire, *8th International* Congress on Mercury as a Global Pollutant, T-125.

Biswas, A., Blum J., Keeler, J., 2008. Mercury storage in a central Washington forest and release during the 2001 Rex Creel Fire. STOTEN-D-07-011986. In review

Biswas, A., Blum, J.D., Klaue, B., Keeler, G.J., 2007. Release of mercury from Rocky Mountain forest fires. Global Biogeochem. Cycles, 21, GB1002; doi:10.1029/2006GB002696.

Brunke, E.-G., Labuschagne, C. and Slemr, F. Gaseous mercury emissions from a fire in the Cape Peninsula, South Africa, during January 2000, *Geophys. Res. Lett.*, 28, 1483–1486, 2001.

Carvalho, J.J., Costa, F.S., Gurgel Veras, C.A., Sandberg, D.V., Alvarado, E.C., Serra, A.M. and Santos, J.M. Biomass fire consumption and carbon release rates of rainforest-clearing experimens conducted in northern Mato Grosso, Brazil, *J. Geophys. Res.*, 106(16), 17877–17887, 2001.

Cinnirella, S. and Pirrone, N. Spatial and temporal distribution of mercury emissions from forest fires in Mediterranean region and Russian federation. *Atmos. Environ.*, 40, 7346–7361, 2006.

Cinnirella, S., Pirrone, N., Allegrini, A., Guglietta, D., 2008. Modeling mercury emissions from forest fires in the Mediterranean region, *Environmental Fluid Mechanics*, 8: 129–145.

Conard, S.G. and Davidenko, E.P. Fire in Siberian Forests- Implications for global Climate and Air Quality. USDA Forest Service Gen. Tech. Rep.PSW-GTR-166, 87–94, 1996.

Driscoll, C., Bushey, J T., Nallana, A.G., Selvendiran, P., Choi, H.-Y. and Holsen T.M. Atmosphere-Land Dynamics of Mercury in a Forest Landscape of the Adirondack Region Page of New York, http://nadp.sws.uiuc.edu/meetings/fall2007/post/6-mercury/driscoll.pdf

Duncan, B.N., Martin, R.V., Staudt, A.C., Yevich, R. and Logan, J.L. Interannual and seasonal variability of biomass burning emissions constrained by satellite observations, *J. Geophys. Res.*, 108(D2), 4100, doi:10.1029/2002JD002378, 2003.

Ebinghaus, R., Slemr, F., Brenninkmeijer, C.A.M., vanVelthoven, P., Zahn, A., Hermann, M., Sullivan, D.A. and Oram, D.E. Emission of gaseous mercury from biomass burning in South America in 2005 observed during CARIBIC flights, *Geophys. Res. Lett.*, 34, L08813; doi:10.1029/2006GL028866, 2007.

Engle, M.A., Sexauer Gustin, M., Johnson, D.W., Murphy, J.F., Miller, W.W., Walker, R.F., Wright, J., Markee, M. Mercury distribution in two Sierran forest and one desert sagebrush steppe ecosystems and the effects of fire, *Science of the Total Environment*, 367, 222–233, 2006.

Erickson, J.A., Gustin, M.S., Schorran, D.E., Johnson, D.W., Lindberg, S.E. and Coleman, J.S. Accumulation of atmospheric mercury in forest foliage, *Atmos. Environ.*, 37, 1613–1622, 2003.

Fay, L. and Gustin, M. Assessing the influence of different atmospheric and soil mercury concentrations on foliar mercury concentrations in a controlled environment, *Water, Air and Soil Poll.* 181, 373–384, 2007, doi10.1007/s11270-006-9308-6.

French, N.H.F., Goovaerts, P., Kasischke, E. Uncertainty in estimating carbon emissions from boreal forest fires. *J. Geophys. Res.*, 109, D14S08; doi:10.1029/2003JD003635, 2004.

Frescholtz, T.F., Gustin, M.S., Schorran, D.E. and Fernandez, G.C.J. Assessing the source of mercury in foliar tissue of quaking aspen, *Environ. Toxicol. Chem.*, 22(9), 2114–2119, 2003.

Friedli, H.R., Radke, L.F., Payne, N.J., McRae, D.J., Lynham, T.J. and Blake, T.W. Mercury in vegetation and organic soil at an upland boreal forest site in Prince Albert National Park, Saskatchewan, Canada. *J. Geophys. Res.*, 112, G01004; doi:10.1029/2005JG000061, 2007.

Friedli, H.R., Radke, L.F., Lu, J.Y., Banic, C.M., Leaitch, W.R. and MacPherson, J.I. Mercury emissions from burning of biomass from temperate North American forests: Laboratory and airborne measurements, *Atmos. Environ.*, 37, 253–267, 2003a.

Friedli, H.R., Radke, L.F., Prescott, R., Hobbs, P.V. and Sinha, P. Mercury emissions from the August 2001 wildfires in Washington State and an agricultural waste fire in Oregon, and atmospheric mercury budget estimates, *Global Biogeochem. Cycles*, 17(2), 1039, doi:10.1029/2002GB001972, 2003b.

Friedli, H.R., Radke, L.F., Lu, J.Y. Mercury in smoke from biomass fires. *Geophys. Res. Lett.* 28 (17), 3223–3226, 2001.

Friedli, H.R. et al. Mercury emissions from experimental burns of Southern African vegetation, unpublished MS, 2008.

Fromm, M.D. and Servranckx, R. Transport of forest fire smoke above the tropopause by supercell convection, *Geophys. Res. Lett.*, 30(10), 1542, doi:10.1029/2002GL016820, 2003.

Giglio, L., van der Werf, G.R., Randerson, J.T., Giglio, L., Collatz, G.J. and Kasibhatla, P.S. Global estimation of burned area using MODIS active fire observations, *Atmos. Chem. Phys.*, 6, 957–974, 2006.

Grégoire, J.-M., Tansey, K. and Silva, J.M.N. The GBA2000 initiative: Developing a global burned area database from SPOT-VEGETATION imagery, *Int. J. Remote Sensing*, 24(6), 1369–1376, 2002.

Grigal, D.F. Mercury sequestration in forests and peatlands: A review. *J. Environ. Qual.*, 32, 393–405, 2003.

Grigal, D.F. Inputs and output of mercury from terrestrial watersheds: a review. *Environ. Rev.* 10, 1–39, 2002.

Guild, L.S., Kauffman, J.B., Ellingson, L.J., Cummings, D.L., Castro, E.A., Babbitt, R.E. and Ward, D.E. Dynamics associated with total aboveground biomass, C, nutrient pools, and biomass burning of primary forest and pasture in Rondonia, Brazil during SCAR-B, *J. Geophys. Res.*, 103 (D24), 32091–32100, 1998.

Gustin, M.S., Lindberg, S.E., Wesiberg, P. An update of our understanding of the role of sources and sinks of the biogeochemical cycle of mercury, 2008, Applied Geochemistry, in press

Hao, W.M. and Liu, M.H. Spatial and temporal distribution of tropical biomass burning, *Global Biogeochem. Cyc.*, 8(4), 495–503, 1994.

Harden, J.W., Neff, J.C., Sandberg, D.V., Turetsky, M.R., Ottmar, R., Gleixner, G., Fries, T.L. and Manies, K.L. Chemistry of burning the forest floor during the FROSTFIRE experimental burn, interior Alaska,1999. *Global Biogeochem. Cycles*, 18, GB3014; doi: 10.1029/2003GB002194, 2004.

Harris, R.C. et al. Whole-ecosystem study shows rapid fish-mercury response to changes in mercury deposition.www.pnas.org/doi:10.1073/pnas.0704186104, 2007

H·ly, C., Alleaume, S., Swap, R.J., Shugart, H.H. and Justices, C.O. SAFARI-2000 characteristics of fuels, fire behavior, combustion completeness, and emissions from experimental burns in infertile grass savannas in western Zambia, *J. Arid Environ*, 54, 381–394, 2003.

Hobbs, P.V., Reid, J.S., Herring, J.A., Nance, J.D., Weiss, R.E., Ross, J.L., Hegg, D.A., Ottmar, R.D. and Liousse, C. Particles and trace gas measurements in smoke from prescribed burns of forest products in the Pacific Northwest, in Biomass Burning and Global Change, vol 1, edited by J.S. Levine, pp. 697–715, MIT Press, Cambridge, Mass., 1996.

Hoelzemann, J.J., Schultz, M.G., Brasseur, G.P., Granier, C. and Simon, M. Global Wildland Fire Emissions Model (GWEM): Evaluating the use of global area burnt satellite data, *J. Geophys. Res.*, 109(D14S04), doi:10.1029/2003JD003666, 2004.

Hoffa, E.A., Ward, D.E., Olbu, G.J. and Baker, S.P. Emissions of CO2, CO, and hydrocarbons from fires in diverse African savanna ecosystems, *J. Geophys. Res.*, 104(D11), 13841–13853, 1999.

Ito, A. and Penner, J.E. Global estimates of biomass burning emissions based on satellite imagery for the year 2000, *J. Geophys. Res.*, 109, D14S05, doi:10.1029/2003JD004423, 2004.

Jain, A.K., Tao, Z., Yang, X. and Gillespie, C. Estimates of global biomass burning for reactive greenhouse gases (CO, NMHCs, and NOx) and CO2, *J. Geophys. Res.*, 111(D06304), doi:10.1029/2005JD006237, 2006.

Kasischke, E.S., Hyer, E.J., Novelli, P.C., Bruhwiler, L.P., French, N.J.F., Sukhinin, A.I., Hewson, J.H. and Stocks, B.J. Influences of boreal fire emissions on Northern Hemisphere atmospheric carbon and carbon monoxide, *Global Biogeochem. Cyc.*, 19, GB1012, doi:10.1029/2004GB002300, 2005.

Kasischke, E.S. and Bruhwiler, L. Emissions of carbon dioxide, carbon monoxide, and methane from boreal forest fires in 1998, *J. Geophys. Res.*, 108 (D1), 8146, doi:10.1029/ 2001JD000461, 2002.

Kasischke, E.S. and Penner, J.E. Improving global emissions of atmospheric emissions form biomass burning, *J. Geophys. Res.*, 109 (D14S01), doi:10.1029/2004JD004972, 2004.

Lavoué, D., Liousse, C., Cachier, H., Stocks, B.J., and Goldammer, J.G. Modeling of carbonaceous particles emitted by boreal and temperate wildfires at northern latitudes, *J. Geophys. Res.*, 105, 26871–26890, 2000.

Lindberg, S.E. Forests and the global biogeochemical cycle of mercury, in *Global and Regional Mercury Cycles: Sources, Fluxes and Mass Balances*, NATO-ASI Ser., vol. 21, edited by W. Baeyens, et al., pp. 359–380, Springer, New York, 1996.

Michelazzo, P.A.M; Fostier, A.H; Magarell, G., Santos, J.C., Carvalho, J.A. Jr. Mercury emissions from forest burning in the region of Alta Floresta. Submitted to Sci. Tot. Environ., 2008.

Müller, J-F. and Stavrakou, T. Inversion of CO and NOx emissions using the adjoint of the IMAGES model, *Atmos. Chem. Phys.*, 5, 1157–1186, 2005.

Obrist, D., Moosmueller, H., Schuermann, R., Antony Chen, L.-w, and Kreidenweis, S.M. Particulate_Phase and Gaseous Elementary Mercury Speciation in Biomass Combustion: Controlling Factors and Correlation with Particulate Matter Emission. Environ. Sci. Technol., 10.1021/es071279n, 2007.

Pétron, G., Grainer, C., Khattatov, B., Yudin, V., Lamarque, J-F., Emmons, L., Gille, J. and Edwards, D.P. Monthly CO surface sources inventory based on the 2000-2001 MOPITT satellite data, *Geophys. Res. Lett.*, 31, L21107, doi:10.1029/2004GL020560, 2004.

Pfister, G., Emmons, L.K., Hess, P.G., Honrath, R.E., Lamarque, J.-F., ValMartin, M., Owen, R.C., Avery, M., Browell, E., Holloway, J., Nedelec, P., Purvis, R., Ryerson, T., Sachse, G., and Schlager, H. (2006), Ozone Production from the 2004 North American Boreal Fires, *J. Geophys. Res.*, 111, D24S07, doi:10.1029/2006JD007695.

Potter, C.S., Randerson, J.T., Field, C.B., Matson, P.A., Vitousek, P.M., Mooney, H.A. and Klooster, S.A. Terrestial Ecosystem Production – A process model based on global satellite and surface data, Global Biogeochem. *Cycles.* 7(4), 811–841, 1993.

Randerson, J.T., Liu, H., Flanner, M.G., Chambers, S.G., Jin, Y., Hess, P.G., Pfister, G., Mack, M.C., Treseder, K.K., Welp, L.R., Chapin, F.S., Harden, J.W., Goulden, M.L., Lyons, E., Neff, J.C., Schuur, E.A.G., and Zender, C.S. The impact of boreal forest fire on climate warming, *Science*. 314, 1130–1132. 2006.

Roberts, G. and Wooster, M.J. New perspectives on African biomass burning dynamics. *EOS*, 88 (18), 369–370, 2007.

Streets, D.G., Yarber, K.F., Woo, J-H., Carmichael, G.R. Biomass burning in Asia: annual and seasonal estimates and atmospheric emissions. *Global Biogeochem. Cycles.* 17(4) 1099, 2003.

Streets, D.G., Hao, J., Wu, Y., Jiang, J., Chan, M., Tian, H., Feng, X. Anthropogenic mercury emissions in China. *Atmos. Environ.* 39, 7789–7806, 2005.

Seigneur, C., Vijayaraghavan, K., Lohman, K., Karamchandani, P., Scott, C. Global source attribution for mercury deposition in the United States, *Environ. Sci. Technol.*, 38, 555–569, 2004.

Seiler, W. and Crutzen, P.J. Estimates of gross and net fluxes of carbon between the biosphere and the atmosphere from biomass burning, *Clim. Change*, 2(3), 207–247, 1980.

Shea, R.W., Shea, B.W., Kauffman, J.B., Ward, D.E., Haskins, C.I. and Scholes, M.C. Fuel biomass and combustion factors associated with fires in savanna ecosystem of South Africa and Zambia, *J. Geophys. Res.*, 101 (D19), 23551–23568, 1996.

Sigler, J.M., Lee, X. and Munger, W. Emission and long-range transport of gaseous mercury from a large-scale Canadian boreal forest fire. *Environ. Sci. Technol.*, 37, 4343–4347, 2003.

Simon, M.S., Plummer, S., Fierens, F., Hoelzemann, J.J. and Arino, O. Burnt area detection at global scale using ATSR-2: The GLOBSCAR products and their qualification, J. Geophys. Res., 109 (D14S02), doi:10.1029/2003JD003622, 2004.

Skyllberg, U., Qian, J., Frech, W., Xia, K., and Bleam, W.F. Distribution of mercury, methyl mercury and organic sulfur species in soil, soil solution and stream of a boreal forest catchment. *Biogeochemistry*, 64, 53–76, 2003.

St. Louis, V.L., Rudd, W.M., Kelly, C.A., Hall, B.D. Rolfhus, K.R., Scott, K.J., Lindberg, S.E. and Dong, W. Importance of the forest canopy to fluxes of methyl mercury and total mercury to a boreal ecosystem, *Environ. Sci. Technol.*, 35, 3089–3098, 2001.

Sukhinin, A.I., French, N.H.F., Kasischke, E.S., Hewson, J.H., Soja, A.J., Csiszar, I.A., Hyer, E.J., Loboda, T., Conrad, S.G., Romasko, V.I., Pavlichenko, E.A., Miskiv, S.I. and Slinkina, O.A. AVHRR-based mapping of fires in Russia: New products for fire management and carbon cycle studies, *Rem. Sens. Environ.*, 96(2), 188–201, 2004.

Turetsky, M.R., Harden, J.W., Friedli, H.R., Flannigan, M., Payne, N., Crook, J., Radke, L.F. Wildfires threaten mercury stocks in northern soils. *Geophy. Res. Lett.*, 33, L16403, doi:10.1029/2005GL025595, 2006.

van der Werf, G.R., Randerson, J.T., Giglio, L., Collatz, G.J., Kasibhatla, P.S. and Arellano, A. Interannual variability in biomass burning emissions from 1997 to 2004, *Atmos. Chem. Phys.*, 6, 3423–3441, 2006.

van der Werf, G.R., Randerson, J.T., Collatz, G.J. and Giglio, L. Carbon emissions from fires in tropical and subtropical ecosystems, *Global Change Bio.*, 9(4), 547–562, 2003.

Veiga, M.M., Meech, J.A. and Ornante, N. Mercury pollution from Deforestation. *Nature*, 368, 816–817, 1994.

Wiedinmyer, C., Quayle, B., Geron, C., Belote, A., McKenzie, D., Zhang, X., O'Neill, S. and Wynne, K.K. Estimating emissions from fires in North America for Air Quality Modeling, *Atmos. Environ.*, 40, 3419–3432, 2006.

Wiedinmyer, C. and Friedli, H. Mercury emission estimates from fires: An initial inventory for the United States, *Environ. Sci. Technol.*, 41, 8092–8098, 2007.

Weiss-Penzias, P., Jaffe, D., Swartzendruber, P., Hafner, W., Chand, D. and Prestbo, E. Quantifying Asian and biomass burning sources of mercury using the Hg/CO ratio in pollution plumes observed at the Mount Bachelor Observatory, *Atmos. Environ.*, 41, 4366–4379, 2007.

Witham, C. and Manning, A. Impact of Russian biomass burning on UK air quality, *Atmos. Environ.* 41, 8075–8090, 2007.

Woodruff, L.G., Harden, J.W., Cannon, W.F., Gough, L.P., 2001. Mercury loss from the forest floor during wildland fire. American Geophysical Union, Fall meeting, abstract # B32B-0117.

Yan, X., Ohara, T. and Akimoto, H., 2006. Bottom-up estimate of biomass burning in mainland China, *Atmos. Environ.*, 40, 5262–5273.

Part II
Spatial Coverage and Temporal Trends of Mercury Measurements

Chapter 9
Spatial Coverage and Temporal Trends of Land-based Atmospheric Mercury Measurements in the Northern and Southern Hemispheres

Ralf Ebinghaus, Catharine Banic, Steve Beauchamp, Dan Jaffe, Hans Herbert Kock, Nicola Pirrone, Laurier Poissant, Francesca Sprovieri, and Peter S. Weiss-Penzias

Summary This chapter presents a review of atmospheric mercury measurements (as total and as speciated mercury) conducted at terrestrial sites during the last decade. A large number of activities have been carried out in different regions of the world aiming to assess the level of mercury in ambient air and precipitation, and its variation over time and with changing meteorological conditions. Recent studies have highlighted that in fast developing countries (i.e., China, India) mercury emissions are increasing in a dramatic fashion due primarily to a sharp increase in energy production from the combustion of coal (Chapter 2 by Street et al.; Chapter 3 by Feng et al. in this report). The large increase in mercury emissions in China over the last decade are not currently reflected in the long-term measurement of total gaseous mercury at Mace Head, Ireland between 1996 to 2006, nor in the precipitation data of the North American Mercury Deposition Network (MDN). There are documented recent increases in the oxidation potential of the atmosphere which might account, at least in part, for the discrepancy between observed gaseous mercury concentrations (steady or decreasing) and global mercury emission inventories (increasing). This chapter provides a detailed overview of atmospheric measurements performed at industrial, remote and rural sites during the last decade with reference to the monitoring techniques and location of monitoring sites in most of the continents.

9.1 Introduction

This chapter provides up to date information of currently available data of mercury concentrations (as total and as speciated mercury) observed at terrestrial sites. As elemental mercury is a semi-volatile contaminant it continuously cycles between the atmosphere, ocean and soil. The biogeochemical cycling can be

N. Pirrone and R. Mason (eds.), *Mercury Fate and Transport in the Global Atmosphere*,
DOI: 10.1007/978-0-387-93958-2_9, © Springer Science + Business Media, LLC 2009

affected by natural and anthropogenic variables and forcing. Mercury is emitted into the atmosphere from a variety of anthropogenic (e.g. power generation facilities, smelters, cement production, waste incineration and many others) (Pirrone et al., 1996; Pirrone et al. 1998; Pirrone et al., 2001) and natural sources (e.g., volcanoes, crustal degassing, oceans) in different chemical and physical forms (Pacyna et al., 2001; Carpi, 1997). Its cycling between different environmental compartments depends on the rate of different chemical and physical mechanisms (i.e., dry deposition, wet scavenging) and meteorological conditions which affect its fate in the global environment. Both source categories, i.e. anthropogenic and natural, contribute to the global atmospheric pool. It has been suggested that due to intensified anthropogenic release of mercury into the atmosphere since the beginning of industrialization this global pool has increased in the past 150 years. Evidence of long-term changes in the atmospheric mercury burden can be derived from chemical analysis of lake sediments, ice cores and peat deposits (Engstrom and Swain, 1997; Bindler et al., 2001; Biester et al., 2002; Lamborg et al., 2002). A growing number of these records from both hemispheres demonstrate about a threefold increase of mercury deposition since pre-industrial times (Lindberg et al., 2007 and references therein).

In principle, an increase in the global atmospheric pool should also be reflected in the background concentration. Since first reliable measurement data were published about 3 decades ago it is extremely difficult to derive a multidecadal global trend estimate based on these spatially and temporally inchoate air concentration data sets. For example, Asian mercury emissions are suggested to be rapidly increasing at least in the past decade however, this is neither reflected in the long-term measurement of TGM at Mace Head, Ireland covering the period between 1996 to 2006, nor in the precipitation data of the North American Mercury Deposition Network (MDN) (Lindberg et al., 2007 and reference therein).

In 1995, Fitzgerald argued for and defined the basic requirements of an Atmospheric Mercury Network (AMNET). This has partly been accomplished on a regional scale within the Canadian Atmospheric Mercury Network (CAMNet) that may be considered as seminal in this respect. Nevertheless, although atmospheric Hg monitoring stations have increased, the database is sparse, especially in remote locations. Fully aware of these constraints, Slemr et al. (2003) attempted to reconstruct the worldwide trend of atmospheric Hg (TGM) concentrations from long-term measurements of known documented quality at 6 sites in the Northern Hemisphere, 2 sites in the Southern Hemisphere, and multiple ship cruises over the Atlantic Ocean made since 1977. The authors interpreted this information to suggest that the TGM concentrations in the global atmosphere had been increasing since the first measurements in 1977 to a maximum in the late 1980s, after which Hg concentrations decreased to a minimum in 1996 and then remained constant at a level of about 1.7 ng m^{-3} in the Northern Hemisphere. It was also hypothesized that the observed temporal profile was primarily the result of the trends in global Hg use, supply, and emissions.

Lindberg et al. (2007) have pointed out a number of reasons to support the null hypothesis (i.e., there has been little change in TGM since 1977). If one particular monitoring station would be excluded from the evaluation then the data suggest there has been little change in TGM levels in the atmosphere between 1977 and 2002. Additional support for the null hypothesis is provided by TGM measurements for the Southern Hemisphere. TGM results for the Southern Hemisphere do not suggest that there has been much change in TGM levels in the global remote atmosphere over the past 25–30 years. Although it may appear that these competing hypotheses on atmospheric TGM levels in recent times would be disconcerting, this situation is not unusual and often aids the development of research strategies. For example, the value of long-term atmospheric Hg monitoring stations and the need for additional sites is obvious, especially in the remote Southern Hemisphere (Lindberg et al., 2007).

General scientific consensus exists about the current global background concentration that refers to the average sea-level atmospheric Hg^0 at remote sites. The background concentration is currently taken as ca. 1.5 to 1.7 ng m^{-3} in the Northern Hemisphere and ca. 1.1 to 1.3 ng m^{-3} in the Southern Hemisphere (Lindberg et al., 2007).

9.1.1 Quality of Data / Field Intercomparisons

Field intercomparisons of atmospheric mercury measurements have been carried out at different locations and with different objectives:

- an intercomparison at Windsor, ON, an urban site in Canada mainly focussing on the comparability of classical manual methods with newly available automated analysers (Schroeder et al., 1995)
- an intercomparison at Mace Head, a marine background site in Ireland including a comprehensive set of coeval methods for the analysis of different mercury species in air and precipitation (Ebinghaus et al., 1999)
- an intercomparison at Sassetta, a rural site in Tuscany, Italy mainly focussing on the comparability of novel techniques for atmospheric mercury species (Munthe et al., 2001)

All 3 intercomparison exercises have revealed that the measured concentrations of TGM showed good agreement between the participating laboratories. At Mace Head it was additionally demonstrated that the comparability of total mercury concentrations in precipitation was satisfactory as well.

During the Mace Head and the Sassetta intercomparisons it could furthermore be shown, that much higher differences are involved when atmospheric mercury species (namely RGM and TPM) are measured however, it was also demonstrated that the repeatability of similar methodologies increased over time. One major conclusion derived from these studies is that good agreement of TGM concentrations in air determined with different techniques, including manual techniques dating back to the 1970s, makes a combination of data sets from different regions of the world feasible (Ebinghaus et al., 1999).

9.2 Measurements of Air Concentrations in North America

9.2.1 Measurements of Air Concentrations in Canada

Currently, most measurements of gaseous atmospheric Hg in Canada are made using a Tekran 2537A Mercury Vapour Analyser (Tekran Inc., Toronto). This instrument concentrates gaseous phase Hg by amalgamation onto gold cartridges, with subsequent thermal desorption and detection of Hg^0 by cold vapour atomic fluorescence spectrophotometry (CVAFS). This automated measurement system compares well with other manual methods operating on the same principal, commonly used to measure atmospheric Hg (Ebinghaus et al. 1999). A Standard Operating Procedure for detection of TGM in ambient air is available (Steffen & Schroeder 1999). The unit used in Canada to express the concentration of mercury measured by this technique is nanograms per standard cubic metre of air, abbreviated ng m^{-3}, where the volume referred to is at the standard temperature and pressure of 0 °C and 1 atm. The Tekran analysers are operated between a 5 and 30 minute integrated sampling interval, dependent on the site.

Definitions: Hg^0: Gaseous Elemental Mercury, RGM: Reactive Gaseous Mercury, i.e. non-elemental gas phase mercury, PM: mercury bound to particles, MDN: mercury in precipitation (mercury deposition network, only wet deposition measured at this time). The above is the coarsest level of "speciation" that we currently work with. TGM: Total Gaseous Mercury.

9.2.1.1 Remote Locations

Much of the area of Canada is remote from anthropogenic mercury sources. Several of the network sites, described in Sections 9.2.4 below, are situated in remote locations, called background sites; measurements are made at these sites to characterize the mercury present in the atmospheric environment far from anthropogenic inputs. In addition to the network sites and the far Arctic site in Alert, measurements have been made at the two additional sites in remote locations, Kuujjuarapik and Mingan.

9.2.1.2 Urban Locations (Including Mining Areas)

During a brief study in summer 2000, the levels of TGM were measured at two locations in downtown Toronto. The TGM concentrations ranged from 1.3 to 50 ng m^{-3}, with values >3 ng m^{-3} seen more frequently than in rural areas. Local sources affect the urban sites with short periods of high air Hg concentrations. More frequent observations of high concentration episodes of Hg in precipitation (>20 ng L^{-1}) were found at sites closer to urban centres, e.g., Reifel Island which is close to Vancouver, Egbert in close proximity to Toronto and Barrie, and St. Anicet which is close to Montreal, than at more remote locations, e.g., Mingan, Cormak and Kejimkujik Park . The increased anthropogenic activity associated with highly populated areas may result in increased air concentrations of RGM and PM which can be more readily incorporated into cloud water and precipitation, as will be discussed below. The Banic et al. (2003) sampling flight compared the TGM in the urban plume from Montreal to the ambient concentrations; concentrations in the urban plume reached 2.5 ng m^{-3} compared with the ambient background of 1.5 ng m^{-3}. The urban plume was observed at altitudes up to 1.25 km.

The TGM concentration was monitored at 2 mine tailing sites in eastern Nova Scotia, Canada (Beauchamp et. al. 2002). The Caribou Mines and Goldenville tailing sites are remnants of gold mining activities which began in the mid 1800's and continued up to the late 1930's. The gold bearing ore was brought to the surface and crushed in stamp mills located on site then gold was extracted using the mercury amalgamation process. Mercury contaminated tailings were then disposed of in lakes and ponds which now form flat expanses of tailings with surface areas up to a square kilometre and depths up to several meters. The concentrations of TGM in near surface ambient air (10 cm above the mine tailings) at the Goldenville site remained above 2 ng m^{-3} throughout the diurnal cycle reaching a maximum 8.1 ng m^{-3} with a 24 hour average concentration of 3.5 ng m^{-3}. Gaseous elemental mercury concentrations in ambient air at the same height above the Caribou Mine tailings averaged 8.4 ng m^{-3} with minimum and maximum concentrations of 3.2 and 23.0 ng m^{-3}, respectively. Ambient TGM concentrations over legacy gold mine tailings remained well above regionally representative background ambient air concentrations of 1.5 ng m^{-3}. Additionally, measurements of air-surface exchange of mercury from tailings from legacy gold mine activity in Nova Scotia which used mercury amalgamation processes have shown emissions 2 orders of magnitude higher than those observed from undisturbed sites.

9.2.1.3 Temporal Trends at Single Locations

In an example of a model study of processes the comparison of the measured concentrations and simulated Hg concentrations (GRAHM model from Dastoor & 2004) are shown for the CAMNet site Egbert. In this study the natural emissions and re-emissions of Hg were set to zero and the anthropogenic emissions were set to a constant rate throughout the year. The meteorology and surface characteristics varied with season (i.e., seasonal circulation patterns and the seasonal cycles in the boundary layer heights, clouds, precipitation and dry deposition characteristics). The results show the extent to which seasonal variation in the TGM concentrations can be driven by meteorological differences

9.2.1.4 Monitoring Networks and Trends

The Canadian Atmospheric Mercury Measurement Network (CAMNet, www.msc.ec.gc.ca/ arqp/ camnet_e.cfm) was established in 1996 to provide accurate, long-term measurements of TGM concentration and the Hg deposition in precipitation (wet deposition) across Canada. A map of the CAMNet sites is shown in Figure 9.1.

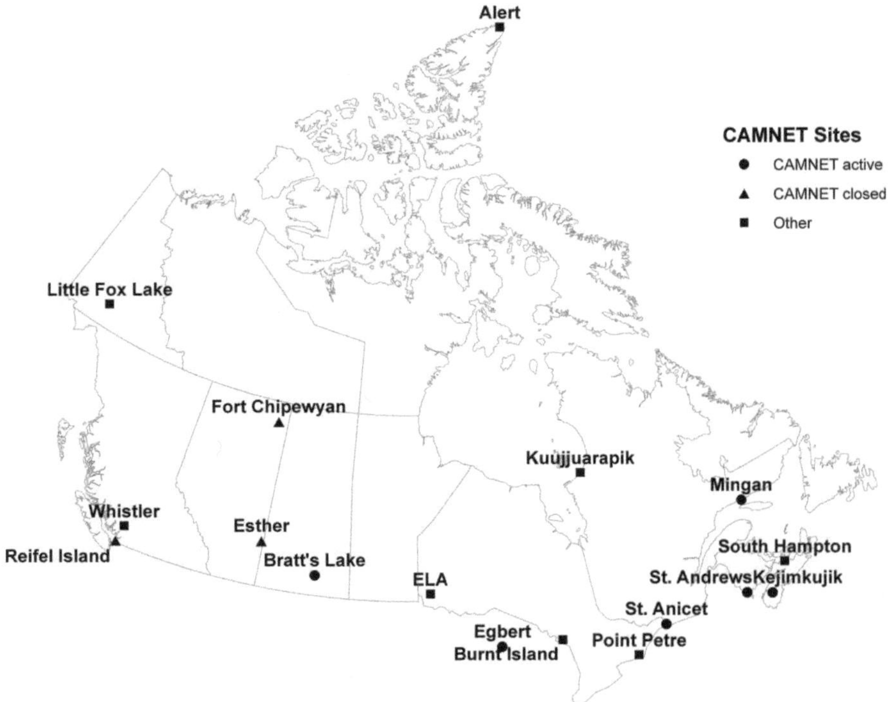

Figure 9.1 Sites in the Canadian Atmospheric Mercury Measurement Network (CAMNet)

The mid-latitude sites will be discussed here, with the Arctic site Alert discussed elsewhere because of different behaviour. The mid-latitude sites are located in background or rural areas, with the latter occasionally impacted by emissions from urban areas. Wet deposition is measured at the CAMNet sites as part of the Mercury Deposition Network (MDN), which includes sites in the United States, Canada and Mexico (http://nadp.sws.uiuc.edu/mdn/). The sampling and analytical methods adhere to the CAMNet Standard Operating Procedure and the USA- National Atmospheric Deposition Network (NADP)-MDN sampling protocol.

An overall average median atmospheric concentration for TGM of 1.60 ± 0.15 ng m^{-3} for the ten Canadian sites was calculated for the years 1997-1999 by averaging together the site medians (Kellerhals et al. 2003). Higher variability of TGM concentrations at the sites in closer proximity to large urban areas appeared to be caused by the alternating exposure of these sites to anthropogenic TGM emissions, depending on wind direction and atmospheric mixing.

Kellerhals et al. (2003) observed that a slight seasonal trend for TGM was seen with higher concentrations observed in winter and spring, and lower concentrations in summer and fall. This is further demonstrated for all years of CAMNet data in Table 9.1. Several factors might contribute to this behaviour (Blanchard et al., 2002), including differences in meteorological conditions and scavenging processes between

Table 9.1 Characteristics of the sampling sites and time periods for the Canadian atmospheric sampling sites

Station	Code	Province	Latitude *(deg)*	Longitude *(deg)*	Altitude (*m* a.s.l.)	Period
Alert	ALT	NU	82.50	-62.33	210	01/95-12/05
Kejimkujik	KEJ	NS	44.43	-65.21	127	01/96-12/04
St. Andrews	STA	NB	45.09	-67.08	80	01/96-12/04
St. Anicet	WBZ	QC	45.12	-74.28	49	01/97-12/05
Point Petre	PPT	ON	43.84	-77.15	75	11/96-12/05
Egbert	EGB	ON	44.23	-79.78	251	12/96-12/05
Burnt Island	BNT	ON	45.81	-82.95	75	05/98-12/05
Bratt's Lake	BRL	SK	50.20	-104.72	577	05/01-12/05
Esther	EST	AB	51.67	-110.20	707	06/98-04/01
Fort Chipewyan	FCH	AB	58.78	-111.12	232	06/00-07/01
Reifel Island	RFL	BO	49.10	-123.17	2	03/99-02/04

a.s.l. above sea level

summer and winter (e.g., reduced mixing heights and higher wind speeds in winter, increased oxidation and larger removal from the atmosphere by wet and dry deposition during warmer months). The seasonal variability observed at the surface will be influenced by changes in the total atmospheric column burden of Hg^0.

Seven of the 10 CAMNet sites experienced a cycle of maximum concentrations near solar noon (1000 to 1400 local standard time (LST)) and minimum concentrations in the early morning hours (0300 to 0700 LST) (Kellerhals et al. 2003). The cycle at the CAMNet sites was attributed to nighttime depletion of TGM in the lowermost atmosphere. Overnight a shallow TGM-depleted layer is formed in the nocturnal inversion layer. Shortly after sunrise there is a rapid increase in near-surface TGM concentration as the nocturnal inversion breaks down and undepleted air is mixed down to the surface. The continued increase in TGM concentration through the morning and into the early afternoon is likely caused by emission of TGM from the surface, which has been observed to increase with increasing solar radiation (Poissant & Casimir 1998).

Long-term monitoring data TGM concentrations from 11 CAMNet sites between 1995 and 2005 were analysed for temporal trends, seasonality and comparability within the network. A statistically significant decreasing trend for TGM concentrations at several rural CAMNet sites was seen for the time period 1995 to 2005 (Table 9.2). The largest declines were observed close to the urban areas of Toronto and Montreal, where levels fell by 17% at Point Petre, and 13% at St. Anicet, respectively. Many of the TGM changes are comparable with the overall trends of total mercury concentrations in precipitation, for similar time periods, at collocated or nearby National Atmospheric Deposition Program's Mercury Deposition Network (NADP-MDN) sites. Results show that these changes are mostly driven by local or regional changes in mercury emissions. Other sites within CAMNet reflect reported changes in hemispherical global background concentrations of airborne mercury, where slight decreases or no statistically significant trend in TGM concentrations exist over the same time period.

Table 9.2 Statistical summary of TGM measurements at CAMNet sites

Station	Days (#)	Mean *(ng m⁻³)*	Median *(ng m⁻³)*	Min *(ng m⁻³)*	Max *(ng m⁻³)*	Lower quartile *(ng m⁻³)*	Upper quartile *(ng m⁻³)*	SD *(ng m⁻³)*
Alert	3603	1.55	1.58	0.03	3.12	1.45	1.73	0.37
Kejimkujik	3168	1.45	1.46	0.54	2.30	1.31	1.59	0.21
St. Andrews	2774	1.42	1.40	0.74	2.46	1.26	1.57	0.23
St.Anicet	3164	1.64	1.60	0.92	16.31	1.44	1.79	0.40
Point Petre	3275	1.78	1.73	0.80	4.26	1.55	1.93	0.34
Egbert	3207	1.67	1.66	0.95	6.90	1.50	1.80	0.27
Burnt Island	2680	1.58	1.58	0.99	2.48	1.43	1.72	0.21
Bratt's Lake	1424	1.53	1.52	0.79	2.68	1.38	1.64	0.24
Esther	878	1.65	1.65	1.19	2.14	1.54	1.75	0.15
Fort Chipewyan	305	1.36	1.35	0.95	1.77	1.28	1.47	0.15
Reifel Island	1642	1.67	1.67	0.91	2.92	1.56	1.79	0.19
Category (median of stations)								
R-W	2612	1.60	1.60	0.91	2.56	1.48	1.71	0.20
R-E	3263	1.43	1.43	0.88	2.09	1.31	1.56	0.19
R-A	3342	1.68	1.67	1.11	2.99	1.52	1.81	0.22
R-C	2680	1.58	1.58	0.99	2.48	1.43	1.72	0.21
ALL	3959	1.58	1.58	0.21	2.75	1.48	1.68	0.17

R-W = RURAL-WEST (RFL, EST, FCH, BRL);
R-E = RURAL-EAST (KEI, STA);
R-A= RURAL-AFFECTED (WBZ, PPT, EGB);
R-C= RURAL-CENTRAL (BNT)

The following statistical analysis was applied using available daily averaged TGM concentrations from all sites. The application of a seasonal decomposition method was used to isolate a long-term systematic trend, regular seasonal effects, perennial irregular (or cyclical) variations and remaining uncertainties. Detailed information about seasonal decomposition can be found in Temme et al. (2004). This method was performed for all sampling sites with a minimum of five complete years of observations. The regression coefficient and the linear slope were tested for significance using the T-test ($p<0.01$) and if $p>0.01$ then the correlation was marked as non-significant. Daily averages from the year 2000 were chosen for Principal Component Analysis (PCA) because that year offered the most complete spatial information, i.e. data from most stations were available for 2000. Bratt's Lake is the only CAMNet site which started its measurements after 2000 and is not included in the PCA. The first two principal components explained more than 50% of the total variance in terms of the spatial differences. The components reflect the seasonality in the annual data (factor 1) and the influence of local sources, which can include anthropogenic, natural or re-emission from water surfaces (factor 2). Analysis isolated the sites expected to be most significantly impacted by nearby sources in the Great Lakes Basin and St. Lawrence River Valley (rural-affected sites). Sites were also divided into eastern (RURAL-EAST) and western sites (RURAL-WEST).

The RURAL-AFFECTED category shows the highest overall median TGM concentration (1.67 ng m^{-3}) and the highest variability of all categories for the entire time period. The three corresponding sites Point Petre, Egbert and St. Anicet also showed substantially higher maximum concentrations (4.26 ng m^{-3}, 6.90 ng m^{-3}, 16.31 ng m^{-3} respectively) than all the other sites. Both the categories RURAL-WEST (RFL, EST, FCH, BRL) and RURAL-AFFECTED (WBZ, PPT, EGB) reveal significantly higher overall median concentrations than the RURAL-EAST (KEJ, STA), RURAL-CENTRAL (BNT).

TGM concentrations at all the CAMNet sites (mean = 1.58 ng m^{-3}) were similar to or slightly lower than those observed at European background sites within a comparable time frame. Seasonal variations of TGM concentrations are observed for all sites. Most sites show higher concentrations in winter and spring, and lower concentrations in summer and fall, which is reflected in the corresponding summer/winter (SUM/WIN) and spring/autumn (SPR/AUT) ratios for each station and category. The time series at Fort Chipewyan does not include enough data to support any conclusions about seasonality. The exception to these findings is observed at Point Petre where the SUM/WIN ratio is > 1; although not statistically significant at Point Petre. The differences between summer and winter median concentration ratios are the highest at the RURAL-EAST (SUM/WIN = 0.88) and RURAL-CENTRAL (SUM/WIN = 0.85) sites where a minimum monthly median TGM concentration was observed in September and a maximum in February.

9.2.1.4.1 Trend Analysis

The seasonal decomposition technique was applied to the original daily averages. The same procedure was applied to all other sites and categories containing a minimum of 5 complete years of observations. For sites (e.g. RURAL-WEST) where less than 5 years of data were available, a simple linear regression was applied to the original time series. The resulting overall and annual rates of change in mercury concentration after seasonal decomposition and linear regression are listed in Table 9.3. Missing data in the time series were replaced by interpolation prior to time series analysis. Therefore the number of days in Table 9.3 often exceeds the number of reported daily averages in Table 9.2. Inter-annual and overall changes are given in ng m^{-3} absolute and as a percentage in respect to the starting intercept, respectively. Statistically significant regression coefficients and slopes, i.e. slope and regression coefficients that were significantly different from 0 ($p<0.01$) are indicated in the last column. Sites are marked with an asterisk when seasonal decomposition was not done or the results were uncertain. Trends for the shorter and sometimes incomplete time series for the individual RURAL-WEST sites can not be determined with the same significance as the other sites and were also marked with an asterisk. This also leads to an uncertain overall change for this category. However these time series were incorporated into the "ALL" category.

From the data considered valid within the constraints of the seasonal decomposition and data coverage, five sites showed significant decreases (between −2.2% and −16.6%) in the TGM concentrations over the corresponding time period. The only exception was the slight positive trend at Kejimkujik where an overall +3.3% change is evident between 1996 and 2004. For the first time since continuous automated

Table 9.3 Results from trend analysis after seasonal decomposition

St.	Start (*dd-mm-yy*)	End (*dd-mm-yy*)	Days (*#*)	Intercept (*ng m*$^{-3}$)	Slope (*ng m*$^{-3}$ *d*$^{-1}$)	Change (*ng m*$^{-3}$ *y*$^{-1}$)	Change (*% year*$^{-1}$)	Change (*ng m*$^{-3}$)	Change (*% overall*)	Signif. ($p<0.01$)
ALT	01-01-1995	31-12-2005	4009	1.58	-0.0000136	-0.0049640	-0.31	-0.05	-3.5	n.s.
KEJ	01-01-1996	31-12-2004	3287	1.43	0.0000144	0.0052560	0.37	0.05	3.3	Yes
STA	01-01-1996	31-12-2004	3288	1.49	-0.0000333	-0.0121545	-0.82	-0.11	-7.4	Yes
WBZ	01 -01-1997	31-12-2005	3286	1.75	-0.0000698	-0.0254770	-1.46	-0.23	-13.1	Yes
PPT	11-05-1996	31-12-2005	3340	2.02	-0.0001000	-0.0365000	-1.81	-0.33	-16.6	Yes
EGB	30-11-1996	31-12-2005	3318	1.69	-0.0000111	-0.0040515	-0.24	-0.04	-2.2	yes
BNT	05-01-1998	31-12-2005	2800	1.62	-0.0000296	-0.0108040	-0.67	-0.08	-5.1	yes
BRL	05-02-2001	31-12-2005	1702							
EST	26-06-1998	22-04-2001	1032							
FCH	17-06-2000	19-07-2001	398							
RFL	03-03-1999	02-11-2004	1807							
R-W	26-06-1998	31-12-2005	2745	1.71	-0.0000755	-0.0275575	-1.61	-0.21	-12.1	n.s.
R-E	01-01-1996	31-12-2004	3288	1.47	-0.0000195	-0.00711750	-0.49	-0.06	-4.4	yes
R-A	11-05-1996	31-12-2005	3343	1.77	-0.0000550	-0.0200750	-1.13	-0.18	-10.4	yes
R-R	05-01-1998	31-12-2005	2800	1.62	-0.0000296	-0.010804	-0.67	-0.08	-5.1	yes
ALL	01-01-1995	31-12-2005	4009	1.61	-0.0000151	-0.0055115	-0.34	-0.06	-3.8	yes

n.s. = not significant, seasonal decomposition not possible (less than 5 years of observations) or questionable
R-W = RURAL-WEST (RFL, EST, FCH, BRL);
R-E = RURAL-EAST (KEI, STA);
R-A = RURAL-AFFECTED (WBZ, PPT, EGB);
R-R = RURAL REMOTE (BNT).

TGM measurements were initiated in North America, Asia and Europe (Kock et al., 2005, Kim et al., 2005), this paper reveals a statistically significant decreasing trend of TGM concentrations from rural locations in Canada between 1995 and 2005. This decreasing trend can be seen with differing intensity at nearly all CAMNet sites and categories, indicating a large spatial distribution of this overall decrease.

9.2.1.4.2 Comparison Between Air Data and Wet Deposition of Mercury in North America

The NADP-MDN network has been operating since 1996 (http://nadp.sws.uiuc.edu/mdn/), with measurements made in Canada as part of CAMNet. The trends of mercury concentration in precipitation from these sites were analyzed using the non-parametric Seasonal Kendall Trend Test (Gilbert, 1987). At the sites that are co-located with, or located in areas near, the CAMNet sites, decreasing trends in MDN concentrations were generally found although there are numerous sites with no statistical significant trends.

Comparisons of the TGM trends and the precipitation concentration trends were made at 4 Canadian sites (St. Andrews, Kejimkujik, St. Anicet and Egbert with site codes NB02, NS01, PQ04 and ON07, respectively) where collocation of TGM and MDN measurements occurs (see Table 9.4). At Kejimkujik where 9 years of MDN data have been collected, concentrations of mercury in precipitation are strongly decreasing (-17.7% overall) whereas TGM is increasing during the same time period (+3.3% overall). At St. Andrews where 7 years of MDN data have been collected, both Hg in precipitation (-13.4%) and TGM (-7.4%) are decreasing. At Egbert, TGM is decreasing slowly, but the co-located and closest MDN sites show no significant change. The co-located MDN site at St. Anicet and nearby MDN stations show good agreement between both the annual and total changes over an 8-year period. St. Anicet has large TGM annual decreases at about -1.5%/year, while three area MDN locations show similar annual changes of between -1.5 to -1.8%. At the other CAMNet sites, there is no co-location with MDN (Burnt Island and Point Petre, Fort Chipewyan) or a data record of less than 5 years (Reifel Island, Bratt's Lake and Esther).

For the MDN sites used in this comparison with significant trends observed (as listed in Table 9.4), the averaged trend is -0.15 ng L^{-1} yr^{-1}. Applying this averaged trend over 10 years (i.e. 1996-2005), with a median total mercury concentration in precipitation of 10 ng L^{-1} (as observed for this time period for these sites), leads to an approximate change of -15%. This observed change in mercury concentration in precipitation is in good agreement with the individual trends in TGM concentrations observed near the major urban areas of Toronto (Point Petre) and Montreal (St. Anicet) for nearly the same time period. We conclude that this agreement indicates that the changes at these stations are most likely driven by local or regional changes in mercury emissions.

In general, the concentration of mercury in precipitation is decreasing at many MDN sites used for comparison in this paper, just as is the TGM at most CAMNet sites. However, some significant differences do exist, particularly at Kejimkujik. At the Canadian MDN sites, the seasonality of Hg concentration in precipitation exhibits an opposite pattern to TGM air concentrations with higher concentrations during the summer months. Two possible factors have been suggested for this seasonal behaviour: increased particle scavenging capacity of rain relative to snow and/or an increase in the oxidation of Hg^0, either in cloud or in the gas phase, during the summer.

Generally, the levels of Hg in precipitation at rural sites in Canada are less than 30 ng L^{-1}, similar to levels observed at border sites in the United States. The lowest median Hg concentrations at the CAMNet sites were found at Cormak (spring, summer, fall and winter concentrations at 4.6, 6.8, 3.0 and 2.9 ng L^{-1}, respectively) and Mingan (summer and winter at 6.6 and 2.9 ng L^{-1}, respectively). Median Hg concentrations in precipitation were higher in the west than in the east in all four seasons. However, sites with the highest observed concentrations do not necessarily have the highest Hg deposition per unit surface area. The deposition is dependent on the precipitation amount as well as on the concentration of Hg in the precipitation. The highest median weekly Hg deposition measured at the Canadian sites on a seasonal basis was at Kejimkujik Park in spring, fall and winter (deposition of 0.120, 0.106 and 0.088 μg m^{-2} $week^{-1}$, respectively), and at St. Anicet in summer (0.166 μg m^{-2} $week^{-1}$). The lowest median weekly deposition was found at Bratt's Lake in the summer and fall (0.083 and 0.036 μg m^{-2} $week^{-1}$, respectively) and at Egbert in winter (0.027 μg m^{-2} $week^{-1}$). For the years 2000 to 2003, the annual wet deposition of mercury at the Canadian sites ranges from 1.9 to 7.9 μg m^{-2} yr^{-1} with the lowest deposition seen at Bratt's Lake and the highest at St. Anicet.

Temporal trends have been developed with the Digital Filtration (DF) technique of Nakazawa for Hg in precipitation at St. Andrews, St. Anicet, Mingan and Kejimkujik Park, each of which have more than 5 years of data. F-tests performed on trends derived by DF have indicated that all correlations are statistically significant with confidence limits of at least 95%. First order half-lives of decline ($t_{1/2}$) were estimated by dividing –ln 2 with the linearly regressed slope of the trend line (Table 9.5).

The half-life is the time required for the concentration or deposition to decline to half its original value and is estimated by assuming first order decline in concentration and deposition. No significant trends were found for the site of Mingan. From Table 9.5, it can be seen that for the three sites, it will require 14 to 22 years for the concentration of Hg in precipitation to decline to half its current value. However, it will only take ~10 to 13 years for the deposition amount to drop to half. This occurs because the annual precipitation rate has generally decreased between 1996 and 2003 at these locations. For all 3 sites, the r^2 for Hg concentration and deposition are approximately 0.4 and 0.5, respectively, with p-values less than 0.01.

Since Canadian measurements began in 1995, mercury levels in the air have shown only a slight decline throughout most of Canada. The greatest decline of airborne mercury in Canada occurred close to the major urban areas of Toronto and Montreal, where levels fell by about -10% between 1996 and 2005. The largest decreases in TGM were seen at Point Petre, on the north shore of Lake Ontario, near Toronto, where levels declined by -17% and at St. Anicet, near Montréal, where levels fell by -13%. This is in good agreement with the overall trend in total mercury concentrations in precipitation observed within the comparable NADP-MDN

Table 9.4 Summary of the trend statistics of total mercury concentrations in precipitation within the MDN network compared with the TGM changes within CAMNet

MDN Site	Total Change(%)	Annual Change(%)	Years		CAMNet site (TGM)	Total Change(%)	Annual Change(%)	Years
NB02	-13.4	-1.9	7	→	St. Andrews	-7.4	-0.83	9
NS01	-17.7	-2.0	9	→	Kejimkujik	3.3	0.37	9
ON07	n.s.		5	→	Egbert	-2.2	-0.04	9
PA30	n.s.							
ME02	-14.7	-1.8	8	→	St. Anicet	-13.1	-1.46	9
ME96	-13.9	-1.7	8					
PQ04	-12.1	-1.5	8					
ME09	n.s.							
ME98	n.s.							
NY20	-14.7	-2.5	6		Point Petre	-16.6	-1.81	9.5
PA90	-12.3	-1.4	9					
ON07	n.s.							
PA30	n.s.							
ON07	n.s.		5	→	Burnt Island	-5.1	-0.67	8
upwind	n.s.							
WA18	-9.7	-1.1	9	→	Reifel Island	-10	-2.03	6
	n.c.				Fort Chipewyan			
	n.c.				Bratt's Lake			
	n.c.				Esther			

notes: n.s. = not significant; n.c. = no comparison possible

Table 9.5 Half lives for decrease in Hg concentration and deposition in precipitation

	t½ (years)	
Sites	Hg concentration in precipitation	Hg deposition due to precipitation
St. Anicet (98-03)	15	9.9
St. Andrews (96-03)	14	7.0
Kejimkujik Park (96-03)	22	13

sites, indicating that these changes are most likely driven by local or regional changes in mercury emissions.

At other sites in rural eastern Canada decreases in TGM concentration up to -7% were seen, with one site, at Kejimkujik National Park in Nova Scotia, recording an increase of 3%. The data record for western Canada is shorter than others and, at times, incomplete. Therefore, a trend with time could not be determined with the same statistical significance as for the other locations. Most sites show higher TGM concentrations in winter and spring, and lower concentrations in summer and fall. It is suggested that the meteorological seasonal variability is the most important factor in the establishment of the observed seasonal cycles of the TGM concentrations. This is consistent with the latest modeling results. Nevertheless further investigations on photochemistry and correlation with meteorological parameters could provide further evidence.

9.2.1.5 Mercury Speciation Analysis

More recently, some sites within CAMNet have been measuring atmospheric Hg-species concentrations in addition to TGM. Reactive gaseous mercury (RGM) includes inorganic Hg compounds such as $HgCl_2$, $HgBr_2$, HgClBr and Hg^0. Current measurements in ambient air cannot distinguish one of these RGM species from another. Concentrations of organic forms of Hg, such as methylmercury or dimethylmercury, are very low in the atmosphere and are not measured in the gas phase, though some measurements have been made in precipitation. In Atlantic Canada, levels of methylmercury have been found to represent 1 to 2% of total Hg in precipitation (Tordon Unpub. Data 2005). Total gaseous mercury (TGM) is the Hg that is detected in particle-free air which has not been deliberately scrubbed of RGM. In most cases, RGM is lost by adsorption to the walls of the inlet system used for the measurement, but some RGM may be detected along with the Hg^0 and this total detected is the TGM.

The current method used consists of a denuder to collect RGM followed by a filter to collect the PM. An impactor which removes particles of aerodynamic diameter greater than 2.5 μm is used at the inlet of the denuder to remove particles in the size range which can be collected by the denuder. Keeler et al. (1995) have shown that the majority of PM is attached to fine particles. A detailed description of this technique is given in Landis et al. (2002). Briefly, a known volume of ambient air is drawn at a set flow rate of 10 L min^{-1} through the following series of collectors: an impactor inlet which removes particles >2.5 μm aerodynamic diameter, a KCl-coated annular denuder of quartz which collects RGM, and a quartz filter housed in a quartz tube which collects p-Hg of aerodynamic diameter <2.5 μm. The typical sample integration period is 2 hours. Analysis for Hg content is carried out by first releasing the Hg species from the collection medium (quartz filter or KCl coating of the denuder) with the medium still contained in the filter housing or denuder. The filter and denuder are heated at separate times (filter before denuder for an automatic integrated system) with a stream of Hg-free air flowing across the collection surface. The filter is heated to 800°C to release the Hg species that are on the filter and to decompose them to Hg^0; the denuder is heated to 500°C, releasing the Hg species from the KCl. The Hg^0 released by this process is collected and analyzed by the method described for Hg^0 above. The fully automated system is produced by Tekran™, Inc. (Model 1130/1135), which in addition to the process above, heats the particle filter to 800°C during the analysis for PM and RGM for complete decomposition of any Hg species eluted from the filter or denuder. There has been one intercomparison study to verify measurement techniques for RGM and PM. Preliminary results indicate significant variability and point to the need for more research (Aspmo et al., 2005).

Continuous measurements of RGM and PM have been made in Quebec, Nova Scotia and Ontario. Poissant et al. (2005) reported values of RGM: 3 ± 11 pg m^{-3} and PM: 26 ± 54 pg m^{-3} at St-Anicet. These values are similar to those found at Point Petre RGM: 5 ± 5 pg m^{-3} and PM: 6 ± 7 pg m^{-3} and at Sterling on the south shore of Lake Ontario, RGM: 6 ± 11 pg m^{-3} (Han, Holsen et al., 2004). Even though RGM and PM constitute a relatively small portion of total Hg in air (0.2 to 1.4%), an evaluation of their role in the atmosphere is essential to understanding the cycle of

Hg. Poissant et al. (2005) reported diurnal and seasonal cycles for both RGM and PM. Additional continuous measurements of RGM and PM are needed to fully assess the seasonality of these species. It should be noted that the values reported for RGM and PM are very often near the detection limit of the measurement method.

Cloud droplets scavenge particles from the air in the nucleation process, and cloud droplets and rain scavenge soluble gases such as RGM. Cloud water concentrations of total Hg were determined in a limited number of samples collected during the summer of 1995 in eastern Canada (Banic et al., 2003).

At the observed cloud liquid water contents in the clouds studied, the median value for the amount of Hg scavenged from the air by the cloud water was 0.02 ng m^{-3}. In south-eastern Canada, concentrations of RGM and p-Hg are near 5 pg m^{-3} and 10 pg m^{-3}, respectively, as given above, suggesting that much of the Hg seen in cloud water can be explained by in-cloud scavenging of these species. Atmospheric mercury speciation measurements consisting of Hg^0, RGM and PM began in January of 2006 in Halifax, Nova Scotia Canada (Figures 9.2–9.4). A summary of the ambient atmospheric mercury species concentrations is given in Table 9.6. The median concentration for Hg^0, RGM and PM were 1.67 ng m^{-3} (0.716 to 46.5 ng m^{-3}), 2.42 pg m^{-3} (detection limit (dl) to 140 pg m^{-3}), and 1.73 pg m^{-3} (dl to 30.8 pg m^{-3}), respectively. The median levels of RGM and PM were a small percentage of the TGM, 0.2% for RGM and 0.1% for PM.

9.2.1.6 Mercury Measurements (incl. air craft) Related to Emissions, and Source Attribution

In the mid- to late-1990s, measurements of Hg^0 were made in three locations in Canada at altitudes up to 7 km (Banic et al., 2003). The data show that, on average, Hg^0 shows a relatively constant distribution with altitude. In the summer in south-eastern

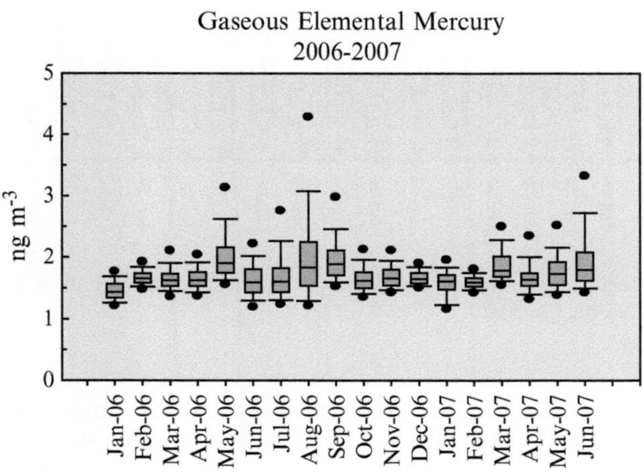

Figure 9.2 Monthly Box-whisker plot trends of Gaseous Elemental Mercury from 2006 and 2007 at Halifax (Nova Scotia, Canada)

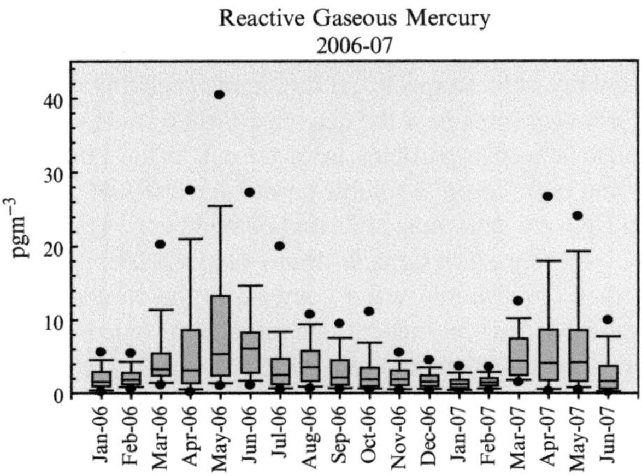

Figure 9.3 Monthly Box-whisker plot trends of Reactive Gaseous Mercury from 2006 and 2007 at Halifax (Nova Scotia, Canada)

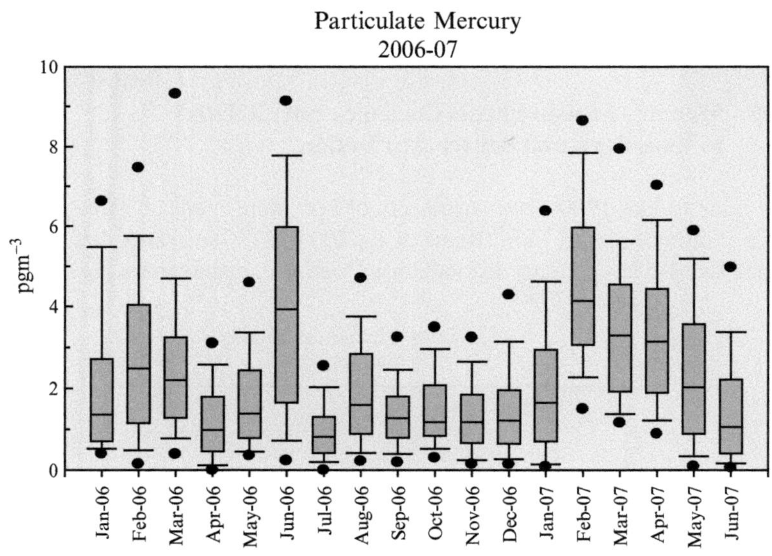

Figure 9.4 Monthly Box-whisker plot trends of Particulate Mercury from 2006 and 2007 at Halifax (Nova Scotia, Canada)

Canada, north-westerly winds bring air with a constant mixing ratio of Hg^0 at altitudes up to 7 km with a concentration near 1.5 ng m^{-3}. In the winter in southern and central Ontario and in spring over the Beaufort Sea and Arctic Ocean, the mixing ratio is still approximately constant at altitudes above 1 km but the concentration is 1.7 ng m^{-3}. Thus, despite differences in the proximity to sources and meteorology,

Table 9.6 Statistical summary of mercury speciation measurements from January 2006 to June 2007 in Halifax (Nova Scotia, Canada)

	Hg^0 (3 hr avg.) *(ng m^{-3})*	RGM *(pg m^{-3})*	PM *(pg m^{-3})*	Hg^0 (5 min.) *(ng m^{-3})*
Mean	1.79	4.62	2.37	1.77
Median	1.69	2.42	1.73	1.67
SD	0.45	7.61	2.30	0.61
Sample #	2835	2835	2835	106283

the observation that there is little difference in Hg^0 aloft over these distance and time scales indicates that there is mixing and uniformity in the Hg^0 concentrations aloft over continental scales. This atmospheric Hg aloft can be drawn down to the Earth's surface by atmospheric mixing processes even in remote regions of the world. The long average atmospheric residence time for Hg^0 is sufficient for this species to become relatively well mixed throughout the troposphere in the northern and southern hemispheres. Thus, since most of the Hg in the gas phase is in the elemental form, TGM concentrations should be of similar magnitude at a wide variety of background sites. Table 9.7 shows differences in Hg^0 below 1 km in the Ontario winter and Arctic spring. The measurements made at altitudes less than 1 km in Ontario demonstrate input of Hg^0 due to anthropogenic sources and over the sea ice in the Arctic demonstrate depletion of Hg^0.

Summary of Measurement Locations

Table 9.8 presents an overview of the measurement efforts that have occurred in the contiguous USA since high-precision measurements have been made (since early 1990s). Generally, a mean and standard deviation is presented, or a range of means from difference subsets of the data.

9.2.2 *Measurements of Air Concentrations in the United States*

Measurements of airborne species of mercury (Hg) [total gaseous (TGM), elemental (Hg^0), reactive gaseous (RGM), and particulate (PM)], as well as dissolved mercury in precipitation have been measured in many geographical areas of the USA including the northeast (New York, Connecticut, Vermont, Maine), north-central (Michigan, Wisconsin, Ohio, Indiana, Illinois), south-eastern (Georgia, Alabama, Tennessee, Florida), eastern seaboard (Maryland and the Atlantic Ocean), the mountain west (Nevada), coastal California, west coast free troposphere (Mt. Bachelor, Oregon), and west coast marine boundary layer (Cheeka Peak, Washington). These locations are shown on a map of the contiguous USA in Figure 9.5. Regions of the USA where there is a lack of published mercury measurements include the middle of the country from the northern plains south to Texas, most of the west (with the exception of Nevada), and the Southwest including southern California. In addition to great

Table 9.7 Elemental mercury for different altitude ranges over Canada. The units are *ng* m^{-3}, with the cubic metre referenced to 0°C and 1 atm in all cases (from Banic et al., 2003)

	< 1 km		1 – 3 km		> 3 km	
Time, Location of Measurement	Mean	Median	Mean	Median	Mean	Median
Summer 1995 Nova Scotia	1.4	1.3	1.5	1.4	1.4	1.4
Summer 1997 Eastern Ontario	1.5	1.5	1.5	1.5	1.5	1.5
Winter 97-98 Southern Ontario	1.7	1.6	1.6	1.6	1.5	1.5
Spring 1998 Arctic Ocean and Beaufort Sea	1.2	1.4	1.7	1.7	1.7	1.7

geographic and climate diversity in the USA, the spatial distribution of mercury emissions are not uniform; the vast majority of point sources lie in the eastern half of the country and are dominated by coal-fired power plants, waste incinerators, and manufacturing facilities. There are some large mercury sources in the West, dominated by mining and metal processing facilities and a few coal-fired power plants, but these emitters are spread out across wide regions and nowhere in the West are there the density of point sources as is found in the East. Thus, the majority of research to date has taken place at locations in the eastern half of the USA, and these studies are largely focused on determining the influences of local (< ~50 km) and regional (< ~500 km) mercury sources on air concentrations of mercury species and mercury concentration in precipitation and its subsequent depositional loadings to the environment. Additionally, some research in the West has determined that mercury emissions from distant locations such as industrial regions in East Asia and boreal fires in Alaska and Siberia contribute to enhanced air concentrations of Hg in the western USA, and that in pristine locations at certain times of the year, this transport may play a dominant role in the overall mercury burden in the environment.

9.2.2.1 Remote Locations

Because most studies in the USA have focused on the impact of regional or local emissions on air concentrations and deposition, there have been few measurements at locations that are truly remote. Figure 9.5 gives an overview on locations in the United States from which there are published measurements of gaseous and particle mercury species.

Perhaps the most pristine location in the USA is Cheeka Peak on the Washington State coast at an elevation of 500 m. At this location when the wind direction is between 160°-315°, the air can be generally classified as "marine" with no influence from continental sources at least 7 days prior to arrival to the site (Weiss-Penzias et al., 2003).

Table 9.8 Summary of Hg^0, RGM and Hg(p) measurements made at remote, rural and urban locations in the United States. NR means "not reported"

Location	Site details	Duration of study	Hg^0 mean *($ng\ m^{-3}$)*	RGM mean *($pg\ m^{-3}$)*	Hg(p) mean *($pg\ m^{-3}$)*	Reference
Remote Sites						
Mount Bachelor, Oregon	2.8 km elevation, mountain top site	1.5 yr, 4 mth	1.4-1.8	39-60	4.4	Weiss-Penzias et al., 2007; Swartzendruber et al., 2006
Cheeka Peak, Washington	500 m elevation, marine boundary layer	1 yr	1.45-1.55	0-2.7	0-2.9	Weiss-Penzias et al., 2003
Ship, between Bermuda and Barbados	Subtropics, marine boundary layer	2 mth	1.63 ± 0.08	5.9	NR	Laurier and Mason, 2007
Rural Sites						
Chesapeake Bay Laboratory, Maryland	70 km south of Baltimore	7 mth	1.7-1.8	6-13	NR	Laurier and Mason, 2007
Look Rock, Tennessee	Smokey Mountains. at 813 m asl	2 mth	1.65	5	7	Valente et al., 2007
Salmon Creek Falls Reservoir, Idaho	SW Idaho (1510 m asl) 100-300 km from large gold mines and industrial plants	1.25 yr	1.3-1.6	1-10	NR	Abbott et al., 2007
Great Mountain Forest, Connecticut	Relatively remote area of NW Connecticut	5 yr	1.4-1.6	NR	NR	Sigler and Lee, 2006
Cove Mountain, Tennessee	Smokey Mountains at 1243 m asl	40 days	3.2	16.4	9.7	Gabriel et al., 2005
Dexter, Michigan	80 km west of Detroit	4 mth, 6 mth	1.49-1.51	2-3	12 ± 5.2	Lynam and Keeler, 2005a; Gildemeister et al., 2005
Potsdam, Stockton, and Sterling, New York	Western New York State	3 yr	1.84-2.59	NR	NR	Han et al., 2004
Pompano Beach, Florida	50 km northeast of Miami	1 mth	1.6-2.0	1.6-4.9	3.5 ± 2.8	Malcom et al., 2003
Stillpond, Maryland	42 km east of Baltimore	1 yr	1.7 ± 0.5	21± 22	42 ± 50	Sheu et al., 2002

(continued)

Table 9.8 (continued)

Location	Site details	Duration of study	Hg^0 mean (*ng m⁻³*)	RGM mean (*pg m⁻³*)	Hg(p) mean (*pg m⁻³*)	Reference
Walker Branch Watershed, Tennessee	20 km away from large coal fired power plants	6 studies over 3 yr	2.2	92 ± 60	NR	Lindberg and Stratton, 1998
Urban, Industrial, Mining, or Fire Sites						
North-central Nevada	1300-1800 m asl, close proximity to enriched substrates, and ore processing facilities	1 mth	2.5-3.0	7-13	9-13	Lyman et al., 2008
Detroit, Michigan	Close proximity to coal-fired power plants and metal smelting plants	1 yr	2.2 ± 1.3	17.7 ± 28.9	20.8 ± 30.0	Liu et al., 2007
Detroit, Michigan	Close proximity to coal-fired power plants and metal smelting plants	2 mth	NR	NR	1-39	Lynam and Keeler, 2005b
Desert Research Institute, Reno, Nevada	1340 m asl, close proximity to enriched substrates and urban emissions	3 yr, 3 mth	2.1-2.5	37 ± 28	7 ± 9	Stamenkovic et al., 2005
Chicago, Illinois	Heavy industrial development 10-20 km away from site	15 mth	3.6 ± 2.9	NR	70 ± 67	Landis et al., 2002
Baltimore, Maryland	Roof of Maryland Science Center	2 yr	4.4 ± 2.7	89 ± 150	74 ± 197	Sheu et al., 2002
Athens, Georgia	Within 350 m of a mercury cell chlor-alkali plant	5 days	3.9-8.7	9-129	NR	Landis et al., 2004
Tuscaloosa, Alabama	Small mercury sources are < 5 km from site	1 mth	4.05	16.4	16.4	Gabriel, et al., 2005
Earlham College, Richmond, Indiana	60-120 km away from large coal-fired power plants	6 studies over 3 yr	4.1	104 ± 57	NR	Lindberg and Stratton, 1998

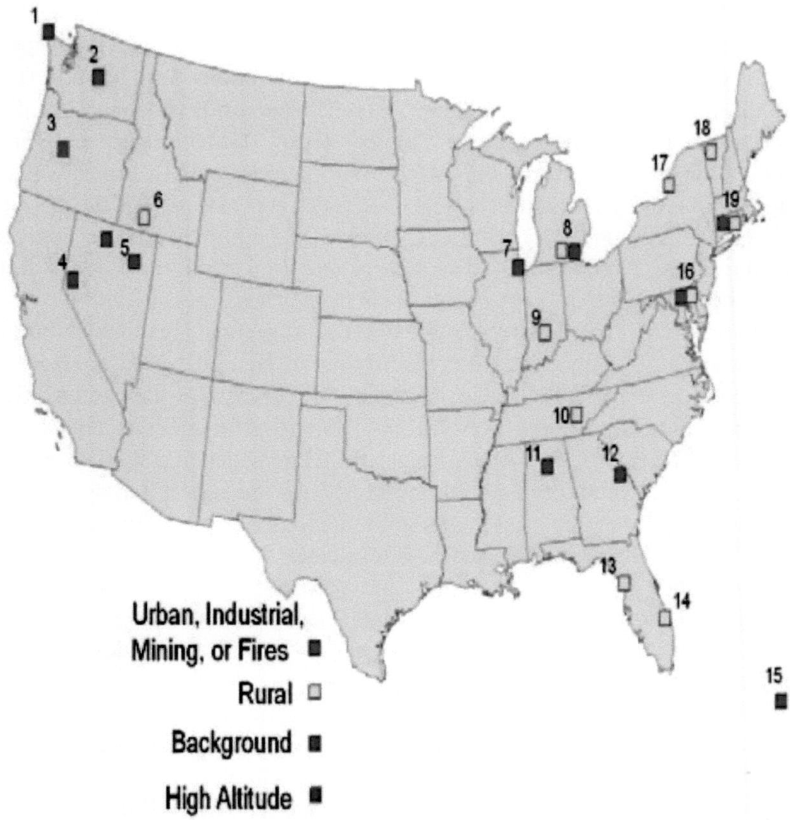

1 Weiss-Penzias et al., 2003
2 Friedli et al., 2003
3 Weiss-Penzias et al., 2006; 2007; Swartzendruber et al., 2006
4 Stamenkovic et al., 2005
5 Lyman and Gustin, 2008
6 Abbott et al., 2007
7 Landis et al., 2002
8 Liu et al., 2007; Lynam and Keeler, 2005a, b; Gildemeister et al., 2005
9 Lindberg and Stratton, 1998
10 Valente et al., 2007; Gabriel et al., 2005; Lindberg and Stratton, 1998
11 Gabriel et al., 2005
12 Landis et al., 2004
13 Pollman et al., 1995; Gill et al., 1995
14 Malcom et al., 2003
15 Laurier and Mason, 2007
16 Laurier and Mason, 2007; Sheu et al., 2002
17 Han et al., 2004
18 Burke et al., 1995
19 Sigler and Lee, 2006; Chen et al., 2004; Nadim et al., 2001

Figure 9.5 Map of locations in the United States from which there are published measurements of gaseous and particle mercury species

Here, TGM averaged 1.45-1.55 ng m^{-3} over an annual cycle in 2001-2002, which is on the low end of the range of commonly accepted northern hemisphere background concentrations (1.5-1.7 ng m^{-3}). Concentrations were slightly (5%) higher in the summer and slightly higher (5%) when the winds were from the direction of the ocean and not the continent. Another set of TGM measurements in a pristine location were made from a ship in the subtropical Atlantic Ocean between Bermuda and Barbados. These data showed that TGM was also within the northern hemispheric background value, averaging 1.63 ± 0.08 ng m^{-3} in the late summer of 2003 (Laurier and Mason, 2007). The lack of temporal variation in these data suggest that there was little influence from local or regional sources.

Other studies at remote and rural sites reveal average Hg^0 concentrations that are within the 1.5-1.7 ng m^{-3} range or slightly above. Hg^0 at Chesapeake Bay Laboratory (CBL), a rural site in the marine boundary layer, was 10% elevated relative to open ocean value (Laurier and Mason, 2007). Short-term spikes in Hg^0 at CBL never exceeded 2.5 ng m^{-3}. Two deployments at CBL revealed average Hg^0 concentrations of 1.8 ng m^{-3} in the fall of 2002 and 1.7 ng m^{-3} in the fall-spring 2003-2004. Speciated mercury measurements were made at Pompano Beach, Florida during June of 2000. The winds alternated from a clean marine sector that produced an Hg^0 concentration of 1.6 ± 0.06 ng m^{-3} and a mixed flow regime that included some continental flow, which produced an Hg^0 concentration of 2.0 ± 0.4 ng m^{-3}.

Gaseous and particle mercury species along with some additional trace gases were measured at Look Rock during two field studies totaling 84 days in the spring and summer of 2004 (Valente et al., 2007). Look Rock is located just west of Smokey Mt. NP at 813 m a.s.l. Hg^0 averaged 1.65 ng m^{-3} over sampling period (spring/summer of 2004). Concentrations were slightly higher in spring. Hg measurements were made at a rural site in Michigan (Dexter, MI) during selected times from 1999 to 2002 (Lynam and Keeler, 2005a). Hg^0 measurements at Dexter (80 km west of Detroit) show an average concentration of 1.51 ng m^{-3} during the winter and 1.49 ng m^{-3} in the spring 2000-2001. Maximum concentration during winter was 4.39 ng m^{-3} and 1.94 ng m^{-3} in spring. TGM was measured at three locations in rural western New York State during the summers of 2000-2001 (Han et al., 2004). The northern (Potsdam) and southern (Stockton) sites revealed equivalent concentrations of 1.84 ng m^{-3}, but the central site was much higher (2.59 ng m^{-3}). The authors suggest that the central site is closer to Hg point sources.

Furthermore, Hg^0 was measured at a rural site about 42 km to the east of Baltimore during 1997-1998 (Sheu et al., 2002). It essentially showed background values (1.7 ± 0.5 n g m^{-3}) albeit with relatively high variability, representative of the regional signal. Not much seasonality was observed. Between 1992 and 1995, TGM was measured at a rural site in eastern Tennessee (Walker Branch Watershed) and produced an average value of 2.2 ng m^{-3} (Lindberg and Stratton, 1998). Two large coal-fired power plants are located approximately 20 km away from the upland forested site, which suggests that there is a higher background signal at this site relative to others in the region. TGM was measured at Underhill Vermont during a complete annual cycle in 1993 (Burke et al., 1995). The average TGM concentration was 2.0 ng m^{-3} with a maximum of 4.2 ng m^{-3}, suggesting that local sources were influencing these measurements.

9.2.2.2 *Urban Locations (including mining areas)*

Many studies have been carried out in urban and mining locations with the aim of assessing the impact of urban, industrial, and metal processing activities on mercury air concentrations and deposition in the local environment. There have been measurements made in several major USA cites (Chicago, Detroit, Baltimore), some smaller cities (Tuscaloosa, Alabama), sites near industrial facilities (Georgia), and mining areas (central Nevada). One of the major questions is to address the great spatial and temporal heterogeneity that is commonly observed in atmospheric mercury when measured in and around these areas.

The emission of RGM from a mercury cell chlor-alkali plant cell building and the impact on near field (100 km) dry deposition was investigated from sites inside and around the facility over a 5 day study (Landis et al., 2004). Measurements in the cell building roof vent showed that RGM constituted 2.1 ± 0.7% of the concurrently measured Hg^0. The percentage of RGM/Hg^0 at ambient monitoring sites 350 m (1.5%) and 800m (1.3%) away while being impacted by cell building emissions suggests the rapid deposition of RGM species. Previous estimates of the RGM% from the emissions in this sector done by the US EPA in the 1997 Mercury Report to Congress were much higher (30%) compared to the 2% observed in this study. Even so, the authors calculate that the magnitude of dry deposition from this plant at a 10 km radius is around 50% (4.6 mg m^{-2} yr^{-1}) of wet deposition (9.8 mg m^{-2} yr^{-1}). A study confirmed that there can be large changes in the near-field environment on short time scales (Gildemeister et al., 2005). TGM values in Detroit were 3-11 times higher than the rural site in Dexter Michigan (Lynam and Keeler, 2005a). TGM varied greatly and generally depended on meteorology conditions that favoured transport to the site. Another study in the Chicago/Gay urban area was investigated to determine its impact on atmospheric Hg concentrations and wet deposition in the Lake Michigan basin (Landis et al., 2002). Over the 1 ½ year period from 1994-1996, the TGM average concentration was 3.6 ± 2.9 ng m^{-3} and the maximum was 22 ng m^{-3} at the downtown Chicago site. These values are elevated and more variable relative to observations at other sites across the Lake Michigan Basin (rural site TGM = 2.2 ± 0.7 ng m^{-3}). Using meteorological clustering, the authors concluded that proximity to the industrial sources in the Chicago/Gary area were the cause of the elevated values. Elevated TGM concentrations were detected at a sampling site in downtown Baltimore (4.4 ± 2.7 ng m^{-3}) (Sheu et al., 2002). When the air came from the SE, S and SW directions, the urban sampling site tended to be impacted by the local emission sources, with higher THg and PM concentrations detected. TGM measurements were also made in smaller cities, but TGM values at these sites are often as high if not higher than in the largest cities. TGM measurements at Tuscaloosa, Alabama averaged 4.05 ng m^{-3} (maximum of 12 ng m^{-3}) during the summer (Gabriel et al., 2005). As with the other studies, short periods of high concentration for all mercury species were found. TGM measurements in Richmond Indiana between 1992-1994 averaged 4.1 ng m^{-3}. Richmond is a small city on the Ohio border and situated nearby (60-120 km) several major coal-fired power plants (Lindberg and Stratton, 1998).

TGM measurements were made at two arid-land rural sites in north-central Nevada, which is an area of diverse natural and anthropogenic mercury sources that include undisturbed and mining-disturbed enriched substrates, coal-fired power plants, ore processing facilities, and industrial facilities (Lyman et al., 2008). The TGM concentration averaged over all campaigns was 3.1 ± 1.7 ng m^{-3} at NV02 and 2.57 ± 3.1 ng m^{-3} at NV99. The highly variable temporal behaviour of TGM was found to be influenced by both local substrate emission and transport from regional source areas.

In summary, the behaviour of TGM in the urban or near-source environment is quite different from its behaviour in the rural, remote, or pristine environment. At near-source areas the concentrations can be between 2-10 times higher than areas beyond about 40-120 km from the emission source, which can be considered the spatial extent of the term "local".

9.2.2.3 Temporal Trends at Single Locations

Measurements of TGM and Hg in precipitation over at least two annual cycles can identify the temporal variation on annual and seasonal timescales. Since Hg is of special concern as an air pollutant and there have been efforts to remove Hg from stack emissions from some facilities, it is of great interest to know if emissions reductions cause an annual decrease in TGM concentrations and Hg deposition at sites downwind of the emissions sources. Also, determining the magnitude and the timing of seasonal changes in TGM and Hg in precipitation can constrain the Hg budget and help to identify natural processes that affect sources and sinks.

TGM concentrations were measured at Desert Research Institute in Reno, Nevada from 2002 to 2005 and averaged 2.3 ng m^{-3} (0.9-8.6 ng m^{-3}) (Stamenkovic et al., 2007). The slope of the daily TGM concentrations at this site over the three year period showed a decrease of ~6% per year, although the concentrations are highly variable suggesting near-field emissions sources (Figure 9.6).

Three main causes of short-term TGM enhancements were identified: 1) after rain events the soil and foliar emissions increase, 2) when the site intercepts an urban plume with high concentrations of co-pollutants, and 3) marked change in meteorological conditions (e.g. inversions). Seasonally averaged concentrations were found to be spring = 2.2 ng m^{-3}, summer = 2.2 ng m^{-3}, fall = 2.1 ng m^{-3}, and winter = 2.5 ng m^{-3}. These data agreed with the patterns for CO and NOx, but this does not necessarily mean that the major sources of TGM are from combustion. The likely cause of this pattern is the greater presence of inversions as well as

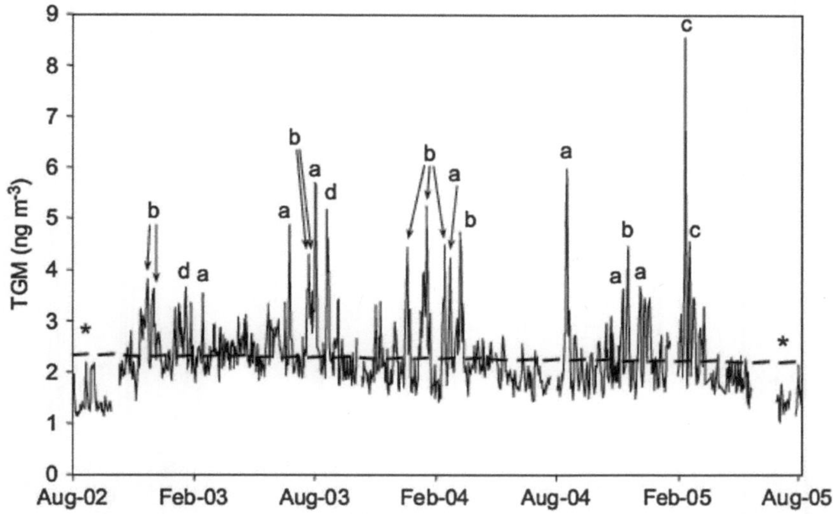

Figure 9.6 Time series plot of average daily concentrations of TGM measured at Desert Research Institute (Reno, Nevada) from 2002 to 2005. Fitted trend equation for all data: y = 2.3–0.00013t (in days). Letters indicate potential sources/causes of elevated TGM (a) following precipitation), atmospheric signature of pollution plumes (b), elevated NOx, CO or PM10 concentrations), marked change in meteorological conditions (c), change in barometric pressure), or no specific indicator (d). Asterisks denote periods of unusually low TGM (Stamenkovic et al., 2005)

weakening atmospheric sinks for all species in winter. There was also significant diurnal variability in TGM (2.1-2.6 ng m^{-3}), with a peak in the morning and gradually decreasing through the afternoon. A decreasing TGM annual average was also observed at a rural site (Chesapeake Bay Laboratory - CBL) near Baltimore/ Washington in the marine boundary layer (Laurier and Mason, 2007). TGM was measured in 2003/2004 and also in 1998/1999. The mean for the earlier time period was 1.9 ng m^{-3} and 1.7 ng m^{-3} for the later time period and no seasonal variation was evident. TGM concentrations never exceeded 2.5 ng m^{-3} at this site, and showed little diurnal pattern. Hg in wet deposition was measured continuously for 11 years (1993-2003) in the Lake Champlain basin at Underhill Vermont (Keeler et al., 2005) (Figure 9.7). There was considerable sample-to-sample and year-to-year variability due amount to variation in precipitation, temperature, and source receptor relationships. As a result of large variability, no increasing or decreasing trend was observed in either annual means of deposition or concentration. This is in contrast to what is observed for the record of sulphate deposition at Underhill, VT, which shows a decreasing trend due to emissions reductions of SO_2 from coal-fired power plants. Hg in wet deposition also displayed a strong seasonal cycle (higher by 4-6 times in May-September), which suggests that atmospheric removal processes play a large part in determining the variability of Hg in wet deposition, as opposed to seasonally varying sources of TGM.

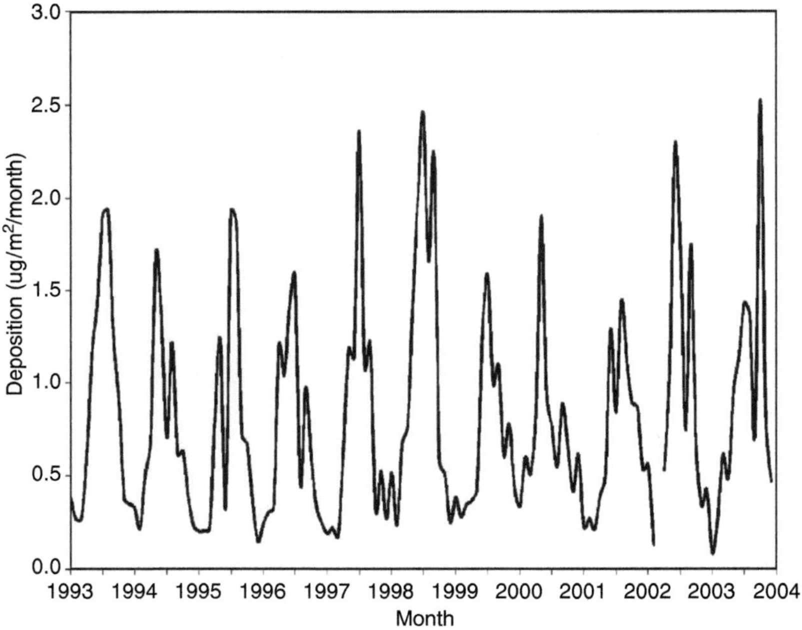

Figure 9.7 Monthly Total Hg wet deposition at Underhill, VT (Keeler et al., 2005)

In summary, there is some evidence that TGM at a rural location on the eastern seaboard downwind of major urban and industrial centers is decreasing and this decrease is likely due to cutbacks in the USA emissions. At more remote sites, there appears to be little change, and Hg is likely controlled by the northern hemispheric pool. However, the temporal dynamics of TGM and Hg in wet deposition are complex, with the magnitude of diurnal and seasonal changes often being larger than annual changes.

9.2.2.4 Monitoring Networks and Trends

Recognizing that TGM and Hg in wet deposition are spatially heterogeneous, several studies have aimed to set up monitoring networks in order to compare trends between sites in the same region, between regions, and to determine the influence of local and regional emissions sources. There is also interest in understanding the processes that contribute to Hg variability on a diurnal, weekly, seasonal, and annual basis. The largest, most ambitious network of sites is the MDN, which has sites across the USA that collect weekly precipitation samples and measure dissolved Hg (Figure 9.8). Some sites are co-located with Nation Trends Network (NTN) sites which measure concentrations of the major ions in precipitation (SO_4^{2-}, NO_3^-, Na^+, Cl^-, etc.) (Figure 9.9).

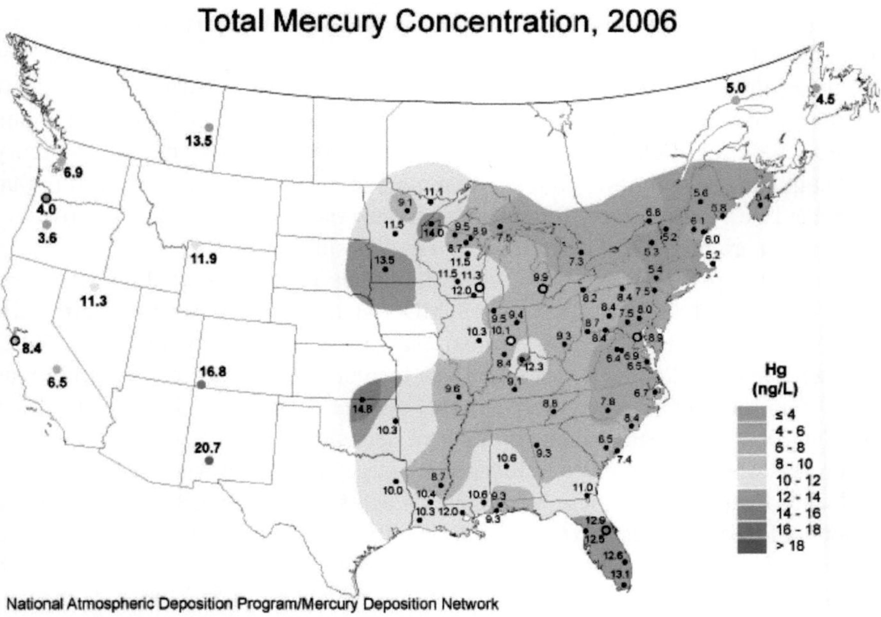

Figure 9.8 Total mercury concentration from the Mercury Deposition Network in 2006

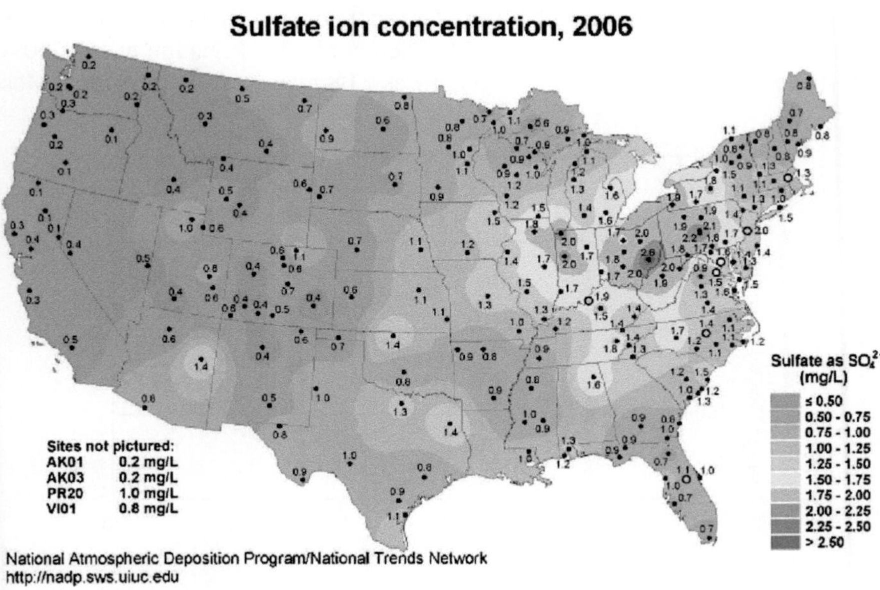

Figure 9.9 Sulfate concentration from the National Trends Network in 2006

The spatial pattern of Hg concentrations in wet deposition has some aspects that have been difficult to reconcile based on known sources and removal processes. For example, there are high mercury concentrations in Florida, where there are few mercury sources, and there are relatively low concentrations in Pennsylvania and Ohio, where there are many coal-fired power plants. Clearly the total mercury and sulphate concentration patterns do not agree in these regions, suggesting that sulphate is not a good proxy for Hg and that removal processes (not only proximity to sources) are important in determining Hg deposition patterns.

One study using MDN Hg wet deposition data in which spatial and temporal trends were evaluated, focused on 13 MDN stations in the north eastern USA (1996-2002) and the Underhill Vermont event-based monitoring site (1993-2002) (Vanarsdale et al., 2005). It was found that mercury deposition and concentrations are not uniform across the region, and that distinct depositional sub-regions could be identified. These sub-regions in order of decreasing deposition are: sites to the south and west that are more influenced by large urban/industrial areas, maritime influenced sites, and northern inland sites. Each site was characterized by a large amount of week to week and seasonal variability with the highest concentration and deposition generally occurring in the spring and summer. Because of this high variability, no increasing or decreasing trend in annual mean deposition or concentration was evident at any site. "Enhanced" mercury deposition weeks (> 250 ng m^{-2}), which may occur only a few times during the year, were found to contribute a large portion of the annual mercury loading (20-60%) at most sites (Figure 9.10). Thus, Hg deposition in this region can be thought of as being episodic, although whether these episodes are driven by sink processes (oxidation/scavenging) or the influence of pollution plumes is not well understood.

In the state of Connecticut, USA, a TGM monitoring campaign was conducted from January 1997 to December 1999 (Chen et al., 2004; Nadim et al., 2001). The three-year mean TGM concentration was 2.08 ng m^{-3} across all 8 sites. Most

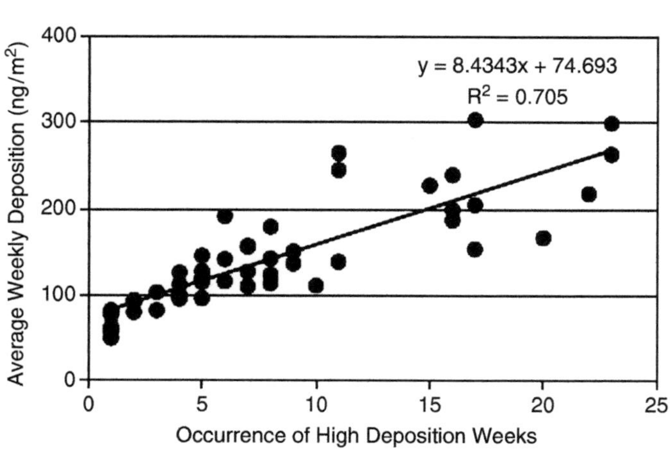

Figure 9.10 The relationship between the annual occurrence (sum) of high mercury deposition weeks (>250 *ng* m^{-2}) and the average annual deposition for those weeks for 13 MDN monitoring sites located in northeast North America (Van Arsdale et al., 2005)

sites displayed mean concentrations slightly above 2 ng m^{-3}, but one site (Waterbury) had a mean close to 4 ng m^{-3}. This anomaly is assumed to be caused by the close proximity of industrial Hg sources at Waterbury. At all sites, the annual mean concentration had no significant differences among the three years of measurements. In the state of Connecticut, USA, a TGM monitoring campaign was conducted from January 1997 to December 1999 (Chen et al., 2004; Nadim et al., 2001).

In another study, TGM and Hg in wet deposition were measured at several sites around Lake Michigan in order to determine the influence of emissions in the Chicago area (Landis et al., 2002). The largest difference in Hg concentration in wet deposition was between the downtown Chicago site (IIT) and remote northern Lake Michigan site (SBD).

The IIT mean summer Hg concentration was ~23 ng L^{-1} while the SBD summer mean was ~12 ng L^{-1}. The IIT site was elevated relative to all other sites in all seasons except winter, when inter-site differences were small. SBD (northern Lake Michigan) was much lower than all sites in every season except spring. Particulate Hg was much higher at IIT (70 pg m^{-3}) relative to other sites (12-24 pg m^{-3}). TGM concentrations were 2.0-2.2 for all sites and 3.6 ng m^{-3} at IIT. The impact of anthropogenic emissions from the Chicago/Gary area on wet deposition flux of mercury to Lake Michigan was modelled and is shown in Figure 9.11. Here it can be seen

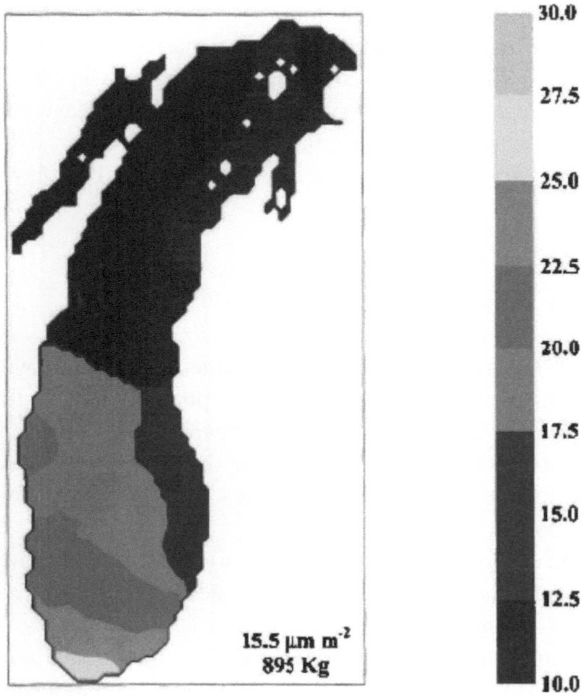

Figure 9.11 Estimated over-water wet deposition flux (July 1, 1994-October 31, 1995) (Landis and Keeler, 2002). The annual mean wet deposition value and total Hg deposited to the lake during the study period is shown at the bottom of the figure

that there is a factor of 3 difference between fluxes near Chicago and at the central/northern end of the lake.

An early study employing sites across Florida and in the Everglades measured TGM and Hg in rainfall between 1992-1994 (Pollman et al., 1995; Gill et al., 1995). The amount of Hg wet deposited shows a strong seasonal pattern with 2-3 times higher concentrations and 4-6 higher deposition in April-September. This pattern along with the lack of a spatial gradient in TGM or rainfall Hg, suggests that deposition is driven by large-scale regional processes as opposed to local emission/deposition processes. Chemical and/or photochemical processes in the atmosphere likely produce the high levels of summer time Hg in rainfall. The pattern of TGM concentrations was consistent with a lack of strong point sources. TGM across all sites had a mean concentration of 1.64 ng m^{-3} (n=191), and no spatial or temporal gradients were observed.

In summary, it seems that not enough attention has been focused on monitoring networks in the USA due to the difficulty and cost in carrying out sensitive measurements at many sites simultaneously. The MDN is essential for understanding the spatial and temporal patterns of Hg in wet deposition, but without at least TGM measurements (and speciated Hg measurements would be the best) in conjunction with precipitation measurements, little can be inferred about the processes responsible for controlling Hg deposition. It is not well understood, for example, which sites would respond most quickly to emissions reduction in various industrial sectors (e.g. coal-fired power plants, incineration, mining, manufacturing, etc.), or if enhanced deposition events are independent of local or regional sources and supplied predominantly by the global Hg pool being caused by natural variations in photochemistry and meteorology. The few studies that have both speciated Hg measurements as well as Hg in wet deposition suggest that Hg in deposition can vary by 2-fold between an urban and a rural site, TGM might vary by 50-75%, and PM may vary 5-10 fold between sites.

9.2.2.5 Mercury Speciation Analysis

In most cases, TGM and Hg in wet deposition measurements provide only part of the story concerning the advection of Hg from point sources and the natural air-surface cycling of Hg. Speciated Hg measurements include the longer lived Hg^0, and the shorter lived RGM, and PM, and these data can provide a wealth of information on the locality of sources and sinks, boundary-layer/free-tropospheric dynamics, and rapid air-sea exchange.

Since the late 1990s it has been possible to make accurate measurements of RGM and PM, and since the early 2000s, automated instruments have been available that can be operated unattended at remote sites.

Some of the important questions that have been investigated are 1) what is the importance of RGM formation in the marine boundary layer as a way of cycling Hg between the air and sea? 2) What is the diurnal pattern of RGM and PM? 3) What are the spatial distributions of these species, especially in close proximity to urban/

mining areas? 4) How do clouds, boundary layer dynamics and exchange with the free troposphere affect the concentration RGM? 5) Are there detectable longer-term changes in RGM or PM that suggest reductions in local emissions?

At two sites in Maryland and on a cruise in the subtropical Atlantic Ocean, the behaviour of RGM and PM in three environments was investigated during 2002-2003: open ocean, coastal, and urban (Laurier and Mason, 2007). RGM showed a distinct diurnal cycle at all sites, especially over the open ocean, where it is believed there is active recycling of Hg between air and water (oxidation by OH and halogens followed by rapid dry deposition of RGM, reduction to Hg^0 and subsequent evasion to replenish the MBL with Hg^0). Open ocean RGM ranged between 0-25 pg m^{-3}, and was generally negatively correlated with O_3, that is, the conditions that destroyed O_3 were responsible for producing RGM (assumed to be the presence of halogens and/or OH). At the rural coastal site (CBL), RGM showed a daytime average of 13 pg m^{-3} and a night time average of 6 pg m^{-3} (maximum value = 113 pg m^{-3}). These values are higher than what was observed over the ocean. The RGM daily peak at CBL generally occurred with the solar maximum under low wind speed conditions, suggesting the importance of in situ production. RGM and Hg^0 were seen to be occasionally elevated simultaneously at CBL, which was never seen over the open ocean. This is likely due to the occasional advection of polluted air to the site. High O_3 days generally produced elevated RGM, although the correlation was weak. No significant peaks of RGM were associated with clean marine air. RGM was very low during precipitation events. RGM variability at the coastal site is largely explained by photochemical production and dry deposition, with other factors such as advection from sources and changes in boundary layer height making minor contributions. Urban RGM concentration averaged 16.9 pg m^{-3} (down from 89 pg m^{-3} in 1997/98), suggesting a long-term reduction in point source emissions. RGM reached concentrations of > 200 pg m^{-3} at the urban site, occurring at night, probably a result of local inputs into a shallow nocturnal boundary layer. Another interesting observation was that RGM was generally depleted to zero during precipitation events at CBL but not at the urban site. This can be explained by the fact that higher concentrations at the urban site are not completely scavenged by precipitation, but that the lower concentrations at CBL are effectively drawn down to near zero.

Semi-continuous measurements of Hg^0, RGM, and PM were made throughout 2003 in Detroit, Michigan, where significant seasonal and diel cycles were observed. Hg^0 and RGM were elevated in the warm months, whereas PM was higher in the winter. RGM displayed a diel cycle that had a daytime maximum and Hg^0 and PM were higher at night, generally being inversely correlated with RGM (Figure 9.12). Multivariate and probability function analysis of Hg spikes with other parameters (ozone, SO_2, meteorology) revealed that both local and regional sources, in addition to boundary layer dynamics, were important in controlling mercury behaviour. The finding that each Hg species had their distinct seasonal and diurnal patterns suggests that the dry and wet deposition of Hg in Detroit and the surrounding area may also vary significantly across different seasons and throughout the day, highlighting the importance of conducting long-term highly time-resolved

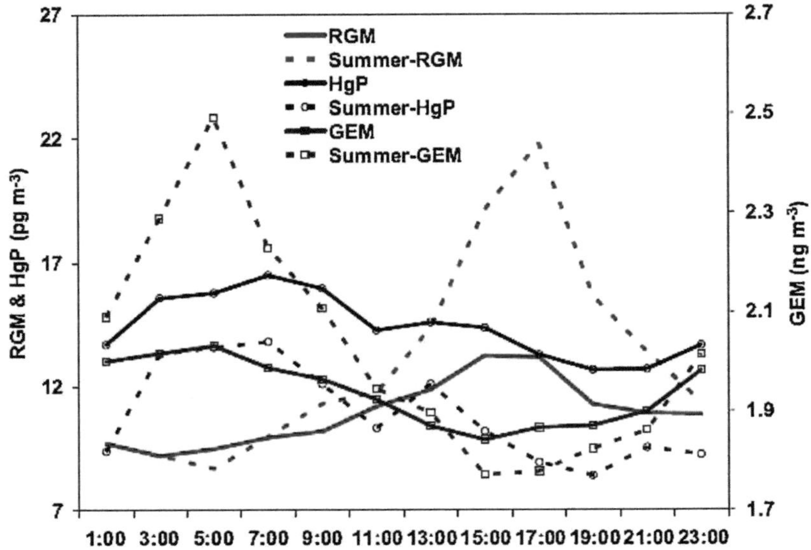

Figure 9.12 Diurnal variation of Hg species (median concentrations) in Detroit (2003) (Liu et al., 2007) Note the inverse relationship between RGM and Hg^0. Hg(p) has the least amount of diel variability

measurements in urban areas that potentially have more Hg source contributions to the surrounding areas.

Earlier efforts by the Keeler group made automated speciated Hg measurements at a rural (Dexter, Michigan) and an urban (Detroit, Michigan) site between 1999 and 2002 (Lynam and Keeler, 2005a). Median RGM at the rural site was 2-3 pg m^{-3} (max = 38.7 pg m^{-3}), was not correlated with Hg^0 in winter and weakly negatively correlated with Hg^0 in the spring. Poor correlations between Hg^0 and RGM at both sites, suggest multiple sources and/or different sinks for these species. Particulate Hg showed a range of 5-60 pg m^{-3} in Detroit. Precipitation and dew caused a decrease in concentration of both species, which was also observed in the Chesapeake Bay-Baltimore studies (Laurier and Mason, 2007). The diurnal cycle in Detroit is much less pronounced than Dexter, with significant amounts of RGM and PM present during the night time hours (presumably due to the trapping of nocturnal emissions and transport).

Higher RGM/PM ratios occurred under drier upper altitude air from the east with fewer point sources and urban areas, whereas lower RGM/PM ratios occurred with wetter low level air from the west, which picked up pollutants from the Chicago area. There was some evidence that high ozone episodes in the Detroit/Dexter region caused enhanced RGM (simultaneous afternoon peak in both species). Additionally, there was a slight negative correlation between Hg^0 and O_3, suggesting photochemical depletion of Hg^0, which produced RGM. However, high ozone conditions (> 80 ppb) also produce a positive sampling bias in the PM measurements on KCl denuded filters versus undenuded filters (Lynam and Keeler, 2005).

The marine boundary layer was determined to not be a source of Hg^0 or RGM at Pompano Beach, Florida, during June of 2000 (Malcom et al., 2003). Higher mercury concentrations were correlated with HNO_3 and SO_2, suggesting urban and industrial sources on land. Hg(p) concentrations were higher than can be explained by sea spray alone suggesting that gaseous Hg is diffusing into the sea salt aerosol. This suggests that deposition of aerosols enriched in Hg via this process may constitute a significant global mercury flux to the oceans.

RGM and PM measurements were made in two contrasting south-eastern USA airsheds (Cove Mountain, Tennessee, and Tuscaloosa, Alabama) during the summers of 2002 and 2003 (Gabriel et al., 2005). A comparison of mercury species concentration shows that the urban site has much higher mean PM concentrations (16.4 vs. 9.73 pg m^{-3}), but mean RGM concentrations between the two sites are roughly equivalent (16.4 pg m^{-3}). There was a stronger dependence between RGM and wind speed at Cove Mtn. suggesting a greater role of boundary layer dynamics and free tropospheric exchange in controlling RGM. The maximum RGM concentration at Tuscaloosa was much higher (162 pg m^{-3}, with higher variability overall) than Cove Mtn. (58.1 pg m^{-3}), suggesting that local point sources of Hg were impacting the urban site. At Cove Mt. there was a small diurnal cycle in RGM with a maximum in the midday, similar to other remote sites. No diurnal cycle was apparent at Tuscaloosa, similar to other urban sites. At another site in the south eastern USA (Look Rock, Tennessee) in the spring and summer of 2004, Valente et al. (2007) observed RGM and Hg(p) means of 5 and 7 pg m^{-3}, respectively. These are lower concentrations than what were seen at other sites in the region (Gabriel et al., 2005; Lindberg and Stratton, 1998). $Hg^{(II)}$ species were found to account for < 1% of TGM.

Urban and rural sites were sampled for RGM and PM along the coast of Chesapeake Bay and in Baltimore (Sheu et al., 2002). One year of data collected at Stillpond (rural) showed that RGM averaged 24 pg m^{-3} in 1997 and 21 pg m^{-3} in 1998. PM at Stillpond for 1997 was 39 pg m^{-3} and for 1998 was 61 pg m^{-3}. There was no seasonal pattern in RGM at Stillpond, but for PM there were higher values in the colder months, suggesting that reactive mercury species favour the condensed phase under colder temperatures. One site in the study (Curtis Creek) displayed the highest mean RGM concentration (385 pg m^{-3}), mean PM concentration (715 pg m^{-3}), and mean Hg^0 (6.1 ng m^{-3}), evidently reflecting near field contamination. RGM + PM as a percentage of TGM varied between 2-19% among sites. Because of the high concentration of RGM, annual dry deposition fluxes at the most polluted site was calculated to be as high as wet depositional fluxes.

Speciated mercury measurements were made in the Pacific Northwest in the marine boundary layer at Cheeka Peak, Washington (Weiss-Penzias et al., 2003), and in the free troposphere at Mount Bachelor, Oregon (Swartzendruber et al., 2006). These locations show very different behaviours of RGM and PM. At Cheeka Peak RGM and Hg(p) were very low (0 - 2.7 and 0 - 2.9 pg m^{-3}, respectively), mostly due to frequent cloudiness and precipitation. Occasional wind-shifts to continental directions (downwind of Seattle and Vancouver) would cause RGM and PM to increase to

10-20 pg m^{-3}, much lower than what is seen downwind of eastern USA cities. At Mount Bachelor, RGM is quite elevated under certain meteorological conditions that bring high-pressure subsidence under high UV conditions (summer). These episodes were clearly marked by dry air at night generally accompanied by higher O_3 and a decrease in Hg^0. There is a nearly quantitative shift in speciation with an increase in RGM and a loss of Hg^0 under subsidence conditions. RGM concentrations as high as 600 pg m^{-3} were observed. Mean RGM concentration at night (n = 203) was 60 pg m^{-3} and the daytime mean (n = 527) was 39 pg m^{-3}, in contrast to other studies that show an RGM maximum at solar noon. PM concentrations for day and night at Mount Bachelor were equivalent (~4.4 pg m^{-3}), indicating that RGM was being formed and unable to condense to particles under these dry air conditions.

In summary, given the available knowledge of speciated mercury, it can be concluded that RGM in situ formation is very important in both the marine boundary layer under cloud-free conditions, and in the free troposphere, where concentrations reach those of the most polluted urban atmosphere. Additionally, because of the diurnal nature of photochemistry and boundary layer/free tropospheric exchange, RGM and to a lesser extent Hg^0 and PM generally have prominent diel cycles. Superimposed on the natural variability is the episodic nature of advection of more or less polluted air masses to a sampling site from urban/industrial/mining sources. With regard to long term changes in RGM or Hg(p) concentrations, the data do not go back far enough or have sufficient spatial coverage to really address this question.

9.2.2.6 Mercury Measurements Related to Emissions and Source Attribution

In an effort to understand the relative importance of anthropogenic and natural emissions of airborne Hg, several studies have attempted to calculate Hg fluxes from source regions. This is primarily done by directly measuring fluxes (usually from enriched surfaces) using flux chambers, or by correlating Hg enhancements in plumes to other tracers whose emissions are known (CO). Source attribution for Hg in precipitation has been done using variations on principal component analysis using other elements as tracers of Hg emission types. Other studies, which are less quantitative, have used back trajectories and meteorological cluster analysis to provide weight-of-evidence for the causes of Hg enhancements in gaseous and precipitation samples.

In two recent studies, total airborne mercury and CO were measured in 22 pollution transport "events" at Mount Bachelor (2800 m a.s.l.) between March 2004 and September 2005 (Weiss-Penzias et al., 2006; 2007) (Figure 9.13). East Asian industrial events yielded a ΔTGM/ΔCO enhancement ratio of ~0.005 ng m^{-3} ppb_v, whereas plumes from western USA anthropogenic sources and from biomass burning in the Pacific Northwest and Alaska gave a ratio of ~0.001 ng m^{-3} ppb_v. Thus, the ΔTGM/ΔCO ratio is an important distinguishing feature of Asian long-range transport. Scaling these ratios with estimated emissions of CO from China and global biomass burning, an emission of 620 Mg yr^{-1} is calculated for total mercury emissions from Chinese anthropogenic sources and 670 Mg yr^{-1} for global biomass burning.

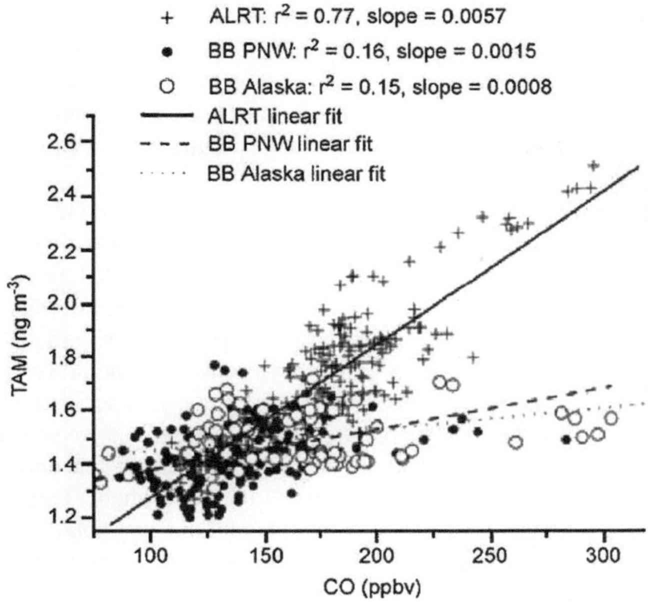

Figure 9.13 Scatter plot of total airborne mercury (TAM) vs. CO measured during 22 pollution events at Mt. Bachelor Observatory, Oregon during 2004-2005. The events are categorized at Asian long-range transport (ALRT), and biomass burning from Alaska or the Pacific Northwest. (Weiss-Penzias et al., 2007)

In a related study, the Hg^0/CO molar enhancement ratio was observed in pollution plumes at Okinawa, Japan and produced a value of 6.2×10^{-7} mol/mol (0.0056 ng m^{-3} ppb_v) (Jaffe et al., 2005). These plumes were identified to have originated in the industrialized region of eastern China, and they produced a similar ratio to those observed at Mt. Bachelor. This implies that there should be the same molar ratio of Hg^0 and CO emissions in China. However, recent Chinese emissions inventories are a factor of two lower than the ratio in the plumes. Likely explanations for this discrepancy are 1) Chinese Hg emissions have been underestimated, 2) there are large natural sources of Hg that are not accounted for and, 3) Hg^0 emissions in China make up larger fraction of the inventory and the shorter lived RGM and PM are less important. Speciated measurements were also made at Okinawa and revealed no correlation between CO (a marker of Asian outflow) and RGM or PM. This suggests that the marine boundary layer and the presence of clouds and precipitation at Okinawa effectively scavenges these reactive Hg species. Thus, the question remains as to the spatial extent of RGM and PM transport from point sources.

Another study also employed the Hg/CO molar enhancement ratio, this time in smoke plumes from large temperate forest wildfires and a wheat stubble fire which burned in August 2001 as sampled by the University of Washington's Convair 580 research aircraft (Friedli et al., 2003). TGM concentrations as high as 7.5 ng m^{-3} were observed in smoke plumes. Hg was in its elemental form (95%) with the remaining fraction in the particulate form. TGM was linearly correlated with CO and using known emission factors (EF) for CO, an EF for TGM was calculated to be 113 μg Hg kg^{-1} of fuel. From these data, an estimate of 3.7 Mg of Hg per year from North American temperate forests was generated based on burn area estimates between 1997-2001. Agricultural waste burning is estimated to contribute 20 Mg yr^{-1} worldwide, but this number is highly uncertain.

Five years of wintertime TGM measurements at a background site in Connecticut were combined with measurements of CO2 as a regional combustion tracer in order to investigate regional Hg emissions (Sigler and Lee, 2006). A 20% decrease in Hg emissions between 1999/2000 and 2001/2002 was found to be not fully explainable by climatological changes in air mass transport, but rather a significant correlation was seen with interannual changes in the emissions from the power sector in the region. The Hg flux from the power sector was calculated to account for 47-75% of the atmospheric flux, yet if 50% or more of the Hg emitted by power sources is RGM and PM and deposits locally, then the power sector can only account for 23-40% of the observed elemental Hg flux to the atmosphere. This suggests that there are other important unaccounted for elemental mercury sources in the region, or that there is significant reduction of $Hg^{(II)}$ in the power plant plumes.

Beginning in the fall of 2002, measurements were made in Steubenville, Ohio as part of a multi-year comprehensive mercury monitoring and source apportionment study to investigate the impact of local and regional coal combustion sources on atmospheric mercury deposition in the Ohio River Valley (Keeler et al., 2006). Using multi-variate component analysis involving numerous trace metals, coal combustion was found to be the source of ~70% of Hg in wet deposition in Steubenville Ohio. The volume-weighted mean Hg concentration over a 2 year period was 13.7 ng L^{-1}, 2-3 times higher than what is observed in Wisconsin and the north-eastern USA.

Source-receptor relationships for Hg and other trace elements wet deposited in south Florida were investigated using daily event precipitation samples collected concurrently at 17 sites from August 6 to September 6, 1995 (Dvonch et al., 2005). A multivariate receptor modeling approach found municipal waste incineration and oil combustion sources to account for 71 ± 8 % of the Hg wet deposited at five Florida Everglades sites. A similar analysis of a year-long record of event samples (June 22, 1995-June 21, 1996) collected at Davie, FL, found 73 ± 6% of the Hg wet deposited to be accounted for by local anthropogenic sources.

In another study, size segregated Hg(p) measurements were made at several urban sites in Detroit, Michigan, in 1996 (Gildemeister et al., 2005). One site (LIV) is located in heavily industrialized section of Detroit near coal-fired power plants, iron and steel smelting, and coke production. This site observed the highest mean concentrations of TGM (2.8 ng m^{-3}) and PM (54 pg m^{-3}) compared to less industrialized sites. PM in the coarse mode made a median daily contribution of 37% of total Hg(p) at the LIV site compared to a median daily contribution of 20% at the rural site. This suggests that industrial sources are impacting the local environment and the impact lessens with a greater distance away from the source. Other studies have been less quantitative but are able to show using back trajectories and cluster analysis that higher concentrations are observed in gaseous or precipitation samples that have been impacted by anthropogenic Hg emissions regions.

Event wet-only precipitation, total particulate, and vapour phase samples were collected for Hg, and trace element determinations from five sites around Lake Michigan from July 1994 through October 1995 as part of the Lake Michigan Mass Balance Study (LMMBS) (Landis et al., 2002). The cluster modelling results indicate that urban and industrial sources in the Chicago area were contributing to enhanced Hg in precipitation and PM concentrations across the entire Lake Michigan basin. Rain water measurements at marine and inland sites in central California show that higher Hg concentrations were associated with back trajectories that stayed the 20-40°N range across the Pacific, suggesting a contribution from Asian sources (Steding and Flegal, 2002).

The Nevada Study and Tests of the Release of Mercury From Soils (STORMS) project focused on the measurement of mercury emissions from a naturally enriched area (Gustin et al., 1999). Hg^0 measurements were as high as 200 ng m^{-3}, indicating large local sources. Fluxes were quite variable due to site heterogeneity, but ranged from 50 to 360 ng m^{-2} hr^{-1}. This represents an increase in the magnitude of natural soil flux by 10-100 times over previously held values.

Finally, to assess the sources, transport and deposition of atmospheric Hg in Michigan, a multi-site network was implemented in which Hg concentrations in event precipitation and ambient samples (vapor and particulate phases) were determined (Hoyer et al., 1995). Elevated Hg concentrations at all rural sites were associated with air mass transport from the west, southwest, south, and southeast. The most northern site (Pellston, MI) had the lowest concentration of vapor phase Hg and precipitation Hg.

9.2.3 Measurements of Air Concentrations in Mexico

Short term field measurements of TGM concentration in ambient air were undertaken at 4 locations in Mexico in 2002 in order to characterize mercury concentration in areas contrasting in mercury sources and exposure. Locations and dates of sampling are provided in Table 9.9. TGM was measured using a Tekran 2537A mercury vapour analyser using a 5 min sampling interval. Mexico City is a large urban/industrial city consisting of a mixture of residential and industrial activities. Zacatecas is a smaller semi-urban centre with a prolonged history of gold and silver mining activities which have traditionally involved mercury amalgamation processes in metal extraction. Mine tailings are a dominant feature in and around Zacatecas and several brick manufacturing facilities are located in the vicinity. Brick manufacturers use the mine tailings as a raw material in the manufacture of bricks.

The other two sites were selected to provide more regionally representative assessments of mercury concentrations in ambient air at locations having little or no known exposure to substantial anthropogenic sources of mercury. Puerto Angel

Table 9.9 Site description and sampling dates for measurements of TGM in air in Mexico

Site	Sampling Date	Lat - Long	Elevation	Description
Zacatecas	Sep 17-20 2002	22° 44'N 102° 28'W	2420 m	mining legacy
Mexico City	Oct 18-21 2002	19° 21'N 99 ° 04'W	2240 m	large urban
Puerto Angel	Oct 16-18 2002	16° 47'N 96 ° 28'W	13 m	rural-remote
Huejutla	Oct 20-23 2002	21° 08'N 98 ° 25'W	172 m	rural-remote

Table 9.10 Location and sampling dates for Puerto Angel and Huejulta MDN sites in Mexico (adapted from NADP-MDN)

Station	Puerto Angel*(OA02)*	Huejutla*(HD01)*
Location	Pochutla County, Oaxaca	Huejutla de Reyes County, Hildago
Dates of Operation	9/28/2004 - 10/11/2005	10/26/2004 - 11/29/2005
Latitude	15.65	21.1583
Longitude	-96.4833	-98.3706
Elevation (m)	110	180

is a very small rural village located on the Pacific coast several hours drive south west of Mexico City. Puerto Angel is a small fishing village and undeveloped beach resort area. Huejutla is located well to the northeast of Mexico City and is a small rural centre. In addition to measurements of TGM, Puerto Angel and Huejulta were also selected as MDN sites to assess regionally representative concentrations of mercury in precipitation. Mexican MDN site descriptions are given in Table 9.10.

9.2.3.1 Remote Locations

TGM concentrations measured at Puerto Angel and Huejutla were consistently at or slightly below the expected hemispheric average concentration (~1.5 ng m^{-3}) and showed very little variability over the brief sampling period of a few days indicating that these two remote sites were not subject to any significant anthropogenic sources of mercury. Slightly lower values may indicate slight TGM depletion through depositional processes since both sites are surrounded by dense forests and subject to frequent moisture through rain and cloud water impaction (Huejulta) and rain and possibly marine advection fog (Puerto Angel). TGM concentrations between Puerto Angel and Huejulta were not statistically different (de la Rosa et al. 2004) (Table 9.11)

Mercury concentration in precipitation at these 2 remote sites had similar annual average precipitation mercury concentrations between 11 and 12 ng L^{-1} (Table 9.12) although concentrations were somewhat more variable between weekly composite samples. These concentrations are well within the range of concentrations observed at other sites in North America.

Table 9.11 Summary of TGM concentrations *(ng m^{-3})* measured at 4 locations in Mexico

Site	Mean	Min	Max	St Dev	n
Zacatecas	72.	0.26	702.	82.	742
Mexico City	9.8	2.8	34.	4.0	1148
Puerto Angel	1.46	0.76	2.45	0.4	546
Huejutla	1.32	1.13	2.93	0.3	701

Table 9.12 Summary of annual mercury concentration (*ng L^{-1}*) in precipitation between Oct 2004 to Oct 2005 at 2 rural-remote locations in Mexico

Site	Mean	Min	Max	St Dev	n
OA02	12.0	9.02	1.33	39.1	10.1
HD01	10.8	8.42	3.21	40.8	7.91

9.2.3.2 Urban Locations (Including Mining Areas)

TGM concentrations measured in Mexico City and Zacatecas stand in direct contrast to those measured at the 2 rural remote sites. Mexico City TGM averaged 9.8 ng m^{-3} with concentrations reaching 34 ng m^{-3}. Higher mean concentrations and variability are clearly related to varying exposures to anthropogenic mercury sources typical of many large, heavily populated urban-industrial centers. However, the largest average and maximum TGM concentrations were observed in Zacatecas which is a much smaller urban area with substantially more limited industrial development compared to Mexico City. There is little doubt that the long history of mining precious metals and the legacy of mercury used in the amalgamation process is reflected in ambient TGM concentrations. Mine tailings are found extensively in the area and the release of mercury into the air may be further exacerbated by the use of the tailings material in the manufacture of bricks which are baked at high temperatures during their manufacture. TGM measurements were not made adjacent to any brick manufacturing facility and would be representative of concentrations found in and around the town. TGM concentrations were significantly different ($p < 0.01$) between sites and significantly different from the two remote sites (de la Rosa et al., 2004).

9.3 Measurements of Air Concentrations in South America

Relatively few observations of atmospheric Hg have been carried out in South America or Mexico. The few observations to date have mostly been carried out near to, or downwind of, major sources. This includes mining, industrial facilities and biomass fires. For nearly all South American observations, measured Hg^0 concentrations were substantially greater than the accepted global background level. Relatively few publications have converted the observed air concentrations into an emission flux. One emission inventory for Hg emissions associated with gold mining has been conducted using estimated gold production and an Hg emission factor. This inventory suggests that gold mining is the largest source of TGM to the atmosphere in South America. At present, there is no information in South America or Mexico that can be used to establish long-term trends.

9.3.1 Urban Locations (including mining areas)

Widescale mining has occurred in the Amazon Basin since at least the 18th century (e.g. Hachiya et al., 1998; Higueras et al., 2005 and references therein). Associated with current and past operations, a number of researchers have examined the TGM air concentrations in mining regions. Hachiya et al. (1998) sampled in urban and rural areas of Brazil near several tributaries of the Amazon river. Samples were collected on gold coated quartz. Hg analysis was carried out in the lab by desorption and UV absorbance. Urban areas sampled included Rio de Janero, Manaus and Brasilia, where concentrations up to 10 ng m^{-3} were found. Adjacent to mining areas concentrations up to 16 ng m^{-3} were found. In some gold workshops, very high levels of TGM were found (max 3.7 ng m^{-3}).

Amouroux et al. (1999) sampled at several sites in two Amazon basins of French Guiana. These sites were also strongly influenced by current and past mining activities. TGM samples were collected on gold coated quartz, with analysis in the lab by desorption and ICP-MS. High TGM concentrations were found and attributed to nearby illegal gold mining. Concentrations were highest at midday and lower at night. The authors attributed this to influence from lake-air exchange. Concentrations were much higher in the Petit Inini River Basin site (mean 15.0 ng m^{-3}), compared to the Petit Saut Lake (2.8 ng m^{-3}), which the authors attributed to mining sources.

De la Rosa et al. (2004) sampled at four sites in Mexico for a few days at each site. The authors measured TGM using a Tekran 2537a instrument and standard calibration methods. High values and high variability were found at the Mexico City and Zacatecas sites, suggesting strong nearby sources. Mean Hg values at Zacatecas were very high at 71.7 ng m^{-3}, whereas the Mexico City site was not as elevated (9.8 ng m^{-3}). At two rural sites, mean TGM values were near accepted global background concentrations (1.46 and 1.32 ng m^{-3}). Brick manufacturing, using mining waste, and mining tailings were attributed by the authors as the most likely source of high Hg at the Zacatecas site.

Higueras et al. (2005) sampled along roads in the Coquimbo region of Northern Chile using a Lumex RA-915+. Their sampling took them past several major ore mining and processing operations for gold and other metals (e.g. Hg, Cu and Mn). Very high concentrations of TGM were found in these current and historical mining regions (some mining back to 16th century). Extreme TGM concentrations, up to nearly 100 µg m^{-3} were observed at some gold recovery operations (milling and amalgamation).

Fostier and Michelazzo (2006) sampled at two sites in Sao Paulo state, Brazil, near the Paulínia industrial area. Hg samples were collected on Au coated quartz sand followed by desorption and CVAFS detection in laboratory. Samples for TPM analysis were also collected on quartz fiber filters. Sampling was conducted for about 5 days at each site during both the rainy and dry seasons. The two sites had an overall mean TGM concentration of 7.0 ng m^{-3}, with no significant difference between the two sites. There was evidence for a diurnal cycle, with higher TGM concentrations during the day. The enhancement in TGM, compared to the global background, was attributed to the wide array of industrial sources in the area. Concentrations of TPM were high, 400 pg m^{-3}. Based on the sampling methodology, this high value likely includes some contribution from RGM sticking to the filters or particulate matter on the filters.

Garcia-Sanchez et al. (2006) measured soil and air concentrations along with some height profiles in the El Callao region of Venezuela using a Lumex RA-915+. The sites included some highly polluted sites due to past mining activities. Extremely high TGM concentrations were observed in some gold processing shops, with several concentrations between 50 and 100 µg m^{-3}. These high values exceed recommended workplace limits (e.g. NIOSH or WHO limits of 50 and 25 µg m^{-3}, respectively).

From the above data, it is clear that past and current gold mining represents a large source of Hg to the atmosphere. Lacerda (1997) estimated global Hg emissions to the environment from gold mining. By this estimate 460 Mg yr^{1} are released to the environment globally, 300 Mg or 65%, of this is released to the atmosphere. Of this total, nearly 60% is released in South America. The atmospheric emissions of Hg in South America by gold mining (179 Mg yr^{1}) calculated by Lacerda (1997) is nearly twice the total Hg emissions from all sources in South America estimated by Pacyna et al. (2006). This estimate was updated by Lacerda (2003) to 107-228 Mg yr^{1}. However it should be noted that the Pacyna et al (2006) inventory does not quantify Hg emissions from South American gold mining nor does it attempt to quantify Hg emissions from illegal gold mining activities. In addition the Pacyna study used a Hg emission factor of 0.5 gram Hg emitted/gram Au mined, whereas Lacerda uses a factor of 1.5. Thus, while emissions of Hg from gold mining in South America are clearly a substantial source to the global atmosphere, there is a significant uncertainty in the actual emissions. Future work on Hg emissions in South America should focus on reducing the large uncertainty in the emissions.

9.3.2 Mercury Measurements (Including Aircraft) Related to Emissions, and Source Attribution

Ebinghaus et al. (2007) observed enhanced CO and TGM on two CARIBIC (*C*ivil *A*ircraft for *R*egular *I*nvestigation of the *A*tmosphere *B*ased on an *I*nstrumented *C*ontainer) flights between São Paolo and Santiago de Chile in 2005. The CARIBIC container is operated monthly onboard a Lufthansa Airbus 340-600 during regular passenger flights. The measured TGM/CO ratio on these two flights, 1.2×10^{-7} and

2.4×10^{-7} mole/mole, respectively, were similar to previous reports of biomass burning plumes, despite significant differences in geographic regions. From these ratios the authors estimate global emissions of TGM from biomass burning in the range of 210-750 Mg yr^{-1}.

9.4 Measurements of Air Concentrations in Europe

9.4.1 Remote Locations

TGM concentrations at remote sites are influenced by the regional and global background pool. Due to emission controls in Europe the regional contribution is declining and gradually less important. As a consequence the TGM background concentration of the Northern hemisphere is dominating at remote European locations. Measurements of TGM have been made over a period of one year using a cold vapour atomic fluorescence absorption technique at Harwell, a rural site in central southern England (Lee et al., 1998). The mean concentration was 1.68 ng m^{-3}, with a maximum hourly mean concentration of 20.5 ng m^{-3} and a minimum hourly mean concentration of 0.26 ng m^{-3}. The data from Harwell show greater variability than those from more remote sites.

Lake Balaton to be between 0.4 and 5.9 ng m^{-3} with higher concentrations during day time. Comparing marine background sites in the Southern Baltic Sea (Preila, Hoburg and Kap Arkona) with a more urban-influenced site in the Gulf of Gdansk, Beldowska et al. (2006) have shown that TGM concentrations and temporal variability at the latter site was much more pronounced. Based on principal component analysis, Beldowska et al. (2006) have separated three air masses reflecting marine, terrigenous and anthropogenic sources. Urba et al. (2000) have identified the southern Baltic Sea and in particular the Gulf of Gdansk as a source region during the summer months.

Another study carried out in Poland explicitly showed the influence of residential heating on air concentrations (Zielonka et al., 2005). TGM concentrations have been measured during one summer campaign (19 - 29 August 2003) and one winter campaign (26 January - 3 February 2004) at a rural Polish site. An average TGM value of 1.63 ± 0.35 ng m^{-3} was obtained in the summer campaign, whereas a 2.5 times higher TGM concentration was found during winter. Since 85% of the houses in this rural environment use low capacity domestic heating units and are fuelled with hard coal during the cold season, Zielonka et al. concluded that residential heating is the most likely explanation for enhanced TGM winter levels.

9.4.2 Urban Locations (Including Mining Areas)

The dominance of the main anthropogenic European mercury source categories varies country wise. However, it appears that the contribution of combustion sources in general is about one third, while industrial emissions make the maximum contribution. Atmospheric mercury levels around the world's largest mining and refining complex (Almaden, Spain) were determined during two field campaigns (September 1993 and February 1994) using both point monitors and LIDAR techniques (Ferrara et al., 1998). High mercury concentrations (0.1 - 5 μg m^{-3}) were measured over the village of Almaden in the prevailing wind direction. At the second largest mercury mine in Idrija, Slovenia five centuries of mining have influenced atmospheric mercury concentrations. Kotnik et al. (2005) have reported that TGM concentrations have decreased significantly in the last decade, from more than 20 ng m^{-3} in the early 1970s to values below 100 ng m^{-3} in the 1980s, and finally reached a level of 10 ng m^{-3} or even lower at the summer of the year 2004.

Total gaseous mercury (TGM) has been monitored at Champ sur Drac, a suburban site of Grenoble in southern east France (Dommergue et al., 2002). TGM measurements have been made over 4 periods of approximately 10 days throughout 1999-2000. The mean TGM concentration was 3.4 ng m^{-3} with maximum hourly mean concentration of 37.1 ng m^{-3}. Although mean TGM concentration was not greatly different from those previously measured in the troposphere, the greater TGM variability as well as the occurrence of high TGM concentration linked to particular wind conditions suggested the strong influence of anthropogenic sources.

9.4.3 *Temporal Trends at Single Locations*

Continuous monitoring data sets exist for the time period 1998 to 2004 for two coastal background sites. At Mace Head, west-Irish Atlantic coast and Zingst peninsula on the southern shore-line of the Baltic Sea, automated TGM measurements have been carried out with the same instrumentation. An intensive evaluation of the two data sets has been published by Kock et al. (2005).

Between 1998 and 2004 the annually averaged TGM concentrations measured at Mace Head (1.74 ng m^{-3}) and Zingst (1.64 ng m^{-3}) remained fairly stable. For both stations higher concentrations were detected during the winter months and lower concentrations during summer, respectively.

Since Mace Head is located at the European inflow boundary and therefore considered to be less influenced by continental emissions an unexpected West to East gradient was observed. Kock et al. found that the overall mean Mace Head TGM concentration was 0.06 ng m^{-3} higher (i.e. between 3 – 4%) than those of Zingst. For the January to June period, the Mace Head TGM values (6-year mean = 1.75 ng m^{-3}) are significantly elevated compared to the Zingst results (6-year mean = 1.64 ng m^{-3}). Since no local anthropogenic mercury sources exist near the Mace Head station, it was concluded that enhanced emission from the sea provide the most probable explanation for the observed differences. During the time period March 1990 to May 1996 total gaseous mercury (TGM) has been monitored at the summit of the Wank mountain (1780 m a.s.l.) in the Bavarian Alps (Slemr and Scheel, 1998, Slemr et al., 1995). This time period covers the probably most drastic changes in the European emission situation, mainly dominated by political and economical changes in East Germany. TGM measurements at Wank showed a linear decrease of 0.169 ± 0.009 ng Hg m^{-3} yr^{-1} i.e. about 7% per year. The frequency of occurrence of extremely high TGM concentrations and the amplitude of the seasonal variation decreased over the observation time. Slemr et al. (1995) concluded that the decrease in TGM concentrations of about 22% in the years between 1990 and 1994 indicate a significant change in the trend of global TGM concentration and that this is most likely the result of reduction in coal consumption and control measures taken in the OECD countries. A decrease in TGM concentrations with time was earlier shown for south-western Scandinavia (Iverfeldt et al., 1995).

Samples for measurements of TGM in air have been collected and evaluated for the time period 1980 to 1992 and show a clear decrease with time. At the Swedish west-coast, yearly average air concentrations and median levels of 3.3 and 3.1 (1980 - 1984), 3.2 and 2.8 (1985 - 1989), and 2.7 and 2.6 ng m^{-3} (1990 - 1992), respectively, were found. Increased average and median winter concentrations were always found, with levels at 3.7 and 3.4, 3.7 and 3.3, and 3.0 and 2.7 ng m^{-3} for the respective time period. Higher winter values were expected due to increased anthropogenic emissions and changes in the mixing height of the atmosphere. As Slemr and Scheel (1998) have substantiated later, a decreased number of episodic events of TGM levels in air, from 1990 and further on is already indicated in the Scandinavian data set (Iverfeldt et al., 1995).

9.4.4 Monitoring Networks and Trends

An extensive evaluation of mercury measurements in air and precipitation at EMEP or OSPAR stations respectively has been carried out by Wangberg et al. (2007). These data were obtained at coastal sites around the North Sea and originate from Ireland, Netherlands, Germany, Norway and Sweden. The observation period is 1995 to 2002 and the two periods 1995 to 1998 and 1999 to 2002 were compared. The reduction in deposition is 10 – 30% when comparing the two periods and the authors relate the decrease to emission controls in Europe. In contrast, no decreasing trend in TGM data could be observed during the same time periods. The authors suggest that a plausible explanation is that TGM concentrations measured in the OSPAR area are to a larger extent dominated by the hemispherical background than before, i.e. European emission reductions may be over-compensated by increasing emissions in other Northern hemispheric regions.

This conclusion is supported by Berg et al. (2006) who found that current mercury levels in surface sediments, surface soils and mosses at background in Norway are substantially affected by long-range atmospheric transport. The project "Mercury species over Europe" (MOE) was aimed at identifying sources, occurrence and atmospheric behaviour of atmospheric mercury species (Pirrone et al., 2001; Munthe et al., 2003). Within MOE simultaneous measurements were carried out on a regional scale as depicted in Figure 9.14.

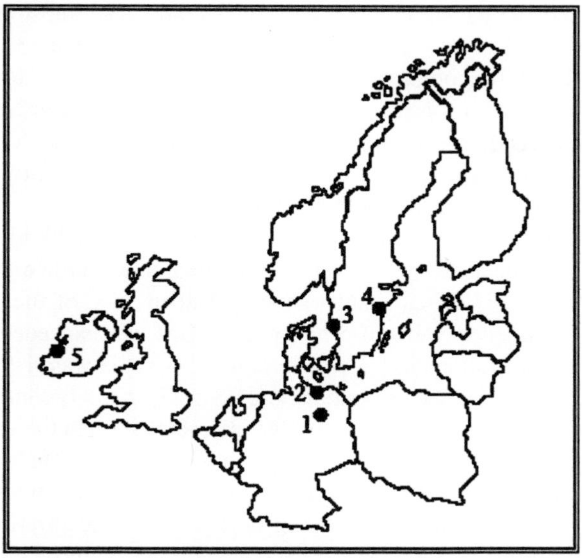

Figure 9.14 MOE measurement sites. (1) Neuglobsow, Germany; (2) Zingst, Germany; (3) Rörvik, Sweden; (4) Aspvreten, Sweden; and (5) Mace Head, Ireland

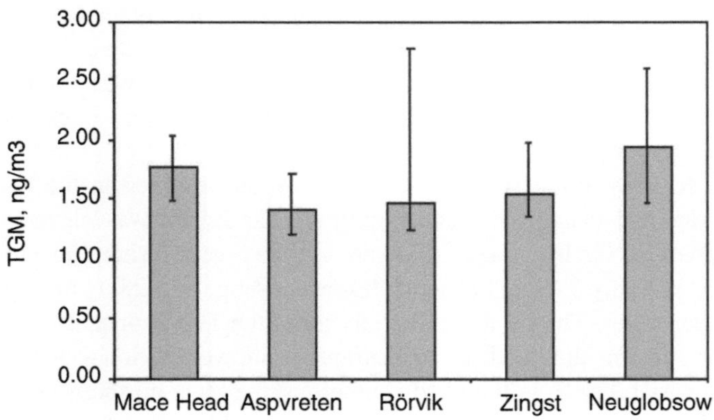

Figure 9.15 Regional differences of TGM concentrations measured during the MOE project Munthe et al., 2003)

As depicted in Figure 9.15 for TGM, the concentration at Mace Head is higher than at the two Swedish stations Rörvik and Aspvreten and more similar to the levels at Zingst on the German Baltic Sea coast. The results from Neuglobsow, Zingst, Rörvik and Aspvreten follow a slightly decreasing trend, which is in line with the location of the main European source areas. There are no local sources of

Hg at Mace Head and the slightly elevated concentrations are most likely caused by re-emissions from the sea surface (Pirrone et al., 2003; Munthe et al., 2003). Wangberg et al. (2001) compared the MOE data with results of a similar project focusing on Southern European sites around the Mediterranean Sea (EU funded project MAMCS) and reported that observed concentrations of TGM (and other species such as TPM and RGM as well) were generally slightly higher in the Mediterranean region compared with Northwest Europe.

Hladíková et al. (2001) reported TGM concentrations from 20 different localities of the Slovak Republic measured in 24-h samples of ambient air in eight times during the period 1996 - 1997. Their results showed that in 34% of the 160 individual cases the WHO guideline value of 5 ng m^{-3} for TGM was exceeded. The range of total mercury concentrations in the ambient air of Slovakia was: 1.13 - 3.98 ng m^{-3} (geom. mean 2.63) in the background area; 2.25-5.27 ng m^{-3} (geom. mean 3.64) in the agricultural areas; 1.73 - 20.53 ng m^{-3} (geom. mean 4.57) in the urban areas; and 1.53 - 39.85 ng m^{-3} (geom. mean 5.28) in the industrial areas. The highest mercury levels occurred in areas with metallurgical industry and coal combustion.

A synoptic view of regionally different background concentrations of TGM in North Central Europe was published by Schmolke et al. (1999). Over a distance of approximately 800 km simultaneous measurements of total gaseous mercury (TGM) were performed at four sampling sites between Stockholm and Berlin. The time resolution of mercury concentration measurements was 5 min. During the sampling period from 26 June to 7 July 1995, in addition to the TGM concentrations, the most common meteorological and air- quality parameters were determined. Comparing the TGM background concentrations at the four sites, a weak but statistically significant south-to-north declining TGM gradient was found. From the most southern sampling site to the northern most site median values of 1.93, 1.78, 1.53 and 1.54 ng m^{-3} TGM were detected.

Compared with the median TGM concentration observed at the two Swedish sites, the regional background concentration near Berlin was elevated by about 25%. Whereas the 0.5 h average TGM varies at the Swedish sampling sites in a very narrow range of only 0.69 ng m^{-3}, a much broader range of 3.28 ng m^{-3} was observed at the southern sites. The short time variability of the TGM concentration measured at the four sites on the south-to-north transect showed regional differences with decreasing variability from the most southern to the most northern site.

In the Mediterranean region recent studies have highlighted the importance of gaseous mercury exchange processes between the atmosphere and surface waters (Sprovieri et al., 2003; Hedgecock and Pirrone, 2004; Horvat et al., 2001; 2003; Gardfeldt et al., 2003; Kotnik et al., 2007; Sprovieri and Pirrone, 2008); the lack of knowledge of the magnitude of these exchange mechanisms is one of the main factors affecting the overall uncertainty associated with the assessment of net fluxes of mercury between the atmospheric and marine environments in the Mediterranean region. Deposition rates of mercury species from the atmosphere to receptors bodies depend on the chemical and physical properties of the species involved and their cycling, from speciated emission, transport, deposition, interaction with biota and possible re-emission to the atmosphere.

In the MBL it has been shown (Pirrone et al., 2000; Hedgecock et al., 2001) that the extremely high chloride ion concentration in sea-salt aerosol is of a great importance in the cycling of RGM species. The sea salt aerosol provides a large excess of complexing ligands for $Hg^{(II)}_{(aq)}$. The sea salt aerosol as well as scavenging Hg^0 produced by the reactions between $Hg^0_{(g)}$ and $OH_{(g)}$, $O_{3(g)}$ and to a lesser extent $H_2O_{2(g)}$ also releases RGM to the gas phase in the form of $HgCl_2$ as a result of the high $Cl^-_{(aq)}$ concentration in the sea-salt aerosol and the fact that the Henry's Law constant for $HgCl_2$ although high is lower than that of HgO (Schroeder and Munthe, 1998). From this point of view, the sea-salt aerosol represents not only a medium by which atmospherically produced $Hg^{(II)}$ could enter the sea, but is also a continuous source of RGM cycling HgO to $HgCl_2$ in the MBL. Photochemical processes in the MBL lead to enhanced oxidation of elemental mercury vapour which would lead to increased concentrations of RGM and TPM via gas-particle interactions. More recently, with the improvement of RGM sampling techniques investigations of the atmospheric chemistry of Hg have been performed across the Mediterranean sea basin (Munthe et al., 2001; Wangberg et al., 2001; 2008; Pirrone et al., 2001).

9.4.5 *Mercury Speciation Analysis*

For TPM, the North-western European distribution pattern has been measured during the MOE project as well. The results are depicted in Figure 9.16. The data represents median values with error bars indicating the 10% and 90% percentiles. Since this data was collected during 4 – 5 campaigns, they include values from all four seasons. Since no direct emissions of particulate mercury were found, a possible explanation for the clear gradient is that TPM is formed after emissions and the measured fractions are actually secondary TPM, i.e. formed in the air mass during transport. A plausible mechanism is adsorption of RGM on existing particles. Reactive gaseous mercury (RGM) is an operationally defined gaseous Hg fraction present in ambient

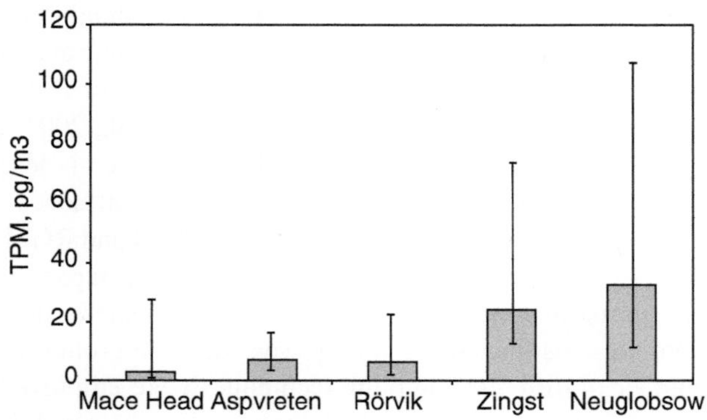

Figure 9.16 Regional differences of TPM concentrations measured during the MOE project (Munthe et al., 2003)

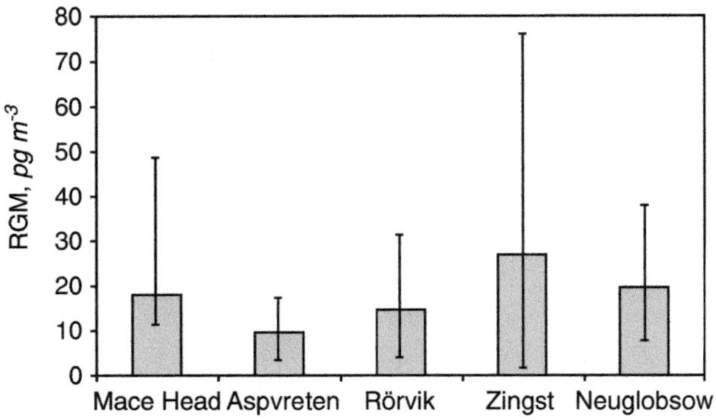

Figure 9.17 Regional differences of RGM concentrations measured during the MOE project (Munthe et al., 2003)

air. However, it is believed that RGM to the most part consist of Hg dichloride ($HgCl_2$), but other divalent Hg species are also possible. MOE results for RGM are given in Figure 9.17.

The results of a number of studies performed first within the MAMCS project (Pirrone et al., 2001; Sprovieri and Pirrone, 2006) and after in MERCYMS project (Hedgecock and Pirrone 2004; Wangberg et al., 2008) highlighted that photochemical processes occur in the MBL of the Mediterranean sea basin, leading to enhanced oxidation of Hg^0 and thus increasing of RGM and PM concentrations via gas-particle interactions.

This suggests that catalytic destruction involving $BrO_{(g)}$ or another Br containing is compound responsible for the sharp increase in the oxidation rate of Hg^0 and the formation of less volatile $Hg^{(II)}$ compounds (Hedgecock and Pirrone, 2001; Hedgecock et al., 2003; 2006). The inclusion of the reactions between halogen molecules and atoms and Hg^0 (Ariya et al., 2002) in a modelling study (Hedgecock and Pirrone, 2004), suggest that the lifetime of Hg^0 may be significantly shorter than 12 months in the MBL. The rate of oxidation of Hg by O_3 could have particular relevance for the Mediterranean MBL because of the consistently high BL concentrations of O_3 throughout the year, and particularly in spring and summer. High $Hg^{(II)}$ concentrations have also been observed in the MBL of the North Atlantic (Bermuda) (Mason et al., 2001) and the Pacific (Laurier et al., 2003) ruling out the possibility of anthropogenic source influences. However in semi-closed sea such as the Mediterranean basin it is possible that they do have an influence.

All methods employed for measurements of TGM, TPM and RGM in the framework of MAMCS project were evaluated in a field intercomparison in Sassetta, Tuscany (Italy) approximately 100 km south of Pisa. The site is situated in a hilly area 25 km from the coastline. All sampling was conducted roughly within a 10m distance. A number of different sampling techniques were employed for different mercury species. A summary of the methodology used is given in Table 9.13. More detailed descriptions of the applied methods are given in Munthe et al., 2001. TGM was measured from the start of the intercomparison exercise using all four methods

Table 9.13 Applied methods for sampling and analysis of atmospheric mercury species

Method	Hg species	Analytical method
Tekran	TGM	CVAFS. semi-continuous
Gardis	TGM	CVAFS. semi-continuous
Manual gold trap	TGM	CVAAS
Charcoal adsorbents-1 NAA	TGM	INAA
Teflon filters	TPM	Acid digestion, $SnCL_2$-CVAFS
Miniature quartz fibre filters	TPM	Thermal desorption, CVAFS
Cellulose acetate filters	TPM	Acid digestion, $SnCL_2$-CVAFS
Glass fibre filters	TPM	Acid digestion, $SnCL_2$-CVAFS
Mist Chamber with 0.1 M HCI	RGM	$SnCL_2$-CVAFS
Tubular KCI-coated denuders	RGM	Thermal desorption, CVAFS
Annular KCL-coated denuders	RGM	Thermal desorption, CVAFS/ Acid rinse-$SnCI_2$-CVAFS

Table 9.14 Average. median and range of observed concentrations of TGM, TPM and RGM in Tuscany, June

	TGM *(ng m^{-3})*	TPM *(ng m^{-3})*	RGM *(ng m^{-3})*
Average	1.98	56	22
Median	1.93	25	22
Maximum	3.38	314	41
Minimum	1.28	13	3

and the results showed that TGM measured with the different techniques are in excellent agreement, provided that careful site selection is made and stringent operational procedures are followed. In Table 9.14, the results of all applied methods for TGM, TPM and RGM are summarised.

The results of RGM and TPM measurements show a larger variability probably due to use of different sampling times but mainly this reflects the analytical difficulties for this operationally defined species. As for TGM, the variability in RGM and TPM results is higher in the beginning of the campaign than in the final three samples.

This indicates that optimisations of the sampling set-up made during the initial phase of the campaign were beneficial for the measurements. Synchronized seasonal field campaigns of 14 days duration were performed during the MAMCS and MOE projects around the Mediterranean and in North Europe from the end of 1998 to 1999; one of the major findings was that TPM and RGM concentrations were generally higher in the Mediterranean area than over northern Europe in spite of the higher density of industrial installations and urban centres in northern compared to southern Europe (Pirrone et al., 2001; Wangberg et al. 2001; Munthe et al. 2001; Pirrone et al. 2001). The most probable interpretation which comes to mind is higher emission rates and/or more active atmospheric transformation processes in the Mediterranean basin. Photochemical processes in the MBL lead to enhanced oxidation of elemental mercury vapour which would lead to increased concentrations of RGM and TPM via gas-particle interactions. The time schedule of each campaign is shown in Table 9.15 and the locations

Table 9.15 MAMCS, MOE multi-sites measurement campaigns

Campaign	Start date	Stop date
MAMCS-1/MOE-1	23-11-1998	06-12-1998
MAMCS-2/MOE-2	15-02-1999	01-03-1999
MAMCS-3/MOE-3	03-05-1999	17-05-1999
MAMCS-4/MOE-4	19-07-1999	02-08-1999
MOE-5	01-11-1999	15-11-1999

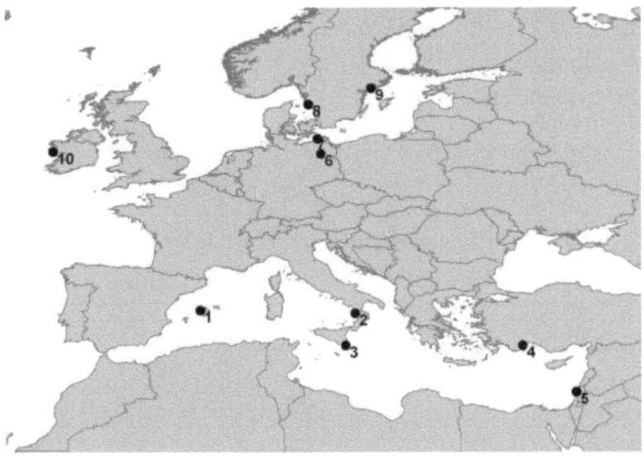

Figure 9.18 MAMCS, MOE measurement sites: 1. Mallorca (39°40'30" N, 2°41'36"E); 2. Calabria (39°25'N, 16°00'E); 3. Sicily (36°40'N, 15°10'E); 4. Turkey (36°28'12"N, 30°20'24"E); 5 Israel (32°40'N, 34°56'E); 6. Germany (53°08'34"N, 13°02'00"); 7. Germany (54°26'14"N, 12°43'30"E); 8. Sweden (57°24'48"N, 11°56'06"E); 9 Sweden (58°48'00", 17°22'54"E); 10. Ireland (53°20'N, 9°54'W)

of the measurement sites are shown in Figure 9.18. Three airborne mercury species, TGM, TPM and RGM were simultaneously measured at all sites along with meteorological parameters.

Average TGM, TPM and RGM values from the MOE and MAMCS campaigns at different seasons, are shown in Figures 9.19 – 9.21. The average TGM concentrations varied between 1.6 and 2.4 ng m^{-3} with no significant seasonal variations mainly due to the relatively stable global/hemispheric background concentration and only occasionally shows higher values for the influence of major sources. Except for the first campaign, the data indicates that the TGM is slightly but significantly higher in the Mediterranean area than in North Europe.

These findings are justified by several reasons: (a) natural emissions both from diffuse sources (Hg enriched minerals) and volcanoes (Ferrara et al., 2000; Pirrone et al., 2001) characterizing the Mediterranean area; (b) enhanced re-emission fluxes of mercury from the sea surface which are partly governed by sunlight and temperature and the warmer climate in the Mediterranean basin.

A similar trend has also been observed for the Mediterranean RGM and TPM concentrations probably due to higher emission rates and/or more active atmospheric transformation processes.

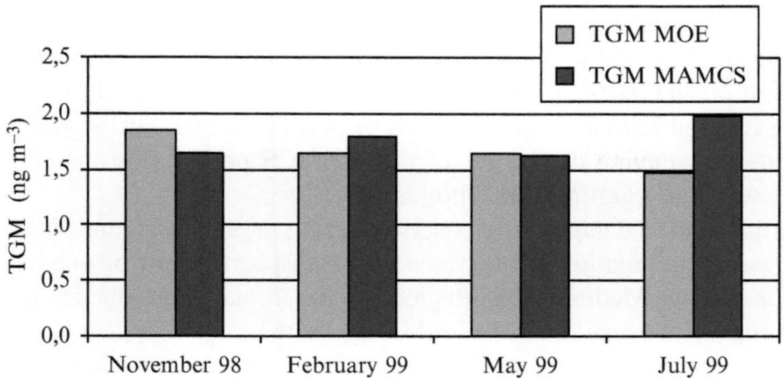

Figure 9.19 Average TGM, values obtained at campaign MOE 1-5 and MAMCS 1-4. The TGM value from the MAMCS campaign 4 should be regarded with some caution since it is based on measurements from two sites only

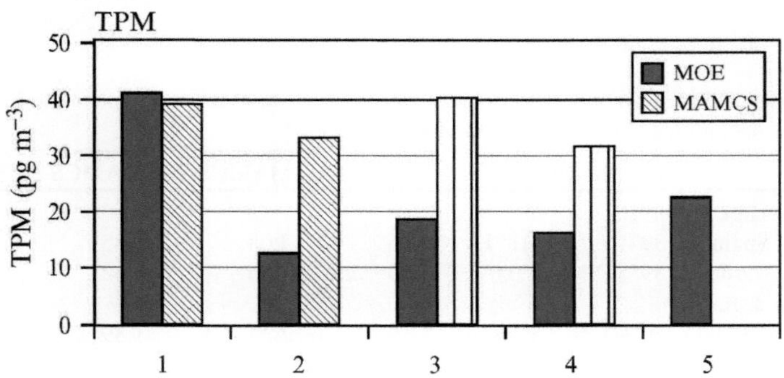

Figure 9.20 Average TPM values obtained at campaign MOE 1-5 and MAMCS 1- 4

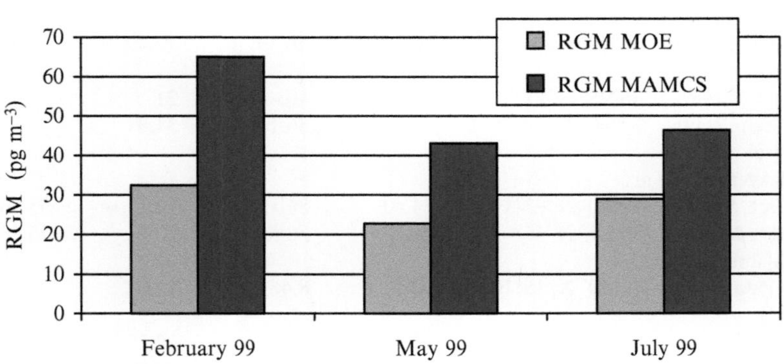

Figure 9.21 Average RGM values obtained at campaign MOE 1- 5 and MAMCS 1- 4

Photochemical processes in the marine boundary layer may lead to enhanced oxidation of elemental mercury vapour which would lead to increased concentrations of RGM and possibly TPM, via gas-particle interactions. Table 9.16 shows the TGM, RGM and TPM average values observed at the five sites in the Mediterranean during the 4 sampling campaigns of the MAMCS project (Pirrone et al., 2001) (http://www.cs.iia.cnr.it/MAMCS/project.htm).

In order to fill the gaps firstly observed during the MAMCS and MOE projects and to assess the relationship between the atmospheric input of mercury and its compounds to the Mediterranean Region and the formation/production of the most toxic forms of mercury (i.e., MeHg, Me_2Hg) in the marine system, five intensive sampling campaigns at five locations around the Mediterranean sea basin have been performed within the MERCYMS project (Pirrone, 2006). The sampling sites are shown in Figure 9.22 (http://www.cs.iia.cnr.it/MERCYMS/project.htm).

The time schedule of measurement campaigns is reported in Table 9.17. The sampling time schedules applied with manual and automatic sampling systems are shown in Table 9.18, whereas the site locations along with mercury species and methods used are shown in Table 9.19. The measurement periods was chosen to

Table 9.16 TGM, RGM and TPM average values observed at the five sites in the Mediterranean during the 4 sampling campaigns of the MAMCS project

Sites	Coordinates	MAMCS-1	MAMCS-2	MAMCS-3	MAMCS-4
TGM Average (*ng m*$^{-3}$)					
Mallorca, Spain	39°40' N, 2°41' E	3.16	3.08	3.85	4.15
Fuscaldo Marina Calabria, Italy	39°25' N, 16°0.0' E	1.30	1.86	1.42	1.09
Porto Palo, Sicily, Italy	36°40' N, 15°10' E	1.34	2.37	1.89	2.18
Antalya, Turkey	36°28' N 30°20' E	1.68	8.71	1.34	---
Neve Yam, Haifa, Israel	32°40' N, 34°56' E	1.83	0.90	1.45	---
RGM Average (*pg m*$^{-3}$)					
Mallorca, Spain	39°40' N, 2°41' E	1.88	99.59	76.02	-----
Fuscaldo Marina Calabria, Italy	39°25' N, 16°0.0' E	40.18	24.84	46.74	35.47
Porto Palo, Sicily, Italy	36°40' N, 15°10' E	90.14	46.39	77.49	29.48
Antalya, Turkey	36°28' N 30°20' E	-----	10.44	21	-----
Neve Yam, Haifa, Israel	32°40' N, 34°56' E	-----	36.14	34.81	-----
TPM Average (*pg m*$^{-3}$)					
Mallorca, Spain	39°40' N, 2°41' E	34.40	86.12	44.11	33.56
Fuscaldo Marina Calabria, Italy	39°25' N, 16°0.0' E	26.32	28.55	22.71	45.50
Porto Palo, Sicily, Italy	36°40' N, 15°10' E	5.57	8.46	11.02	9.11
Antalya, Turkey	36°28' N 30°20' E	14.66	14.39	25.25	65.25
Neve Yam, Haifa, Israel	32°40' N, 34°56' E	115.39	27.3	97.89	4.19

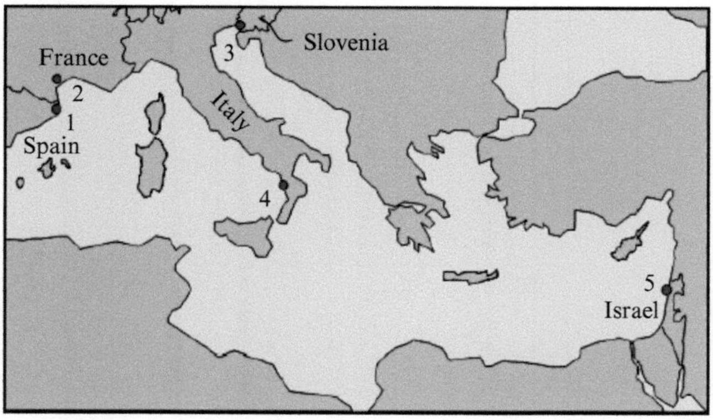

Figure 9.22 MERCYMS coastal measurement sites: 1 Cabo de Creus, 2 Meze, Thau Lagoon, 3 Piran Marine, 4 Fuscaldo, 5 Neve Yam

Table 9.17 Time-schedule of sampling campaigns in the framework of the MERCYMS Project

Coastal campaigns	Start date	Stop date
Autumn 2003	20-10-2003	03-11-2003
Winter 2004	19-01-2004	02-02-2004
Spring 2004	26-04-2004	10-05-2004
Summer 2004	19-07-2004	02-08-2004

Table 9.18 Sampling type and time schedule applied during MERCYMS campaigns

Specie	Manual	Automatic
TGM	Two 12 h average samples per 24 h period. starting 7:30 and 19:30 UTC	Continuous 5 mm average samples yielding 288 samples per 24 h period
RGM	Two 12 h average samples per 24 h period. starting 7:30 and 19:30 LTC	One 2 h sample per every 3 h period, alternatively one 5.25 h sample per every 6 h period
TPM	Two 12 h average samples per 24 h period. starting 7:30 and 19:30 UTC	One 2 h sample per every 3 h period

cover four seasons. Table 9.20 shows the average TGM, RGM and TPM values from coastal stations during four season.

9.4.6 *Mercury Measurements (Including Aircraft) Related to Emissions, and Source Attribution*

Many emission inventories for mercury have been compiled but rarely constrained using observations of ambient air concentrations with a known quality. Slemr et al.

Table 9.19 Site locations and mercury species and methods used during MERCYMS project

Sites	Coordinates	TGM	RGM	TPM
Cabo de Creus, Spain	42°19.2' N, 3°18.9' E	Automatic measurements with 5 min time resolution, Tekran Model 2537A	Manual 12 h samples on KCl annular denuders. Automatic measurements during the spring and summer campaigns	Manual 24 h samples on TPM mini-traps
Mèze,Thau Lagoon, France	43°25' N, 3°35' E	Manual Au traps 2h samples	Manual 12 h samples on KCl annular denuders	Manual 12 h samples on TPM mini-traps
Piran Marine, Slovenia	45°32.9' N, 13°33.0' E	Manual Au traps 12 h samples (24 h during winter)	Manual 12 h samples on KCl annular denuders (24 h during winter)	Manual 12 h samples on TPM mini-traps (24 h during winter)
Fuscaldo/S. Lucido, Calabria, Italy	39°25' N, 16°0.0' E	Automatic measurements with 5 min time resolution, Tekran Model 2537A	Automatic sampling on KCl annular denuders using the Tekran 1130 speciation unit	Automatic sampling of fine particulate mercury, using the Tekran 1130 speciation unit
Neve Yam, Haifa, Israel	32°40' N, 34°56' E		Manual 12 h samples on KCl denuders	Manual 12 h samples on TPM mini-traps

Table 9.20 Average TGM, RGM and TPM values from coastal stations during four seasons

	Fall			Winter			Spring			Summer		
Site	TGM	RGM	TPM	TGM	RGM	TPM	TGM	RGM	TPM	TGM	RGM	TPM
Cabo de Creus, Spain	1.6	2.2	9.6	1.5	0.24	9.1	2.0	1.2	9.5	2.1	1.2	11.2
RGM Automatic		–	–		–			1.9			6.9	
Mèze,Thau Lagoon, France	1.6	8.6	3.0	2.9	41.9	82	1.9	10.4	26.7	3.3	191	662
Piran Marine, Slovenia	-	4.5	-	0.8	1.0	18.7	1.8	2.5	7.4	4.0	15.4	9.4
Fuscaldo/S.Lucido, Italy	1.3	1.6	1.0	1.9	4.2	6.1	1.8	2.1	1.7	1.6		
Neve Yam, Israel	Day 1.19 Night 0.78	33	89	Day 0.80 Night 0.50	2.2	3.9	Day 0.86 Night 0.46	10.7	40.1	Day 1.24 Night 1.21	8.3	22.7

(2006) have derived Hg/CO, Hg/halocarbon, and Hg/CH_4 emission ratios from pollution episodes observed during the long-term mercury monitoring at the Mace Head Atmospheric Research Station in Ireland. The authors conclude that mercury emissions calculated from the emission ratios and the European emissions of the above gases are in reasonable agreement with the estimated anthropogenic total mercury emissions of 250 Mg yr^{-1} in 1995. However, the measurements encompass almost exclusively elemental mercury whose anthropogenic emissions are estimated to be only 152 Mg yr^{-1}. Slemr et al. propose several hypotheses to explain this discrepancy, including natural sources, underestimation of the emissions of elementary mercury, and erroneous speciation of anthropogenic emissions.

One of the very few aircraft measurements in Europe have been carried out on 13 June 1996, during a level flight from Munich to Halle at an altitude of 900 m a.s.l. and back at 2500 m a.s.l. Ebinghaus and Slemr (2000) showed that TGM was horizontally evenly distributed over a distance of 400 km. TGM concentration at an altitude of 2500 m a.s.l. (in the free troposphere) was 1.635 ± 0.094 ng m^{-3} (n=22) slightly lower than at 900 m a.s.l. (in the mixing layer) 1.774 ± 0.101 ng m^{-3} (n=17). Higher TGM concentrations of 2.190 ± 0.255 ng m^{-3} (n=2) during the southernmost part of the flight south of Munich and of 2.321 ± 0.133 ng m^{-3} (n=8) at the summit of the mountain Wank was attributed to a different air mass. Substantially higher TGM concentrations were observed in the mixing layer downwind of a former chlor-alkali plant in East Germany. From the downwind/upwind concentration difference and the prevailing wind conditions, a mercury emission of about 0.4 kg d^{-1} was estimated for the plant area.

9.5 Measurements of Air Concentrations in Asia

9.5.1 Remote Locations

Kim et al. (1996) have reported TGM concentrations from 13 remote mountainous sampling stations in Korea, for the time period October 1987 through February 1993. The TGM concentrations determined during these field campaigns were found to be in the range of 1.48 to 8.00 ng m^{-3}, 75% of them spanning between 2 to 5 ng m^{-3}. Kim et al. concluded that the observed Hg levels and the wide spreadness of the observed data suggests that Hg pollution in the Korean atmosphere may result in generally enhanced levels compared to other Northern hemispheric regions. This finding is supported by Sohn et al. (1993) who reported rural concentrations in Korea to be between 1.0 to 7.0 ng m^{-3} (mean 3.8 ng m^{-3}) for the years 1988 – 1989.

Liu et al. (2002) have reported Chinese data for two remote sampling locations for 1998 in the Beijing area to be between 3.1 – 5.3 ng m^{-3} in winter and 4.1 – 7.7 ng m^{-3} in summer. Wang et al. (2007) also found higher concentrations in summer compared to winter for a remote site on the Mount Waliguan, China to be 1.7 ± 1.1 ng m^{-3} in summer and 0.6 ± 0.08 ng m^{-3} in winter. TGM data for a remote site in the Yangtze Delta were higher and expressed a more pronounced variability (5.4 ± 4.1 ng m^{-3}). Different seasonal variation patterns of TGM concentration were found between urban and remote sites. In Beijing (urban) the highest TGM concentration was in winter and the lowest in summer, while in Mt. Waliguan the TGM concentration in summer was higher than that in winter. These indicated that different processes and factors controlled TGM concentration in urban, regional and remote areas.

TGM monitoring data for Korean GAW station (An-Myun Island) have been published by Nguyen et al. (2007). Measurements were routinely recorded on An-Myun Island off the coast of Korea between December 2004 and April 2006. The mean TGM concentration for the measurement period was 4.61 ± 2.21 ng m^{-3} with a range of 0.10 – 25.4 ng m^{-3}. Analysis of the seasonal patterns indicated TGM concentration levels generally peaked in spring, while reaching a minimum in summer. Nguyen et al. (2007) concluded that Hg concentration levels at An-Myun Island can be affected intensively by trans-boundary input processes over certain periods of time. and that its springtime dominance hence suggests combined effects of various local source processes and the meteorological conditions favourable for the massive air mass transport phenomenon (such as Asian Dust storms).

9.5.2 Urban Locations (including mining areas)

For Seoul, Korea Sohn et al. (1993) have reported TGM data for the years 1988 – 1989 to be in the range from 5.0 to 88.8 ng m^{-3} with a mean of 25.1 ng m^{-3}. For September 1997 and May/June 1998 Kim et al. (2001b) have found mean concentrations of TGM in Seoul (1999–2000) for the two study periods were computed as 3.94 and 3.43 ng m^{-3}, respectively. For January 1999 to August 2000 the mean hourly TGM concentration was 5.26 ± 3.27 ng m^{-3} and when divided seasonally, the highest mean of 6.01 ng m^{-3} was observed during winter. Kim et al. (2001a) suggested that this most likely due to anthropogenic sources such as household heating systems.

Urban data for Beijing, China, show a similar distribution between summer and winter. Liu et al (2002) and Wang et al. (2007) give winter concentration ranges between 8 and 25 ng m^{-3}, and lower summer values between 5 and 13 ng m^{-3}, with autumn and spring concentration in between. Feng et al. (2004) have reported TGM concentration data for Guiyang city, in 2001. The mean TGM concentration at this site is 8.40 ng m^{-3} on the basis of one year observation. Feng et al. concluded that TGM concentrations in Guiyang are significantly elevated compared to the continental global background values and that coal combustion from both industrial and domestic uses is probably the primary atmospheric source. Similar data were obtained earlier (Feng et al., 2003) during 4 measurement campaigns in 2000 and 2001 in Guiyang. The seasonal geometric mean of TGM in Guiyang was 7.45 ng m^{-3} for winter, 8.56 ng m^{-3} for spring, 5.20 ng m^{-3} for summer and 8.33 ng m^{-3} for autumn. The overall average TGM covering the sampling periods was 7.39 ng m^{-3}. Data for Guangzhou were reported to be 13.5 ± 7.1 ng m^{-3} (Wang et al., 2007).

9.5.3 Temporal Trends at Single Locations

In 1995, Nakagawa published data for TGM levels in air that have been monitored over a period of 17-years (1978-1994) at three different stations in Japan. Nakagawa concluded that TGM was at minimum levels at the three sampling stations in 1988 and was corresponding to background levels. Afterwards Nakagawa found TGM levels rising steadily up to the present time and claimed that the current atmospheric concentrations of total mercury inferred from measurements of air sampled at the Japanese stations has an increasing tendency every year in the time period 1978-1994 (Nakgawa, 1995).

9.5.4 Mercury Speciation Analysis

Xiu et al. (2005) have published data of size-fractionated particulate mercury (TPM-species) in Shanghai, China, collected from ambient air between March 2002 to September 2003. Speciation data are operationally defined as (i) volatile particle-phase mercury (VPM), (ii) reactive particle-phase mercury (RPM) and (iii) inert particle-phase mercury (IPM). Concentrations for particulate phase mercury as published by Xiu et al. (2005) are:

- volatile particle-phase mercury (VPM) = 0.058 – 0.252 ng m^{-3}
- reactive particle-phase mercury (RPM) = 0.148 – 0.398 ng m^{-3}
- inert particle-phase mercury (IPM) = 0.233 – 0.529 ng m^{-3}

Fang et al. (2001a; 2001b) have measured particulate phase mercury concentration in Changchun, an urban location in China, Changchun between July 1999 to January 2000. During non-heating season they found PM concentration levels in air between 0.022 to the 0.398 ng m^{-3} with an average of 0.145 ng m^{-3}. During the heating season PM concentration in air were reported to be 0.148-1.984 ng m^{-3} with an average of 0.461 ng m^{-3}. Coal burning and wind-blown soil material were identified to be two important sources of Hg(p) for the Changchun area.

For Guiyang, an urban location in southwest China (Shang et al. (2003) have measured average concentrations for TGM to be 7.09 ng m^{-3}, and RGM as 37.5 pg m^{-3} in March 2002. They concluded that TGM concentrations in Guiyang are significantly elevated comparing to the global background values, whereas RGM concentrations are only slightly higher than the reported values in remote areas in Europe and the USA.

9.5.5 *Mercury Measurements (Including Aircraft) Related to Emissions, and Source Attribution*

In situ mercury emission fluxes from soil in Lanmuchang Hg-Tl mining area, in south-western Guizhou, China, were measured using the dynamic flux chamber method (December 2002 and May 2003). Huge mercury emission fluxes from soil were obtained in the mining area, ranging from 623 to 10,544 ng m^{-2} h^{-1} with the maximal mean Hg flux of 2,283 ± 2,434 ng m^{-2} h^{-1} (Wang et al. 2005).

Tomiyasu et al. (2000) have shown that Hg emissions from localized sources are reflected in ambient air concentrations. For Kagoshima City, Japan, the influence of mercury emitted from Sakurajima has been measured between January 1996 to January 1997. The mercury concentration obtained was in the range 1.2-52.5 ng m^{-3} (mean 10.8 ng m^{-3}, n = 169). Values of 8.1 ± 5.3 ng m^{-3}, 14.8 ± 7.9 ng m^{-3}, 13.9 ± 11.7 ng m^{-3} and 4.4 ± 1.6 ng m^{-3} (mean ± S.D.) were obtained for spring, summer, autumn and winter, respectively. For July 1999 to March 2002 the concentration ranged from 1.24 to 29.4 ng m^{-3} in the daytime and 1.38 to 14.8 ng m^{-3} overnight. The concentration gradient showed a high degree of dependency on solar radiation. These observations suggested that mercury is emitted from the ground with an increase in solar radiation and rise of the temperature in the daytime and deposits with the fall of temperature at nighttime. The influence of rain and volcanic activity on the periodic variation of mercury is also discussed in detail by Tomiyasu et al., (2006).

Kim et al. (2002) have published data from flux measurements over a paddy field in Kang Hwa, Korea, carried out in March 2001. The Hg fluxes over a bare paddy field at Hari, were comparable to those reported in the literature for polluted environments and showed clear diurnal pattern with maximum (< 600 ng m^{-2} h^{-1}) during midday and the minimum (~100 ng m^{-2} h^{-1}) at nighttime. The mean air concentrations of TGM was found to be elevated compared to global background concentrations, i.e. 3.72 ng m^{-3}.

Friedli et al. (2004) reported that during the Aerosol Characterization Experiment (ACE-Asia), research flights the air at all sampled altitudes contained concentrations of atmospheric mercury above the global background. Highest mixing ratios for Hg^0 were found in industrial plumes exiting China (6.3 ng m^{-3}), Korea (3 ng m^{-3}), and Japan (3 ng m^{-3}).

9.6 Measurements of Air Concentrations in Africa

9.6.1 *Monitoring Networks and Trends*

The monitoring of total gaseous Hg (TGM) was established at the Cape Point Global Atmospheric Watch (GAW) station in September 1995. Baker et al. (2002) presented the first data obtained until June 1999. Atmospheric Hg concentrations were found to be fairly homogeneous fluctuating between 1.2 – 1.4 ng m^{-3}. Whilst no significant diurnal variation is detectable, a slight seasonal variation with a TGM minimum in March-May and maximum in June-August was observed. A minimum annual TGM concentration was detected in 1997. The Cape Point GAW station was found to constitute a suitable site for monitoring TGM concentrations in the Southern Hemisphere (SH).

9.6.2 Mercury Measurements (Including Aircraft) related to Emissions, and Source Attribution

During mid-January 2000 the plume from a fire, which destroyed 9000 ha of mixed vegetation in the southern part of the Cape Peninsula, passed over the Cape Point Global Atmosphere Watch station (34°S, 18°E). Measurements of total gaseous mercury (TGM) made during this episode provided Hg/CO and Hg/CO_2 emission ratios of $(2.10 \pm 0.21) \times 10^{-7}$ and $(1.19 \pm 0.30) \times 10^{-8}$ mol/mol, respectively (Brunke et al. 2001).

9.7 Summary and Conclusion

Mercury concentration measurements in ambient air of documented and accepted quality are available since the mid 1970 and concentration data are available for both hemispheres. A significant impulse on international mercury research was achieved by the implementation of new automated analysers in the early 1990's. Long-term monitoring of atmospheric mercury with high time resolution has been started at Alert, Canada (January 1995) and Mace Head, Ireland (September 1995), followed by numerous other sites since then. Field intercomparisons have been carried out and included the comparison of new automated analysers with traditional manual methodologies. One major conclusion derived from these studies is that good agreement of TGM concentration data in air determined with different techniques, including techniques dating back to the 1970's, makes a combination of these data sets with contemporary ones from different regions of the world feasible.

General scientific consensus exists about the current global background concentration of airborne mercury which is taken as ca. 1.5 to 1.7 ng m^{-3} in the Northern Hemisphere and ca. 1.1 to 1.3 ng m^{-3} in the Southern Hemisphere. Seasonal variations of TGM concentrations are observed at almost all sites with sufficient data coverage. Most sites show higher concentrations in winter and spring, and lower in summer and fall. It is suggested that the meteorological seasonal variability is the most important factor in the establishment of the observed seasonal cycles of the TGM concentrations. Temporal dynamics of TGM and also Hg in wet deposition are complex, with the magnitude of diurnal and seasonal changes often being larger than annual changes.

Canadian data show that the seasonality of Hg concentration in precipitation exhibits an opposite pattern to TGM air concentrations with higher concentrations during the summer months. Two possible factors have been suggested for this seasonal behaviour: increased particle scavenging capacity of rain relative to snow and/or increase in the oxidation of Hg, either in cloud or in the gas phase, during summer. However, sites with the highest observed concentrations do not necessarily have the highest Hg deposition per unit and surface area since the deposition is also dependent on the precipitation amount. Regional differences, temporal trends and potential sources and source regions can be identified by monitoring, especially when carried out in networks. In principal, an increase of the global atmospheric pool should also be reflected in the background concentration of mercury in ambient air. Fitzgerald's initiative for the installation of a global AMNET has partly been accomplished on a regional scale within the CAMNet that may be considered as seminal in this respect. CAMNet has revealed a decreasing trend for TGM at a number of rural sites for the time period 1995 to 2005. The largest declines were observed close to urban areas of Toronto and Montreal. This is supported by data from United States giving evidence that TGM concentrations at rural locations on the eastern USA seaboard downwind of major urban and industrial centers are decreasing.

The changes are mostly driven by local or regional changes such as cutbacks in emissions. Many of the TGM changes are comparable with the overall trends of total mercury concentrations in precipitation, again reflecting local or regional emission reductions in decreasing concentration levels. For remote sites in Europe and North America it could be shown that only slight decreases or no statistically significant trend in the TGM concentration exist over the same time period. This is in contrast with emission estimates especially for Europe, where drastic reductions have taken place over the past 10 to 20 years. It has been suggested that TGM concentration in the North-western European atmosphere are to a larger extent dominated by hemispherical background than before, i.e. European emission reductions may be over-compensated by increasing emissions in other Northern hemispheric regions. It should be noted that for the same time period and same set of stations (1995 to 1998 vs. 1999 to 2002) a reduction in deposition of 10 to 30% was found which can possibly be related to emissions controls in Europe.

Deposition networks on a regional scale are essential for understanding the spatial and temporal patterns of Hg in wet deposition, but without at least TGM measurement and additional speciated Hg measurements wherever possible in conjunction with precipitation measurements, little can be inferred about the processes responsible for controlling Hg deposition. It is still not well understood which sites would respond most quickly to emission reductions in certain industrial sectors, or if enhanced deposition events are independent of local or regional sources and supplied predominantly by the global pool being caused by natural variations in photochemistry and meteorology. US studies show that Hg deposition can vary by 2-fold between an urban and a rural site, TGM might vary by 50 – 75%, whereas TPM may vary 5 – 10 fold between sites. Regional differences of individual speciation have also been reported for Europe. Observed concentrations of TGM, TPM and RGM are generally slightly higher in the Mediterranean region than in Northwest Europe.

In general, it may be concluded that RGM in situ formation is very important in both the marine boundary layer under cloud-free conditions, and in the free troposphere, where concentrations reach those of the most polluted urban atmosphere. Additionally, because of the diurnal nature of photochemistry and boundary layer/free tropospheric exchange, RGM, and to a lesser extent Hg^0, and TPM generally have prominent diel cycles. Superimposed on the natural variability is the episodic nature of advection of more or less polluted air masses to a sampling site from urban/industrial/mining sources. With regard to long-term changes in RGM or TPM concentrations, the data do not go back far enough or have sufficient spatial coverage to appropriately address this question.

Asian emissions are considered to be of global importance and are suggested to be rapidly increasing in the past decade. The importance of Asian emissions is obvious from recent emission estimates. Furthermore, experimental data are showing long-range transport across the Pacific and suggest a significant underestimate of Asian mercury emissions. However, potentially increased Asian emissions are neither reflected in the long-term measurement of TGM at Mace Head (1995 – 2007), nor in the precipitation data of the North American MDN. The reason for this is not yet clear however, it was hypothesized that atmospheric mercury cycling is possibly going on a faster rate than previously thought.

In general it can be concluded that monitoring at single locations or in networks is a very useful scientific tool in order to identify regional differences, temporal trends and for source attribution. Monitoring is necessary to evaluate the effectiveness of control measures, as demonstrated on regional scales. In connection with air quality data, mercury monitoring can be used to verify and/or improve emission estimates. Monitoring data can give new insights in the mercury cycling on different temporal and spatial scales, due to "unexpected findings", such as AMDEs as a prominent example. They help us to improve our understanding of the global mercury cycle in order to evolve purposeful regulations on an international scale. The value of long-term atmospheric mercury monitoring and the need for additional sites is obvious, especially in the remote Southern Hemisphere.

References

Abbott, M. C-J. Lin, P. Martian, J. Einerson, 2007 Atmospheric Mercury Near Salmon Falls Creek Reservoir in Southern Idaho, Idaho Department of Environmental Quality report, INL/EXT-06-12048, accessed at http://www.osti.gov/bridge/product.biblio.jsp?osti_id= 923511, May 14, 2008.

Amouroux, D., Wasserman, J.C., Tessier, E., Donard, O.F.X., (1999), Elemental mercury in the atmosphere of a tropical Amazonian forest (French Guiana), Environmental Science and Technology 33 (17), pp. 3044-3048

Ariya, P.A., Khalizov, A., Gidas, A. (2002) Reactions of gaseous mercury with atomic and molecular halogens: kinetics, product studies, and atmospheric implications. J. Phys. Chem. A 106, 7310–7320.

Aspmo Katrine, Pierre-Alexis Gauchard, Alexandra Steffen, Christian Temme, Torunn Berg, Enno Bahlmann, Cathy Banic, Aurelien Dommergue, Ralf Ebinghaus, Christophe Ferrari,

Nicola Pirrone, Francesca Sprovieri and Grethe Wibetoe, (2005), Measurements of atmospheric mercury species during an international study of mercury depletion events at Ny-Ålesund, Svalbard, spring 2003, How reproducible are our present methods? Atmospheric Environment, Volume 39, Issue 39, 7607–7619

Baker, P.G.L., Brunke, E.-G., Slemr, F., Crouch, A.M., (2002), Atmospheric mercury measurements at Cape Point, South Africa, Atmospheric Environment 36 (14), pp. 2459–2465

Banic, C.M., Beauchamp, S.T., Tordon, R.J., Schroeder, W.H., Steffen, A., Anlauf, K.A., Wong, H.K.T., (2003), Vertical distribution of gaseous elemental mercury in Canada, Journal of Geophysical Research D: Atmospheres 108 (9), pp. ACH 6–1 ACH 6-14

Beauchamp, S., R. Tordon, L. Phinney, K. Abraham, A. Pinette, A. MacIntosh, A. Rencz, H. Wong, J. Dalziel, 2002. Air-surface exchange of mercury over natural and impacted surfaces in Atlantic Canada. Geochemistry: Exploration, Environment and Analysis. 2 (2), pp. 157–165.

Bedowska, M., Falkowska, L., Lewandowska, A., (2006), Airborne trace metals (Hg, Cd, Pb, Zn) of the coastal region, Gulf of Gdańsk, Oceanological and Hydrobiological Studies 35 (2), pp. 159–169

Berg, T., Fjeld, E., Steinnes, E., (2006), Atmospheric mercury in Norway: Contributions from different sources, Science of the Total Environment 368 (1), pp. 3–9

Biester, H., Kilian, R., Franzen, C., Woda, C., Mangini, A. and Schöler, H.F. 2002. Elevated mercury accumulation in a peat bog of the Magellanic Moorlands, Chile (538S)—An anthropogenic signal from the Southern Hemisphere. Earth Planet. Sci. Lett. 201, 609–620

Bindler, R., Renberg, I., Appleby, P.G., Anderson, N.J. and Rose, N.L. 2001. Mercury accumulation rates and spatial patterns in lake sediments from west Greenland: a coast to ice margin transect. Environ. Sci. Technol. 35, 736–741.

Blanchard, P., Froude, F.A., Martin, J.B., Dryfhout-Clark, H., Woods, J.T., (2002), Four years of continuous total gaseous mercury (TGM) measurements at sites in Ontario, Canada, Atmospheric Environment 36 (23), pp. 3735–3743

Brunke, E.-G., Labuschagne, C., Slemr, F., (2001), Gaseous mercury emissions from a fire in the Cape Peninsula, South Africa, during January 2000, Geophysical Research Letters 28 (8), pp. 1483–1486

Burke, J. Hoyer, M. Keeler G. and Scherbatskoy. Wet deposition of mercury and ambient mercury concentrations at a site in the Lake Champlain basin. Water Air and Soil Pollution 80 (1995), pp. 353–362.

Carpi, A., 1997. Mercury from combustion sources: a review of the chemical species emitted and their transport in the atmosphere. Water , Air, and Soil Pollution 98, 241–254.

De La Rosa, D.A., Volke-Sepúlveda, T., Solórzano, G., Green, C., Tordon, R., Beauchamp, S., (2004), Survey of atmospheric total gaseous mercury in Mexico, Atmospheric Environment 38 (29), pp. 4839–4846

Dommergue, A., Ferrari, C.P., Planchon, F.A.M, Boutron, C.F., (2002), Influence of anthropogenic sources on total gaseous mercury variability in Grenoble suburban air (France), Science of the Total Environment 297 (1-3), pp. 203–213

Dvonch, J.T., Keeler, G.J., Marsik, F.J., (2005), The influence of meteorological conditions on the wet deposition of mercury in southern Florida, Journal of Applied Meteorology 44 (9), pp. 1421–1435

Ebinghaus, R., Jennings, S.G., Schroeder, W.H., Berg, T., Donaghy, T., Guentzel, J., Kenny, C., (1999), International field intercomparison measurements of atmospheric mercury species at Mace Head, Ireland, Atmospheric Environment 33 (18), pp. 3063–3073

Ebinghaus, R., Slemr, F., (2000), Aircraft measurements of atmospheric mercury over southern and eastern Germany, Atmospheric Environment 34 (6), pp. 895–903

Ebinghaus, R., Slemr, F., Brenninkmeijer, C.A.M., van Velthoven, P., Zahn, A., Hermann, M., O'Sullivan, D.A., Oram, D.E., (2007), Emissions of gaseous mercury from biomass burning in South America in 2005 observed during CARIBIC flights, Geophysical Research Letters 34 (8), art. no. L08813

Engstrom, D.R. and Swain, E.B. 1997. Recent declines in atmospheric mercury deposition in the upper Midwest. Environ. Sci. Technol. 312, 60–67

Fang, F., Wang, Q., Li, D., Zhao, G., Wang, L., (2001a), Atmospheric particulate mercury concentration in Changchun City and its dry deposition flux, Huanjing Kexue/Environmental Science 22 (2), pp. 60–63

Fang, F., Wang, Q., Li, J., (2001 b), Atmospheric particulate mercury concentration and its dry deposition flux in Changchun City, China, Science of the Total Environment 281 (1-3), pp. 229–236

Feng, X., Tang, S., Shang, L., Yan, H., Sommar, J., Lindqvist, O., (2003), Total gaseous mercury in the atmosphere of Guiyang, PR China, Science of the Total Environment 304 (1-3), pp. 61–72

Feng, X., Shang, L., Wang, S., Tang, S., Zheng, W., (2004), Temporal variation of total gaseous mercury in the air of Guiyang, China, Journal of Geophysical Research D: Atmospheres 109 (3), pp. D03303 1–9

Ferrara, R., Maserti, B.E., Andersson, M., Edner, H., Ragnarson, P., Svanberg, S., Hernandez, A., (1998), Atmospheric mercury concentrations and fluxes in the Almaden district (Spain), Atmospheric mercury concentrations and fluxes in the Almaden district (Spain)

Ferrara, R., Mazzolai, B., Lanzillotta, E., Nucaro, E., Pirrone, N., (2000), Temporal trend in gaseous mercury evasion from the Mediterranean Sea Waters. Science of Total Environment, 259, 183–190

Fostier, A.H., Michelazzo, P.A.M., (2006), Gaseous and particulate atmospheric mercury concentrations in the Campinas Metropolitan Region (Sao Paulo State, Brazil), Journal of the Brazilian Chemical Society 17 (5), pp. 886–894

Friedli, H.R., Radke, L.F., Prescott, R., Hobbs, P.V., Sinha, P., (2003), Mercury emissions from the August 2001 wildfires in Washington State and an agricultural waste fire in Oregon and atmospheric mercury budget estimates, Global Biogeochemical Cycles 17 (2), pp. 8–1

Friedli, H.R., Radke, L.F., Prescott, R., Li, P., Woo, J.-H., Carmichael, G.R., (2004), Mercury in the atmosphere around Japan, Korea, and China as observed during the 2001 ACE-Asia field campaign: Measurements, distributions, sources, and implications, Journal of Geophysical Research D: Atmospheres 109 (19), pp. D19S25 1–13

Gabriel, M.C., Williamson, D.G., Brooks, S., Lindberg, S., (2005), Atmospheric speciation of mercury in two contrasting Southeastern US airsheds, Atmospheric Environment 39 (27), pp. 4947–4958

Garcia-Sanchez, A., Contreras, F., Adams, M., Santos, F., (2006), Airborne total gaseous mercury and exposure in a Venezuelan mining area, International Journal of Environmental Health Research 16 (5), pp. 361–373

Gardfeldt, K., Sommar, J., Ferrara, R., Ceccarini, C., Lanzillotta, E., Munthe, J., Wangberg, I., Lindqvist, O., Pirrone, N., Sprovieri, F., Pesenti, E. (2003). Evasion of mercury from coastal and open waters of the Atlantic Ocean and the Mediterranean Sea. Atmospheric Environment. Vol. 37-S1, 73–84

Gildemeister, A.E., Graney, J., Keeler, G.J., (2005), Source proximity reflected in spatial and temporal variability in particle and vapor phase Hg concentrations in Detroit, MI, Atmospheric Environment 39 (2), pp. 353–358

Gill, G.A., Guentzel, J.L., Landing, W.M., Pollman, C.D., (1995), Total gaseous mercury measurements in Florida: The FAMS project (1992- 1994), Water, Air, and Soil Pollution 80 (1-4), pp. 235–244

Gustin, M.S., Lindberg, S., Marsik, F., Casimir, A., Ebinghaus, R., Edwards, G., Hubble-Fitzgerald, C., (…), Zhang, H., (1999), Nevada STORMS project: Measurement of mercury emissions from naturally enriched surfaces, Journal of Geophysical Research D: Atmospheres 104 (D17), pp. 21831–21844

Hachiya, N., Takizawa, Y., Hisamatsu, S., Abe, T., Abe, Y., Motohashi, Y., (1998), Atmospheric mercury concentrations in the basin of the Amazon, Brazil, Environmental Health and Preventive Medicine 2 (4), pp. 183–187

Han, Y.-J., Holsen, T.M., Lai, S.-O., Hopke, P.K., Yi, S.-M., Liu, W., Pagano, J., (…), Andolina, C., (2004), Atmospheric gaseous mercury concentrations in New York State: Relationships with meteorological data and other pollutants, Atmospheric Environment 38 (37), pp. 6431–6446

Hedgecock, I.M. and Pirrone, N. (2001), Mercury and photochemistry in the marine boundary layer-modelling studies suggest the in situ production of reactive gas phase mercury. Atmos. Env. 35, 3055–3062

Hedgecock, I., Pirrone, N., Sprovieri, F., Pesenti, E. (2003) Reactive Gaseous Mercury in the Marine Boundary Layer: Modeling and Experimental Evidence of its Formation in the Mediterranean. Atmos. Environ. 37/S1, 41–50

Hedgecock, I. M., and Pirrone, N., (2004) Chasing Quicksilver: Modeling the Atmospheric Lifetime of Hg^0 (g) in the Marine Boundary Layer at Various Latitudes. Environmental Science and Technology, Vol.38, 69–76

Hedgecock, I.M., Hedgecock, I.M., Trunfio, G.A., Pirrone, N., Sprovieri, F. (2006) MERCYMS Final Technical Report, Annex 2.2: Mercury chemistry in the MBL: Mediterranean case and sensitivity studies using the AMCOTS (Atmospheric Mercury Chemistry over the Sea) model. *Technical report CNR/IIA/2006/10,* Rende, Italy, 16 p

Higueras, P., Oyarzun, R., Lillo, J., Oyarzun, J., Maturana, H., (2005), Atmospheric mercury data for the Coquimbo region, Chile: Influence of mineral deposits and metal recovery practices, Atmospheric Environment 39 (39 SPEC. ISS.), pp. 7587–7596

Hladikova, V., Petrik, J., Jursa, S., Ursinyová, M., Kocan, A., (2001), Atmospheric mercury levels in the Slovak Republic, Chemosphere 45 (6-7), pp. 801–806

Horvat, M., Kotnik, J., Fajon, V., Logar, M., Zvonaric, T., Pirrone, N., (2001), Speciation of mercury in waters of the Mediterranean Sea. Materials and Geoenvironment 48, 24–252.

Horvat, M., Kotnik, J., Fajon, V., Logar, M., Zvonaric, T., Pirrone, N. (2003) Speciation of Mercury in Surface and Deep-Sea waters in the Mediterranean Sea. Atmospheric Environment, Vol. 37/S1 , 93–108

Hoyer, M., Burke, J., Keeler, G., (1995), Atmospheric sources, transport and deposition of mercury in Michigan: Two years of event precipitation, Water, Air, and Soil Pollution 80 (1-4), pp. 199–208

Iverfeldt, A., Munthe, J., Brosset, C., Pacyna, J., (1995), Long-term changes in concentration and deposition of atmospheric mercury over Scandinavia, Water, Air, and Soil Pollution 80 (1-4), pp. 227–233

Jaffe, D., Prestbo, E., Swartzendruber, P., Weiss-Penzias, P., Kato, S., Takami, A., Hatakeyama, S., Kajii, Y., (2005), Export of atmospheric mercury from Asia, Atmospheric Environment 39 (17), pp. 3029–3038

Keeler, G.J., Gratz, L.E., Al-Wali, K., (2005), Long-term atmospheric mercury wet deposition at Underhill, Vermont, Ecotoxicology 14 (1-2), pp. 71–83

Keeler, G.J., Landis, M.S., Norris, G.A., Christianson, E.M., Dvonch, J.T., (2006), Sources of mercury wet deposition in eastern Ohio, USA, Environmental Science and Technology 40 (19), pp. 5874–5881

Kellerhals, M., Beauchamp, S., Belzer, W., Blanchard, P., Froude, F., Harvey, B., McDonald, K., (…), Tordon, R., (2003), Temporal and spatial variability of total gaseous mercury in Canada: Results from the Canadian Atmospheric Mercury Measurement Network (CAMNet), Atmospheric Environment 37 (7), pp. 1003–1011

Kim, K.-H., Kim, M.Y., (1996), Preliminary measurements of atmospheric mercury in mountainous regions of Korea, Journal of Environmental Science and Health - Part A Toxic/Hazardous Substances and Environmental Engineering 31 (8), pp. 2023–2032

Kim, K.-H., Kim, M.-Y., (2001 a), Some insights into short-term variability of total gaseous mercury in urban air, Atmospheric Environment 35 (1), pp. 49–59

Kim, K.-H., Kim, M.-Y., (2001 b), The temporal distribution characteristics of total gaseous mercury at an urban monitoring site in Seoul during 1999-2000, Atmospheric Environment 35 (25), pp. 4253–4263

Kim, K.-H., Kim, M.-Y., Kim, J., Lee, G., (2002), The concentra-tions and fluxes of total gaseous mercury in a western coastal area of Korea during late March 2001, Atmospheric Environment 36 (21), pp. 3413–3427

Kim, K.-H., Ebinghaus, R., Schroeder, W.H., Blanchard, P., Kock, H.H., Steffen, A., Froude, F.A., (…), Kim, J.-H., (2005), Atmospheric mercury concentrations from several observatory sites in the Northern Hemisphere, Journal of Atmospheric Chemistry 50 (1), pp. 1–24

Kock, H.H., Bieber, E., Ebinghaus, R., Spain, T.G., Thees, B., (2005), Comparison of long-term trends and seasonal variations of atmospheric mercury concentrations at the two European coastal monitoring stations Mace Head, Ireland, and Zingst, Germany, Atmospheric Environment 39 (39 SPEC. ISS.), pp. 7549–7556

Kotnik, J., Horvat, M., Dizdarevic, T., (2005), Current and past mercury distribution in air over the Idrija Hg mine region, Slovenia, Atmospheric Environment 39 (39 SPEC. ISS.), pp. 7570–7579

Kotnik, J., Horvat, M., Tessier, E., Ogrinc, N., Monperrus, M., Amouroux, D., Fajon, V., Gibicar, D., Zizek, S., Horvat, N., Sprovieri, F., Pirrone, N. (2007). Mercury speciation in surface and deep waters of the Mediterranean Sea. Marine Chemistry Vol.107, 13–30.

Lacerda LD., (1997), Global mercury emissions from gold and silver mining. Water Air and Soil Pollution 97 (3-4): 209–221

Lacerda L.D., (2003), Updating global Hg emissions from small-scale gold mining and assessing its environmental impacts. Envir Geol. 43, 308–314.

Lamborg, C.H., Fitzgerald, W.F., Damman, A.W.H., Benoit, J.M., Balcom, P.H. and Engstrom, D.R. 2002. Modern and historic atmospheric mercury fluxes in both hemispheres: global and regional mercury cycling implications. Glob. Biogeochem. Cycles 16, 104, doi:10.1029/2001GB1847.

Landis, M.S., Vette, A.F., Keeler, G.J., (2002), Atmospheric mercury in the Lake Michigan basin: Influence of the Chicago/Gary urban area, Environmental Science and Technology 36 (21), pp. 4508–4517

Landis, M. S., G. J. Keeler, K. I. Al-Wali, R. K. Stevens, (2004) Divalent inorganic reactive gaseous mercury emissions from a mercury cell chlor-alkali plant and its impact on near-field atmospheric dry deposition. Atmos. Environ., 38, 613–622.

Laurier, F. J. G., R. P. Mason, L. Whalin, and S. Kato, (2003) Reactive gaseous mercury formation in the North Pacific Ocean's marine boundary layer: A potential role of halogen chemistry. J. Geophys. Res., 108(D17), 4529, doi:10.1029/2003JD003625

Laurier, F., Mason, R., (2007), Mercury concentration and speciation in the coastal and open ocean boundary layer, Journal of Geophysical Research D: Atmospheres 112 (6), art. no. D06302

Lindberg, S.E., Turner, R.R., Meyers, T.P., Taylor Jr., G.E., Schroeder, W.H., (1991), Atmospheric concentrations and deposition of Hg to a deciduous forest at Walker Branch Watershed, Tennessee, USA, Water, Air, and Soil Pollution 56 (SPEC. VOL.), pp. 577–594

Lindberg, S. E., Stratton, W. J., (1998), Atmospheric mercury speciation: concentrations and behaviour of reactive gaseous mercury in ambient air, Environmental Science and Technology, 32, 49–57

Lindberg, S., R. Bullock, R. Ebinghaus, D. Engstrom, X. B. Feng, W. Fitzgerald, N. Pirrone, E. Prestbo, and C. Seigneur (2007), A synthesis of progress and uncertainties in attributing the sources of mercury in deposition, Ambio 36(1), 19–32

Liu, S., Nadim, F., Perkins, C., Carley, R.J., Hoag, G.E., Lin, Y., Chen, L., (2002), Atmospheric mercury monitoring survey in Beijing, China, Chemosphere 48 (1), pp. 97–107

Liu B., et al., (2007) Temporal variability of mercury speciation in urban air. Atmos. Environ., 41, 1911–1923

Lynam, M., G. J. Keeler, (2005b), Artifacts associated with the measurement of particulate mercury in an urban environment: The influence of elevated ozone concentrations, Atmos. Environ., 39, 3081–3088

Malcolm, E. G., G. J. Keeler, and M. S. Landis, (2003) The effects of the coastal environment on the atmospheric mercury cycle, J. Geophys. Res., 108(D12), 4357, doi:10.1029/2002JD003084

Malcolm, E.G., Keeler, G.J., Lawson, S.T., Sherbatskoy, T.D., (2003), Mercury and trace elements in cloud water and precipitation collected on Mt. Mansfield, Vermont

Mason, R.P., Lawson, N.M., Sheu, G.-R., (2001), Mercury in the Atlantis Ocean: factors controlling air-sea exchange of mercury and its distribution in the upper waters. Deep-Sea Res. II. 48, 2829 2853

Munthe, J., Wangberg, I., Iverfeldt, A., Lindqvist, O.Stromberg, D.Sommar, J., Gardfeldt, K., (…), Siemens, V., (2003), Distribution of atmospheric mercury species in Northern Europe: Final results from the MOE project, Atmospheric Environment 37 (SUPPL. 1), pp. S9–S20

Munthe, J., Wangberg, I., Pirrone, N., Iverfeld, A., Ferrara, R., Ebinghaus, R., Feng, R., Gerdfelt, K., Keeler, G.J., Lanzillotta, E., Lindberg, S.E., Lu, J., Mamane, Y., Prestbo, E., Schmolke, S., Schroder, W.H., Sommar, J., Sprovieri, F., Stevens, R.K., Stratton, W., Tuncel, G., Urba, A. (2001) Intercomparison of Methods for Sampling and Analysis of Atmospheric Mercury Species. Atmospheric Environment, 35, 3007–3017

Nguyen, H.L., Leermakers, M., Kurunczi, S., Bozo, L., Baeyens, W., (2005), Mercury distribution and speciation in Lake Balaton, Hungary, Science of the Total Environment 340 (1-3), pp. 231–246

Nguyen, H.T., Kim, K.-H., Kim, M.-Y., Hong, S., Youn, Y.-H., Shon, Z.H. and Lee, J.S., (2007). Monitoring of atmospheric mercury at a Global Atmospheric Watch (GAW) site on An-Myun Island, Korea, Water, Air, and Soil Pollutions 185 , pp. 149–164

Pacyna, E.G., Pacyna, J.M., Pirrone, N. (2001). European emissions of atmospheric mercury from anthropogenic sources in 1995. Atmospheric Environment, 35, 2987–2996

Pacyna, E.G., Pacyna, J.M., Steenhuisen, F., Wilson, S, (2006), Global Anthropogenic Mercury Emission Inventory for 2000, Atmospheric Environment 40, 4048–4063

Pirrone, N., Keeler, G.J., Nriagu, J. (1996). Regional differences in worldwide emissions of mercury to the atmosphere. Atmos. Environ., 17, 2981–2987.

Pirrone, N., Allegrini, I., Keeler, G.J., Nriagu, J.O., Rossmann, R. and Robbins, J.A. (1998) Historical Atmospheric Mercury Emissions and Depositions in North America Compared to Mercury Accumulations in Sedimentary Records. Atmospheric Environment, Vol. 32, 929–940

Pirrone, N., Hedgecock, I., Forlano, L. (2000). The role of the ambient aerosol in the atmospheric processing of semi-volatile contaminants: a parameterised numerical model (GASPAR). Journal of Geophysical Research D105 (8), 9773–9790

Pirrone, N., Costa, P., Pacyna, J.M., Ferrara, R., (2001). Mercury emissions to the atmosphere from natural and anthropogenic sources in the Mediterranean region. Atmos. Env., 35, 2997–3006

Pirrone, N., Sprovieri, F., Hedgecock, I., Gensini, M. (2001) MAMCS final report. *Technical rep ort CNR/IIA/2001/08*, Rende (CS), pp. 253.

Pirrone, N., Pacyna, J.M., Munthe, J., Barth, H. (Editors) (2003) Dynamic Processes of Mercury and Other Atmospheric Contaminants in the Marine Boundary Layer of European Seas. Special Issue of Atmospheric Environment, Elsevier Science Publishers, Amsterdam, Netherlands, Vol. 37/S, 1-178

Pirrone, N. (2006) An integrated approach to assess the mercury cycling in the Mediterranean basin (MERCYMS). Final Technical Report + Annexes + CD-Rom. *Technical report CNR/ IIA/2006/08,* Rende, Italy, 377 p

Poissant, L. and Casimir, A. , (1998), Water-air and soil-air exchange rate of total gaseous mercury measured at background sites, Atmospheric Environment, 32, 883–893

Poissant, L., Pilote, M., Beauvais, C., Constant, P., Zhang, H.H., (2005), A year of continuous measurements of three atmospheric mercury species (Hg0, RGM and Hgp) in southern Québec, Canada, Atmospheric Environment 39 (7), pp. 1275–1287

Pollman, C., Gill, G., Landing, W., Guentzel, J., Bare, D., Porcella, D., Zillioux, E., Atkeson, T., (1995), Overview of the Florida Atmospheric Mercury Study (FAMS), Water, Air, and Soil Pollution 80 (1-4), pp. 285–290

Schmolke, S.R., Schroeder, W.H., Kock, H.H., Schneeberger, D., Munthe, J., Ebinghaus, R., (1999), Simultaneous measurements of total gaseous mercury at four sites on a 800 km transect: Spatial distribution and short-time variability of total gaseous mercury over central Europe, Atmospheric Environment 33 (11), pp. 1725–1733

Schroeder, W.H., Ebinghaus, R., Shoeib, M., Timoschenko, K., Barrie, L.A., (1995), Atmospheric mercury measurements in the northern hemisphere from 56° to 82.5° N latitude, Water, Air, and Soil Pollution 80 (1-4), pp. 1227–1236

Schroeder, W.H., Keeler, G., Kock, H., Roussel, P., Schneeberger, D., Schaedlich, F., (1995), International field intercomparison of atmospheric mercury measurement methods, Water, Air, and Soil Pollution 80 (1–4), pp. 611–620

Schroeder, W.H., Munthe, J., (1998), Atmospheric mercury - An overview, Atmospheric Environment 32 (5), pp. 809–822

Shang, L., Feng, X., Zheng, W., Yan, H., (2003), Preliminary study of the distribution of gaseous mercury species in the air of Guiyang city, China, Journal De Physique. IV : JP 107 (II), pp. 1219–1222

Sheu, G.-R., Mason, R.P., Lawson, N.M., (2002), Speciation and distribution of atmospheric mercury over the northern Chesapeake Bay, ACS Symposium Series 806, pp. 223–242

Sigler, J. M., and X. Lee (2006), Recent trends in anthropogenic mercury emission in the northeast United States, J. Geophys. Res., 111, D14316, doi:10.1029/2005JD006814

Slemr, F., Junkermann, W., Schmidt, R.W.H., Sladkovic, R., (1995), Indication of change in global and regional trends of atmospheric mercury concentrations, Geophysical Research Letters 22 (16), pp. 2143–2146

Slemr F, Brunke EG, Ebinghaus R, Temme C, Munthe J, Wängberg I. (2003). Worldwide trend of atmospheric mercury since 1977. Geophys Res Lett., 30 Art No 1516

Slemr, F., Ebinghaus, R., Simmonds, P.G., Jennings, S.G., (2006), European emissions of mercury derived from long-term observations at Mace Head, on the western Irish coast, Atmospheric Environment 40 (36), pp. 6966–6974

Sohn, D.H., Jung, W.T., Kim, D., (1993), Distribution of total mercury in the ambient atmosphere of Seoul and its diurnal, monthly and altitudinal variations, Japanese Journal of Toxicology and Environmental Health 39 (6), pp. 582–588

Sprovieri, F., Pirrone, N., Gärdfeldt., K. and Sommar, J. (2003) Mercury speciation in the marine boundary layer along a 6000 km cruise path around the Mediterranean Sea, Atmos. Environ., 37, S6371

Sprovieri, F., Pirrone, N. (2006) MERCYMS Final Technical Report, Annex 3.2: Evaluation of the MERCYMS coastal and Mediterranean Sea measurement data base concerning TGM. *Technical report CNR/IIA/2006/11*, Rende, Italy, pp. 6

Sprovieri, F. and Pirrone, N. (2008) Spatial and Temporal Distribution of Atmospheric Mercury Species over the Adriatic Sea. Environmental Fluid Mechanics, 8(2), 117–128.

Stamenkovic, J., Lyman, S., Gustin, M.S., (2007), Seasonal and diel variation of atmospheric mercury concentrations in the Reno (Nevada, USA) airshed , Atmospheric Environment 41 (31), pp. 6662–6672

Steding, D., Flegal, A., (2002), Mercury concentrations in coastal California precipitation: Evidence of local and trans-Pacific fluxes of mercury to North America, Journal Geophysical Research, 107, D24, 4764

Steffen, S. and Schroeder, B., (1999), Standard Operating Procedure Manual for Atmospheric Mercury Measurements, Environment Canada, Toronto

Swartzendruber P. C., D. A. Jaffe, E. M. Prestbo, P., Weiss-Penzias, N. E. Selin, R. Park, D. Jacob, S. Strode, L. and Jaegl, (2006), Observations of Reactive Gaseous Mercury in the Free-Troposphere at the Mt. Bachelor Observatory , Journal of Geophysical Research, 111(D24), D24301, 2006

Temme, C., Blanchard, P., Steffen, A., Banic, C., Beauchamp, S., Poissant, L., Tordon, R., Wiens, B., (2007), Trend, seasonal and multivariate analysis study of total gaseous mercury data from the Canadian atmospheric mercury measurement network (CAMNet), Atmospheric Environment 41 (26), pp. 5423–5441

Tomiyasu, T., Nagano, A., Sakamoto, H., Yonehara, N., (2000), Background levels of atmospheric mercury in Kagoshima City, and influence of mercury emission from Sakurajima Volcano, Southern Kyushu, Japan, Science of the Total Environment 259 (1-3), pp. 231–237

Tomiyasu, T., Eguchi, M., Sakamoto, H., Anazawa, K., Imura, R., (2006), Seasonal change and vertical movement of atmospheric mercury at Kagoshima City in relation with Sakurajima Volcano, Japan, Geochemical Journal 40 (3), pp. 253–263

Urba, A., Kvietkus, K., Marks, R., (2000), Gas-phase mercury in the atmosphere over the southern Baltic Sea coast, Science of the Total Environment 259 (1–3), pp. 203–210

Valente, R.J., Shea, C., Lynn Humes, K., Tanner, R.L., (2007), Atmospheric mercury in the Great Smoky Mountains compared to regional and global levels, Atmospheric Environment 41 (9), pp. 1861–1873

Vanarsdale, A., J. Weiss, G. Keeler, E. Miller, G. Boulet, R. Brulotte, and L. Poissant (2005), Patterns of mercury deposition and concentration in northeastern North America (1996-2002), Ecotoxicology, 14(1-2), 37–52

Wang, S., Feng, X., Qiu, G., Wei, Z., Xiao, T., (2005), Mercury emission to atmosphere from Lanmuchang Hg-Tl mining area, Southwestern Guizhou, China, Atmospheric Environment 39 (39 SPEC. ISS.), pp. 7459–7473

Wang, Z., Chen, Z., Duan, N., Zhang, X., (2007), Gaseous elemental mercury concentration in atmosphere at urban and remote sites in China, Journal of Environmental Sciences 19 (2), pp. 176–180

Wangberg, I., Munthe, J., Pirrone, N., Iverfeldt, A., Bahlman, E., Costa, P., Ebinghaus, R., (…), Tuncel, G., (2001), Atmospheric mercury distribution in Northern Europe and in the Mediterranean region, Atmospheric Environment 35 (17), pp. 3019–3025

Wangberg, I., Munthe, J., Berg, T., Ebinghaus, R., Kock, H.H., Temme, C., Bieber, E., (…), Stolk, A., (2007), Trends in air concentration and deposition of mercury in the coastal environment of the North Sea Area, Atmospheric Environment 41 (12), pp. 2612–2619

Wangberg, I., Munthe, J., Amouroux, D., Andersson, M. E., Fajon, V., Ferrara, R., Gårdfeldt, K., Horvat, M., Mamane, Y., Melamed, E., Monperrus, M., Ogrinc, N., Yossef, O., Pirrone, N., Sommar, J., Sprovieri, F. (2008) Atmospheric Mercury at Mediterranean Coastal Stations. Environmental Fluid Mechanics, 8 (2), 101–116

Weiss-Penzias, P., Jaffe, D., McClintick, A., Prestbo, E., Landis, M., (2003), Gaseous Elemental Mercury in the Marine Boundary Layer: Evidence for Rapid Removal in Anthropogenic Pollution, Environmental Science and Technology 37(17), 3755 –3763, 2003

Weiss-Penzias P., D. A. Jaffe, P. Swartzendruber, W. Hafner, D., Chand, and E. Prestbo, (2007), Quantifying Asian and biomass burning sources of mercury using the Hg/CO ratio in pollution plumes observed at the Mount Bachelor Observatory , Atmospheric Environment 41, 4366–4379

Weiss-Penzias P., D. A. Jaffe, P. Swartzendruber, W. Hafner, D., Chand, and E. Prestbo, (2007 a), Observations of Asian air pollution in the free troposphere at Mount Bachelor Observatory during the spring of 2004, Atmospheric Environment 41, 4366–4379

Xiu, G.L., Jin, Q., Zhang, D., Shi, S., Huang, X., Zhang, W., Bao, L., (…), Chen, B., (2005), Characterization of size-fractionated particulate mercury in Shanghai ambient air, Atmospheric Environment 39 (3), pp. 419–427

Zielonka, U., Hlawiczka, S., Fudala, J., Wängberg, I., Munthe, J., (2005), Seasonal mercury concentrations measured in rural air in Southern Poland: Contribution from local and regional coal combustion, Atmospheric Environment 39 (39 SPEC. ISS.), pp. 7580–75

Chapter 10
Spatial Coverage and Temporal Trends of Atmospheric Mercury Measurements in Polar Regions

Aurélien Dommergue, Christophe P. Ferrari, Marc Amyot, Steve Brooks, Francesca Sprovieri, and Alexandra Steffen

Summary The discovery of atmospheric mercury depletion events (AMDEs) in the Canadian Arctic at Alert in 1995 initiated the intense study of atmospheric Hg processes. Mercury has unique characteristics that include long-range atmospheric transport to regions like the Arctic and the Antarctica, the transformation to more toxic methylmercury compounds and the ability of these compounds to biomagnify in the aquatic food chain. Following the discovery of AMDEs, studies have been conducted throughout Polar Regions where the same phenomenon was observed. Since then, many scientific projects have focused on studying the mechanisms related to AMDEs. Additionally, special attention is paid to the consequences of AMDEs in terms of contamination of Polar Regions because AMDEs rapidly convert atmospheric gaseous mercury into reactive and water-soluble forms that may potentially become bioavailable. Finally, the contribution of this unique reactivity occurring in polar atmospheres to the global budget of atmospheric mercury and the role played by snow and ice surfaces of these regions are important issues. This chapter presents a review of atmospheric measurements conducted both in the Arctic and the Antarctic since 1995 (continuous data). Our current understanding of AMDEs in these regions is presented.

10.1 Introduction

Polar Regions used to be considered pristine environments. Indeed, the Arctic is relatively far from industrial centers located at mid-latitude in the northern hemisphere and is less populated than other parts of the world- indigenous peoples of the Arctic are estimated to comprise nearly 650 000 individuals, most of whom live in northern Russia (AMAP, 2003). In the Southern Hemisphere, Antarctica is even less populated and impacted by anthropogenic activities, except on a local scale by a few scientific stations. However, due to a combination of long-range transport associated with a specific climatology, the Arctic and, to a lesser extent the Antarctic, are affected on a large scale by pollutants originating from the mid-latitudes of the northern hemisphere. It is now established that several pollutants

N. Pirrone and R. Mason (eds.), *Mercury Fate and Transport in the Global Atmosphere*,
DOI: 10.1007/978-0-387-93958-2_10,

including Persistent Organic Pollutants (POPs) and heavy metals are present in several compartments of the polar ecosystems (AMAP, 2003).

For example, certain Arctic species, particularly those at the upper end of the marine food chain as well as sea birds, carry high levels of PCBs and organo-chlorine pesticides (Braune et al., 2005). POPs are transported to the Arctic by regional and global physical processes and are then subjected to biological mechanisms that lead to the high levels found in certain species. Organo-chlorine pesticides are also reported present in marine mammals from the southern hemisphere. Levels are, however, lower than those observed in marine mammals from the northern hemisphere, yet the presence of these contaminants in the Antarctic food web indicates transport from uses in some developing countries within the southern hemisphere (Miranda et al., 2007).

In addition to organic compounds, heavy metals such as mercury (Hg) can be dispersed and contaminate Polar Regions. Mercury is transported far from its emission sources, mainly fossil fuel combustion, industry and mining, to Polar Regions. A recent review paper (Steffen et al., 2008); provides a comprehensive assessment of the state of the Hg science in the context of Atmospheric Mercury Depletion Events (AMDEs) in Polar Regions since 1995. Unlike other metals, Hg predominantly exists in the atmosphere as a gaseous compound, namely elemental gaseous mercury (Hg^0 or GEM). This species has a relatively long life-time (more than 6 months) making it subject to long range transport. The substantial different geographical distribution of landmasses around both poles influences the Hg^0 annual mean concentration observed in the Arctic (~ 1.6 ng m^{-3}) and Antarctica (~1.0 ng m^{-3}). The Antarctic region is remote from human activities and far from landmasses; in contrast, the Arctic region is surrounded by northern North America, northern Europe and northern Asia, therefore, anthropogenic influences are stronger. Atmospheric mercury concentrations, however, are several times lower than Hg levels measured at mid-latitudes sites. After its oxidation to divalent species ($Hg^{(II)}$), which are more water soluble and reactive than the elemental form, Hg is deposited and has the potential to be converted into organo-metallic forms such as methylmercury (MeHg), although currently there are no measurements to confirm this link.

Both $Hg^{(II)}$ and MeHg bioaccumulate and biomagnify in the polar ecosystem and can be found at high and increasing levels in marine species. In most mammals, the level of mercury increases with their age and high concentrations of more than 10 ppm (w/w) have been found in ringed seals from Canadian Arctic (Braune et al., 2005). In the Antarctic, Hg levels are much lower in the food chain; however, data are limited. Still, increasing trends of Hg in marine (coastal) food web can be observed (Riva et al., 2004). Mercury exposure poses a health risk to animals and some people in the Arctic. For instance, Hg levels in the liver and kidney of polar bears in East Greenland have been observed in levels which may lead to toxic effects (Sonne et al., 2007). Of concern is the fact that people in Greenland or in Canada, who have a traditional marine diet (fish, seabird, seals and whales), exhibit high blood concentration of Hg. At times, the intake of mercury exceeds acceptable daily intake level (Johansen et al., 2007). In Greenland, the intake of mercury from food is 10 times higher than in Denmark (Johansen et al., 2007). In Canada, Hg levels from Inuit are higher in both maternal and cord blood than in the rest of other ethnic groups, indicating that the consumption of marine species is potentially a source of mercury. Because in some cases levels of MeHg in the cord samples exceed some dose limit (such as EPA's), the incidence of adverse outcome could be expected (Walker et al., 2006).

It is believed that anthropogenic activities have led to a 3-times increase in the global deposition of Hg in ecosystems since pre-industrial times (Lindberg et al., 2007). It is also believed that more recent increases (Lockhart et al., 1998) can be observed occasionally (Lockhart et al., 2005; Wagemann et al., 1996) in marine food-chains. However, there is a great discrepancy in assessing recent changes in the biota because of regional differences and difficulties in combining data sets for different species of differing location or age. Despite a global decrease in Hg emissions since the 70's, pressure is still exerted on arctic ecosystems and native populations are more exposed to Hg contamination than populations outside Arctic regions (AMAP, 2005).

To date, the reason for the contamination of mercury in Arctic is still unclear. However, the discovery made in Alert (Canada) in 1995 (Schroeder et al., 1998) which revealed that Hg^0 is oxidized and deposited onto polar environmental surfaces more rapidly than anywhere else due to a phenomenon called Atmospheric Mercury Depletion Events (AMDEs) sparked considerable interest in the research community to further investigate the mercury situation in the north. AMDEs occur during the spring (of the Northern Hemisphere) after the polar sunrise in high latitude and coastal areas. The atmospheric drop in Hg^0 can last several hours to several days and is strongly correlated with tropospheric ozone depletion events. Numerous studies suggest that sea-salt initiates a photochemically-initiated and autocatalytic release of reactive halogen (Br and BrO radicals). The conversion of Hg^0 by bromine radicals is certainly fast, even if the kinetics is poorly studied, and leads to the formation of divalent species of Hg. These species are more water-soluble and reactive and thus can be deposited by wet and dry processes on surfaces. Significant increases in oxidized levels of Hg have been measured in the snow following AMDEs (Lindberg et al., 2002). A fraction of Hg deposited by AMDEs has been reported as bioavailable (Scott, 2001) yet the link to the methylation of Hg in this system is still unknown.

AMDEs have been observed in several arctic and sub-arctic sites (section 10.2.2) and in Antarctica (section 10.2.3) as well. Quantification of Hg deposition due to these events is today limited. The magnitude of Hg deposition to the Arctic ranges between 50-200 Mg per year but some uncertainties exist in assessing the contribution AMDEs play in the global Hg cycle. No estimate has yet been made for Antarctica. Several uncertainties remain in the understanding of AMDEs. Firstly, the origin of AMDEs is unknown, it could be a recent phenomenon enhanced for example by the increasing occurrence of open sea waters. Secondly, several studies have shown that a part of deposited mercury can be reemitted back to the atmosphere by photoreduction processes (Steffen et al., 2008) thus adding uncertainty to the net deposition. Thirdly, the bioavailability of deposited Hg is poorly known, thus the question of how mercury enters the food web is still unsolved. Finally, the mechanisms that produce MeHg - which is likely the most relevant form that biomagnifies - in these cold environments are to date unsolved even though many pathways have been suggested (Barkay and Poulain, 2007).

Polar Regions, like other regions of the planet, are impacted by man-made emission of Hg, and therefore studying the cycling of Hg in Polar Regions is necessary to understand and follow the extent of the contamination within these ecosystems. Additionally, the ice and snow cover play an important role in the reactivity of the overlying atmosphere. The rapid changes observed in Polar Regions (temperature, sea-ice extent, precipitation) during the last years may complicate our understanding of how Hg cycles within these regions. The following chapter is a current state of Hg measurements in the polar atmospheres. Most of the research activities are today located in the northern hemisphere with long-term data for only a few sites. The Antarctic regions have not been extensively monitored yet and only sporadic measurements have been made. However, for both regions, an effort has been made to study the processes of AMDEs. The two first sections of this chapter present the current research in the Arctic and in Antarctica. A third part is devoted to the influence of snow surfaces on the global balance of atmospheric Hg.

10.2 Results and Discussion

10.2.1 Methods

10.2.1.1 Definitions

Gaseous Elemental Mercury (Hg^0 or GEM), Reactive Gaseous Mercury (RGM) and Particle associated mercury (PM) are the most commonly measured and monitored fractions in the atmosphere of Polar Regions. Hg^0 is the only gaseous Hg component that is easily and accurately measured in the field. RGM and PM are operationally defined and thus measurements from different sites may be complex to inter-compare. In some cases, Total Gaseous mercury (TGM) may be provided. It generally refers to the sum of Hg^0 and RGM.

10.2.1.2 Atmospheric Measurements in Cold Regions

Polar mercury speciation, mercury fluxes measurements, and snow pack sampling methods are similar to methods conducted around the world with exceptions made for the extreme cold, the blowing snow layer, the high altitude of the polar ice caps, and the high magnitude of mercury fluxes in and out of the surface snow. Care must be taken to (1) ensure that flow volumes and residency times are appropriate for the speciation of mercury into the 3 components, (2) prevent unintended mercury absorption in the sampling stream, and (3) ensure near 100% collection efficiency onto the pre-concentrating gold cartridges. Atop the high-altitude polar plateau item (1) requires matching the volume flow to a 0.1 second residency time over the KCl coated annular denuder. At the foggy coastal sites item (2) requires that dry air must be used to flush the system, otherwise the iodated carbon canisters (used to trap mercury from the flush unintentionally) can potential introduce iodine into the flush stream, where it can unintentional oxidize Hg^0. In all locations item (3) requires a high purity inert carrier gas, and a sampling location (such as a clean air sector) where unintended contaminates are not introduced.

Under very cold conditions the heated sample lines should be kept fully external to the climate controlled area, otherwise the temperature change between interior and exterior portions will induce hot/cold zones and mercury absorption/desorption at the tubing walls. The exterior front-end cases and the exterior sampling stream should have robust insulation and heating systems that will not significantly vary the set temperatures regardless of weather conditions. The inlet position must be placed sufficiently above the blowing snow layer, but remain within the lowest 10% of the atmospheric boundary layer, which may be as shallow as a few 10's of meters. Snow pack sampling must take care to use a sun shield to prevent photoreduction during pit excavation and sampling.

10.2.2 Atmospheric Mercury in the Arctic

10.2.2.1 Introduction

Atmospheric mercury is of concern in the Arctic because it has been anthropogenically mobilized during the last century and it has the ability to bioaccumulate and biomagnify Arctic marine wildlife (Macdonald and Bewers, 1996). Hg^0 has unique characteristics that include long-range atmospheric transport to regions like the Arctic. Thus, Hg^0 can be transported globally while oxidized forms of Hg are more reactive and travel much shorter distances before they are scavenged or deposited. How pollutants such as mercury are in the Arctic results from an accumulation from a combination of robust stratification, resulting from strong surface temperature inversions inhibiting turbulent transport, and the atmospheric transport of pollutants from mid-latitudes. This pole ward transport of pollutants is due to the geographic position of a meteorological phenomenon known as blocking (Iversen and Joranger, 1985). Polluted air masses can reach the Arctic troposphere within 2 to 10 days (e.g. Raatz and Shaw, 1984). Once atmospheric contaminants reach the Polar Regions, their lifetime in the troposphere is dictated by local removal processes.

10.2.2.2 Atmospheric Mercury Depletion Events in the Arctic

The discovery of atmospheric mercury depletion events (AMDEs) in the Canadian Arctic at Alert in 1995 (Schroeder et al., 1998) initiated almost a decade of intense study of atmospheric Hg processes. The reactive forms of Hg (e.g. RGM and some PM) have short lifetimes in the atmosphere and are deposited from the atmosphere close to emission sources and thus would not be expected in the Arctic which is far from these emission sources. However, the existence of reactive Hg in a particular air sample does not necessarily imply the existence of a local emission source but can be the

Table 10.1 Atmospheric mercury measurements conducted in arctic and sub-arctic sites (based on the table presented in Steffen et al. (2008))

Location	Analyte	Period	Analytical method/ Instrumentation	Reference
Alert, Canada	Hg^0 PM, RGM TAM TFM, TPM	Continuous data for Hg^0 since 1995	Tekran 2537A Tekran 2537A/1130/1135 Tekran 2537A/ CRPU Manual TPM/ TFM Minisampler	(Ariya et al., 2004; Banic et al., 2003; Cobbett et al., 2007; Lu and Schroeder, 2004; Lu et al., 2001; Schroeder et al., 1998; Slemr et al., 2003; St. Louis et al., 2005; Steffen et al., 2003a; Steffen et al., 2003b; Steffen et al., 2002; Steffen et al., 2005)
Amderma, Russia	Hg^0	Continuous data since 2001	Tekran 2537A	(Steffen et al., 2005)
Barrow, Alaska, USA	Hg^0, RGM PM, RGM RGM	Continuous data for GEM since 1998	Tekran 2537A, manual denuders Tekran 2537A/1130/1135	(Brooks et al., 2006; Lindberg et al., 2001; Lindberg et al., 2002; Skov et al., 2006; Tackett et al., 2007)
Ny-Ålesund, Svalbard, Norway	Hg^0, PM, RGM TAM	Continuous data for Hg^0 since 2000	Tekran 2537/1130/1135, CRPU TGM (manual gold trap) Gardis RGM manual, KCl coated denuders, PM manual filters	(Aspmo et al., 2005; Berg et al., 2003; Fain et al., 2006; Ferrari et al., 2005; Gauchard et al., 2005b; Sommar et al., 2007; Sprovieri et al.,2005a; Sprovieri et al., 2005b; Wangberg et al., 2003)
Resolute, Canada	Hg^0	May-June 2003	Tekran 2537A	(Lahoutifard et al., 2005)
Station Nord, Greenland, Denmark	Hg^0	1998-2002	Tekran 2537A Gardis	(Ferrari et al., 2004a; Ferrari et al., 2004b; Skov et al., 2004)
Summit, Greenland, Denmark	Hg^0	July 2005 and May-June 2006	Tekran 2537A	(Faïn et al., 2007)
North Atlantic Ocean	Hg^0, RGM, PM	June-August 2004	Tekran 2537A/1130/1135	(Aspmo et al., 2006)
Churchill, Canada	Hg^0, RGM, PM	March-August 2004	Tekran 2537A/1130/1135	(Kirk et al., 2006)
Kuujjuarapik, Canada	Hg^0, RGM	Continuous data for Hg^0 since 1999	Tekran 2537A/1130/1135 Gardis	(Dommergue et al., 2003a; Dommergue et al., 2003b; Gauchard et al., 2005a; Kirk et al., 2006; Lahoutifard et al., 2006; Poissant and Pilote, 2003; Steffen et al., 2005)

result of atmospheric chemical reactions involving Hg^0 transported from distant sources (Bottenheim and Chan, 2006; Gauchard et al., 2005b; Lindberg et al., 2007). Experimental evidence has demonstrated the presence of RGM and PM at remote locations ranging from Polar Regions to the open oceans (Steffen et al., 2008)

As summarized in Table 10.1 and in Figure 10.1 atmospheric mercury measurements have been collected on continuous and/or within field campaigns at several sites throughout the Arctic and sub-arctic regions. These sites include: Alert, Canada 82°28'N 62°30'W (Schroeder et al., 1998); Ny-Ålesund, Svalbard 78°54'N 11°53'E (Berg et al., 2003); Pt. Barrow, USA 71°19'N 156°37'W (Lindberg et al., 2001); Station Nord, Greenland 81°36'N 16°40E (Skov et al., 2004); Kuujjuarapik, Canada 55°16'N 77°45'W (Poissant and Pilote, 2003), Amderma, Russia 69°45'N 61°40'E (Steffen et al., 2005), the North Atlantic Ocean (Aspmo et al., 2006), Resolute, Canada 74°42'N 94°58W, (Lahoutifard et al., 2005) Churchill, Canada 58°8'N 94°1W(Kirk et al., 2006), and Summit, Greenland 72°6' N 38°5'W (Faïn et al., 2007).

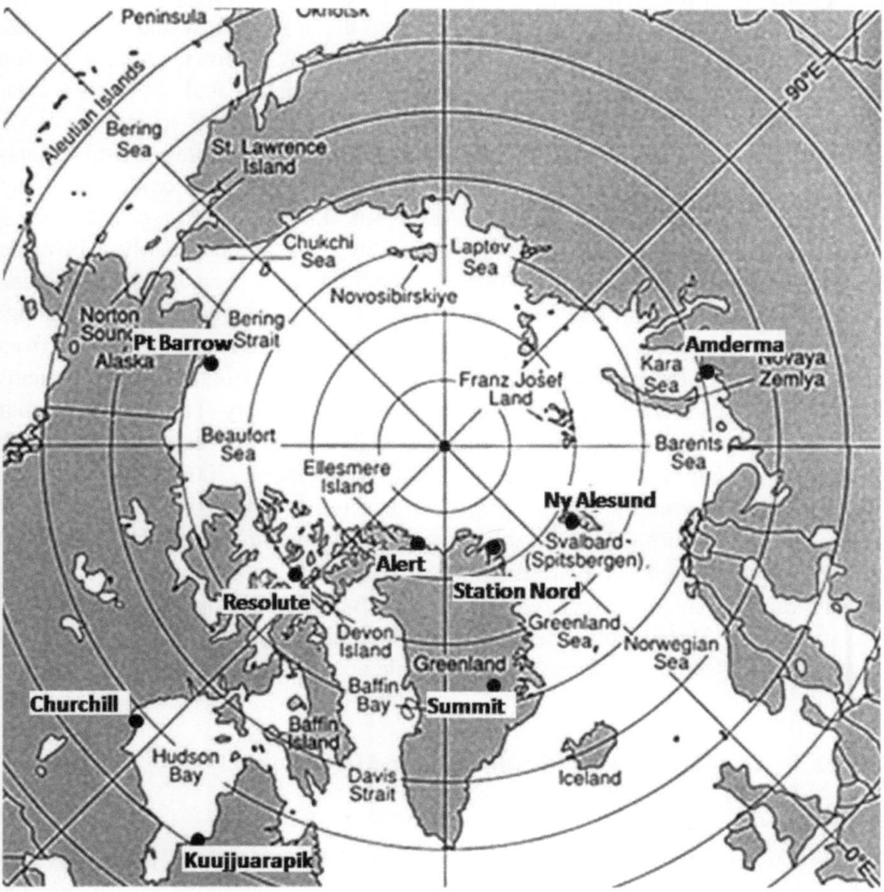

Figure 10.1 Measurement site for atmospheric mercury in the Arctic

The first arctic annual time series of high-resolution atmospheric Hg vapor data was collected at Alert in 1995 (Schroeder et al., 1998). It was found that after sunrise the Hg^0 concentrations underwent extraordinary fluctuations, decreasing at times from values approximately 1.7 ng m^{-3} to less than 0.1 ng m^{-3} within periods of 24 hours. This behavior runs counter to what is expected for an air pollutant characterized by a long atmospheric residence time (Schroeder and Munthe, 1995). This depletion of Hg^0 from the Arctic atmosphere was confirmed at several locations throughout the Arctic (Steffen et al., 2008). In 1998, Lu et al. (2001) and Lu and Schroeder (2004) reported high levels of PM during AMDEs that anti-correlated with measured gas phase Hg. They suggested that Hg^0 was being converted to PM and RGM when AMDEs occurred. This hypothesis that RGM is produced during AMDEs was confirmed in 2000 through direct measurements by Lindberg et al. (2001) at Barrow, Alaska, USA. Steffen et al. (2002) measured total atmospheric mercury at Alert in 2000 and showed that during depletion events other forms of Hg species exist in the air besides Hg^0. This study also demonstrated that, during depletion events, not all of the converted Hg^0 remains in the air and it was proposed that the remainder of the converted Hg is deposited onto the nearby snow and ice surfaces. It is now thought that the chemistry that causes the well known ozone depletion events (Barrie et al., 1988; Bottenheim et al., 1986; Simpson et al., 2007) is similar to what drives AMDEs (Ariya et al., 2002; Lindberg et al., 2001; Lindberg et al., 2002) in that Hg^0 is oxidized by reactive halogens, namely Br atoms or BrO radicals (Ariya et al., 2004; Goodsite et al., 2004; Skov et al., 2004). The oxidation of Hg^0 with these reactive halogens yields inorganic RGM. While there are mechanisms and theoretical calculations that suggest that RGM is predominantly a bromide compound (Calvert and Lindberg, 2004), its identity has not been directly elucidated and thus RGM is operationally defined. The reactive halogen species oxidizing Hg^0 are assumed to be generated from open water regions such as leads or polynyas from refreezing sea ice (nilas) forming on open waters and UV radiation. Measurements have shown that AMDEs occur right at the snow surface (Berg et al., 2003; Sommar et al., 2007; Sprovieri et al., 2005b; Steffen et al., 2002) and are limited to the surface up to a maximum of 1 km (Banic et al., 2003; Tackett et al., 2007). Lindberg et al. (2001, 2002) reported the first and highest measured concentration levels of RGM (up to 900 pg/m^3) during AMDEs at Barrow and showed a strong correlation between RGM production and UV-B irradiation and with an increase in surface snow Hg concentrations. The discovery of AMDEs has revolutionized our understanding of the cycling of Hg in Polar Regions while stimulating a significant amount of research to understand its impact to this fragile ecosystem.

10.2.2.3 Temporal Trends of Atmospheric Mercury and Comparisons Between Sites

Continuous long-term TGM measurements in the Arctic using highly time-resolved automatic monitors have been carried out at several observatory sites within the Northern Hemisphere. Hg^0 measurements have been collected at Ny-Ålesund, Norway

(1994-2000 [manual samples]; 2000-present [automated samples]) and Alert, Canada (1995-present). Other long term measurements of Hg^0 in the Arctic include Amderma Russia (2001-present) and Kuujjuarapik Canada (1999-present). Trend analysis was conducted on these measurements and has showed some temporal and spatial trends. An illustration of some of these measurements is provided on Figure 10.2.

The data sets that have been collected for more than 5 years were subjected to robust statistical trend analysis (Temme et al., 2007) and both of these time series showed no evidence of annual long-term trends during each respective monitoring period. In the springtime, highly variable Hg^0 concentrations as well as the lowest

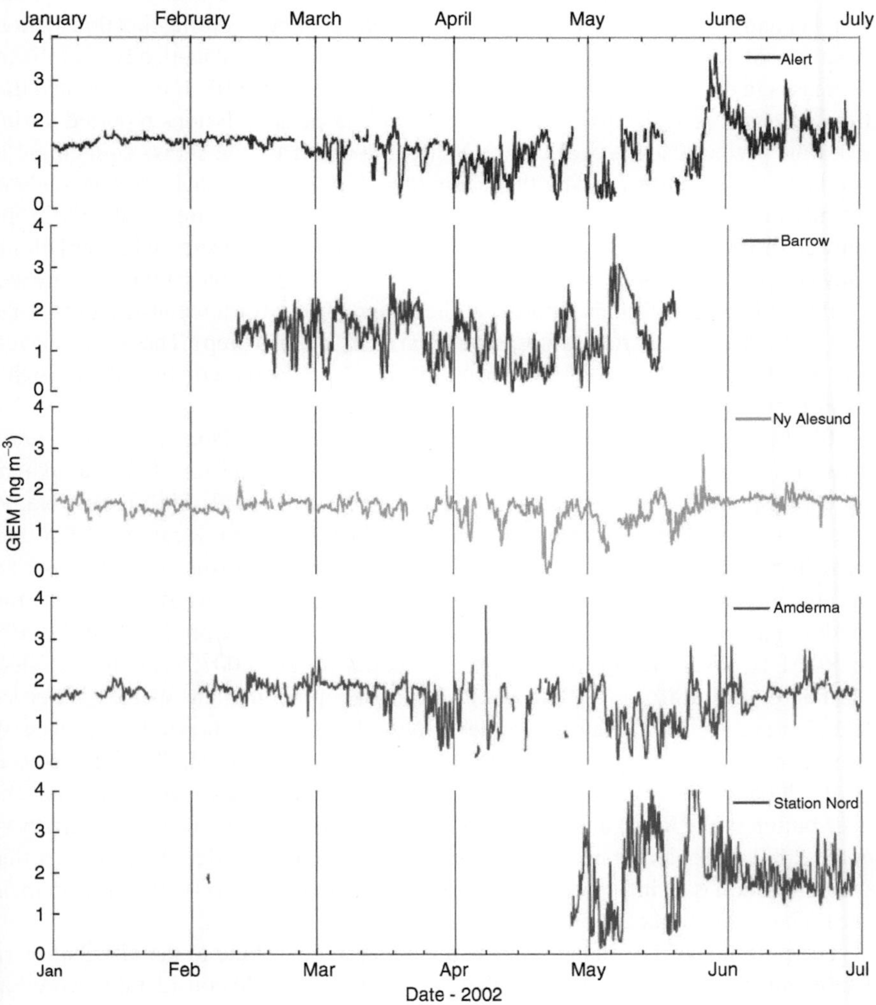

Figure 10.2 Temporal trends of Hg^0 measurements conducted in the Arctic in 2002

median concentrations of all the seasons were found and are a result of AMDEs that are known to occur in these regions. While the low springtime median concentrations at Alert revealed no significant trend (95% CI) from 1995 to 2002, the summer Hg^0 concentrations indicated a statistically significant (95% CI) decrease from 1995 to 2002. Mercury concentrations measured in the summer were higher than the springtime at Alert perhaps due to the emission of Hg from tundra and snow surfaces (Steffen et al., 2005).

This decreasing summer trend in Hg^0 concentration is in contrast to a more recent report of a trend at Alert, between 1995 and 2005, where it is shown that no statistically significant trend for each season was found (Temme et al., 2007). The authors hypothesize that this change in trends may be due to higher re-emission from the oceans coupled with effects from rising air temperatures during Arctic summer and effects from decreasing European emission rates during that time period. Analysis of the measurements from Ny-Ålesund reveals statistically significant decreases in the fall and winter months between 2000 and 2005. It is speculated that this decrease is a reflection of the decrease in European emissions reported during that time period (Pacyna et al., 2006; Steffen, 2007). This site shows an increase in summer concentrations during that same time period. Some analysis was made of the measurements from Alert, Amderma and Kuujjuarapik (Steffen et al., 2005) and showed distinct repeating seasonal patterns for all sites yet the sites at higher latitude showed higher seasonal amplitude. Longer time series of measurements are not available; however, firn air records in Greenland allows for a reconstruction of the past Hg^0 signal in the Arctic atmosphere (Faïn et al. 2008, in prep). These data suggest that Hg^0 levels were the highest during the 70's and decreased since then. Such a trend is similar to the worldwide mercury production.

To the best of the authors' knowledge, only Alert, Canada has long term measurements of RGM and PM in the Arctic. However, several field campaigns have measured these species during AMDEs (see Steffen et al., 2008). The relative distribution of these two atmospheric species differs between locations. RGM can exist in the gas phase but will be readily sorbed onto aerosols present in the air because of its hygroscopic properties (Ariya et al., 2004). At Alert, the overall predominant species in spring is PM but a clear shift from the predominance of PM to RGM is observed during the spring (Cobbett et al., 2007; Kirk et al., 2006; Steffen et al., 2008). At Barrow, RGM is the predominant species observed (Lindberg et al., 2002). Several studies at Ny-Ålesund have shown that, in general, there is no predominance of either RGM or PM (Gauchard et al., 2005b; Sprovieri et al., 2005a; Sprovieri et al., 2005b). Some researchers have suggested that the distribution of the RGM and PM is an indication of the age of an air mass (Lindberg et al., 2002; Sprovieri et al., 2005a; Steffen et al., 2003a) while others suggest that the distribution is an indication of local versus transported events (to the measurement site) (Gauchard et al., 2005b; Wangberg et al., 2003).

The presence of UV radiation is also thought to contribute to the distribution of RGM and PM as suggested by Lindberg et al. (2002). While considerable work has been done to study mercury in the Arctic, research continues to fully understand the behaviour of mercury over time, the transport of mercury to this area. The processes involved in its conversion between differing media and the impact that this contaminant has on this vulnerable region needs more study. It is important to understand all of these issues with a changing climate occurring at such a rapid pace in the Arctic region.

10.2.3 Atmospheric Mercury in the Antarctic

10.2.3.1 Introduction

Antarctica and the Southern Ocean are located in a remote region, with no indigenous human population and no industrial activity. Human activity is minimal and localized. Human presence in the region largely consists of scientific investigations and logistical operations in support of these investigations. The greatest human impact can be expected where research is carried out at long-term stations yet these typically have populations of fewer than 100 people. The overwhelming majority of anthropogenic Hg loading to the environment and biota derives from global rather than local input. Antarctica is characterized by a vast, cold, dry, high-altitude polar plateau, and a coastal region where the seasonal freezing and melting of sea ice surrounding the continent is the Earth's largest seasonal energy exchange event. This vast freezing of sea ice liberates sea salt bromine. Far from anthropogenic emissions, and isolated by the circumpolar vortex, only the longest-lived of the global atmospheric contaminants, such as GEM, make their way to the Antarctica polar plateau.

Similar to the Arctic (see section above), atmospheric mercury and ozone depletion events are most noticeable along Antarctic coastlines where polynyas and coastal, or flaw, leads provide frequently freezing sea ice surfaces as a source of atmospheric bromine. Far from the coasts, the main source of bromine is photochemical recycling between the cold surface snow pack and the atmosphere (Piot and von Glasow, 2008). Evidence of Br_2 emissions from snow surfaces was first reported by Foster et al. (2001) and enhancements of BrO above the snow pack were reported by Avallone et al. (2003). Recent chemical modeling studies, including Saiz-Lopez et al. (2007) and Piot and von Glasow (2008), show that bromine recycling in the surface snow is required to match air column chemical observations. Most recently, Dibb et al. (2008) directly observed photochemical bromine recycling from the surface snow, along with correlated ozone and mercury depletion events at an icecap location >3000 m in altitude and >1000 km from the nearest coast.

10.2.3.2 Atmospheric Measurements on the Antarctic Region

Few field mercury experiments have been performed in Antarctica compared to those carried out in the Arctic. Mercury measurements performed at different locations of the Antarctic region are reported in Table 10.2 and in Figure 10.3. The Hg^0

Table 10.2 Summary of atmospheric mercury measurements performed at different Antarctic locations from 1985 to 2005. TPM is measured by manual mini-traps

Measurement Sites	Period	Measurement type	Techniques	Statistical Parameters ($ng\ m^{-3}$) Mean ± SD	Min	Max	References
Lake Vanda 77°33'S 161°37'E	Dec. 1985	TGM	Manual-silvered/gilded sand collectors;	0.23 ± NA	NA	NA	(De Mora et al., 1991; De Mora et al., 1993)
Scott Base 90°00'S 139°16'W	1987	TGM	Manual-silvered/gilded sand collectors;	0.52 ± 0.14	0.16	0.83	(De Mora et al., 1991; De Mora et al., 1993)
	1988	TGM		0.60 ± 0.40	0.02	1.85	
Arrival Heights 77°11'S 166°40'E	1989	TGM	Manual-silvered/gilded sand collectors;	0.52 ± 0.16	0.11	0.78	(De Mora et al., 1991; De Mora et al., 1993)
Neumayer 70°39'S 08°15'W	2000-2001	TGM	Tekran 2537A; 1130 and KCl-Coated Annular Denuders; AESmini-Traps;	1.08 ± 0.29	0.27	2.34	(Ebinghaus et al., 2002; Temme et al., 2003)
		Hg^0		0.99 ± 0.27	0.16	1.89	
		RGM		NA	5×10^{-3}	$\sim 300 \times 10^{-3}$	
		TPM		NA	15×10^{-3}	120×10^{-3}	
Terra Nova Bay 74°41'S 164°07'E	1999-2001	TGM	Tekran 2537A; 1130 and KCl-Coated Annular Denuders; Gold-mini Traps;AE-TPM Traps	0.81 ± 0.1	0.5	0.9	(Sprovieri and Pirrone, 2000; Sprovieri et al., 2002)
		Hg^0		0.9 ± 0.3	0.29	2.3	
		RGM		$(116 \pm 78) \times 10^{-3}$	$\sim 11 \times 10^{-3}$	334×10^{-3}	
		TPM		$(12 \pm 6) \times 10^{-3}$	$\sim 4 \times 10^{-3}$	20×10^{-3}	
S. Pole 90°00'S	2003- 2005	Hg^0	Tekran 2537A; 1130, 1135	0.54± 0.19	0.24	0.82	(Brooks et al., 2008a)
		RGM		$(344 \pm 151) \times 10^{-3}$	95×10^{-3}	705×10^{-3}	
		PM		$(224 \pm 119) \times 10^{-3}$	71×10^{-3}	660×10^{-3}	
McMurdo 77°13'S 166°45'E	2003	Hg^0	Tekran 2537A; 1130, 1135	1.20 ± 1.08	BDL	11.16	(Brooks et al., 2008b)
		RGM		$(116 \pm 45) \times 10^{-3}$	29×10^{-3}	275×10^{-3}	
		PM		$(49 \pm 36) \times 10^{-3}$	5×10^{-3}	182×10^{-3}	

NA: data not available; BDL: Below detection limit

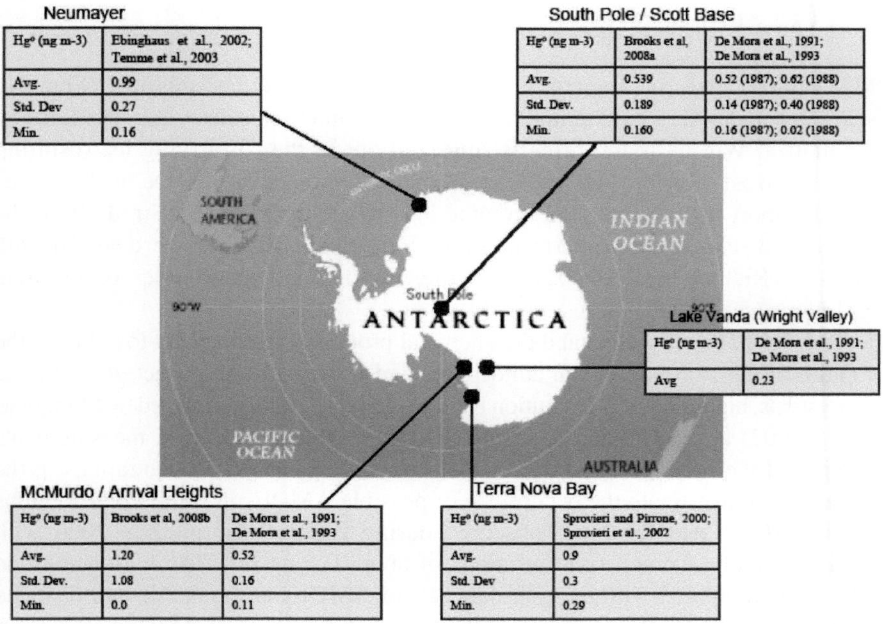

Figure 10.3 Measurement sites for atmospheric mercury in Antarctica

levels are far below the concentrations observed in the Arctic due to the remoteness from anthropogenic sources.

The first baseline data for the concentration and speciation of atmospheric mercury in Antarctica were reported by De Mora et al. (1993). Mercury measurements were carried out at three sampling location throughout 1985 and 1988. In particular, a preliminary study was carried out on the frozen surface of Lake Vanda (77°33'S, 161°37'E) in the Wright Valley during December 1985. While obviously limited, the data were interesting and suggested that TGM concentrations in Antarctica were substantially lower than those observed elsewhere (0.23 ng m^{-3}). Therefore, further studies were conducted throughout 1987 and 1988 at Scott Base (90°00'S, 139°16'W) and during 1989 at Arrival Heights (77°11'S, 166°40'E) on Ross Island. The mean TGM for 1987 was 0.52 ± 0.14 ng m^{-3} whereas the corresponding 1988 value was 0.60 ± 0.40 ng m^{-3}. At the third site, mean TGM value was 0.52 ± 0.16 ng m^{-3}. The first data reporting and referring to AMDEs were obtained at Neumayer (70°39'S, 8°15'W) a coastal location in Antarctica (Ebinghaus et al., 2002).

10.2.3.3 Temporal Trends of Atmospheric Mercury in Antarctica

To the best knowledge of the authors, there are no long-term data sets in Antarctica. Therefore, only measurements conducted during field campaigns are reported.

10.2.3.3.1 Coastal Sites

Mercury processes in Antarctica begin with marine bromine emissions. Freezing sea water under very cold temperatures traps bromine sea salts within the forming ice matrix. Within hours, brine is squeezed out of the solidifying ice resulting in briny frost flowers, which both dramatically increase the ice surface area and transport the concentrated bromine ions to the air interface. Similarly to the Arctic, the related atmospheric bromine compound, BrO, can be detected with satellite (Richter et al., 2002) indicating regions and magnitudes of bromine emissions (see Figure 10.4).

In order to better understand the chemical processes that may act to enhance the capture of Hg from the global atmosphere and its deleterious impact on Antarctic ecosystems, high-temporal-resolution Hg measurements were performed by Ebinghaus et al. (2002) at the German Research Station at Neumayer. These measurements comprised the first annual time series of ground-level TGM concentrations in the Antarctic to investigate the occurrence of possible AMDEs in south Polar Regions (Figure 10.5). AMDEs were observed during Antarctic springtime 2000 with minimum daily average concentrations of about 0.1 ng m^{-3}. The high-resolution data were compared with existing data sets of AMDEs in the Arctic and revealed similarities between the temporal and quantitative sequence of AMDEs after polar sunrise. TGM and O_3 were positively correlated (Figure 10.5) as in the Arctic boundary layer after polar sunrise (Sprovieri and Pirrone, 2000; Sprovieri et al., 2005a; Steffen et al., 2008), even if the ozone depletion events at Neumayer are less frequent, and shorter (Lehrer, 1999).

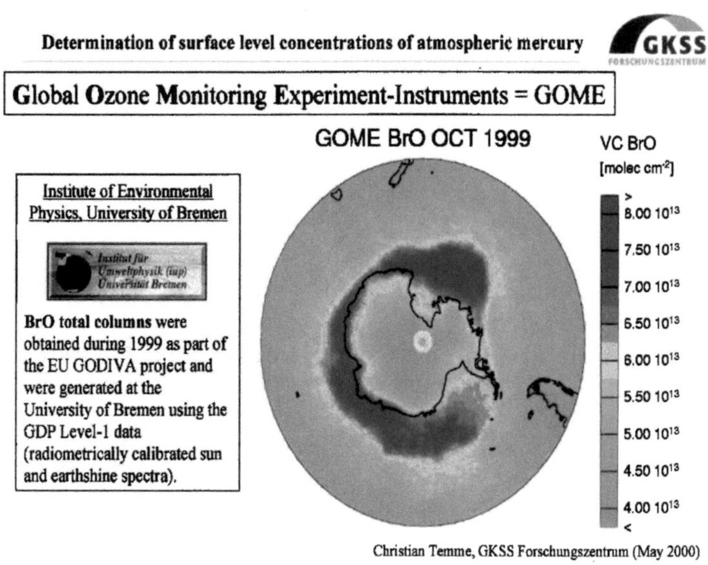

Figure 10.4 Distribution of BrO around the Antarctic Continent (Richter et al., 2002)

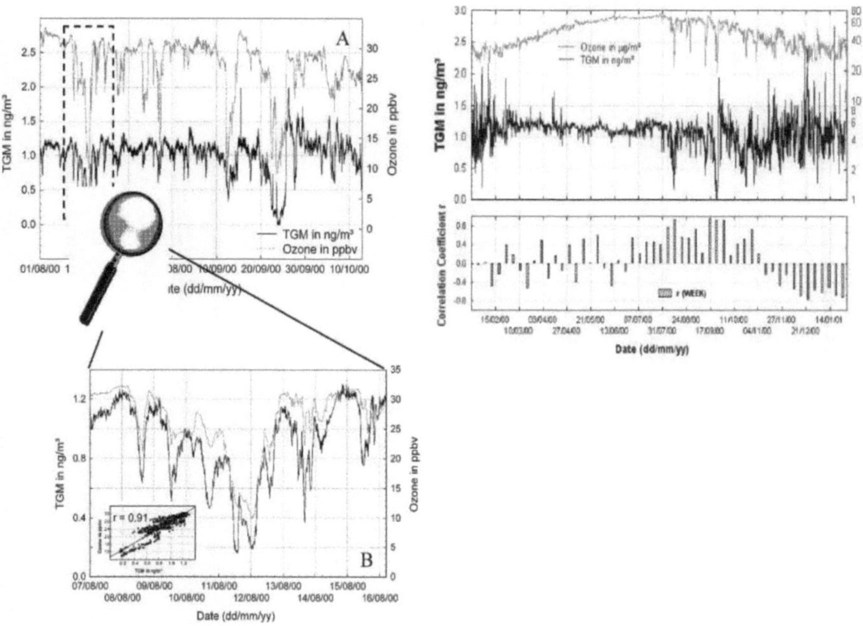

Figure 10.5 Ozone and TGM concentrations during the AMDEs observed at Neumayer, Antarctica from August to October 2000 (Reprinted with permission from Ebinghaus et al., 2002 and Temme et al., 2003. Copyright 2002 and 2003 American Chemical Society)

Friess (2001) detected enhancements of BrO in the lower troposphere, during the same period at Neumayer, using DOAS. Ebinghaus et al. (2002) also found that AMDEs coincided with enhanced column densities of BrO from measurements by the satellite borne GOME instrument over the sea ice around the Antarctic continent after polar sunrise. Air masses at ground level coming from the sea ice surface, accompanied by BrO enhancements, could be a necessary condition for the AMDEs at Neumayer. Therefore, the sea ice is a possible place where the photochemical reaction of ozone and Br atoms and/or the reaction of BrO radicals and Hg^0 can take place in the Antarctic during the springtime.

Figure 10.6 shows, in particular, simultaneous Hg^0, RGM and O_3 ambient concentrations for the period November 26th to December 31th, 2000 (Sprovieri et al., 2002) observed at Terra Nova Bay (74°41'S, 164°70'E). RGM concentrations were high and comparable to those directly observed by anthropogenic Hg sources. Interestingly, these high levels were measured in the absence of simultaneous ozone and mercury depletion events. Similar RGM results (Table 10.2) have been reported by Temme et al. (2003). The very high RGM concentrations at both Antarctic coastal sites could be influenced by the local production of oxidized gaseous mercury species over the Antarctic continent or by shelf ice during polar summer. The Hg^0 depletions recorded in January show no significant correlation to any additional parameters that were measured (Temme et al., 2003). This suggests that the oxidation of Hg^0 to RGM,

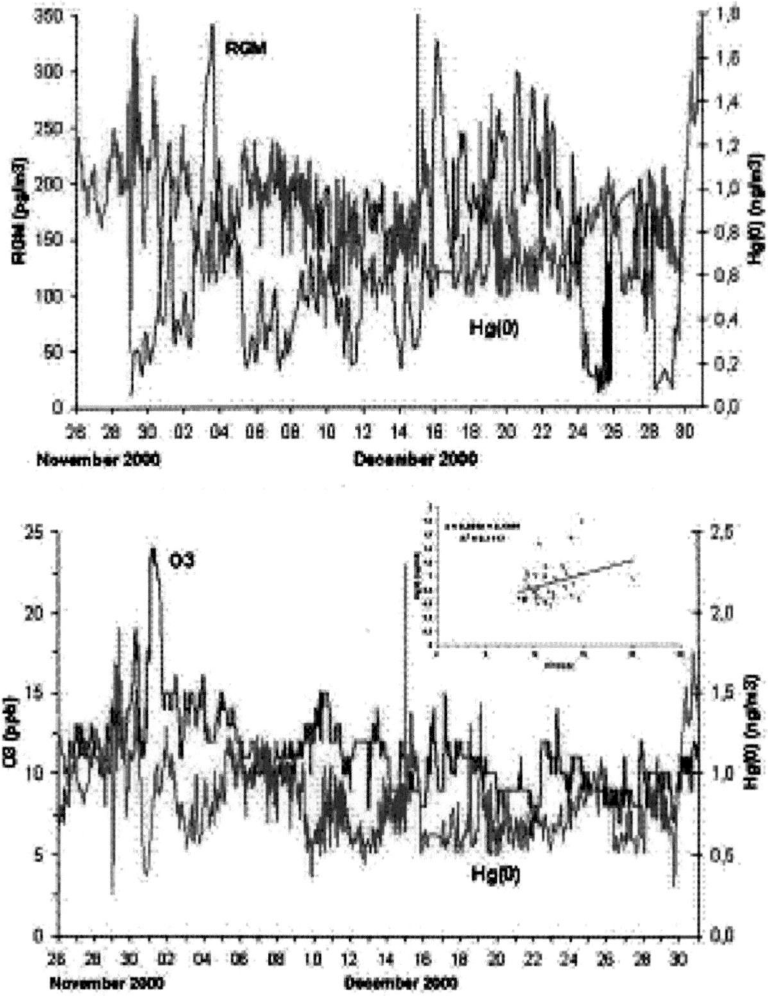

Figure 10.6 Two-hourly mean concentrations of Hg^0 and RGM measured at Terra Nova Bay, Antarctica from November to December, 2000 (Source: Sprovieri et al., 2002)

and a concurrent production of O_3, has already occurred before the air parcels were advected to the sampling site. The authors proposed a gas-phase oxidation of Hg^0 by potential oxidants (i.e. ·OH, $HO\cdot_2$, $NO\cdot_3$) associated with high levels of NO. These oxidants result from photo-denitrification processes in the snow-pack (Zhou et al., 2001) which may maintain the high RGM concentrations that were observed. Therefore, additional atmospheric measurements of potential precursor compounds and isentropic trajectory calculations are required to potentially ascertain the reaction mechanism and origin of the air masses reaching the measurements locations where these high RGM levels are observed during the Antarctic summer.

10.2.3.3.2 Sites on the Polar Plateau

On the Antarctic Polar Plateau where the snowpack is perennial and the bromine process decoupled by distance from the original freezing sea ice sources, polar sunrise (~September 21 at the South Pole) heralds negligible mercury oxidation. Mercury oxidation rates only begin to peak around the summer solstice with maximum values ~February 1. This reflects the bromine transport times from the coastal sea ice and the photochemical processes (Brooks et al., 2008a).

Oxidized mercury species were first reported by Arimoto et al. (2004) from high volume filter results at the S. Pole station clean air sector. More recently (Brooks et al., 2008a) combined mercury concentrations in snow and air, with vertical mercury flux measurements at the South Pole. These observations showed atmospheric oxidized mercury depositing to the snow pack, subsequent photo reduction, and emissions of Hg^0 from the surface. High oxidized mercury concentrations were measured in the near-surface air (e.g., RGM and fine PM; 100–1000 pg m^{-3}). From these concentrations and flux rates, it was calculated that the Antarctic polar plateau sequesters ~60 Mg of Hg annually. The seasonal cycling of atmospheric mercury oxidation, deposition, and re-emission via photoreduction is on the order of ~500 Mg annually (see section 2.4 for explanation). This dynamic mercury cycle on the polar plateau is driven by the surrounding sea ice as a vast bromine source, Southern Hemisphere Hg emissions, the sun, and the cold Spring/Summer temperatures.

A major global obstruction to the formation of atmospheric $Hg^{(II)}$ is believed to be the fast thermal decomposition of the $Hg^{(I)}$ radical, HgBr (Holmes et al., 2006). This fast thermal decomposition rate dominates chemistry above 0° C, but the rate decreases by half with every 6° C drop in temperature below 0° C (Goodsite et al., 2004; Holmes et al., 2006). The mercury in the air over the polar plateau (the coldest place on Earth), unlike any other location, is predominately $Hg^{(II)}$ in Spring and Summer (Brooks et al., 2008b). While Arctic and Antarctic coastal sites experience episodic mercury depletion events which occur predominantly in the late winter and early spring, the polar plateau experiences nearly-constant mercury events, peaking in the summer (Figure 10.7).

While the effects of mercury oxidation and deposition in coastal and marine environments have been postulated (AMAP, 2005) the effects of the observed ice cap mercury cycle are less clear. Given the dry conditions of the Antarctic polar plateau (burial/snowfall rate is ~10 cm/year) only ~10% of the deposited mercury is buried (sequestered), resulting in some 60 Mg Hg annually. The much smaller Greenland ice cap (1.8 million square km) has a burial/snowfall rate of ~70 cm/year, and similarly sequesters ~40 Mg Hg annually (Dibb et al., 2008).

10.2.3.4 Antarctica vs Arctic

A comparison between the two Polar Regions is problematic because both spatial and temporal coverage of Hg measurements are limited. The behaviour of mercury species may be associated with a number of reactive chemicals and reactions that take

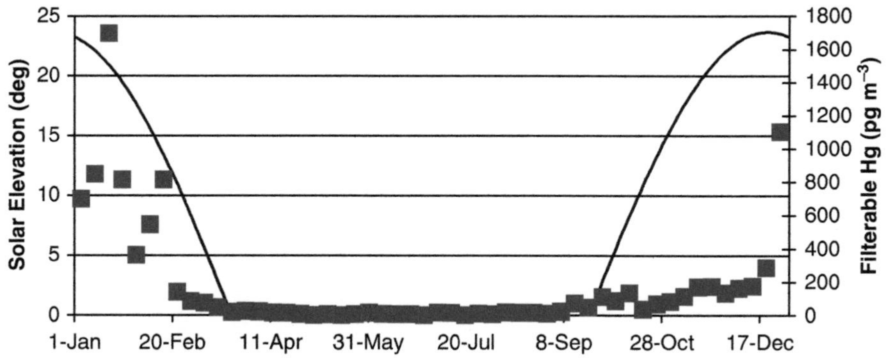

Figure 10.7 Weekly averages of total filterable mercury concentrations (the sum of RGM and PM) collected as Hg on high volume filters, and the annual solar elevation angles at South Pole Station. High volume filters allow GEM is pass but collect PM and a significant portion of the RGM. Most notably, filterable Hg concentrations (RGM + PM) are totally absent during the dark fall and winter seasons, implying that sunlight is a requirement to produce oxidized mercury species, RGM and PM. The figure also shows that filterable Hg (RGM+PM) concentration in the near-surface air much higher in the summer compared to the spring, indicating a delay between the re-emerging sunlight and, the GEM transport and bromine snow pack recycling, the drives the atmospheric chemical production of oxidized mercury species, RGM and PM. This delay could be due to the requirement of "seed" reactive halogens to drive the recycling of Br from the surface snow (Simpson et al., 2007; Piot and von Glascow, 2008)

place in the atmosphere after polar sunrise. The tropospheric chemistry of the polar areas is distinctly different than in the other parts of the Earth due to natural differences of meteorological and solar radiation conditions. During the winter months, in total lack of solar radiation, temperature and humidity conditions are very low, so the vertical mixing of the lower stratified Antarctic troposphere is hindered. The direct consequence is that the abundance of photo chemically labile compounds will rise, while the level of photochemical products will be low. During spring and summer, solar radiation is present 24 hours a day and under sunlight conditions, the elevated concentrations of reactants present in the Antarctic atmosphere can initiate a sequence of atmospheric chemical transformations often different than other latitudes.

It can be anticipated that in the polar troposphere, free radical precursors that build up in the darkness of the polar winter begin to photo dissociate and the resulting gas phase radicals may play a fundamental role in the elemental gas phase mercury decrease seen in Antarctica and in Arctic. Although in the Arctic the highest RGM concentrations were found during AMDEs, elevated concentrations were found at Barrow during snowmelt. Snowmelt is more limited in the Antarctic, even at coastal sites, than it is in the Arctic at sites such as Barrow, which suggests that the snowpack is directly involved in maintaining high RGM concentrations. The higher Hg^0 concentrations observed in the Arctic when compared to the Antarctica clearly indicate the different chemical composition of the troposphere as a result of the location of the measurements areas. In fact, the Arctic is surrounded by populated continents from which pollution is released and transported to the north. In contrast, the Antarctic is entirely surrounded by the Pacific Ocean and is far from any anthropogenic emissions. In particular, fluxes of mercury to the atmosphere, mainly from anthropogenic and continental sources in the Northern Hemisphere (particularly from Eurasian and North America in late winter and spring), are greater than those in the Southern Hemisphere, and higher atmospheric concentrations are found in the North than the South.

The dynamic transformations of atmospheric mercury species during the polar spring illustrate the complexity of photochemical reactions in Polar Regions and have revealed the limitations in our understanding of the chemical cycling of mercury and other atmospheric constituents/contaminants in remote regions with seasonally variable sea-ice coverage. More research and investigation on possible reaction mechanisms and chemical kinetics of these phenomena are required to successfully improve our understanding of chemical-physical processes involved in the mercury cycle in order to assess the resulting net input into the polar biosphere.

10.2.4 The Role of Snow Surfaces on Atmospheric Hg Trends

Snow surfaces are well recognized as important sites of chemical transformations in Polar Regions for many organic compounds. It has recently been established that mercury can also undergo redox transformation at the snow/air interface (for a review, see Steffen et al., 2008). Since the reduced form of Hg, Hg^0, is volatile, these interfacial transformations may lead to changes in the evasion and deposition of Hg in cold regions. However, there have been debates on the importance of these snow processes on atmospheric Hg trends. In this section, the available information regarding Hg emission and deposition processes at the snow/air interface will be reviewed followed by a discussion on the potential impact of these processes on atmospheric Hg levels.

10.2.4.1 Role of Snow in Emission and Deposition Processes

Snowpacks can alter atmospheric Hg concentrations by two types of processes. On one hand, they can promote the deposition of Hg^0 by, for instance, favouring its conversion into oxidized species through heterogeneous processes. On the other hand, newly deposited $Hg^{(II)}$ can be transformed within the snowpack into Hg^0, leading to snow-to-air transfer of Hg. In the latter case, oxidative processes within the snowpack may hinder this transfer. The balance between these evasional and depositional processes will change temporally (on seasonal and daily scales) and spatially (e.g. coastal vs. inland snowpacks).

10.2.4.1.1 Snowpacks as Promoters of Atmospheric Hg Deposition

Snow can influence Hg deposition through physical and/or chemical processes. First, snow packs can act as a source of halogens to the lower troposphere, thereby contributing to the halogen-assisted atmospheric oxidation of Hg^0 into RGM in

Polar Regions. This phenomenon is particularly likely to occur in snow over sea ice and in coastal snow packs, where halogen levels in snow are high (Brooks et al., 2006; Lindberg et al., 2002; Simpson et al., 2007). This impact of snow on atmospheric processes is partly supported by data showing that RGM formation is occurring mostly at the snow/air interface (Steffen et al., 2002).

Once RGM is formed in the atmosphere, snow can act as an efficient surface for the sorption of newly formed RGM. Douglas et al. (2005) hypothesized that active growth of snow and ice crystals from the vapour phase near leads readily scavenged available RGM, leading to the highest Hg levels ever reported in snow and ice. However, the sorptive properties of snow towards Hg species are still very poorly documented and represent a fruitful field of future research.

Once Hg deposition has occurred, oxidative processes within the snowpack may promote the retention of Hg on the ground and its eventual transport to receiving ecosystems. In particular, photooxidation of Hg by UV-A radiation has been shown to be substantial in waveband exclusion experiments, particularly in snow containing high chloride concentrations (e.g. Poulain et al., 2004; Poulain et al., 2007a). It is believed that chloride stabilizes intermediate $Hg^{(I)}$ species during a two-step oxidation of Hg^0 toward $Hg^{(II)}$. Alternatively, halides found in coastal snow can form radicals able to oxidize Hg within the snowpack (e.g. Ariya et al., 2004; Fain et al., 2006). Hydrogen peroxide, formed through photoreactions, has also been shown to be a possible Hg oxidant in snow, under acidic conditions (Lahoutifard et al., 2006). Overall, coastal and sea ice snowpacks can be considered as important sites promoting the deposition and retention of $Hg^{(II)}$ on the ground, leading to higher risk of contamination for neighboring ecosystems, when compared to inland snowpacks.

10.2.4.1.2 Snowpacks as Promoters of Hg Evasion

Redox reactions within the snowpack may also lead to the formation of Hg^0, and its release to the atmosphere. This reduction of Hg can either be driven by photochemical or biological processes, although the latter have received less attention until recently. According to a recent study conducted over a 80 m deep snowpack in Greenland, these redox transformations are limited to the first meters of the snowpack, since Hg^0 levels in the deep firn were constant between 15 and 70 m (Faïn et al., 2007).

Photochemical reactions are generally believed to be the main drivers of $Hg^{(II)}$ reduction. Field and laboratory polychromatic experimentations have shown that UV-B bands were particularly favorable to $Hg^{(II)}$ reduction in snow (Dommergue et al., 2007). The importance of the reductive process has been demonstrated in the Arctic by rapid Hg losses in the irradiated surface snow layers (Poulain et al., 2004). Observations of diel cycles of Hg^0 in interstitial air also suggest an important evasional flux of Hg driven by $Hg^{(II)}$ reduction (Dommergue et al., 2003a; Dommergue et al., 2003b).

The relative importance of Hg oxidation vs. reduction is likely to depend on seasonal and daily changes in: 1) light intensity and quality (as influenced by solar irradiation and cloud cover); 2) snow composition; 3) atmospheric and snow oxidants and reductants; 4) Hg speciation in snow. In fact, it has been suggested that following AMDEs, Hg in snow undergoes cycles of reduction and oxidation at the snow/air interface and within the superficial snow layer (Steffen et al., 2002). It should be noted that these photochemical processes will be limited to the upper centimeters of the superficial snow, due to limited light penetration within the snowpack (King and Simpson, 2001).

Mechanistically, we still have a very limited understanding of Hg reduction in snow. The exact substrates (Hg species) involved are not well characterized. Potential reductants have been proposed (e.g., HO_2), although their role has not been confirmed by thermodynamical experiments (Gardfeldt and Jonsson, 2003). Direct photolysis of some Hg complexes has also been proposed, but not demonstrated in snow. Ferrari et al. (2005) have suggested that the quasi-liquid layer surrounding snowflakes may be the main site of Hg redox transformations in snowpacks, and that temperature therefore plays an important role in dictating the occurrence and magnitude of these processes. However, this hypothesis needs to be supported by additional experimental data.

Biological processes may also be involved in the formation of Hg^0 in snowpacks. Indeed, data from Poulain et al. (2007b) have recently established the presence and expression of genes coding for mercury resistance in High Arctic bacteria. Since one of the main axes of resistance in bacteria is the reduction of $Hg^{(II)}$ into the less toxic Hg^0, these results call for additional studies on microbiological Hg redox transformations in snow. Additionally, snow microorganisms are likely to influence heterogeneous redox processes by acting as sorbents for Hg, or by generating exudates. Microbial processes are likely to be particularly significant during snowmelt and, possibly, in the quasi-liquid layer of snowflakes.

10.2.4.2 Potential Influence on Local, Regional and Global Mercury Levels

10.2.4.2.1 Local Scale

These redox processes can ultimately have an impact on atmospheric Hg levels. At the local scale, aircraft measurements, micrometeorological and flux chamber studies have shown that the snow/air interface was a dynamic layer where both evasional and depositional fluxes occurred. Few studies have established the height of this dynamic layer over the snow/air interface. Banic et al. (2003), using aircraft measurements made between 0.1 and 7 km, observed depletion of atmospheric Hg in air masses up to 1 km in height, but not above this limit. However, it is not possible to ascertain which portion of these depletion events can be attributed to snow related processes. Steffen et al. (2002) provided detailed profiles of Hg^0 between 0 and 2 m over snow and were able to demonstrate the occurrence of evasional fluxes of Hg^0 from snow during Arctic Spring. It is therefore likely that snow processes partly influence Hg^0 atmospheric levels in the atmospheric layer immediately overlying snowpacks.

The influence of snow on atmospheric Hg levels is not constant seasonally. Recently, Cobbett et al. (2007) presented Hg^0 fluxes measurements over Arctic night and day, and over the transition period between both periods. They reported no significant net Hg^0 flux during or soon after AMDEs, and after snowmelt; the largest Hg^0 depositional fluxes were measured during the Arctic night, and were attributed to concentration gradients. The maximum Hg^0 evasional fluxes were recorded when the tundra was first visible during snowmelt. This suggests that Hg^0 production in snowpacks may have only limited influence on atmospheric Hg^0 levels, except during the snowmelt period, when their influence is more pronounced. Indeed, Poulain et al. (2007a) have measured high $Hg^{(II)}$ reduction rates in snowmelt water, supporting the notion of a significant Hg^0 emission during the snowmelt period. However, it should be noted that other studies have reported significant Hg^0 fluxes immediately following AMDEs (e.g. Kirk et al., 2006); therefore, at least at some polar locations, snowpacks influence local atmospheric Hg^0 levels both in early and late spring.

10.2.4.2.2 Regional and Global Scales

In order to assess the influence of snow processes on regional and global atmospheric Hg levels, it is necessary to include these processes in large scale models. There are currently two hemispheric Hg models including the Danish Eulerian Hemispheric Model (DEHM) and the Meteorological Synthesizing Centre-East hemispheric model (MSCE-Hg-Hem) (Travnikov and Ryaboshapko, 2002) and one global model, the Global/Regional Atmospheric Heavy Metals Model (GRAHM; (Dastoor and Larocque, 2004). However, these models do not include heterogeneous chemistry on surface snow and snow physics because of the lack of laboratory data (Steffen et al., 2008). Re-emission from snow is not considered in the DEHM and the MSCE-Hg-Hem model. In the GRAHM model, re-emission is treated as Hg^0 as a fraction of the annual Hg deposition (Ariya et al., 2004) instead of a mechanistic representation including photochemical processes, due to a lack of understanding of these processes.

Using the GRAHM model, Ariya et al. (2004) concluded that re-volatilization was not significant over the scale of weeks, except in the late Spring; these model results are coherent with field reports from Cobbett et al. (2007) discussed above, which showed significant Hg^0 emission from snowpacks in late Spring. Finally, Brooks et al. (2006) calculated a mass balance for the springtime arctic environment and concluded that photo-induced emission of Hg^0 from snowpack represented 59% of depositional fluxes. It is therefore likely that snow processes are impacting Hg^0 levels in the lower atmosphere at least at the regional scale (i.e. the Arctic), but only during springtime.

10.3 Gaps of Knowledge, Future Research and Policy Implications

The observations seen in the polar regions, thus constitute direct evidence of a link between sunlight assisted Hg^0 oxidation, greatly enhanced atmospheric $Hg^{(II)}$ wet and/or dry deposition, and elevated Hg concentrations in the polar snow-pack in spring. It has been thought, in fact, that the AMDEs are, probably, recent phenomena due to the climate global changes in the Polar Regions and, in general, to the global warming of the planet. This last leading to a decreasing trend in multi-year ice coverage, earlier timing of snowmelt, increasing ocean temperature and increasing atmospheric circulation can impact the dynamics of the AMDEs (Lindberg et al., 2002). These climate changes have, in fact, increased atmospheric transport of photo-oxidants and production of reactive halogens (Br/Cl) in the Polar Regions enhancing Hg oxidation reactions. The decreasing, in addition, in total column ozone amounts over the polar areas and the subsequent increasing of the incident solar UV-B which influence the production of reactive halogen species could lead to increase of the Hg^0 oxidation processes and Hg accumulation in polar ecosystems. Further studies are necessary to explain the reaction mechanism and the kinetics of the AMDEs and the RGM production identified during measurement campaigns in the Polar Regions during springtime depletion periods. It is also important to combine the results observed with trajectory calculations in combination with sea ice maps to investigate the origin of the depleted air masses and the actual places where the chemical reactions involving ozone, reactive bromine species, hydroxyl radical and Hg^0 occur.

Moreover, it is clear that current models are limited in their integration of snow processes by a lack of sufficient mechanistic studies detailing physical, chemical and biological interactions. In particular, there is a need to assess the rates of scavenging/adsorption of Hg^0 and $Hg^{(II)}$ species by snow and ice crystals, both inland and on the sea ice. In order to reach such an objective, a more detailed knowledge of Hg speciation in air and snow is also required. Laboratory studies coupled to field ones are also needed to elucidate heterogeneous redox transformation at the snow/air and ice/air interface. Additional studies should assess the role of the quasi-liquid layer as a potential key site of chemical reactions for Hg on snow. At this point in time, photochemical processes seem to dominate all others, and the balance between photo reduction and photo oxidation is considered to be dependent on snow chemistry (i.e. halide levels) and light quality (UV-A vs. UV-B). However, microbial processes have received too little attention both in the Arctic and Antarctica, and may prove to be significant, particularly in late Spring when metabolic activity is rapidly rising and, incidentally, when observed and modeled late spring Hg^0 snow-to-air fluxes are maximal.

The Arctic is currently undergoing rapid and dramatic changes including warming which is changing the timing and extent of sea ice and its coverage (Serreze et al., 2002; Stroeve et al., 2005) and it is affecting the seasons with winter coming later and spring melt coming earlier. As well, coal and fossil fuel combustion in Asia, a major global source of Hg, is expected to increase up to 350% between 1990 levels and 2020 (van Aardenne et al., 1999). The effects of these increasing emissions on AMDEs processes and the long term deposition of Hg to the Polar Regions will only be discernible if long term measurements are collected at numerous locations. In Antarctica, gross mercury input is controlled by the South Hemisphere emissions. While Northern Hemisphere mercury emissions have been decreasing over the last couple decades, Southern Hemisphere mercury emissions increased from 1990 to 1995 and have stayed roughly constant since 1995. From 1990 to 1995 Africa emissions increased from 200 to 400 Mg yr^{-1}, Australia from 50 to 100 Mg yr^{-1}, and South America from 55 to 80 Mg yr^{-1} (Lindberg et al., 2007; Pacyna et al., 2006). A warmer and wetter environment will have positive and negative effects. On the plateau the formation of $Hg^{(II)}$ will decrease with the enhanced thermal decomposition of the intermediate $Hg^{(I)}$ radical, HgBr. However, a wetter environment will increase snowfall rates and bury (sequester) a greater proportion of the deposited $Hg^{(II)}$.

Long term measurements of Hg^0 and other atmospheric Hg species in the Polar Regions are very limited and need to be increased. These types of measurements can yield critical information to better understand the processes involved in the cycling of Hg in the polar atmosphere and thus the deposition of this pollutant to this fragile environment. Long term measurements of Hg in the polar atmosphere must be put into place so that the effects of these changes to Hg distribution in this environment can be monitored and scrutinized.

References

AMAP, 2003. AMAP Assessment 2002: Human Health in the Arctic. Arctic Monitoring and Assesment Programme (AMAP). Oslo, Norway.

AMAP, 2005. AMAP Assessment 2002: Heavy Metals in the Arctic. Arctic Monitoring and Assessment Programme (AMAP). Oslo, Norway.

Arimoto, R., Schloesslin, C., Davis, D., Hogan, A., Grube, P., Fitzgerald, W., Lamborg, C., 2004. Lead and mercury in aerosol particles collected over the South Pole during ISCAT-2000. Atmospheric Environment 38, 5485–5491.

Ariya, P.A., Khalizov, A., Gidas, A., 2002. Reactions of Gaseous Mercury with Atomic and Molecular Halogens: Kinetics, Product Studies, and Atmospheric Implications. Journal of Physical Chemistry A 106, 7310–7320.

Ariya, P.A., Dastoor, A.P., Amyot, M., Schroeder, W.H., Barrie, L., Anlauf, K., Raofie, F., Ryzhkov, A., Davignon, D., Lalonde, J., Steffen, A., 2004. The Arctic: a sink for mercury. Tellus Series B-Chemical and Physical Meteorology 56, 397–403.

Aspmo, K., Gauchard, P.-A., Steffen, A., Temme, C., Berg, T., Bahlmann, E., Banic, C., Dommergue, A., Ebinghaus, R., Ferrari, C., Pirrone, N., Sprovieri, F., Wibetoe, G., 2005. Measurements of atmospheric mercury species during an international study of mercury depletion events at Ny-Alesund, Svalbard, spring 2003. How reproducible are our present methods? Atmospheric Environment 39, 7607–7619.

Aspmo, K., Temme, C., Berg, T., Ferrari, C., Gauchard, P.A., Fain, X., Wibetoe, G., 2006. Mercury in the atmosphere, snow and melt water ponds in the North Atlantic Ocean during Arctic summer. Environmental Science and Technology 40, 4083–4089.

Avallone, L.M., Toohey, D.W., Fortin, T.J., McKinney, K.A., Fuentes, J.D., 2003. In situ measurements of bromine oxide at two high-latitude boundary layer sites: Implications of variability. Journal of Geophysical Research-Atmospheres 108.

Banic, C., Beauchamp, S.T., Tordon, R.J., Schroeder, W.H., Steffen, A., Anlauf, K.A., 2003. Vertical distribution of gaseous elemental mercury in Canada. Journal of Geophysical Research 108, 4264.

Barkay, T., Poulain, A.J., 2007. Mercury (micro)biogeochemistry in polar environments. Fems Microbiology Ecology 59, 232–241.

Barrie, L.A., Bottenheim, J.W., Schnell, R.C., Crutzen, P.J., Rasmussen, R.A., 1988. Ozone destruction and photo-chemical reactions at polar sunrise in the lower Arctic atmosphere. Nature 334, 138–141.

Berg, T., Sekkesæter, S., Steinnes, E., Valdal, A.K., Wibetoe, G., 2003. Springtime depletion of mercury in the European Arctic as observed at Svalbard. Science of the total environment 304, 43–51.

Bottenheim, J.W., Gallant, A.G., Brice, K.A., 1986. Measurements of NOy species and O3 at 82oN latitude. Geophysical Research Letters 22, 599–602.

Bottenheim, J., Chan, H.M., 2006. A trajectory study into the origin of spring time Arctic boundary layer ozone depletion. Journal of geophysical research 111,

Braune, B.M., Outridge, P.M., Fisk, A.T., Muir, D.C.G., Helm, P.A., Hobbs, K., Hoekstra, P.F., Kuzyk, Z.A., Kwan, M., Letcher, R.J., Lockhart, W.L., Norstrom, R.J., Stern, G.A., Stirling, I., 2005. Persistent organic pollutants and mercury in marine biota of the Canadian Arctic: An overview of spatial and temporal trends. Science of the Total Environment 351, 4–56.

Brooks, S.B., Saiz-Lopez, A., Skov, H., Lindberg, S.E., Plane, J.M.C., Goodsite, M.E., 2006. The mass balance of mercury in the springtime arctic environment. Geophysical Research Letters 33, L13812.

Brooks, S., Arimoto, R., Lindberg, S., and Southworth, G., 2008a. Antarctic polar plateau snow surface conversion of deposited oxidized mercury to gaseous elemental mercury with fractional long-term burial, Atmospheric Environment vol. 42, no. 12, 2877–2884.

Brooks, S., Lindberg, S., Southworth, G., and Arimoto, R., 2008b. Springtime atmospheric mercury speciation in the McMurdo, Antarctica coastal region. Atmospheric Environment vol. 42, no. 12, 2885–2893.

Calvert, J.G., Lindberg, S.E., 2004. The potential influences of iodine containing compounds on the chemistry of the troposphere in the polar spring II. Atmospheric Environment 38, 5105–5116.

Cobbett, F.D., Steffen, A., Lawson, G., Van Heyst, B.J., 2007. GEM fluxes and atmospheric mercury concentrations (GEM, RGM and HgP) in the Canadian Arctic at Alert, Nunavut, Canada (February-June 2005). Atmospheric Environment 41, 6527–6543.

Dastoor, A.P., Larocque, Y., 2004. Global circulation of atmospheric mercury: a modelling study. Atmospheric Environment 38, 147–161.

De Mora, S.J., Bibby, D.M., Patterson, J.E., 1991. Baseline concentration and speciation of atmospheric mercury at Baring Head (41"S), New Zealand. Environmental Technology 12, 943–946.

De Mora, S.J., Patterson, J.E., Bibby, D.M., 1993. Baseline atmospheric mercury studies at Ross Island, Antarctica. Antarctic Science 5, 323– 326.

Dibb, J.E., Stutz, J., Huey, L.G., Brooks, S.B., Lefer, B.L., von Glasow, R., Tanner, D.J., 2008. Bromine and mercury air-snow exchange on the Greenland ice cap. in prep.

Dommergue, A., Ferrari, C.P., Poissant, L., Gauchard, P.-A., Boutron, C.F., 2003a. Chemical and photochemical processes at the origin of the diurnal cycle of gaseous mercury within the snow-pack at Kuujjuarapik, Québec. Environmental Science & Technology 37, 3289–3297.

Dommergue, A., Ferrari, C.P., Gauchard, P.-A., Boutron, C.F., Poissant, L., Pilote, M., Jitaru, P., Adams, F., 2003b. The fate of mercury species in a sub-arctic snow-pack during snowmelt. Geophysical Research Letters 30, doi: 10.1029/2003GL017308.

Dommergue, A., Bahlmann, E., Ebinghaus, R., Ferrari, C., Boutron, C., 2007. Laboratory simulation of Hg° emissions from a snowpack. Analytical and Bioanalytical Chemistry 388, 319–327.

Douglas, T.A., Sturm, M., Simpson, W.R., Brooks, S., Lindberg, S.E., Perovich, D.K., 2005. Elevated mercury measured in snow and frost flowers near Arctic sea ice leads. Geophysical Research Letters 32, -.

Ebinghaus, R., Kock, H.H., Temme, C., Einax, J.W., Löwe, A.G., Richter, A., Burrows, J.P., Schroeder, W.H., 2002. Antarctic Springtime Depletion of Atmospheric Mercury. Environmental Science and Technology 36, 1238 –1244.

Fain, X., Ferrari, C.P., Gauchard, P.-A., Magand, O., Boutron, C.F., 2006. Fast depletion of elemental gaseous mercury in the kongsvegen Glaciersnowpack in Svalbard. Geophysical Research Letters 33, L06826, doi:06810.01029/02005GL025223.

Faïn, X., Ferrari, C., Dommergue, A., Albert, M., Battle, M., Arnaud, L., Barnola, J.-M., Cairns, W.R.L., Barbante, C., Boutron, C., 2007. Mercury in the snow and firn at Summit Station, Central Greenland, and implications for the study of past atmospheric mercury levels. Atmospheric Chemistry and Physics Discussions 7, 18221–18268.

Ferrari, C.P., Dommergue, A., Boutron, C.F., 2004a. Profiles of mercury in the snow pack at Station Nord, Greenland shortly after polar sunrise. Geophysical Research Letters 31, L03401, doi: 03410.01029/02003GL018961.

Ferrari, C.P., Dommergue, A., Boutron, C.F., Skov, H., Goodsite, M., Jensen, B., 2004b. Nighttime production of elemental gaseous mercury in interstitial air of snow at Station Nord, Greenland. Atmospheric Environment 38, 2727–2735.

Ferrari, C.P., Gauchard, P.A., Aspmo, K., Dommergue, A., Magand, O., Bahlmann, E., Nagorski, S., Temme, C., Ebinghaus, R., Steffen, A., Banic, C., Berg, T., Planchon, F., Barbante, C., Cescon, P., Boutron, C.F., 2005. Snow–to–air exchanges of mercury in an Arctic seasonal snow pack in Ny–Alesund, Svalbard. Atmospheric Environment 39, 7633–7645.

Foster, K.L., Plastridge, R.A., Bottenheim, J.W., Shepson, P.B., Finlayson-Pitts, B.J., Spicer, C.W., 2001. The role of Br-2 and BrCl in surface ozone destruction at polar sunrise. Science 291, 471–474.

Friess, U., 2001. Spectroscopic Measurements of Atmospheric Trace Gases at Neumayer-Station, Antarctica. Ph.D. Thesis, University of Heidelberg, Heidelberg, Germany.

Gardfeldt, K., Jonsson, M., 2003. Is bimolecular reduction of Hg(II) complexes possible in aqueous systems of environmental importance. Journal of Physical Chemistry A 107, 4478–4482.

Gauchard, P.A., Ferrari, C.P., Dommergue, A., Poissant, L., Pilote, M., Guehenneux, G., Boutron, C.F., Baussand, P., 2005a. Atmospheric particle evolution during a nighttime atmospheric mercury depletion event in sub-Arctic at Kuujjuarapik/Whapmagoostui, Quebec, Canada. Science of the Total Environment 336, 215–224.

Gauchard, P.A., Aspmo, K., Temme, C., Steffen, A., Ferrari, C.P., Berg, T., Ström, J., Kaleschke, L., Dommergue, A., Bahlmann, E., Magand, O., Planchon, F., Ebinghaus, R., Banic, C., Nagorski, S., Baussand, P., Boutron, C.F., 2005b. Study of the origin of atmospheric mercury depletion events recorded in Ny-Ålesund, Svalbard, spring 2003. Atmospheric Environment 39, 7620–7632.

Goodsite, M.E., Plane, J.M.C., Skov, H., 2004. A theoretical study of the oxidation of Hg-0 to HgBr2 in the troposphere. Environmental Science & Technology 38, 1772–1776.

Holmes, C.D., Jacob, D.J., Yang, X., 2006. Global lifetime of elemental mercury against oxidation by atomic bromine in the free troposphere. Geophysical Research Letters 33, -.

Iversen, T., Joranger, E., 1985. Arctic air pollution and large scale atmospheric flows. Atmospheric Environment 19, 2099–2108.

Johansen, P., Mulvad, G., Pedersen, H.S., Hansen, J.C., Riget, F., 2007. Human accumulation of mercury in Greenland. Science of the Total Environment 377, 173–178.

King, M.D., Simpson, W.R., 2001. Extinction of UV radiation in Arctic snow at Alert, Canada (82 degrees N). Journal of Geophysical Research 106, 12499–12507.

Kirk, J.L., St. Louis, V.L., Sharp, M.J., 2006. Rapid reduction and reemission of mercury deposited into snow packs during atmospheric mercury depletion events at Churchill, Manitoba, Canada. Environmental Science & Technology 40, 7590–7596.

Lahoutifard, N., Sparling, M., Lean, D., 2005. Total and methyl mercury patterns in Arctic snow during springtime at Resolute, Nunavut, Canada. Atmospheric Environment 39, 7597–7606.

Lahoutifard, N., Poissant, L., Scott, S.L., 2006. Scavenging of gaseous mercury by acidic snow at Kuujjuarapik, Northern Quebec. Science of the Total Environment 355, 118–126.

Lehrer, E., 1999. Polar Tropospheric Ozone Loss. Ph.D. Thesis, University of Heidelberg, Heidelberg, Germany.

Lindberg, S.E., Brooks, S.B., Lin, C.J., Scott, K., Meyers, T., Chambers, L., Landis, M., Stevens, R.K., 2001. Formation of reactive gaseous mercury in the Arctic: evidence of oxidation of Hg^0 to gas-phase $Hg^{(II)}$ compounds after arctic sunrise. Water Air and Soil Pollution 1, 295–302.

Lindberg, S.E., Brooks, S., Lin, C.J., Scott, K.J., Landis, M.S., Stevens, R.K., Goodsite, M., Richter, A., 2002. Dynamic Oxidation of Gaseous Mercury in the Arctic Troposphere at Polar Sunrise. Environmental Science and Technology 36, 1245–1256.

Lindberg, S., Bullock, R., Ebinghaus, R., Engstrom, D., Feng, X.B., Fitzgerald, W., Pirrone, N., Prestbo, E., Seigneur, C., 2007. A synthesis of progress and uncertainties in attributing the sources of mercury in deposition. Ambio 36, 19–32.

Lockhart, W.L., Wilkinson, P., Billeck, B.N., Danell, R.A., Hunt, R.V., Brunskill, G.J., Delaronde, J., St. Louis, V., 1998. Fluxes of mercury to lake sediments in central and northern Canada inferred from dated sediment cores. Biogeochemistry 40, 163–173.

Lockhart, W.L., Stern, G.A., Wagemann, R., Hunt, R.V., Metner, D.A., DeLaronde, J., Dunn, B., Stewart, R.E.A., Hyatt, C.K., Harwood, L., Mount, K., 2005. Concentrations of mercury in tissues of beluga whales (Delphinapterus leucas) from several communities in the Canadian Arctic from 1981 to 2002. Science of the Total Environment 351, 391–412.

Lu, J.Y., Schroeder, W.H., Barrie, L.A., Steffen, A., Welch, H.E., Martin, K., Lockhart, L., Hunt, R.V., Boila, G., Richter, A., 2001. Magnification of atmospheric mercury deposition to polar regions in springtime: the link to tropospheric ozone depletion chemistry. Geophysical Research Letters 28, 3219–3222.

Lu, J.Y., Schroeder, W.H., 2004. Annual time-series of total filterable atmospheric mercury concentrations in the Arctic. Tellus 56B, 213–222.

Macdonald, R.W., Bewers, J.M., 1996. Contaminants in the arctic marine environment: priorities for protection. ICES Journal of Marine Science 53, 537–563.

Miranda, K.C., Metcalfe, T.L., Metcalfe, C.D., Robaldo, R.B., Muelbert, M.M.C., Colares, E.P., Martinez, P.E., Bianchini, A., 2007. Residues of persistent organochlorine contaminants in Southern elephant seals (Mirounga leonina) from Elephant Island, Antarctica. Environmental Science & Technology 41, 3829–3835.

Pacyna, E.G., Pacyna, J.M., Steenhuisen, F., Wilson, S., 2006. Global anthropogenic mercury emission inventory for 2000. Atmospheric Environment 40, 4048–4063.

Piot, M., von Glasow, R., 2008. The potential importance of frost flowers, recycling on snow, and open leads for ozone depletion events. Atmospheric Chemistry and Physics 8, 2437–2467.

Poissant, L., Pilote, M., 2003. Time series analysis of atmospheric mercury in Kuujjuarapik/Whapmagoostui (Quebec). Journal de Physique IV 107, 1079–1082.

Poulain, A.J., Lalonde, J.D., Amyot, M., Shead, J.A., Raofie, F., Ariya, P.A., 2004. Redox transformations of mercury in an Arctic snowpack at springtime. Atmospheric Environment 38, 6763–6774.

Poulain, A.J., Garcia, E., Amyot, M., Campbell, P.G.C., Raofie, F., Ariya, P.A., 2007a. Biological and chemical redox transformations of mercury in fresh and salt waters of the high arctic during spring and summer. Environmental Science & Technology 41, 1883–1888.

Poulain, A.J., Ni Chadhain, S.M., Ariya, P.A., Amyot, M., Garcia, E., Campbell, P.G.C., Zylstra, G.J., Barkay, T., 2007b. Potential for mercury reduction by microbes in the high arctic. Applied and Environmental Microbiology 73, 2230–2238.

Raatz, W.E., Shaw, G.E., 1984. Long-range tropospheric transport of pollution aerosols into the Alaskan arctic. Journal of Climatology and Applied Meteorololgy 7, 1052–1064.

Richter, A., Wittrock, F., Ladstatter-Weissenmayer, A., Burrows, J.P., 2002. GOME measurements of stratospheric and tropospheric BrO. Remote Sensing of Trace Constituents in the Lower Stratosphere, Troposphere and the Earth's Surface: Global Observations, Air Pollution and the Atmospheric Correction 29, 1667–1672.

Riva, S.D., Abelmoschi, M.L., Magi, E., Soggia, F., 2004. The utilization of the Antarctic environmental specimen bank (BCAA) in monitoring Cd and Hg in an Antarctic coastal area in Terra Nova Bay (Ross Sea-Northern Victoria Land). Chemosphere 56, 59–69.

Saiz-Lopez, A., Plane, J.M.C., Mahajan, A.S., Anderson, P.S., Bauguitte, S.J.B., Jones, A.E., Roscoe, H.K., Salmon, R.A., Bloss, W.J., Lee, J.D., Heard, D.E., 2007. On the vertical distribution of boundary layer halogens over coastal Antarctica: implications for O3, HOx, NOx and the Hg lifetime. Atmospheric Chemistry and Physics Discussions 7, 9385–9417.

Schroeder, W.H., Munthe, J., 1995. Atmospheric mercury - an overview. Atmospheric Environment 32, 809–822.

Schroeder, W.H., Anlauf, K.G., Barrie, L.A., Lu, J.Y., Steffen, A., Schneeberger, D.R., Berg, T., 1998. Arctic springtime depletion of mercury. Nature 394, 331–332.

Scott, K.J., 2001. Bioavailable mercury in arctic snow determined by a light-emitting mer-lux bioreporter. Arctic 54, 92–95.

Serreze, M.C., Maslanik, J.A., Scambos, T.A., Fetterer, F., Stroeve, J., Knowles, K., Fowler, C., Drobot, S., Barry, R.G., Haran, T.M., 2002. A record minimum arctic sea ice extent and area in 2002. Geophysical Research Letters 30, 1110.

Simpson, W., Von Glasow, R., Riedel, K., Anderson, P., Ariya, P.A., Bottenheim, J., Burrows, J.P., Carpenter, L., Freisse, U., Goodsite, M., Heard, D., Hutterli, M., Jacobi, H.-W., Kaleschke, L.,

Neff, B., Plane, J., Platt, U., Richter, A., Roscoe, H., Sander, R., Shepson, P.B., Sodeau, J., Steffen, A., Wagner, T., Wolff, E., 2007. Halogens and their role in polar boundary-layer ozone depletion. Atmospheric Chemistry and Physics 7, 4375–4418.

Skov, H., Christensen, J.H., Heidam, N.Z., Jensen, B., Wahlin, P., Geernaert, G., 2004. Fate of elemental mercury in the Arctic during atmospheric depletion episodes and the load of atmospheric mercury to the Arctic. Environmental Science & Technology 38, 2373–2382.

Skov, H., Goodsite, M.E., Lindberg, S.E., Meyers, T.P., Landis, M., Larsen, M.R.B., McConville, G., 2006. The fluxes of Reactive Gaseous mercury measured with a newly developed method using relaxed eddy accumulation. Atmospheric Environment 40, 5452–5463.

Slemr, F., Brunke, E., Ebinghaus, R., Temme, C., Munthe, J., Wängberg, I., Schroeder, W.H., Steffen, A., Berg, T., 2003. Worldwide trend of atmospheric mercury since 1977. Geophysical Research Letters 30, 23–21.

Sommar, J., Wangberg, L., Berg, T., Gardfeldt, K., Munthe, J., Richter, A., Urba, A., Wittrock, F., Schroeder, W.H., 2007. Circumpolar transport and air-surface exchange of atmospheric mercury at Ny-A°lesund (79° N), Svalbard, spring 2002. Atmospheric Chemistry and Physics 7, 151–166.

Sonne, C., Dietz, R., Leifsson, P.S., Asmund, G., Born, E.W., Kirkegaard, M., 2007. Are liver and renal lesions in East Greenland polar bears (Ursus maritimus) associated with high mercury levels? Environmental Health 6, -.

Sprovieri, F., Pirrone, N., 2000. A preliminary assessment of mercury levels in the Antarctic and Arctic troposphere. Journal of Aerosol Science 31, 757–758.

Sprovieri, F., Pirrone, N., Landis, M., Stevens, R.K., 2005a. Oxidation of gaseous elemental mercury to gaseous divalent mercury during 2003 polar sunrise at Ny-Alesund. Environmental Science & Technology 39, 9156–9165.

Sprovieri, F., Pirrone, N., Landis, M., Stevens, R.K., 2005b. Atmospheric mercury behaviour at different altitudes at Ny Alesund during Spring 2003. Atmospheric Environment, 39, 7646–7656.

Sprovieri, F., Pirrone, N., Hedgecock, I.M., Landis, M.S., Stevens, R.K., 2002. Intensive atmospheric mercury measurements at Terra Nova Bay in Antarctica during November and December 2000. Journal of Geophysical Research-Atmospheres 107

St. Louis, V.L., Sharp, M.J., Steffen, A., May, A., Barker, J., Kirk, J.L., Kelly, D.J.A., Arnott, S.E., Keatley, B., Smol, J.P., 2005. Some Sources and Sinks of Monomethyl and Inorganic Mercury on Ellesmere Island in the Canadian High Arctic. Environmental Science & Technology 39, 2686–2701.

Steffen, A., Schroeder, W.H., Bottenheim, J., Narayan, J., Fuentes, J.D., 2002. Atmospheric mercury concentrations: measurements and profiles near snow and ice surfaces in the Canadian Arctic during Alert 2000. Atmospheric Environment 36, 2653–2661.

Steffen, A., Schroeder, W.H., Edwards, G., Banic, C., 2003a. Mercury throughout polar sunrise 2002. Journal de Physique IV 107, 1267–1270.

Steffen, A., Schroeder, W.H., Poissant, L., MacDonald, R., 2003b. Mercury in the arctic atmosphere. Indian and Northern Affairs Canada, Ottawa.

Steffen, A., Schroeder, W.H., Macdonald, R., Poissant, L., Konoplev, A., 2005. Mercury in the arctic atmosphere: an analysis of eight years of measurements of GEM at Alert (Canada) and a comparison with observations at Amderma (Russia) and Kuujjuarapik (Canada). Science of the total environment 342, 185–198.

Steffen, A., 2007. Mercury Measurements at Alert. QS-8602-060-EE-A1, Ministry of Indian Affairs and Northern Development, Ottawao.

Steffen, A., Douglas, T., Amyot, M., Ariya, P., Aspmo, K., Berg, T., Bottenheim, J., Brooks, S., Cobbett, F., Dastoor, A., Dommergue, A., Ebinghaus, R., Ferrari, C., Gardfeldt, K., Goodsite, M.E., Lean, D., Poulain, A., Scherz, C., Skov, H., Sommar, J., Temme, C., 2008. A synthesis of atmospheric mercury depletion event chemistry in the atmosphere and snow. Atmospheric Chemistry and Physics 8, 1445–1482.

Stroeve, J.C., Serreze, M.C., Fetterer, F., Arbetter, T., Meier, W., Maslanik, J., Knowles, K., 2005. Tracking the Arctic's shrinking ice cover: Another extreme September minimum in 2004. Geophysical Research Letters 32,

Tackett, P.J., Cavender, A., Shepson, P.B., Bottenheim, J.W., Morin, S., Deary, J., Steffen, A., 2007. A Study of the vertical scale of halogen chemistry in the Arctic troposphere during polar sunrise at Barrow, AK. Journal of geophysical research 112, D07306.

Temme, C., Einax, J.W., Ebinghaus, R., Schroeder, W.H., 2003. Measurements of Atmospheric Mercury Species at a Coastal Site in the Antarctic and over the South Atlantic Ocean during Polar Summer. Environmental Science & Technology 37, 22–31.

Temme, C., Blanchard, P., Steffen, A., Beauchamp, S.T., Poissant, L., Tordon, R.J., Weins, B., 2007. Trend, seasonal and multivariate analysis study of total gaseous mercury data from the Canadian Atmospheric Mercury Measurement Network (CAMNet). Atmospheric Environment 41, 5423–5441.

Travnikov, O., Ryaboshapko, A., 2002. EMEP/MSC-E report 6/02.

van Aardenne, J.A., Carmichael, G.R., Levy, H., Streets, D., Hordijk, L., 1999. Anthropogenic NOx emissions in Asia in the period 1990-2020. Atmospheric Environment 33, 633–646.

Wagemann, R., Innes, S., Richard, P.R., 1996. Overview and regional and temporal differences of heavy metals in Arctic whales and ringed seals in the Canadian Arctic. Science of the Total Environment 186, 41–66.

Walker, J.B., Houseman, J., Seddon, L., McMullen, E., Tofflemire, K., Mills, C., Corriveau, A., Weber, J.P., LeBlanc, A., Walker, M., Donaldson, S.G., Van Oostdam, J., 2006. Maternal and umbilical cord blood levels of mercury, lead, cadmium, and essential trace elements in Arctic Canada. Environmental Research 100, 295–318.

Wangberg, I., Sommar, J., Berg, T., Gardfeldt, K., Munthe, J., 2003. Interpretation of mercury depletion events observed at Ny-Alesund, Svalbard during spring 2002. Journal de Physique IV 107, 1353–1356.

Zhou, X.L., Beine, H.J., Honrath, R.E., Fuentes, J.D., Simpson, W., Shepson, P.B., Bottenheim, J.W., 2001. Snowpack photochemical production of HONO: a major source of OH in the Arctic boundary layer in springtime. Geophysical Research Letters 28, 4087–4.

Chapter 11
Spatial Coverage and Temporal Trends of Over-Water, Air-Surface Exchange, Surface and Deep Sea Water Mercury Measurements

Francesca Sprovieri, Nicola Pirrone, Robert P. Mason, and Maria Andersson

Summary The world's oceans and seas are both sources and sinks of mercury, and although it appears that the atmosphere is the major transport/distribution medium for mercury, because most Hg emissions are to the atmosphere, oceans and seas also play an important role. The transformations of Hg and its compounds which take place in marine water are of crucial importance to the understanding of the way in which mercury released to the atmosphere is eventually incorporated into biota, thereby becoming a risk to human and ecosystem well being. This chapter provides an overview of where and when measurements of atmospheric and aquatic concentrations of mercury and its compounds have been made in the marine environment. These measurements cover – in part obviously – the Pacific, Atlantic, Southern and Arctic oceans, North and Baltic Seas and the Mediterranean,. There are relatively few direct measurements of the air-sea exchange of mercury, however simultaneous measurements of Dissolved Gaseous Hg (DGM) and Hg in air, when combined with measurements of the sea and air temperature and wind speed, can be used to estimate the evasion and deposition fluxes. The magnitude of these fluxes is one of the indispensable parameters in compartmental and atmospheric Hg models. There remains some uncertainty as a result of the, so far, limited spatial and temporal coverage of the measurements.

11.1 Introduction

Mercury occurs naturally in the Earth's crust principally as the ore, cinnabar, HgS and it is quite different from other metals in several respects: (i) it is the only metal that is liquid at room temperature; (ii) it is the only metal that boils below 650° C; (iii) it is quite inert chemically, having a higher ionization potential than any other electropositive element with the sole exception of hydrogen; (iv) it exists in oxidation states of 0 (Hg^0) and 1 ($Hg^{(I)}$) in addition to the expected state of 2 ($Hg^{(II)}$). Hg is unique among metals in possessing appreciable volatility in its elemental state (Hg^0) at ambient temperatures and pressures. This volatility means that a significant quantity of Hg can exist in the gas phase. Hg enters the atmosphere either through

N. Pirrone and R. Mason (eds.), *Mercury Fate and Transport in the Global Atmosphere*,
DOI: 10.1007/978-0-387-93958-2_11,

discharge to natural ecosystems followed by volatilization, or by direct emission to the air. In the latter case, the metal can be released as either Hg^0 or a variety of oxidized Hg forms, $Hg^{(II)}$. The oxidized (mercuric) forms, which may account for as much as half of the emitted Hg, have short residence times in the atmosphere, and are deemed responsible for most of the Hg contamination close to point sources (Nriagu, 1989, Pirrone et al., 1996; Pirrone et al. 1998; Pirrone et al., 2001). Elemental Hg (Hg^0), on the other hand, is thought to have an average atmospheric residence time of approximately one year (Hall, 1995, Shia et al., 1999), although there is considerable uncertainty on this point, with estimates ranging from 0.2 to 2 years. Nevertheless, because of this time scale and its consequent ability to travel long distances prior to deposition, most of the directly emitted Hg^0 enters the global Hg cycle (Hedgecock and Pirrone, 2004; Hedgecock et al., 2006).

Historical as well as present variations in anthropogenic influence in the atmosphere are undoubtedly due to geological and climatic variability. The effect of anthropogenic activities is to redistribute naturally-occurring Hg from its original matrix (in which it may be highly immobile) to other environmental compartments, and to alter its speciation, often towards more mobile and toxic forms. The task of distinguishing anthropogenic from natural contributions to the global cycle is a very difficult one, since the so-called "natural" sources include a large component of reemitted previously-deposited Hg. For example, only 20-30% of oceanic emissions are thought to arise from Hg originally mobilized from natural sources (Pirrone et al., 2001; Pacyna et al., 2001; Mason and Sheu, 2002). To some extent then, today's "natural" emissions are thus yesterday's anthropogenic emissions, which includes remnants of Hg used in gold mining centuries ago. The amount of Hg presently in the atmosphere is believed to exceed the pre-industrial level by a factor of between two and five (Expert Panel on Hg Atmospheric Processes. 1994; Mason et al., 1994). Likewise, it has been suggested that Hg deposition rates have increased by a factor of 2-3 over the pre-industrial rate, based on analysis of sediment records (Engstrom, et al. 1994). Since the early 1960s, the growing awareness of environmental Hg pollution (e.g. the Minamata tragedy resulting from MeHg poisoning) has stimulated the development of more accurate, precise and efficient methods of determining Hg and its compounds in a wide variety of matrices.

Research on atmospheric emissions, transport and deposition mechanisms to terrestrial and aquatic receptors, chemical transformations of Hg^0 to more toxic species (i.e. MeHg), studies on the bioaccumulation of Hg in the aquatic food web, and exposure and risk assessments have driven the scientific and political communities to consider this toxic element as a pollutant of global concern (i.e. Nriagu and Pacyna, 1988; Iverfeldt, 1991a, 1991b; Mason et al., 1994; Pirrone et al., 1996, 2001; Schroeder and Munthe, 1998; Petersen et al., 1998; Schroeder et al., 1998; Ebinghaus et al., 1999). International and national organizations and programmes have been involved in assessing the current status of environmental contamination by Hg and in developing strategies and policies to reduce anthropogenic emissions of this pollutant (i.e. LRTAP, OSPAR, Barcelona Convention). There have also been research programmes carried out at European and national levels that have investigated different aspects of Hg cycling in European ecosystems (i.e. EU-MAMCS/MOE project, EU-MERCYMS project) focused among others on the identification and possible quantification of the magnitude of major chemical and physical processes/mechanisms involved in the transfer of Hg to terrestrial and aquatic receptors (Pirrone et al., 2003; 2005; Pirrone, 2006). In recent years, new analytical techniques have become available and have been used in environmental studies and consequently the understanding of Hg chemistry in natural systems has improved significantly.

Field measurements carried out in the framework of the MAMCS and MOE programmes did not show significant spatial gradients with ambient concentrations of Total Gaseous Mercury (TGM) in the range of 1.6 – 3.2 ng m^{-3} during the four measurement campaigns carried out between November 1998 and July 1999. On the contrary, Reactive Gaseous Mercury (RGM) levels were found to show significant spatial gradients during the four seasons over the entire domain, with ambient concentrations in the range 35 – 65 pg m^{-3} at coastal sites in the Mediterranean Sea region and between 10 and 32 pg m^{-3} at North European sites (Wangberg et al., 2001; 2008; Munthe et al., 2001). Total Particulate Mercury (TPM) concentrations showed a similar trend to that observed for RGM with seasonal averages in the range of 30 – 40 pg m^{-3} (Wangberg et al., 2001). The major environmental and health problems associated with Hg pollution result from high concentrations of MeHg in fish. Changes in speciation from inorganic to methylated forms are the first steps in the aquatic bioaccumulation processes. These processes are considered to occur in both the water column and sediments. The mechanism of synthesis of MeHg is not very well understood. Although MeHg is the dominant form of Hg in higher organisms, it represents only a very small amount of the total Hg in aquatic ecosystems and in the atmosphere. Methylation-demethylation reactions are assumed to be widespread in the environment and each ecosystem attains its own steady state equilibrium with respect to the individual species of Hg. However, owing to the bioaccumulation of MeHg, methylation is more prevalent in the aquatic environment than demethylation.

Once MeHg is formed, it enters the food chain by rapid diffusion and tight binding to proteins in aquatic biota and attains its highest concentrations in the tissues of fish at the top of the aquatic food chain due to biomagnification through the trophic levels. The main factors that affect the levels of MeHg in fish are the diet/trophic level of the species, age of the fish, microbial activity and MeHg concentration in the upper layer of the local sediment, and its bioavailability, which is influenced by dissolved organic carbon content, salinity, pH, and redox potential. At the scale of aquatic systems, one can classify the processes responsible for the high Hg concentration in fish into three categories: (i) accumulation of inorganic Hg in aquatic systems; (ii) methylation of inorganic Hg; (iii) MeHg bioaccumulation and magnification in the food chain.

In most systems, the primary cause of high fish MeHg concentrations is thought to be elevated atmospheric inputs of Hg to the water bodies and their watersheds (Fitzgerald et al., 1998; Horvat et al., 2001; 2003; Lindberg et al., 2007), mostly by wet deposition with over 60% of the Hg input from direct deposition (Mierle, G. 1990), although in some cases, geologic sources of Hg may also be important (Rasmussen, P. E. 1994; Cossa et al., 1997). Hg in atmospheric deposition is predominantly inorganic Hg ($Hg^{(II)}$). In oceanic waters, after it undergoes a complex set of chemical and biological transformations, most of the deposited Hg is reduced to Hg^0 and returned to the atmosphere (Pirrone et al., 2000a; Hedgecock et al., 2001; Mason et al., 2001; Gardfeldt et al., 2003; Andersson et al., 2007); only a small fraction is permanently exported to the sediments. Thus the Hg inventories in the atmosphere and surface seawater are tightly coupled by an effective precipitation/volatilisation cycle driven by oxidation/reduction reactions (Hedgecock and Pirrone, 2004; Hedgecock et al., 2006). In this context, factors that influence the reduction and oxidation of Hg represent a knowledge gap that must be addressed. At present, it is thought that Hg reduction proceeds mostly via photochemical processes (Amyot et al., 1997) although in some systems, biologically-mediated reduction may be significant (Mason et al., 1995). Once in an aquatic environment, inorganic Hg is transformed into MeHg through microbial activity. MeHg is the most toxic and most bioavailable form of Hg (Xun et al., 1987; IARC, 1994; US-EPA, 1997; Horvat et al., 1993a, 1993b, 1997; Horvat et al., 2003). Levels of MeHg may build up in the food chain to a point that poses a risk to wildlife and humans consuming fish. Therefore, one of the keys to understanding the risk from Hg is to determine the link between atmospheric inputs of Hg, methylation and bioaccumulation.

11.2 Over-Water Mercury Measurements

The biogeochemical cycles of Hg involves the transfer of quantities of this metal between atmospheric, aquatic and biological reservoirs. This transfer is facilitated by the mobility of Hg, in part due to its volatility and long atmospheric lifetime. Global scale models predict that atmospheric transport and deposition are the principal pathways for Hg to the surface open ocean. Further, these models indicate the important role which oceanic processes, especially air-sea exchange, play in distributing Hg across the Earth's surface (Mason et al, 1994). The evasion of Hg from the surface ocean recycles both natural and anthropogenic contributions to deposition back to the atmosphere. Atmospheric deposition to surface waters supplies Hg for important reactions such as reduction to Hg^0, methylation and particulate scavenging. Biotic/abiotic production of Hg^0 in surface waters results in supersaturation and thereby a wind-induced efflux of Hg^0 from the surface mixed layer. This evasion process serves to remobilise inputs from the atmosphere, recycling Hg from aquatic systems (Mason et al., 1994) giving the major mechanism within the oceanic Hg bio-geochemical cycle established. Aspects of the atmospheric Hg cycle in open-ocean environments have been investigated with a focus principally on determining concentrations of species and their variations with latitude and atmospheric transport (Mason and Fitzgerald, 1991) as well as atmospheric Hg chemistry and the biogeochemical controls on Hg reduction and evasion (Laurier et al., 2004).

Long-term monitoring of Hg in the atmosphere would provide valuable information about the impact of emission controls on the global budget of atmospheric Hg, and their observance (Fitzgerald, 1995). In addition, systematic Hg assessment could also provide an insight into the global Hg cycle, especially into the ratio of anthropogenic and natural emissions which is currently poorly defined (Ebinghaus et al., 1999b). The need for such measurements was recognized in the mid-1990s (Fitzgerald, 1995) and several monitoring stations have been brought into operation since (e.g. Schroeder et al., 1998; Slemr and Scheel, 1998; Ebinghaus et al., 2002a, 2002b). A complementary approach to measurements at a few stationary sites for long periods are campaign measurements from moving platforms such as ships or aircraft covering large areas of the globe. TGM measurements on board ships proved to provide valuable complementary information to measurements from the ground based monitoring network. This information consists of a snapshot of large-scale geographical distribution. With proper quality control to ensure comparability and a relatively low measurement uncertainty, the combination of intermittent shipboard and long-term ground measurements can provide information about the worldwide distribution and trend of atmospheric Hg. Occasional shipboard measurements should thus be a part of the global monitoring network for atmospheric Hg.

11.2.1 Atlantic Ocean

The first measurements made on board ships during north–south traverses of the Atlantic Ocean were made between 1977–1980 (Slemr et al., 1981,1985) and the cruises were repeated in 1990 (Slemr and Langer, 1992) and 1994 (Slemr et al., 1995).

In 1996, Lamborg et al. (1999) performed Hg measurements in the south and equatorial Atlantic Ocean during a cruise from Montevideo, Uruguay to Barbados (Figure 11.1) on the research vessel R/V *Knorr*. TGM ranged between 1.17 and 1.99 ng m^{-3} with a weighted mean of 1.61 ± 0.09 ng m^{-3}. The results obtained by Lamborg in 1996 compared well with Pacific Ocean data and earlier results from the Atlantic (see Figure 11.3). The open-ocean samples recorded a distinctive inter-hemispheric gradient, which is consistent with a long-lived trace gas emitted to a greater extent in the Northern than in the Southern Hemisphere (Lamborg et al., 2002). In the same year, Slemr's cruise (Slemr and Langer, 1992; Slemr et al., 1995) for the Atlantic Ocean found values very similar to those reported by Lamborg (1999).

From 1996 to 2001 atmospheric Hg measurements were carried out on board the RV Polarstern during three cruises (Temme et al., 2003). The first cruise was from Bremerhaven to Punta Quilla in October and November 1996 (cruise ANT XIV); the second from Bremerhaven to Cape Town and then Antarctica and back to Cape Town in December 1999–March 2000 (cruises ANT XVII/1 and XVII/2), and the third from Antarctica to Punta Arenas in February 2001 (cruise ANT XVIII). Figure 11.2 shows the tracks of the Polarstern cruises in 1996 and 1999–2001. The 1999/2000 cruise followed a route near to the European and African coasts of the Atlantic Ocean up to western Africa and then took a direct course to Cape Town in South Africa. The cruise then continued to Antarctica. The 2001 cruise then proceeded from the Neumayer station to Punta Arenas in Chile. During all these cruises Hg was measured quasi-continuously using an automatic Hg vapour analyser (Model 2537A, manufactured by Tekran Inc., Toronto, Canada).

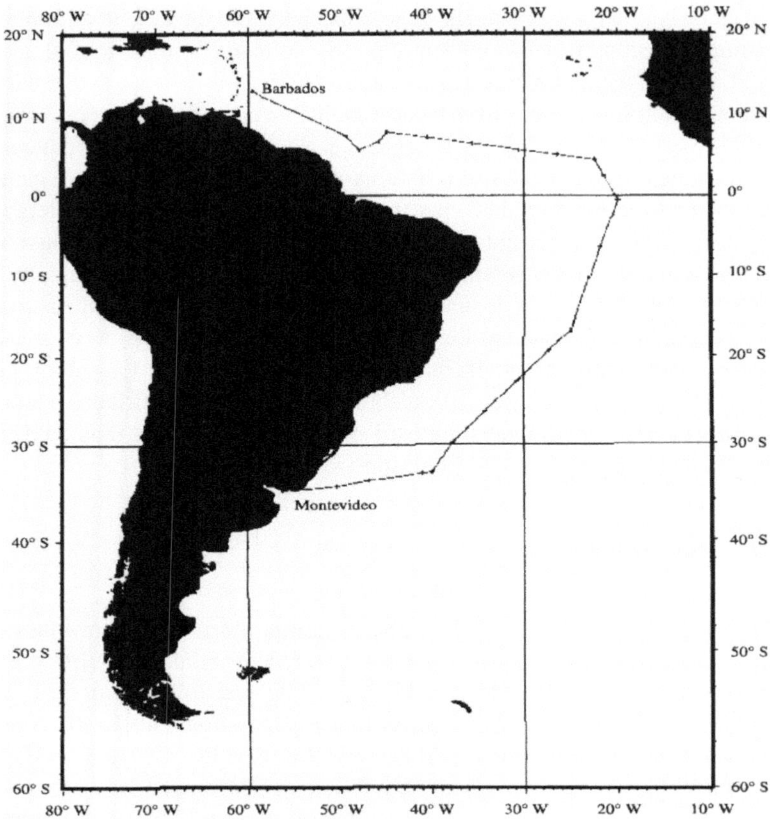

Figure 11.1 The 1996 South Atlantic cruise track from Montevideo to Barbados
Source: Lamborg et al., (1999)

The results of all cruises made over the Atlantic Ocean by these investigators are summarised in Table 11.1 in statistical terms. In the northern hemisphere (NH) the medians are almost always smaller than averages pointing to an asymmetric distribution with a few higher TGM concentrations in the vicinity of Hg sources. In the southern hemisphere the medians and the averages are mostly equal. A rather homogeneous distribution of TGM in the southern hemisphere (SH) was observed during the previous cruises (Slemr et al., 1981,1985,1995; Slemr and Langer, 1992); higher median TGM concentrations in both hemispheres were observed in 1990. The median TGM concentrations in each hemisphere in 1977–1980 are comparable with those observed in 1994–2000. Temme et al. (2003) data for 1977–1980 in both hemispheres are comparable to the respective TGM measurements reported by Fitzgerald (1995) for the Pacific Ocean in 1980–1983. When latitude is taken into account, the TGM concentrations measured on board a ship are also comparable to measurements at remote coastal sites such as Mace Head (Ireland), Cape Point (South Africa), and Lista (Norway) (Figure 11.3). The agreement shows that combina-

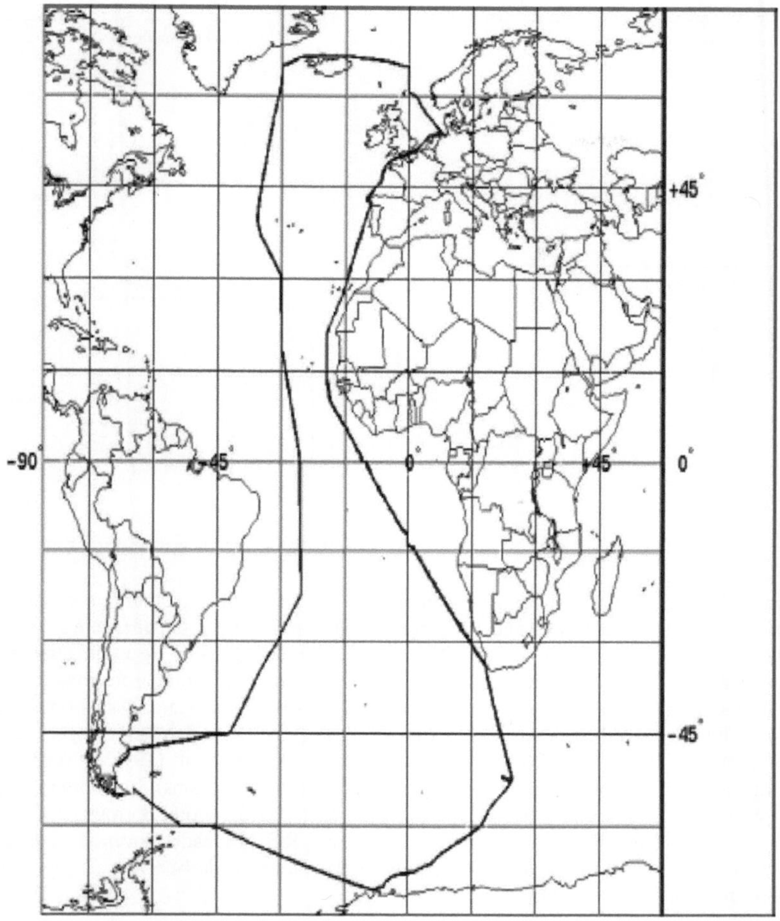

Figure 11.2 Tracks of the Polarstern cruises from Bremerhaven to Punta Quilla (Argentina) in 1996, from Bremerhaven over Cape Town (South Africa) to Antarctica in 1999–2000, and from Antarctica to Punta Arenas (Chile) in 2001
Source: Temme et al., (2003)

tion of long-term measurements at several sites with snapshots of latitudinal distribution obtained by ship measurements is feasible and may provide information about the worldwide trends of atmospheric Hg. All cruises show a pronounced concentration gradient between the hemispheres (Figure 11.3). The average NH/SH ratio of TGM hemispheric medians of all Polarstern cruises is 1.49 ± 0.12 (n = 8). This ratio is almost the same as 1.45 used by Slemr et al. (1985) to estimate the atmospheric residence time of Hg, approximately 1 yr. This atmospheric residence time is also consistent with the variability of the TGM concentrations of about 21% in the northern and about 8% in the southern hemispheres.

The gradient and the higher variability in the northern hemisphere suggest that the majority of emissions and re-emissions are located in the northern hemisphere.

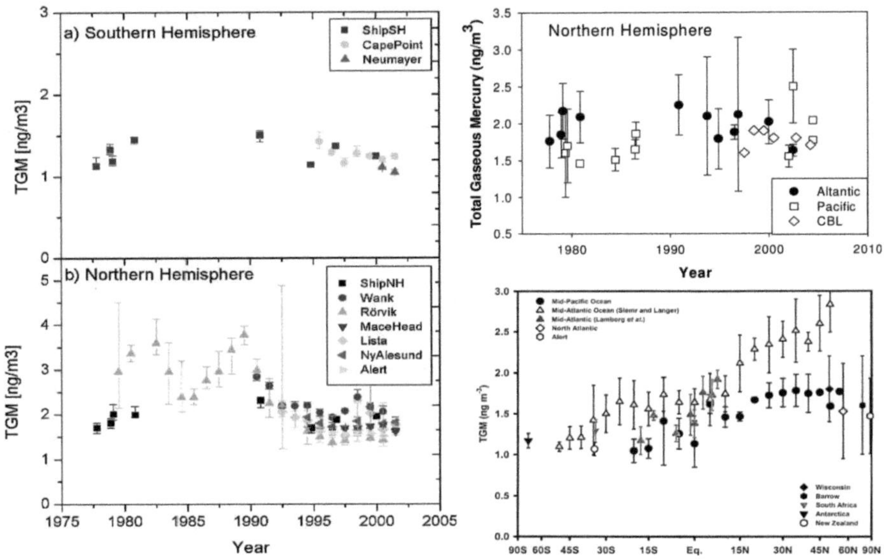

Figure 11.3 Total Gaseous Mercury concentrations over the ocean and other remote locations as reported by a) Slemr et al. (2003) and b) Laurier and Mason (2007). Both papers are a compilation of data from numerous sources. c) Latitudinal gradient in total gaseous Hg in the atmosphere at several locations. Notice the discernible inter-hemispheric gradient, resulting from greater emissions of Hg to the atmosphere in the more industrialized Northern Hemisphere. Data from Fitzgerald (1995) - Mid-Pacific (filled circles); Slemr and Langer (1992) - Mid-Atlantic (open triangles); Lamborg et al. (1999) - Mid-Atlantic (shaded triangles); Mason et al. (1998) - North Atlantic (open diamonds); Schroeder et al. (1998) - Alert, N.W.T. (open hexagons); Lamborg et al. (1995) - rural Wisconsin, U.S.A. (filled diamonds); Lindberg et al. (2000) – Point Barrow, Alaska, U.S.A. (filled hexagons); Ebinghaus et al. (2000) - Cape Town, R.S.A. (shaded inverted triangles) and Antarctica (filled inverted triangles); Fitzgerald (1989) - Ninety Mile Beach, N.Z. (open circles). Taken from Lamborg et al., (2002)

The TGM concentrations observed over the Atlantic Ocean in 1996 and 1999–2001 were comparable with those measured in 1977–1980 but they were lower than those measured in 1990. Similar concentrations have been found in other studies (Fitzgerald, 1995; Lamborg et al., 1998; Lamborg et al., 2002; Mason et al., 2001; Laurier and Mason, 2007). There have been suggestions of a possible global decrease of TGM concentrations between 1990 and 1996 (Slemr et al., 2003), which could be linked with changes and/or a decrease of anthropogenic emissions, although there is no consensus within the scientific community of the degree to which remote ocean BL concentrations have changed over time (see also Figure 11.3). Several processes might have contributed to this reduction; such as a reduction of emissions from waste incineration, and closure of large emitters and introduction of desulphurisation in the power plants in the countries of the former Eastern Block (Slemr and Scheel, 1998). The inter-hemispherical gradient with higher TGM concentrations in the northern hemisphere remained nearly constant over the years. The former estimate of the TGM atmospheric residence time (Slemr et al., 1985), based

Table 11.1 Summary of the measurements of Total Gaseous Mercury (TGM) over the Atlantic Ocean

Cruise	Latitude	Range (*ng m*$^{-3}$)	Mean (*ng m*$^{-3}$)	SD (*ng m*$^{-3}$)	No. of samples	VK [a]	Median (*ng m*$^{-3}$)
Northern hemisphere[b]							
October 1977 [c]	11-33°N	1.0-2.6	1.763	0.362	62	14.9	1.70
Nov./Dec. 1978	5-51°N	1.42-2.70	1.849	0.306	89	15.4	1.81
Jan./Feb. 1979	21-53°N	1.63-3.06	2.169	0.382	52	16.6	2.01
Oct./Nov. 1980	12-54°N	1.41-3.41	2.085	0.351	100	15.8	1.99
Oct./Nov. 1990	7-54°N	1.41-3.41	2.247	0.409	117	17.3	2.31
Oct./Nov. 1994	6-54°N	1.31-3.18	1.788	0.410	101	22.3	1.69
Oct./Nov. 1996	8-67°N	0.44-15.95	2.120	1.035	636	48.8	1.88
Dec.1999/Jan.2000	6-54°N	1.44-3.73	2.022	0.300	715	14.8	1.95
Southern hemisphere[d]							
October 1977[c]	32°S-11°N	080-1.70	1.187	0.249	64	15.3	1.13
Nov./Dec. 1978	23°S-3°N	0.86-1.85	1.350	0.207	63	14.2	1.33
Jan./Feb. 1979	2°S-4°N	1.07-2.09	1.259	0.216	37	16.2	1.19
Oct./Nov. 1980	34°S-11°N	1.10-1.89	1.453	0.157	82	9.1	1.45
Oct./Nov. 1990	48°S-7°N	086-2.44	1.497	0.295	158	18.8	1.50
Oct./Nov. 1994	46°S-6°N	0.82-2.13	1.180	0.167	165	12.9	1.14
Oct./Nov. 1996	37°S-8°N	0.95-2.26	1.391	0.131	423	9.4	1.36
Dec.1999/Jan.2000	71°S-3°N	0.54-1.84	1.266	0.093	1570	7.3	1.25
Feb./March 2000	71°S-34°N	0.24-1.30	1.001	0.116	861	11.6	1.02
01/02/01	71°S-54°N	0.75-1.42	1.066	0.100	968	9.4	1.07

Measurements made during Walther Herwing cruises in 1977 and 1978, Meteor cruises in 1979 and 1980, and Polarstern cruises in 1990, 1994, 1996, and 1999-2001.

[a] Variation coefficient after correction for the reproducibility of the analytical method; $VK^2=VK^2$ (measured) $-VK^2$ (analytical) where VK (analytical)=14.5% during the Walther Herwing cruise in 1977 and 5.8% for cruises until 1994. No correction was made for cruises with Tekran measurements, i.e. since 1996.

[b] North of ITCZ.

[c] Corrected for systematic error in the analytical method used during the Walther Herwing cruise in 1977.

[d] South of ITCZ.

Source: Temme et al., 2003.

on the inter-hemispherical TGM concentration difference thus does not need any revision, although there is evidence, and atmospheric models predict, that the residence time may be lower and a value between 0.5-1 year appears more reasonable.

Measurements of Hg species over water have also been performed during the BATS cruises by Mason et al., 2001 and on a follow-up cruise in August 2003. The TGM concentrations were expected for this latitude (Fitzgerald and Mason, 1997 and references therein), and are shown as with other open ocean TGM measurements in Figure 11.3. While the particulate Hg measurements are limited over the Atlantic, the values found by Mason et al. (2001) were at or below the detection limit (<2 pg m^{-3}). Using much longer collection times, Lamborg et al. (1998) found a time-weighted mean value of 2 pg m^{-3} for the South and equatorial Atlantic, and studies in other remote ocean locations (i.e. the equatorial Pacific; Mason et al., 1992) have found similar values. RGM measurements by Mason et al (2001) represented the first reported measurements in the remote ocean marine boundary layer. In September, measured concentrations of RGM were elevated overall compared to December and March, and to the average value measured in coastal Maryland at CBL, It is likely that the September BATS values do not represent the long-term average for the marine boundary layer at this location as they were collected within days of Hurricane Gert, and it is possible that some factor associated with the passage of the hurricane lead to enhanced $Hg^{(II)}$ production in the atmosphere during the cruise. Preliminary aircraft measurements in Florida and elsewhere (Prestbo, 1996) have shown higher RGM concentrations at higher elevations and mixing of these air masses during the hurricane could account for the elevated values.

Speciation measurements ($Hg^{(II)}$ and Hg^0) were made by Laurier and Mason (2007) on a cruise in August 2003. Data from the cruise, collected south of Bermuda are shown in Figure 11.4 and $Hg^{(II)}$ showed a clear and consistent diurnal cycle with maxima in concentration in the afternoon and minima at night. The $Hg^{(II)}$ maxima ranged up to 27 pg m^{-3}, and the minima were often at the instrument detection limit. Overall, the average concentration was 5.9 pg m^{-3}; the average for the log-transformed data was 4.9 pg m^{-3}. Similarly, a diurnal fluctuation in RGM, with afternoon maxima, was found during the North Pacific cruise (Laurier et al., 2003), although concentrations were much higher, up to 100 pg m^{-3}. For the subtropical North Pacific, the average

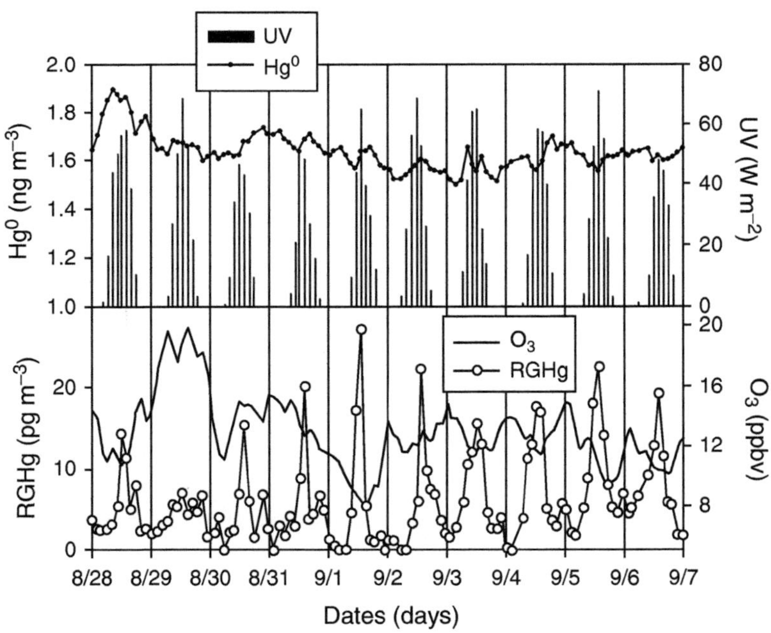

Figure 11.4 Concentrations of reactive Gaseous Mercury (RGHg in figure), Hg^0 and ozone, as well at the UV radiation, measured during the cruise in the North Atlantic in August 2003. Taken from Laurier and Mason (2007)

Table 11.2 Concentrations of TGM and RGM for samples collected using the filter pack method (Sheu and Mason, 2001) from the Weatherbird II in the vicinity of BATS

Sampling details	TGM *(pmol m^{-3})*	RGM *(pmol m^{-3})*
9/24-25/99; On 18:08, Off 2:15	10.95	3.56
9/25-26/99; On 19:09, Off 4:20	9.05	6.86
9/26-27/99; On 19:12, Off 7:05	10.30	2.54
12/13-14/99; On 23:49, Off 7:45	-	2.68
12/15/99; On 0:53, Off 6:35	-	0.54
12/15-16/99; On 19:15, Off 5:15	10.25	1.07
3/7-8/00; On 18:30, Off 6:00	12.40	0.23
3/8-9/00; On 18:00, Off 5:00	-	<0.05
3/9-10/00; On 17:45, Off 6:00	6.50	<0.05
Bermuda average	9.91 ± 2.0	1.38 ± 1.30
CBL average 1997/1998	9.50 ± 4.0	0.19 ± 0.33
Baltimore city average 1997/1998	21.0 ± 9.5	0.39 ± 0.72

Souce: Mason et al., (2001).

RGM concentration was 11.8 pg m^{-3}, which was somewhat higher than that found for the North Atlantic, but showed the same degree of variability. The diurnal change in the UV radiation is also shown in Figure 11.4 and it is clear that the changes in UV and $Hg^{(II)}$ occur temporally in a similar fashion, suggesting that the processes leading to a $Hg^{(II)}$ build-up are likely photo-chemical and there appears to be an anti-correlation with ozone (O_3), in that there was a measurable decrease in O_3 during the day, and an increase at night, and the decrease in O_3 coincided with the increase in RGM. This is consistent with the hypothesis that processes that destroy O_3 are those responsible for Hg^0 oxidation, and therefore for the build up of the product, RGM. This inverse correlation can be seen if the maximum afternoon RGM concentration is plotted against the corresponding O_3 concentration (Figure 11.5). In addition to the data from the North Pacific (Laurier et al., 2004) and North Atlantic (Laurier and Mason, 2007), unpublished data collected during a study off the coast of New Hampshire, USA, is also plotted. There is a strong non-linear relationship that shows that the highest measured concentrations occur under conditions of very low O_3, less than 10 ppbv. There is consistency in all these data in terms of their relationship with O_3.

Comparison of the speciation and concentration of Hg in the atmosphere and in precipitation at CBL compared to Bermuda reinforces the notion that there must be RGM in the marine atmosphere. At CBL, particulate and RGM concentrations are typically < 40 pg m^{-3}, and Hg in precipitation averages around 14–16 ng L^{-1} (Mason et al., 2000). In the Atlantic, wet deposition concentrations are about a factor of 3–5 times less, but particulate concentrations are 10–20 times less. Something besides scavenging of atmospheric particulate matter must be contributing to Hg in open-ocean rain. Guentzel et al. (1995) reached a similar conclusion in their studies in Florida. They concluded that particulate scavenging could not support the measured values for Hg in rain and that $Hg^{(II)}$ must be an important contributor to wet deposition. Recent measurements suggest that dry deposition of $Hg^{(II)}$ is important in Florida with both local (Dvonch et al., 1999) and long-range transport as potential

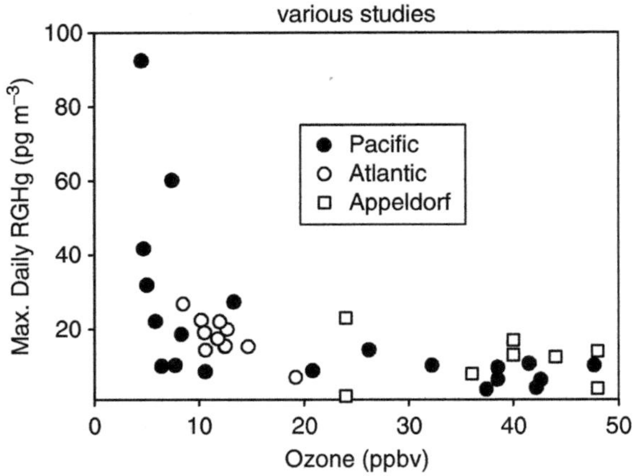

Figure 11.5 Relationship between the maximum measured daily RGM concentration (2 hr average) and the midday ozone concentration from a number of different measurements over the ocean. Taken from Laurier and Mason (2007)

sources for the $Hg^{(II)}$ (Guentzel et al., 2001). While in-cloud processes leading to Hg^0 oxidation can contribute to Hg in wet deposition, these processes occur over both the land and the ocean. Clearly, the disparity between the precipitation concentrations and the atmospheric particulate concentrations at CBL and over the ocean argues for the presence and formation of $Hg^{(II)}$ in the marine boundary layer. Further data, from other oceans and from the Mediterranean, further support this notion.

11.2.2 Pacific Ocean

Measurements made over the Pacific Ocean in the early 1980s have been summarized in a number of publications and will be briefly reviewed here (Fitzgerald, 1995). As shown in Figure 11.3, there has been little change in concentration over time for the North Pacific, based on the samples collected (Lamborg et al., 2002; Laurier and Mason, 2007). There is insufficient data from the recent measurements, given their relative wide distribution latitudinally, and the differences in the seasons of sampling, to conclude that the apparent recent increase in concentrations in the North Pacific boundary layer atmosphere are statistically significant.

More recent studies have included speciation measurements, particularly during a cruise between Japan and Hawaii in May/June 2002 that sampled both the surface waters and the atmosphere. The atmospheric results are shown in Figure 11.6 (Laurier et al., 2003). As discussed above for the Atlantic speciation data, there is clear evidence for a diurnal trend in $Hg^{(II)}$ concentration, especially in the latter part of the cruise where the ship was in a lower ozone region, and there was higher UV radiation and higher temperatures, and often lower wind. All these factors would enhance the photochemical production of RGM and also lead to an increase in the atmospheric concentration (maximum production, minimum removal).

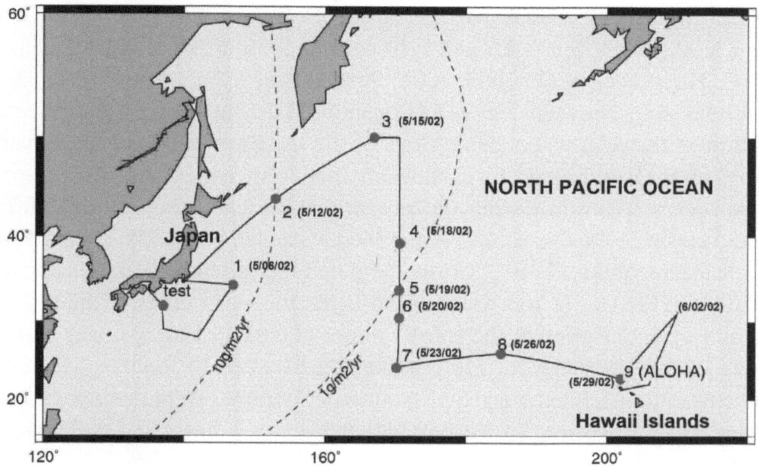

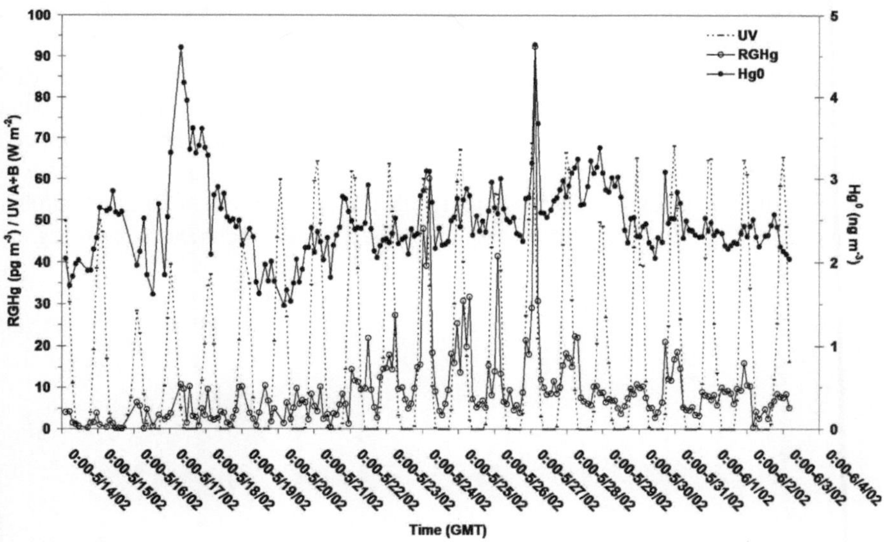

Figure 11.6 The cruise track for the 2002 North Pacific cruise from Japan to Hawaii. Note that the Hg data cover the period between May 14 and June 4, which includes sampling in a north-south transect and sampling in a west-east transect. The UV radiation is also shown. Taken from Laurier et al. (2003)

11.2.3 Mediterranean Sea

Within the framework of the MED-OCEANOR project funded by the Italian National Research Council (CNR) an in-depth investigation was carried out from 2000 to 2007 by group of research institutes to quantify and possibly explain spatial and temporal patterns of Hg and its species concentrations in air, surface and deep water samples, and gaseous Hg exchange rates at the air–water interface along paths of a

6000 km cruise routes around Mediterranean Sea basin aboard the *RV Urania* (Sprovieri et al., 2003; Sprovieri and Pirrone 2008; Gardfeldt et al., 2003; Horvat et al., 2003; Hedgecock et al., 2005; 2006; Pirrone et al., 2003; Kotnik et al., 2007).

The relationship between $Hg^0_{(g)}$, RGM and TM is one which requires careful examination as the relative concentrations of the three species depends in varying proportions on local emissions, local chemistry and long range transport. Figure 11.7 shows the routes followed during the oceanographic campaigns during different seasons and covering the two sectors of the Mediterranean basin. The overall strategy for the cruise campaigns was to perform integrated measurements of the atmosphere (specifically the MBL), the top-water micro-layer, the water column and sediments with the aim of understanding the role of major chemical, physical and biological parameters on the exchange of Hg compounds between different environmental compartments of the Mediterranean Sea. Atmospheric species were monitored continuously along routes covering the whole Mediterranean sea basin and both water and sediment samples have been collected at selected stations including Western Basin, Thyrrenian Sea, Ionian Sea, Levantine Basin, Aegean Sea and principal shelf areas (Gulf of Lyon, Adriatic Sea and Strait of Sicily) (see Figure 11.7).

The associated measurement programme for all MED-OCEANOR campaigns is detailed in Table 11.3

A full set of atmospheric measurements were performed with automated and manual methods with a time resolution ranged betwee 5min and 24hours. Ambient concentrations of Hg^0, $Hg^{(II)}$, PM were measured by an integrated Tekran automated systems with a frequency of 5min for Hg^0 and 2-hrs for $Hg^{(II)}$ and PM. TGM was also measured by a Tekran 2537A system with a time resolution of 5min. A number of ancillary measurements were also performed which includes meteorological

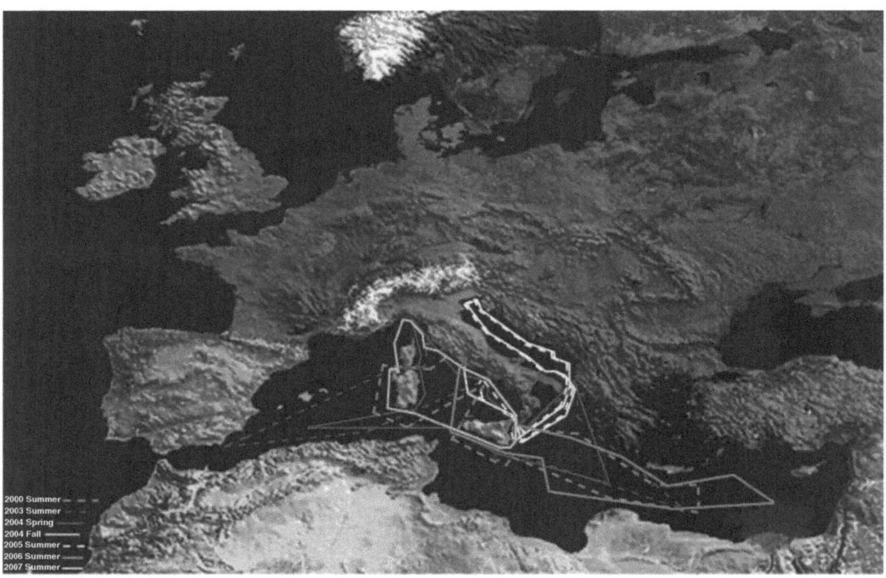

Figure 11.7 Routes followed during the oceanographic campaigns in the Mediterranean Sea during different seasons and covering two sectors (Eastern and Western) from 2000 to 2007

Table 11.3 Mercury measurements programme carried out during the cruises over the Mediterranean Sea from 2000 to 2007

Atmospheric Measurements	TGM	Hg^0	RGM	PM	$PM_{2,5}$-PM_{10}	SO_2	O_3	CO_2	Halogens	
Water Measurements (Hydrocast)	DGM	GSM	TM,TDM, TPM	DIM, TIM	DIM, TIM	RDM	DOC; Particulate organic C, N; Particulate 13C, 15N	Nutriens; Rn, O2	Phytoplankton speciation; Zooplankton speciation; Zooplankton biomass (C,N); Chlorophyll pigments	Hg methylation potential; Hg reduction potential; Hg species uptake rate (phyto)
Sediment Measurements	Bulk Hg (TM, TMM)	Porewater Hg (PWM, PWMM)	Bulk C,N,S,S	Bulk 13C, 15N	Bulk heavy metals, Pb isotopes	Porewater DOC, DIC, S-2	Porewater Fe, Mn, O2	Hg methyl, potential		
Meteorological Parameters	Air Temp	Wind speed	Wind direction	Relative Humidity	Solar radiation	Atmospheric pressure				
Water Parameters	Temp	Conductivity	Salinity							

data, and ozone concentrations with a time resolution of 5min. Table 11.4 contains a summary of the sampling/analytical techniques employed to assess the level of TGM/Hg^0, RGM and PM in the MBL during the cruise campaigns.

Measurements of Hg^0, $Hg^{(II)}$, and PM ambient concentrations were performed using an integrated automated system combining the analytical capability of the Tekran (Toronto, Canada) Model 2537A Cold Vapour Atomic Fluorescence Spectrometer (CVAFS) with the Tekran Model 1130 speciation unit and the Tekran Model 1135 Hg^0 unit (Landis et al., 2002).

A statistical summary of the TGM, Hg^0, $Hg^{(II)}$ and PM concentrations observed during the cruise campaigns performed in the Mediterranean sea basin is reported in Tables 11.3–11.6. Tables 11.5 and 11.6 show the spatial and temporal distribution of atmospheric Hg concentrations in the East and West sectors of the Mediterranean

Table 11.4 Sampling/analytical methods used to assess atmospheric Hg species in the MBL of the Mediterranean Sea basin during the cruise campaigns (2000–2007)

Species	Sampling Methods	Analisys	Frequency
TGM (Aut)	Tekran 2537A	Tekran/ CVAFS	Continuous
TGM (Manual)	Gold Traps	CVAFS	12/24 hours
TGM2	Gardis	Gardis	Continuous
Hg(p) (Manual)	AES-Mini-traps	Thermal-desorb.-CVAFS	24 hours
PM (Aut)	Tekran 1135 unit/Quartz-Annular denuders	CVAFS	2 hours
$Hg^{(II)}$ (Manual)	Quartz-Annular denuders	Thermal desorb.-CVAFS	24 hours
$Hg^{(II)}$ (Aut)	Tekran 1130 Unit	CVAFS	2 hours
Meteorological data	Meteorological station		Continuous

Table 11.5 Main statistical parameters for atmospheric Hg species concentrations observed over the East sector of the Mediterranean Sea Basin during the MED-OCEANOR campaigns from 2000 to 2006

Mediterranean East-sector	Year	Statistical Parameters	TGM (*ng m^{-3}*)	Hg^0 (*ng m^{-3}*)	$Hg^{(II)}$ (*pg m^{-3}*)		PM (*pg m^{-3}*)
Medoceanor (Summer)	2000	Max.	9.5	---	Aut. 8.6	Man. 11.1	16.4
		Min.	0.2	---	1.1	0.1	1.9
		Mean	1.9	---	3.8	4.9	7.3
		Std Dev	0.5	---	2.0	3.4	4.5
Medoceanor (Summer)	2003	Max.	15.7	11.4	22.5		10.1
		Min.	0.7	0.2	2.8		0.1
		Mean	1.6	1.3	9.1		1.8
		Std Dev	0.5	0.7	5.3		2.0
Medoceanor (Spring)	2004	Max.	2.0	1.9	9.7		5.7
		Min.	1.0	1.1	0.6		1.9
		Mean	1.6	1.6	3.9		3.6
		Std Dev	0.2	0.1	2.5		1.1
Medoceanor (Summer)	2006	Max.	---	2.8	76		14.8
		Min.	---	0.4	0.4		0.1
		Mean	---	1.2	14.4		4.4
		Std Dev	---	0.5	16.4		2.6

Table 11.6 Main statistical parameters for atmospheric Hg species concentrations observed over the West sector of the Mediterranean Sea Basin during the MED-OCEANOR campaigns from 2000 to 2007

Mediterranean West-Sector	Year	Statistical Parameters	TGM *(ng m^-3^)*	Hg^0 *(ng m^-3^)*	$Hg^{(II)}$ *(pg m^-3^)*	PM *(pg m^-3^)*
Medoceanor (Summer)	2000	Max.	11.1	---	Aut. 30.1 Man. 8.7	17
		Min.	0.1	---	0.2 2.6	4.8
		Mean	1.7	---	11.6 5.0	9.6
		Std Dev	0.8	---	9.8 1.9	3.2
Medoceanor (Summer)	2003	Max.	32	2.8	13.1	7.1
		Min.	0.1	0.8	1.0	0.3
		Mean	2.2	1.2	6.3	1.4
		Std Dev	1.5	0.2	4.4	1.7
Medoceanor (Spring)	2004	Max.	8.6	4.4	25.3	11.9
		Min.	1.0	0.5	0.1	0.2
		Mean	1.8	1.7	6.2	2.6
		Std Dev	0.3	0.3	5.5	2
Medoceanor (Summer)	2007	Max.	---	116.9	97.8	77.5
		Min.	---	0.2	0.1	0.4
		Mean	---	2.2	8.2	11.2
		Std Dev	---	4	10.4	10.1

Table 11.7 Main statistical parameters for atmospheric Hg species concentrations observed over the Adriatic Sea during the MED-OCEANOR campaigns from 2004 to 2005

Adriatic-Sea	Statistical Parameters	TGM *(ng m^-3^)*	Hg^0 *(ng m^-3^)*	$Hg^{(II)}$ *(pg m^-3^)*	PM *(pg m^-3^)*
Medoceanor 2004 (Fall)	Max.	4.0	4	63	51
	Min.	0.7	0.7	0.04	0.04
	Mean	1.6	1.5	6.7	4.5
	Std Dev	0.5	0.4	12	8
Medoceanor 2005 (Summer)	Max.	---	5.4	40	9.1
	Min.	---	0.1	0.8	0.04
	Mean	---	2.0	8.2	2.9
	Std Dev	---	0.7	8.1	2

sea, respectively, whereas Table 11.7 shows the temporal and spatial distribution of atmospheric Hg concentrations over the Adriatic Sea, a blind alley of the Mediterranean basin, with a very slow change of waters, separated from the rest of the Mediterranean by the Ionian Sea. Table 11.8, finally, shows a statistical summary of the atmospheric Hg species observed in the Mediterranean Sea from 2000 to 2007.

11.3 Air-Water Mercury Exchange

Open-ocean investigations have demonstrated the importance of air–sea exchange in controlling the Hg concentration in the atmosphere, and ultimately, in determining the long-term fate of Hg released to the atmosphere from both natural and anthropogenic sources. Hg exists as a dissolved gas in ocean waters, both as Hg^0

Table 11.8 Main statistical parameters for atmospheric Hg species concentrations observed over the Mediterranean Sea Basin during the MED-OCEANOR campaigns from 2000 to 2007

Med-Oceanor campaigns	Statistical Parameters	TGM (*ng m*$^{-3}$)	Hg^0 (*ng m*$^{-3}$)	$Hg^{(II)}$ (*pg m*$^{-3}$)	PM (*pg m*$^{-3}$)
Medoceanor 2003	Max.	31.9	11.4	22.5	10.1
(Summer)	Min.	0.1	0.2	1.0	0.13
	Mean	1.9	1.3	8.2	1.7
	Std Dev	1.2	0.6	5.1	1.9
Medoceanor 2004	Max.	8.6	4.4	25.3	11.9
(Spring)	Min.	1.0	0.5	0.1	0.2
	Mean	1.7	1.7	5.8	2.8
	Std Dev	0.3	0.3	5.2	1.8
Medoceanor 2004	Max.	4.0	4	63	51
(Fall)	Min.	0.7	0.7	0.04	0.04
	Mean	1.6	1.5	6.7	4.5
	Std Dev	0.5	0.4	12	8
Medoceanor 2005	Max.	---	5.4	40	9.1
(Summer)	Min.	---	0.1	0.8	0.04
	Mean	---	2.0	8.2	2.9
	Std Dev	---	0.7	8.1	2
Medoceanor 2006	Max.	---	2.8	76	14.8
(Summer)	Min.	---	0.4	0.4	0.1
	Mean	---	1.2	14.4	4.4
	Std Dev	---	0.5	16.4	2.6
Medoceanor 2007	Max.	---	116.9	97.8	77.5
(Summer)	Min.	---	0.2	0.1	0.4
	Mean	---	2.2	8.2	11.2
	Std Dev	---	4	10.4	10.1

and as dimethylmercury (Me_2Hg) (Kim and Fitzgerald, 1988; Mason and Fitzgerald, 1990; Mason et al., 1995a; Cossa et al., 1997; Mason and Sullivan, 1999; Lamborg et al., 1998). However, in the upper ocean, Hg^0 appears to be the dominant form of the Dissolved Gaseous Hg (DGM) fraction. In most studies, Hg^0 concentrations in surface waters have been found to be saturated relative to the atmosphere, and thus the resultant flux of Hg^0 is from the ocean to the atmosphere. Flux estimates have been made on cruises in both the Atlantic (Mason et al., 1998; Cossa et al., 1997; Mason and Sullivan, 1999; Lamborg et al., 1998) and Pacific Oceans (Kim and Fitzgerald, 1988; Mason and Fitzgerald, 1993) as well as in the Mediterranean Sea basin (Gardfeldt et al., 2001; 2003; Andersson et al., 2007; Kotnik et al., 2007). The results of the more recent studies suggest that the evasional flux, as estimated based on DGM concentrations in the surface waters, is greater than the estimated inputs from both atmospheric and riverine inputs.

The ocean receives 90% of its Hg through wet and dry deposition, and a significant fraction of the Hg is in the oxidised form (Mason et al., 1994a). Oxidised Hg may be transformed by several processes in the aqueous phase and reemitted to the atmosphere. Gas exchange of Hg between the surface water and the atmosphere is considered the major mechanisms driving Hg from the seawater to the atmosphere (e.g., Mason et al., 1994; Schroeder and Munthe, 1998). The major components of total mercury (Hg-tot) in seawater are mercuric chloride complexes, mercuric ions associated with dissolved organic carbon (DOC) (Munthe, 1991) and suspended particles. Some of these Hg forms can be reduced to Hg^0 both through biotic (that is enzymatically catalysed by microorganisms) (Mason et al., 1995) and abiotic processes (Allard and Arsenie, 1991; Xiao et al., 1994, 1995; Costa and Liss, 1999). In fact, these processes may significantly contribute to the super saturation of DGM found in many natural waters and thus to the evasion of Hg to the atmosphere (Schroeder and Munthe, 1998). It has been found that a majority of aquatic environments are supersaturated with respect to dissolved DGM (Schroeder and Munthe, 1998). Part of this Hg^0 may be emitted to the atmosphere and Mason et al. (1994a) estimated the emission (mostly Hg^0) from the water surfaces to account for 30% (2000 Mg per year) of the total emission of Hg to the atmosphere. The total global emission from the sea surface has been re-evaluated (Mason and Sheu, 2002) to 2600 Mg per year, however the error in this estimation could, according to the authors, be as high as a factor of 5. The global flux of Hg from ocean evasion, and the potential variability and error, is discussed in detail in Chapter 7 (Sunderland and Mason, 2007).

Evasion occurs at the air-water interface, and the top water micro layer probably plays an important role in this process, although it is difficult to sample this layer. The efficiency of these processes depends upon the intensity of the solar radiation, the ambient temperature of the air parcel above the seawater, and the water temperature. It could be assumed that the temperature of the micro layer is influenced by solar radiation in the same way as that of the air, and this could explain the correlation observed between the temporal trend of Hg^0 evasion and air temperature. Ferrara et al. (2003) report Hg depth profile measurements in the Mediterranean Sea showing that DGM losses may also be a result of oxidation in the upper thermo cline. The influence of sunlight on both the reduction of mercuric complexes and the oxidation of DGM in the aqueous phase has been observed by many scientists and was investigated during an Atlantic Ocean cruise as reported by Mason et al. (2001) and others (Whalin and Mason, 2006; Whalin et al., 2007). Additionally, the major DGM species Hg^0 can be formed from microbial degradation of methyl mercury (Mason et al., 1995). It has been shown that the photoinduced processes in open-ocean surface waters could result in either a net oxidation or net reduction of Hg species.

There are two general methods for estimating Hg fluxes: i) flux chamber techniques (Xiao et al., 1991; Kim and Lindberg 1995; Capri and Lindberg, 1998; Poissant and Casimir, 1998; Ferrara and Mazzolai, 1998; Gårdfeldt et al., 2001, 2003; Ferarra et al., 2000, 2001; Amyot et al., 2004) and ii) gas exchange models (Lindberg et al., 1995a; Kim et al., 1995; Gårdfeldt et al., 2001; Rolfhus and Fitzgerald, 2001; Poissant et al., 2000; Wängberg et al., 2001a, 2001b; Baeyens et al. 1991; Baeyens and Leermakers, 1998; Cossa et al., 1997). Both approaches have advantages and disadvantages. The chamber technique is often used because of it simplicity and good detection limits, but it disrupts the local environment by covering the study area, and affects the meteorological parameters and the concentration gradient of Hg. The models on the other hand include separate measurements of Hg^0 in air and water, wind speed and water temperature. These measurements do not disrupt the environment and meteorological parameters are included. A limitation with these methods is the need for multiple samples in order to get a proper estimate. When comparing the methods it has been shown that the flux chamber technique consistently measures lower values than the calculations (Lindberg et al., 1995a, b; Gustin et al., 1999).

Surveys of Hg evasion from European waters have been reported by Cossa et al. (1997), and Ferrara et al. (2001) for confined areas in the Mediterranean Sea and by Gardfeldt et al. (2001), Andersson et al., 2007; and Wangberg et al. (2001b) in the North Sea, the Skagerrack and the open Baltic Sea, respectively. Measurements have also been carried out in the Atlantic Ocean and the Arctic Ocean (e.g. Mason et al., 1995b, 1998; Gill and Fitzgerald 1985, 1987, 1988; Dalziel 1992; Daziel and Yeats 1985; Cossa et al., 1992, 1994; Mason and Sullivan 1999; Temme et al., 2005; Andersson 2008; Leitch et al., 2007; St Louis et al., 2007; AMAP 2005 and references therein; Coquery et al., 1995; Sommar et al., 2007; Andersson et al., 2008).

11.3.1 Atlantic Ocean

A few studies have been carried out to study Hg speciation in the Atlantic Ocean (e.g. Mason et al., 1995b, 1998; Gill and Fitzgerald 1985, 1987, 1988; Dalziel 1992; Daziel and Yeats 1985; Cossa et al., 1992, 1994; Mason and Sullivan 1999; Temme et al., 2005; Andersson 2008), but most of these studies have focused on total and reactive Hg.

Only a few of these studies focused on measurements of DGM at offshore sites (Mason et al., 1998; Mason and Sullivan 1999; Temme et al., 2005; Andersson 2008). Mason et al. (1998) carried out measurements in the North Atlantic Ocean, while Mason and Sullivan (1999) measured DGM in the equatorial and South Atlantic Ocean. Measurements were carried out by discrete sampling methods, i.e. each sample was collected separately and analysed by the purge and trap technique within hour after sample collection. On the other hand, Temme et al. (2005) and Andersson (2008) carried out continuous measurements, resulting in a much large amount of data. However, only limited areas were sampled even though continuous sampling was performed.

Mason et al. (1998) collected discrete samples of DGM in the North Atlantic in August 1993. The expedition covered the area from Newfoundland to Reykjavik, passing around Iceland. Samples were collected both at stations using Go-Flo bottles and between stations using a surface "fish" sampler. In total 15 samples were collected. The average concentration measured was 130 ± 80 pg L^{-1} with a range from 30 pg L^{-1} north of Iceland to 300 pg L^{-1} south of Iceland. The flux was estimated to be 15.9 ±

10.8 ng m^{-2} h^{-1}. The authors presented a simple mass balance calculation and concluded that the concentration must represent a non-steady state situation, since the concentrations and the calculated flux can neither be supported by deposition nor diffusion of Hg from depth. An estimated monthly flux based on the data from this expedition was made by Mason and Gill (2005) to be 12,000 ng m^{-2} $month^{-1}$.

In May/June 1996, Mason and Sullivan (1999) measured DGM along a cruise track in the eastern South Atlantic, from Montevideo, Uruguay, to Bridgetown, Barbados (Figure 11.1). Six deep stations where sampled and surface samples were collected in-between stations. The average concentration from the expedition was 240 ± 160 pg L^{-1}, with a range of 100 to 800 pg L^{-1}. It is suggested that the concentrations reflect a net accumulation of Hg^0 in the surface water during the summer, rather than the annual average. During the expedition other Hg species were collected, i.e. total, reactive, dimethyl, monomethyl, and particulate Hg. For most of the samples collected, the reactive and DGM were in the same range, i.e. all reactive Hg was in the form of DGM, except from about 30 to 22°S where the concentration of reactive Hg increased while the DGM was detected at lower levels than previously. This potentially indicated atmospheric input of reactive Hg from continental sources. During the expedition airborne Hg species was measured and by using the TGM concentration, the flux was calculated using the flux model developed by Wanninkhof (1992) as 80 ng m^{-2} h^{-1}. The flux was extrapolated to an annual flux of 700 ng m^{-2} yr^{-1}, a value that could not balance the atmospheric input. Therefore, it was suggested that the DGM cycling in the surface was not in balance with the atmospheric input (Lamborg et al. 1999). Temme et al. (2005) measured DGM continuously in the North Atlantic Ocean, from June to August 2004, along the west coast of Norway (62°N) past Spitsbergen up to 85°N. The concentration range in the North Atlantic was 11-14 pg L^{-1}. By using the airborne gaseous mercury concentration measured by Aspmo et al. (2006) during the same expedition, and the annual average wind speed, 9.5 m s^{-1}, and water temperature, 281 K (NOAA data base), an estimated flux, according to Nightingale et al. (2000), of 750 ng m^{-2} $month^{-1}$ was estimated.

Andersson (2008) measured DGM continuously in the North Atlantic in July 2005 along a transect from north of the British Isles to Cap Farwell, south of Greenland. The concentrations in the sampled water were in the range of 5.6 to 17.9 pg L^{-1} with an average of 11.6 ± 2 pg L^{-1}. The variation observed in the data-set was not diurnal, and was considered to be random. In parallel, measurements of TGM was carried out and by using the flux model developed by Nightingale et al. (2000) the flux was estimated. Due to the high resolution in the measurements, high resolution fluxes could be calculated, and the average flux was calculated to be 0.42 ± 0.36 ng m^{-2} h^{-1}, ranging from -0.64 ng m^{-2} h^{-1}, i.e. deposition of Hg^0 to 2.5 ng m^{-2} h^{-1}, i.e. evasion of Hg^0. These results show that the waters sampled was both undersaturated and super-saturated with respect to Hg^0. An annual flux was estimated to be 460 ng m^{-2} $month^{-1}$, by using the NOAA data base for the annual wind speed and water temperature, 9.5 m s^{-1} and 281 K respectively.

Table 11.9 Summary of the results from the Atlantic Ocean

Location	DGM *(pg L^{-1})*	Flux *(ng $m^{-2}h^{-1}$)*	Reference
North	130 ± 80	15.9 ± 10.8	Mason et al., 1998
Equatorial and Southern	240 ± 160	80	Mason and Sullivan, 1999
North	11-14		Temme et al., 2005
North	11.6 ±2	0.42± 0.36	Andersson, 2008

11.3.2 Pacific Ocean

Measurements of DGM in ocean waters were first made in the equatorial Pacific Ocean. (Kim and Fitzgerald, 1986; 1988; Fitzgerald, 1995). These studies showed that there was a broad relationship between DGM concentration and chlorophyll, suggesting that there was a link with primary productivity. However, other factors, such as light levels which vary at the same time, could also be important as the reactions leading to net Hg^0 formation can be photo chemically induced (e.g. Whalin and Mason, 2006). Also, as noted below, upwelling waters could be a source of reactive Hg that is then reduced. For example, on a transect at 85°W and across the equator, DGM varied from 10 to 27 pg L^{-1}, and waters were saturated relative to the atmosphere. In contrast, measurements in the North Pacific gyre region showed lower concentrations. The data from these earlier measurements are summarized in Table 11.10. Measurements of DGM were also obtained in the equatorial region during a cruise in 1990/91 (Mason et al., 1991; Mason and Fitzgerald, 1993). Mass balance estimates, based on these data, indicate that the concentrations measured, and the evasional fluxes estimated, cannot be sustained purely via mixing of Hg^0 from deeper waters into the mixed layer. As a result, it was concluded that formation of Hg^0 in the mixed layer of the ocean must be occurring (Mason et al., 1994b; Mason and Fitzgerald, 1993). Mass balance calculations suggest that reduction rates of the order of 1-10% per day were required to balance estimated evasion rates in open ocean waters. For the equatorial Pacific Ocean, the estimated evasion rates ranged between 12 and 230 ng m^{-2} day^{-1} (Table 11.10). These evasion rates exceed the rate of atmospheric deposition estimated for the region (16 ng m^{-2} day^{-1}; Mason et al., 1994b) and another source of inorganic Hg to the mixed layer is required to maintain the estimated average evasion. It has been suggested that equatorial upwelling of Hg into the thermo cline provides this additional Hg source and this Hg is supplied to the equatorial thermocline by meridional circulation of Hg deposited at mid-latitudes (Mason and Fitzgerald, 1993; Mason et al., 1994b). In the gyre region, estimated evasion rates are less that the depositional flux.

The mass balance calculations for Hg^0 formation and evasion in the equatorial Pacific, and similar data from the North Atlantic, suggest that the rate of formation of Hg^0 in ocean regions is constrained, in productive regions, by the rate of supply of Hg to the ocean mixed layer (Mason and Fitzgerald, 1993; Mason et al., 1998).

Table 11.10 Concentrations of Hg^0 measured in various ocean regions and the associated estimated evasional flux to the atmosphere. (References: 1: Mason and Fitzgerald (1993), 2: Kim and Fitzgerald (1986), 3: Kim (1987))

Region	Conc. *(pM)*	Hg^0 *(%)*	Potential flux *(ng $m^{-2}day^{-1}$)*	Reference
EPO 1990	0.05 - 0.36	14 ±10	32 - 230	1
EPO 1984	0.03 - 0.23	5 ± 3	12 - 230	2
150°W, 10°N-12°S	0.04 - 0.09	-	0.1 - 0.35	3
North Pacific	0.03	-	0.03	3

This notion has been further supported by more recent data for DGM, collected on a cruise in the North Pacific in May/June 2002. Surface waters were sampled using a "fish surface sampler" and discreet measurements of total and DGM were made along the cruise track at individual stations (Laurier et al., 2003; Laurier et al., 2004). The data obtained are shown in Figure 11.8 and were higher for both total Hg and DGM closer to Asia, decreasing offshore.

From this data, the evasional flux was estimated, based on the average wind speed, surface water temperatures and the gas exchange equation of Wanninkhof (1992), modified for Hg. Although there was some variability, the results show that on average the evasional flux is higher in the tropical water and particularly during high wind speed events, which occurred during the last week of the cruise (Laurier et al., 2003). Based on both DGM concentrations in the surface water and atmospheric Hg^0 measurements, the estimated evasional flux was equivalent to -0.5 up to 57.6 % (average of 12.7 ± 16.1%) of the Hg^0 present in the MBL with the highest fluxes corresponding to the highest wind speeds (Table 11.11).

11.3.3 Mediterranean Sea

In the Mediterranean area there are several cinnabar deposits (e.g. Almadén (Spain), Idrjia (Slovenia) and Monte Amiata (Italy)), where Hg has been released by both natural and anthropogenic activities. These characteristics along with favourable climate conditions represent synergic factors that yield significant evasional fluxes of Hg from the surface water during most of the annual period (Ferrara et al., 2003; Lanzillotta et al., 2001; Hedgecock et al., 2006).

Measurements by Ferrara et al. (1998; 2000) were performed in optimal weather conditions, in the absence of wind and clouds, and with a smooth sea. Wind action is of particular importance, since a speed of up to 8 m s^{-1} the water-sea exchange of gases increases with a linear trend. At higher winds, the subsequently-formed waves lead to an even bigger increase, due to extension of the water surface and the bubbles produced by breaking waves. Ferrara et al. (1998; 2000) performed Hg flux measurements using a floating flux chamber and a semi-automatic sampling device linked to the modified GARDIS 1A Mercury Analyzer to allow the

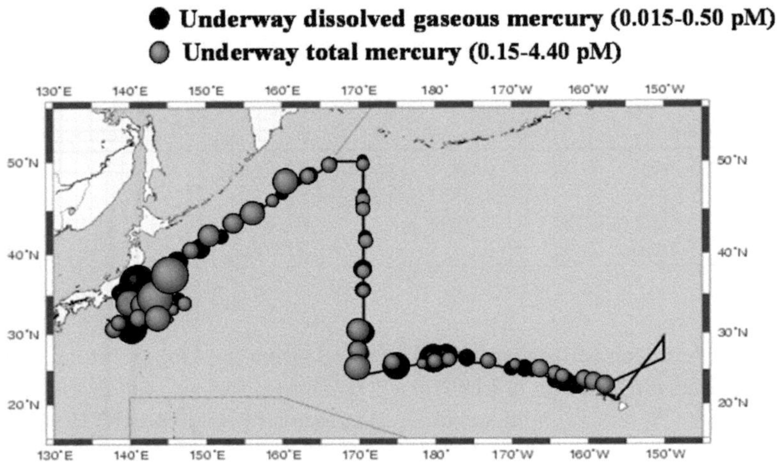

Figure 11.8 Concentrations of total mercury and dissolved gaseous mercury for surface samples collected using the fish sampler during the may/June 2002 cruise. Adapted from Laurier et al. (2004). Size of the circle illustrates the relative concentrations between the maximum and minimum values shown in the leg

Table 11.11 Mean calculated reactive gaseous mercury ($Hg^{(II)}$ dry deposition, total mercury (Hg) in wet deposition and dissolved gaseous mercury (DGM) evasion fluxes. Adapted from Laurier et al. (2003)

Sampling Period (month/day)	5/14 - 5/20	5/21 - 5/28	5/29 - 5/31
RGM dry deposition flux (*ng* m^{-2} d^{-1})	6.2 ± 5.8	19.9 ± 21.4	12.2 ± 6.3
Hg wet deposition flux (*ng* m^{-2} d^{-1})	no data	12.6 ± 9.5	58.3 ± 23.3
DGM evasional flux (*ng* m^{-2} d^{-1})	20.9 ± 18.4	26.0 ± 31.1	59.8 ± 50.3

[a]Standard deviation (±).

determination of low fluxes of Hg from water (Ferrara et al., 1998) (see Figure 11.9). The measurements of the mercury evasional fluxes were carried out at three sites of the northern Tyrrhenian Sea during 1998. Two sites were located in a unpolluted and a polluted (near chlor-alkali plant) coastal area, and the third was an offshore site. The Hg^0 flux observed near the polluted coastal site was almost three times higher then that observed at the other two sites, which is consistent with the higher degree of Hg contamination in the area whereas mercury fluxes at the offshore site were very similar to those observed at the unpolluted coastal site. The evasional flux showed a typical daily trend, being highest at midday when the ambient temperature and solar radiation were at a maximum, and lowest, near to zero, during the night, suggesting that solar radiation is one of the major driving factors affecting the release of Hg^0 from surface waters. In addition, a seasonal trend was also observed, with minimum values during the winter period and maximum values during the summer, probably due to higher water temperature that may have facilitated

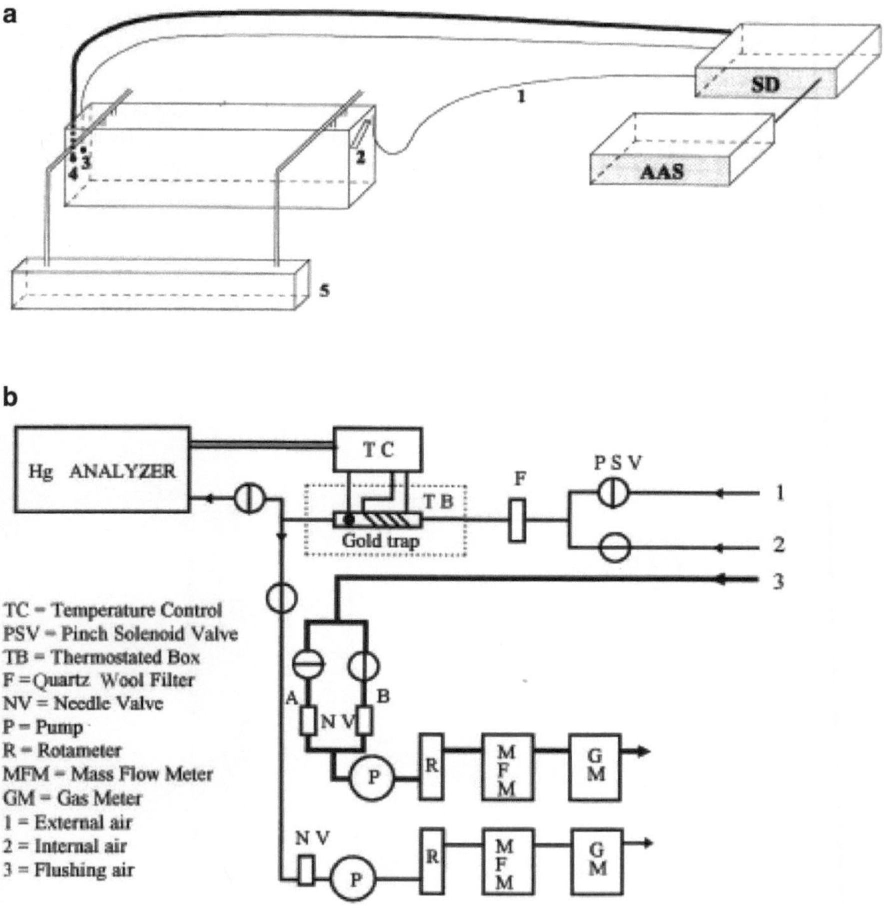

Figure 11.9 **(a)** Floating flux chamber. 1) Teflon tube for external air sampling; 2) inlet port; 3) and 4), outlet ports; 5) floating PVC foam bar; the second floating bar is not shown in the figure; AAS, atomic absorption spectrometer; SD, sampling device. **(b)** Semi-automatic air sampling device for measurements of mercury degassing rate
Source: Ferrara et al., 1998

biotic and abiotic processes in the water column. The daily average for emission rates from the surface water are reported in Table 11.12.

DGM concentrations were measured in Atlantic coastal water, west of Ireland, and in the Mediterranean Sea (Gardfeldt et al., 2003) in order to estimate mercury emissions using a gas-exchange model developed by Wanninkhof (1992). The measurements in the Mediterranean Sea were carried out during the MED-OCEANOR campaign on board of the Research Vessel *Urania* from July 14 to August 9 2000. Flux chamber measurements were performed at near shore sites during 2000 MED-OCEANOR cruise over the western sector of the Mediterranean Sea basin (see Figure 11.10).

Table 11.12 Concentration of dissolved mercury $Hg^{(D)}$, mercury associated with particulate matter Hg(p) and emission from the seawater surface of three selected sites during the summer season

Site	Date	$Hg^{(D)}$ *(ng L⁻¹)*	Hg(p) *(ng L⁻¹)*	Hg emitted in sunshine hours *(ng m⁻²)*	Hg emitted in night-time hours *(ng m⁻²)*	Hg emitted in a day *(ng m⁻²)*
Unpolluted coastal zone	11-08-1998	6.3	2.4	40	12	52
Polluted coastal zone	17-08-1998	15.0	9.0	135	29	164
Offshore	26-08-1998	6.0	0.8	30	14	44

Source: Ferrara et al., 1998.

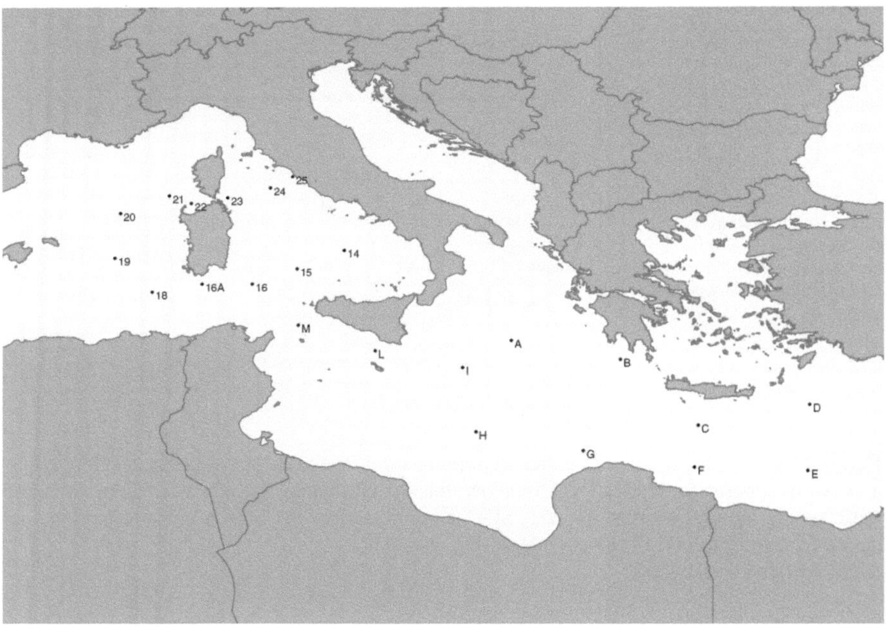

Figure 11.10 Hg measurements performed at near shore sites during 2000 MED-OCEANOR cruise over the western sector of the Mediterranean sea basin (see Table 11.13 for details). Source: Gardfeldt et al., 2003

The Mediterranean Sea is a source of atmospheric Hg^0 that can influence the European domain. Table 11.13 shows the observations from the Mediterranean Sea which have been sorted into four regional categories: Western, Tyrrhenian, Strait of Sicily and Eastern Mediterranean Sector. The average evasion from the western Mediterranean Sea was lower than the eastern sector. As shown in Table 11.13, the mean wind speed was nearly the same at the eastern and western Mediterranean

Table 11.13 Some physical parameters at the sampling sites with corresponding mercury evasion estimations provided by the gas exchange model

Position and salinity	Sampling point	Date and time *(yyyy-mm-dd hh:mm)*	DGM *(pg L^-1^)*	TGM *(ng m^-3^)*	Water temperature *(°C)*	Saturation (-)	Wind Speed $u_{10}(ms^{-1})$	Hg evasion *(ng m^-2^hr^-1^)*
Atlantic coastal water at the Irish west coast: 35 PSU								
59°19N 9°53W		1999-08-31 16:20	40.3	1.54	15.0	6.4	7.0	6.0
59°19N 9°53W		1999-09-01 16:20	18.7	1.68	15.0	2.7	6.8	2.1
59°19N 9°53W		1999-09-02 16:20	20.0	1.64	15.8	3.0	7.1	2.6
59°19N 9°53W		1999-09-03 16:20	21.5	1.79	16.1	3.0	4.8	1.3
59°19N 9°53W		1999-09-04 16:20	19.4	1.70	16.1	2.9	3.0	0.4
59°19N 9°53W		1999-09-05 16:20	15.8	1.64	16.6	2.5	8.7	2.8
59°19N 9°53W		1999-09-06 12:10	13.4	1.57	16.0	2.1	8.1	1.9
59°19N 9°53W		1999-09-07 12:10	21.9	1.26	16.0	4.4	8.6	4.6
Average			21.4	1.6	15.8	3.4	6.8	2.7
Eastern Mediterranean Sea: 38.4-39.2 PSU								
36°47N 18°37E	A	2000-07-17 13:30	35.1	1.20	25.7	10.7	7.2	7.6
36°11N 22°04E	B	2000-07-18 10:30	54.2	1.92	23.8	9.8	3.6	2.7
34°05N 24°32E	C	2000-07-19 06:20	22.0	2.37	23.3	3.2	4.9	1.6
34°44N 28°03E	D	2000-07-20 03:10	22.6	1.42	24.0	5.6	11.0	9.8
32°40N 28°00E	E	2000-07-20 17:45	32.2	2.20	25.4	5.3	7.2	6.2
32°47N 24°24E	F	2000-07-21 13:15	34.8	1.94	24.8	6.4	10.8	15.2
32°18N 20°54E	G	2000-07-22 09:20	43.1	1.90	25.7	8.3	6.5	7.4
33°53N 17°30E	H	2000-07-23 10:00	36.4	1.77	26.6	7.7	9.6	13.7
35°55N 17°05E	I	2000-07-23 23:30	33.6	1.50	26.2	8.3	7.1	6.9
Average			34.9	1.8	25.0	7.2	7.5	7.9
Strait of Sicily: 37.3-37.6 PSU								
36°27N 14°18E	L[a]	2000-07-25 00:30	32.5	1.5	25.8	7.9	4.2	2.3
37°16N 11°52E	M[a]	2000-07-25 15:00	86.6	1.6	24.1	19	10.6	40.5

(continued)

Table 11.13 (continued)

Position and salinity	Sampling point	Date and time *(yyyy-mm-dd hh:mm)*	DGM *(pg L^{-1})*	TGM *(ng m^{-3})*	Water temperature *(°C)*	Saturation (-)	Wind Speed $u_{10}(ms^{-1})$	Hg evasion *(ng $m^{-2}hr^{-1}$)*
Tyrrenian Sea: 38.1-38.2 PSU								
39°40N 13°20E	14	2000-07-29 05:00	26.5	1.88	25.7	5.2	10.0	9.9
39°05N 11°50E	15	2000-07-29 17:00	22.1	1.62	25.4	4.9	6.7	3.6
38°35N 10°25E	16	2000-07-30 08:00	13.8	1.58	23.5	3.0	13.5	7.3
38°35N 08°50E	16A	2000-08-01 00:00	19.5	2.71	24.8	2.6	1.6	0.1
41°21N 09°39E	23	2000-08-08 06:15	17.8	1.44	23.7	2.9	6.2	2.3
41°40N 10°60E	24	2000-08-08 13:00	18.2	1.42	24.0	4.4	9.0	5.0
42°03N 11°42E	25	2000-08-08 16:45	17.0	1.58	24.1	3.5	4.8	1.2
Average			19.3	1.73	24.4	3.8	7.4	4.2
Western Mediterranean Sea: 37.3-37.9 PSU								
38°20N 07°15E	18	2000-08-02 08:30	14.8	1.49	24.3	3.5	3.2	0.5
39°25N 06°05E	19	2000-08-02 18:30	19.0	2.37	25.6	2.3	6.1	2.1
40°51N 06°15E	20	2000-08-03 08:00	15.6	1.58	24.4	4.3	7.5	2.8
41°25N 07°48E	21	2000-08-03 17:30	15.2	2.07	23.0	2.6	8.1	2.6
41°10N 08°30E	22	2000-08-03 21:00	12.0	1.36	24.8	4.0	11.0	4.5
Average			15.3	1.8	24.4	3.3	7.2	2.5

Source: Gardfeldt et al., 2003

sites, therefore, the higher Hg^0 evasion from the eastern region is, consequently, explained by the higher mean degree of Hg^0 saturation in the east compared to the west (see Table 11.13). The average Hg evasion value for the Tyrrhenian Sea, calculated by Ferrara et al. (2000), is consistent with a suggested gradient from west to south-east. The Hg^0 evasion reported by Gardfeldt et al., 2003 from the western Mediterranean and the Tyrrhenian Sea is of the same order of magnitude as that estimated by Ferrara et al. (2000) from unpolluted and offshore water in the Tyrrhenian sea (2.2 and 1.8 ng m^{-2} hr^{-1}, 24 h average). Cossa et al. (1997) estimated a mass-budget for the western Mediterranean based on measurements of several mercury species as well as literature data, based on a global model (Cossa et al., 1997, and references therein) and the estimated evasion is also is of the same order of magnitude as that obtained by Gardfeldt et al. (2003).

A summary of observed Hg^0 fluxes is presented in Table 11.14. The maximum DGM concentration and corresponding mercury evasion from the Mediterranean

Table 11.14 Mercury evasion from some aquatic environments reported in the literature including this study. For a more detailed description on averages and methods the reader is may refer to the original article

Location	Evasion[a] *(ng $m^{-2}hr^{-1}$)*	Method[b]	Author
Open waters			
Baltic Sea summer average	1.6	GEM	Wangberg et al., (2001a)
Baltic Sea winter average	0.8	GEM	Wangberg et al., (2001b)
North Sea	1.6-2.5	GEM	Cossa et al., (1996)
	0.49-9.25	GEM	Baeyens and Leermakers (1998)
The mid-Atlantic Bight adjacent the East Coast of North America	2.5	GEM	Mason et al., (2001)
Open waters, Mediterranean sites			
North West Mediterranean sites	1.2	GEM [c]	Cossa et al., (1997)
Western Mediterranean Sea	2.5	GEM	Gardfeldt et al. (2003)
Tyrrenian Sea	4.2	GEM	Gardfeldt et al. (2003)
Tyrrenian Sea	1.8	FC	Ferrara et al., (2000)
Strait of Sicily	2.3-40.5[d]	GEM	Gardfeldt et al., (2003)
Eastern Mediterranean Sea	7.9	GEM	
Coastal waters			
Skagerack part the North Sea Summer average	0.8	FC	Gardfeldt et al., (2001)
Atlantic water at the Irish west coast	2.7	GEM	Gardfeldt et al., (2003)
The Tyrrenian Sea, polluted coastal zone	6.8	FC	Ferrara et al., (2000a)
Mediterranean Sea, near shore sites round Sardinia	3.8	FC	

[a]Average values from the corresponding sampling periods.
[b]GEM: gas exchange models. FC: flux camber.
[c]Not based on DGM measurements due to a high detection limit (26 *pg* L^{-1}). Literature data was to estimate Hg evasion from the sea surface according to a global model, Cossa et al., (1997) and references therein.
[d]Range from the corresponding sampling period.
Source: Gardfeldt et al., 2003.

Sea was found in the Strait of Sicily. High DGM concentrations were also found at two locations south of Greece.

Past or present tectonic activity may contribute to the high DGM concentrations found at these positions. DGM data combined with an empirical gas-exchange model (Wanninkhof, 1992) data suggested that about 66 Mg of Hg^0 are released to the atmosphere from the Mediterranean Sea during the summer (Gardfeldt et al., 2003). This emission is considerable in comparison to European anthropogenic emissions and should thus be taken into account in regional atmospheric modelling and assessment. During three seasons (August 2003, March–April 2004 and October–November 2004) Andersson et al. (2007) performed measurements of DGM, TGM as well as meteorological and waters parameters on board the RV *Urania* in the Mediterranean Sea Basin, Figures 11.11, 11.12 and 11.13 show the selected measurement stations along the cruise tracks. The first two cruises performed in August 2003 and March–April 2004 covered parts of the Eastern and Western Mediterranean basin, (Figures 11.11 and 11.12), while the last cruise performed in October–November 2004, covered the Adriatic Sea (Figure 11.13). The flux from the sea surface was calculated using the gas exchange model developed by Nightingale et al. (2000).

In Table 11.15, data for DGM/saturation, TGM, wind speed and temperature are listed. The data is the average value from each station and sorted into eight different sections: (1) Western Mediterranean, (2) Tyrrhenian Sea, (3) Ionian Sea, (4) Adriatic Sea, (5) Northern Adriatic Sea, (6) Strait of Sicily, (7) Strait of Messina, (8) Strait of Otranto, as each section contained the stations from these parts of the cruises.

Table 11.16 summaries the seasonal average concentrations of TGM, DGM and saturation from each section. The DGM concentrations varied within the Mediterranean Sea. Two possible explanations for this might be geographical location and daily variation. The DGM night concentration during spring, autumn and summer did not vary significantly. The daytime DGM concentration was clearly

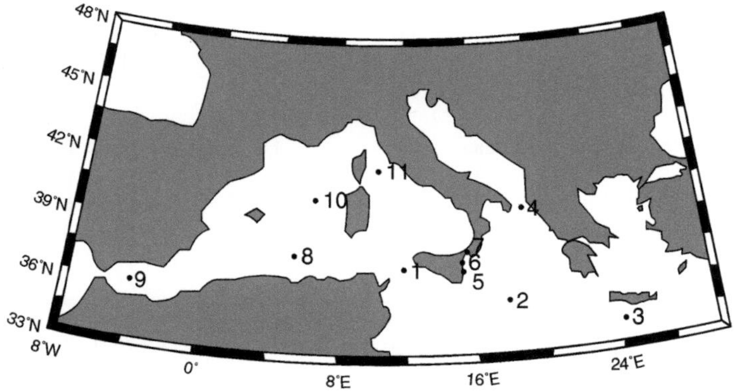

Figure 11.11 Cruise 1, Summer 2003
Source: Andersson et al., 2007.

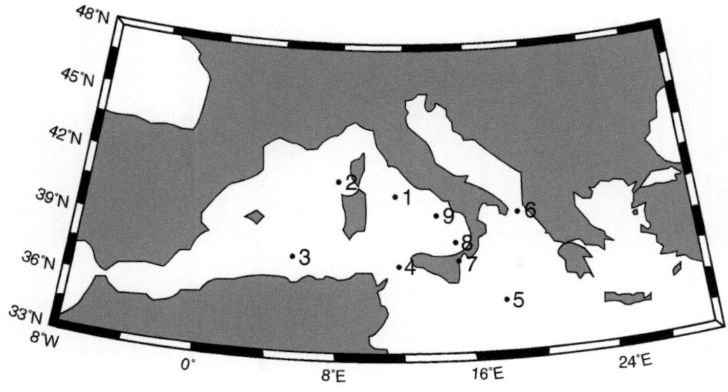

Figure 11.12 Cruise 2, Spring 2004
Source: Andersson et al., 2007.

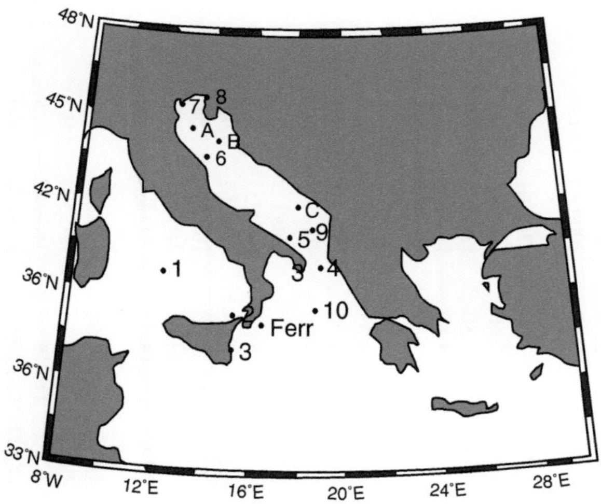

Figure 11.13 Cruise 3, Fall 2004
Source: Andersson et al., 2007.

influenced by sunlight; the difference between day and night during the summer can be as large as 130 fM.

The degree of saturation (S = DGM*H' / TGM, where H' is the dimensionless Henry's law constant) varied between the different seasons with the highest average degree of saturation during the summer cruise, 850%, while the autumn cruise had an average degree of saturation of 740% and the spring cruise 320%. The degree of saturation showed the highest variation during the summer cruise (600–1150%). During the autumn cruise a variation in degree of saturation was also observed (500–3260%). The variations in saturation between the seasons might be explained

Table 11.15 Results for DGM/saturation, TGM, wind speed and temperature from each section of the Mediterranean Sea

Sample site	Position	Date and time *(year-month-day h:min)*	DGM *(fM)/(%)*	TGM *(fM)*	Temperature *(°C)*	Wind Speed $u_{10}(ms^{-1})$	Hg flux *(pmol* $m^{-2}hr^{-1}$*)*	N
(1) Weast Mediterranean								
1:8	37°52N5°21E	2003-08-20 07:40 to 03:40[b]	200/689	10.5	28.5	6.6	31.0	199
1:9	35°56N -4°0E	2003-08-23 07:50 to 16:10	200/648	10.0	25.0	5.4	20.5	76
(2) Tyrrenian Sea								
1:11	41°53N10°18E	2003-08-27 02:20 to 07:50	170/601	9.5	27.7	1.9	3.5	67
2:8	38°39N15°7E	2004-04-02 08:55 to 17:40	100/258	9.5	14.6	3.6	3.5	106
2:9	39°55N13°59E	2004-04-03 06:10 to 21:50	120/334	9.0	15.1	2.2	2.0	189
3:1	39°41N12°8E	2004-10-28 00:30 to 08:40	150/603	7.5	22.1	4.6	11.0	97
3:2	38°18N15°14E	2004-10-29 03:30 to 21:55	300/1070	8.5	23.2	3.4	20.5	216
(3) Ionian Sea								
1:2	35°45N17°55E	2003-08-08 08:30 to 11:05	330/1146	10.0	28.7	4.1	24.0	26
1:3	34°19N24°20E	2003-08-10 06:35 to 14:25	180/728	8.0	25.3	4.9	15.5	77
1:5	37°12N15°17E	2003-08-14 19:30 to 09:45[b]	300/1082	17.5	27.8	2.5	10.5	160
1:6	37°37N16°16E	2003-08-15 14:10 to 21:55	250/1124	8.0	28.4	1.7	5.0	94
2:5	35°75N17°92E	2004-0.3-29 09:10 to 20:25	80/273	7.5	15.9	6.1	6.0	135
2:7	37°37N15°15E	2004-04-01 02:40 to 18:55	140/362	9.5	14.4	3.1	4.0	196
3:3	37°11N15°17E	2004-10-30 12:30 to 21:30	18/683	8.0	23.4	4.9	16.5	55
3:Ferr	38°0N16°30E	2004-10-31 06:55 to 14:20	240/921	8.0	23.5	2.4	6.5	82
3:10	38°32N18°46E	2004-11-11 09:10 to 14:45	160/790	6.0	21.8	8.2	33.0	63
(4) Adriatic Sea								
3:5	40°59N17°37E	2004-11-02 04:35 to 12:50	160/659	7.0	20.3	6.7	23.0	99
3:6	43°36N13°45E	2004-11-04 08:55 to 20:15	180/504	9.5	19.3	8.0	31.0	69
3:B	44°10N14°11E	2004-11-07 10:20 to 16:05	160/802	5.5	20.0	6.5	22.0	66
3:C	42°0N18°0E	2004-11-09 09:25 to 19:00	150/600	7.0	19.8	4.3	10.0	113
3:9	41°12N18°32E	2004-11-10 05:45 to 14:35	210/886	6.5	19.0	8.6	48.5	54
(5) North Adriatic Sea								
3:A	44°32N13°2E	2004-11-05 05:15 to 12:50	350/553	17.0	19.2	-	-	92
3:7	44°22N12°26E	2004-11-05 19:45 to 23:45	350/1716	10.0	18.0	7.4	118.5	29
3:8	45°40N13°34E	2004-11-06 06:15 to 12:25	1030/3260	8.0	18.0	7.1	165.5	75

(6) Strait of Sicily								
1:1	37°19N11°50E	2003-08-06 13:00 to 21:55	140/772	6.0	26.7	5.6	17.5	76
2:4	37°29N11°29E	2004-03-26 21:50 to 18:50[b]	100/307	8.0	14.6	3.8	3.5	253
(7) Strait of Messina								
1:7	38°5N15°34E	2003-08-16 02:10 to 03:00	220/907	8.0	27.1	9.1	62.0	11
(8) Strait of Otranto								
1:4	40°0N19°0E	2003-08-12 16:35 to 17:50[b]	180/776	7.5	26.2	4.7	16.0	256
2:6	40°0N19°0E	2004-03-30 20:55 to 04:05[b]	130/364	8.5	15.3	3.1	4.0	87
3:4	39°59N19°0E	2004-11-01 05:25 to 17:45	160/585	7.5	21.6	4.6	10.5	147

Source: Andersson et al., 2007

Table 11.16 Average concentrations and saturation from the different parts of the Mediterranean Sea

	Summer			Spring			Autumn		
Section	DGM (*fM*)	TGM (*fM*)	Sat. (%)	DGM (*fM*)	TGM (*fM*)	Sat. (%)	DGM (*pM*)	TGM (*fM*)	Sat (%)
(1) Western Mediterranean	200	10.5	669	-	-	-	-	-	-
(2) Tyrrenian Sea	170	9.5	601	110	9.5	296	220	8.0	837
(3) Ionian Sea	270	11.0	1020	110	8.5	318	140	7.5	798
(4) Adriatic Sea	-	-	-	-	-	-	170	7.0	690
(5) North Adriatic Sea	-	-	-	-	-	-	670	11.5	1843
(6) Strait of Sicily	140	6.0	772	-100	8.0	307	-	-	-
(7) Strait of Messina	220	8.0	907	-	-	-	-	-	-
(8) Strait of Otranto	180	7.5	776	130	8.5	364	160	7.5	585
Average from the stations	190	8.8	790	120	8.6	321	190[a]	8.3	728[a]
Average from the entire cruise		9.5			8.0			8.0	

[a] Average without the North Adriatic Sea

Source: Andersson et al., 2007

Table 11.17 Average wind speed and flux from the stations in the different parts of the Mediterranean Sea

Section	Summer		Spring		Autumn	
	Wind speed μ_{10} (*ms*$^{-1}$)	Hg flux (*pmol m*$^{-2}$ *h*$^{-1}$)	Wind speed μ_{10} (*m s*$^{-1}$)	Hg flux (*pmol m*$^{-2}$ *h*$^{-1}$)	Wind speed μ_{10} (*m s*$^{-1}$)	Hg flux (*pmol m*$^{-2}$ *h*$^{-1}$)
(1) Western Mediterranean	6.0	25.8	-	-	-	-
(2) Tyrrenian Sea	1.9	20.5	2.9	3.7	4.0	15.8
(3) Ionian Sea	3.3	13.8	4.6	5.0	5.2	18.7
(4) Adriatic Sea	-	-	-	-	6.8	26.9
(5) North Adriatic Sea	-	-	-	-	4.8	94.7
(6) Strait of Sicily	5.6	17.5	3.8	3.5	-	-
(7) Strait of Messina	9.1	62.0	-	-	-	-
(8) Strait of Otranto	4.7	16.5	3.1	4.0	4.6	10.5
Average from the stations	5.1	26.0	3.6	4.1	5.1	18.0[a]
Average from the entire cruise	5.9	22.3	5.8	7.6	7.0	24.6

[a]Average without the North Adriatic Sea

Source: Andersson et al., (2007).

by differences in water temperature, but also by the clear diurnal variation observed in the DGM concentrations measured during the summer cruise. However, the variation in water temperature between different seasons is probably the dominant factor for the variation in average saturation, since the night DGM concentrations between the different seasons did not vary substantially.

In Table 11.17, the wind speed and the flux, calculated using the formula of Nightingale et al. (2000), are tabulated. The wind speed differed between the different sections and between the different stations and the flux also varied. In order to obtain a good estimate of the flux from the Mediterranean Sea, average concentrations of DGM and TGM were used, and, for each season, the average wind speeds from the entire cruise were used. Taking these assumptions into account the estimated summer flux would be 22.3 pmol m^{-2} h^{-1} which is comparable to the value estimated by Gårdfeldt et al. (2003) (24.3 pmol m^{-2} h^{-1}), calculated using the gas exchange model developed by Wanninkhof (1992). Ferrara et al. (2000) determined comparable fluxes in the Tyrrhenian Sea for the summer season using a flux chamber (16.1 pmol m^{-2} h^{-1}). Estimated fluxes by Andersson et al. (2007) for the spring cruise were 7.6 pmol m^{-2} h^{-1} which are comparable to the value Ferrara et al. (2000) in the Tyrrhenian Sea at an unpolluted site in May 1998, 7.3 pmol m^{-2} h^{-1}. The estimated flux for the autumn was the highest of the measured seasons, 24.6 pmol m^{-2} h^{-1}, this value is twice as high as Ferrara et al. (2000), measured at the unpolluted site in the Tyrrhenian Sea (11.9 pmol m^{-2} h^{-1}). Cossa et al. (1997) estimated an average evasion from the Western Mediterranean of 5.7 pmol m^{-2} h^{-1} using the global MFMs model (Mason et al., 1994a).

In order to calculate the total evasion from the Mediterranean Sea, the winter season had to be estimated. This was done using a TGM concentration of 8.0 fM, a DGM concentration of 120 fM, and an estimated wind speed of 7 m s^{-1} and water temperature of 12 °C, giving an average flux of 12.2 pmol m^{-2} h^{-1} for this season

The total Hg^0 evasion from the Mediterranean Sea surface was calculated to be 77 Mg per year. Using the estimations of the total oceanic evasion by Mason and Sheu (2002) (2600 Mg per year) and Mason et al. (1994a) (2000 Mg per year) the Mediterranean Sea would account for approximately 3–4% of the total oceanic evasion. Taking into consideration that the Mediterranean Sea represents 0.8% of the oceanic surface, the evasion from the Mediterranean Sea is almost 4–5 times higher compared to the global average sea surface evasion. However, as pointed out by the authors, Mason and Sheu (2002), there are great uncertainties in the estimation of the global Hg emissions from sea surfaces.

11.3.4 Other Ocean Waters

11.3.4.1 Arctic Ocean

The greatest source of Hg to the Arctic environment has been suggested to be long range transport. In the winter about two-thirds of the Hg in the Arctic area originates from regions outside the area. The projected main source is emissions from

Asia as, based on models, 5 to 10 % of the total emissions from that region end up in the high Arctic. Another significant source to the Arctic, or more correctly to the Arctic Ocean, has been suggested to be river input. There are several large rivers on both the Russian and American sides of the Arctic Ocean, which may contribute to the Hg load into the Arctic Ocean (AMAP, 2005).

Most measurements in the Artic have focused on atmospheric Hg measurements, which are presented in Chapter 10. Measurements of total Hg and MeHg have been conducted in the Mackenzie area (Leitch et al. 2007) and around Ellesmere Island (St Louis et al., 2007). Moreover, measurements of total mercury have been carried out in Laventiya, the Gulf of Ob, the Laptev Sea, at the coast in the Pechora Sea and at coastal sites along the Northern Siberian coast (AMAP 2005 and references therein; Coquery et al. 1995). DGM measurements have been carried out by Sommar et al. (2007) in Kongsfjorden, Spitzbergen, Temme et al. (2005) from 60°N 0°E to around 85°N 0°E, St Louis et al. (2007) around Ellesmere Island and Andersson et al. (2008) covering large parts of the Arctic Ocean.

Sommar et al. (2007) reports measurements of DGM in Kongsfjorden over a period of 7 days in May 2002. Discrete samples were collected close to shore during daytime. A clear diurnal variation was observed with concentrations as high as 70 pg L^{-1} during day-time and concentrations as low as 12 pg L^{-1}. A close relation between the measured DGM concentration and solar radiation was observed. All waters measured were found to be supersaturated with respect to Hg, and evasion between 0.1 and 7 ng m^{-2} h^{-1} was calculated by the flux model developed by Wanninkhof (1992). It was suggested that this evasion was overestimated due to the fjord being partly ice covered, thus hindering the wave field.

As mentioned in section 3.1, Temme et al (2005) conducted measurements of DGM along the Norwegian coast (62°N) past Spitsbergen up to 85°N. The DGM concentration observed in the North Atlantic Ocean (62°N to 74°N) was between 11 and 14 pg L^{-1}, and increased by a factor of 3 to 35 pg L^{-1} north of 74°N and/or in ice-covered areas. The authors suggest that deposited Hg is transported into the water column where part of it accumulates due to the ice-cover.

St Louis et al. (2007) carried out measurements of DGM at two ice-covered locations off of Ellesmere Island. The average concentration was 129 ± 36 pg L^{-1}, which corresponds to super-saturation. The average flux was calculated according to Wanninkhof and McGillis (1999) to be 5.4 ± 1.2 ng m^{-2} hr^{-1}. The authors suggest that an even higher evasion would be expected during Hg depletions events, when more Hg is deposited into the Arctic Ocean and during ice-break-up and melt each spring.

Andersson et al. (2008) carried out continuous measurements of DGM along the west coast of Greenland, into the Canadian archipelago, along the Alaskan coast into Russia around the Wrangel Island and finally crossing the Arctic Ocean, from Barrow, Alaska across the North Pole to Spitsbergen. The overall average concentration measured in the water sampled was 45 ± 22 pg L^{-1}, however a wide range of concentrations were measured - from 5 pg L^{-1} in the Canadian archipelago to 134 pg L^{-1} north of Alaska. Several specific features were highlighted in the data set. For example, when passing the Mackenzie River delta the concentrations of DGM increased from 40 to 100 pg L^{-1}, indicating Hg was species being discharged into the Arctic Ocean. In the Straight of Bering the measured concentrations were in the same range as in the North Atlantic Ocean (14 pg L^{-1}). Enhanced DGM concentrations were observed close to the North Pole, which might be related to river water run-off. Measurements were carried out in both ice-covered and non-ice-covered areas, and the DGM concentration increased up to 80 % between non-ice-covered and ice-covered areas. During transit through ice-covered areas, enhanced TGM concentrations were observed, and it was speculated that the sea ice may act as a barrier for the evasion of Hg^0 from the sea surface, however when the ship broke the ice evasion became possible. Since most of the waters sampled were in ice-covered areas, the evasion of Hg^0 may be restricted. The authors, however, calculated according to Nightingale et al. (2000) the Hg^0 flux for the open waters. Both deposition (-1.6 ng m^{-2} h^{-1}) and evasion (98 ng m^{-2} h^{-1}) of Hg^0 was estimated.

11.3.4.2 North Sea

Offshore in the North Sea only a few studies have been carried out for Hg speciation (Coquery and Cossa 1995, Baeyens and Leermakers 1998). Both these studies include measurements of DGM in the surface water and estimation of the flux from the area. Coquery and Cossa (1995) carried out measurements within the North Sea, both offshore and at coastal stations. The average DGM concentration for the entire study was 52 ± 22 pg L^{-1}, however the concentrations measured at the offshore sites were in the range from undetectable to 70 pg L^{-1}. Most samples demonstrated that the water was super-saturated with respect to mercury, giving a net evasion of mercury from the sea surface of 0.9 - 1.8 ng m^{-2} h^{-1}. Baeyens and Leermakers (1998) carried out measurements of mercury at off shore station and costal stations. The concentration measured at the off shore station was 12 pg L^{-1}. This concentration corresponds to a net evasion to the atmosphere of 0.5 ng m^{-2} h^{-1}. The authors concluded that the evasion was in the same range as the deposition of mercury to this area.

11.3.4.3 Baltic Sea

Six expeditions have been carried out in the Baltic Sea. Wangberg et al. (2001b) conducted two expeditions in summer 1997 and late winter 1998, in the south Baltic Sea. During both expeditions similar DGM concentrations were observed; 17.6 and 17.4 pg L^{-1} respectively. The flux was calculated using the flux model developed by Wanninkhof (1992). Higher flux was observed during the summer, 1.6 ng m^{-2} h^{-1}, than the winter, 0.8 ng m^{-2} h^{-1} (see Table 11.18).

Kuss and Schneider (2007) carried out measurements in the south of the Baltic Sea during all seasons in 2006, with continuous measurements of 7 to 10 days each time. The lowest DGM concentrations were measured during winter, 10 - 17 pg L^{-1}, and autumn, 11 - 14 pg L^{-1}. In the summer, the highest DGM concentrations were observed, 19 - 32 pg L^{-1} and during the spring a DGM concentration range of 15 - 20 pg L^{-1} was measured. Fluxes were calculated using the flux model developed for the

Table 11.18 Measurements conducted in the Baltic Sea

Season	DGM (*pg* L^{-1})	Flux (*ng* $m^{-2}h^{-1}$)	Reference
Summer	18	1.6	Wangberg et al. (2001b)
Winter	18	0.8	Wangberg et al. (2001b)
Winter	10-17	-02-0.2	Kuss and Schneider (2007)
Spring	15-20	1.0-2.1	Kuss and Schneider (2007)
Summer	19-32	3.1-6.2	Kuss and Schneider (2007)
Autumn	11-14	0.8-2.1	Kuss and Schneider (2007)

Baltic Sea by Weiss et al. (2007). The lowest flux was observed during the winter season, when deposition of Hg was observed, -0.2 to 0.2 ng m^{-2} h^{-1}. During the spring and autumn expeditions similar fluxes were calculated, 1.0-2.1 and 0.8-2.1 ng m^{-2} h^{-1} respectively. The highest flux was calculated for the summer expedition, 3.1-6.2 ng m^{-2} h^{-1} (Table 18). The authors state that the annual evasion calculated for the Baltic Sea cannot be compensated by deposition of mercury.

11.4 Surface and Deep Sea Water Mercury Measurements

The chemical form of Hg in aquatic systems is strongly influenced by redox and pH conditions, as well as the concentrations of inorganic and organic complexing agents. The proportion of $Hg^{(II)}$ bound to humic substances is lower in seawater than in freshwater environments, due to chloride ion competition. In coastal environments, however, a substantial portion of traditionally defined dissolved Hg is composed of Hg bound to organic colloids (Mason et al., 1993; Leermakers et al., 1995; Stordal et al., 1996). The methylation of Hg in the marine environment is usually lower than in freshwater environments (Compeau and Bartha, 1987), which has generally been attributed to salinity effects and the presence of sulphide and chloride complexes. It has been demonstrated that uncharged $HgCl_2$ is more bio-available than the negatively charged $HgCl_4^{2-}$, because the negative charge may reduce the availability for methylating bacteria (Barkayet al., 1997).

Reducing conditions and high salinity promote demethylation processes of MeHg (Hines et al., 2000; 2001). Other sea-salts may also effect Hg speciation, i.e. bicarbonate has a negative influence on Hg methylation under aerobic and anaerobic conditions (Campeau and Bartha, 1987). MeHg is kinetically inert towards decomposition (Stumm and Morgan, 1996); however, it might be efficiently degraded by photochemical and/or microbial actions (Whalin et al., 2007). But, methylation in the ocean is not confined to low oxygen zones, which indicates that there must be additional mechanisms for methylation/demethylation processes. Seasonal variations of Hg speciation in oceanic waters are relatedto temperature, redox conditions, as well as to seasonal changes in productivity and the bio-availability of nutrients. Me_2Hg has been reported as the dominant methylated species in deep ocean waters (Mason and Fitzgerald, 1990; Cossa et al., 1994; Mason et al., 1995a, 1995b), where the governing process for its production is not very clear. In surface oceanic waters no Me_2Hg has been detected, this might be due to that Me_2Hg is readily lost from the aquatic surface by evaporation and/or photolytic degradation and it is therefore not considered to be available for accumulation in aquatic organisms (Morel et al., 1998). DGM consists mostly of Hg^0 with a small percentage of Me_2Hg in the surface water (Mason and Fitzgerald, 1991). The production of DGM is considered to be a function of abiotic and biotic processes. The potential ability of humic substances and the importance of solar radiation in the reduction of $Hg^{(II)}$ to Hg^0 in aquatic systems has been proven (Allard and Arsenie, 1991). One of the main mechanisms of Hg^0 production involves the $Hg^{(II)}$ complexing capacity of the dissolved organic matter (DOM) and the electron transfer from DOM to $Hg^{(II)}$ after absorbing a photon produced via solar radiation (Nriagu, 1994). Microorganisms and phytoplankton may also play a role in the reactions mechanisms of DGM production (Mason et al., 1995b). In the water column Hg^0 formation, is a consequence of the reduction of reactive forms of mercury, which in turns reduces the pool of $Hg^{(II)}$, leading to the species no longer being available for methylation.

11.4.1 Atlantic Ocean

Dalziel and Yeats (1985) carried out deep profile measurements of reactive Hg in the waters near the Mid Atlantic Ridge south of the Azores in July 1982, and in the Sargasso Sea in April 1983. The concentration profile for the station near the Mid Atlantic Ridge varied between 0.28 to 0.72 ng L^{-1} with a mean of 0.52 ± 0.12 ng L^{-1}. Measurements were also carried in bottom water, averaging at 0.58 ± 0.2 ng L^{-1}. In the Sargasso Sea concentrations in the same range were measured 0.26 to 0.7 ng L^{-1} with an average of 0.44 ± 0.12 ng L^{-1}. Dalziel (1992) carried out measurements of reactive Hg on the Scotian Shelf and in the northwest Atlantic Ocean. Concentrations ranged from 0.24 to 0.68 ng L^{-1} at the two stations on the Scotian Shelf. For both stations similar average values were observed, 0.46 ± 0.16 ng L^{-1} and 0.38 ± 0.12 ng L^{-1}. For the two offshore stations the concentration ranged from 0.18 to 0.76 ng L^{-1}, with averages of 0.48 ± 0.16 ng L^{-1} and 0.46 ± 0.16 ng L^{-1}. The measurements carried out by Dalziel and Yeats (1985) and Dalziel (1992) were treated with acid prior to the analysis, therefore these measurements are more likely to be comparable to the measurements carried out for total Hg in more recent papers. Gill and Fitzgerald (1988) carried out measurements of deep profiles at two locations in the Sargasso Sea. The total Hg concentration and the distribution at both stations were similar, with surface water concentration of 0.8 ng L^{-1} and a maximum of 2 ng L^{-1} just below the main thermocline, and lower concentrations of Hg were detected in the underlying water.

Cossa et al. (1992) carried out measurements at the ICES (International Council for the Exploration of the Sea) reference station in the eastern North Atlantic; one deep profile was carried out at 14 depths. The total Hg concentration ranged from 0.08 to 2 ng L^{-1}. The authors suggested that the shape of the profile, with a broad peak between 150 and 1500m, could be a result of remobilisation of Hg during mineralization of organic material combined with scavenging in deeper waters.

Cossa et al. (1994) carried out measurements of Me_2Hg at three stations in the Alboran Sea and the Strait of Gibraltar in February 1992. The concentrations ranged from the detection limit (6 pg L^{-1}) up to 58 pg L^{-1}. the highest concentration was detected in the deep water at the station in the eastern part of the Alboran Sea, the concentration than decreased towards the west, with the lowest concentrations observed in the Atlantic Ocean.

Mason et al. (1995) presented the preliminary results from an expedition in the North Atlantic Ocean conducted in August 1993; further details were then described in Mason et al. (1998). During the expedition several mercury species were measured, i.e. total Hg, reactive Hg, DGM, Me_2Hg and MeHg. The reactive mercury concentration ranged from the detection limit (70 pg L^{-1}) to 410 pg L^{-1} (uncorrected for DGM). All the profiles for reactive Hg were consistent with each other, the authors suggested that there were measurable differences between different water masses and that each had a signature of their source region and/or any mercury input due to mineralization. The fraction of DGM in the reactive mercury changed with depth, with the highest fraction observed in the surface water and then decreasing downwards. However, the concentration of DGM was similar throughout the water column. Me_2Hg increased downwards, with almost no detectable Me_2Hg in the surface waters. This was explained, by the authors, as Me_2Hg being unstable in the warm surface water, but building up to steady concentrations with time in the deeper waters. The MeHg concentrations observed were low, around the detection limit (0.1 ng L^{-1}) in the surface waters and then increased in the intermediate and deep waters, varying from 0.1 to 0.36 ng L^{-1}. The average total mercury concentration was 0.48 ± 0.32 ng L^{-1}, there was a concentration difference of less then 2 for the measured average concentration between different locations. Mason et al. (1995) suggest that there was a trend of higher total mercury in the "younger water".

Mason and Sullivan (1999) carried out deep profile measurements in the South and Equatorial Atlantic Ocean, for Me_2Hg, MeHg, DGM and total Hg. The Me_2Hg for the top 1500 m could be correlated to the apparent oxygen utilization implying that microbial processes were the primary factors controlling its formation. The concentration ranged from 2 to 22 pg L^{-1} at the stations measured. For MeHg in the deep water the observed concentration were at the detection limit (10 pg L^{-1}) or lower. The total mercury concentration was similar through out the water column and the different stations had average concentrations ranging from 0.16 to 0.48 ng L^{-1}. The authors concluded that the mercury in the deep waters of the South Atlantic Hg primary consists of DGM with a small fraction present as methylated species.

11.4.2 Pacific Ocean

Measurements of reactive Hg were made in the North Pacific on various cruises and at various locations in the early 1980's by Fitzgerald's research group (Gill and Fitzgerald, 1987; 1988). No strong offshore gradients were found and the authors concluded that riverine inputs were not important sources, as more recent data, mass balances and modeling have confirmed (Mason et al., 1994; Sunderland and Mason, 2007). These early studies set the baseline for the concentration and distribution of Hg in these waters and were mostly of the same order, but lower, than those of the Atlantic Ocean (Fitzgerald. 1989). Most measurements prior to 1980 must be discarded due to contamination either during sampling, storage or analysis. The earlier studies have been summarized in a number of reviews and papers and will not be detailed here. Overall, concentrations range from 0.64 ± 0.26 pM for the North Pacific to somewhat higher values in the equatorial regions (1 - 2 pM) and 1.2 ± 0.3 pM for the deep Pacific (Mason and Fitzgerald, 1990; 1993; Gill and Fitzgerald, 1988; Laurier et al., 2004). These measurements reflect the different sources of Hg to the various Pacific Ocean regions.

The earliest measurements of DGM were made on these cruises, and mostly in the equatorial Pacific Ocean (Kim and Fitzgerald, 1986; 1988; Fitzgerald et al., 1984). Studies focused on the equatorial Pacific Ocean in the late 1980's/early 90's as it was found that elevated concentrations of methylated Hg species, principally Me_2Hg, but also MeHg, were present in the suboxic low oxygen waters of the equatorial Pacific (Kim and Fitzgerald, 1986; Mason et al., 1990; 1991). Both MeHg and Me_2Hg were identified in these waters and it was shown that the highest concentrations were found in the sub-thermocline, sub-oxic waters of the equatorial region, and the concentration and distribution reflected to some degree surface ocean productivity and the apparent oxygen utilization in these sub-thermocline waters – higher methylated Hg concentrations in the most oxygen depleted waters (Mason and Fitzgerald, 1990; 1993). Similar distributions of the methylated Hg species were found during a cruise in the South and equatorial Atlantic Ocean, suggesting formation by similar mechanisms (Mason and Sullivan, 1999).

Recently, there was an assessment to compare the concentration and distribution of Hg in North and central Pacific Ocean over time through the examination of the data collected during a cruise in 2002 with measurements of 20 years earlier (Laurier et al., 2004; Figure 11.14). Vertical Hg distributions, for 3 different cruises (N. Pacific (1980), VERTEX (1986-87) and IOC (2002)) show similar values (Laurier et al., 2004 and references therein). During the 2002 cruise, total Hg averaged 1.15 ± 0.86 pM with the highest concentrations found in Japanese coastal waters. The overall upper-water Hg concentration average for this study (0.64 ± 0.26 pM) was similar to the earlier VERTEX cruise (0.58 ± 0.37 pM) but was lower compared to the North Pacific cruise (1.40 ± 0.34). Variance in Hg concentrations in the upper water, generally within or close to the main thermocline, was observed among several stations, and for both VERTEX and IOC campaigns. Factors that can account for the observed distribution of Hg in the upper water include horizontal advection of water along isopycnals, vertical mixing, ventilation, the presence of a thermocline and the associated remobilization of Hg as a result of remineralization of settling particles, and the influence of diagenetic processes from continental margin sediments and seasonal stratification of the euphotic zone. Seasonal impacts on the upper water column concentration are shown by data collected over a season at one of the VERTEX sites (Figure 11.15).

The higher summer concentrations are likely reflective of inputs from rainfall during this period coupled to the strong seasonal thermocline in the region (Laurier

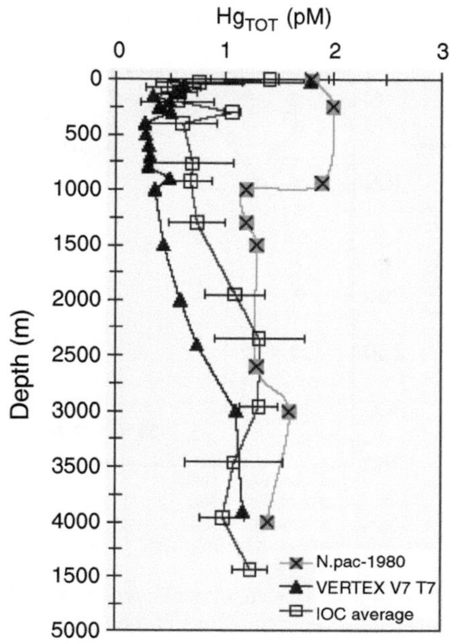

Figure 11.14 Concentrations of total mercury (reactive mercury for the earlier studies) measured on the three cruises discussed in the text: the 1980 North Pacific cruise, the VERTEX cruise data and the 2002 IOC cruise. Taken from Laurier et al. (2004)

et al., 2004). Winter overturning and deep water mixing due to high winter winds and storms erases this temporary upper ocean signal. Thus, comparison between collections must take such variability into account and multiple measurements are required to ascertain a clear signal in changing concentration over time. The deep and bottom waters of the North Pacific Ocean averaged 1.10 ± 0.31 pM and were characterized by comparable Hg concentrations among the different cruises. These data reinforce the homogeneous characteristics of these water masses, and suggest that their concentration has changed little in the last 20 years. In addition, a comparison between deep water in the Atlantic and the Pacific shows that Hg concentrations in the North Pacific are three to six times lower than in the deep Atlantic (Laurier et al., 2004). Even for the upper waters, there is little clear evidence for a substantial increase in Hg concentration over time.

11.4.3 Mediterranean Sea

The circulation and hydrography of the Mediterranean Sea waters are driven by the net fresh water loss and heat loss to the atmosphere and the exchange of salinity and heat through the Strait of Gibraltar. The loss by evaporation exceeds the input by precipitation, the river runoff and Black Sea exchange. The Mediterranean Sea

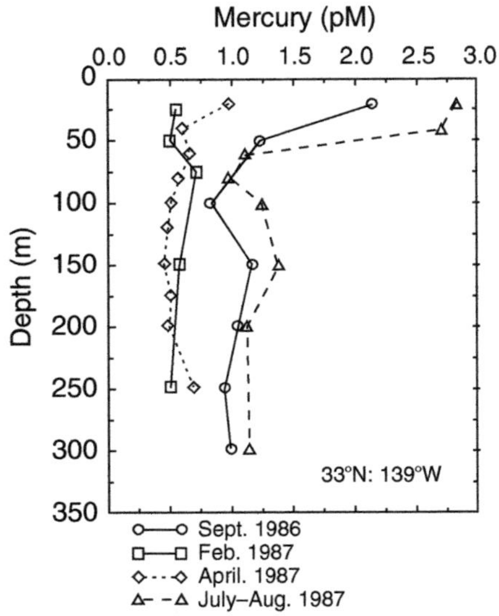

Figure 11.15 Concentrations of reactive mercury measured at the VERTEX site over a season in 1986/87 showing the changes in the concentrations in the upper waters over time. Taken from Laurier et al. (2004)

water circulation has been the subject of numerous studies and is relatively well understood and modeled (e.g. Katz, 1972; Zavatarelli and Mellor, 1995; Gvirtzman et al., 2001; Horvat et al., 2003).

An interesting feature of Hg biogeochemistry in the Mediterranean Sea is that in several fish species higher concentrations of Hg in their tissues have been observed, compared to the same species in the Atlantic Ocean (FAO, 1986), although the concentrations of Hg in the open waters of both oceans are in the same range (Cossa et al., 1997; Mason and Sullivan, 1999; Mason et al., 1998, 2001). It has been suggested that the higher Hg levels noted in many larger pelagic fish species in the Mediterranean Sea are not related to anthropogenic inputs, but rather due to the higher average of natural environmental levels of the metal, originating from the Mediterranean Hg anomaly. Although this was an interesting hypothesis, it conflicts with certain facts relating to the levels and distribution of Hg in the Mediterranean Sea. Elevated Hg levels have been noted in environmental matrices, from the Mediterranean regions, adjacent to known mercury anomalies. The data do not clearly indicate that the effects of these anomalies have been transmitted to open waters or to lower trophic level species living in these waters. Cossa and Martin (1991) presented a vertical profile of Hg concentrations for the Gulf of Lyons waters, which is similar to the Atlantic Ocean profiles (Cossa et al., 1992; Mason et al., 1995a, 1995b). Cossa et al., (1997) observed the distribution and cycling of Hg in the Western Mediterranean Sea and calculated a mass balance. While the total Hg exchanges at the straits are not balanced, Hg enters the western

Mediterranean as inorganic Hg and is exported to the Atlantic partially as methylated species. At the Strait of Sicily the Hg budget is balanced (Cossa et al., 1997).

A summary of data recently obtained for Hg speciation (reactive Hg, total Hg, MeHg), dissolved gaseous mercury (DGM) and Me_2Hg in open and coastal waters of the Mediterranean Sea was compiled by Horvat et al. (2003). The majority of the results were obtained during an oceanographic cruise aboard the research vessel *Urania* from July 14 to August 9, 2000, as part of the MED-OCEANOR Project, funded by the National Research Council of Italy.

The results are compared to the results obtained in contaminated coastal environments of the Adriatic Sea (The Gulf of Trieste and Kastela Bay) and uncontaminated coastal waters of the eastern Adriatic coast in 1998. A summary of the results for Hg and its species, temperature, salinity and oxygen in surface water samples during the *Urania* research cruise is given in Table 11.19. In addition, Tables 11.20 and 11.21 summarize data for the eastern Adriatic Sea and other oceanic waters. In the Hg contaminated coastal environment of the Gulf of Trieste, total Hg and total MeHg values were much higher than at other stations within the Adriatic Sea, due to the influence of the drainage area of the Hg mine, at Idrija in Slovenia, and due to the chloralkali industry in the area. During the study carried out in 1995 by Cossa et al. (1997) in the western part of the Mediterranean Sea values observed for total Hg were, on average, significantly higher than those observed by Horvat et al., 2003. A direct comparison between the mentioned studies is not really possible, because of differences in sampling time and location. The only stations that could be compared directly were the ones in the Strait of Sicily. Cossa et al. (1997) observed concentrations of total Hg in the range between 1.2 and 2.5 pM as compared to 0.42 and 0.30 pM found by Horvat et al. (2003). Westward from the isles of Sardinia and Corsica, Cossa et al. (1997) reported a value of 3.4 pM compared to 0.27–0.30pM observed by Horvat (2003). One possible explanation for the differences in the data sets obtained from the studies above, could be due to differences in the sampling period; Horvat et al., (2003) performed measurements during the summer season characterized by low productivity and low precipitation; this results in a loss of Hg from warm surface waters either due to evaporation and/or scavenging of Hg bound to particulate matter. The concentration of total Hg observed by Horvat et al., (2003) were lower than the observations made in the Atlantic and Pacific Oceans (Mason et al., 1998; Mason and Sullivan, 1999; Mason and Fitzgerald, 1990, 1993). Reactive Hg data observed by Horvat et al. (2003) for the western part of the Mediterranean Sea were higher than those obtained for the same area by Cossa et al. (1997), but were in good agreement with data obtained for the Atlantic Ocean (Mason et al., 1998; Dalziel, 1995) and higher values were obtained in the South and equatorial Pacific (Mason and Sullivan, 1999). Hg depth profiles were also performed by Horvat et al., (2003) during the MED-OCEANOR cruise campaign in summer 2000 in the eastern Mediterranean to a maximum depth of about 2500 m and in the western part. In the western part a constant increase of DGM with depth until about 1000 m was observed and then the concentration remained at approximately the same level. Significantly higher values of DGM were obtained with depth in the western part (up 0.44 pM) as compared to the

Table 11.19 Concentrations of mercury in surface and deep water of the Mediterranean Sea (expressed in pM) Source: Horvat et al., (2003)

Location	Depth *(m)*	Position	Day time of the sampling	Water Temp *(°C)*	Salinity *(psu)*	Oxygen *(µM)*	Reactive *Hg (pM)*	Total Hg *(pM)*	Dissolved Hg *(pM)*	Total MeHg *(pM)*	Dissolved MeHg *(pM)*	DGM *(pM)*	Me_2Hg *(fM)*
(a) In the eastern basin of the Mediterranean Sea													
A	0	36°47'N18°37'E	13:30	25.016	38.420	197	0.91	1.97	1.77	0.24	0.15	0.14	<LOD
B	0	36°11'N22°04'E	10:00	23.484	38814	204	0.82	1.34	0.89	0.20	0.15	0.29	n.d.
C	0	34°05'N24°32'E	6:00	23.100	38.976	207	1.14	1.35	0.60	0.19	0.16	0.07	n.d.
D	0	34°44'N28°03'E	3:00	23.600	39.238	203	0.46	1.07	0.81	0.33	0.16	0.14	<LOD
E	0	32°40'N28°00'E	18:00	25.596	39.191	190	0.52	1.53	1.16	0.19	0.16	0.13	<LOD
F	0	32°47'N24°24'E	13:30	24.856	38.788	192	0.51	1.55	0.50	0.26	0.15	0.32	<LOD
G	0	32°18N20°54'E	10:00	25.609	38.584	191	0.48	2.33	1.08	0.27	0.16	0.21	<LOD
H	0	33°53'N17°30'E	10:00	26.417	38.656	189	0.68	1.55	1.32	0.26	0.19	0.24	<LOD
H	20	33°53'N17°30'E	10:00	26.396	38.652	190	0.68	1.50	contaminant	0.26	0.20	0.12	1.05
H	100	33°53'N17°30'E	10:00	15.092	38.573	233	0.34	1.55	0.86	0.26	0.20	0.17	1.05
H	500	33°53'N17°30'E	10:00	13.786	38.772	184	0.34	1.00	0.83	0.32	0.15	0.20	1.05
H	1000	33°53'N17°30'E	10:00	13.612	38.718	185	0.39	2.64	0.98	0.27	0.15	0.18	1.85
H	2000	33°53'N17°30'E	10:00	13.737	38.726	192	-	-	-	-	-	-	--
H	2500	33°53'N17°30'E	10:00	13.371	38.717	194	0.39	1.28	1.02	0.33	0.15	0.28	1.05
I	0	35°05'N14°16'E	23:30	26.661	38.469	184	0.66	2.08	1.98	0.25	0.16	0.16	n.d.
I	4000	35°05'N14°16'E	23:30	13.266	38.689	194	0.53	1.39	1.32	0.27	0.17	0.24	1.6
L	0	37°16'N11°54'E	0:30	25.167	37.629	199	0.75	1.48	1.20	0.32	0.20	0.21	<LOD
M	0	37°16'N11°52'E	14:30	24.006	37.329	202	0.66	2.12	1.15	0.27	0.19	0.52	<LOD
(b) In the western basin of the Mediterranean Sea													
14	0	39°40'N13°20'E	5:00	25.722	38.233	196.15	1.11	2.07	1.89	0.26	0.19	0.16	<LOD
15	0	39°05'N11°50'E	17:00	25.415	38.253	195.75	0.84	5.55[a]	1.94	0.30	0.16	0.21	<LOD
16	0	38°35'N10°25'E	8:00	23.548	38.078	205.64	0.63	1.31	0.91	0.31	0.20	0.08	<LOD
16A	0	38°15'N08°50'E	14:00	23.904	38.328		1.05	1.12	0.58	0.35	0.21	0.18	n.d.
18	0	38°20'N07°15'E	8:30	24.327	37.908		0.66	1.23	1.17	0.39	0.26	0.07	<LOD

19	0	39°25'N06°05'E	18:30	25.526	37.332	196.66	0.87	1.35	0.88	0.31	0.20	0.07	<LOD
19	20	39°25'N06°05'E	18:30	23.478	37.512		0.79	1.54	1.05	0.30	0.34	0.12	1.00
19	100	39°25'N06°05'E	18:30	14.649	37.696	231.38	0.35	1.34	0.84	0.28	0.15	0.24	2.40
19	500	39°25'N06°05'E	18:30	13.391	38.560	171.25	0.52	1.65	1.31	0.28	0.18	0.44	4.50
19	1000	39°25'N06°05'E	18:30	13.208	38.518	185.00	0.57	1.72	1.20	0.33	0.21	0.33	11.35
19	2780	39°25'N06°05'E	18:30	13.252	38.455	198.14	0.66	1.93	1.67	0.29	0.20	0.40	7.65
20	0	40°51'N06°15'E	8:00	24.241	37.617	201.02	0.74	1.36	0.83	0.27	0.19	0.06	n.d.
21	0	41°25'N07°48'E	17:30	23.054	37.952		0.46	1.49	1.10	0.27	0.20	0.09	n.d.
22	0	41°10'N08°30'E	21:00	24.166	38.239		1.05	1.14	0.75	0.30	0.21	0.06	1.00
23	0	41°21'N09°39'E	6:15	23.191	38.240		0.94	0.96	0.85	0.29	0.17	0.03	n.d.
24	0	41°40'N10°60'E	13:00	23.294	38.311		0.68	0.82	0.59	0.32	0.21	0.02	<LOD
25	0	42°03'N11°42'E	16:45	23.889	38.285		0.98	1.04	0.98	0.32	0.22	0.03	<LOD

n.d.: not determined; <LOD: below the limit of detection.

[a] Possible contamination.

Table 11.20 Summary table for mercury analysis and speciation in surface water samples during the Urania cruise (results are expressed in *pM* concentrations of Hg). Source: Horvat et al., (2003)

		Reactive Hg	Total Hg	Dissolved Hg	Total MeHg	Dissolved MeHg	DGM
Eastern Basin	Min. Value	0.46	1.07	0.50	0.19	0.15	0.07
(n = 11)	Max. Value	1.91	2.33	1.98	0.33	0.20	0.52
	Mean	0.78	1.67	1.13	0.25	0.17	0.22
	St. dev.	0.42	0.40	0.45	0.05	0.02	0.12
	Median	0.66	1.54	1.15	0.26	0.16	0.21
Western Basin	Min. Value	0.45	0.81	0.57	0.26	0.16	0.02
(n = 12)	Max. Value	1.11	2.06	1.93	0.39	0.26	0.21
	Mean	0.83	1.26	1.03	0.31	0.20	0.09
	St. dev.	0.20	0.33	0.44	0.04	0.02	0.06
	Median	0.85	1.22	0.89	0.30	0.20	0.07
All results	Min. Value	0.45	0.81	0.50	0.19	0.15	0.02
(n = 12)	Max. Value	1.91	2.33	1.98	0.39	0.26	0.52
	Mean	0.81	1.46	1.08	0.28	0.18	0.15
	St. dev.	0.32	0.41	0.44	0.05	0.03	0.11
	Median	0.74	1.35	0.97	0.27	0.19	0.14

Table 11.21 Comparison of results for mercury speciation in surface ocean waters (Results are expressed in pM concentrations of Hg). Source: Horvat et al., (2003)

Study area/reference	Total Hg	Reactive Hg	DGM (Hg^0)	MeHg (n.f.w.)	$MeHg_{diss}$
Mediterranean Sea 2000 surface water (Horvat et al., 2003)	1.46 ± 0.41 (0.81-2.33)	0.81 ± 0.32 (0.45-1.91)	0.15 ± 0.12 (0.020-0.518)	0.28 ± 0.05 (0.19-0.39)	0.19 ± 0.03 (0.15-0.26)
Western Mediterranean (Cossa et al., 1997)	2.54 ± 1.25 (0.8-6.4)	0.38 (0.02-0.97)	<LOD-39	<0.15	n.d.
North Atlantic (Mason et al., 1998)	2.4 ± 1.6	0.80±0.44	0.48 ± 0.31	1.04 ± 1.08	n.d.
South and Equatorial Atlantic (Mason and Sullivan, 1999)	2.9 ± 1.2	1.7±1.2	1.2 ± 0.8	<0.05	n.d.
North Atlantic and Antarctic bottom water	n.d	0.48-1.34	n.d.	n.d.	n.d.
Mediterranean outflow water (Dalziel, 1995)		0.67-1.25 0.86-1.06			
Equatorial Pacific (Mason and Fitzgerald, 1990,1993)	n.d.	0.47-1.85	0.04-0.325	<LOD-0.28	n.d.

n.d.: not determined; n.f.w.: non filtered water

eastern Mediterranean (up to 0.28 pM). The profiles for DGM and Me_2Hg were found to be similar, but differing in the concentration levels. The concentrations of Me_2Hg were very low in the profiles performed. In general, the values are of the same order of magnitude as found at some stations in the western Mediterranean

(Cossa et al., 1997), but are significantly lower compared to data reported for the Pacific and Atlantic Oceans (Mason and Fitzgerald, 1990, 1993).

The major sources of DGM in the Mediterranean may be due to biological processes in deep and surface waters, as well as due to tectonic activity at the sea bottom of the Mediterranean (Horvat et al., 2003). In surface waters photo reduction and photo oxidation are two important driving forces for the presence of net DGM concentrations (mainly Hg^0). The sharp decrease of DGM in the surface waters of the open Mediterranean may also be due to oxidation of Hg^0 due to a strong oxidizing environment at the surface (e.g. the presence of chlorine/bromine and UV light). This hypothesis is further supported by the presence of highly reactive Hg in surface waters. The profile of Hg distribution with depth is related to various processes. At the surface biological and photochemical transformation processes seem to be the most important driving forces, while at depth biologically mediated transformation mechanisms are possible, but in addition, other factors such as hydrodynamic movement of oceanic waters and their mixing as well as tectonic activities should be considered (Gvirtzman and Nur, 2001; Kahle and Mueller, 1998). The differences that are observed may also be explained by several factors, among which differences in biogeochemical behaviour and Hg source terms are the most important. One should also not forget the intensive tectonic characteristics of the seafloor, in particular, in the western leg of the *Urania* cruise, which maybe the major supply of DGM from the depths as evidenced by higher concentrations in deep waters. The decrease of DGM in the surface layer may be a result of photochemical oxidation of DGM diffusing from the depths.

Mercury speciation and its distribution in surface and deep waters of the Mediterranean Sea were also studied during two oceanographic cruises on board the Italian research vessel *Urania* in summer 2003 and spring 2004 as part of the MED-OCEANOR and MERCYMS projects. The study is presented in Kotnik et al., (2007) and includes deep water profiles of dissolved gaseous Hg (DGM), reactive Hg (RHg), total Hg (THg), monomethyl Hg (MeHg) and dimethyl Hg (Me_2Hg) in open ocean waters. The data obtained by Kotnik et al., 2007 complement the measurement campaigns performed in 1995 by Cossa et al. (1997), in 2000 by Horvat et al. (2003), and by Cossa and Coquery (2005) adding more data on Hg speciation profiles in deep and surface Mediterranean waters. The results obtained by Kotnik et al., 2007 were in good agreement with the results measured in a previous study performed in the Western and Eastern Mediterranean Sea (Horvat et al., 2003). A slight differences in the data could reflect the fact that sampling was performed during different seasons and at different sampling locations. Total Hg concentrations were relatively low during both cruises, with average values of 1.46 ± 0.62 and 1.21 ± 0.32 pM in summer and spring, respectively. Spatial and temporal distribution of reactive Hg in the Mediterranean Sea was dependent on the presence of volcanic and geotectonic activity such as demonstrated by the higher values found in the Strait of Sicily and in the Tyrrhenian Sea, and seasonal changes, such as demonstrated by high values during The biologically active spring period.

The data obtained for reactive Hg are in agreement with data reported by Cossa et al. (1997) for The Western Mediterranean and slightly lower than those reported by Horvat et al. (2003) for west and east Mediterranean and for the Atlantic Ocean (Mason et al., 1998; Dalziel, 1995). Much higher values were found for the south and equatorial Pacific Ocean (Mason and Sullivan, 1999). Since no significant dimethyl Hg (Me_2Hg) was found in surface water it was assumed that at the surface the value for DGM mostly represents the dissolved Hg^0 concentration, whereas towards the bottom a significant fraction of Me_2Hg can be found in the lower oxygen zone, representing up to 5% of DGM in deeper layers, as found in other studies (Cossa et al., 1997; Tessier et al., 2004).

In the summer 2003 cruise Me_2Hg was found in measurable concentrations at only two locations (south western Mediterranean and the Strait of Gibraltar). The relatively high concentrations and fractions of DGM and MeHg indicate high reactivity of Hg in open marine waters. The MeHg summer values obtained are in agreement with those reported by Horvat et al. (2003) for the Eastern and Western Mediterranean and by Mason and Fitzgerald (1990, 1993) for the Equatorial Pacific, but significantly higher than those reported by Cossa et al. (1997) for the western Mediterranean and those reported by Mason and Sullivan (1999) for the South and Equatorial Atlantic. In the vertical profile, the observed increase of MeHg towards the bottom could be the consequence of photochemical degradation and/or microbial action in surface waters and microbiologically mediated methylation in deeper waters. Particulate scavenging also removes MeHg from surface waters. In deeper waters particulate dissolution releases MeHg and inorganic Hg into solution (Cossa et al., 1997). However, there must be additional mechanisms for methylation and demethylation processes since methylation in the ocean is not confined to the low oxygen zone. Methylating and demethylating processes were also studied in the MERCYMS project and are presented in more detail elsewhere (Monperrous et al., 2007).

Vertical profiles also show that DGM generally increases with depth, suggesting the presence of a source of volatile Hg in deeper waters. One possible source of DGM in deeper and bottom waters could be the intensive tectonic activity of the seafloor, as indicated by higher concentrations and fractions of DGM near the bottom at locations with strong tectonic activity (Alboran Sea, Strait of Sicily, Tyrrhenian Sea, Ionian Sea). Rajar et al. (2007) calculated total natural underwater emission of Hg in the Mediterranean Sea to be 11 to 20 Mg y^{-1} with a mean estimate of 15 tons per year. In order to test this hypothesis measurements of radon were carried out (Vaupoti et al., 2007), confirming the emission of this gas at tectonically more active areas. A second possible source could be bacterial activity that might produce DGM (Ramamoorthy et al., 1983; Ferrara et al., 2003). From certain profiles the increase of DGM corresponds to a decrease in dissolved oxygen levels, suggesting that bacteria produce DGM in the oxygen minimum zone, confirming the findings of Kim and Fitzgerald (1988) for the tropical Pacific Ocean and Ferrara et al. (2003) for the Western Mediterranean. Loss of DGM in surface waters could be due to its emission to the atmosphere (Ferrara et al., 2000; Gardfeldt et al., 2003; Ferara et al., 2003) and/or oxidation of DGM due to strong UV irradiation (Hedgecock et al., 2005) and the presence of chlorine and/or bromine (Horvat et al., 2003) and photolytically produced hydroxyl radicals (Gårdfeldt et al., 2001b; Mason et al., 2001).

References

Allard, B., Arsenie, I., 1991. Abiotic reduction of mercury by humic substances in aquatic systems-an important process for the mercury cycle. Water, Air and Soil Pollution 56, 457–464.

AMAP, 2005. AMAP assessment 2002: Heavy metals in the Arctic. Arctic Monitoring and Assessment program (AMAP), Oslo, Norway. Xvi + 265 pp.

Amyot, M. et al. 1997. Environmental Science and Technology; 31: 3606–3611.

Amyot, M., Southworth, G., Lindberg, S., Hintelmann, H., Lalonde, J., Ogrinc, N., Poulain, A., Sandilands, K., 2004. Formation and evasion of dissolved gaseous mercury in large enclosures amended with $^{200}HgCl^2$. Atmospheric Environment 38, 4279–4289.

Andersson, M., Gardfeldt, K., Wangberg, I., Sprovieri, F., Pirrone, N., O. Lindqvist (2007) Seasonal and daily variation of mercury evasion at coastal and off-shore sites at the Mediterranean Sea. Marine Chemistry, Vol.104, pp.214–226.

Andersson, M. 2008, Transport of Mercury Species in the Environment, Exchange between Oceanic Waters and the Atmosphere. ISBN 978-91-628-7391–2.

Andersson, M.E., Sommar, J., Gardfeldt, K., Lindqvist, O., 2008, Enhanced concnetrations of dissolved gaseous mercury in the surface waters of the Arctic Ocean. In press, Marine Chemistry, doi:10.1016/j.marchem.2008.04.002.

Aspmo, K., Temme, C., berg, T., Ferrari, C., Gauchard, P.-A., Fain, X., Wibetoe, G. 2006. Mercury in the Atmosphere, Snow and Melt Water Ponds in the North Atlantic Ocean during Arctic Summer. Environmental Science and Technology, 40, 4083–4089

Baeyens, W., Leermakers, M., Dedeurwaerder, H., Lansens, P., 1991. Modelization of the mercury fluxes at the air–sea interface. Water, Air and Soil Pollution 56, 731–744.

Baeyens, W. and Leermakers, M., 1998. Elemental mercury concentrations and formation rates in the Scheldt estuary and the North Sea. Marine Chemistry, 60, 257–266.

Barkay, T., Gillman, M., Turner, R.R., 1997. Effects of dissolved organic carbon and salinity on bioavailability of mercury. Applied Environmental Microbiology 63, 4267–4271.

Capri, A., Lindberg, S., 1998. Application of a Teflon dynamic flux chamber for quantifying soil mercury fluxes: test and results over background soil. Atmospheric Environment 32, 873–882.

Compeau, G.C., Bartha, R., 1987. Effects of salinity on mercury methylating activity of sulphate-reducing bacteria in estuarine sediments. Applied Environmental Microbiology 53, 261–265.

Coquery, M., Cossa, D., 1995. Mercury speciation in surface waters of the North Sea. Netherlands Journal of Sea Research, 34, 245–257.

Coquery, M., Cossa, D., Martin, J.M., 1995. The distribution of dissolved and particulate mercury in three Siberian estuaries and adjacent Arctic coastal waters. Water, Air and Soil Pollution, 80, 653–664.

Cossa, D., Michel, P., Noel, J., Auges, D., 1992, Vertical Profile of total mercury in relation to arsenic, cadmium and copper distributions at the Easter North Atlantic ICES reference station. Oceanologica Acta 15, 603–308.

Cossa, D., Martin, J.-M., Sanjuan, J., 1994, Dimethylmercury formation in the Alboran Sea. Marine Pollution Bulletin, 28, 381–384.

Cossa, D., Martin, J.-M., Takayanagi, K., Sanjuan, J., 1997. The distribution and cycling of mercury species in the western Mediterranean. Deep-Sea Research II 44, 721–740.

Cossa, D., Coquery, M., 2005. The Mediterranean mercury anomaly, a geochemical or a biological issue. Hdb. Env. Chem., vol. 5. Springer-Verlag, Heidelberg.

Costa, M., Liss, P., 1999. Photoreduction of mercury in sea water and its possible implications for Hg0 air–sea fluxes. Marine Chemistry 68, 87–95.

Dalziel, J.A., Yeats, P., 1985, Reactive mercury in the central North Atlantic. Marine Chemistry, 15, 357–361.

Dalziel, J.A., 1992, Reactive mercury on the Schotian Shelf and in adjacent northwest Atlantic Ocean. Marine Chemistry, 37, 171–178.

Dalziel, J.A., 1995. Reactive mercuryin the eastern North Atlantic and southern Atlantic. Marine Chemistry 49, 307–314.

Dvonch, J.T., Graney, J.R., Marsik, F.J., Keeler, G.J., Stevens, R.K., 1999. An investigation of source–receptor relationships for mercury in south Florida using event precipitation data. Science of the Total Environment 213, 95–108.
Ebinghaus, R., Jennings, S.G., Schroeder, W.H., Berg, T., Donaghy, T., Guentzel, J., Kenny, C., Kock, H.H., Kvietkus, K., Landing, W., Munthe, J., Prestbo, E.M., Schneeberger, D., Slemr, F., Sommar, J., Urba, A., Wallschlager, D., Xiao, Z., 1999. International field intercomparison measurements of atmospheric mercury species at Mace Head, Ireland. Atmospheric Environment 33, 3063–3073.
Ebinghaus, R., Tripathi, R.M., Wallschlager, D., Lindberg, S.E., 1999b. Natural and anthropogenic mercury sources and their impact on the air-surface exchange of mercury on regional and global scales. In:Ebinghaus, R., Turner, R.R., de Lacerda, L.D., Vasiliev, O. (Eds.),Mercury Contaminated Sites. Springer, Berlin, pp.1–50.
Ebinghaus R., Temme C. H., Kock H. H., Loewe A., and Schmolke S. R. (2000). Determination of surface level concentrations of atmospheric mercury in Antarctica. Proceedings of International Conference on Heavy Metals in the Environment, Ann Arbor, MI, August 2000.
Ebinghaus, R., Kock, H.H., Coggins, A.M., Spain, T.G., Jennings, S.G., Temme, Ch., 2002b. Long-term measurements of atmospheric mercury at Mace Head, Irish west coast, between 1995 and 2001. Atmospheric Environment 36, 5267–5276.
Engstrom, D. R. et al. 1994. In: Environmental Chemistry of Lakes and Reservoirs. L. A. Baker (ed.). American Chemical Society. pp. 33–66.
Expert Panel on Hg Atmospheric Processes. 1994. Electric Power Research Institute Report No. TR-104214.
FAO, 1986. FAO Fisheries Report No. 325, Supplement, FIPL/R325, Meeting on the Biogeochemical Cycle of Mercury in the Mediterranean, Sienna, Italy, 1984. FAO, Rome, 1986.
Ferrara, R., Mazzolai, B. (1998) A dynamic flux chamber to measure mercury emission from aquatic systems. Sci Total Environ, 215: 5–57.
Ferrara, R., Mazzolai, B., Lanzillotta, E., Nucaro, E., Pirrone, N. (2000) Temporal trends in gaseous mercury evasion from the Mediterranean seawaters. Sci Total Environ, 259: 183–190.
Ferrara, R., Lanzillotta, E., Ceccarini, C. (2001) Dissolved gaseous mercury concentration and mercury evasional flux from seawater in front of a chlor-alkali plant. Environ. Technol, 22: 971–978.
Ferrara, R., Ceccarini, C., Lanzillotta, E., Gardfeldt, K., Sommar, J., Horvat, M., Logar, M., Fajon, V., Kotnik, J., 2003. Profiles of dissolved gaseous mercury concentration in Mediterranean Sea water. Atmospheric Environment 37 (S1), S85–S92.
Fitzgerald, W.F., Gill, G.A. and Kim, J.P. (1984) An equatorial Pacific Ocean source of atmospheric mercury. Science 224, 591–599.
Fitzgerald, W.F., 1989. Atmospheric and oceanic cycling of mercury. In:Riley, J.P., Chester, R. (Eds.),Chemical Oceanography, vol. 10. Academic Press, London, pp.151–186.
Fitzgerald, W.F., 1995. Is mercury increasing in the atmosphere? The need for an atmospheric mercury network (AMNET). Water, Air and Soil Pollution 80, 245–254.
Fitzgerald, W.F., Mason, R.P., 1997. Biogeochemical cycling of mercury in the marine environment. In: Sigel, A., Sigel, H. (Eds.), Mercury and its effects on environment and biology. Marcel Dekker, New York, pp.53–111.
Fitzgerald, W. F. et al. 1998. Environmental Science and Technology; 32: 1–7.
Gardfeldt, K., Feng, X., Sommar, J., Lindqvist, O., 2001a. Total gaseous mercury exchange between air and water over lake and sea surfaces. Atmospheric Environment 35, 3027–3038.
Gardfeldt, K., Sommar, J., Stromberg, D., Feng, X., 2001b. Oxidation of atomic mercury by hydroxyl radicals and photoinduced decomposition of MeHg species in the aqueous phase. Atmospheric Environment 35, 3039–3047.
Gardfeldt, K., Sommar, J., Ferrara, R., Ceccarini, C., Lanzillotta, E., Munthe, J., Wangberg, I., Lindqvist, O., Pirrone, N., Sprovieri, F., Pesenti, E. (2003) Evasion of mercury from coastal and open waters of the Atlantic Ocean and the Mediterranean Sea. Atmospheric Environment. Vol.37-S1, 73–84.

Gill, G.A., Fitzgerald, W.F., 1985, Mercury sampling of open ocean waters at the picomolar level. Deep-Sea Research, 32, 287–297.

Gill, G.A., Fitzgerald, W.F., 1988, Vertical mercury distribution in the Oceans. Geochimica et Cosmochimica Acta, 52, 1719–1728.

Gill, G.A., Fitzgerald, W.F., 1987, Mercury in the surface waters of the open ocean. Global Biogeochemical Cycles, 3, 199–212.

Guentzel, J.L., Landing, W.M., Gill, G.A., Pollman, C.D., 1995. Atmospheric deposition of mercury in Florida: the FAMS project (1992–1994). Water, Air and Soil Pollution 80, 393–402.

Guentzel, J.L., Landing, W.M., Gill, G.A., Pollman, C.D., 2001. Processes influencing rainfall deposition of mercury in Florida: the FAMS project (1992–1996). Environmental Science and Technology 35, 863–873.

Gvirtzman, Z., Nur, A., 2001. Residual topography, lithospheric structure and sunken slabs in the central Mediterranean. Earth and PlanetaryScience Letters 187, 117–130.

Hall, B. 1995. The phase oxidation of elemental mercury by ozone. Water, Air and Soil Pollution, 80, 301–315.

Hines, M.E., Horvat, M., Faganeli, J., Bonzongo, J.-C.J., Barkey, T., Major, E.B., Scott, K.J., Bailey, E.A., Warwick, J.J., Lyons, W.B., 2000. Mercury biogeochemistry in the Idrija River, Slovenia, from above the Mine into the Gulf of Trieste. Environmental Research 83, 129–139.

Hines, M.E., Horvat, M., Faganeli, J., 2001. MeHg formation and degradation in sediments of the Gulf of Trieste. RMZ—Materials and Geoenvironment 48, 157–164.

Hedgecock, I.M. andPirrrone, N. 2001. Mercury and photochemistry in the marine boundary layer-modelling studies suggest the in situ production of reactive gas phase mercury. Atmos. Env. 35, 3055–3062.

Hedgecock, I.M. and Pirrrone, N. (2004) Chasing Quicksilver: Modeling the Atmospheric Lifetime of $Hg^0(g)$ in the Marine Boundary Layer at Various Latitudes, Env. Sci. Technol., 38, 69–76.

Hedgecock, I.M. Tunfio, G.A. Pirrone N. and Spuovieni F. (2005) Mercury chemistry in the MBL: Mediterranean case and sensitivity studies using the AMCOTS (Atmospheric Mercury Chemistry over the Sea) model, Atmos. Environ., 39, 7217–7230.

Hedgecock, I.M., Pirrone, N.,Trunfio, G.A. and Sprovieri, F. (2006) *Integrated mercury cycling, transport, and air-water exchange (MECAWEx) model.* Journal of Geophysical Research, 111 (D20302), doi: 10.1029/2006JD007117.

Horvat, M., Bloom, N.S., Liang, L., 1993a. Comparison of distillation with other current isolation methods for the determination of methyl mercury compounds in lowlevel environmental samples. Part I: sediments. Analytica Chimica Acta 281, 135–152.

Horvat, M., Bloom, N.S., Liang, L., 1993b. Comparison of distillation with other current isolation methods for the determination of methyl mercury compounds in lowlevel environmental samples. Part II: water. Analytica Chimica Acta 282, 153–168.

Horvat, M., Liang, L., Azemard, S., Mandic, V., Coquery, M., Villeneuve, J.-P., 1997. Certification of total mercury and MeHg concentrations in Mussel Homogenate (Mytilus edulis) reference material. Fresenius Journal of Analytical Chemistry 358, 411–418.

Horvat, M., Kotnik, J., Fajon, V., Logar, M., Zvonaric, T., Pirrone, N., 2001. Speciation of mercury in waters of the Mediterranean Sea. Materials and Geoenvironment 48, 24–252.

Horvat, M., Kotnik, J., Fajon, V., Logar, M., Zvonaric, T., Pirrone, N. (2003) Speciation of Mercury in Surface and Deep-Sea waters in the Mediterranean Sea. Atmospheric Environment, Vol.37/S1 ,93–108.

Kahle, H.-G., Mueller, S., 1998. Structure and dynamics of the Eurasian–African/Arabian plate boundary system: objectives, tasks and resources of the WEGENER group. Journal of Geodynamics 25, 303–325.

Katz, E.J., 1972. The Levantine Intermediate water between the straight of Sicilyand the straight of Gibraltar. Deep-Sea Research II 19, 507–520.

Kotnik, J., Horvat, M., Tessier, E., Ogrinc, N., Monperrus, M., Amouroux, D., Fajon, V., Gibicar, D., Zizek, S., Horvat, N., Sprovieri, F., Pirrone, N. (2007). Mercury speciation in surface and deep waters of the Mediterranean Sea. Marine Chemistry Vol.107, 13–30.

Kim, J.P., Fitzgerald, W.F., 1986. Air portioning of mercury in the Tropical Pacific Ocean. Science 23, 1131–1133.

Kinder, T.H., Parrilla, G., 1987. Yes, some of the Mediterranean outflow does come from great depth. Journal of Geophysical Research 92, 2901–2906.

Kim, J., Fitzgerald, W., 1988. Gaseous mercury profiles in the tropical Pacific Ocean. Geophysical Research Letters 15 (1), 40–43.

Kim, K.H., Lindberg, S., 1995. Design and initial tests of dynamic enclosure chamber for measurements of vapour-phase mercury fluxes over soils. Water, Air and Soil Pollution 80, 1059–1068.

Kim, K.H., Lindberg, S.E., Meyers, T.P., 1995. Micrometeorological measurements of mercury vapour fluxes over background forest soils in eastern Tennessee. Atmospheric Environment 29, 267–282.

Kuss, J., Schneider, B. 2007. Variability of the Gaseous Elemental Mercury Sea–Air Flux of the Baltic Sea. Environmental Science and Technology, 41, 8018–8023

IARC, 1994. Monographs on the Evaluation of Carcinogenic Risk to Humans, Mercury and Mercury Compounds. Vol. 58. Lyon.

Iverfeldt, A., 1991a. Mercury in forest canopy throughfall water and its relation to atmospheric deposition. Water, Air, and Soil Pollution 56, 553–564.

Iverfeldt, A., 1991b. Occurrence and turnover of atmospheric mercury over the Nordic Countries. Water Air and Soil Pollution 56, 251–265.

Lamborg, C.H., Fitzgerald, W.F., Vandal, G.M., Rolfhus, K.R., 1995. Atmospheric mercury in northern Wisconsin: sources and species. Water, Air and Soil Pollution 80, 189–198.

Lamborg, C.H., Rolfhus, K.R., Fitzgerald, W.F., Kim, G., 1998. The atmospheric cycling and air–sea exchange of mercury species in the South and equatorial Atlantic. Deep-Sea Research II 46, 957–977.

Lamborg, C.H., Rolfhus, K.R., Fitzgerald, W.F., Kim, G., 1999. The atmospheric cycling and air-sea exchange of mercury species in the south and equatorial Atlantic Ocean. Deep Sea Research II, 46, 957–977.

Lamborg, C.H., Fitzgerald, W.F., Donnell, J. and Toagersen, T. (2002) A non-steady state compartmental model of global-scale mercury biogeochemistry with interhemispheric atmospheric gradients, Geochim. Cosmochim. Acta, 66, 1105–1118.

Landis, M.S., Stevens, R.K., Schaedlich, F., Prestbo, E.M., 2002. Development and characterisation of an annular denuder methodology for the measurement of divalent inorganic reactive gaseous mercury in ambient air. Environmental Science and Technology 36, 3000–3009.

Lanzillotta, E., Ferrara, R., 2001. Daily trend of dissolved gaseous mercury concentration in coastal seawater of the Mediterranean basin. Chemistry 45, 935–940.

Laurier, F.J.G., Mason, R.P., Whalin, L., Kato, S., 2003. Reactive gaseous mercury formation in the North Pacific Ocean's marine boundary layer: a potential role of halogen chemistry. Journal of Geophysical Research 108 (D17) (art. no. 4529).

Laurier, F.J.G., Mason, R.P., Gill, G.A., and Whalin, L. (2004), Mercury distributions in the North Pacific Ocean: 20 years of observations, Mar. Chem., 90, 3–19.

Laurier, F. and R.P. Mason. 2007. Mercury concentration and speciation in the coastal and open ocean boundary layer. JGR-Atmos. 112, D06302, doi: 10.1029/2006JD007320.

Leermakers, M., Meuleman, C., Baeyens, W., 1995. Mercury speciation in the Scheldt estuary. Water, Air and Soil Pollution 80, 641–652.

Leitch, D.R., Carrie, J., Lean, D., Macdonald, R.W., Stern, G.A., Wang, F., 2007. The delivery of mercury to the Beaufort Sea of the Arctic Ocean by the Mackenzie River. Science of the Total Environment, 373, 178–195.

Lindberg, S.E., Kim, K.-H., Meyers, T.P., Owens, J.G., 1995a. Micrometeorological gradient approach for quantifying air– surface exchange of mercury-vapour tests over contaminated soils. Environmental Science and Technology 29, 126–135.

Lindberg S. E., Brooks S., Lin C.-J., Meyers T., and Chambers L. (2000) The Barrow Arctic Mercury Study (BAMS): Recent measurements of the production of reactive gaseous mercury durino mercury depletion events at Point Barrow, Alaska. Proceedings of International Conference on Heavy Metals in the Environment, Ann Arbor, MI, August 2000.

Lindberg, S.E., Bullock, R., Ebinghaus, R., Engstrom, D., Feng, X., Fitzgerald, W., Pirrone, N., Presto, E. and Seigneur, C. (2007). A Synthesis of Progress and Uncertainties in Attributing the Sources of Mercury in Deposition. Ambio, Vol.36, 19–32.

Mason, R.P., Fitzgerald, W.F., 1990. Alkylmercury species in the equatorial Pacific. Nature 347, 457–459.

Mason, R.P., Fitzgerald, W.F., 1991. Mercury speciation of open ocean waters. Water, Air, and Soil Pollution 56, 779–789.

Mason, R.P., Fitzgerald, W.F., 1993. The distribution and biogeochemical cycling of mercury in the equatorial Pacific Ocean. Deep-Sea Research II 40 (9), 1897–1924.

Mason, R.P., Fitzgerald, W.F., Morel, M.M., 1994. The biological cycling of elemental mercury: anthropogenic influences. Geochimica et Cosmochimica Acta 58, 3191–3198.

Mason, R.P., Fitzgerald, W.F., Morel, F.M.M., 1994a. The biogeochemical cycling of elemental mercury: anthropogenic influences. Geochimica et Cosmochimica Acta 58 (15), 3191–3198.

Mason, R., O'Donnell, J., Fitzgerald, W., 1994b. Elemental mercury cycling within the mixed layer of the Equatorial Pacific Ocean. In:Watras, C., Huckabee, J. (Eds.), Pollution Integration and Synthesis. Lewis Publishers, Boca Raton, pp.83–97.

Mason, R.P. et al. 1995. Water, Air and Soil Pollution; 80: 775–787.

Mason, R.P., Morel, F.M.M., Hemond, H.F., 1995a. The role of microorganisms in elemental mercury formation in natural waters. Water, Air and Soil Pollution 80, 775–789.

Mason, R.P., Rolfhus, K.R., Fitzgerald, W.F., 1995b. Methylated and elemental mercury in the surface and deep ocean waters of the North Atlantic. Water, Air and Soil Pollution 80, 665–677.

Mason, R.P., Rolfhus, K.R., Fitzgerald, W.F., 1998. Mercury in the North Atlantic. Marine Chemistry, 61, 37–53.

Mason, R.P., Sullivan, A.R., 1999. The distribution and speciation of mercury in the South and Equatorial Atlantic. Deep Sea Research II, 46, 937–956.

Mason, R.P., Lawson, N.M., Sheu, G.-R., 2000. Annual and seasonal trends in mercury deposition in Maryland. Atmospheric Environment 34, 1691–1701.

Mason, R.P., Lawson, N.M., Sheu, G.-R. 2001. Mercury in the Atlantis Ocean: factors controlling air-sea exchange of mercury and its distribution in the upper waters. Deep-Sea Res. II. 48, 2829–2853.

Mason, R.P., Sheu, G.R., 2002. Role of the ocean in the global mercury cycle. Global and Biochemistry Cycles 16 (4) (art. no.-1093).

Mason, R.P., Gill, G.A., 2005. Mercury in the marine Environment, in: Parsons, M. B., Percival, J. B. (Eds.), Mineralogical Association of Canada Short Course 34, Mercury : Sources, Measurements, Cycles and Effects, Halifax, Chapter 10 pp 179–216.

Mierle, G. 1990. Environ. Toxicol. Chem.; 9: 843–851.

Monperrus, M., Tessier, E., Amouroux, D., Leynaert, A., Huonnic, P., Donard, O., 2007. Mercury methylation, demethylation and reduction rates in coastal and marine surface waters of the Mediterranean Sea. Marine Chemistry 107, 49–63.

Morel, F.M.M., Kraepiel, A.M.L., Amyot, M., 1998. The chemical cycle and bioaccumulation of mercury. Annual Review of Ecologyand Systematics 29, 543–566.

Munthe, J., 1991. The redox cycling of mercury in the atmosphere. Ph.D. Thesis, Department of Inorganic Chemistry, University of Goteborg, Sweden.

Munthe, J., Wangberg, I., Pirrone, N., Iverfeld, A., Ferrara, R., Ebinghaus, R., Feng, R., Gerdfelt, K., Keeler, G.J., Lanzillotta, E., Lindberg, S.E., Lu, J., Mamane, Y., Prestbo, E., Schmolke, S., Schroder, W.H., Sommar, J., Sprovieri, F., Stevens, R.K., Stratton, W., Tuncel, G., Urba, A. (2001) Intercomparison of Methods for Sampling and Analysis of Atmospheric Mercury Species. Atmospheric Environment, 35, 3007–3017.

Nightingale, P.D., Malin, G., Law, C.S., Watson, A.J., Liss, P.S., Liddicoat, M.I., Boutin, J. andUp-still-Goddard, R.C., 2000. In situ evaluation of air-sea gas exchange parameterization using novel conservative and volatile tracers. Global Biogeochemical Cycles, 14(1), 373–387.

NOAA, database, http://www.cdc.noaa.gov/cdc/data.ncep.reanalysis.derived.surface.html

Nriagu, J.O., Pacyna, J.M., 1988. Quantitative assessment of worldwide contamination of air, water, and soils by trace metals. Nature 333, 134–139.

Nriagu, J.O., 1989. A global assessment of natural sources of atmospheric trace metals. Nature 338, 47–49.

Nriagu, J.O. Mechanistic steps in the photoreduction of mercury in natural waters (1994) Sci Total Environ, 154: 1–8.

Pacyna, E.G., Pacyna, J.M., Pirrone, N. (2001). European emissions of atmospheric mercury from anthropogenic sources in 1995. Atmospheric Environment, 35, 2987–2996.

Petersen, G., Munthe, J., Pleijel, K., Bloxam, R., Vinod Kumar, A., 1998. A comprehensive eulerian modeling framework for airborne mercury species: development and testing of the tropospheric chemistry module (TCM). Atmospheric Environment 29, 829–843.

Pirrone, N., Keeler, G.J., Nriagu, J.O., 1996. Regional differences in worldwide emissions of mercury to the atmosphere. Atmospheric Environment 30, 2981–2987.

Pirrone, N., Allegrini, I., Keeler, G.J., Nriagu, J.O., Rossmann, R., Robbins, J.A., 1998. Historical atmospheric mercury emissions and depositions in North America compared to mercury accumulations in sedimentary records. Atmospheric Environment 32, 929–940.

Pirrone, N., Hedgecock, I., Forlano, L., 2000a. The role of the ambient aerosol in the atmospheric processing of semivolatile contaminants: a parameterised numerical model (GASPAR). Journal of Geophysical Research 105 (D8), 9773–9790.

Pirrone, N., Costa, P., Pacyna, J.M., Ferrara, R., 2001. Mercury emissions to the atmosphere from natural and anthropogenic sources in the Mediterranean Region. Atmospheric Environment 35, 2997–3006.

Pirrone, N., Ferrara, R., Hedgecock, I.M., Kallos, G., Mamane, Y., Munthe, J., Pacyna, J.M., Pytharoulis, I., Sprovieri, F., Voudouri, A., Wangberg, I. (2003) Dynamic Processes of Mercury Over the Mediterranean Region: results from the Mediterranean Atmospheric Mercury Cycle System (MAMCS) project. Atmospheric Environment .Vol.37-S1, 21–39.

Pirrone, N., Sprovieri, F., Hedgecock, I.M., Trunfio, A., Cinnirella, S. (2005) Dynamic Processes of Atmospheric Mercury in the Mediterranean Region. In: Dynamics of Mercury Pollution on Regional and Global Scales, N. Pirrone and K. Mahaffey (Editors), Springer Verlag Publishers, Norwell, MA, USA. Chapter 23, pp.541–579.

Pirrone, N. (2006) An integrated approach to assess the mercury cycling in the Mediterranean basin (MERCYMS). Final Technical Report + Annexes + CD-Rom. *Rapporto Tecnico CNR/IIA/2006/08*, Rende, Italy, pp. 377.

Poissant, L., Casimir, A., 1998. Water–air and soil–air exchange rate of total gaseous mercury measured at background sites. Atmospheric Environment 32, 883–893.

Poissant, L., Amyot, M., Pilote, M., Lean, D., 2000. Mercury water– air exchange over the Upper St. Lawrence River and Lake Ontario. Environmental Science and Technology 34, 3069–3078.

Prestbo, E., 1996. Mercury speciatin in the boundary layer and free troposphere advected to South Florida: phase I reconnaissance. Report submitted to Florida DEP, Tallahassee, FL.

Rajar, R., Žagar, D., Horvat, M., Cetina, M., 2007. Mass balance of mercury in the Mediterranean Sea. Mar. Chem. 107, 89–102.

Ramamoorthy, S., Cheng, T.C., Kushner, D.J., 1983. Mercury speciation in waters. Canadian Journal of Fisheries and Aquatic Sciences 40, 1795–1798.

Rasmussen, P.E. 1994. Environ. Sci. Technol.; 28: 2233–2241.

Rolfhus, K.R., Fitzgerald, W.F. (2001) The evasion and spatially temporal distribution of mercury species in Long Island Sound, CTNY. Geochim Cosmochim Acta, 65(3): 407–418.

Schroeder, W.H., Anlauf, K.G., Barrie, L.A., Lu, J.Y., Steffen, A., Schneeberger, D.R., Berg, T., 1998. Arctic springtime depletion of mercury. Nature 394, 331–332.

Schroeder, W.H., Munthe, J., 1998. Atmospheric mercury: An overview. Atmospheric Environment 29, 809–822.

Shia, R.L., Seigneur, C., Pai, P., Ko, M. andSze, N.D. 1999. Global simulation of atmospheric mercury concentrations and deposition fluxes. J. Geophys. Res., 104 (D19), 23760.

Slemr, F., Seiler, W., Schuster, G., 1981. Latitudinal distribution of mercury over the Atlantic Ocean. Journal of Geophysical Research 86, 1159–1166.

Slemr, F., Schuster, G., Seiler, W., 1985. Distribution, speciation, and budget of atmospheric mercury. Journal of Atmospheric Chemistry 3, 407–434.

Slemr, F., Langer, E., 1992. Increase in global atmospheric concentrations of Hg inferred from measurements over the Atlantic Ocean. Nature 355, 434–437.

Slemr, F., Junkermann, W., Schmidt, R.W.H., Sladkovic, R., 1995. Indication of change in global and regional trends of atmospheric mercury concentrations. Geophysical Research Letters 22, 2143–2146.

Slemr, F., Scheel, H.E., 1998. Trends in atmospheric mercury concentrations at the summit of the Wank mountain, Southern Germany. Atmospheric Environment 32, 845–853.

Slemr, F., Brunke, E.-G., Ebinghaus, R., Temme, Ch., Munthe, J., Wangberg, I., Schroeder, W., Steffen, A., Berg, T., 2003. Worldwide trends of atmospheric mercury since 1977. Geophyscial Research Letters, 30, NO.10, 1516, doi:10.1029/2003GL016954.

Sprovieri, F., Pirrone, N., Gärdfeldt, K. andSommar, J. (2003) Mercury speciation in the marine boundary layer along a 6000 km cruise path around the Mediterranean Sea, Atmos. Environ., 37, S6371.

Sprovieri, F. and Pirrone, N. (2008) Spatial and Temporal Distribution of Atmospheric Mercury Species over the Adriatic Sea. Environmental Fluid and Mechanics (in press).

Sommar, J., Wängberg, I., Berg, T., Gårdfeldt, K., Munthe, J., Richter, A., Urba, A., Wittrock, F., Schroeder, W.H., 2007. Circumpolar transport and air-surface exchange of atmospheric mercury at Ny-Ålesund (79° N), Svalbard, spring 2002. Atmospheric Chemistry and Physics, 7(1), 151–166.

Stordal, M.C., Gill, G.A., Wen, L.S., Santschi, P.H., 1996. Mercury phase speciation in the surface waters of three Texas estuaries: importance of colloidal forms. Limnology and Oceanography 41, 52–61.

St Louis, V., Hintelmann, H., Graydon, J., Kirk, J.L., Barker, J., Dimock, B., Sharp, M.J., Lehnherr, I., 2007. Methylated Mercury Species in Canadian High Arctic Marine Surface Waters and Snowpacks. Environmental Science and Technology, 41, 6433–6441.

Stumm, W., Morgan, J.J. (Eds.) (1996). Aquatic Chemistry - Chemical Equilibria and Rates in Natural Waters, 3rd Edition. WileyInterscience , New York, 1996 (Chapter 10).

Sunderland, E.M. and Mason, R. P. 2007 Human impacts on open ocean mercury concentrations. Global Biogeochemical Cycles 21, GB4022, doi:10.1029/2006GB002876.

Temme, C., F. Slemr, R. Ebinghaus, and J. W. Einax (2003), Distribution of mercury over the Atlantic Ocean in 1996 and 1999–2001, Atmos. Environ., 37, 1889–1897.

Temme, C., Bakau, J., Schneider, B., Aspmo, K., Fain, X., Ferrari, C., Gauchrd, P.-E., Ebinghaus, R. 2005. Air/water exchange of mercury in the North Atlantic Ocean during arctic summer. Extended Abstract for the XIII International Conference on Heavy Metals in the Environment, Rio de Janeiro, June 2005.

Tessier, E., Monperrus, M., Amouroux, D., Pinaly, H., De Wit, R., Leynaert, A., Donard, O.F.X., 2004. Mercury species distribution in the water column of the Mediterranean Sea during summer 2003. RMZ—Materials and Geoenvironment. 7th International Conference on Mercury as a Global Pollutant, Ljubljana, J, vol. 51, pp. 1408–1411.

US-EPA, 1997. Mercury Study Report to Congress. US Environmental Protection Agency, EPA-452/R-97-003.

Xun, L., Campbell, N.E.R., Rudd, J.W.M., 1987. Measurements of specific rates of net MeHg production in the water column and surface sediments of acidified and circumneutral lakes. Canadian Journal of Fisheries and Aquatic Science 44, 750–757.

Vaupoti , J., Kotnik, J., Repinc, U., Kobal, I., Horvat, M., Pirrone, N. 2007. Selected radionuclides and DGM in coastal and deep sea waters in the Mediterranean. Marine Chemistry, 107.

Wängberg, I., Munthe, J., Pirrone, N., Iverfeldt, Å., Bahlman, E., Costa, P., Ebinghaus, R., Feng, X., Ferrara, R., Gårdfeldt, K., Kock, H., Lanzillotta, E., Mamane, Y., Mas, F., Melamed, E., Osnat, Y., Prestbo, E., Sommar, J., Schmolke, S., Spain, G., Sprovieri, F., Tuncel, G. (2001) Atmospheric Mercury Distributions in North Europe and in the Mediterranean Region. Atmospheric Environment, 35, 3019–3025.

Wängberg, I., Schmolke, S., Schager, P., Munthe, J., Ebinghaus, R., Iverfeldt, Å., 2001a. Estimates of air–sea exchange of mercury in the Baltic Sea. Atmospheric Environment 35, 5477–5484.

Wangberg, I., Schmolke, S., Schager, P., Munthe, J., Ebinghaus, R. andIverfeldt, A., 2001b. Estimates of air-sea exchange of mercury in the Baltic Sea. Atmospheric Environment, 35, 5477–5484.

Wängberg, Munte, J., Amouroux, D., Andersson, M. E., Fajon, V., Ferrara, R., Gårdfeldt, K., Horvat, M., Mamane, Y., Melamed, E., Monperrus, M., Ogrinc, N., Yossef, O., Pirrone, N., Sommar, J., Sprovieri, F. (2008) Atmospheric Mercury at Mediterranean Coastal Stations. Environmental Fluid Mechanics. (in press).

Wanninkhof, R., 1992. Relationship between wind speed and gas exchange over the ocean. Journal of Geophysical Research, 97(C5), 7373–7382.

Wanninkhof, R. and McGillis, W.R., 1999. A cubic relationship between air-sea CO2 exchange and wind speed. Geophysical Research Letters, 26(13), 1889–18892.

Weiss, A., Kuss, J., Peters, G., Schneider, B. Evaluating transfer velocity–wind speed relationship using a long-term series of direct eddy correlation CO2 flux measurements. Journal of Marine Systems. 2007, 66 (Special Issue), 130–139.

Whalin, L.M. and R. P. Mason. 2006. A new method for the investigation of mercury redox chemistry in natural waters utilizing deflatable Teflon® bags and additions of isotopically labeled mercury. Anal. Chim. Acta, 558: 211–221.

Whalin, Lindsay; Kim, Eun-Hee; Mason, Robert (2007) Factors influencing the oxidation, reduction, methylation and demethylation of mercury species in coastal waters Marine Chemistry Volume:107, Issue:3, December 1, 2007, pp.278–294.

Xiao, Z., Munthe, J., Schroeder, W., Lindqvist, O., 1991. Vertical fluxes of volatile mercury over forest soil and lake surfaces in Sweden. Tellus 43B, 267–379.

Xiao, Z.F., Munthe, J., Stromberg, D., Lindqvist, O., 1994. Photochemical Behaviour of Inorganic Mercury Compounds in Aqueous Solution. Lewis Publishers, Boca Raton (Chapt. VI.6).

Xiao, Z.F., Strömberg, D., Lindqvist, O., 1995. Influence of humic substances on photolysis of divalent mercury in aqueous solution. Water, Air, and Soil Pollution 80, 789–798.

Zavatarelli, M., Mellor, G.L., 1995. A numerical study of the Mediterranean Sea circulation. Journal of Physical Oceanography 25, 1384–1414.

Chapter 12
Monitoring and Modeling the Fate of Mercury Species in Japan

Noriyuki Suzuki, Yasuyuki Shibata, and Koyo Ogasawara

Summary Japan has started a monitoring program that will provide background air monitoring data of mercury and other heavy metals to contribute to the understanding of their atmospheric long-range transport. For this purpose, Japan started a pilot project in 2007 to monitor levels of mercury and obtain information on the long range transport of mercury and other heavy metals in the Asia-Pacific region, and to develop monitoring methodologies and contribute to international efforts in ambient atmospheric monitoring. The project will also develop fate modelling methodology that helps understand the global cycling of mercury in the atmosphere and environment, by expanding the information from the monitoring outputs. Pilot monitoring is conducted at the "Cape Hedo Atmosphere and Aerosol Monitoring Station" operated by the National Institute for Environment Studies, in Okinawa. Mercury speciation such as gaseous elemental mercury (Hg^0), divalent reactive gaseous mercury (RGM), and particulate mercury (PM) are continuously measured with a Tekran mercury speciation system. Hazardous heavy metals in particles and in precipitation are also measured. Japan will continue the monitoring at Cape Hedo station to understand the background levels of mercury and other toxic substances and contribute to the better understanding of the current status of mercury in the environment by joining in the international efforts in ambient atmospheric monitoring. In addition to the monitoring program, fate analysis using modelling methodology is also studied in the project framework.

12.1 Introduction

During the high economic growth period in Japan, Minamata disease, caused by toxic organic mercury effluent, inflicted serious damage to human health and the environment. In addition, Minamata disease has imposed considerable damage for the whole community of the region. To never repeat such a tragedy, Japan continues to convey and share the experiences of Minamata disease both at home and abroad. Furthermore, by using the lessons learned, the Japanese government is making many efforts to prevent human health hazards from toxic chemicals in order to

N. Pirrone and R. Mason (eds.), *Mercury Fate and Transport in the Global Atmosphere*,
DOI: 10.1007/978-0-387-93958-2_12,

establish a society in which people can safely live and protect the environment. From this point of view, Japan has made considerable efforts to reduce mercury releases to the environment.

Japan started a monitoring program that will provide background air monitoring data of mercury and other heavy metals to contribute to the understanding of their atmospheric long-range transport. In this chapter, firstly we address Japan's current effort to monitor the background levels of mercury and other heavy metals at the remote background area in Okinawa. Some preliminary measurement results of mercury are reported. Secondly the comprehensive fate model of mercury species in a multimedia environment is presented. Some progress in development and verification of the model is reported.

12.2 Monitoring Project for Ambient Atmospheric Mercury and Other Heavy Metals in a Remote Background Location

The Ministry of the Environment started a pilot project at a emote background location in Okinawa in 2007. The objectives of the pilot project are to:

- Monitor current levels of mercury and other heavy metals in air, airborne particles, and precipitation and obtain information to assess the long-range transportation of hazardous trace elements in the Asia-Pacific region;
- Develop monitoring methodologies and contribute to the international efforts in ambient atmospheric monitoring.

12.2.1 Project Site for the Field Measurement

Pilot monitoring is conducted at the Cape Hedo Atmosphere and Aerosol Monitoring Station (CHAAMS) operated by the National Institute for Environment Studies, in Okinawa. Cape Hedo is located on the north end of the island of Okinawa as shown in Figure 12.1. The Cape Hedo Station has been used for many years to study the outflow of pollution from East Asia and the Asian continent and was assigned as one of the major sites in Japan for the UNEP/ABC (Atmospheric Brown Clouds) project. Therefore the data for the assessment of background levels of toxic trace elements across Japan, including the contribution from the Asian continent and other sources, are most likely to be obtained.

12.2.2 Methods of Sampling and Analysis

Mercury speciation such as gaseous elemental mercury (Hg^0), divalent reactive gaseous mercury (RGM), and particulate mercury (PM) are continuously measured with Tekran's mercury speciation system coupled to a cold vapor atomic fluorescence spectroscopy (CVAFS) detector. Airborne particles are collected on a

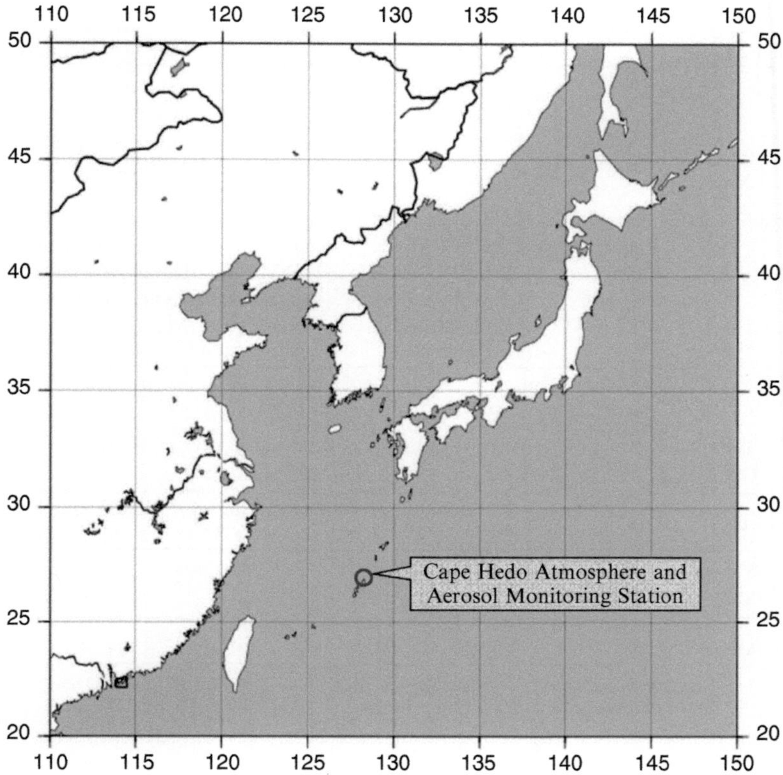

Figure 12.1 Project site: Cape Hedo Atmosphere and Aerosol Monitoring Station (CHAAMS) in Okinawa

polytetrafluoroethylene filter using a low-volume sampler. Sampling is performed at a flow rate of twenty liters per minute for seven days, and measurement is conducted once a week. Heavy metals including Pb, Cd, Cu, Zn, As, Cr, V, Ni, etc. in particles are analyzed with an inductively-coupled plasma mass spectrometer (ICP/MS) (ME, 2008). Precipitation samples are collected using an automatic wet-only sampler for a month, and heavy metals are measured every month. Heavy metals and their analytical methods are the same as those of particulates (ME, 2008) other than Hg. The concentration of total mercury is determined by following EPA method 1631 (EPA, 2002). Measurement items, methods of sampling and analysis are summarized in Table 12.1.

12.2.3 Measurement Results

Measurements of mercury species and other heavy metals started in February 2007. Figure 12.2 shows the preliminary measurement results of gaseous elementary mercury (Hg^0) from 16 October 2007 to 31 January 2008. Table 12.2 shows the statistical

Table 12.1 Measurement items, sampling, and analytical methods

Component		Measurement items	Sampling and analytical methods
Atmosphere	Mercury	Mercury speciation (Hg^0, RGM, TPM) Hg^0	Continuous measurement with Tekran mercury speciation system 24 hours continuous sampling by gold trap amalgamation method and analysis by ICP/MS
Precipitation	Particulate matter Hg and the same	Pb, Cd, Cu, Zn, As, Cr, V, Ni, Se, Sb, Ba, Co, Mn, Sn, Te, Tl, Be, Al, Fe, Ca, Na, K items as particulate matters	Seven days continuous sampling by the low-volume sampler and analysis by ICP/MS[1)] Sampling by the automatic wet-only sampler and analysis of heavy metals by ICP/MS, Hg by gold trap amalgamation[2)]

Table 12.2 Monthly statistics of gaseous elementary mercury at CHAAMS from October 2007 to January 2008, compared with Jaffe et al. (2005)

Hg^0*(ng m-3)*	October 2007a	November 2007	December 2007	January 2008	23 Mar-2 May 2004
Mean	1.3	1.4	1.5	1.7	2.04
Median	1.2	1.3	1.5	1.5	1.99
Min	1.0	0.8	1.1	1.4	1.37
Max	1.8	2.6	1.3	4.0	4.74

[a.] Average concentrations of October were calculated based on the data from 16th to 31st

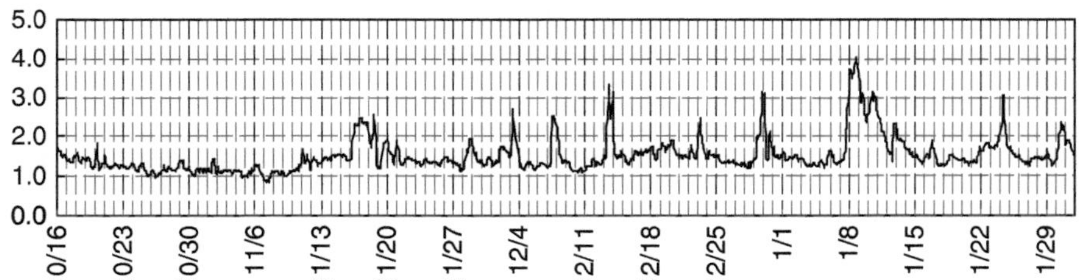

Figure 12.2 Observation of gaseous elemental Hg^0 at CHAAMS from October 2007 to January 2008

information of Hg^0 at Cape Hedo station. Previous observation from 23 March to 2 May 2004, reported by Jaffe et al. (2005), are also summarized in Table 12.2. Monthly mean concentrations of Hg^0 from October 2007 to January 2008 were approximately 1.3 to 1.7 ng m^{-3}, which were slightly lower than the spring observation in 2004. Measurement data of RGM and PM are now under careful data quality consideration. The occasions where the values of Hg^0 exceeded 2.0 ng m^{-3} appear to indicate significant events.

From Figure 12.2, we are able to identify nine substantial peaks of Hg^0 concentration which exceeded 2.0 ng m^{-3}, at around 16 - 19 Nov. (#1), 3 - 4 Dec. (#2), 7 - 8 Dec. (#3), 12 - 14 Dec. (#4), 22 - 23 Dec.(#5), 28 - 31 Dec. (#6), 6 - 15 Jan.(#7),

21 - 26 Jan. (#8), 29 - 31 Jan. (#9). The average Hg^0 concentration is generally constant around 1.3 ng m^{-3}. Using the NOAA HYSPLIT model, we conducted the backward trajectory analysis to overview the most likely source regions of the nine episodes (#1 to #9).

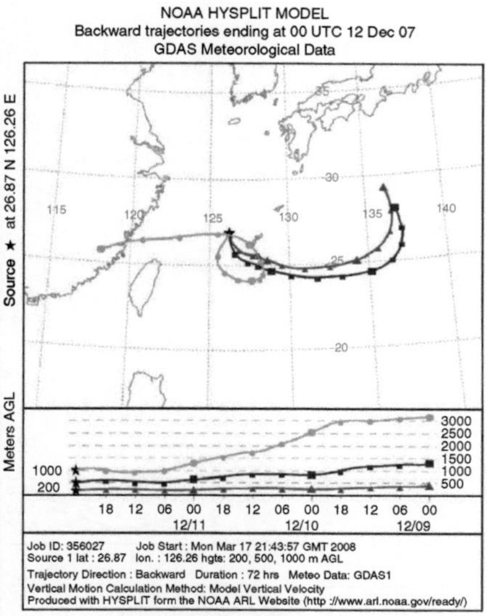

(a)12 Dec 2007, JST hour 9
Hg^0 at CHAAMS: 1.3 *ng m*$^{-3}$

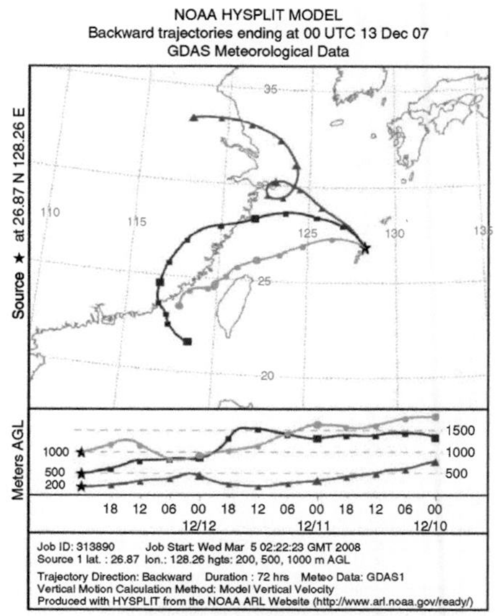

(b)13 Dec 2007, JST hour 9
Hg^0 at CHAAMS: 2.6 *ng m*$^{-3}$

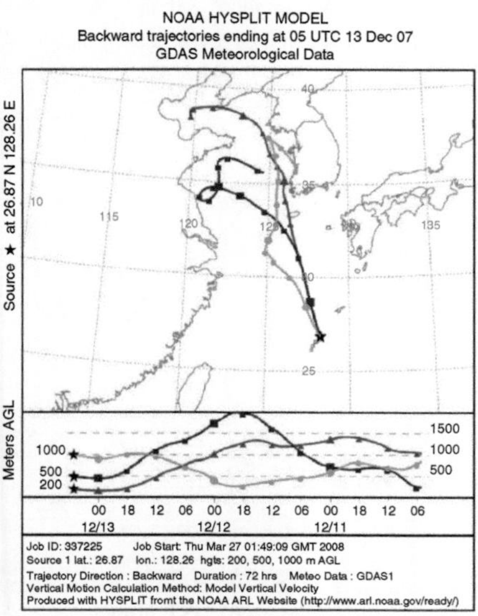

(c) 13 Dec 2007, JST hour 14
Hg^0 at CHAAMS: 3.3 *ng m*$^{-3}$

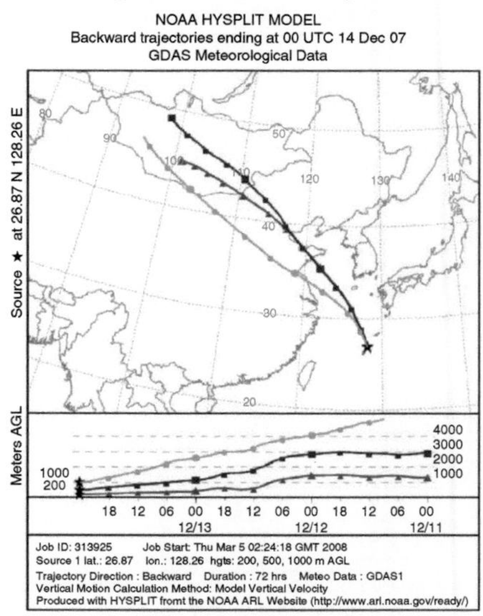

(d) 14 Dec 2007, JST hour 9
Hg^0 at CHAAMS: 1.4 *ng*$^{-3}$

Figure 12.3 NOAA HYSPLIT backward trajectory for Episode #2 Japanese Standard Time (JST) = Coordinate Universal Time (UTC) + 9 hours

For example, Figure 12.3 shows the trajectories for Episode #2 from just before the beginning of the peak at 09:00 JST on 12 December 2007 to the end of the peak at 09:00 JST 14 December 2007. The peak Hg^0 concentration of 3.3 ng m^{-3} was observed at 14:00 JST on 13 December 2007 and its trajectory is shown in Figure 12.3(c). Also Figure 12.4 shows the trajectories for Episode #3 from the beginning of the peak at 09:00 JST on 28 December 2007 to the end of the peak at 09:00 JST 30 December 2007.

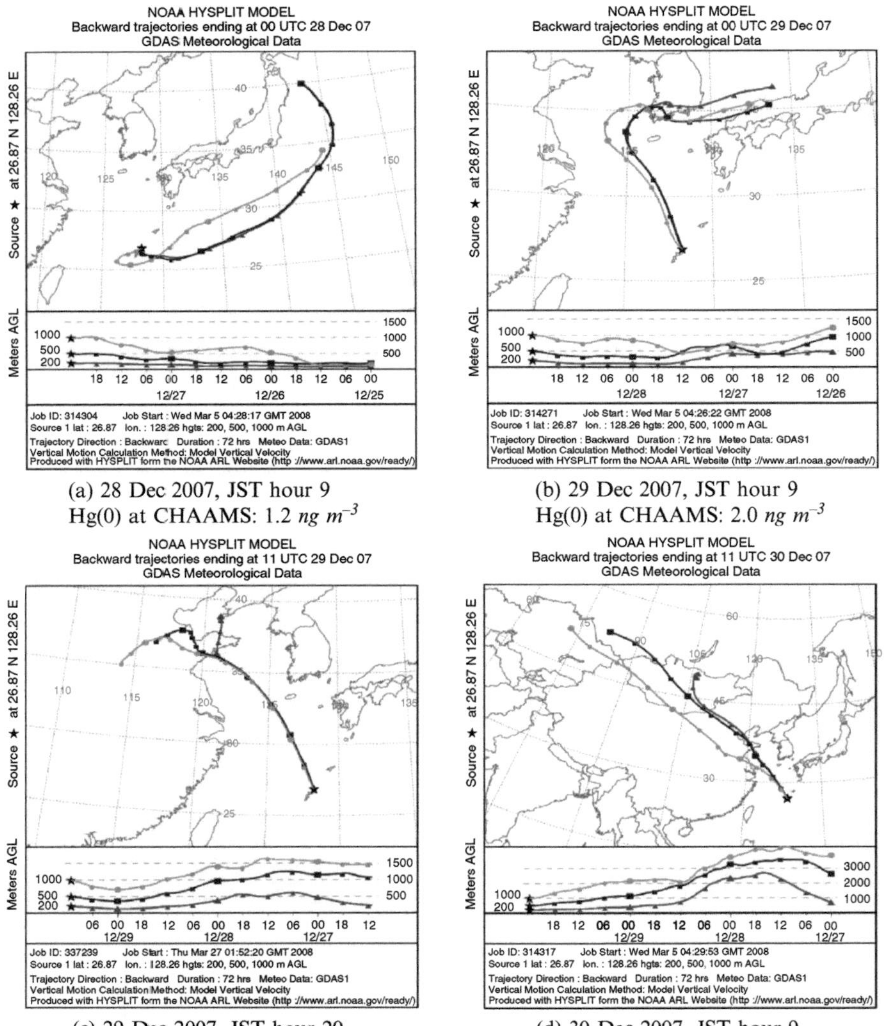

(a) 28 Dec 2007, JST hour 9
Hg(0) at CHAAMS: 1.2 $ng\ m^{-3}$

(b) 29 Dec 2007, JST hour 9
Hg(0) at CHAAMS: 2.0 $ng\ m^{-3}$

(c) 29 Dec 2007, JST hour 20
Hg(0) at CHAAMS: 3.1 $ng\ m^{-3}$

(d) 30 Dec 2007, JST hour 9
Hg(0) at CHAAMS: 1.4 $ng\ m^{-3}$

Figure 12.4 NOAA HYSPLIT backward trajectory for Episode #3 Japanese Standard Time (JST) = Coordinate Universal Time (UTC) + 9 hours

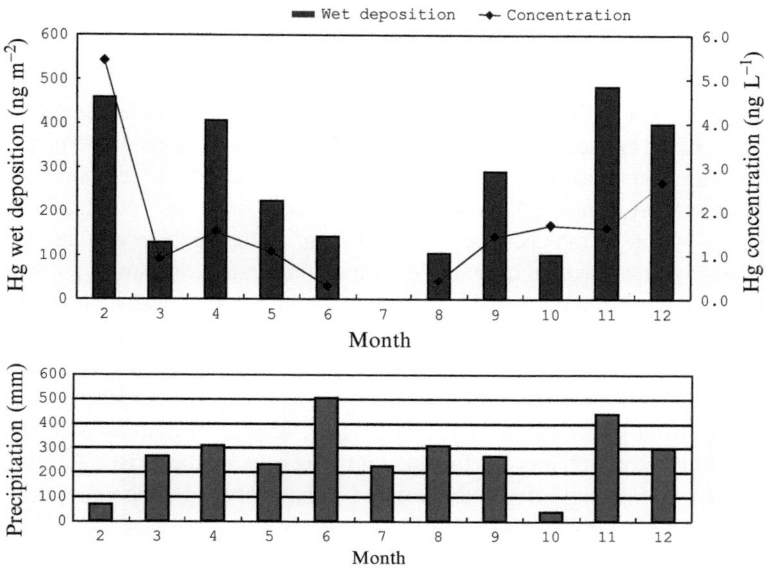

Figure 12.5 Monthly mercury wet deposition flux, concentration and precipitation observed at CHAAMS in 2007. (Data of July was unobtainable due to the typhoon)

The peak of Hg^0 concentration of 3.1 ng m^{-3} was observed at 20:00 JST on 30 December 2007 as shown in Figure 12.4 (c). According to these trajectories, in both cases, before the beginning of the increase of Hg^0 concentration, the air mass came from various regions other than the Asian continent. However, the trajectories for the hour of the Hg^0 peak indicate that the air mass flow is from the Asian continent and East Asia region.

Monthly changes in precipitation, wet deposition flux and concentration of total mercury at Cape Hedo Station in 2007 are shown in Figure 12.5. Total mercury concentration in precipitation was approximately 1 to 5 ng L^{-1} and wet deposition flux was approximately 100 to 500 ng m^{-2} per month in 2007. The highest wet deposition fluxes were observed in February, April, November and December. These data are under examination by experts to understand their relation to meteorological conditions and the outflow from the Asian continent.

12.3 Fate Analysis of Mercury Species for the Monitoring Data Using a Multimedia Environmental Fate Model

Long-range transport of mercury species has been simulated by a number of atmospheric transport and chemistry modeling frameworks. Although atmospheric transport and resultant deposition are believed to be the major source of entry into the surface environment, inter-media exchange processes between air and surface media including water, soil etc. may not necessarily be sufficiently well described in an existing modelling framework.

The National Institute for Environmental Studies has developed a multimedia-modelling framework to assess the inter-media transport of mercury species through media-boundaries based on the multimedia-modelling framework for organic chemicals, which mainly focuses on transport across media boundaries explicitly. By combining existing chemical/transport atmospheric modelling experience with the inter-media transport simulation, a more comprehensive fate modelling, including both air and terrestrial/aquatic environments, would be possible for more integrated assessment purposes.

The objective of the pilot project is to develop inter-media transport schemes and process descriptions for mercury species by expanding the multimedia modeling frameworks from the monitoring outputs. The multimedia fate model G-CIEMS (Suzuki et al.,, 2004) is used as the basis of the study, which is under the POP model inter-comparison study by MSC-E/EMEP. Hg^0, RGM, Particulate and MeHg are the first set of target chemicals for the study. Results of the existing and our new atmospheric monitoring information are to be considered in the process description and validation of simulation results.

12.3.1 *Outline of the Model*

The study is based on the multimedia fate model G-CIEMS (Suzuki et al., 2004) which describes the inter-media transport of contaminants among air, water (river and ocean), sediments and soil including several sub-compartments within each main compartment. Transport in air and water compartments are also described in the model. The model consists of compartments including air (vapor/particulate and wet/dry depositions), water (soluble/particulate and sedimentation), soil (several land-cover information and leaf part of forests, for both river and oceanic regions) and sediment (beneath all aquatic environments). The detailed scheme of the model is described in the literature (Suzuki et al., 2004). This model for organic chemicals is applied for the simulation of mercury species by introducing newly developed mercury modules.

Hg^0, RGM, particulate and MeHg are considered as the target species and transformations between species are formulated as shown in Figure 12.6. Although part of the parameters are established based on the case study of PCB (polychlorinated biphenyl), many of the chemical-specific or related parameters are now being tentatively assigned for mercury and its species. The domain of the model is global with 2.5 by 2.5 degree resolution.

12.3.2 *Results and Comparison to the Monitoring Outputs*

Results of the simulation of the atmosphere are shown in Figure 12.7. The air concentration at the Cape Hedo location is roughly within the order of magnitude compared to our monitoring data, although the RGM from the model is substantially lower than

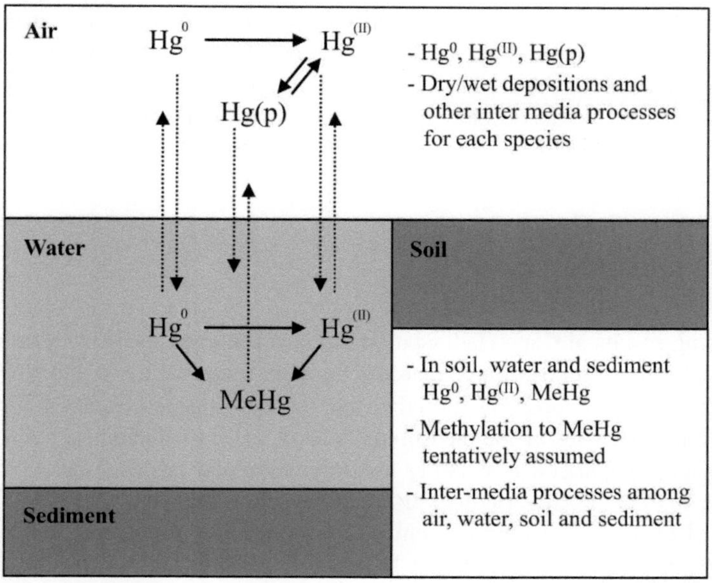

Figure 12.6 Transformation scheme of mercury species in the modeling study

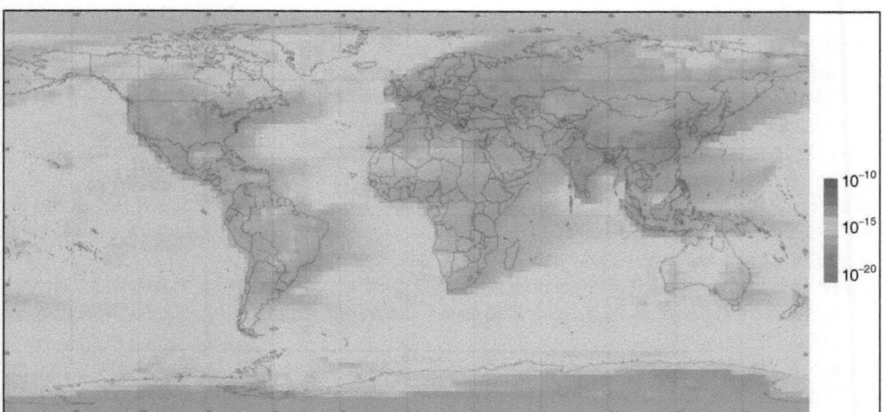

Simulated air concentration at cape Hedo in Okinawa, Japan

	Hg^0(ng/m^3)	Hg^{2+}(pg/m^3)	Hg(p) (pg/m^3)	Total (ng/m^3)
Conc.	1.1	0.83	1.1	1.1

Figure 12.7 Preliminary results of atmospheric concentration and comparison to the simulated air concentration at the Cape Hedo location

the measured levels. This shows the need of the further development of the model. However, on the other hand, the model can be considered to represent the general fate processes in the air and related compartments. As the modelling is currently at a preliminary stage, the model development and validation studies are necessary for further analysis of the environmental fate of mercury in the air and related compartments.

12.4 Future Directions

Monitoring results at Cape Hedo station are now under analysis in terms of measurement, levels, composition of mercury species, temporal trend and fluctuations, and relation to the deposition and other fate mechanisms and events that may control the characteristics of the measured status in the East Pacific region, including the detailed comparison to existing measurements and prior studies (Jaffe et al., 2005; Selin et al., 2007; Friedli et al., 2004) and fate analysis.

The Ministry of the Environment will continue the monitoring at Cape Hedo Station (CHAAMS), within a budget, to understand the background levels of mercury and other heavy metals and contribute to the better understanding of the current status of mercury in the environment by joining the international efforts in ambient atmospheric monitoring.

References

Friedli, H.R., Radke, L.F., Prescott, R., Li, P., Woo, J.-H., and Carmichael, G.R. Mercury in the Atmosphere around Japan, Korea and China as Observed during 2001ACE-Asia Field Campaign: Measurements Distributions Sources, and Implications. *Journal of Geophysical Research*, 109(D19), S25 (2004)

Jaffe, D., Swartendruber, P., Weiss-Penzias, P., Kato, S., Takami, A., Hatakeyama, S. and Kajii, Y. Export of Atmospheric Mercury from Asia, *Atmos. Environ.* 39, 3029–3038 (2005)

Ministry of the Environment, Japan, Manual for the Measurement Methods for Hazardous Air Pollutants, - The Measurement Method for Atmospheric Heavy Metals with Simultaneous Multielement Analysis (2008)

United States Environmental Protection Agency (US-EPA), Method 1631, Revision E: Mercury in Water by Oxidation, Purge and Trap, and Cold Vapor Atomic Fluorescence Spectrometry (2002)

Selin, N.E., Jacob, D.J., Yantosca, R.M., Strode, S., Jaegle, L. and Jaffe, D. Chemical Cycling and Deposition of Atmospheric Mercury, *Journal of Geophysical Research*, 112(D2), D02308 (2007)

Suzuki, N., Murasawa, K., Sakurai, T., Nansai, K., Matsuhashi, K., Moriguchi, Y., Tanabe, K., Nakasugi, O. and Morita, M. Geo-Referenced Multimedia Environmental Fate Model (G-CIEMS). Model formulation and comparison to the generic model and monitoring approaches, *Environ. Sci. Technol.* 38, 5682–5693 (2004)

Chapter 13
The Need for a Coordinated Global Mercury Monitoring Network for Global and Regional Models Validations

Gerald J. Keeler, Nicola Pirrone, Russel Bullock, and Sanford Sillman

Summary Currently, there is not a coordinated observational network for mercury (Hg) that could be used by the modelling community or for establishing recommendations for protecting human and environmental health on a global scale. Current national networks are inadequate as they lack (1) observations of all forms of Hg in the ambient air and in both wet and dry deposition; (2) long-term measurements of Hg and other air pollutants; (3) comprehensive monitoring sites in the free-troposphere; and (4) measurement sites that permit a careful investigation of inter-hemispheric transport and trends in background concentrations. Programs such as the World Meteorological Organization's Global Atmosphere Watch have made substantial efforts to establish data centers and quality control programs to enhance integration of air quality measurements from different national and regional networks, and to establish observational sites in under-sampled, remote regions around the world. Similarly, the International Global Atmospheric Chemistry project (of the International Geosphere-Biosphere Programme) has strongly endorsed the need for international exchange of calibration standards and has helped coordinate multinational field campaigns to address a variety of important issues related to global air quality. Following the lead of these programs and incorporation of a well-defined Hg monitoring component into the existing network sites would be the most expeditious and efficient approach. Close coordination of the global modelling community with the global measurement community would lead to major advances in the global models and advance our understanding of the Hg science while decreasing the uncertainties in global assessments for Hg.

13.1 Introduction

Mercury in the atmosphere exists in both gaseous phase and particulate phase with three possible oxidation states: zero oxidation state, also called elemental Hg (Hg^0); monovalent state (Hg^I), which is not found in the atmosphere; and divalent state ($Hg^{(II)}$) (Schroeder and Munthe, 1998; Lin and Pehkonen, 1999). For operational

N. Pirrone and R. Mason (eds.), *Mercury Fate and Transport in the Global Atmosphere*,
DOI: 10.1007/978-0-387-93958-2_13, © Springer Science+Business Media, LLC 2009

reasons, based on methods of quantification, atmospheric Hg is classified as Hg^0, gaseous $Hg^{(II)}$ or reactive (divalent) gaseous Hg (RGM), and particulate mercury (PM). Particulate mercury consists of Hg^0 and $Hg^{(II)}$ that are associated with atmospheric particulates. Mercury in the atmosphere is dominated (~98%) by Hg^0 (Schroeder and Munthe, 1998) which is relatively insoluble in water and unreactive, and its atmospheric lifetime (>30 d) allows for global-scale transport. In contrast, RGM in the atmosphere is water-soluble and is efficiently removed through both wet and dry deposition processes. Elevated levels of RGM are typically associated with direct emissions from localized anthropogenic sources, but can also be produced by photochemical conversion from Hg^0 (e.g. Laurier and Mason, 2007). The importance of the atmosphere in the global redistribution of Hg is well-known. Over the last two decades a significant effort has been made to understand the atmospheric transport, transformation and deposition processes of Hg around the globe.

More recently, developments in measurement and analytical techniques for Hg have led to a much improved understanding of the sources and cycling of Hg in the environment. However, there are many atmospheric chemical and physical processes involving Hg that are still not well understood (see Chapters 14 & 15). For example, Bergan and Rodhe (2001) noted that an important step in our ability to model the global redistribution of Hg involves reducing the uncertainty in our understanding of the removal of gaseous Hg^0 from the atmosphere via deposition and atmospheric chemical transformation (Schroeder and Munthe, 1998; Selin et al., 2007). Measurements of the key Hg species are critical in our developing a better understanding of these processes, and how they may have changed over time.

It has been suggested that due to intensified anthropogenic release of Hg into the atmosphere, the global pool of Hg has increased in the past 150 years (Mason and Sheu, 2002). Evidence of long-term changes in the atmospheric Hg burden can be derived from a variety of methods including chemical analysis of lake sediments, ice cores and peat deposits (Fitzgerald et al., 2007; Krabbenhoft et al., 2007). Measurements from both hemispheres demonstrate at least a threefold increase of Hg deposition since pre-industrial times (mason et al., 1994; Lamborg et al., 2002; Fitzgerald et al., 2007). An increase in the global atmospheric pool should also be reflected in the background Hg concentration. However, to date, it has been extremely difficult to derive a multi-decadal global trend estimate based on the existing spatially and temporally uncoordinated air concentration data sets (e.g. Slemr et al., 2003). The estimated large increases in Hg emissions in China (see Chapters 2 & 3) over the last decade are not currently reflected in the long-term measurement of total gaseous Hg at Mace Head, Ireland between 1996 to 2006, nor in the precipitation data of the North American Mercury Deposition Network (MDN), as noted in Chapter 9. Total gaseous Hg (TGM) measurements performed in the Southern Hemisphere also do not suggest that a significant change in TGM levels in the global remote atmosphere over the past two or three decades (Slemr et al., 2003).

Given these observations, and the apparent contradictions, it has become clear that there is a need for global mercury monitoring network incorporating existing long-term atmospheric Hg monitoring stations, such as Mace Head, but with the incorporation of a number of additional sites to obtain a globally representative picture of atmospheric Hg in the troposphere. While atmospheric Hg models have had some success in predicting the levels and trends in ambient Hg levels, the scarcity of global measurement data available for the comparisons make the exercise and results less significant. Efforts to improve our understanding of atmospheric Hg chemistry and inter-hemispheric transport will require a comprehensive research framework that integrates observations covering a wide range of temporal and spatial scales with modeling and process studies. Hence, there is a critical need for a coordinated global Hg monitoring network designed to support the development of global and regional scale Hg models relied upon in the policy making process. A well-planned international network is required to provide a consistent, standardized set of long-term data on the concentrations and deposition of atmospheric Hg in its various forms, together with trace gases, particles, and physical parameters at strategic sites that are globally distributed.

Such a global Hg network should leverage its efforts by collocating with other existing monitoring programs such as the World Meteorological Organization's Global Atmosphere Watch sites, US and Canadian Monitoring sites, and UN-ECE's European Monitoring and Evaluation Programme (EMEP) sites. In addition, as noted in Chapter 9, there are already aspects of such a monitoring program in Europe, Canada and the USA. In the USA, there has been a recent concerted effort to develop and design a monitoring network that would be able to evaluate the ecosystem response to changing concentrations of Hg in the atmosphere and in deposition, and to examine how this impacts levels of methylmercury (MeHg) in fish (Mason et al., 2005; Harris et al., 2007). While this program focuses on all aspects of Hg biogeochemical cycling, the conclusions of Driscoll et al. (2007) have relevance to the approach to designing a global Hg monitoring network. These authors suggest that a successful network would consist of a relative small number of "intensive" sites, where the full range of measurements are made (e.g. atmospheric Hg speciation and dry deposition estimation, event-based wet deposition and flux, and the measurement of required ancillary parameters and detailed meteorology), and a larger number of "cluster" sites where only weekly wet deposition is collected. The cluster sites would allow for integration between the intensive sites, and examine the effects of local and regional conditions, while the intensive sites would provide the detailed information needed to calibrate and test global and regional Hg models (Driscoll et al., 2007; Saltman et al., 2007). This approach is one model of how such a network would be constructed.

Clearly, the over-arching benefit of a coordinated global Hg monitoring network (CGMMN) would be the universal availability of high-quality measurement data that are desperately needed to develop, refine, and validate models on different spatial and temporal scales. The data from the set of coordinated monitoring sites would support the evaluation and ground truthing of models as research and management tools to evaluate our understanding of the global cycling of Hg and deposition to sensitive ecosystems. This chapter will specifically highlight the need for a global Hg monitoring designed to support the development of global and regional scale Hg models that should be considered as the scientific basis of the policy making process.

To reiterate, the principal goals of a global Hg monitoring network are:

- To study the temporal and spatial variability of atmospheric Hg and atmospheric composition;
- To provide long-term monitoring of changes in the physical and chemical state of Hg in the lower atmosphere; in particular to provide the means to discern andunderstand the causes of such changes;
- To establish the links between changes in atmospheric Hg, tropospheric chemistry and climate;
- To support intensive field campaigns focusing on specific Hg processes occurring at various latitudes and seasons, and in highly sensitive ecosystems;
- To produce verified data sets for testing and improving global and regional models for atmospheric Hg and those coupled to aquatic and terrestrial ecosystem models.

13.2 Existing Global Monitoring Programs

Currently a global monitoring network for Hg does not exist. There are a number of state and national programs that are collecting atmospheric Hg data but the parameters monitored, the locations of the monitoring sites, and the methods employed may prohibit their utility in assessing global trends and changes. In spite of this, there is a great deal of recent information available on the levels of Hg in the atmosphere around the globe. Chapters 9-12 of this book provide a detailed compilation of the available atmospheric Hg data in the literature. Many of the issues discussed in these chapters, regarding data quality and the need for Hg speciation data at background and source impacted areas, are relevant to the development of a global network as well. It is important to stress that the measurement of Hg by itself is not sufficient for us to improve our understanding of Hg sources and impacts. Measurements of other key atmospheric constituents at the global monitoring sites are necessary for us to develop a better understanding of the global redistribution of Hg and to further refine our model parameterizations of the key processes. The co-location of Hg measurements with ongoing global programs is the most efficient approach to initiating this coordinated monitoring network.

13.2.1 Ambient Measurements

Current air monitoring programs for air pollutants such as ozone, sulfur dioxide, and nitrogen compounds include regulatory monitoring networks that report daily air quality changes at sites located primarily in urban or populated areas; global and regional networks designed to measure background atmospheric composition at selected remote sites or at regionally representative sites; remote-sensing (satellite) instruments that provide global-scale observations of selected atmospheric species; and a variety of radiosonde and aircraft based instrument programs focused on specialized measurement campaigns. These different observational approaches vary widely in their scope and degree of analytical quality. Initial efforts are underway to develop such monitoring capabilities for Hg in the Northern Hemisphere, as detailed in earlier chapters. In the USA, there is currently an effort underway to expand the MDN network to include measurements of Hg speciation in air, and for estimating dry deposition (http://nadp.sws.uiuc.edu/mdn/). Coordination of such efforts globally will help advance the development of the CGMMN.

Programs such as the World Meteorological Organization's Global Atmosphere Watch (GAW) have established data centers and quality control programs to push for the integration of air quality measurements from different national and regional networks, and to establish observational sites in data spare and remote regions of the globe. The International Global Atmospheric Chemistry project (of the International Geosphere-Biosphere Programme) has strongly endorsed the need for international exchange of calibration standards and has helped coordinate multinational field campaigns to address a variety of important issues related to global air quality.

13.2.2 Mercury Measurements at Altitude

Most air quality monitoring networks rely entirely upon ground-based sites that sample within the boundary layer (the lowest portion of atmosphere in contact with the surface). Addressing global air quality problems such as Hg contamination, however, will require observations that are made at higher altitudes above the boundary layer. Studies have shown that transport of pollution including Hg between Asia and the United States occurs primarily through the middle and upper troposphere, and because of the highly episodic nature of this transport, there can be significant inhomogeneity in the air masses reaching the continental United States Thus, networks that only sample air masses within the boundary layer would not allow a quantitative determination of long-range pollutant fluxes. While sampling with aircraft can provide detailed information about Hg in the upper atmosphere (Banic et al., 2003; Ebinghaus et al., 2000; Friedli et al., 2004; Swartzendruber et al., 2008; Talbot et al., 2007), in terms of long-term monitoring, the use of aircraft ahs obvious limitations. The preferred approach it to use mountain-top monitoring sites, that are frequently in the free-troposphere, and which located around the globe are essential to understanding the global transport of Hg and other pollutants (Jaffe et al. 2003). Currently, there are a number of such sites in existence, including Mt. Batchelor in the western USA (Jaffe et al., 2003), Mona Loa in Hawaii (Landis et al., 2005), Wank Mt. in Germany (Slemr et al., 2003), and the Lulin station in Taiwan (Sheu et al., 2007).

13.2.3 Episode-based Measurement Intensives

In addition to long-term monitoring efforts, investigations of both ozone and aerosols have included numerous measurement intensives. These have typically featured measurements of a more extensive range of species that are related to photochemistry, including radicals such as OH and HO_2 (e.g. Ren et al., 2006). Intensive field campaigns have occurred in urban, rural and remote locations and have frequently included both surface-based and aircraft measurements (e.g. Daum et al., 2004, Lei et al., 2007). Such intensives have also occurred for Hg (see Chapters 9-12) but in many cases have not included sufficient ancillary information that the details of the formation and removal of various Hg species can be determined in detail (See Chapters 14 & 15).

13.2.4 Meteorological Measurements

Meteorological processes operating over a wide range of spatial scales play a central role in the air quality and deposition of all atmospheric pollutants including Hg. The emissions, chemical transformations, deposition, and re-emission of Hg are strongly affected by meteorological parameters such as temperature, cloud cover, humidity, mixing height, and wind speed and direction. In addition, dynamical forces related to vertical temperature profiles control the dispersion of pollution within the urban/local boundary layer and the release of pollutants from the boundary layer into the free troposphere. Finally, the long-range transport of pollutants is influenced by atmospheric high- and low-pressure systems occurring on synoptic scales (100s to 1000s km). Thus, the meteorological context of atmospheric chemical measurements must be established in order to accurately assess impacts of Hg loadings and to develop the models used for studying long-range Hg transport and climate-Hg interactions.

13.2.5 Atmospheric Deposition

The atmosphere provides the main environmental pathway for redistribution of Hg around the globe, and therefore, quantifying the transfer of Hg from the air to the earth's surface via wet and dry deposition is critically important. Like ambient Hg, there is currently not a globally coordinated network of atmospheric Hg deposition sites. However, there are currently a few coordinated networks in certain regions of the world including: North American Mercury Deposition Network (MDN) that was initiated in the early 1990s as part of the National Atmospheric Deposition Program (NADP); EMEP in Europe; and networks in Japan as well as other parts of Asia. Coordination of the national networks in the various regions and formation of a standard set of measurement objectives and techniques will be important at the onset of the project. While wet deposition networks are currently in-place for Hg, the measurements challenges involved in quantifying Hg dry deposition have thus far prevented these measurements from becoming routine. Dry deposition measurement techniques have been developed using both surrogate surface approaches (Keeler and Dvonch, 2005), and using inferential techniques that measure the various forms of Hg in the atmosphere as well as meteorological parameters to model the dry deposition flux at the measurement site (as discussed in Driscoll et al., 2007). Inferential methods have been used in both US EPA and Environment Canada Networks for acid rain species and may hold the most promise for the global network as well. They key to estimating the dry deposition flux will be the accurate measurement of atmospheric Hg in the gaseous forms and on size-fractionated particulate matter. There is evidence that particulate Hg is bimodal in the atmosphere and undergoes reversible gas-particle partitioning during transport from source to receptor.

13.3 Measurements and Model Development

Models for chemistry and transport of atmospheric species are based on input with a significant range of uncertainty. This is especially true for models for Hg, because essential features of atmospheric cycling and photochemical transformation of Hg involve more uncertainty relative to other atmospheric species. The major role of measurements in model development is to provide a basis for evaluating the model accuracy in simulating ambient conditions. Comparisons between predicted and observed ambient values are the central test for establishing the accuracy of models (e.g. Selin et al., 2007; Strode et al., 2008; Hedgecock et al., 2006). When model predictions show significant differences from measured values, this is frequently regarded as evidence of errors in the models, and prompts investigation and modification of model input assumptions or spatial resolution. As such, ensembles of measured values provide a basis for constraining the assumptions used by models. Chemistry/transport models also use ambient meteorological measurements as direct input. Meteorological input is typically provided by prognostic models with assimilated data in order to generate simulations for specific time periods. The methods of prognostic meteorological modeling and data assimilation have been studied extensively and are used in chemistry/transport models for all atmospheric species. Here we will describe the role of ambient measurements that are used for model evaluation, through comparison with model predicted values, rather than as a source of model input. This type of model evaluation is an essential part of the model development process. Model-measurement comparisons provide a basis for constraining the range of assumptions used by models, especially with regard to emission rates and photochemical reaction sequences. Model-measurement comparisons can also be used to provide evidence on issues that are relevant to policy concerns.

13.3.1 General Evaluation of Model Transport and Photochemistry

Chemistry transport models for ozone and its precursors and for sulfate and nitrate aerosols have been the subject of extensive evaluation over the past 20 years at both the urban/regional and global scales (e.g. TF-HTAP, 2007). Many of the approaches used to evaluate these models are applicable to Hg as well, and thus Hg has been included under the umbrella of HTAP, and many existing models for atmospheric species (e.g. Geos-Chem) have been adapted to include Hg. As with ozone, initial model evaluations must focus on model accuracy and the ability of the model to reproduce peak concentrations and diurnal variations (especially for urban/regional models) and on the vertical distribution throughout the troposphere (for global models) (e.g. Logan, 1999, Oltmans, 2006). As with models for sulfate and nitrate aerosols, many Hg models have focused initially on reproducing the observed wet deposition of the MDN and other networks. As model evaluation continues, there will be an increased emphasis on comparing measurements of a wider variety of species and evaluations that could be used to test the accuracy of specific model components. Efforts for reactive species and Hg have included the following: 1) Correlations with CO as a basis for identifying transport from polluted regions (e.g. Parrish, et al., 1993, Weiss-Penzias et al., 2006; 2007; Jaffe et al., 2005); 2) Correlations between O_3 and NO_x reaction products (HNO_3, various organic nitrates) as a basis for providing constraints on the ratio between the rates of production of O_3 and other species and evaluating chemistry (e.g. Trainer et al., 1993; Sillman et al., 1998; Roberts et al., 2004; Horowitz et al., 2007; Hynes et al., 2008 chapter 14)); 3) Direct measurement of OH and HO_2 radicals as a basis for evaluating model chemistry (e.g. Ren et al., 2006, Olson et al., 2006, Lei et al., 2007); 4) A combination of episodic and long-term measurements designed to measure continental outflow and intercontinental transport for O_3 and aerosols (e.g. Prospero et al., 2003, Savoie et al., 2002), and Hg (e.g. Jaffe et al., 2005); and 5) Model evaluations based on a wide range of primary and secondary organic species and reactive nitrogen (e.g. Lei et al., 2007; Bey et al., 2001).

As noted, the evaluation of models for Hg has initially focused on two benchmarks: ambient TGM and the observed wet deposition of total Hg. There have also been limited comparisons with ambient RGM (Selin et al., 2007; Strode et al., 2008). Wet deposition is of interest in particular because it is linked to policy concerns. In addition, the spatial variation of wet deposition within the U.S. provides a basis for evaluating the ability of models to represent the variety of factors (including emissions, transport and chemistry) that affect deposition rates. Comparisons with the observed wet deposition have been a common basis for model evaluation (Figure 13.1) (see Chapter 16). However, although wet deposition will likely remain the most important basis for evaluating model performance, it also has significant drawbacks. Wet deposition alone does not provide sufficient information for evaluating the impact of local versus global emission sources. It also does not provide a basis for identifying photochemical transformations that affect Hg. Significant improvements in model accuracy may be obtained by an expanded range of model-measurement comparisons. Figure 13.2 illustrates the large differences

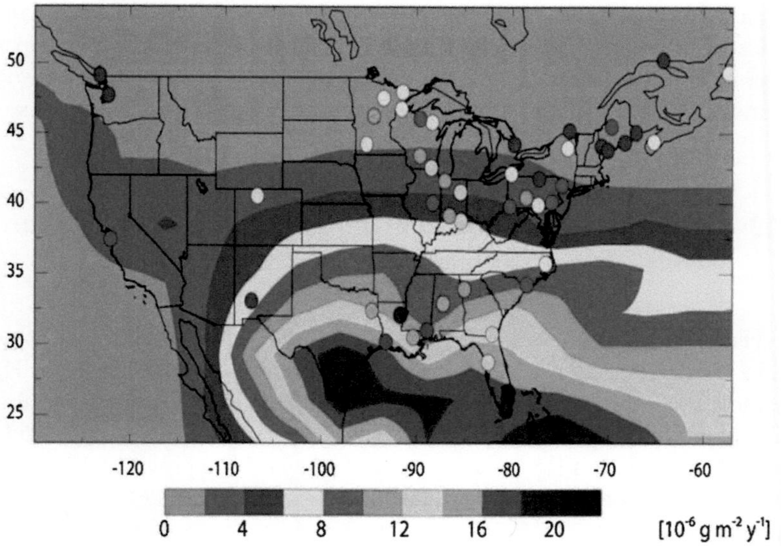

Figure 13.1 Annual mercury wet deposition fluxes over the United States for 2003–2004. Observations from the Mercury Deposition Network (circles) are compared to model results (background). From Selin et al. (2007)

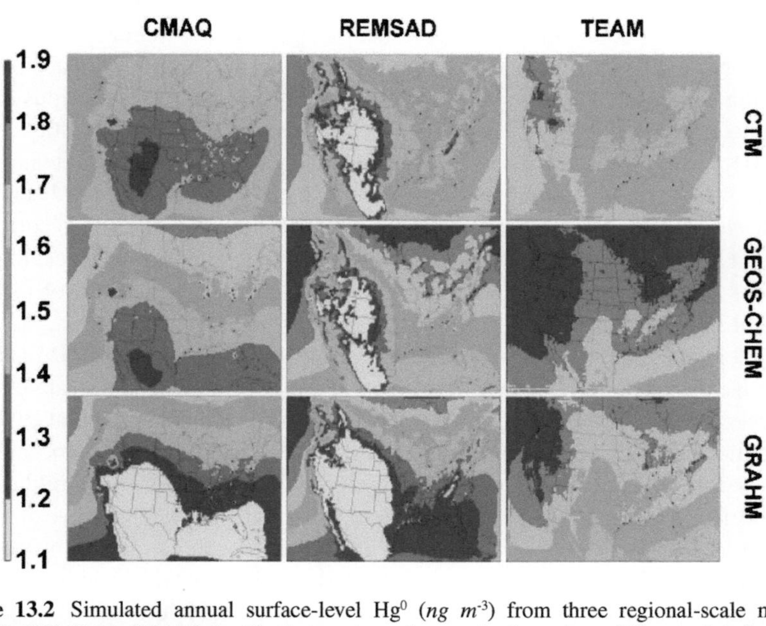

Figure 13.2 Simulated annual surface-level Hg^0 (*ng m*$^{-3}$) from three regional-scale models (CMAQ, REMSAD, TEAM) for North America using results from three different global models (CTM, GEOS-CHEM and GRAHM) as boundary conditions. From Bullock *et al.* (2007)

in ambient Hg^0 and RGM predicted by different regional and global-scale models. The size of these differences identifies the importance of model intercomparisons as a basis for identifying the effects of differing model assumptions. Networks of measurements should provide information for resolving these differences.

13.3.2 Source-receptor Relationships: Long-range Transport Versus Local Sources

The relative impact of local emissions and long-range transport is important because it determines the effectiveness of proposed control strategies for Hg wet deposition. It is especially important to develop methods for testing the accuracy of model predictions in this area and revising models in response to results. Evidence for episodic transport of Hg from Asia to the U.S. has been provided through measurements at sites in the western U.S. (Weiss-Penzias et al., 2006).

As shown in Figure 13.4, measurements showed episodes with elevated ambient TGM, O_3, particulates and CO coinciding with circulation that suggested possible transport from Asia. Because CO has a lifetime of >2 months it is viewed as a marker for long-range transport and for anthropogenic origin. The observed ratio between CO, and TGM provides a basis for evaluating the accuracy of model representations of emissions and transport. Weiss-Penzias et al. (2007) report values for ΔTGM/ΔCO plumes of Asian origin as opposed to North American origin and found significant differences (0.0045-0.0048 ng m^{-3} ppb^{-1} in Asian plumes vs. 0.0013 ng m^{-3} ppb^{-1} for North American sources). Models should reproduce these ratios. Recent modeling efforts suggest that the agreement is better when non-point source emissions in Asia are accounted for (Strode et al., 2008). Evidence for local emission sources of Hg have largely been based on statistical correlations or relationships between deposited Hg and trace elements or stable isotopic ratios, e.g. Pb 206/204 in rainwater or aerosols (Dvonch et al. 2005; Fitzgerald et al., 2007) (Figure 13.3). This relationship between trace metals and Hg has not been tested yet in comparison with deterministic models. Because the statistical correlation between Hg and trace metals is closely associated with emission sources, this ratio provides a basis for evaluating the accuracy of model representations of local and regional sources.

Model results have suggested that correlations between ambient RGM and Hg^0, or the lack of such correlations, may also provide evidence for the relative impact of local emissions versus long-range transport. Model results predict that RGM and Hg^0 anticorrelate in situations where RGM has been produced photochemically from the global pool of Hg^0 (e.g. Jaffe et al., 2005) (Figure 13.4), while directly emitted RGM is predicted to positively correlate with Hg^0 (Sillman et al., 2007) (Figure 13.5). More extensive measurements of these species would allow an evaluation of the predicted versus measured correlation, which would provide an indirect evaluation of model representations of local versus global sources.

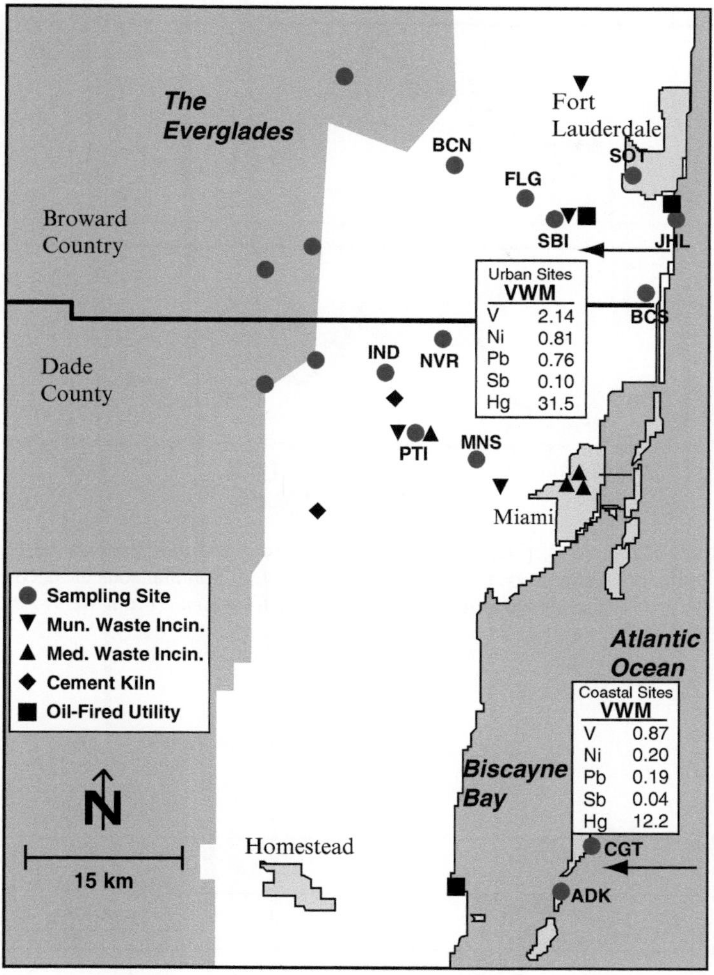

Figure 13.3 Volume-weighted mean concentrations of V, Ni, Pb, and Sb (*μg L*$^{-1}$) and Hg (*ng L*$^{-1}$) in precipitation arriving from the east at SoFAMMS urban (BCN, BCS, FLG, IND, MNS, NVR, PTI, SBI, and SOT) and coastal (ADK, CGT, and JHL) sites. (After Dvonch et al. 2005).

13.3.3 Evaluation of Photochemical Processes

The photochemical transformation between Hg^0, RGM and PM represents one of the major uncertainties in the current understanding of the cycle of atmospheric Hg. Hg^0 is converted to RGM through gas-phase reaction with OH and O_3, and through reactions with reactive halogen species, but the rates of those reactions are uncertain (see Chapter 14). RGM may be converted to Hg^0 through aqueous reaction with HO_2,

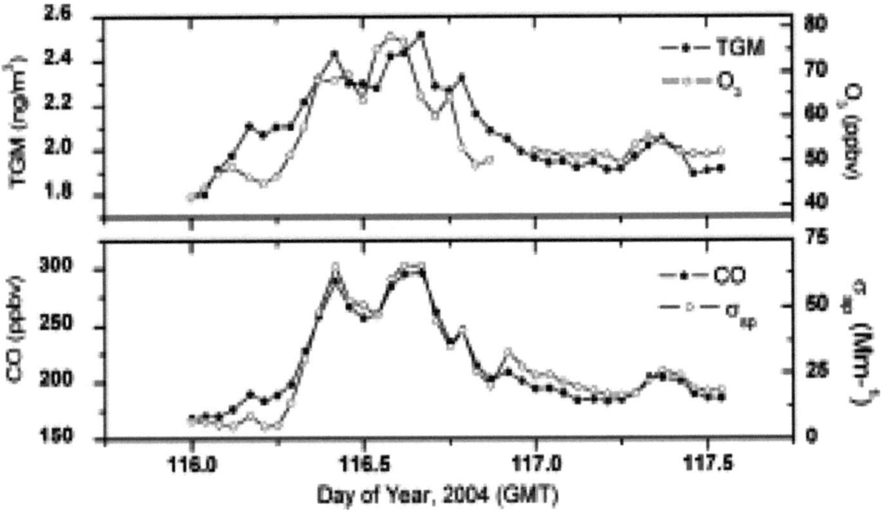

Figure 13.4 Measured O_3, CO, particulate scattering (σ_{sp}) and total gaseous mercury (TGM) during a pollution transport event at Mt. Bachelor, Oregon. Simultaneous elevated concentrations is interpreted as evidence of transport, most likely from Asia. (From Weiss-Penzias et al., 2006)

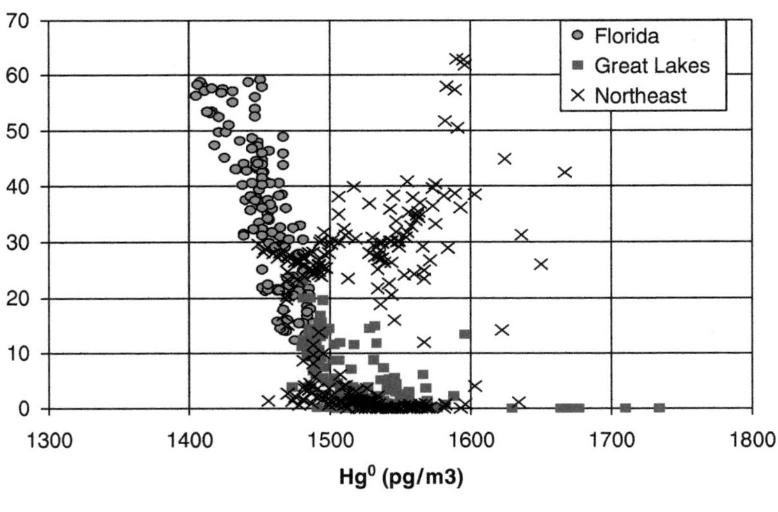

Figure 13.5 Model correlation between Hg^0 and RGM in *pg m*$^{-3}$ for Florida (circles), the northeast corridor(X's) and Great Lakes (squares). Negative correlations identify photochemical production of RGM from global background Hg^0. Positive correlations identify directly emitted RGM and Hg^0. (Source: Sillman et al., (2007)

but this reaction is also uncertain (Gartfeldt and Jonnson, 2003). Other mechanisms for the reduction of RGM have also been proposed. Lin et al. (2006; 2007) and Pongprueksa et al. (2007) tested the viability of these proposed reactions and reaction rates in a regional-scale simulation. They used measured wet deposition rates and ambient Hg^0 as constraining factors, proposing in effect that an ensemble of photochemical reaction rates must result in Hg^0 and wet deposition that is consistent with measurements. Pongprueksa et al. (2007) reported that the agreement between model and measured wet deposition is improved somewhat if a reduction reaction between RGM and CO is used in place of aqueous reduction via HO_2. This type of investigation might be improved if measured ambient RGM were more widely available.

Another approach to evaluating photochemistry is to use correlations between ambient RGM and O_3. Model results predict a strong correlation between RGM and O_3, due to enhanced photochemical activity during pollution events with high O_3 (Sillman et al., 2007). Measurements also indicate a correlation between RGM and O_3 (Keeler and Dvonch, 2005). Evaluation of this correlation can also provide an indirect evaluation of model photochemistry. Selin et al. (2007) used the diurnal variation of RGM as a basis for proposing rapid uptake of RGM by sea-salt aerosols.

13.3.4 Evaluation of Gas-particle Partitioning

Studies have shown that PM is primarily from combustion processes and is found mainly in fine fraction particles with aerodynamic diameters below 2.5 μm (Keeler et al., 1995; Pirrone et al., 1996). PM associated with coarse particles (> 2.5 μm) were also observed in studies conducted in urban/industrial areas (e.g. Keeler et al., 1995). These coarse fractions PM were thought to be formed through near source adsorption of the two gaseous Hg species (Hg^0 and RGM) into the existing aerosols in the air. Since PM has a greater solubility and reactivity, but lower volatility, than Hg^0, PM produced through adsorption is more likely to be derived from RGM than Hg^0 (Pleijel and Munthe, 1995; EC, 2001). The lifetime of PM in the atmosphere depends on aerosol characteristics such as particle size and the ability of aerosols to be scavenged by precipitation and clouds (Schroeder and Munthe, 1998).

Coarse particles settle more rapidly than fine particles and may play a dominant role in atmospheric dry deposition, especially near the sources; while fine particles can distribute uniformly in the air during their longer atmospheric residence time and contribute substantially to the total atmospheric dry deposition on a larger spatial area (Keeler et al., 1995). In the absence of precipitation, PM has a lower removal rate and longer residence time (days to months) than RGM (hours to days) (Schroeder and Munthe, 1998). Like RGM, high concentrations of PM are often found near emission sources (Keeler et al., 1995). While RGM and PM constitute a small percentage of the total atmospheric Hg, they contribute a significant portion of the deposition of this toxic metal. Studies conducted in the U.S. and in Europe have also shown observing significant spatial gradients in Hg deposition near urban/industrial areas where large anthropogenic Hg sources are located, pointing to the importance of local anthropogenic influences (Dvonch et al., 1998; Schroeder and Munthe, 1998; Landis and Keeler, 2002; Keeler and Dvonch, 2005). Measurements are needed that can help to validate the model parameterizations used in current Hg models that incorporate gas-particle partitioning which may play an important role in the deposition and transport of PM on global scales. This is particularly important in the upper troposphere where elevated RGM levels would likely result in elevated PM levels at altitude as well.

13.3.5 Evaluation of Emission Inventories

Emission inventories represent a major source of uncertainty in models for Hg, especially on the global scale. Errors and emissions in emission inventories can sometimes be suggested by a careful comparison between model TGM an ensembles of measured values (Selin et al., 2007). Indeed, recent estimates in south Africa suggest that previous inventories are substantially higher than is supported by data and modeling (Chapter 5; Dabrowski et al., 2008; Selin et al., 2008). Evaluations of emission inventories based on ambient measurements have been made based on inverse modeling techniques (Bergamaschi et al., 2000). These approaches have frequently used satellite measurements and have been used to derive emission estimates for CO (e.g. Bergamaschi et al., 2000; Streets et al., 2006), NO_x (Müller and Stavrakou, 2005) and biogenics (Fu et al., 2007). A summary of these studies can be found in TF-HTAP (2007). Although this type of evaluation typically requires satellite measurements that are not available for Hg, it is possible to derive estimates by using ratios of measured TGM to CO (Weiss-Penzias et al., 2006; 2007). Constraints on the emission rates for CO can be used to develop constraints for Hg emissions.

13.3.6 Evaluation of Past/future Changes and Effectiveness of Control Strategies

Evaluation of past changes in pollutant levels based on a historical record of measurements provides a strong basis for evaluating processes related to pollution control. The historical record can identify changes in emission rates and global ambient concentrations that might otherwise be overlooked. A comparison between model and measured ambient concentrations and/or wet deposition for a historical period (e.g. Selin et al., 2008; Sunderland and Mason, 2007) also provides a direct evaluation of changes in response to control strategies, and an evaluation of the model accurately in representing the change.

Significant information has been obtained from the historical record for pollutants other than Hg. Measurements in remote locations suggest a significant increase in the global background O_3 in the Pacific (Jaffe et al., 2003; Parrish et al., 2004) (see Figure 13.6), and possibly in the Atlantic (Simmonds et al., 2004; Derwent et al., 2007). The observed increase has amounted to 6 ppb (20%) over a 15-year period (Jaffe

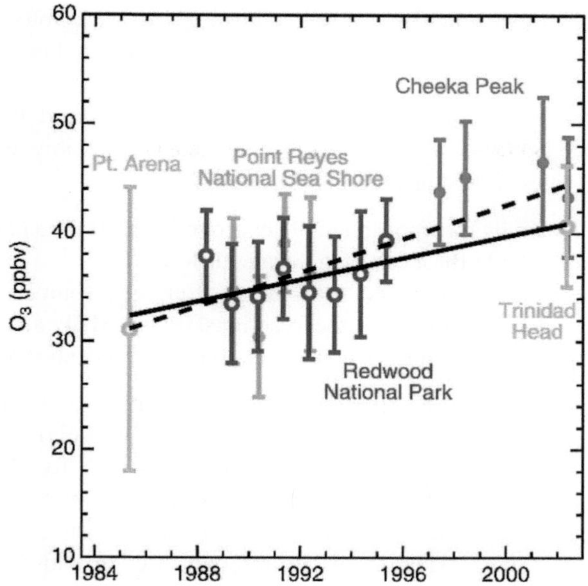

Figure 13.6 Spring mean mixing ratio ±1 standard deviation for background O_3 at 5 sites representing the marine boundary layer along the U.S. Pacific coast, with linear regression lines. The linear fit to the data from the 4 sea level sites (solid line), yields a slope and year 2000 O_3 mixing ratio (with 95% confidence intervals) of 0.50 ± 0.36 ppbv/year, 39.9 ± 3.3 ppbv, and an r^2 of 0.44. If the higher altitude and latitude Cheeka Peak site is included (dashed line), the calculated slope and year 2000 mixing ratio become 0.78 ± 0.28 ppbv/yr^{-1}, 42.9 ± 2.4 ppbv, and the r^2 is 0.68. (Jaffe et al., 2003)

et al., 2003) and may have a significant impact on air quality in the USA. It is uncertain whether models can identify the reason for the observed increase. Current research also proposes to use measurements and models to evaluate reductions in NO_x emissions in the USA between 2000 and 2004 and their effect on ambient O_3 (Frost et al., 2006). This type of historical evaluation requires sets of measurements that extend over time periods of 10 years or more. Such measurements are rarely available for Hg, especially in terms of its atmospheric chemistry.

13.3.7 Proposed Measurements to Enhance Model Development

A consistent set of measurements made on a global scale would dramatically improve our ability to test and validate global and regional scale atmospheric Hg models. The models would benefit from measurements at surface based sites performed as part of the coordinated network but would also benefit from closely linked intensive aircraft studies. Below is a brief list of some of the most critical measurement needs for model development:

1. A network for monitoring Hg wet deposition at global background sites far from anthropogenic sources as well as sites strategically located in/downwind of various source areas.
2. Surface sites for continuous monitoring of Hg^0, RGM and PM, along with fundamental gas-phase species: CO, O_3, particulate and possibly reactive nitrogen and sulfate. Sites should include the following:

   a. Remote locations for tropospheric background condition (e.g. Mt. Bachelor)
   b. Locations suitable for measuring Asian outflow
   c. Locations in the USA subject to influence from local sources
   d. Locations that can examine the reactions in the MBL and the reactions occurring in polar regions, where oxidation of Hg^0 is enhanced

3. Aircraft-based studies to identify the vertical distribution of Hg and correlations between Hg and other atmospheric species (CO, O_3, aerosols).
4. Aircraft-based studies to evaluate the evolution of Hg^0, RGM and Hg(p) in plumes downwind of major emission sources, in combination with measurements of gas-phase species (O_3, CO, NO_x, SO_x). Such measurements can be used to resolve the rate of oxidation of Hg^0 through reaction with O_3 and OH, and evidence for reduction of RGM through reaction with CO or photolysis.
5. Aircraft-based studies to evaluate the effect of cloud processing on ambient Hg, as a basis for assessing the possible reduction of RGM through aqueous reactions.

13.4 Establishment of the Coordinated Global Mercury Monitoring Network (CGMMN)

13.4.1 Key Components of a Coordinated Long-term Network

All global networks must document quality and consistency of routine operations and data treatment through development of protocols for primary and auxiliary measurements, data processing, instrument comparisons, analysis, and data validation. Establishment of a coordinated global Hg monitoring network would benefit greatly from the experience and lessons learned from other global measurement networks, and from the on-going design of regional and national Hg networks (see Chapter 9 and Driscoll et al., 2007). Additionally, climate change research and studies of atmospheric CO_2 and carbon cycling have taught us valuable lessons about what is needed to initiate and maintain effective, reliable networks for observing the chemical state of the atmosphere. As suggested by Driscoll et al. (2007), some of the keys in establishing a successful global network include:

1. Commitment to long-term Hg data collection. Maintaining high-quality, long-term measurement programs for atmospheric Hg requires vision, governmental and international support. There is a need to ensure that the support of observations

is adequate to prevent breaks in the data record. Similarly, mechanisms are needed to provide support for development and validation of new instrumentation and its deployment in the field, which can often take many years.

2. Calibration. Inadequate calibration of instruments in a measurement program limits the value of observations for understanding global atmospheric changes. Absolute calibration is critical for calculating the budgets and lifetimes of different chemical species. Relative calibration is essential in applications that require the use of observations from multiple measurement stations or networks. The goal should be to calibrate each species with an absolute accuracy approaching the analytical precision of the measurement technique.
3. Measurement/Analytical Quality Control/Quality Assurance. If a quality assurance/ quality control structure has not been established or is inadequately funded for a particular measurement program, then the uncertainties associated with the measurements may render them unusable for trend analysis and model evaluation, and thus the collection efforts are wasted. Standardized quality assurance criteria are especially important in efforts to integrate data from multiple observational programs.
4. Collaboration among researchers with different missions but similar data needs. Each type of measurement has limitations and is most useful when it can be meaningfully combined with other types of data. Effective coordination among programs with related observational needs can avoid redundant data collection efforts or data gaps that occur when individual programs lack the resources to adequately support continued observational efforts. For example, coordination of ambient Hg measurements with plant uptake or terrestrial cycling can aid both the air quality and biological cycling research communities.
5. Institutional and Personnel Requirements. Experience has shown that maintaining and advancing long-term observational science in atmospheric chemistry depends primarily on highly qualified and dedicated individuals. Attracting such individuals requires strong educational programs and promising career opportunities. Having more than one laboratory striving for the same goal is all the most effective way to assure that the highest quality observations will be made (i.e., all key species should be measured by more than one research group). The complexity in making speciated Hg measurements mandates that well-trained operators be identified for this task.

13.4.2 Mercury Measurement Methods

There are a relatively small number of reliable methods that have been developed for the measurement of atmospheric Hg in its various forms. Examples of the sampling and analysis approaches for total gaseous Hg and particulate bound Hg can be found in the literature (e.g. Keeler et al. 1995; Landis et al. 2002). These manual approaches give reliable Hg data but require more operational time and laboratory analysis. More recent advancements have led to the development of field instruments that sample and analyze the concentration of Hg in the gaseous phase and that fraction associated with fine particulate material. These methods, while not quite at the stage of being routine field instruments, have added greatly to our fundamental knowledge of the behavior of Hg species and its deposition. A brief description of various measurement and analytical methods can be found in Driscoll et al. 2007. Here, a brief description of the type of the ambient Hg instrumentation that would be needed for a global Hg network is described. Continuous instruments that provide equivalent information on the time-scale of minutes to hours are important for a detailed study of Hg transport and chemistry.

Simultaneous measurements of Hg^0, RGM and PM (2.5 μm) can be performed using a Tekran® 2537A/1130/1135 automated Hg measurement system (Tekran Inc., Ontario, Canada). Studies have reported that this system provides reliable data in several intercomparison studies on Hg measurements (Munthe et al., 2001; Landis et al., 2002). The Tekran system measures all three Hg species in a semi-continuous fashion with a total of 12 sample cycles performed within a 24-hour period. Each complete single sample cycle consists of a 1-hour sampling period and a 1-hour desorbing period. The four major parts of the system: Model 2537A analyzer, Model 1130 pump module, Model 1130 denuder module, and Model 1135 particulate module, work in concert through a program installed in the Model 1130 controller.

During sampling, ambient air drawn at a flow rate of 10 liters per minute (L min^{-1}) first passes through a denuder housed in the Model 1130 denuder module. RGM in the air stream is collected by the potassium chloride (KCl) coating of the denuder with a diffusion coefficient > 0.1 cm^2 s^{-1}(Poissant et al., 2005). The air stream continuously travels to a regenerable particulate filter (RPF) assembly located within the Model 1135. Particulate Hg (2.5 μm) in the air stream is captured by a unique quartz filter disk, which resides in the RPF assembly. Any Hg^0 left in the air stream is then directed into the Model 2537A analyzer at a rate of 1 L min^{-1}, while the remaining 9 L min^{-1} is vented out through the Model 1130 pump module.

The Model 2537A is equipped with a pair of gold cartridges which allow alternate sampling and detection of Hg using Cold Vapor Atomic Florescence Spectrometry (CVAFS) at a wavelength of 253.7 nm. During the sampling period, Hg^0 is measured and detected every five minutes while RGM and PM are only collected but not measured. A total of twelve readings of Hg^0 concentrations are made in the sampling hour. During the desorbing period, PM was detected in three consecutive 5-minute intervals. The sum of these three 5-minute readings is the total PM collected during the sampling hour. Similarly, the sum of the three 5-minute readings during the denuder heating is the total RGM collected during the sampling hour.

The 10 L min^{-1} flow rate at the inlet corresponding to a standard temperature (0 °C) and pressure of 1 atmosphere (STP) was drawn together by the Model 2537A (1 L min^{-1}) and the Model 1130 denuder pump module (9 L min^{-1}). The flow rate and the design of the inlet insure that only particles with aerodynamic diameters 2.5 μm continue to the next stage of the sample system, while the coarse fraction (>2.5 μm) of the particles are removed by aerodynamic impaction onto a glass impactor plate housed inside the inlet glass-ware (Landis et al., 2002). A heating boot wrapped around the inlet keeps the inlet at 50 °C during the sampling period and at 65 °C during the desorbing period to prevent condensation problems. The RPF-assembly and the denuder are both kept at 50 °C during the sampling period, and are heated to 800 °C and 500 °C during the desorbing period, respectively. Throughout the whole 2-hour sample cycle, the sample air line and the zero air line are kept at 50 °C, while the case of Model 1130 denuder module and Model 1135 module are kept at 35 °C.

The measurement of Hg associated with coarse fraction (>2.5 m) particles has been performed to date only using manual methods and over longer time intervals than that available using a system such as the Tekran speciation system. In source regions and in coastal regions the measurement of size-segregated particulate eg is critical to understanding the loading of Hg to terrestrial and coastal ecosystems. Since the deposition velocities of the larger fraction Hg are an order of magnitude greater than the fine fraction Hg, the dry deposition of these large particles is likely to dominate the Hg flux to the surface. More research and development is required before a continuous approach for coarse particle Hg could be deployed in monitoring network. Initially, it is recommended that manual approaches be employed to develop the first order estimate of the importance of the coarse particles in the chemistry and deposition of Hg at different sites around the globe.

13.5 Coordinated Monitoring and Modelling

Integrated measurement programs form the core component of the overall research framework needed for the study of global air quality change. Models also represent a key component of a global research framework, as they are able to play many important roles such as: helping to determine the optimal locations for long-term observational sites and short-term process studies, in order to maximize the usefulness of the data collected; assimilating the observational data acquired from different monitoring systems and helping to place isolated measurements in a larger context; and providing simulations of future Hg trends, fluxes, and coupling with water quality changes. Careful integration of various types of models, modeling tools, and observations will be extremely important in the future. Atmospheric process models can be incorporated into regional-scale models to provide detailed representation of critical chemical and surface processes. Similarly, regional models are embedded in global models to increase the spatial and temporal resolution in areas with sensitive watersheds. Chapter 16 addresses some of these issues in more detail.

Coordinating measurements among various networks, with airborne platforms, and into other special studies, and integrating these measurements with detailed modeling studies presents an immensely complex research challenge. The CGMMN must go beyond model evaluation via comparison with observations/measurements. Model-observation comparisons can elucidate which models might be in error but do not provide information on which model processes are causing the error in large scale simulations. In addition, uncertainties in current Hg measurements can sometimes make results inconclusive. Therefore a comprehensive comparison to multiple types of observations/measurements is needed, and these comparisons should be defined a priori to provide information about the performance of processes in models. This is often difficult as the modeling community often may not understand the technical details of the Hg monitoring and analysis, or what data quality entails and uncertainties actually mean.

Currently, there is a lack of a centralized, standardized observational database that includes information on Hg data quality, data uncertainties and other measurement details. Matching the spatial and temporal scales of atmospheric model estimates to the measurements will ensure that meaningful comparisons can be performed. Coordination of the monitoring network with the global and regional modeling community can help to enhance the model-measurement comparisons and help identify key parameters that need to be measured in the network and at what spatial scales.

13.5.1 Importance of Establishing Boundary Conditions

The North American Mercury Model Intercomparison Study (NAMMIS) was conducted to provide guidance to the research community regarding which scientific process uncertainties are contributing most to the observed discrepancies in model simulations of Hg deposition. As a starting point each regional-scale model used the same inputs for initial (IC) and boundary conditions (BC), meteorology and emissions. The three IC/BC data sets were each developed from simulations of one of three global-scale atmospheric Hg models, the Chemical Transport Model (CTM) adapted for Hg (Shia et al., 1999; Seigneur et al., 2001), the GEOS-Chem model adapted for Hg (Yantosca, 2005; Selin et al., 2007), or the Global-Regional Atmospheric Heavy Metal (GRAHM) model (Dastoor and Larocque; 2004; Ariya et al., 2004). In addition, each regional-scale model uses the same horizontal modeling domain.

The study was conducted to, in part, separate the effects of input data and scientific process treatments within each model so they can be better understood. The model intercomparison showed that the models were sensitive to changes in Hg species concentrations at the boundary of the grid, and the lack of observations of Hg species severely limits our ability to evaluate the accuracy of the global model estimates for the boundary conditions, which varied significantly from one model to another. The observed model sensitivity also clearly suggests that global-scale transport is critical in regional-scale assessments. However, the magnitudes on the estimated wet deposition across North America varied significantly using the boundary conditions estimated using the three different models. It was clear from the study (see Figure 13.7) observations of Hg^0, RGM and Hg(p) are needed at all levels of the troposphere, especially along the western boundary of the North American continent. The model evaluations performed were thought to be largely subjective due to the lack of observation of the Hg species, especially aloft (Bullock, 2007).

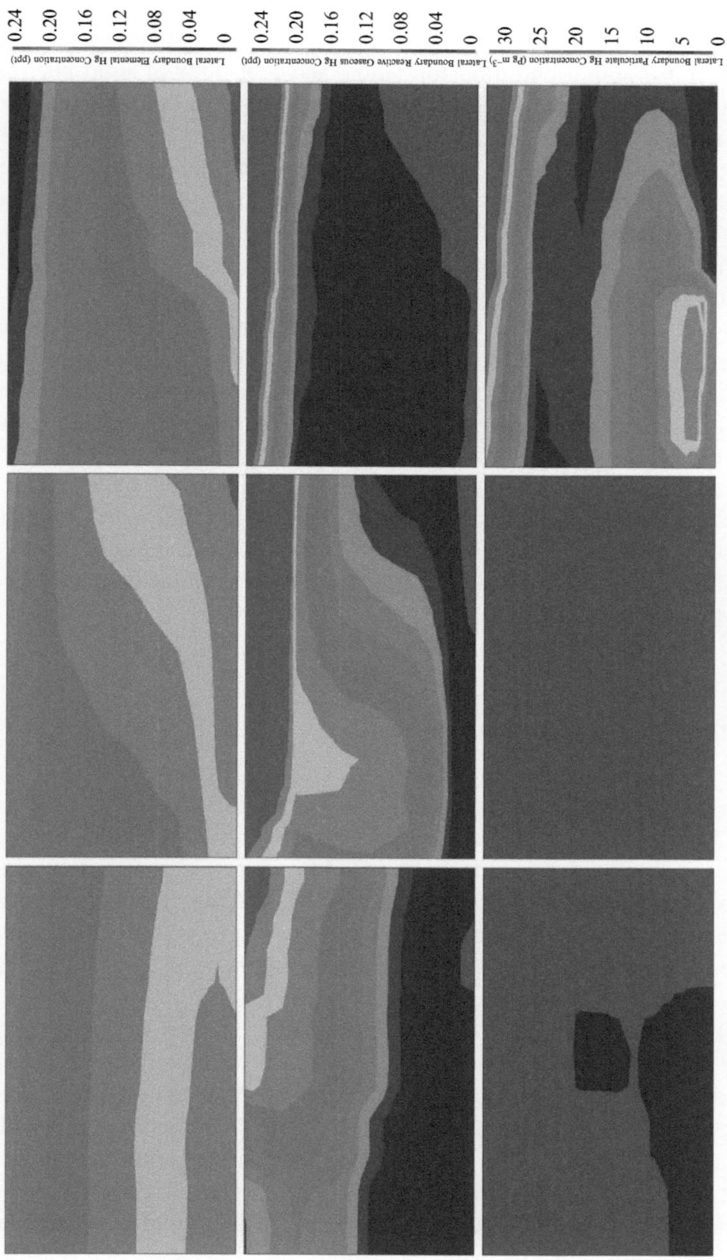

Figure 13.7 Annual Average Western Boundary Values for the three mercury species determined by the three Global Mercury Models for the NAMMIS

13.5.2 Identification of Key Measurements Parameters and Species

Reliably modeling the transport, transformation, and deposition of atmospheric Hg and elucidating the importance of local, regional, and global emission sources is currently limited because of uncertainties in its atmospheric chemistry (Sillman et al. 2007; Chapter 14). While Hg in the atmosphere is dominated by Hg^0 (Schroeder and Munthe, 1998), it is the RGM which is thought to be most important for wet and dry deposition. Elevated levels of RGM are typically associated with direct emissions from localized anthropogenic sources, but can also be produced by photochemical conversion from Hg^0 (Liu et al. 2007).

Ambient gaseous Hg species are affected by gas phase and aqueous photochemical reactions that involve a wide range of species (O_3, OH, Cl, Br and sulfates). Modeling the transport and transformation of Hg in the atmosphere is a challenge because it involves processes on widely different spatial and temporal scales. Deposition of Hg is affected by localized convective events, and processing by small-scale convective clouds can also affect photochemistry. Photochemical conversion from Hg^0 to RGM also results in the formation of TPM, which frequently occurs as part of multi-species conglomerates. Particulate bound Hg in the atmosphere has a bi-modal distribution with a significant fraction of the mass in the >2.5 μm size range (Keeler et al. 1995)

13.5.3 Four-D Data Assimilation

Development must continue on data assimilation methods for weather and climate prediction. They have led to remarkable progress in estimating global water and energy fluxes. Applying the same techniques to biogeochemistry can yield quantitative data for variables that have heretofore been unavailable. Significant progress has been made in validating physical models and in analyzing how calibration can improve their performance. Improvements in modeling have also been directed to problems of water management (e.g., Wood et al., 1993).

Air quality monitoring networks provide extensive ambient measurements of gaseous and particulate pollutant species. Most air quality monitoring sites are sited in populated areas, and are designed to determine whether or not the area is in compliance with national ambient air quality standards. There has been a lack of air monitoring sites across the globe that allow detailed investigation of long-range transport or that provide trends in background concentrations. Trends in pollutants such as ozone in particular are difficult to assess because changes due to anthropogenic forcing are often confounded by natural variations.

Figures 13.8 show comparison between model ambient RGM and measured values from the aircraft flights described in section 2.6. Figure 13.8 shows the variation of RGM with altitude in the model for the 5 d that correspond with measurements

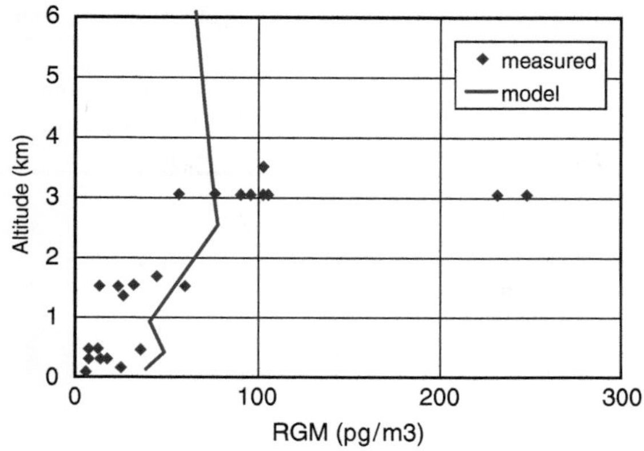

Figure 13.8 Measured RGM (*pg/m*³) versus altitude (*km*) from aircraft measurements over the Atlantic Ocean off the coast of south Florida during June 2000 (points). The line represents model RGM versus altitude, based on an average of model results during the afternoon on the 5 d (9, 12, 14, 25 and 26 June) that coincide with measurements (After Sillman et al. 2007)

(9, 12, 14, 25 and 26 June, always at 1700 LT). Figure 13.8 also shows measured RGM versus altitude for the full ensemble of measurements during June 2000, including days not represented by the model.

The full set of measurements is included here in order to show a complete picture of the observed variation with height. Results show that the model is consistent with measurements in many aspects, although there are also significant discrepancies. RGM increases with altitude from 0 to 3 km in both the model and in the measured ensemble. The rate of increase versus altitude is steeper for the ensemble of measurements than for the model, but the comparison is not extensive enough to show whether this is a consistent trend. Individual vertical profiles of RGM in the model sometimes show a complex layered pattern, reflecting cloud layers at various elevations, but the measurements represent flight path averages and cannot show this type of detail.

13.5.4 Observation-based Apportionment Methods for Emission Inventory Reconciliation

Table 13.1 presents data from the U.S. EPA emissions inventory for point sources within the SoFAMMS study domain developed from a national Hg emissions database (USEPA, 1998). Municipal waste incineration was estimated to be the largest Hg emission source type in south Florida. Municipal waste incineration was found to be the dominant source of Hg wet deposited to Everglades sites during the intensive study. An evaluation of the robustness of the Hg source apportionment was reported

Table 13.1 Mercury emissions inventory for Broward and Dade County, Florida USA

		U.S. EPA estimated Hg emission *(kg yr^{-1})*	Revised Hg emission (from this study) *(kg yr^{-1})*	Receptor model Hg source apportionment *(% of total)*
Municipal waste	Dade	1176.35		
	Broward	1125.64		
	total	2301.99	508 ± 101	57 ± 7
Medical waste	Dade	22.62		
	Broward	2.08		
	total	24.70	268 ± 127	30 ± 14[a]
Utility	Dade	3.25		
	Broward	7.36		
	total	10.61		
Boilers	Dade	67.50		
	Broward	43.70		
	Total	111.20	130 ± 43[b]	14 ± 5[b]

[a]Determined from emissions reconciliation
[b]Includes contributions from utility, industrial, and residential oil boilers.

after a detailed reconciliation between the source based emission inventory, the SoFAMMS stack testing data, and the receptor modeling results.

Stack testing was performed at a municipal waste incinerator in Dade County and this incinerator was estimated by the U.S. EPA to account for 50% of total Hg emissions from all municipal waste incinerators in Dade and Broward Counties (USEPA, 1998). Source testing performed during SoFAMMS indicated that emissions of Hg from this facility may have been overestimated in the U.S. EPA emissions inventory by a factor of 4. Although it was estimated in the emission inventory that 1156.1 kg yr^{-1} Hg were emitted from this municipal waste incinerator, source measurements determined this value to be much lower at 255.0 (50.5 kg yr^{-1} (Stevens et al. 1996)). As a means of re-assessing emissions from all municipal waste incinerators in Dade and Broward Counties, the factor of four over-estimation was assumed to be consistent for the source category. This factor was applied to all four municipal waste incinerators in Dade and Broward Counties, which lowered the total estimate for municipal waste Hg emissions from 2302 to 508 (101 kg yr^{-1}).

Stack emissions testing was performed at the largest medical waste incinerator in Dade County which accounted for 37% of total Hg emissions from all medical waste incinerators in Dade and Broward Counties. Stack testing indicated that emissions of Hg from this facility was likely underestimated in the U.S. EPA emissions inventory by a factor of 10 for this category in south Florida. Although it was estimated in the emission inventory that 9.2 kg yr^{-1} Hg were emitted from this medical waste incinerator, source measurements determined this value to be much higher at 100.4 (47.6 kg yr^{-1}) (Stevens et al. 1996). Again, assuming this factor to be consistent for the source category, this underestimation factor was applied to all 12 medical waste incinerators in Dade and Broward Counties. This brought the total estimate for medical waste Hg emissions up from 25 to 268 (127 kg yr^{-1}).

The use of observations by Dvonch et al. 1999 revealed that 57% of the Hg wet deposited at the Everglades sites was accounted for by municipal waste incineration. Since nearly all of the Hg emitted from waste incineration sources was measured during the study in a highly soluble RGM form, these emissions should be directly proportional to the measured wet deposition. Therefore, the 57% Hg contribution from municipal waste incineration emissions would correspond to the 508 kg yr^{-1} Hg estimated to be emitted from municipal waste incinerators within the SoFAMMS study domain. Using this approach, the 268 kg yr^{-1} Hg estimated to be emitted from medical waste emissions would correspond to a 30 (14%) contribution of Hg wet deposited at the Everglades sites (with the uncertainty estimate derived from the variability in the SoFAMMS medical waste incinerator stack data). It was previously determined that 29% of the Hg measured at the Everglades sites was left unaccounted for by the PCA/MLR receptor modeling approach. Therefore, from the previous analysis, it was estimated that Hg emissions from medical waste incineration sources within the SoFAMMS study domain may account for the 29% of Hg wet deposition at the Everglades sites that was previously left unexplained.

As evidenced by the large degree of uncertainty associated with the apportionment of Hg to medical waste incineration, it was acknowledged that additional sources within the study domain as well as regional sources associated with long-range transport may have contributed to the Hg deposited and could be responsible for a portion of the deposition unaccounted for by the PCA/MLR model. Although the multivariate model did not identify any additional significant source(s) of the remaining (29%) Hg, a number of smaller sources that did not significantly add to the variance in the data set may have contributed to the measured deposition. However, like the medical waste incinerators, they were not traceable using the methods employed in this study.

Mercury stack testing was not performed at oil combustion sources during SoFAMMS. However, the emission/deposition relationship established above was also used to predict Hg emissions from utility, industrial, and residential boilers within the SoFAMMS domain. The 14% of Hg in precipitation at Everglades sites determined from PCA/MLR analysis to be contributed from oil combustion sources would equate to an emission of 130 (43 kg yr^{-1}) Hg from oil boilers. This estimated value of Hg emitted is in agreement with the 122 kg yr^{-1} Hg that was calculated by the U.S. EPA to be emitted from oil combustion associated with utility, industrial, and residential boilers within the SoFAMMS study domain. However, since previous source tests of fossil fuel combustion sources within the United States have indicated the ratio of $Hg^{(II)}$/Hg(total) in emissions to be much less than 0.5, these source reconciliation results suggest that either the oil combustion sources in south Florida are emitting a much greater amount of total Hg or that they are simply emitting a much higher percentage of $Hg^{(II)}$ than previous estimates suggested. Lastly, it was estimated that 1.8 kg day^{-1} Hg, on average, is wet deposited to the Florida Everglades during the spring and summer seasons. The revised Hg emissions inventory, calculated using the actual emissions measured at sources in south Florida during SoFAMMS, indicated that 2.5 (0.5 kg day^{-1}) Hg was emitted from municipal waste, medical waste, and oil combustion sources in Dade and Broward Counties alone.

Therefore, enough Hg was emitted in these two counties alone, which was measured during the study in a soluble $Hg^{(II)}$ form that is easily incorporated into cloud droplets, to account for the Hg wet deposited to the entire Florida Everglades.

13.5.5 Case Study of Coordinated Measurement/Modeling in the Mediterranean

Both box model and regional model studies have been carried out for the Mediterranean region to study the Hg cycle between the sea surface and the atmosphere. The models were used to elucidate the chemical mechanisms which oxidise Hg in the marine boundary layer (MBL) and to estimate both Hg deposition and re-emission. These modeling studies made use of data obtained during measurement campaigns performed as part of three EU funded projects (MAMCS, MOE and MERCYMS). The measurement campaigns of two and four weeks were performed in each project (one each season). This relatively sparse database of measurements was supplemented by both atmospheric and aquatic measurements performed aboard the Italian CNR *R.V.* Urania. More recently two monitoring stations, one in the Calabria mountains, and the other on the Tyrrhenian coast of Calabria, have come on-line and provide continuous measurement data for Hg and ancillary pollutants. Using the MECAWEx model (Hedgecock, et al. 2006) the twelve-month period during which the four MAMCS measurement campaigns (one each season) were performed was simulated. The simulation results were in reasonable to good agreement

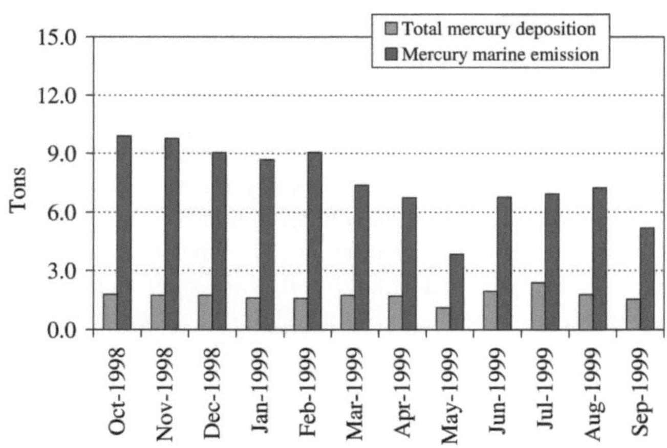

Figure 13.9 Calculated monthly total deposition and evasion of mercury (1 ton=1Mg) to and from the surface of the Mediterranean Sea during the MAMCS campaign (Hedgecock et al. 2006)

with the total Hg observed at the five measurement sites, and allowed estimates to made of the annual deposition to and emission from the surface of the Mediterranean Sea (Figures 13.9 and 13.10).

Further studies were performed comparing MECAWEx output with the results obtained from research cruises aboard the R.V. *Urania*, (the Med-Oceanor campaigns) which added another dimension to the quality of the comparison between model and measurements because $Hg^0_{(g)}$, RGM and TPM were measured. The other benefit that ship board measurements have is that the measurements are performed in what in the model is the first layer of the atmosphere and that which is directly influenced by the parametrisation employed to calculate the evasive flux from the sea surface, Figure 13.11.

The only measured data available for Hg deposition within the modelling domain at present is from some of the EMEP stations in northern Europe. These data, which are mostly available as monthly totals, were used to validate the model results, Figure 13.12. It would be extremely useful however for modellers to have Hg in precipitation data available on an event basis as this really would illustrate the model's skill in reproducing Hg deposition fields.

The results from the oceanographic cruises were also studied using a photochemical box model, which uses a much more complex chemical scheme than MECAWEx, including as it does both sea salt and non-sea-salt sulphate aerosol phases in order to simulate the release of reactive halogen containing compounds from aerosol particles, (Hedgecock, et al. 2005). The studies suggested that the cycling of Hg emitted from the sea surface could be quite rapid given the appropriate meteorological conditions.

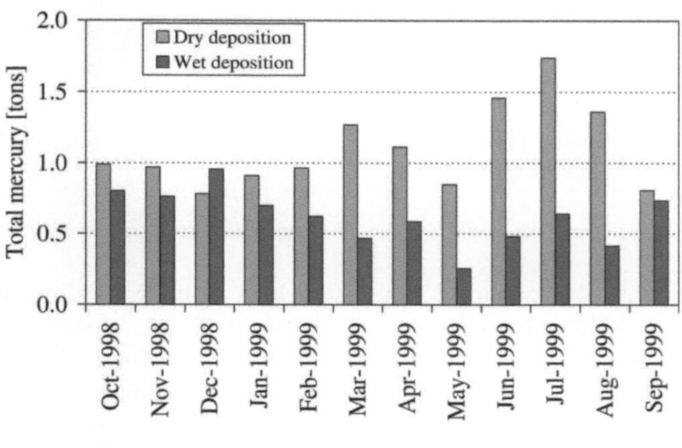

Figure 13.10 Calculated monthly dry and wet deposition (1 ton=1Mg) to the surface of the Mediterranean Sea during the MAMCS campaign (Hedgecock et al. 2006)

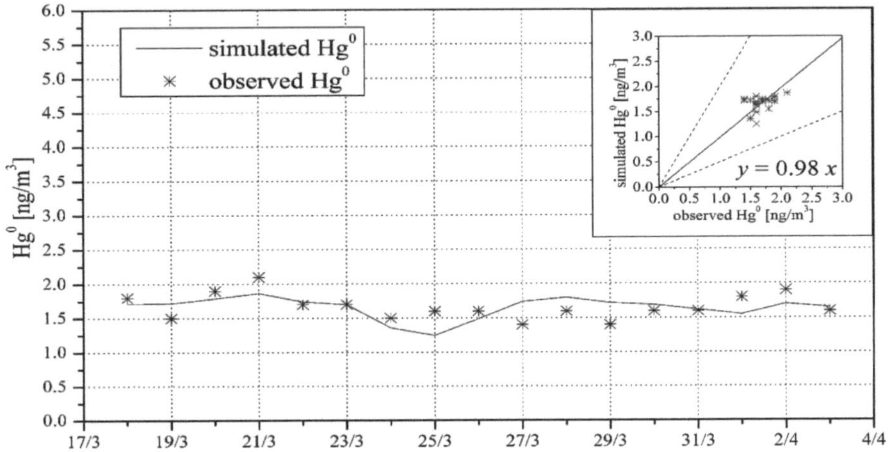

Figure 13.11 The measured and calculated $Hg^0_{(g)}$ concentration during the Med-Oceanor 2004 oceanographic campaign (Hedgecock et al. 2006)

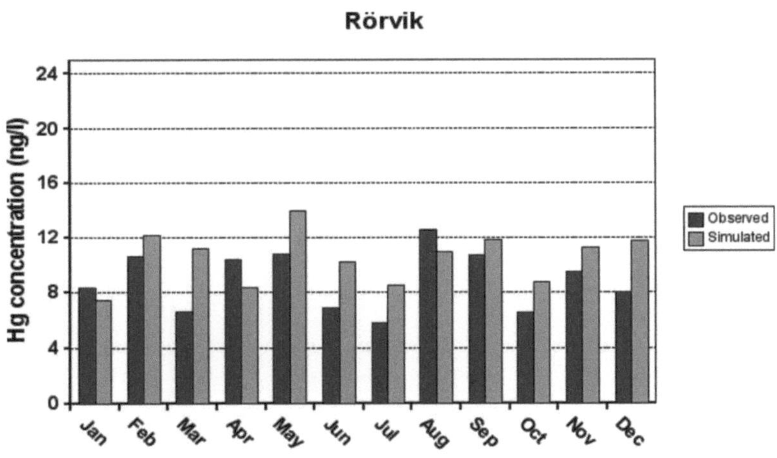

Figure 13.12 Comparison of monthly average Hg concentration in rain for the EMEP site at Rorvik during 2000 (Pirrone et al., 2008)

However, notwithstanding the progress made in the understanding of Hg cycling in the MBL and the ability to reproduce reasonably TGM values over a region as large as the Mediterranean, there remain some uncertainties which cannot be addressed using a measurement campaign approach. The hourly, daily and monthly variation in TGM does not actually cover a very wide range of values; this makes it difficult not only to be certain that the model is reproducing atmospheric transport and transformation phenomena accurately, but it also makes statistical analyses of the simulated vs. observed results less easy to interpret. As regional Hg models have improved over the years it has become clear that global models are required to provide time-dependent boundary conditions, as discussed earlier in this chapter. The results obtained with MECAWEx showed that, even over a domain which covers Europe, the whole of the Mediterranean Sea and a substantial part of N. Africa, the simulated concentration fields at the centre of the domain can still be sensitive to the boundary conditions when transport into the model domain is relatively rapid, for example when there are strong westerlies. During typical Mediterranean anticyclonic summer conditions the boundary conditions play a less significant role due to restricted transport.

Estimating annual deposition and emission fluxes in the case of the Mediterranean is also fraught with uncertainty for two fundamental reasons. Firstly, the model validation over the long term was based on the results from four two week measurement campaigns which measured TGM and not speciated Hg concentrations. Admittedly this situation was improved by the results obtained from the cruises, but the second point of uncertainty is that there are no measurements of deposition fluxes for the Mediterranean region, neither wet nor dry. Dry deposition of Hg is extremely difficult to quantify but Hg in precipitation is regularly measured in some parts of the world.

A majority of the problems discussed above could be resolved by the establishment of a monitoring network, with some monitoring stations measuring speciated Hg compounds and Hg in precipitation. If this were to be done, then with the computing power which is currently available it would be possible to use models much more extensively. Combining models and monitoring data it would provide a more solid base than currently exists for assessing the reliability of emission inventories, assessing atmospheric chemistry parameterisations, quantifying the influence of regional and global sources and therefore also for evaluating the effectiveness of pollution control initiatives and strategies on both local and global scales in combating the impact on ecosystems of Hg deposition.

References

Ariya P, Dastoor A, Amyot M, Schroeder W, Barrie L, Anlauf K, Raofie F, Ryzhkov A, Davignon D, Lalonde J, Steffen A. 2004. Arctic: A sink for mercury. Tellus, 56B: 397–403.

Banic, S.T. Beauchamp, R.J. Tordon, W.H. Schroeder, A. Steffen, K.A. Anlauf and H.K.T. Wong, Vertical distribution of gaseous elemental mercury in Canada. J. Geophys. Res. 108 D9 (2003), p. 4264.

Bergan et al. (1999). L. Gallardo and H. Rodhe, Mercury in the global troposphere: a three-dimensional model study. Atmos. Environ. 33 (1999), pp. 1575–1585.

Bergan T. and H. Rodhe, Oxidation of elemental mercury in the atmosphere; constraints imposed by global scale modeling, Journal of Atmospheric Chemistry 40 (2001), pp. 191–212.

Bergamaschi, P., R. Hein, M. Heimann, and P.J. Crutzen (2000), Inverse modeling of the global CO cycle 1. Inversion of CO mixing ratios, J. Geophys. Res., 105(D2), 1909–1927.

Bey, I., D. Jacob, R. Yantosca, J. Logan, B. Field, A. Fiore, Q. Li, H. Liu, L. Mickley, and M. Schultz (2001), Global modeling of tropospheric chemistry with assimilated meteorology: Model description and evaluation, J. Geophys. Res., *106*(D19), 23,073–23,096.

Bullock and Brehme (2002). Atmospheric mercury simulation using the CMAQ model: formulation, description and analysis of wet deposition results. Atmos. Environ. 36 (2002), pp. 2135–2146.

Bullock, R. Jr., Benjey W.G. and Keating, M.H. (1997) The modeling of regional scale atmospheric mercury transport and deposition using RELMAP. In: J.E. Baker, Editor, Atmospheric Deposition of Contaminants to the Great Lakes and Coastal Waters, SETAC, Pensacola, FL (1997), pp. 323–347.

Bullock O.R., Brehme KA. 2002. Atmospheric mercury simulation using the CMAQ model: formulation, description, and analysis of wet deposition results. Atmospheric Environment; 36: 2135–2146.

Bullock O. R., (2007) Braverman T., Chapter 2.2 Application of the CMAQ mercury model for U.S. EPA regulatory support Developments in Environmental Sciences, Vol.6 pp. 85–95

A.P. Dastoor, Cloudiness parameterization and verification in a large-scale atmospheric model. Tellus 46A (1994), pp. 615–634.

Dastoor AP, Larocque Y. 2004. Global circulation of atmospheric mercury: A modeling study. Atmos. Environ. 38, 147–161.

Daum, P. H., L. I. Kleinman, S. R. Springston, L. J. Nunnermacker, Y.-N. Lee, J. Weinstein-Lloyd, J. Zheng, and C. M. Berkowitz (2004), Origin and properties of plumes of high ozone observed during the Texas 2000 Air Quality Study (TexAQS 2000), J. Geophys. Res., 109, D17306, doi:10.1029/2003JD004311.

Derwent, R. G., P.G. Simmonds, A.J. Manning and T.G. Spain, Trends over a 20-year period from 1987 to 2007 in surface ozone at the atmospheric research station, Mace Head, Ireland, Atmos. Environ., 41, 9091–9098, 2007.

Driscoll, C.T., Han, Y.J., Chen, C.Y., Evers, D.C., Lambert, K.F., Holsen, T.M., Kamman, N. Dvonch, J.T., Graney, J.R., Marsik, F.J., Keeler, G.J., Stevens R.K. (1998) An investigation of source–receptor relationships for mercury in south Florida using event precipitation data. The Science of The Total Environment, Vol. 213, Issues 1-3, Pages 95–108.

Dvonch, J.T., Graney, J.R., Keeler, G.J., and Stevens, R.K. Utilization of Elemental Tracers to Source Apportion Mercury in South Florida Precipitation. Environmental Science and Technology 33, 4522–4527. 1999.

Dvonch, J.T., Marsik, F.J. and Keeler, G.J. The Use of WSR-88D Radar Data for Source-Apportionment of Wet-Deposition Measurements from the 1995 SoFAMMS. Journal of Climate and Applied Meteorology. 1421–1435, 2005.

Ebinghaus R, Slemr F. 2000. Aircraft measurements of atmospheric mercury over southern and eastern Germany. Atmos. Environ., 34(6): 895–903.

Ebinghaus, R, R.M. Tripathi, D. Wallshlager and S.E. Lindberg, Natural and anthropogenic mercury sources and their impact on the air-surface exchange of mercury on regional and global scale. In: R. Ebinghaus et al. *Mercury Contaminated Sites*, Springer, New York (1999), pp. 1–50.

Fitzgerald, W.F., Lamborg, C.H. and Hammerschmidt, C.R., 2007. Marine biogeochemical cycling of mercury. Chemical Reviews, 107(2): 641–662.

Friedli, H.R. et al., 2004. Mercury in the atmosphere around Japan, Korea, and China as observed during the 2001 ACE-Asia field campaign: Measurements, distributions, sources, and implications. J. Geophys. Res.-Atmospheres, 109(D19).

Frost G. J., et al. (2006), Effects of changing power plant NO x emissions on ozone in the eastern United States: Proof of concept, J. Geophys. Res., 111, D12306, doi:10.1029/2005JD006354.

Fu, T.-M., D. J. Jacob, P. I. Palmer, K. Chance, Y. X. Wang, B. Barletta, D. R. Blake, J. C. Stanton, and M. J. Pilling (2007), Space-based formaldehyde measurements as constraints on volatile organic compound emissions in east and south Asia and implications for ozone, J. Geophys. Res., 112, D06312, doi:10.1029/2006JD007853.

Gardfeldt, K., and M. Jonsson, Is Bimolecular Reduction of Hg(II) Complexes Possible in Aqueous Systems of Environmental Importance, J. Phys. Chem. A. 107 (22); 4478–4482, 2003.

Gildemeister, A.E., Keeler, G.J. and Graney, J.R. Source proximity reflected in spatial and temporal variability in particle and vapor phase Hg concentrations in Detroit, MI. Atmos. Environ. 38, 5227–5236. 2005.

Hedgecock, I. M., Trunfio, A., Pirrone, N., Sprovieri, F. (2005) Mercury Chemistry in the MBL: Mediterranean Case and Sensitivity Studies Using the AMCOTS (Atmospheric Mercury Chemistry Over the Sea) Model. Atmospheric Environment, 39, 7217–7230.

Hedgecock, I.M., Pirrone, N., Trunfio, G., Sprovieri, F. (2006) Integrated mercury cycling, transport, and air-water exchange (MECAWEx) model. Journal of Geophysical Research, 111 (D20302), doi: 10.1029/2006JD007117.

Horowitz, L. W., A. M. Fiore, G. P. Milly, R. C. Cohen, A. Perring, P. J. Wooldridge, P. G. Hess, L. K. Emmons, and J.-F. Lamarque (2007), Observational constraints on the chemistry of isoprene nitrates over the eastern United States, J. Geophys. Res., 112, D12S08, doi:10.1029/2006JD007747.

Hoyer M, Burke J, Keeler G. 1995. Atmospheric sources, transport and deposition of mercury in Michigan: two years of event precipitation. Water Air Soil Poll., 80: 199–208.

Jaffe, D., H. Price, D. Parrish, A. Goldstein, and J. Harris (2003), Increasing background ozone during spring on the west coast of North America, Geophys. Res. Lett., 30(12), 1613, doi:10.1029/2003GL017024.

Keeler, G.J., Glinsorn, G., Pirrone, N., 1995 Particulate mercury in the atmosphere: Its significance, transport, transformations and sources. Water Air Soil Poll. 80, 159–168.

Keeler, G.J., Hoyer, M. 1997. Recent measurements of atmospheric mercury in the Great Lakes region. In: Atmospheric Deposition of Contaminants to the Great Lakes and Coastal Waters. (Baker JE, editor), SETAC Press, Pensacola, FL, USA, 477pp.

Keeler, G.J. and Dvonch, J.T. Atmospheric Mercury: A Decade of Observations in the Great Lakes. In: Dynamics of Mercury Pollution on Regional and Global Scales: Atmospheric Processes and Human Exposures around the World. N. Pirrone and K. Mahaffey Eds. Kluwer Ltd. 2005.

Keeler, G.J., Gratz, L. and Al-Wali, K. Influences on the Long-term Atmospheric Mercury Wet Deposition at Underhill, Vermont. Ecotoxicology, 14, 71–83. 2005.

Kellerhals, M., S. Beauchamp, W. Belzer, P. Blanchard, F. Froude, B. Harvey, K. McDonald, M. Pilote, L. Poissant, K. Puckett, B. Schroeder, A. Steffen and R. Tordon, Temporal and spatial variability of total gaseous mercury in Canada: results from the Canadian Atmospheric Mercury Measurement Network (CAMNet). Atmos. Environ. 37 7 (2003), pp. 1003–1011.

Lamborg, C.H. et al., 2002. Modern and historic atmospheric mercury fluxes in both hemispheres: Global and regional mercury cycling implications. Global Biogeochemical Cycles, 16(4): art. no.-1104.

Landis MS, Keeler, GJ. 1997. A critical evaluation of an automatic wet-only precipitation collector for mercury and trace element determinations. Environ. Sci. Technol. 31: 2610–2615.

Landis, M.S., Stevens, R.K., Schaedlich, F., Prestbo, E.M., 2002. Development and characterization of an annular denuder methodology for the measurement of divalent inorganic reactive gaseous mercury in ambient air. Environ. Sci. Technol. 36, 3000–30009.

Lei, W., B. de Foy, M. Zavala, R. Volkamer, and L. T. Molina Characterizing ozone production in the Mexico City Metropolitan Area: a case study using a chemical transport model, Atmos. Chem. Phys., 7, 1347–1366, 2007.

Li, Q., D.J. Jacob, I. Bey, P.I. Palmer, B.N. Duncan, B.D. Field, R.V. Martin, A.M. Fiore, R.M. Yantosca, D.D. Parrish, P.G. Simmonds, and S.J. Oltmans, Transatlantic transport of pollution and its effects on surface ozone in Europe and North America, J.Geophys. Res., 107, doi: 10.1029/2001JD001422, 2002.

Lin, C., Pehkonen, S.O., 1999. The chemistry of atmospheric mercury: a review. Atmos. Environ. 33, 2067–2079.

Lin, C-J., P. Pongprueksa, S. E. Lindberg, S. O. Pehkonen, D. Byun, C. Jang, Scientific uncertainties in atmospheric mercury models I: Model science evaluation, Atmos. Environ. 40 (2006) 2911–2928.

Lin, C-J., P. Pongprueksa, O.R. Bullock, S.E. Lindberg, S.O. Pehkonen, C. Jang, T. Braverman and T.C. Ho, Scientific uncertainties in atmospheric mercury models II: sensitivity analysis in the CONUS Domain, Atmos. Environ. 41 (2007), pp. 6544–6560.

Liu, B., Keeler, G. J., Dvonch, J. T., Barres, J. A., Lynam, M. M., Marsik, F. J., and Morgan, J.T.: Temporal variability of mercury speciation in urban air, Atmos. Environ., 41, 1911–1923, 2007.

Logan L.A., et al., (1999), Trends in the vertical distribution of ozone: A comparison of two analyses of ozonesonde data, J. Geophys. Res., 104, D21, 26373–26399.

Lynam, M.M. and Keeler, G.J. Automated speciated mercury measurements in Michigan, Environ. Sci. Technol. 39, 3289–3299, 2005.

Mason, R.P., Fitzgerald, W.F. and Morel, F.M.M., 1994. The aquatic biogeochemistry of elemental mercury. Geochim. Cosmochim. Acta, 58: 3191–3198.

Mason, R.P. and Sheu, G.R., 2002. Role of the ocean in the global mercury cycle. Global Biogeochemical Cycles, 16(4): art. no.-1093.

Mason, R.R. et al., 2005. Monitoring the response to changing mercury deposition. Environ. Sci. Technol., 39(1): 14A–22A.

Miller, E.K., Van Arsdale, A., Keeler, G. J., Chalmers, A. Poissant, L. and Kammen, N. Estimation and Mapping of Wet and Dry Mercury Deposition Across Northeastern North America. Ecotoxicology, 14, 53–70. 2005.

Müller, J.F. and T Stavrakou (2005), Inversion of CO and NOx emissions, Atmos. Chem. Phys., 5, 1157–1186.

Munthe, J., Wangberg, I., Pirrone, N., Iverfeld, A., Ferrara, R., Ebinghaus, R., Feng., R., Gerdfelt, K., Keeler, G.J., Lanzillotta, E., Lindberg, S.E., Lu, J. (2001). Intercomparison of Methods for Sampling and Analysis of Atmospheric Mercury Species. Atmospheric Environment.Vol. 35, 3007–3017.

Munson, C., R.K., 2007. Mercury contamination in remote forest and freshwater ecosystems in the northeastern U.S.: sources, transformations and management options. BioScience 57 (1)

Olson, J. R., J. H. Crawford, G. Chen, W. H. Brune, I. C. Faloona, D. Tan, H. Harder, and M. Martinez (2006). A reevaluation of airborne HOx observations from NASA field campaigns, J. Geophys. Res., 111, D10301, doi:10.1029/2005JD006617.

Oltmans, S.J., et al., (2006), Long-term changes in tropospheric ozone, Atmos. Environ., 40, 3156–3173.

Parrish, D. D., et al. (2004), Changes in the photochemical environment of the temperate North Pacific troposphere in response to increased Asian emissions, J. Geophys. Res., 109, D23S18, doi:10.1029/2004JD004978.

Pai, P. P. Karamchandani and C. Seigneur, Simulation of the regional atmospheric transport and fate of mercury using a comprehensive Eulerian model. Atmos. Environ. 31 (1997), pp. 2717–2732.

Parrish, D. D., J. S. Holloway, M. Trainer, P. C. Murphy, G. L. Forbes, and F. C. Fehsenfeld. Export of North American ozone pollution to the North Atlantic Ocean. Science, 259, 1436–1439, 1993.

Pirrone, N., G. J. Keeler, I. Allegrini (1996). Particle size distributions of atmospheric mercury in urban and rural areas. Journal of Aerosol Science,Vol. 27, Suppl. 1,1996, pp.S13–S14.

Pleijel K. and J. Munthe, Modelling the atmospheric mercury cycle-chemistry in fog droplet, Atmospheric Environment 29 (1995), pp. 1441–1457.

Pongprueksa, P., C-J. Lin, S.E. Lindberg, C. Jang, T. Braverman, O.R. Bullock, T.C. Ho and H-W. Chu, Scientific uncertainties in atmospheric mercury models III: Boundary and initial conditions, model grid resolution, and $Hg^{(II)}$ reduction mechanism, Atmos. Environ., 2007, in press, doi:10.1016/j.atmosenv.2007.11.020.

Poissant L., Pilote, M., Beauvais, C., Constant, P., Zhang, H.H., 2005. A year of continuous measurements of three atmospheric mercury species (GEM, RGM, and Hgp) in southern Quebec, Canada. Atmos. Environ. 39, 1275–1287.

Prospero, J. M., D. L. Savoie, and R. Arimoto, (2003), Long-term record of nss-sulfate and nitrate in aerosols on Midway Island, 1981–2000: Evidence of increased (now decreasing?) anthropogenic emissions from Asia, Journal Geophysical Research 108(D1),4019, doi:10.1029/ 2001JD001524.

Rea, A.W., S.E. Lindberg and G.J. Keeler, Dry deposition and foliar leaching of mercury and selected trace elements in deciduous forest throughfall. Atmos. Environ. 35 (2001), pp. 3453–3462.

Ren, X., et al. (2006), OH, HO2, and OH reactivity during the PMTACS–NY Whiteface Mountain 2002 campaign: Observations and model comparison, J. Geophys. Res., 111, D10S03, doi:10.1029/2005JD006126.

Roberts, J. M. et al., (2004), Measurement of peroxycarboxylic nitric anhydrides (PANs) during the ITCT 2K2 aircraft intensive experiment, J. Geophys. Res., 109(D23S21), doi: 10.1029/2004JD004960.

Savoie, D.L., R. Arimoto, W.C. Keene, J.M. Prospero, R.A. Duce and J.N. Galloway, (2002), Marine biogenic and anthropogenic contributions to non-sea-salt-sulfate in the marine boundary layer over the North Atlantic, J. Geophys. Res., 107, 4356, doi:1029/2001JD000970.

Schroeder, W.H., Munthe, J., 1998. Atmospheric mercury --- an overview. Atmos. Environ. 32, 809–822.

Schroeder, W.H. K.G. Anlauf, L.A. Barrie, J.Y. Lu, A. Steffen, D.R. Schneeberger and T. Berg, (1998) Arctic springtime depletion of mercury. Nature 394 (1998), pp. 331–332.

Seigneur C, Karamchandani P, Lohman K, Vijayaraghavan K, Shia R.-L. 2001. Multiscale modeling of the atmospheric fate and transport of mercury. J. Geophys. Res. 106(D21), 27,795–27,809.

Selin, NE, Jacob DJ, Park RJ, Yantosca RM, Strode S, Jaegle L, Jaffe D. 2007. Chemical cycling and deposition of atmospheric mercury: Global constraints from observations. . J. Geophys. Res , 112: D02308, doi:10.1029/2006JD007450.

Selin, N.E. et al., 2008. Global 3-D land-ocean-atmosphere model for mercury: present-day vs. preindustrial cycles and anthropogenic enhancement factors for deposition. Global Biogeochem. Cycles, Accepted.

Shia RL, Seigneur C, Pai P, Ko M, Sze ND. 1999. Global simulation of atmospheric mercury concentrations and deposition fluxes. . J. Geophys. Res , 104(D19), 23,747–23,760.

Sheu G, Lee C, Lin N. 2007. Measurements of atmospheric mercury at a high elevation site (Lulin Atmospheric Background Station, LABS). Abstract A53C-1350, AGU Fall meeting, San Francisco, 2005.

Sillman, S., D. He, M. Pippin, P. Daum, L. Kleinman, J. H. Lee and J. Weinstein-Lloyd. Model correlations for ozone, reactive nitrogen and peroxides for Nashville in comparison with measurements: implications for VOC-NOx sensitivity. J. Geophys. Res. 103, 22629–22644, 1998.

Sillman, S., F. J. Marsik, K. I. Al-Wali, G J. Keeler, and M. S. Landis (2007), Reactive mercury in the troposphere: Model formation and results for Florida, the northeastern United States, and the Atlantic Ocean, J. Geophys. Res., 112, D23305, doi:10.1029/2006JD008227.

Simmonds P., R. Derwent, A. Manning, and G. Spain, (2004), Significant growth in surface ozone at Mace Head, Ireland, 1987–2003, Atmos. Environ., 38(28), 4769–4778.

Slemr, F. et al., 2003. Worldwide trend of atmospheric mercury since 1977. Geophys. Res. Lett., 30(10).

Slemr (1992). F. Slemr and E. Langer, Increase in global atmospheric concentrations of mercury inferred from measurements over the Atlantic Ocean. Nature 355 (1992), pp. 434–437.

Slemr et al (1995). F. Slemr, W. Junkermann, R.W.H. Schmidt and R. Sladkovic, Indication of change in global and regional trends of atmospheric mercury concentrations. Geophysical Research Letters 22 (1995), pp. 2143–2146.

Stevens, R. K.; Zweidinger, R.; Edgerton, E.; Mayhew, W.; Kellog, R.; Keeler, G. Source Characterization in Support of Modeling the Transport of Mercury Emissions in South Florida. Presented at Measurement of Toxic and Related Air Pollutants Symposium, May 7-9, Research Triangle Park, NC, 1996. (25) Dzubay, T.; Stevens, R.; Lewis, C.; Hern, D.; Courtney, W.; Tesch,

Streets, D.G., Q. Zhang, L. Wang, K. He, J. Hao, Y. Wu, Y. Tang, and G.R. Carmichael (2006), Revisiting China's CO emissions after the Transport and Chemical Evolution over the Pacific (TRACE-P) mission: Synthesis of inventories, atmospheric modeling, and observations, J. Geophys. Res., 111, D14306.

Strode, S.A. et al., 2008. Trans-Pacific transport of mercury. J. Geophys. Res.-Atmospheres, 113 (D15).

Sunderland, E.M. and Mason, R.P., 2007. Human impacts on open ocean mercury concentrations. Global Biogeochem. Cycles, 21: GB4022, doi:10.1029/2006GB002876.

Swartzendruber, P.C. et al., 2008. Vertical distribution of mercury, CO, ozone, and aerosol scattering coefficient in the Pacific Northwest during the spring 2006 INTEX-B campaign. J.Geophys. Res.-Atmospheres, 113 (D10).

Talbot, R., Mao, H., Scheuer, E., Dibb, J. and Avery, M., 2007. Total depletion of Hg degrees in the upper troposphere-lower stratosphere. Geophys. Res. Lett., 34 (23).

Task Force on Hemispheric Transport of Air Pollution (TF-HTAP), Hemispheric Transport of Air Pollution 2007, United Nations Economic Commission for Europe, Air Pollution Studies No. 16, Report number ECE/EB.AIR/94, 2007. Available at www.htap.org/activities/2007 _ interim_report/HTAP 2007 EB version.pdf

Trainer, M., D. D. Parrish, M. P. Buhr, R. B. Norton, F. C. Fehsenfeld, K. G. Anlauf, J. W. Bottenheim, Y.Z. Tang, H.A. Wiebe, J.M. Roberts, R.L. Tanner, L. NewmanCorrelation of ozone with NOy in photochemically aged air. J. Geophys. Res., 98, 2917–2926, 1993.

U.S. Environmental Protection Agency. Mercury Study Report to Congress; EPA-452/R-97-003; Office of Air Quality Planning and Standards, Office of Research and Development, U.S. Government Printing Office: Washington, DC, 1998.

VanArsdale, A., Weiss, J., Keeler, G.J. and Miller, E. Patterns of mercury deposition in northeastern North America (1996-2002). Ecotoxicology, 14, 84–101. 2005.

Weiss-Penzias, P., D. A. Jaffe, P. Swartzendruber, J. B. Dennison, D. Chand, W. Hafner, and E. Prestbo, Observations of Asian air pollution in the free troposphere at Mt. Bachelor Observatory in the spring of 2004, J. Geophys. Res., 110, D10304, doi:10.1029/2005JD006522, 2006.

Weiss-Penzias, P., D. A. Jaffe, P. Swartzendruber, W. Hafner, D. Chand, and E. Prestbo, Quantifying Asian biomass burning sources of mercury using the Hg/CO ratio in pollution plumes observed at the Mount Bachelor Observatory. Article In-Press, Atmos. Environ., February 2007.

Woods JS, Martin MD, Naleway CA, Echeverria D. Urinary porphyrin profiles as a biomarker of mercury exposure: studies in dentists with occupational exposure to mercury vapor. J Toxicol Environ Health. 1993;40:235–246.

Yantosca B. 2005. GEOS-Chem v7-03-06 User's Guide, Atmospheric Chemistry Modeling Group, Harvard University, Cambridge, MA, posted 8 November 2005 at www.as.harvard.edu/chemistry/trop/geos/doc/man/index.html.

Part III
Understanding Atmospheric Mercury on Hemispheric and Global Scales

Chapter 14
Our Current Understanding of Major Chemical and Physical Processes Affecting Mercury Dynamics in the Atmosphere and At the Air-Water/Terrestrial Interfaces

Anthony J. Hynes, Deanna L. Donohoue, Michael E. Goodsite, and Ian M. Hedgecock

Summary The predictions of atmospheric chemical models are limited by the accuracy of our understanding of the basic physical and chemical processes that underlie the models. In this work we review the current state of our knowledge of the chemical processes that transform atmospheric mercury species via gas and aqueous phase reactions and the physical processes of deposition. We concur with the conclusions of other recent reviews that our understanding of the basic chemistry that controls mercury is incomplete and the experimental data either limited or non-existent. In spite of this recent experimental and theoretical studies of mercury reaction kinetics have clarified some issues. Observations in Polar Regions suggest that Hg^0 can undergo fast oxidation in the presence of elevated levels of bromine compounds. Both experimental and theoretical studies suggest that the recombination of Hg^0 with Br atoms is sufficiently fast to initiate this oxidation process. However there is a large uncertainty in the value of the rate coefficient for this recombination reaction and in the fate of the reaction product, HgBr. Most global mercury models incorporate reactions of Hg^0 with OH and O_3. Based on the most recent high level ab-initio calculations of the stability of HgO it appears that neither of these reactions is likely to play a significant role in mercury oxidation. The most important aqueous oxidation for Hg^0 appears to be reaction with O_3 however that there has only been one determination of the Hg + O_3 reaction rate constant in the aqueous phase. Aqueous phase reduction of oxidized mercury via reaction with HO_2 is the only significant reduction reaction in current models but now seems unlikely to be significant. Again this suggests that the chemistry controlling mercury transformation in current models requires significant modification.

14.1 Introduction

Over the past decade our understanding of the chemical cycling of mercury has been changed, particularly by the recognition that homogeneous gas phase chemistry may play a significantly more important role in this cycling than was previously recognized.

N. Pirrone and R. Mason (eds.), *Mercury Fate and Transport in the Global Atmosphere*, 427
DOI: 10.1007/978-0-387-93958-2_14,

In this chapter we have attempted to review the current state of mercury science with regard to the important chemical and physical processes that affect the atmospheric transformation of mercury and its rate of deposition. If we view the progress that has been made in our overall understanding of chemical cycling in the troposphere and stratosphere we see iterative interactions between three branches of atmospheric chemistry. The three branches are typically compartmentalized between modeling, field measurements and laboratory measurements. Laboratory measurements include both experimental and theoretical studies of the elementary processes that are important in atmospheric transformation and include both measurements and calculations of reaction rate coefficients, thermodynamics and spectroscopic properties of atmospheric species. These processes form the basic input to atmospheric models that are then used to predict the spatial and temporal distributions of atmospheric species. The output from models is then compared with field measurements and gives a measure of our detailed understanding of the processes that are taking place. We consider separately homogeneous gas phase transformation, liquid phase transformation and atmospheric deposition processes.

14.2 Homogeneous Gas Phase Transformation

Our understanding of the importance of gas phase chemistry in the biogeochemical cycling of mercury has gone through a complete transformation over the past decade. This has been driven by atmospheric observations and we briefly review these in the context of our overall understanding of tropospheric chemical cycling.

14.2.1 Field Observations

The seminal observation that led to a reassessment of mercury chemistry was provided by Schroeder and coworkers during a study of polar ozone depletion events (Schroeder et al., 1998). Observations of springtime ozone depletion have been known for over two decades and are linked to halogen chemistry, with bromine containing radicals thought to be the most important species. In 1998 Schroeder et al. reported the results of 1995 field observations at Alert, Canada that showed that for three months following polar sunrise there were frequent episodic depletions of gaseous elemental mercury which correlated strongly with depletions of ozone. Subsequently Lindberg et al. (2002) published a series of observations from Barrow, Alaska demonstrating the same phenomenon. They also incorporated measurements of RGM using a newly developed denuder technique. Again they observed a strong correlation between mercury and ozone depletion events and found that the depletion events were correlated with high column densities of BrO as measured by the GOME satellite. They also found that that RGM showed a strong diurnal profile indicating photochemical production of the species responsible for Hg^0 oxidation. They proposed an oxidation mechanism based on Hg^0 reaction with bromine and chlorine atoms and the halogen monoxides. Subsequently these AMDEs (Atmospheric Mercury Depletion Events) have been observed by several groups in the springtime in both the Arctic and Antarctic.

Recently Peleg et al. (2007) reported the observation of diurnal cycles of mercury, ozone, and BrO behavior based on short-time resolution measurements near the Dead Sea, Israel, in Summer 2006. The results showed that mercury depletion events occurred almost daily, accompanied always by the presence of BrO and concurrent ozone destruction. The intensity of the MDE corresponded to increasing BrO levels. Mercury depletions of more than 40% were observed when BrO levels rose above 60 - 70 ppt.

In addition to mercury transformation that is clearly consistent with a halogen mediated oxidation mechanism a number of observations have suggested that elevated levels of RGM are present in the upper troposphere. Landis et al. (2005) monitored both Hg^0 and RGM in a series of flights in southern Florida in June 2000. During the flights, they monitored Hg^0 and RGM between 0.06 and 3.5 km and observed a small negative trend with altitude in Hg^0 and a more pronounced positive trend between RGM and increasing altitude. They observed RGM concentrations of 100 pg m^{-3} at the highest altitudes. The Landis et al. data set is the only aircraft data that has monitored both Hg^0 and RGM. Murphy and coworkers have obtained and analyzed the mass spectra of individual particles using a Particle Analysis by Laser Mass Spectroscopy (PALMS) instrument (Murphy et al., 2006a, 2006b). The extensive PALMS data set revealed that at the tropopause a large fraction of the observed particles in both the tropics and mid-latitudes contain Hg. In addition a single sample was collected in the lower stratosphere for analysis by Scanning Transmission Electron Microscopy. The distribution of Hg onto up to 70% of sampled particles, even small (20nm) particles is indicative of a local Hg(p) source rather than transport of Hg-rich aerosols from surface sites. The peak in aerosols containing Hg is centered at the tropopause with almost none of the particles sampled below 5 km containing Hg. This rapid decline could be due the re-volatilization of Hg or RGM from the aerosol once in the warmer troposphere or the rapid washout of Hg(p) and RGM during precipitation events. As the PALMS instrument measures both positive and negative ions it is of interest that Hg ions and Br^- follow a similar profile, indicating that they are present in the same aerosol species even if they are not chemical bound to each other.

A limited number of observations at elevated sampling sites at Mona Loa, Hawaii (Landis et al., 2005), and Mount Batchelor, Oregon (Swartzendruber et al., 2006) also show evidence for elevated RGM in the free and upper troposphere. Swartzendruber et al. (2006) saw elevated levels of RGM of up to 500 pg m^{-3} in subsiding air at Mount Batchelor. These air masses contained decreased levels of elemental gaseous mercury suggesting that total mercury was conserved and the RGM was a result of local conversion of Hg^0 to RGM in the upper troposphere.

These observations are clearly indicative of relatively rapid transformation of Hg^0 to RGM and suggest that gas phase chemistry is responsible. The observations of mercury depletion in polar regions and the Dead Sea are accompanied by ozone depletion and elevated levels of bromine radicals and are clearly suggestive of halogen chemistry. Transformation in the upper troposphere carries no "chemical signature" which is suggestive of the chemical mechanism and oxidation via reaction with ozone, OH and Br atoms have all been considered as possible mechanisms. We now consider the potential oxidation chemistry that might be responsible for this transformation.

14.2.2 Kinetic of Homogeneous Gas Phase Reactions

In spite of the clear evidence for rapid, homogeneous gas chemistry involving reaction of gas phase elemental mercury with other atmospheric reactants, the identification of the specific chemistry remains elusive. As we review the limited amount of experimental and theoretical data available we try and apply two criteria 1) are the observations consistent with our understanding of basic physical chemistry and 2) are they consistent with field observations. As we discuss below recent calculations on the stability of diatomic gas phase HgO suggest that homogeneous gas phase oxidation of Hg^0 by O_3 is not a viable atmospheric oxidation pathway for Hg^0. Both calculations and observations suggest that weakly bound complexes of mercury may play a significant role, at least in laboratory experiments. Since an understanding of the kinetics, particularly of weakly bound complexes, can be difficult and may not be treated clearly in the mercury literature, we provide some background on basic kinetics, thermochemistry and atmospheric chemistry.

14.2.2.1 Terminology

We briefly review some of the basic terminology associated with this chemistry giving examples from tropospheric chemistry and some mercury specific examples. The bimolecular rate coefficient for the reaction of two species A and B to give products C and D:

$$A + B \rightarrow C + D$$

is defined as

$$-d[A]/dt = -d[B]/dt = d[C]/dt = d[D]/dt = k[A][B]$$

where k is the rate coefficient. The rate coefficient is sometimes referred to as a "rate constant" but k will typically vary as a function of temperature and sometimes as pressure and hence rate coefficient is a preferred term. For the simplest type of bimolecular chemical reaction involving an abstraction reaction, k is typically independent of pressure and has a positive activation energy, E_a, as defined by the Arrhenius equation.

$$k\ (T) = A \exp\ (-E_a/RT)$$

where A is the pre-exponential factor.
The major oxidant in the troposphere is the hydroxyl radical and one of its major sinks is reaction with atmospheric methane. This is a simple H atom abstraction reaction to form water and methyl radicals.

$$CH_4 + OH \rightarrow CH_3 - H_2O \quad \Delta H^0 = -57.8\ \text{kJ} \cdot \text{mol}^{-1}$$

The rate coefficient is well described by an Arrhenius expression 1.85×10^{-12} $\exp(-1690/T)$ cm^3 $molecule^{-1}$ s^{-1} over the temperature range 200-300 K, which gives a room temperature rate coefficient of $k = 6.4\times10^{-15}$ cm^3 $molecule^{-1}$ s^{-1} at 298 K (Sander et al., 2006, Atkinson et al., 2006). Because this is a simple abstraction reaction this rate coefficient shows no dependence on pressure or bath gas composition. If this rate coefficient is measured at one atmosphere in air, or one Torr in He, by either relative or absolute methods, the same result should be obtained.

The rate coefficient can exhibit more complicated pressure and temperature dependencies for unimolecular and termolecular reactions and particularly complex behavior if reaction proceeds via formation of a weakly bound addition complex. Since it appears likely that HgO may form this type of a complex we describe an example of such a reaction and the type of behavior observed. Both carbon disulfide and the alkyl sulfides react with OH via the formation of weakly bound addition complexes that undergo further reaction with molecular oxygen (Hynes et al., 1988, Williams et. al., 2007). In the case of the alkyl sulfides this addition process proceeds in parallel with an abstraction channel, but for CS_2 reaction proceeds only via addition so this case will be identical to behavior we might expect from Hg^0. The reaction of CS_2 with OH is described by the following sequence of reactions (Hynes et al., 1988).

$$OH + CS_2 + M \leftrightarrow OHCS + M \qquad (2, -2)$$

$$OHCS_2 + O_2 \rightarrow \text{products} \qquad (3)$$

In such a system the results of an experimental investigation of the rate coefficient at any particular temperature will produce results that depend on the pressure, bath gas composition *and the temporal resolution of the experiment* !! If we attempt to measure the rate coefficient for this reaction in the absence of O_2 in an experiment with sub-microsecond time resolution or via a relative rate technique we will see no evidence for chemical reaction. If we measure using a direct technique with fast temporal resolution we can directly observe the equilibration (2,-2) (Hynes et al., 1988) and by analyzing the observed biexponential decays we can measure the rate coefficients for reactions 2 and -2 and hence the equilibrium constant. By measuring the equilibrium constant as a function of temperature we can the obtain the bond energy of the adduct. If we perform the same experiments in the presence of O_2 under conditions in which the concentration of the adduct is much smaller than that of OH we will observe a net loss of OH. If reaction is performed with the concentration of CS_2 in large excess with respect to OH then we will see pseudo-first order decays of OH and we can define an “effective rate coefficient”, k_{obs} which is given by:

$$k_{obs} = \frac{(k_3 / k_2) * k_2 * [O_2]}{1 + (k_3 / k_{-2}) * [O_2]}$$

The key point here is that k_{obs} is not a true elementary rate coefficient but rather a composite which is a function of the three elementary rate coefficients, the total pressure and the specific O_2 concentration. Since the three elementary rate coefficients show very different temperature dependencies the overall temperature dependence is complex, but, at least for CS_2 and the alkyl sulfides, is dominated by k_{-2}. As the temperature decreases k_{-2} decreases rapidly, hence for any particular conditions of total pressure and O_2 concentration, the effective rate coefficient increases rapidly as the temperature decreases! As we noted above, if we perform experiments at 298 K we will obtain a different value of k_{obs} each time we change the pressure or O_2 concentration. However if we are working under conditions in which the OH radical concentration is much greater than that of the adduct, conditions which apply in the real atmosphere, we will obtain the same value of k_{obs}, irrespective of how we actually perform the measurement, i.e. direct and relative rate techniques will give exactly the same result. This, of course, assumes that there are no artifacts due to secondary chemistry or other systematic errors.

14.2.2.2 Thermodynamics

The overall rate of a reaction is governed by an interaction between kinetics, a rate process, and thermodynamics, which considers the energetics of a process. Processes that are exothermic and increase the entropy of the system favor reaction. The balance between enthalpy, ΔH, and entropy, ΔS, is given by the Gibbs Free Energy.

$$\Delta G = \Delta H - T\Delta S$$

If Gibbs Free Energy is negative a reaction will proceed spontaneously from a thermodynamic perspective and we can calculate the equilibrium constant, K_p from:

$$\ln (K_p) = -\Delta G^0/RT$$

and the ratio of the forward and reverse rate coefficients from:

$$K_c = k_f/k_r$$

In many cases a reaction that is thermodynamically favorable will not proceed at a measurable rate because a large endothermic barrier to reaction exists. In attempting to evaluate the available kinetic data on mercury chemistry it is useful to place this in the context of kinetic data evaluation. Typical models of atmospheric chemistry contain hundreds of chemical reactions and for the most critical reactions there will be many independent determinations of the rate coefficients by multiple experimental techniques. Even though these experiments may have been published in the peer reviewed literature the rate coefficients that have been obtained can differ significantly and, as a consequence, two independent review panels were established to evaluate the published literature and recommend the value of the rate coefficient that would be appropriate for use in models together with an estimate of the uncertainty associated with the recommendation (Sander et al. 2006, Atkinson et al. 2006). In this context, for the most important reactions the range of uncertainty is less than a factor of two, based on multiple independent determinations by a variety of different experimental techniques. In contrast there are very few independent determinations of any reactions of Hg^0 and in most cases very large discrepancies exist in the published data. At the moment neither of the two international data panels includes mercury chemistry in their evaluations.

14.2.2.3 Experimental Approaches

Experimental approaches to the study of mercury kinetics can be broadly divided between absolute and relative rate methods. In direct measurements the rate of loss one reactant is monitored in the presence of a large excess of a second reactant. For very slow reactions, such as the reaction of mercury with ozone, chemistry can be initiated by physical mixing. For fast reactions, photolysis, using a laser or flash lamp is used. In relative rate methods the loss of mercury is measured relative to that of a reference compound. The rate coefficient for the reaction of the reference compound with the reactant of interest should be well known.

14.2.2.4 Ab-Initio Thermochemistry

The challenges with using ab-initio quantum mechanics to calculate the thermochemistry of mercury compounds has been discussed by Ariya and Peterson (2005). They note, in particular, the challenge in dealing with relativistic effects for a large element like mercury with eighty electrons, a result of the large nuclear charge. The ability of such calculations to calculate thermodynamic properties with "chemical accuracy" normally implies an uncertainty of ± 1 kcal mol^{-1}, the accuracy required to calculate rate coefficients within a factor of two or three of the actual rate. At this point it appears that Peterson and coworkers have been able to achieve something that certainly approaches "chemical accuracy" and, as we discuss below, this has profound implications for our evaluation of potential mercury oxidation reactions.

14.3 Specific Reaction Systems

Ariya and Peterson (2005) reviewed the kinetic data on mercury reactions and discussed some of the problems associated with measurements. In this review we have focused on a limited number of reactions that are currently used in models of mercury oxidation or which may be of importance. In each case we detail our best estimate of the thermo-chemistry if such an estimate is feasible.

14.3.1 $Hg^0 + O_3$

$$Hg^0 + O_3 \rightarrow HgO + O_2 \qquad \Delta H = 93 \text{ kJ mol}^{-1} \qquad (4a)$$

$$Hg^0 + O_3 + M \rightarrow HgO_3 + M \qquad (4b)$$

There have been three recently published studies of this reaction. Hall (1995) studied the reaction of Hg^0 with O_3 as a function of temperature in a static configuration in Teflon reactors. He also investigated the effects of changes in the surface to volume (s/v) ratio and the effects of sunlight. Experiments were performed at atmospheric pressure in a gas mixture that was composed of N_2, with 2-10 % O_2, and with trace levels of Hg^0 and O_3 (20-1500 ppm). The decrease of Hg^0 and the formation of $Hg^{(II)}$ on the walls were monitored. Monitoring Hg^0 loss in an excess of O_3, i.e. under essentially pseudo first order conditions, the Hg^0 decays showed good pseudo first order behavior. Essentially quantitative conversion of Hg^0 to $Hg^{(II)}$ was observed and calculated rate coefficients were based on an assumed mechanism that included gas phase and wall reaction. Based on the change in rate coefficient observed when the s/v was changed Hall concluded that reaction was occurring in both the gas phase and on the reactor walls and a gas phase rate coefficient of $3\pm2\times10^{-20}$ cm^3 $molecule^{-1}$ s^{-1} was derived. However the pseudo-first order rates did not depend linearly on $[O_3]$ and gave a reaction order of 0.8 with respect to O_3. Experiments were performed at two elevated temperatures and the rate coefficient was found to increase with temperature. Expressing the data in Arrhenius form gave an A factor of 2.1×10^{-18} cm^3 $molecule^{-1}$ s^{-1} and an activation energy of 10 kJ mol^{-1}, however the large uncertainty in the value of the heterogeneous component of the rate produces a very large uncertainty in these values which is not discussed in the paper. The rate coefficient showed no dependence on relative humidity but it was found to increase in the presence of sunlight, a result that might be due to either gas phase or surface reactions. No in-situ diagnostics were employed to confirm that reaction products were being formed in the gas phase.

Pal and Ariya (2004) used both relative and absolute measurement approaches to study this reaction. They noted that the kinetics appeared to be very sensitive to the nature of the wall and coated their reaction vessels with halocarbon wax to minimize surface chemistry. The relative rate studies monitored the decay of Hg^0 relative to that of propone or 1-butene in an excess of N_2 buffer gas at 750 Torr total pressure. They reported five rate coefficients between 15 and 7×10^{-19} cm^3 $molecule^{-1}$ s^{-1} and concluded that the reactions of the reference molecules with O_3 produced OH radicals that initiated secondary chemistry. In two experiments trimethylbenzene was added to reaction mixtures to scavenge the OH radicals giving rate coefficients of 7×10^{-19} cm^3 $molecule^{-1}$ s^{-1}. In the absolute measurements Hg^0 decays were monitored in an excess of O_3 at several surface to volume ratios. The decays showed a complex behavior and the authors attributed the initial component of the decays to a simple gas phase reaction, the second component to a surface enhanced reaction and they obtained a rate coefficient of $(7.5\pm0.9)\times10^{-19}$ cm^3 molecule^{-1} s^{-1}. No in-situ diagnostics were employed to confirm that reaction products were being formed in the gas phase. Analysis of volatile surface reactants, using chemical ionization mass spectroscopy, produced evidence for a HgO containing species. As we discuss below Tossell (2003) has reported calculations that suggest that oligomers of HgO are much more strongly bound than the diatomic molecule. Such oligomers would be stable in the gas phase and would likely react in a chemical ionization experiment to form $HgOH^+$. A less extensive absolute rate study of the reaction has also been reported by Sumner et al. (2005) who monitored Hg^0 removal in excess O_3. They estimated that 30-40% of the reacted Hg^0 was deposited to the wall and obtained a rate coefficient of $6.4\pm2.3\times10^{-19}$ cm^3 molecule^{-1} s^{-1}.

Accurate thermochemistry is critical in assessing this reaction and its potential role in atmospheric transformation of mercury. Hall reported a large negative free energy change for this reaction based on JANAF thermochemistry that assumes that HgO is bound by 268 kJ mol^{-1} and this thermochemistry is also cited by Pal and Ariya. However recent calculations on the electronic structure of HgO suggest that in fact it is a very weakly bound molecule. Tossell (2003) reported calculations on a series of possible reactions of HgO and found that HgO was unbound with respect to reactants while Shepler and Peterson (2003) reported that HgO was bound by approxinately17 kJ mol^{-1}. Shepler and Peterson discuss the prior experimental and theoretical calculations and make a convincing argument that the prior experimental evidence for a strongly bound diatomic HgO molecule is unconvincing. Using this new thermochemistry the atom transfer reaction (4a) becomes endothermic by approximately 90 kJ mole^{-1} with a large positive free energy for any reasonable standard entropy of HgO. We can consider the implications by calculating the free energy change and equilibrium constant based on the new thermochemistry. Using an HgO bond distance of 1.91Å we calculate a standard entropy of $S^0 = 239.0$ J K^{-1} mol^{-1} for HgO which is close to the JANAF value of 239.3 J K^{-1} mol^{-1}. We can use these values to calculate the free energy change and equilibrium constant for reaction (4a).

$$\Delta G^0 = \Delta H^0 - T\Delta S^0$$

Using $\Delta H^0 = 93$ kJ mol^{-1} and $\Delta S^0 = 30$ J K^{-1} mol^{-1} we obtain $\Delta G^0 = 84$ kJ mol^{-1}

$$K = \exp(-\Delta G^0/RT) = 2.4\times10^{-15} \text{ at } 300\text{ K}$$

Using the principle of detailed balance we can calculate the rate coefficient for reaction (-4a) as follows assuming the rate coefficient reported by Pal and Ariya (2004) is correct.

$$k_r = k_f/K = 7.5\times10^{-19}/2.4\times10^{-15} = 3.1\times10^{-4}\text{ cm}^3\text{ molecule}^{-1}\text{ s}^{-1}$$

This is at least six orders of magnitude faster than any physically reasonable rate coefficient, hence it is very unlikely that the oxidation of Hg by O_3 proceeds as a homogeneous gas phase reaction. Indeed HgO would have to be bound by at least 50 kJ mol^{-1} for the above rate coefficient to be consistent with a physically realistic value for reaction (-4a). There is some level of uncertainty associated with the quantum mechanical calculations because of the large number of electrons and the issue of incorporating relativistic effects, nevertheless it seems inconceivable that the error could be large enough to make the O atom transfer energetically favorable. Calvert and Lindberg (2005) suggest that reaction might proceed via an addition reaction to produce a weakly bound molecule with a binding energy of ~16 kJ mole^{-1}. Such an addition complex would have no exothermic dissociation routes but it might diffuse to the reaction walls. In more recent work Tossell (2006) has shown that, in contrast to the monomer, ring-type oligomers, i.e. $(HgO)_n$, are stable and their formation on surfaces should be facile. Tossell (2006) states that a stable van der Waals complex of Hg and O_3 exists but gives no value for a binding energy and also notes that the isomerization of this species to $OHgO_2$, as suggested by Calvert and Lindberg (2003), is unlikely since it is endothermic but again quantitative thermodynamic data is not provided. If the reaction does proceed in the gas phase it must be via the formation of a weakly bound complex that is likely reversible since no exothermic decomposition pathways are accessible. In the laboratory experiments such a complex could diffuse to the reactor surface to form solid mercuric monoxide, possibly via oligomer formation. Such a process would be energetically favorable because of the exothermicity of oligomer formation and the lattice energy associated with the formation of the solid. If the concentration of the complex were sufficiently high it might also undergo self-reaction in the gas phase to form $(HgO)_2$ and higher oligomers but this is unlikely if the complex is bound by less than 20 kJ mol^{-1}. The overall loss kinetics initiated by such a reversible addition process are analogous to those described above for the reaction of OH with CS_2 and will be extremely complex since the observed rate coefficient will be a composite of the elementary rate coefficients involved in the formation and decomposition of the adduct and other irreversible loss processes which in this case is most likely surface loss. If reaction is initiated by a reversible addition process it appears that the adduct does not react with water molecules since the rate coefficient showed no dependence on relative humidity in the study by Hall. The combination of complex gas phase kinetics coupled with contributions from heterogeneous reactions could certainly explain the very large differences in the rate coefficients obtained in the recent studies of this reaction. However oxidation based on such a mechanism is unlikely to be of any significance in the atmosphere. At typical atmospheric concentrations a very small fraction of the Hg^0 would be present as adduct and the chances of heterogeneous removal would not be significant. If we assume that quantum calculations are not correct and the rate coefficient measured by Pal and Ariya (2004) does produce HgO it has significant implications for global models and would imply that ozone is the main oxidant of Hg^0 in the atmosphere. We can calculate the rate of transformation of Hg^0 by O_3 in the following manner: If we take the rate coefficient measured by Pal and Ariya (2004), 7.5×10^{-19} cm^3 molecule^{-1} s^{-1}, we require concentrations of both Hg^0 and O_3 in molecule cm^{-3}. Ozone concentrations are typically given as a mixing ratio in ppb (parts per billion), with a typical value of 40 ppb at sea level over much of the continental US. At one atmosphere pressure and 300 K, 40 ppb corresponds to an absolute concentration of approximately 10^{12} molecules cm^{-3}. Hg^0 concentrations are typically quoted as absolute values, but in ng m^{-3}, a unit that is not typically used by atmospheric chemists to designate trace gas concentrations. An Hg^0 concentration of 2 ng m^{-3} corresponds to approximately 6×10^6 molecules cm^{-3}. The ozone concentration is much larger than the Hg^0 concentration and the reaction will not perturb the O_3 concentration. We can define an Hg^0 "half life", $t_{1/2}$, with respect to reaction with O_3, which is the time required to reduce Hg^0 to 50% of its original concentration. This is given by $\ln(2)/(k^*[O_3]) = \ln(2)/(7.5\times10^{-19} \times 1\times10^{12}) = 9.2\times10^5$ s, or approximately 10.6 days. It should be noticed that this number does not depend on the Hg^0 concentration so the time required to reduce the concentration to 25% of its original concentration is 2 $t_{1/2}$. Although photochemically active, ozone does not have a strong diurnal variation as is observed with the production of, for example, OH. However the limited observations of RGM with hourly temporal resolution suggest that its production shows a strong diurnal variation. Several recent global modeling studies have concluded that field observations are inconsistent with this reaction being the major oxidant of Hg^0. Bergan and Rohde (2001) used a global model to examine the sensitivity of global mercury concentrations and their seasonal variation to changes in the rate coefficient for this reaction. They concluded that the rate coefficient of Hall (1995) is too slow to explain observed atmospheric concentrations of Hg^0. They also examined the effect of using a faster rate coefficient and making O_3 the sole oxidant of Hg^0. Using a rate coefficient that produced realistic Hg^0 concentrations they found latitudinal and seasonal variations that were not consistent with observations. More recently Seigneur et al. (2006) examined the effect of using the Pal and Ariya (2004) rate coefficient in their global model and found that it gave surface concentrations that ranged from 0.3 to 1.2 ng m^{-3}. The half-life of Hg^0 was reduced from 9 months to approximately 11 days which appears unrealistic unless there are competing reactions which reduce $Hg^{(II)}$ compounds in the atmosphere, which are as yet unidentified.

14.3.2 Hg^0 + OH

Again our evaluation of potential pathways for this reaction is dependent on the thermodynamics of the HgO molecule. A direct atom transfer is endothermic using either the old or new binding energy of HgO.

$$Hg + OH \rightarrow HgO + H \quad \Delta H^0 = 415 \text{ kJ mol}^{-1} \tag{5a}$$

$$Hg + OH + M \leftrightarrow HgOH + M \quad \Delta H^0 = -40 \text{ kJ mol}^{-1} \tag{5b}$$

However it does appear that HgOH is weakly bound, with both Tossell (2003) and Goodsite et al. (2004) calculating binding energies of ~30 and

40 kJ mole^{-1} respectively and, if so, it will rapidly redissociate to reactants at atmospheric temperatures. Goodsite et al. (2004) calculate a forward addition rate of 3.2×10^{-13} cm^3 molecule^{-1} s^{-1} and a dissociation rate of ~3200 s^{-1} at 300 K. Again the kinetic situation is similar to that of the reaction of CS_2 with OH where both the forward and reverse rate coefficients are about a factor of ten larger than those calculated for HgOH but the equilibrium constant is almost the same. Hence, in the absence of further reaction of HgOH, we expect the reactants and products will form an equilibrium mixture with no permanent removal of reactants.

Using the old (Chase, 1998) binding energy for HgO, reaction with O_2 would be energetically favorable generating HgO and HO_2.

$$HgOH + O_2 \rightarrow HgO + HO_2 \quad \Delta H^0 = 210 \text{ kJ mol}^{-1}$$

However, using the more recent binding energies of Shepler and Peterson (2003) or Tossell (2006), this reaction becomes strongly endothermic, with $\Delta H = 210$ kJ mol^{-1} based on the binding energy of Shepler and Peterson (2003).

There have been three experimental studies of this reaction utilizing both absolute and relative rate approaches. Sommar et al. (2001) used a relative rate technique monitoring the loss of Hg^0 relative to cyclohexane in the presence of OH generated by the photolysis of methyl nitrate. They reported four experiments in which mercury consumption was 3 to 5% of the starting concentration and reported a rate coefficient of $(8.7 \pm 2.8) \times 10^{-14}$ cm^3 molecule^{-1} s^{-1}. Bauer et al. (2003) used a pulsed laser photolysis - pulsed laser induced fluorescence approach to monitor the decay of OH in a pseudo-first order excess of Hg^0 at room temperature and atmospheric pressure in both air and He buffer gases. They saw no evidence for reaction and reported an upper limit of $(1.2) \times 10^{-13}$ cm^3 molecule^{-1} s^{-1}. These experiments have the time resolution to directly observe an equilibrium formation of HgOH and based on their failure to observe such an equilibrium they concluded that the equilibrium constant, K_c, for this reaction must be less than 5×10^{-16} cm^3 molecule^{-1}.

Pal and Ariya (2004) used a relative rate technique generating OH by photolysis of isopropyl nitrate and monitoring the loss of mercury relative to five reference compounds. Measurements relative to cyclohexane, n-butane and 2-methylcyclopropane gave rate coefficients of approximately 13×10^{-14} cm^3 molecule^{-1} s^{-1}, while experiments relative to ethane and cyclopropane gave rate coefficients of approximately 9.5×10^{-14} cm^3 molecule^{-1} s^{-1}. The authors suggest that the lower values are appropriate because the rate coefficients of the unknown and reference compounds are of similar magnitude. This should improve the precision of the measurements since the consumption of mercury is larger, however it cannot explain the difference between the sets of measurements, such differences may well be indicative of kinetic complications. Pal and Ariya (2004) made product measurements that were similar to those employed in their study of the reaction of ozone with Hg^0 and observed $HgOH^+$ in chemical ionization experiments.

As we have noted, thermodynamic considerations suggest this reaction must proceed via a complex mechanism involving reversible addition and unless the weakly bound HgOH reacts to form stable products then no reaction will be observed.

$$HgOH + X \rightarrow \text{products}$$

Since reaction with O_2 to form HgO is highly endothermic loss might occur via reaction with another radical species or wall loss. Sommar et al., (2001) suggest that "X" is O_2 and cite an endothermicity of ~ 30 kJ mol^{-1}, they suggest that the forward rate is measurable because of the low ratio of $[HO_2] / [O_2]$, however this estimate of the endothermicity appears unrealistic. Pal and Ariya (2004) suggest that "X" may be O_2 or OH but do not discuss the implications of an addition mechanism for their measurements. Goodsite et al. (2004) have calculated rates for the forward and reverse reactions based on a binding energy of 39 kJ mol^{-1}. Calvert and Lindberg (2005) have discussed the measurements of Pal and Ariya (2004) and Sommar et al. (2001) and suggest that "X" is the sum of the potential reactants, {X = OH, HO_2, RO, RO_2, NO, NO_2} in the experiments. They assign a rate coefficient of 2.5×10^{-10} cm^3 molecule^{-1} s^{-1} to reaction, however in their simulations X is primarily the sum of [NO] plus $[NO_2]$ which build up to significant concentrations and there is no experimental evidence or theoretical calculations to support such a rate or indeed to show that NOHgOH or NO_2HgOH are stable molecules. Calvert and Lindberg (2005) conclude that this reaction is probably unimportant in the actual atmosphere because the concentration of reactants that can remove HgOH will be low, making this an ineffective loss pathway.

Bergan and Rodhe (2001) used the rate coefficient reported by Sommar et al. (2001) as the sole oxidant for Hg^0 in their global model and concluded that it was a factor of three too large to be consistent with observations. However Seigneur et al. (2006) note that this model did not contain the aqueous phase reduction of $Hg^{(II)}$ by HO_2 and conclude that observations are consistent with a faster OH rate coefficient if the reduction path is occurring. Selin et al. (2007) also use the faster rate coefficient in their global model and conclude that this reaction is the dominant sink for Hg^0. None of these models incorporated bromine chemistry.

This reaction is the major route for the oxidation of Hg^0 in most current models so developing a clear understanding of the detailed mechanism of this reaction is critical. It seems clear that neither a direct atom transfer, nor an addition followed by reaction with O_2 would be energetically feasible. However, if the Goodsite et al.(2004) dissociation rate for HgOH is correct it has a lifetime of less than 1 ms at room temperature and reaction with O_2 would be the only feasible route for permanent loss of HgOH. It should be recognized that no reasonable, i.e. consistent with known thermochemistry, chemical mechanism has been proposed for the OH initiated oxidation of Hg^0 and it seems unlikely that this reaction contributes to the oxidation of Hg^0 in the atmosphere. Direct observation of the equilibration process and measurements of the reactivity of HgOH are key in resolving the importance of this reaction. Relative rate studies using an OH source that is less susceptible to secondary chemistry and does not generate NO_x would also be very useful.

14.3.3 Halogen Reactions

14.3.3.1 Hg^0 + Cl

$$Hg^0 + Cl + M \rightarrow HgCl + M \quad (6)$$

This reaction involves the three-body recombination of Hg^0 and Cl atoms and the rate coefficient would be expected to be pressure dependent and to exhibit a negative activation energy if expressed in Arrhenius form. This type of atom recombination involves the formation of an energized HgCl* molecule which, in the absence of collisions will redissociate to reactants. If the energized molecule undergoes a collision which removes enough energy to give a molecule below the dissociation threshold it will thermalize forming a stable HgCl molecule.

$$Hg + Cl \leftrightarrow HgCl^*$$

$$HgCl^* + M \rightarrow HgCl + M$$

This reaction mechanism predicts that the reaction will show a distinct pressure dependence with an effective rate coefficient which is third order at low pressures, i.e. $-d[Cl]/dt = k[Hg][Cl][M]$ and becomes pressure independent at high pressures. The transition from third to second order behavior is known as the fall-off region. For a typical atom-atom recombination we expect the reaction to be in the low pressure, third order regime at atmospheric pressure and below. The effective rate coefficient also depends on the identity of the "third body", M, as different sizes of M will differ in their efficiency in deactivating HgCl*. We expect atoms to be least efficient, and then deactivation efficiency to increase as we move from diatomic to more complex polyatomic molecules. Since the effective rate of reaction depends on the rate of deactivation we expect that the rate coefficient will increase as the temperature is lowered. As we lower the temperature the energy associated with translational motion of the reactants decreases hence less energy has to be removed from HgCl* to bring it below the dissociation threshold.

There have been three recent experimental studies of this reaction. Ariya et al. (2002) reported a relative rate study of this reaction. In initial experiments Cl atoms were produced by photolysis of trichloroacetylchloride in air. Five different reference molecules were used obtaining results, which differed by a factor of 270 in the measured relative rates together with a strong non-linearity of the relative rate plot. They concluded that the variation was caused by the presence of a secondary reaction between the reference molecules and OH. The buffer gas was switched from air to nitrogen to eliminate oxygen chemistry, giving an overall reduction in the observed rate. However, the variation in the measured relative rate between the reference molecules was still a factor of 30 and the non-linearity remained. Ultimately, a series of 8 measurements were made using 1,3-dichloropropane as the reference molecule with the addition of 835 ppm of benzene as an OH scavenger. The reported rate coefficient, $(1.0 \pm 0.2) \times 10^{-11}$ cm^3 $molecules^{-1}$ s^{-1}, was determined from this sub-set of the data.

Spicer et al. (2002) reported a relative rate study utilizing the photolysis of Cl_2 as a source of Cl atoms. They attempted to use both toluene and dimethyl sulfide as reference molecules but obtained inconsistent results with toluene. The experiments were performed in air containing 29 ppbv NO. The reason for the addition of NO was not specified. They derived a rate coefficient of 6.4×10^{-11} cm^3 $molecules^{-1}$ s^{-1}. These experiments were performed in air and, as we discuss below, reaction of Cl atoms with O_2 could cause kinetic complications.

Donohoue et al. (2005) used a pulsed laser photolysis-pulsed laser induced fluorescence approach to measure the rate coefficient over a wide range of temperature and pressure. This type of system has been used extensively for the measurement of elementary rate coefficients that are important in atmospheric and combustion systems. Such experiments are typically performed under pseudo-first order conditions that require that one reactant is present in at least a factor of 10 excess in concentration relative to the other reactant. Normally the stable reactant in a radical molecule reaction is in excess. In such a configuration only the absolute concentration of the reactant that is present in excess is required to obtain a rate coefficient. The relative change in the concentration of the minor reactant is then monitored as a function of time to obtain the rate coefficient. The configuration of choice would be to monitor the variation in Cl atom as a function of time in an excess of mercury. The mercury concentration can be monitored photometrically with high precision and the variation in the Cl atom concentration can be monitored using laser-induced fluorescence. Using this approach Donohoue et al. (2005) were unable to see any evidence for reaction at room temperature. The low vapor pressure of mercury limits such an approach to fast reactions and the problem is exacerbated for studies that are attempting to measure the rate coefficient at low temperature. As an alternative approach, Donohoue et al. (2005) generated a large excess of Cl atoms by photolysis and monitored the decay of Hg^0. This approach makes the experiments more difficult to implement and complicates the data analysis. Since Cl atoms react via a recombination reaction it requires the temporal profile of both species to be monitored using LIF. Since the Cl atom concentration is changing as a function of time, the Hg^0 profiles do not show a simple exponential decay and rate coefficients are obtained by numerical integration to fit the observed Hg^0 temporal profiles. In addition the approach requires that the absolute concentration of Cl atoms is known accurately. This cannot be measured directly and has to be calculated from the concentration of Cl_2 and the photolysis laser flux. The advantage of this approach is that it allows the rate coefficient to be measured over a large temperature and pressure range and the effect of a change in the bath gas can be tested. In this case the results were consistent with expectations and the observed second order rate coefficients showed a linear dependence on pressure, a slightly negative temperature dependence and a significant difference in deactivation efficiency with N_2 and He as third bodies. Donohoue et al. (2005) give an expression that can be used to calculate the recombination rate at any temperature and pressure in the troposphere. Their expression gives a rate coefficient of 5.4×10^{-13} cm^3 $molecule^{-1}$ s^{-1} at 298 K and one atmosphere pressure in air.

The main uncertainty in this approach is the calculation of the absolute [Cl] and the authors discuss this in detail. One potential complication that Donohoue et al. (2005) did not discuss relates to the potential reaction of the HgCl product of the reaction with excess Cl atoms. They assumed this reaction proceeds via another three body recombination to form $HgCl_2$. Balabanov et al. (2005) have suggested that the analogous Br reaction can proceed via an alternative bimolecular route reforming Hg^0 atoms. For Cl this reaction would be:

$$HgCl + Cl \rightarrow HgCl_2 \tag{7a}$$

$$\rightarrow Hg + Cl_2 \tag{7b}$$

If reaction (7b) were significant then Hg^0 atoms would be regenerated and the experiments which monitored Hg^0 loss would underestimate the rate coefficient. Such a channel would not impact the experiments which were performed with Hg^0 in excess and this suggests (7b) is not significant.

Donohoue et al. (2005) discuss an additional issue that is relevant when these experiments are performed in air. Cl reacts with molecular oxygen to form a weakly bound adduct, ClO_2. This produces potential complications due to the reaction of Cl atoms with ClO_2. Donohoue et al. (2005) saw a large increase in the rate of loss of Cl atoms in the presence of O_2 that was consistent with the following chemistry.

$$Cl + O_2 + M \leftrightarrow ClO_2 + M$$
$$Cl + ClO_2 \rightarrow Cl_2 + O_2$$
$$\rightarrow 2ClO$$

A significant enhancement in the apparent recombination rate coefficient of mercury and chlorine atoms was also observed in the presence of air. If this loss were solely due to the mercury chlorine recombination reaction it would have resulted in an increase in the recombination rate coefficient by a factor of 4 relative to the rate coefficient obtained in N_2. Again this is not consistent with the expected relative three body efficiencies of nitrogen and oxygen and they chose not to report a recombination rate coefficient for $M = O_2$. This observation suggests an additional loss of mercury atoms due to the secondary chemistry described above. Based on this chemistry the only plausible candidates are ClO or ClO_2. The most likely candidate is reaction with ClO_2.

$$Hg + ClO_2 \rightarrow HgCl + O_2$$

This reaction requires further investigation since the reaction is endothermic and, unlike the recombination reaction, no third body is required to stabilize the HgCl product.

Khalizov et al. (2003) calculated the recombination rate coefficient. They used electronic structure calculations to obtain both molecular parameters and the capture rate or high-pressure limit. They determined a pressure dependent rate coefficient by assuming strong collisional deactivation. This, typically unrealistic, assumption should produce the maximum possible recombination rate coefficient at any particular pressure. The value they obtained, 2.8×10^{-12} cm^3 $molecule^{-1}$ s^{-1} at 298 K, 760 Torr, is a factor of three smaller than the rate coefficient reported by Ariya et al., (2002) a factor of twenty smaller than that reported by Spicer et al. (2002) and a factor of five faster that the rate coefficient reported by Donohoue et al. (2005). In the atmosphere chlorine atoms react rapidly with methane which limits their concentration. As a consequence concentrations are much lower than those of bromine atoms and this reaction is not significant in global models. Based on current models it also appears to be unimportant in polar mercury depletion events, but it may be a source of RGM in the marine boundary layer. An accurate determination of this rate is required for modeling purposes and the discrepancies in the experimental database need resolution. In addition the potential role of ClO_2 should be resolved as this could dramatically enhance the effective rate of the Cl reaction, especially at low temperatures where the stability of ClO_2 increases.

14.3.3.2 Hg^0 + Br

$$Hg^0 + Br + M \rightarrow HgBr + M \quad (8)$$

This reaction has been studied in three recent experimental systems using both direct and relative rate techniques. As with the Cl atom reaction this is a standard three body recombination reaction. Ariya et al. (2002) performed a relative rate study using a single reference reaction, Hg^0 with 1- butene, to measure the relative rate of reaction. They obtained a rate coefficient of 3.2×10^{-12} cm^3 $molecules^{-1}$ s^{-1} Kinetic studies on the reaction of Hg^0 and 1-butene are limited, with only one study referenced. This study was also a relative rate study and the observed rate coefficient depended on the O_2 partial pressure, indicating a complex reaction mechanism. In the Ariya et al. (2002) study the rate observed depended on the concentration of the reference molecule, the concentration of the OH scavenger, and the identity of the buffer gas. The observed rate coefficient for reaction varied by a factor of 3 as the buffer gas was varied between nitrogen and air. In order to obtain linear relative rate plots they added large amounts of an OH scavenger (cyclohexane) leading to an enhancement in the absorption of reactant on the cell walls. Ariya et al. (2002) reported that the primary complication to their system was enhanced removal of the reference compound by reaction with OH or loss on the cell walls.

Spicer et al. (2002) report a relative rate study monitoring Hg^0 loss relative to DMS and propene. They obtained rate coefficients of 3×10^{-13} cm^3 $molecule^{-1}$ s^{-1} relative to DMS and 9×10^{-13} cm^3 $molecule^{-1}$ s^{-1} relative to propene with the latter considered more reliable.

Donohoue et al. (2006) used a pulsed laser photolysis-pulsed laser induced fluorescence approach to measure the rate coefficient as a function of temperature and pressure. As described above for their Cl atom study, they generated a large excess of Br atoms by photolysis and monitored the decay of Hg^0 and the same complications ensue. Both Br and Hg^0 profiles were monitored using LIF and rate coefficients were obtained by numerical integration to fit the observed Hg^0 temporal profiles. Again the approach requires that the absolute concentration of Br atoms is known accurately. The rate coefficient was measured over a large temperature and pressure range and the effect of a change in the bath gas was measured. The observed second order rate coefficients showed a linear dependence on pressure, a slightly negative temperature dependence and a significant difference in deactivation efficiency with N_2 and He as third bodies. Donohoue et al. (2006) give an expression that can be used to calculate the recombination rate at any temperature and pressure in the troposphere. Their expression gives a rate coefficient of 3.6×10^{-13} cm^3 $molecule^{-1}$ s^{-1} at 298 K and one atmosphere pressure in air. As noted above, a significant reaction between HgBr and Br to regenerate Hg^0 could produce an underestimation of the rate coefficient in these experiments.

There have been three computational studies of this reaction, Khalizov et al. (2003) calculated the collision frequency or "capture rate" to form an energized HgBr* and assumed that every collision stabilizes the molecule forming HgBr. This "strong collision" assumption should give an upper limit to the rate coefficient since, in practice, not every collision will remove enough energy to deactivate HgBr*. They obtained a rate coefficient of 2.07×10^{-12} cm^3 $molecule^{-1}$ s^{-1} at 298 K, 760 Torr. Goodsite et al. (2004) performed a more physically realistic calculation and obtained a rate coefficient of 1.1×10^{-12} cm^3 $molecule^{-1}$ s^{-1} at 298 K, 760 Torr. They used a master equation approach and assumed an average energy transfer per collision of 400 cm^{-1}. However this energy transfer parameter is an estimate and is not based on experiment or a detailed calculation of the HgBr-N_2 potential. Shepler et al. (2007) used high level calculations to construct a global potential energy surface for HgBr + Ar and calculated the collisional dissociation rate using quasi-classical trajectories. The HgBr recombination rate was then calculated from detailed balance and they obtained a rate coefficient of 9.8×10^{-13} cm^3 $molecule^{-1}$ s^{-1} at 298 K in 760 Torr of Ar buffer.

There have been *two* modeling studies that have examined the impact of the inclusion of this model in studies. Holmes et al. (2006) used kinetic data from Donohoue et al. (2006) and Goodsite et al. (2004) together with the global Br concentrations from the model of Yang et al. (2005) to model the global mean atmospheric lifetime of Hg^0. They concluded that reaction with Br atoms is a major and possibly the dominant global sink for Hg^0, with most of the oxidation taking place in the middle and upper troposphere. They also that noted that lack of kinetic data on the fate of HgBr is a major source of uncertainty in the model.

It is likely that this is a key reaction both in polar mercury depletion events and in the global Hg^0 oxidation cycle. Overall we see that the level of uncertainty is high relative to a typical atmospheric reaction but two very different experimental approaches and theoretical calculations give results that span less than an order of magnitude. For the purposes of atmospheric modeling we would suggest that values between 1×10^{-12} and 3×10^{-13} are used to examine the effects of the uncertainty in this rate coefficient on model predictions. Additional studies are needed to reduce the uncertainty in this rate coefficient and to characterize the fate of HgBr, another key parameter in models.

14.3.3.3 Hg + BrO

$$Hg + BrO \rightarrow HgBr + O \ \Delta H^0 = 166 \text{ kJ mol}^{-1} \quad (9a)$$

$$\rightarrow HgO + Br \ \Delta H^0 = 219 \text{ kJ mol}^{-1} \quad (9b)$$

There have been two experimental determinations of the rate coefficient using relative rate techniques. Spicer et al. (2002) used photolysis of Br_2 in the presence of O_3 to produce BrO and monitored the loss of Hg^0 relative to the loss of dimethyl sulfide. They reported the results of three experiments which ranged from 3.0 - 6.4 $\times 10^{-14}$ cm^3 $molecule^{-1}$ s^{-1}. Raofie and Ariya (2003) also report a relative rate measurement using photolysis of bromine and dibromomethane in the presence of O_3 to produce BrO. The monitored Hg^0 loss relative to three reference molecules, propane, butane and DMS and reported upper and lower limits of 1×10^{-13} and 1×10^{-15} cm^3 $molecule^{-1}$ s^{-1} for the rate coefficient. They also performed a product study again using photolysis of bromine and dibromomethane in the presence of O_3 to produce BrO (Raofie and Ariya, 2004). They reported observation of HgBr, HgO, and HgOBr / HgBrO. In all of these experiments BrO would be photolyzed and there would be a steady state concentration of Br atoms which would react with Hg^0. Calculation of the Br atom concentration requires a knowledge of the photolysis rate but this is a potential source of secondary chemistry in these studies. It is clear that the thermodynamics of this reaction are not consistent with an abstraction reaction to form HgBr or HgO, however calculations suggest that the addition complex HgBrO is bound by 84 kJ mol^{-1}. It seems unlikely that this reaction can play a role in mercuy transformation via reactions 9a or 9b. Studies on the stability of the addition complex and its potential reactions are necessary to assess the importance of this channel.

14.3.3.4 $Hg^0 + NO_3$

$$Hg + NO_3 \rightarrow HgO + NO_2 \quad \Delta H^0 = 195 \text{ kJ mol}^{-1} \quad (10)$$

There has been a single reported measurement of this reaction by Sommar et al. (1997) using a discharge flow technique. The authors report an exothermicity of -12 kJ mol^{-1} for this reaction based on the old HgO thermochemistry. Again the new thermochemistry makes this reaction sufficiently endoergic that the possibility of the atom transfer can be dismissed. The reported experiments measured a heterogeneous wall reaction at two surface-to-volume (S/V) ratio's, 7.6 and 4.0 × 10^{-14} cm^3 $molecule^{-1}$ s^{-1} and then extrapolated to S/V = 0 to extract a gas phase component of 4.0 ×10^{-15} cm^3 $molecule^{-1}$ s^{-1}, a factor of ten slower than the slowest measured rate coefficient. They acknowledged that "In practice, the stated rate coefficient should be taken as an upper limit for the reaction between Hg and NO_3". The endothermicity of this reaction is such that it cannot proceed at a measurable rate in the gas phase and it will not be significant in the atmosphere, even at night in polluted atmospheres.

14.4 Gas Phase Oxidation: Issues and Uncertainties

It is clear that that the limited amount of laboratory data on mercury reaction kinetics makes it difficult to provide a definitive evaluation of published rate coefficients. No data exists on reaction with I or IO although again the calculations of Goodsite et al. (2004) suggest that HgI is weakly bound and will rapidly dissociate to Hg^0 and I. Based on this evaluation of the laboratory data, thermodynamics and models we would suggest that reaction of Hg^0 with Br and possibly BrO are likely to explain both the localized dramatic loss of Hg^0 during mercury depletion events and the global oxidation of Hg^0 in the troposphere. Currently there are no published data on the reactions of HgBr and this is a key uncertainty. To constrain models independent measurements of key rates that agree within a factor of two or better are required. In addition field measurements showing the vertical profiles of both Hg^0 and RGM together with speciated measurements of RGM would do much to resolve these issues.

14.5 Mercury Chemistry in the Atmospheric Aqueous Phase

The aqueous phase chemistry of Hg in the atmosphere involves redox reactions, complexation and phase exchange equilibria and while occurring mostly in cloud droplets, can also take place in deliquesced aerosols. In terms of the amount of water present (atmospheric liquid water content) the most important atmospheric aqueous phase is cloud water. The liquid water content associated with deliquesced aerosol particles is orders of magnitude lower, but their specific chemical composition and the high concentrations of some chemical species make them relevant in terms of the atmospheric aqueous phase chemistry of Hg.

14.5.1 Redox reactions

Elemental mercury is not highly soluble it has a Henry's Law constant of 1.1 ×10^{-1} M atm^{-1} (Sanemasa, 1975) and its concentration in the atmospheric aqueous phase reaches a stable value in a matter of minutes. Once in the aqueous phase Hg can be oxidized by a number of compounds present in the droplet. The most important of these in cloud water is ozone (Munthe, 1992; Iverfeldt and Lindqvist, 1986) on account of its relative abundance and the rate constant for the $Hg^0_{(aq)} + O_{3(aq)}$ reaction. Other aqueous phase reactions which oxidize Hg^0 are those with the hydroxyl radical (OH), (Lin and Pehkonen, 1997; Gardfeldt et al., 2001), $HOCl/OCl^-$ (Lin and Pehkonen, 1998), and the most recently identified, the reaction with $Br_2/HOBr/OBr^-$ (Wang and Pehkonen, 2004). The aqueous phase reactions of Hg which have been studied to date are shown in Table 14.1.

The reduction of oxidized Hg compounds ($Hg^{(II)}$) depends on the complex which Hg has formed in solution. Munthe et al. (1991) first investigated the reduction of Hg via complexation with the sulphite ion, SO_3^{2-}, and obtained a first order rate constant of 0.6 s^{-1}. This reaction was studied again by van Loon et al. (2000), who found a rate constant more than an order of magnitude slower, 0.0106 s^{-1}. Photolytic

Table 14.1 Aqueous phase reactions of Hg

Mercury Reactions and Equilibria	k or K (298K)	Reference
$Hg^0_{(aq)} + O_{3(aq)} \rightarrow HgO_{(aq)}$	$4.7\times10^7\ M^{-1}s^{-1}$	Munthe (1992)
$HgO_{(aq)} + H^+_{(aq)} \rightarrow Hg^{++}_{(aq)} + OH^-_{(aq)}$	$1\times10^{10}\ M^{-1}s^{-1}$	Pleijel and Munthe (1995)
$Hg^{++}_{(aq)} + OH^-_{(aq)} \leftrightarrow HgOH^+_{(aq)}$	$3.9\times10^{10}\ M^{-1}$	Pleijel and Munthe (1995)
$HgOH^+_{(aq)} + OH^-_{(aq)} \leftrightarrow Hg(OH)_{2(aq)}$	$1.6\times10^{11}\ M^{-1}$	Pleijel and Munthe (1995)
$HgOH^+_{(aq)} + Cl^-_{(aq)} \leftrightarrow HgOHCl_{(aq)}$	$2.7\times 10^7\ M^{-1}$	Pleijel and Munthe (1995)
$Hg^{++}_{(aq)} + Cl^-_{(aq)} \leftrightarrow HgCl^+_{(aq)}$	$5.8 \times 10^6\ M^{-1}$	Pleijel and Munthe (1995)
$HgCl^+_{(aq)} + Cl^-_{(aq)} \leftrightarrow HgCl_{2(aq)}$	$2.5 \times 10^6\ M^{-1}$	Pleijel and Munthe (1995)
$HgCl_{2(aq)} + Cl^-_{(aq)} \leftrightarrow HgCl^-_{3\,(aq)}$	$6.7\ M^{-1}$	Clever et al. (1985)
$HgCl^-_{3\,(aq)} + Cl^-_{(aq)} \leftrightarrow HgCl^{--}_{4\,(aq)}$	$13\ M^{-1}$	Clever et al. (1985)
$Hg^{++}_{(aq)} + Br^-_{(aq)} \leftrightarrow HgBr^+_{(aq)}$	$1.1 \times 10^9\ M^{-1}$	Clever et al. (1985)
$HgBr^+_{(aq)} + Br^-_{(aq)} \leftrightarrow HgBr_{2(aq)}$	$2.5 \times 10^8\ M^{-1}$	Clever et al. (1985)
$HgBr_{2(aq)} + Br^-_{(aq)} \leftrightarrow HgBr^-_{3\,(aq)}$	$1.5 \times10^2\ M^{-1}$	Clever et al. (1985)
$HgBr^-_{3\,(aq)} + Br^-_{(aq)} \leftrightarrow HgBr^{--}_{4\,(aq)}$	$23\ M^{-1}$	Clever et al. (1985)
$Hg^{++}_{(aq)} + SO^{--}_{3\,(aq)} \leftrightarrow HgSO_{3(aq)}$	$2.1 \times 10^{13}\ M^{-1}$	Van Loon et al. (2001)
$HgSO_{3(aq)} + SO^{--}_{3\,(aq)} \leftrightarrow Hg(SO_3)^{--}_{2\,(aq)}$	$1.0 \times 10^{10}\ M^{-1}$	Van Loon et al. (2001)
$HgSO_{3(aq)} \leftrightarrow Hg^0_{(aq)}$ + products	T×exp((31.971×T)-12595)/T s-1	Van Loon et al. (2001)
$Hg^0_{(aq)} + OH_{(aq)} \rightarrow Hg^+_{(aq)} + OH^-_{(aq)}$	$2.0 \times 10^9\ M^{-1}\ s^{-1}$	Lin and Pehkonen (1997)
$Hg^+_{(aq)} + OH_{(aq)} \rightarrow Hg^{++}_{(aq)} + OH^-_{(aq)}$	$1.0 \times 10^{10}\ M^{-1}\ s^{-1}$	Lin and Pehkonen (1997)
$Hg^{II}_{(aq)} + O^-_{2\,(aq)} \rightarrow Hg^+_{(aq)} + O_{2(aq)}$	$1.1\times 10^4\ M^{-1}\ s^{-1}$	Pehkonen and Lin (1997)
$Hg^{II}_{(aq)} + HO_{2(aq)} \rightarrow Hg^+_{(aq)} + O_{2(aq)} + H^+_{(aq)}$	$1.1 \times 10^4\ M^{-1}\ s^{-1}$	Pehkonen and Lin (1997)
$Hg^+_{(aq)} + O^-_{2\,(aq)} \rightarrow Hg_{(aq)} + O_{2(aq)}$	fast	Pehkonen and Lin (1997)
$Hg^+_{(aq)} + HO_{2(aq)} \rightarrow Hg^0_{(aq)} + O_{2(aq)} + H^+_{(aq)}$	fast	Pehkonen and Lin (1997)
$Hg^{II}_{(aq)} + O^-_{2\,(aq)} \rightarrow Hg^+_{(aq)} + O_{2(aq)}$	0	Gårdfeldt, and Jonsson (2003)
$Hg^{II}_{(aq)} + HO_{2(aq)} \rightarrow Hg^+_{(aq)} + O_{2(aq)} + H^+_{(aq)}$	0	Gårdfeldt, and Jonsson (2003)
$Hg_{(aq)} + HOCl_{(aq)} \rightarrow Hg^{++}_{(aq)} + Cl^-_{(aq)} + OH^-_{(aq)}$	$2.09 \times 10^6\ M^{-1}\ s^{-1}$	Lin and Pehkonen (1999)
$Hg_{(aq)} + ClO^-_{(aq)} \rightarrow Hg^{++}_{(aq)} + Cl^-_{(aq)} + OH^-_{(aq)}$	$1.99 \times10^6\ M^{-1}\ s^{-1}$	Lin and Pehkonen (1999)
$Hg_{(aq)} + HOBr_{(aq)} \rightarrow Hg^{++}_{(aq)} + Br^-_{(aq)} + OH^-_{(aq)}$	$0.279\ M^{-1}\ s^{-1}$	Wang and Pehkonen (2004)
$Hg_{(aq)} + OBr^-_{(aq)} \rightarrow Hg^{++}_{(aq)} + Br^-_{(aq)} + OH^-_{(aq)}$	$0.273\ M^{-1}\ s^{-1}$	Wang and Pehkonen (2004)
$Hg_{(aq)} + Br_{2(aq)} \rightarrow Hg^{++}_{(aq)} + 2Br^-_{(aq)} + OH^-_{(aq)}$	$0.196\ M^{-1}\ s^{-1}$	Wang and Pehkonen (2004)
$Hg^0_{(g)} \leftrightarrow Hg^0_{(aq)}$	$0.13\ M\ atm^{-1}$	Schroeder and Munthe (1998)
$HgO_{(g)} \leftrightarrow HgO_{(aq)}$	$2.69 \times 10^{12}\ M\ atm^{-1}$	Schroeder and Munthe (1998)
$HgCl_{2(g)} \leftrightarrow HgCl_{2(aq)}$	$2.75 \times 10^6\ M\ atm^{-1}$	Schroeder and Munthe (1998)
$HgBr_{2(g)} \leftrightarrow HgBr_{2(aq)}$	$2.75 \times 10^6\ M\ atm^{-1}$	Schroeder and Munthe (1998)

reduction of $Hg(OH)_2$ also occurs in solution, but not at a rate which is important in the atmospheric aqueous phase, (Xiao et al., 1994; Lin and Pehkonen, 1998b). The reduction of $Hg^{(II)}$ (independent of speciation) via a two step mechanism involving HO_2 was proposed by Pehkonen and Lin (1998), with the first reduction step to Hg^{I} having a rate constant of 1.1×10^4 M^{-1} s^{-1}, in the presence of chloride ions (otherwise 1.7×10^4 M^{-1} s^{-1}). However, whether this mechanism can occur in the atmosphere was called into question by Gardfeldt and Jonsson (2003), who suggested that the Hg^{I} formed in the first step of the reduction mechanism would be rapidly oxidized by dissolved O_2 ($k >> 10^9$ M^{-1} s^{-1}) before the second reduction step to elemental Hg could take place. This question is discussed more fully below.

14.5.2 Does Reduction Actually Occur in the Aqueous Phase?

The reaction between $Hg^{(I)}$ and dissolved O_2 (Nazhat and Asmus, 1973) is the reaction which is missing in the vast majority of atmospheric Hg models. It is potentially fundamental for two reasons: firstly if the oxidation of Hg^0 by OH proceeds as a two step reaction, as proposed by Pehkonen and Lin (1998), then there is no need to consider the second step as a rapid reaction between $Hg^{(I)}$ and OH, because long before $Hg^{(I)}$ encounters a second OH radical it will have already been oxidized by O_2. In fact in some reaction schemes this reaction is considered as Hg^0 + OH leading directly to $Hg^{(II)}$, see Lin et al. (2006), which would appear to be a reasonable approximation. Secondly, the proposed reduction of all $Hg^{(II)}$ complexes by HO_2 proceeds via the formation of $Hg^{(I)}$ (Pehkonen and Lin, 1998), which then needs to react again with yet another HO_2 radical in order to be further reduced to Hg^{0}. If this is the case then the reaction is of absolutely no relevance whatsoever under atmospheric conditions because the chances of $Hg^{(I)}$ encountering a second HO_2 before it encounters a dissolved O_2 molecule are so small as to be irrelevant. Having said this, for the most part Hg models, from zero dimensional box models, to regional and global models include the reduction of $Hg^{(II)}$ by HO_2 in their reaction mechanisms without including the oxidation of $Hg^{(I)}$ by O_2. In many cases the reaction is assumed to proceed very rapidly after the initial reduction to $Hg^{(I)}$, so that some reaction schemes include this reaction as a direct reduction of $Hg^{(II)}$ to Hg^0, which, apart from being kinetically inaccurate leads to an overestimation of the reduction rate. If the reaction between $Hg^{(I)}$ and O_2 is included then the $Hg^{(II)}$ + HO_2 reduction reaction becomes redundant in reaction schemes because the amount of $Hg^{(II)}$ reduced by sequential reactions with HO_2 is negligible.

14.5.3 Speciation

$Hg^{(II)}$ in solution can form complexes with a large number of ligands present in the atmospheric aqueous phase. The equilibrium constants for complex formation are shown in table 14.1, which complexes will actually be present in solution depends on the abundance of the appropriate ligands and the pH of the droplets. The composition of droplets has been studied using both kinetic (where the equilibria are represented as forward and backward reactions) and equilibrium models, (Lin and Pehkonen, 1997; 1998; Pirrone et al. 2000). It is generally agreed that organic ligands are in too short supply to have any influence, and that in the presence of chloride the major part of the $Hg^{(II)}$ present in droplets is present as $HgCl_2$. Some concentrated solutions, such as occur when soluble atmospheric aerosols reach deliquescence may not follow this generalization, however they represent a very small proportion of the atmosphere's liquid water content. The marine sea salt aerosol is however potentially important in the atmospheric Hg cycle and is discussed separately below.

14.5.4 Is Aqueous Phase Chemistry Important Compared to Gas Phase Oxidation?

Measurements of Hg in precipitation have made it clear that wet deposition is in many areas the most important route by which atmospheric Hg enters terrestrial and aquatic ecosystems. It is also well known that Hg once oxidized in the gas phase is readily scavenged by cloud droplets, because the two compounds which are usually assumed to be formed, HgO and $HgCl_2$ are much more readily soluble that elemental Hg, the Henry's Law constant for $HgCl_2$ is 1.4×10^6 M atm^{-1} (Lindqvist and Rodhe, 1985), and that of HgO 2.75×10^6 M atm^{-1} (Schroeder and Munthe, 1998). The exception to this are the Marine Boundary Layer (MBL) and Arctic / Antarctic mercury depletion events, where a significant part of the oxidized Hg in the gas phase probably contains Br. Whether the oxidized Hg present in rain and cloud water is predominantly due to gas phase oxidation followed by scavenging or to Hg oxidation in the aqueous phase can be evaluated using photochemical box models.

The AMCOTS box model (Hedgecock et al., 2005) has been used to investigate the proportion of gas phase to aqueous phase oxidation of Hg in clouds. The model, which has been used up to now to study the influence of non-sea-salt sulphate and sea salt aerosol on Hg chemistry in the MBL has been adapted to include just one atmospheric aqueous phase, the cloud droplets, and the definition of the aqueous phase in terms of liquid water content, droplet radius and droplet number density changed to reflect the properties of different types of clouds. Simulating clean and polluted continental cumulus and marine stratus and cumulostratus clouds it was found that aqueous phase oxidation accounted for between 2 and 30% of all Hg oxidation. The major factors which influenced the proportion of Hg oxidized in the droplets are temperature, droplet number, cloud optical depth and time of day. The lower the temperature the higher the solubility of the gas phase species and therefore higher concentrations of reactants are found in the drops. The influence of droplet number was shown by comparison of polluted continental cumulus and maritime stratus clouds, in both cases a liquid water content of 0.3 g m^{-3} was used.

The maritime cloud has fewer and larger droplets (ND = 80) compared to the continental cloud (ND = 1300), the percentage of total oxidation which took place in cloud water was almost double in the continental cloud. The period during which aqueous phase oxidation is found to be most important is the night, and when the optical depth of the cloud is such that O_3 photolysis is severely reduced restricting the production of OH. Photolytically produced OH is responsible for a large part of the total oxidation during the day, but clearly has little influence during the hours of darkness. Simulating a polluted continental cumulus cloud ozone accounted for 65% of all the oxidation occuring, 35% of the total occuring in the gas phase and 30% in the aqueous phase.

Clearly then Hg chemistry in clouds is important and oxidized Hg compounds in rain come from both scavenging of RGM and in-droplet oxidation. Interestingly the proportion of in-droplet oxidation due to the reaction with O_3 never dropped below 95% in these simulations, and was usually between 98 and 99.9%. It would be worth investigating this finding further in regional models as the simplification of the aqueous phase reaction scheme could lead to savings in calculation times.

14.5.5 The Sea Salt Aerosol

Marine aerosols have been studied primarily because as they become acidified by the uptake of acidic gases they can release reactive halogen containing compounds to the atmosphere. This process is of interest to atmospheric chemists in general, particularly due to these halogen compounds potential for reaction with ozone and volatile organic compounds. It is also of potential significance in the cycling of Hg in the atmosphere as this process in a sense mimics the 'bromine explosion' seen in the Arctic at polar dawn. The concentrations of reactive halogens in the MBL does not reach the levels seen in the Arctic, but given that two thirds of the Earth's surface is water the spatial scale of this phenomena is important. Reactive halogen compounds, particularly those containing Br readily oxidize Hg, see above, and as oxidized Hg compounds are scavenged by droplets there is potential for the sea salt aerosol to contain relatively high concentrations of Hg. On top of this there is the composition of the aerosol themselves to be considered, the very high concentrations of chloride allow Hg to form complexes which are not formed in other atmospheric aqueous phases, such as $HgCl_3^-$ and $HgCl_4^{2-}$ (Hedgecock and Pirrone, 2001; 2004). The ability of the aerosol to 'carry' Hg increases dramatically as its chloride content increases (Pirrone et al., 2000). Modeling studies have also shown that the aerosol could process Hg which has been oxidized by O_3 or OH to give HgO. HgO is very readily scavenged and once in the aqueous phase reacts with H^+ to give an $Hg^{(II)}$ ion which in turn complexes with available ligands. The concentration of the chloride ion is such that the concentration of $HgCl_2$ in solution can increase until it is no longer in equilibrium with the concentration of $HgCl_2$ in the gas phase and therefore outgases from the aerosol.

14.6 The Uncertainty due to Hg Chemistry in Atmospheric Models

The uncertainty which results from the atmospheric processes which are included in atmospheric Hg modeling studies has been studied by Lin et al. (2006). Their article discusses the uncertainties relating to gas and aqueous phase chemistry as well as the other important factors such as emission speciation and dry and wet deposition parameterizations. They conclude that:

> *"Of the greatest importance for reducing model uncertainties is the accurate quantification of the reaction kinetics of mercury transformation mechanisms in both gas and aqueous phases. Thecurrent understanding of atmospheric mercury chemistry and related kinetics are based on the extrapolation of limited laboratory investigations. The appropriateness of such extrapolation has been questioned (e.g., Calvert and Lindberg, 2005, Gardfeldt and Jonsson, 2003). In addition, some of the reaction kinetics and products are not clearly defined. Since chemistry is the most important driving force for mercury deposition in regions away from the anthropogenic sources, further experimental investigations addressing these kinetic and product uncertainties will greatly improvemodel performance in predicting both dry and wet depositions."*

They continue, mentioning the lack of data relating to adsorption isotherms for Hg and its compounds both for particulate matter in the gas phase and that included in cloud and rain droplets.

It is probably safe to say that the uncertainty in Hg chemistry lies less with the aqueous phase than with the gas phase and the adsorption of Hg and its compounds to solid particulates, especially in light of the box modeling tests mentioned above, from which it appears that by far the most important aqueous oxidant is O_3.

It must be pointed out however that there has only been one determination of the Hg + O_3 reaction rate constant in the aqueous phase and that the reduction reaction generally considered to be the most important might not actually occur at all. The last point raises a number of difficulties because observations quite clearly indicate a relatively long lifetime for atmospheric elemental Hg, but if indeed the reduction of $Hg^{(II)}$ by HO_2 is not pertinent to atmospheric aqueous phase chemistry there are no known reactions which reduce $Hg^{(II)}$ in the atmosphere at a rate which could offset all the oxidation which is believed to be occurring, so that the observed lifetime of Hg is essentially unexplainable in terms of the known reactions which occur in the atmosphere.

14.7 Deposition Processes

After emission mercury is subject to physical and chemical processes in the atmosphere which may lead to conversion into the RGM form which is more soluble, adsorption to particles or ultimately some type of deposition, where its low volatility can lead to re-emission. The enhanced solubility of RGM compared to GEM will lead to relatively quicker deposition, generally closer to sources, whereas GEM may be transported globally with a lifetime of approximately 1 year on average. GEM, RGM and TPM are eventually deposited by wet or dry deposition. What is known with relative certainty is that divalent mercury is much efficiently removed from the troposphere via wet deposition than elemental mercury due to its solubility. Gaseous and particulate forms of mercury are much more likely removed via dry deposition processes.

The deposition of RGM and GEM have been measured over various surfaces with chamber and micrometeorological techniques (Edwards, et al., 2001; Skov et al., 2006; Cobbett and Van Heyst, 2007) with varying results. Deposition has also been modeled (Steffen et al., 2007). Cobbett et al. (2007), summarize their own flux measurements of GEM, RGM and TPM over an arctic sampling site and review the published flux data for the ten years previous to their study. None of the studies have yet documented the causal effect that the observations of mercury deposited end up in the Arctic food chain. Though a reasonable hypothesis, observational data shows that under proper conditions, the mercury could be reemitted to the atmosphere. More studies are needed to ultimately describe the mass balance and fate of mercury in the Arctic.

Is it possible to identify regions that are most sensitive? Mercury has been investigated in all regions of the world, see previous chapters. There is general agreement in the literature that similar trends are noted in the northern hemisphere. Data from the southern hemisphere is more limited but note similar depositional trends and similar preanthropogenic natural background levels when compared to the northern hemisphere (Steffen et al., 2007).

Peatlands under the proper conditions can be used to investigate the environmental record of mercury deposition (Madsen, 1981). In general, the "natural" or pre anthropogenic rate of deposition for mercury, based on a rough approximation of all various environmental archives for example: ice cores, lake and marine sediments and peat sediments is on the order of 2 $\mu g\ m^{-2}$ per yr^{-1}, depending on the archive and site (Givelet et al., 2003). This amount generally increases with a factor of three to ten through the anthropogenic peak, and has been falling since then. The peak has been seen in the 1990s or in the 1950s. What will happen to the mercury in the peatlands with increased warming and the hydrological changes with accompanying biological activity this will entail is pure conjecture. Never the less, peatlands represent significant natural sinks of mercury that could be mobilized in the next 50 years.

Swain et al. (1992) investigated whole-basin mercury fluxes, determined from lake-wide arrays of dated sediment cores. The analyses indicated that the annual deposition of atmospheric mercury increased from 3.7 to 12.5 $\mu g\ m^{-2}$ (1990) since 1850, with the last decades showing a decline. 25 percent of atmospheric mercury deposition to the terrestrial catchment was exported to the lake. The deposition increase was similar among sites, implying regional or global sources for the mercury entering these lakes.

Peat records from the Faroe Islands in the North Atlantic had average rate of atmospheric Hg accumulation from 1520 BC to AD 1385 of $1.27 \pm 0.38\ \mu g\ m^{-2}\ yr^{-1}$.

Br and Se concentrations and the background Hg/Br and Hg/Se ratios were used to calculate the average rate of *natural* Hg accumulation for the same period of 1.32 ± 0.36 μg m^{-2} yr^{-1} and 1.34 ± 0.29 μg m^{-2} yr^{-1}, respectively (Shotyk et al., 2005).

The total mercury fluxes were similar to the preanthropogenic rates obtained using peat cores from Switzerland, southern Greenland, southern Ontario, Canada, and the northeastern United States (Givelet et al., 2003; 2004; Shotyk et al., 2003; 2005; Roos-Barraclough and Shotyk, 2003; Roos-Barraclough et al., 2006). Values on the Faroe Islands were representative of those measures at other sites on the northern hemisphere in similar studies. On the Faroe Islands, the maximum rate of Hg accumulation was 34 μg m^{-2} yr^{-1}. The greatest fluxes of anthropogenic Hg accumulation calculated using Br and Se, respectively, were 26 and 31 μg m^{-2} yr^{-1}. The rate of atmospheric Hg accumulation in 1998 of 16 μg m^{-2} yr^{-1} was comparable to the values recently obtained by atmospheric transport modeling for Denmark, the Faroe Islands, and Greenland (Shotyk et al., 2005).

Environmental archives in the above studies already show a decreasing load of mercury in the past decade, likely due to reductions in emissions from the power industry (due to better emission controls for many other types of pollutants that likely inhibit much of the emission of mercury), though many more experiments are needed in this area as well, and fall in post eastern block industrial activity decreased emissions. The actual reduction in loadings have not been calculated or modelled as it is not a simple linear relationship, since warming climate will likely increase mercury emissions from previously deposited mercury.

Although many archives have been investigated, there is still poor geographic resolution in both hemispheres. The transfer functions which describe the mass transfer between the atmosphere and the surface of the archive needs to be further studied for all archives as do possible physical, chemical and biological perturbations which might be affecting the accuracy of the records. The records of mercury deposition are only as good as their chronology and advances in high resolution dating should continue to be made. Archives that have been investigated for mercury should be reanalyzed to establish a climate proxy and these records should be compared, as relatively few records exist that both contain historical mercury and climate proxy records.

References

Ariya, P. A., Alexei Khalizov, and Alexios Gidas, 2002, Reactions of Gaseous Mercury with Atomic and Molecular Halogens: Kinetics, Product Studies, and Atmospheric Implications, J. Phys. Chem. A. 106, 7310–7320; (Article) DOI: 10.1021/jp020719o

Ariya P. A. and K. A. Peterson in Pirrone, N., and Mahaffey, K.R., eds., Dynamics of Mercury Pollution on Regional and Global Scales: Atmospheric Processes and Human Exposures Around the World: Springer Science and Business Media, Inc. 2005

Atkinson, R., Baulch, D. L., Cox, R. A., Crowley, J. N., Hampson, R. F., Hynes, R. G., Jenkin, M. E., Rossi, M. J., Troe, J., and IUPAC Subcommittee, :Evaluatedkinetic and photochemical data for atmospheric chemistry: Volume II – gas phase reactions of organic species, Atmos. Chem. Phys., 6, 3625–4055, 2006

Balabanov N. B., B. C. Shepler, and K. A. Peterson, 2005, Accurate Global Potential Energy Surface and Reaction Dynamics for the Ground State of HgBr2, J. Phys. Chem. A, 109, 8765–8773, DOI: 10.1021/jp0534151

Bauer, D., L. D'Ottone, P. Campuzano-Jost, A. J. Hynes, 2003, Gas phase elemental mercury: a comparison of LIF detection techniques and study of the kinetics of reaction with the hydroxyl radical, Journal of Photochemistry and Photobiology A, 157, 247–256

Bergan T. and H. Rohde, 2001, Oxidation of Elemental Mercury in the Atmosphere; Constraints Imposed by Global Scale Modelling, J. Atmos. Chem., 40, 191–212, DOI 10.1023/A: 1011929927896

Calvert J. G. and S. E. Lindberg, Mechanisms of mercury removal by O3 and OH in the atmosphere Atmos. Environ. 39, 3355, 2005.

Chase, M.W. NIST-JANAF Thermochemical Tables, Fourth Edition: Journal of Physical Chemical Reference, Monograph 9, 1998.

Clever, H.L., Johnson, S.A., Derrick, M.E. The solubility of Mercury and Some Sparingly Soluble Mercury Salts in Water and Aqueous Electrolyte Solutions. J. Phys. Chem. Refer. Data, 14, 631–680, 1985.

Cobbett F. D. and B.J. Van Heyst, 2007, Measurements of GEM fluxes and atmospheric mercury concentrations (GEM, RGM and Hgp) from an agricultural field amended with biosolids in Southern Ont., Canada (October 2004–November 2004), Atmospheric Environment, 41, 2270–2282

Cobbett, F. D. A. Steffen, G. Lawson, B. J. Van Heyst, 2007, GEM fluxes and atmospheric mercury concentrations (GEM, RGM and Hgp) in the Canadian Arctic at Alert, Nunavut, Canada (February–June 2005), Atmospheric Environment, 41, 6527–6543

Donohoue D. L. , D. Bauer, and A. J. Hynes, 2005, Temperature and Pressure Dependent Rate Coefficients for the Reaction of Hg with Cl and the Reaction of Cl with Cl: A Pulsed Laser Photolysis-Pulsed Laser Induced Fluorescence Study, J. Phys. Chem. A, 109, 7732–7741, DOI: 10.1021/jp0513541

Donohoue D. L., D. Bauer, B. Cossairt, and A. J. Hynes, 2006, Temperature and Pressure Dependent Rate Coefficients for the Reaction of Hg with Br and the Reaction of Br with Br: A Pulsed Laser Photolysis-Pulsed Laser Induced Fluorescence Study, J. Phys. Chem. A, 110, 6623–6632, DOI: 10.1021/jp054688j

Edwards, G., P. Rasmussen, W. Schroeder, R. Kemp, G. Dias, C. Fitzgerald-Hubble, E. Wong, L. Halfpenny-Mitchell, and M. Gustin (2001), Sources of variability in mercury flux measurements, J. Geophys. Res., 106(D6), 5421–5435.

Gårdfeldt K., J. Sommar, D. Strömberg and X. Feng, 2001, Oxidation of atomic mercury by hydroxyl radicals and photoinduced decomposition of methylmercury in the aqueous phase, Atmospheric Environment, 35 (17), 3039–3047

Gårdfeldt, K. Jonsson, M. Is bimolecular reduction of Hg(II) complexes possible in aqueous systems of environmental importance. J. Phys. Chem. A, 107, 4478–4482, 2003.

Givelet, N., Roos-Barraclough, F. and Shotyk, W. (2003) Rates and predominant anthropogenic sources of atmospheric Hg accumulation in southern Ontario recorded by peat cores from three bogs: comparison with natural "background" values (past 8,000 years). Journal of Environmental Monitoring 5, 935–949

Givelet, N.; Roos-Barraclough, F.; Goodsite, M. E.; Cheburkin, A. K.; Shotyk, W., 2004, Atmospheric Mercury Accumulation Rates Between 5900 and 800 Calibrated Years BP in the High Arctic of Canada Recorded by Peat Hummocks, Environ. Sci. Technol., 38, 4964–4972. DOI: 10.1021/es0352931

Goodsite, M. E., Plane, J. M. C., and Skov, H.: A theoretical study of the oxidation of Hg-0 to HgBr2 in the troposphere, Environ. Sci. Technol., 38(6), 1772–1776, 2004.

Hall, B. D.: The gas-phase oxidation of elemental mercury by ozone, Water Air and Soil Pollution, 80, 301–315, 1995.

Hedgecock, I. and Pirrone, N. (2001) Mercury and Photochemistry in the Marine Boundary Layer – Modelling Studies suggest the in situ Production of Reactive Gas Phase Mercury. Atmospheric Environment. Vol. 35, 3055–3062.

Hedgecock, I. M., and Pirrone, N., (2004) Chasing Quicksilver: Modeling the Atmospheric Lifetime of Hg0 (g) in the Marine Boundary Layer at Various Latitudes. Environmental Science and Technology, Vol.38, 69–76.

Hedgecock, I. M., Trunfio, G.A., Pirrone, N., Sprovieri, F. (2005) Mercury chemistry in the MBL: Mediterranean case and sensitivity studies using the AMCOTS (Atmospheric Mercury Chemistry over the Sea) model. Atmospheric Environment, Vol. 39, 7217–7230.

Hynes, A.J., P. H. Wine, and J. M. Nicovich, 1988, Kinetics and mechanism of the reaction of hydroxyl with carbon disulfide under atmospheric conditions, J. Phys. Chem., 92, 3846–3852; DOI: 10.1021/j100324a034

Holmes C. D., D. J. Jacob, X. Yang (2006), Global lifetime of elemental mercury against oxidation by atomic bromine in the free troposphere, Geophys. Res. Lett., 33, L20808, doi:10.1029/2006GL027176.

Iverfeldt, A and O. Lindqvist, 1986, Atmospheric oxidation of elemental mercury by ozone in the aqueous phase, Atmospheric Environment, 20 (8), 567–1573)

Khalizov A. F., B. Viswanathan, P. Larregaray, and P. A. Ariya, 2003, A Theoretical Study on the Reactions of Hg with Halogens: Atmospheric Implications, J. Phys. Chem. A., 107, 6360–6365, DOI: 10.1021/jp0350722

Landis M. S. et al., in Pirrone, N., and Mahaffey, K.R., eds., Dynamics of Mercury Pollution on Regional and Global Scales: Atmospheric Processes and Human Exposures Around the World: Springer Science and Business Media, Inc. 2005

Lin, C-J., Pehkonen, S.O. Aqueous Phase Reactions of Mercury with Free Radicals and Chlorine: Implications for Atmospheric Mercury Chemistry. Chemosphere, 38, 1253–1263, 1999.

Lin, C-J., Pehkonen, S.O. Two-Phase Model of Mercury Chemistry in the Atmosphere. Atmos. Environ., 32, 2543–2558, 1998b.

Lin, C.-J., Pehkonen, S.O. Aqueous free radical chemistry of mercury in the presence of iron oxides and ambient aerosol. Atmos. Environ., 31, 4125–4137, 1997.

Lin, C.-J., Pehkonen, S.O. Oxidation of elemental mercury by aqueous chlorine (HOCl/OCl-): implication for tropospheric mercury chemistry. J. Geophys. Res., 103, 28093–28201, 1998.

Lin, C.J, P. Pongprueksa, S.E. Lindberg, S.O. Pehkonen, D. Byun ,C. Jang, Scientific uncertainties in atmospheric mercury models. SI: Model science evaluation, Atmospheric Environment 40, 2911–2928, 2006.

Lindberg, S. E., Brooks, S., Lin, C.-J., Scott, K. J., Landis, M. S., Stevens, R. K., Goodsite, M., and Richter, A.: Dynamic oxidation of gaseous mercury in the Arctic troposphere at polar sunrise, Environ. Sci. Technol., 36, 1245–1256, 2002.

Lindqvist, O., Rodhe, H. Atmospheric mercury—a review. Tellus, 37B, 136–159, 1985.

Madsen, P. P., Peat bog records of atmospheric mercury deposition, Nature 293, 127, 1981.

Munthe, J., 1992, The aqueous oxidation of elemental mercury by ozone, Atmos. Environ., 26A, 1461–1468

Munthe, J., Xiao, Z.F. and Lindquist, O., The aqueous reduction of divalent mercury by sulphite. Water, Air and Soil Pollution 56, 621–630, 1991.

Murphy, D. M.; Cziczo, D. J.; Froyd, K. D.; Hudson, P. K.; Matthew, B. M.; Middlebrook, A. M.; Peltier, R. E.; Sullivan, A.; Thomson, D. S.; Weber, R. J. 2006a Single-particle mass spectrometry of tropospheric aerosol particles J. Geophys. Res., Vol. 111, No. D23, D23S32 10.1029/2006JD007340

Murphy, D. M.; Hudson, P. K.; Thomson, D. S.; Sheridan, P. J.; Wilson, J. C., 2006b Observations of Mercury-Containing Aerosols Environ. Sci. Technol. 40 3163–3167. DOI: 10.1021/es052385x

Nazhat, N. B., K.-D. Asmus, Reduction of mercuric chloride by hydrated electrons and reducing radicals in aqueous solutions. Formation and reactions of HgCl, J. Phys. Chem., 77(5), 614–620, 1973.

Pal B. and P. A. Ariya, 2004, Gas-Phase HO-Initiated Reactions of Elemental Mercury: Kinetics, Product Studies, and Atmospheric Implications, Environ. Sci. Technol., 38, 5555–5566, DOI: 10.1021/es0494353

Pal, B. and Ariya, P. A.: Kinetics and mechanism of O3-initiated reaction of Hg0: atmospheric implication, J. Phys. Chem.-Chem. Phys., 6, 752, 2004.

Pehkonen, S.O., Lin, C.-J. Aqueous photochemistry of mercury with organic acids. J. A. W. M. A.,. 48, 144–150, 1998.

Peleg, M.; Matveev, V.; Tas, E.; Luria, M.; Valente, R. J.; Obrist, D., Mercury Depletion Events in the Troposphere in Mid-Latitudes at the Dead Sea, Israel, Environ. Sci. Technol.; (Article); 2007; 41(21); 7280–7285. DOI: 10.1021/es070320j

Pirrone, N., Hedgecock, I., Forlano, L. The Role of the Ambient Aerosol in the Atmospheric Processing of Semi-Volatile Contaminants: A Parameterised Numerical Model (GASPAR). Journal of Geophysical Research, 105, D8, 9773–9790, 2000.

Pleijel, K., Munthe, J. Modelling the Atmospheric Mercury Cycle - Chemistry in Fog Droplets. Atmos. Environ., 29, 1441–1457, 1995.

Raofie F. and P. A. Ariya, 2003, Kinetics and products study of the reaction of BrO radicals with gaseous mercury, J. Phys. IV, p. 1119

Raofie F. and P.A. Ariya, 2004, Product Study of the Gas-Phase BrO-Initiated Oxidation of Hg0: Evidence for Stable Hg1+ Compounds, Environ. Sci. Technol., 38, 4319–4326, DOI: 10.1021/es035339a

Roos-Barraclough F. and W. Shotyk, 2003, Millennial-Scale Records of Atmospheric Mercury Deposition Obtained from Ombrotrophic and Minerotrophic Peatlands in the Swiss Jura Mountains Environmental Science and Technology 37, 235–244. DOI: 10.1021/es0201496

Roos-Barraclough, F.; Givelet, N.; Cheburkin, A. K.; Shotyk, W.; Norton, S. A., 2006

Sander S. P. et al. Chemical Kinetics and Photochemical Data for Use in Atmospheric Modeling, JPL 06-2, Jet Propulsion Laboratory, Pasadena, CA, 2006.

Sanemasa, I. (1975) The solubility of elemental mercury vapour in water, Bull. Chem. Soc. Jpn., 48, 1795–98.

Schroeder, W. H., Anlauf, K. G., Barrie, L. A., Lu, J. Y., Steffen, A., Schneeberger, D. R., and Berg, T.: Arctic springtime depletion of mercury, Nature, 394, 331–332, 1998.

Schroeder, W.H., Munthe, J., Atmospheric mercury - an overview. Atmos. Environ., 32, 809–822, 1998.

Seigneur C., K. Vijayaraghavan, K. Lohman (2006), Atmospheric mercury chemistry: Sensitivity of global model simulations to chemical reactions, J. Geophys. Res., 111, D22306, doi:10.1029/2005JD006780.

Selin, N.E., D.J. Jacob, R.J. Park, R.M. Yantosca, S. Strode, L. Jaegle,and D. Jaffe, Chemical cycling and deposition of atmospheric mercury: Global constraints from observations, J. Geophys. Res, 112, DO2308, doi:10.1029/2006JD007450, 2007.

Shepler B. C, N. B. Balabanov, and K. A. Peterson, 2007, Hg+Br HgBr recombination and collision-induced dissociation dynamics, J. Chem. Phys. 127, 164304 (2007); DOI:10.1063/1.2777142

Shepler B. C. and K. A. Peterson, 2003, Mercury Monoxide: A Systematic Investigation of Its Ground Electronic State, J. Phys. Chem. A, 107, 1783–1787; DOI: 10.1021/jp027512f

Shotyk, W., M. E. Goodsite, F. Roos-Barraclough, R. Frei, J. Heinemeier, G. Asmund, C. Lohse, T. S. Hansen, 2003, Anthropogenic contributions to atmospheric Hg, Pb and As accumulation recorded by peat cores from southern Greenland and Denmark dated using the 14C "bomb pulse curve", Geochimica et Cosmochimica Acta, 67, 21, 3991–4011

Shotyk, W., M.E. Goodsite, F. Roos-Barraclough, N. Givelet, G. Le Roux, D. Weiss, A.K. Cheburkin, K. Knudsen, J. Heinemeier, W.O. van Der Knaap, S.A. Norton, C. Lohse, 2005, Accumulation rates and predominant atmospheric sources of natural and anthropogenic Hg and Pb on the Faroe Islands, Geochimica et Cosmochimica Acta, 69, 1–17

Skov, H., S. B. Brooks, M. E. Goodsite, S. E. Lindberg, T. P. Meyers, M. S. Landis, M. R.B. Larsen, B. Jensen, G. McConville, J. Christensen, 2006, Fluxes of reactive gaseous mercury measured with a newly developed method using relaxed eddy accumulation, Atmospheric Environment, 40, 5452–5463

Sommar, J., M. Hallquist, 1997, On the Gas Phase Reactions Between Volatile Biogenic Mercury Species and the Nitrate Radical, Journal of Atmospheric Chemistry, 27, 233, 1997

Sommar, J. K. Gårdfeldt, D. Strömberg, X. Feng, 2001, A kinetic study of the gas-phase reaction between the hydroxyl radical and atomic mercury, Atmospheric Environment, 35, 3049–3054

Spicer C. W. et al., Kinetics of Gas-Phase Elemental Mercury Raction with Halogen Species, Ozone, and Nitrate Radical under Atmospheric Conditions: Tallahassee, FL, Florida Department of Environmental Protection, 2002

Steffen, A., T. Douglas, M. Amyot, P. Ariya, K. Aspmo, T. Berg, J. Bottenheim, S. Brooks, F. Cobbett, A. Dastoor, A. Dommergue, R. Ebinghaus, C. Ferrari, K. Gardfeldt, M. E. Goodsite, D. Lean, A. Poulain, C. Scherz, H. Skov, J. Sommar, C. Temme, A synthesis of atmospheric mercury depletion event chemistry linking atmosphere, snow and water, Atmos. Chem. Phys. Discuss. 2007, 7, 10837.

Sumner A. L. in Pirrone, N., and Mahaffey, K.R., eds., Dynamics of Mercury Pollution on Regional and Global Scales: Atmospheric Processes and Human Exposures Around the World: Springer Science and Business Media, Inc. 2005

Swain, E. B., D. R. Engstrom, M. E. Brigham, T. A. Henning, and P. L. Brezonik, 1992, Increasing Rates of Atmospheric Mercury Deposition in Midcontinental North America, Science, 257, 784–787.

Swartzendruber, P. C., D. A. Jaffe, E. M. Prestbo, P. Weiss-Penzias, N. E. Selin, R. Park, D. J. Jacob, S. Strode, and L. Jaeglé (2006), Observations of reactive gaseous mercury in the free troposphere at the Mount Bachelor Observatory, J. Geophys. Res., 111, D24301, doi:10.1029/2006JD007415.

Tossell J. A. 2003 Calculation of the Energetics for Oxidation of Gas-Phase Elemental Hg by Br and BrO, J. Phys. Chem. A., 107, 7804–7808 DOI: 10.1021/jp030390m

Tossell, J. A. 2006, Calculation of the Energetics for the Oligomerization of Gas Phase HgO and HgS and for the Solvolysis of Crystalline HgO and HgS, J. Phys. Chem. A., 110, 2571–2578, DOI: 10.1021/jp056280s

F. Roos-Barraclough, N. Givelet, Andriy K. Cheburkin, W. Shotyk, and S. A. Norton. Use of Br and Se in Peat To Reconstruct the Natural and Anthropogenic Fluxes of Atmospheric Hg: A 10000-Year Record from Caribou Bog, Maine, Environmental Science and Technology 40, 3188–3194. DOI: 10.1021/es051945p.

Van Loon, L., Mader, E., Scott, S.L. Reduction of the aqueous mercuric ion by sulfite: UV spectrum of HgSO3 and its intramolecular redox reaction. J. Phys. Chem. A, 104, 1621–1626, 2000.

Van Loon, L. L., Mader, E. A., Scott, S. L. Sulfite Stabilization and Reduction of the Aqueous Mercuric Ion: Kinetic Determination of Sequential Formation Constants, J. Phys. Chem. A., 105, 3190–3195, 2001.

Wang, Z., Pehkonen, S.O., 2004. Oxidation of elemental mercury by aqueous bromine: atmospheric implications. Atmospheric Environment 38, 3675–3688.

Williams, M. B., P. Campuzano-Jost, B. M. Cossairt, A. J. Hynes, and A. J. Pounds, 2007, Experimental and Theoretical Studies of the Reaction of the OH Radical with Alkyl Sulfides: 1. Direct Observations of the Formation of the OH-DMS Adduct-Pressure Dependence of the Forward Rate of Addition and Development of a Predictive Expression at Low Temperature, J. Phys. Chem. A, 111, 89–104, DOI: 10.1021/jp063873+

Xiao Z F. et al. In: Watras, C.J., Huckabee, J.W. (Eds.), Mercury as a Global Pollutant-Integration and Synthesis. Lewis Publishers, New York, 581–592, 1994.

Yang X., R. A. Cox, N. J. Warwick, J. A. Pyle, G. D. Carver, F. M. O'Connor, N. H. Savage (2005), Tropospheric bromine chemistry and its impacts on ozone: A model study, J. Geophys. Res., 110, D23311, doi:10.1029/2005JD006244.

Chapter 15
Mercury Chemical Transformation in the Gas, Aqueous and Heterogeneous Phases: State-of-the-art Science and Uncertainties

Parisa A. Ariya, Kirk Peterson, Graydon Snider, and Marc Amyot

Summary Mercury is a persistent, toxic and bio-accumulative pollutant of global interest. This element is assumed to exist predominantly in the atmosphere, as elemental mercury, undergoing chemical reactions in the presence of atmospheric oxidants. The oxidized mercury can further deposit on the Earth's surface and may potentially be bioaccumulative in the aquatic food chain, through complex, but not yet well understood, mechanisms. Since the atmosphere plays a significant role as a medium for chemical and physical transformation, it is imperative to understand the fundamentals of the kinetics and thermodynamics of the elementary and complex reactions of $Hg^0_{(g)}$ and oxidized mercury not only in the atmosphere as gas phase, but also the reactions in the aqueous and heterogeneous phases at atmospheric interfaces such as aerosols, fogs, clouds, and snow-water-air interfaces. In this chapter, we compile a comprehensive set of theoretical, laboratory and field observations involving mercury species in the course of homogeneous and heterogeneous reactions. We herein describe the state-of-the-knowledge in this domain and put forward the open questions and future direction of research.

15.1 Introduction

Atmospheric chemical processes of mercury promoted by interfaces have been largely overlooked until recently, although heterogeneously-catalyzed chemical reactions in the stratosphere have been well established for several decades. As a fluid metal, mercury is a liquid at room temperature and has the lowest known critical temperature of any metal (*c.a.*, 1478 °C) Although the dominant form of atmospheric mercury is gaseous elemental mercury ($Hg^0_{(g)}$), traces of oxidized mercury in aquatic and heterogeneous systems are expected (Seiler et al., 1980; Slemr et al.,1985; Lindqvist et al., 1985). Figure 15.1 illustrates a simplified schematic of mercury transformation in the atmosphere and at atmospheric interfaces.

N. Pirrone and R. Mason (eds.), *Mercury Fate and Transport in the Global Atmosphere*,
DOI: 10.1007/978-0-387-93958-2_15,

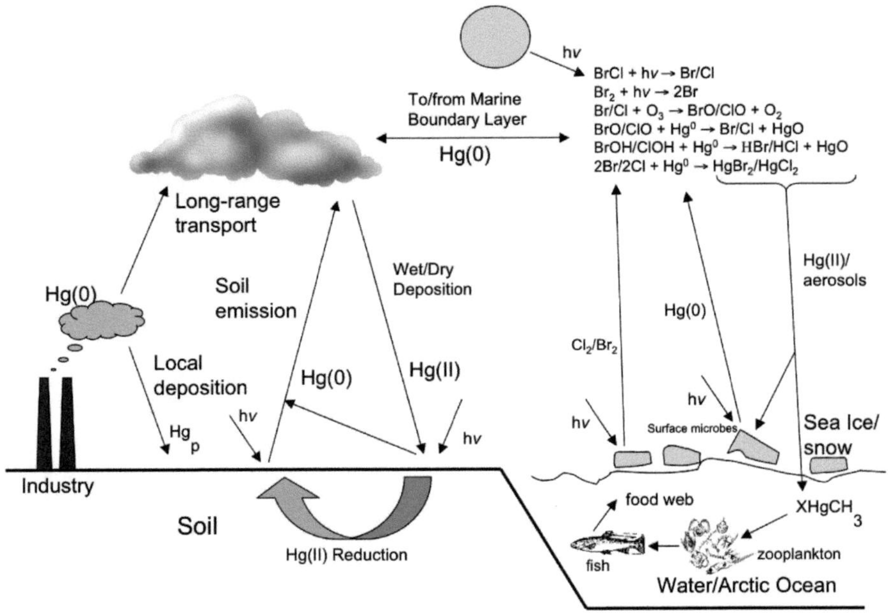

Figure 15.1 A simplified schematic of mercury transformation in the Earth's environment (inspired by Ariya, et al., 2004; Macdonald et al., 2005; Lindberg et al., 2007)

Elemental mercury (Hg^0) exists in ambient air, both in the vapour and particle phase associated with aerosols. As pointed out in the earlier chapters, mercury in the atmosphere is predominantly anthropogenic in origin, such as fuel and coal combustion and waste incineration. Natural emissions, including those from volcanic eruptions, soils, lakes, open water and forest fires, contribute less significantly than anthropogenic sources (c.a. *40%*). However, there are significant uncertainties on natural emission inventories (Mason et al., 1994; Gardfeldt and Jonsson, 2003). Atmospheric chemical transformations of mercury can indeed play an important role in the global cycling of this toxic element, as the atmosphere is the fastest moving fluid in the Earth's ecosystem. A major interest in the understanding of atmospheric transformation stems from its potential impact on mercury bioaccumulation. Mercury speciation in the atmosphere has a significant influence on its deposition on environmental surfaces. Solubility and deposition of elemental mercury is quite distinct from $Hg^{(II)}$ (Ariya and Peterson, 2005), and thus deposition rates on the Earth's surface vary substantially. Amongst all mercury species, methyl mercury has been considered to be bio-magnified in fish. The extent of methylation depends on a constant supply of inorganic mercury from the atmosphere (Mason and Sheu, 2002). Indeed, atmospheric deposition is considered to be a major source of mercury in most remote aquatic systems (Mason et al., 1994; Nriagu, 1994). It is noteworthy that the chemical-biological processes that dictate the bioaccumulation of mercury in the food chain have yet to be fully characterized (Morel, 1998). Consequently, the extent of incorporation of oxidized mercury produced via atmospheric chemical reactions into the food chain has yet to be evaluated. Figure 15.1 depicts a simplified schematic of mercury cycling in the Earth's ecosystem in atmosphere and at environmental interfaces.

Hg species are removed from atmosphere through dry and wet deposition processes (Lindqvist and Rodhe, 1985). Interestingly, in the high-Arctic region (Schroeder et al., 2002), Arctic (Lindberg et al., 1998), and sub-Arctic, the rapid depletion of mercury has been observed. Nearly complete depletions of ozone in the boundary layer occurred over large areas, and evidence of reactive halogens have been observed during most mercury depletion events (MDEs) (Ariya et al., 1998; 1999). Upon reaction with atmospheric oxidants, elemental mercury can be transformed to its oxidized forms, which are also more bio-accumulative than elemental mercury (Gardfeldt et al., 2001). Observed ozone depletion events at the ground are suggested to be driven by sunlight and bromine atoms derived from reactions of atmospheric reactive halogens with marine sea salt in surface snow and ice (Ariya et al., 1998; 1999; Gardfeldt et al., 2001). Soon after, mercury depletion was found to be wide spread. Such depletion events have also been observed in the Antarctic (Ebinghaus et al., 2002), where they are influenced by the photochemical oxidation of elemental mercury in the troposphere involving sea salt on snow/icepack or aerosols (Schroeder and Munthe, 1998).

The mechanism of the volatilization of gaseous elemental mercury from surfaces, chemical transformation in gas and condensed phases (liquid/solid/ heterogeneous), and deposition mechanisms are not well-defined processes. For instance, not much is known about chemical reactions occurring in the snow, especially catalytic and heterogeneous reactions occurring at the surface of snow grains and removal of $Hg^0_{(g)}$ over fly ash, but field observations support the importance of such surfaces in mercury cycling (Dommergue et al., 2007; Pavlish et al., 2003). Pure gaseous oxidation of mercury is mechanistically difficult to explain as well, and in some cases can be explained via heterogeneous phase chemistry (Raofie and Ariya, 2003). Inconsistencies between kinetic and thermodynamic data describing the homogeneous gas phase oxidation of mercury such as in case of one of the most predominant atmospheric oxidant, ozone or its well reactions (Calvert and Lindberg, 2005, Hall et al., 1995). Attempts to more clearly understand reduction of $Hg^{(II)}_{(s,\, aq)}$ to $Hg^0_{(g)}$ (or the reverse oxidation) are motivated by uncertainties in the Hg chemistry of the Arctic and in finding suitable surface catalysts for $Hg^0_{(g)}$ emission reduction in coal fire combustion. Noting the lack of detailed *mechanistic* understanding of mercury redox reactions, we herein strive to examine what changes alter surface reactivity, including the presence of water, various trace surface impurities, photochemistry, temperature, or other competing reactions. Surfaces can act as reactive sites for chemical reactions, active sites for catalysis, and as a platform for exchange between different planetary ecosystem compartments such as air-snow, air-water (lake/ocean), vegetation-air, water-soil and air-soil. However, due to the complexity of the nature of surfaces, its variability, its sensitivity towards environmental variables, its temporal and spatial heterogeneity, environmental surfaces studies are one of the major scientific domain of uncertainty that will face environmental scientists in this new century.

A significant part of the deposited mercury is photo-reduced and re-emitted as GEM (Ferrari et al., 2003; 2004; Brooks et al., 2006; Aspmo et al., 2006). There have been several excellent review articles on mercury transformation in atmosphere (Schroeder, 1991; Lin and Pehkonen, 1999; Steffen, et al., 2007), particularly on its properties, sources, sinks, and fluxes of mercury. As such, in the light of recent laboratory and computational studies, we will attempt to focus instead on a comprehensive review of the kinetic, product studies and thermochemical calculations of mercury redox reactions in the homogeneous and heterogeneous phases. We will discuss the importance of environmental interfaces and environmentally relevant (or potentially relevant) carbon surfaces (such as fly ash, charcoal). We will outline major gaps and some future research directions.

15.2 Atmospheric Oxidation and Reductions

Most atmospheric oxidation reactions are likely to react via multi-step reaction mechanisms. Tables 15.1-15.3 illustrate the detailed kinetic and theoretical studies, as well as estimation from field studies on various atmospheric reactions in gaseous and aqueous phases as well as reactions on surfaces i.e., heterogeneous phases. To be concise, we only focus on a few reactions of significant atmospheric relevance in the text, as the Tables are very detailed and full of information on conditions where these data were taken. Please note that in calculation of atmospheric lifetime for mercury, in contrast with many other chemical in the atmosphere that they can be irreversibly transformed to products, mercury can be oxidized, and oxidized mercury can be reduced by various atmospheric reductants in aerosols, fog, clouds and interfaces (see Tables 15.1 and 15.2). Hence, the oxidation alone represents merely one aspect of its transformation, can not yield to proper calculation of the lifetime.

15.2.1 Kinetic and Product Studies

The rate of the atmospheric chemical transformation of elemental mercury towards a given oxidant is dependent on two factors. The first factor is the reactivity of mercury toward a given oxidant at environmentally relevant conditions, such as temperature, pressure, oxygen concentration, and relative humidity. Since in our previous book chapter in 2004 edition (Ariya and Peterson, 2005), we have described in detail, the techniques used, as well as the importance of environmental condition, we will herein not describe them again. The second factor is the concentration (or mixing ratio) of the oxidant. The existing laboratory studies of mercury kinetic reactions have been obtained using steady state reaction chamber or fast flow tubes and a single study has been carried on the analysis on field data. Both relative and absolute techniques were used in these studies. Both absolute and relative techniques have advantages and disadvantages.

Table 15.1 Compilation of known gas-phase kinetics of mercury

Gas Phase Reaction	Reaction Type [1]	Gas Type, Press	Temp *(K)*	Rate $(cm^3 molec^{-1} s^{-1})^2$ $[R] = J\ K^{-1} mol^{-1} [T] = Kelvin$	Reference
$Hg(6^3P_1) + N_2 \rightarrow Hg(6^3P_0)$	Abs	N_2, 1 atm	298	$3.9\text{-}7.1 \times 10^{-12}$	Hall et al., 2005
$Hg(6^3P_0) + N_2 \rightarrow Hg(6^3P_0)$				6.1×10^{-15}	
$Hg + I_2 \rightarrow HgI_2 + HgOI$	Abs	N_2, 1 atm	296 ± 1	$< 1.27 \times 10^{-19}$	Raofie et al., 2008
$Hg + Br_2 \rightarrow HgBr_2$	Abs	Air, N_2, 1 atm	298 ± 1	$< 9 \times 10^{-17}$	Ariya et al., 2002
	Abs	air, 1 atm	~298	No reaction detected	Sumner et al., 2005
$Hg + BrO \rightarrow HgBrO$	Rel rate	N_2, 1 atm	298	$10^{-15} < k < 10^{-13}$	Raofie and Ariya, 2003
$Hg + Br \rightarrow HgBr$	Ab initio	N/A, 1 atm	Range?	$1.01 \times 10^{-12}\ e^{1738/RT}$	Khalizov et al., 2003
			298	2.1×10^{-12}	
	RRKM/ B3LYP	N/A, 1 atm	180-400	$1.1 \times 10^{-12}\ (T/298)^{-2.37}$	Goodsite et al., 2004
			298	1.1×10^{-12}	
	Abs	He, 0.26-0.79 atm	243-293	$(1.46 \pm 0.36) \times 10^{-32}$ $[cm^6/molecule^2\ s^{-1}]$ $(T/298)^{(-1.86 \pm 1.49)}$	Donohoue et al., 2006
		1 atm	298	$(3.6 \pm 0.9) \times 10^{-13}$	
	CCSD(T)	Ar, 1 atm	260	1.2×10^{-12}	Shepler et al., 2007
	Rel	Air, N_2, 1 atm	298 ± 1	$(3.2 \pm 0.3) \times 10^{-12}$	Ariya et al., 2002
	Abs, fitted equation	CF_3Br, 0.26 atm	397	$\sim 3 \times 10^{-16}\ molec^{-1}\ s^{-1}$	Greig, G., et al., 1970
$HgBr + Br \rightarrow HgBr_2$	Abs	CF_3Br, 0.26 atm	397	$7 \times 10^{-17}\ molec^{-1}\ s^{-1}$	Greig, G., et al., 1970
$HgBr + HgBr \rightarrow (HgBr_2 + Hg)$ or Hg_2Br_2	Abs, fitted equation	CF_3Br, 0.26 atm	397	$(2.1 \pm 1.3) \times 10^{-14}\ molec^{-1}\ s^{-1}$	
$Hg + Cl_2 \rightarrow HgCl_2$	Abs	Air, N_2, 1 atm	298 ± 1	$(2.6 \pm 0.2) \times 10^{-18}$	Ariya et al., 2002
	Abs	Humid air	273	$0.4(dry) - 2.5(80\%\ RH) \times 10^{-15}$	Menke and Wallis, 1980; Seigneur et al., 1994
	Abs	Air	293-973	5.6×10^{-15}	Sliger, et al., 2000
	Abs	air, 1 atm	~298	$(2.5 \pm 0.9) \times 10^{-18}$	Sumner et al., 2005
	Abs		470 – 870	$0.386(units?)\ exp^{(-607/RT)}$	Agarwal et al., 2007
$Hg + HCl \rightarrow \ldots \rightarrow HgCl_2(g)$	Abs	N_2, 1 atm	673 - 1173	$3.65 \times 10^{-14}\ exp^{(-45400/RT)}$	Widmer, et al., 1998
			673	4.8×10^{-15}	

(continued)

Table 15.1 (continued)

Gas Phase Reaction	Reaction Type [1]	Gas Type, Press	Temp (*K*)	Rate ($cm^3\ molec^{-1} s^{-1}$)[2] $[R] = J\ K^{-1} mol^{-1} [T] = Kelvin$	Reference
$HgCl + HCl \rightarrow HgCl_2 + H$	Ab initio QCISD/1997	1 atm	298-2000	$3.2\times10^{-15} \exp^{(-104640/RT)}$	Wilcox et al., 2003
			298	1.5×10^{-33}	
$HgCl + M \rightarrow Hg + Cl + M$	Ab initio QCISD/1997	1 atm	298-2000	$9.0\times10^{-11} \exp^{(-89800/RT)}$	
			298	1.6×10^{-26}	
$Hg + Cl \rightarrow HgCl$	Ab initio	N/A, 1 atm		$1.38 \times 10^{-12} e^{1729/RT}$	Khalizov et al., 2003
			298	2.8×10^{-12}	
	Abs	He	243-293	$(2.2 \pm 0.5) \times10^{-32} e^{[(680 \pm 400)(1/T\ 1/298)]}$ cm^6 $molecule^{-2} s^{-1}$	Donohoue et al. 2005
	Abs, fitted equation	Ar, 0.93 atm	383-443	$(3.2 \pm 1.7) \times 10^{-11}$	Horne et al., 1968
	Abs	Air, N_2, 1 atm	298 ± 1	$(1.0 \pm 0.2) \times10^{-11}$	Ariya, et al., 2002
	Abs	Ar, 0.95 atm	397	1.5×10^{-14} $molec^{-1} s^{-1}$	Greig, et al., 1970
	Abs	?	> 873	$30 - 2.5 \times10^{-11}$	Senior, et al., 2000
$HgCl + HgCl \rightarrow Hg_2Cl_2$	Abs	Ar, 0.95 atm	397	$(3 \pm 2) \times10^{-13}$ $molec^{-1} s^{-1}$	Greig et al., 1970
$Hg + F_2 \rightarrow Prod$	Abs	air, 1 atm	~298	$(1.8 \pm 0.4) \times10^{-15}$	Sumner et al., 2005
$Hg + F \rightarrow HgF$	Ab initio	N/A		$9.2\times10^{-13} e^{1720/RT}$	Khalizov, et al., 2003
			298	1.9×10^{-12}	
$Me_2Hg + OH \rightarrow MeHgOH + Me$	Rel rate	Air, 0.92 atm	~298	$(1.9 \pm 0.2) \times10^{-11}$	Niki et al., 1983
				2.72×10^{-13}	Lin and Pehkonen, 1999
$Me_2Hg + O_3 \rightarrow ?$				$< 10^{-21}$	
$Me_2Hg + Cl \rightarrow CH_3HgCl$				$(2.75 \pm 0.30) \times10^{-10}$	
$Me_2Hg + NO_3 \rightarrow Hg^0HgO(s)$				$(7.4 \pm 2.6) \times10^{-14}$	
$Me_2Hg + NO_3 \rightarrow O(^3p)\ HgO$				$(2.5 \pm 0.20) \times10^{-11}$	
$Me_2Hg + F \rightarrow ?$				$(4.7 \pm 0.5) \times10^{-10}$	
$Hg + OH \rightarrow HgOH$	Rel rate	N_2, 1 atm	283-353	$3.55\times10^{-14} e^{-(2440\pm130)/RT}$	Pal and Ariya, 2004
			298	$(9.0\pm1.3)\times10^{-14}$	
	Abs	He, 1 atm	298	$<1.2 \times10^{-13}$	Bauer et al., 2003
	RRKM/ B3LYP	N/A, 1 atm	180-400	$3.2\times10^{-13} \times (T/298)^{-3.06}$	Goodsite et al., 2004
	Rel.	Air, 1atm	298	$(8.7\pm2.8)\times10^{-14}$	Sommar et al., 2001

$Hg^0(g) + h\nu\ ½O_2 \rightarrow Hg^O(g/s)$	Abs, 3-step reaction	Air, 1 atm	293	1×10^{-23}	Hall et al., 1995
			973	4×10^{-23}	
$Hg + O_3 \rightarrow HgO(s) + O_2$	Abs., extrapolated	N_2, 1 atm	283-323	$8.43\times10^{-17}\ e^{-(11700\pm270)/RT}$	Pal and Ariya, 2004
			298	$(7.5\pm0.9)\times10^{-19}$	
	Abs., S/V extrapolated	N_2/O_2, 1 atm	293	$(3\pm2)\times10^{-20}$	Hall, 1995
	Abs.	air, 1 atm	~298	$(6.4\pm2.3)\times10^{-19}$	Sumner, et al., 2005
	Abs., re-plotted data	1 atm	293	4.9×10^{-18} (?)	Schroeder et al., 1991, P'yankov, 1949
			303	8.4×10^{-18} (?)	
			293	4.2×10^{-19} (?)	Slemr, 1985, P'yankov, 1949
	Abs.,	Air, 1 atm	293	$\sim1.7\times10^{-18}$	Iverfeldt, and Lindqvist, 1986
	QCIS(T)//MP2; TST theory	1 atm	~298	$1.2\times10^{-9}\ e^{-176,500/RT}$	Xu eal., 2008
			298	1.4×10^{-40} (?)	
$Hg + NO_2 \rightarrow$	Abs.	N_2, 1 atm	293	$(2.8\pm0.5)\times10^{-35}$ $cm^6\ molec^{-2}\ s^{-1}$, 2nd order in $[NO_2]$	Hall et al., 1995
$Hg + NO_3 \rightarrow HgO + NO_2$	S/V Extrapolation	N_2, 5-10 $\times10^{-3}$ atm	294 ± 2	$< 4\times10^{-15}$	Sommar, et al., 1997
	Rel	air, 1 atm	~298	$< 7\times10^{-15}$, $< 1.3\times10^{-14}$, 3×10^{-14}	Sumner et al., 2005
$Hg + H_2O_2 \rightarrow ?$	Abs	N_2,	293	$< 8.5\times10^{-19}$	Tokos et al., 1998

See Lee et al., (2004) for recent discussion of theoretical and experimental mercury kinetics gas phase reactions.
[1]ab initio, semi- empirical, rel., abs., extrapolated
[2]unlessotherwise stated

Table 15.2 Liquid (water)-phase kinetics of mercury

Liquid	Type of study	Environment *pH*	Temp *(K)*	Rate *($M^{-1}s^{-1}$, unless otherwise stated)*	Reference
$Hg^{(II)} + hv \rightarrow Hg^0(aq)$	Field study	lake water, daylight radiation	ambient	2.4 – 9 times $Hg^0(aq)$ production than dark. 1.6%/h	Amyot et al., 1994, 1997
$Hg^0(aq) + O_3 + H_2O \rightarrow Hg^{2+} + OH^- + O_2$	Field	Water, pH = 4.5-9.5	~298 (T-ind)	$(4.7 \pm 2.2) \times 10^7$	Munthe, 1992
$Hg^{0(II)} + H_2O + hv \rightarrow Hg(OH)_2+$	Field	pH = 7	ambient	$1.2 \times 10^{-4}\ s^{-1}$	Zhang, 2006
$Hg^0(aq) + HOCl + H^+ \rightarrow Hg^{2+} + Cl^- + H_2O$	Lab	Water	Ambient	$(2.09 \pm 0.06) \times 10^6$	Lin and Pehkonen, 1998
$Hg^0(aq) + OCl^- + H_2O \rightarrow Hg^{2+} + Cl^- + 2OH^-$	Lab	Water	Ambient	$(1.99 \pm 0.05) \times 10^6$	
$HgCl_2 + e^- \rightarrow Cl^- + HgCl$	Avg'd Lit Cit.	pH = 5	Ambient	4.0×10^{10}	
$HgBr_2 + e^- \rightarrow Br^- + HgBr$	Avg'd Lit Cit.		Ambient	3.7×10^{10}	
$HgI_2 + e^- \rightarrow I^- + HgI$	Avg'd Lit Cit.		Ambient	3.0×10^{10}	
$Hg(CN)_2 + e^- \rightarrow CN^- + HgCN$	Avg'd Lit Cit.		Ambient	1.4×10^{10}	
$Hg(SCN)_2 + e^- \rightarrow SCN^- + HgSCN$	Avg'd Lit Cit.		Ambient	4.5×10^{10}	
$Hg(EDTA)^{2-} + e^- \rightarrow$ Products	Avg'd Lit Cit.		Ambient	$2.1, 5.1 \times 10^9$	Buxton et al., 1998
$Hg(CN)_2 + e^- \rightarrow CN^- + HgCN$	Avg'd Lit Cit.		Ambient	1.4×10^{10}	
$Hg^{(II)} + e^- \rightarrow Hg(I)$	Avg'd Lit Cit.		Ambient	7.1×10^9	Zhang, 2006
$H + Hg_2^{2+} \rightarrow H^+ + Hg_2^+$	Lit. Cit.	pH = 1.0	Ambient	4.7×10^9	Buxton et al., 1988
$H + Hg^{2+} \rightarrow H^+ + Hg^+$	Lit. Cit.	pH = 1.5	Ambient	2.0×10^9	
$H + Hg(OH)_2 \rightarrow H_2O + HgOH$	Lit. Cit.	pH = 7	Ambient	2.4×10^9	
$H + HgCl_2 \rightarrow HCl_{aq} + HgCl$	Lit. Cit.	pH = 1	Ambient	$1.0, 1.5 \times 10^9$	
$H + HgI_2 \rightarrow HgI_2(H)$	Lit. Cit.		Ambient	1.5×10^9	
$OH + HgCl \rightarrow OH^- + HgCl^+$	Lit. Cit.	pH = 5	Ambient	$\sim 1 \times 10^{10}$	Buxton et al., 1988
$OH + HgBr_2 \rightarrow Br + HgBrOH$	Lit. Cit.		Ambient	$> 9 \times 10^8$	
$OH + HgCN \rightarrow$ products	Lit. Cit.		Ambient	3.1×10^9	

$Hg^0(aq) + OH \rightarrow Hg^+ + OH^-$	Rel.	Water	298	$(2.4 \pm 0.3) \times 10^9$	Gardfeldt et al., 2001
	Lab expt	Water, pH = 3.9	298 ± 2	2.0×10^9	Lin and Pehkonen, 1997
	Field			5.5×10^9	Hines and Brezonik, 2004
$Hg^+(aq) + OH \rightarrow Hg^{2+} + OH^-$	Lab		Ambient	1×10^{10}	Zhang, 2006
$Hg^0(aq) + OH \rightarrow HgOH$	Lab			2.4×10^{9}	
$HgOH + OH \rightarrow Hg(OH)_2$	Lab				
$HgOH + H_2O + O_2 \rightarrow Hg(OH)_2 + H^+ + O_2^-$	Lab			1×10^{10}	
				10^9	
$HO_2 + Hg^{2+} \rightarrow Hg^+ + O_2 + H^+$	Lab		Ambient	1.7×10^4	Pehkonen and Lin, 1998
				1.1×10^4 (chloride present)	
$Hg(HSO_3)^-_{aq} + h\nu \rightarrow Hg^0(aq) + S(VI)$				$4 \times 10^{-6}\ s^{-1}$	Lin and Pehkonen, 1997
$Hg(HSO_3)_2^{2-} + h\nu \rightarrow Hg^0(aq) + S(VI)$				$< 10^{-4}\ s^{-1}$	
$HgSO_3(aq) + h\nu \rightarrow Hg^+ + SO_3^- \rightarrow Hg^0(aq)$	Lab expt			$0.6\ s^{-1}$	Lin and Pehkonen, 1999
$Hg(OH)_{2,aq} \rightarrow Hg^0(aq) + prod.$	Expt			$3 \times 10^{-7}\ s^{-1}$	
$Hg^0(aq) + Cl^- \rightarrow Hg^{2+} + Cl^-$	Lab study	Sea water	298	$0.28 - 1.1 \times 10^{-4}\ s^{-1}$	Yamamoto, 1996
$Hg^0(aq) + Cl^- + h\nu(UV) +$ benzo-quinone $\rightarrow Hg^{(II)}$	Lab study	Pure water spiked	298	$1.7 \times 10^{-4}\ s^{-1}$	Lalonde et al., 2001
$Hg^0(aq)$ + bacteria exudates + Cl $\rightarrow Hg^{(II)}$	Lab study	Spiked water	298	Dark reactions; 40% decrease in $Hg^0(aq)$	Poulain et al., 2007
$HgMe + OH \rightarrow$ products	Lab expt	pH = 6.2 – 8.2	293	1.2×10^9	Zepp et al., 1987
$Hg^{(II)} + h\nu \rightarrow Hg^0(aq)$	Field expt	Lake water	Ambient	UV-A: $7.76 \times 10^{-5}\ s^{-1}$	O'Driscoll et al., 2006
		River water		UV-B: $8.91 \times 10^{-5}\ s^{-1}$	
				UV-A: $1.78 \times 10^{-4}\ s^{-1}$	
				UV-B: $1.81 \times 10^{-4}\ s^{-1}$	

(continued)

Table 15.2 (continued)

Liquid	Type of study	Environment *pH*	Temp (K)	Rate *($M^{-1}s^{-1}$, unless otherwise stated)*	Reference
$Hg^0(aq) \rightarrow Hg^{(II)}$	Lab expt; oxid^n isolated from red.	Coastal water, dark	Ambient	$2.6 - 5.3 \times 10^{-4}\ s^{-1}$	Whalin and Mason, 2006
		Fresh water, dark		$8 - 15 \times 10^{-4}\ s^{-1}$	
				$10 - 20 \times 10^{-4}\ s^{-1}$	
				$0.28 \times 10^{-4}\ s^{-1}$	
				$0.17 \times 10^{-4}\ s^{-1}$	
$Hg^0(aq)$ + fulvic acid $\rightarrow Hg^{(II)}$		Artificial water		$1.4 - 1.6 \times 10^{-4}\ s^{-1}$	
$Hg^0(aq)$ + semiquinones + $Cl^- \rightarrow Hg^{(II)}$		Artificial water, 0.5 M HCl		$0.7 \times 10^{-4}\ s^{-1}$	
$Hg^0(aq)$ + saline water $\rightarrow Hg^{(II)}$		Saline water, dark		$1.6 \times 10^{-4}\ s^{-1}$	
$Hg^0(aq) + h\nu \rightarrow Hg^{(II)}$		Saline water, UV	Ambient		Whalin and Mason, 2006
$Hg^0(aq)$ + UV + DOC $\rightarrow Hg^{(II)}$		Lake water			
$Hg^0(aq) + KCl(aq) \rightarrow Hg^{(II)}$		Lake water			
$Hg^0(aq) \rightarrow Hg^{(II)}$		Freshwater			
$Hg^0(aq) + h\nu \rightarrow Hg^{(II)}$		Freshwater, Vis light			
$Hg^0(aq) + h\nu \rightarrow Hg^{(II)}$		Freshwater, UV light			
				$<0.1 \times 10^{-4}\ s^{-1}$	Whalin and Mason, 2006; Lalonde, et al., 2004
$Hg^0(aq) + h\nu \rightarrow Hg^{(II)}$		Sea water + UV		$0.25 \times 10^{-4}\ s^{-1}$	
$Hg^{(II)} + h\nu \rightarrow Hg^0(aq)$		Sea water + UV		$1.6 - 1.9 \times 10^{-4}\ s^{-1}$	
$Hg^0(aq) + Fe(III) \rightarrow Hg^{(II)}$		Freshwater, dark (prev. UV expos.)			
$Hg^{(II)} + Fe(III) + h\nu \rightarrow Hg^0(g)$		Freshwater, UV exposed		$0.3 - 3.9 \times 10^{-4}\ s^{-1}$	Poulain et al., 2007
				$1.2 - 4.4 \times 10^{-4}\ s^{-1}$	
				$0.6 - 0.8 \times 10^{-4}\ s^{-1}$	Zhang and Lindberg, 2001; Whalin and Mason, 2006
				470% increase in Hg^0 production in 4h; $0.3 - 0.6 \times 10^{-4}\ s^{-1}$	Zhang and Lindberg, 2001

$Hg^{(II)}$ + MerA bacteria → $Hg^0(aq)$	pH = 7.0	MerA, intact cells	Ambient	4 - 18 nmol Hg^{2+} min^{-1} mg $protein^{-1}$	Philippidis et al., 1991
$Hg^0(aq) + h\nu \rightarrow Hg^{(II)}$		Natural water, UV exposure	River water	Hg^0 loss: $0.6 \times 10^{-4}\ s^{-1}$ $Hg^{(II)}$ production: $0.9 \times 10^{-4}\ s^{-1}$	
			Sea Water	Hg^0 loss: $1.2 \times 10^{-4}\ s^{-1}$ $Hg^{(II)}$ production: $1.2 \times 10^{-4}\ s^{-1}$	
$Hg^0(aq) + h\nu \rightarrow Hg^{(II)}$ (HgO?)		Lake water + Hg lamp	Ambient Temp	$1.1 - 2.1 \times 10^{-4}\ s^{-1}$ $1.4 \times 10^{-4}\ s^{-1}$	Hines and Brezonik, 2004; Whalin and Mason, 2006
$Hg^0(aq) + NO_3^- + h\nu \rightarrow Hg^{(II)}$		Lake water		$15.3 \times 10^{-4}\ s^{-1}$	
$Hg^0(aq) + Cl^- + h\nu \rightarrow Hg^{(II)}$		Lake water		$0.006 \times 10^{-4}\ s^{-1}$	
$Hg^0(aq) \rightarrow Hg^{(II)}$		HPLC water, dark Lake water, dark		$0.06 \times 10^{-4}\ s^{-1}$	
$Hg^0(aq) + NO_3^- + h\nu \rightarrow Hg^{(II)}$		Q-water	Ambient temp	$8.7 \times 10^{-4}\ s^{-1}$	Hines and Brezonik, 2004; Mason et al., 2001
$Hg^0(aq) + Cl^- + NO_3^- + h\nu \rightarrow Hg^{(II)}$		Q-water		$46 \times 10^{-4}\ s^{-1}$	
$Hg^0(aq) + Br_2 \rightarrow P$	Lab expt	pH = 2	296 - 298	0.20 ± 0.03	Wang and Pehkonen, 2004
$Hg^0(aq) + HOBr \rightarrow P$		pH = 6.8		0.28 ± 0.02	
$Hg^0(aq) + OBr^- \rightarrow P$		pH = 11.7		0.27 ± 0.04	
$Hg^0(aq) + H_2O_2 \rightarrow HgO(s) + Hg^{2+} + H_2O$	Lab			6.0	Munthe and Mcelroy, 1992
$HgSO_3(aq) \rightarrow Hg^+ + SO_3^-(aq)$	Lab			$0.6\ s^{-1}$	Munthe et al., 1991

Table 15.3 Inter-phase (heterogeneous/surface) kinetics and emission rates of mercury

Surfaces	Type of expt	Interface	Temp (K)	Rates/rate constants/results; No general units	Reference
$Hg^0(g) \rightarrow Hg^0(ads)$	Abs	N_2, 1 atm, N_2/Teflon wall (s/v = 0.58 cm^{-1})	293	$4.5 \times 10^{-6}\ s^{-1}$	Hall, 1995
			323	$1.7 \times 10^{-5}\ s^{-1}$	
			348	$3.0 \times 10^{-5}\ s^{-1}$	
$Hg^0(g) \rightarrow Hg^0(ads)$	Abs	air, 1 atm, air/carbon	293	$90\text{-}120 \times 10^{-4}\ s^{-1}$	Hall et al., 1995
			423-523	$1.3\text{-}5.0 \times 10^{-4}\ s^{-1}$	
			573	$\sim 0\ s^{-1}$	
	Abs	air, 1 atm, air/fly ash	293	$81 \times 10^{-4}\ s^{-1}$	
			423-523	$11.2\text{-}27.2 \times 10^{-4}\ s^{-1}$	
			573	$6.8\text{-}7.7 \times 10^{-4}\ s^{-1}$	
$Hg^0/HgF_2/HgNO_3\ (aq) + h\nu \rightarrow Hg(l)$	Lab, Hg lamp	Water/TiO_2 surface	293	90% conversion after 40 min irradiation. Lower pH = less reduction.	Habibi et al., 2003
$Hg(l) + \Delta \rightarrow Hg^0(g)$		TiO_2 surface/N_2 flow	423	Removes $Hg^{(l)}$ in ~1 hour	
$Hg^0(g) + H_2O(g) + O_2(g) + h\nu \rightarrow HgO(s)$	Lab	Air/TiO_2 surface	297-408	$k = Ae^{-(Es-\lambda)/RT}$ $d[Hg]/dt = k[Hg]^{(1.4\pm0.1)}[J_{UV}]^{(0.35\pm0.05)}$	Lee et al., 2004
$HgCl_2$ + dodecyl sulfate(DS) + $h\nu \rightarrow Hg^{(l)}$	Lab	Water/TiO_2 surface	Ambient	97% reduction in 6 min	Zhang, 2006,
$HgCl_2 + h\nu$ + cetyltrimethylammonium $(CTA^+) \rightarrow Hg^{(l)}$	Lab		Ambient	99% reduction in 25 min	Horvath et al., 2005
$HgCl_2 + h\nu$ + arginine $\rightarrow Hg^{(l)}$	Lab	Water/TiO_2 surface	Ambient?	arginine binds $Hg^{(II)}$ to TiO_2, facilitates charge transfer	Skubal and Meshkov, 2002
$HgCl_2(aq) + UV \rightarrow Hg^0(aq) +$ 2Cl+-	Lab	Water/TiO_2 semiconductor	293 ± 5	E = 0.41V, poss. catalytic TiO_2 action, rate oxid = $0.051\sqrt{[Hg]}$	Prairie et al., 1993
$Hg^0(g) + H_2O + UV \rightarrow HgO(s)$	Lab	Humid Air + TiO_2 surface	298	Surface mechanism developed, remov. rate incr. with $[H_2O]$, Hg removal < 80%, > 30%.	Rodríguez et al., 2004
$Hg^0(g) + H_2O + UV \rightarrow HgO(s)$	Lab	Humid Air+ SiO_2/TiO_2 surface	298	L-H mech., remov. rate decreases with $[H_2O]$, up to 95% removal when dry.	Li and Wu, 2007
$Hg^0(g) \rightarrow Hg^0(ads)$	Model/field data	Air/Forest floor	Ambient	$0.12\ cm\ s^{-1}$	Lindberg et al., 1992
				$0.006\ cm\ s^{-1}$	

$Hg^0(g) \rightarrow Hg^0(ads)$	Field study	Air/ground dry deposition	Ambient	0.5 cm s^{-1} (particulate; Hg_p) 0.1 cm s^{-1} ($Hg^0(g)$)	Schroeder et al, 1998
		Air/(snow/ barren ground)	Ambient	12.5 ± 2.5 pmol m^{-2} h^{-1}	
Hg(II)/humic acid + hv $\rightarrow$ Hg(g)	Lab study, xenon radiation	Water/air,	Ambient	2×10^{-2} s^{-1}	Xiao et al., 1995
$Hg(OH)_2 + h\nu \rightarrow Hg(g)$				1.2×10^{-4} s^{-1}	
$Hg^0(g) + H_2O_2(ads) + hv \rightarrow Hg^{(II)}$	Field study	Air/snow pack		5-fold increase in $Hg^0(g)$ deposition with H_2O_2-spiked snow	Lahoutifard et al., 2006
$Hg^0(I/II)?_g \rightarrow Hg^0(I/II)_{ads}$	Field study	Air/snow; dry dep.	Ambient	1 cm s^{-1}	Lindberg et al., 2002
$Hg^{2+}(ads) \rightarrow Hg^{2+}(snowpack)$	Field study	Snow/Snow vertical diffusion	273	5.8 – 7.0 pg m^{-2} h^{-1}	Ferrari et al., 2004
$Hg^0(ads, II) + h\nu \rightarrow Hg^0(g)$	Field study	Snow/air	273?	> 20% reduction loss in 3h	Ariya et al., 2004
	Field study	Snow/air	273?	> 40% reduction in 24h	Lalonde et al., 2003
	Field study	Snow/air,	269	0.24 h^{-1}, 45 pmol m^{-2} h^{-1}. Reduc enhanced by $H_2O(l)$	Fain et al., 2007
$Hg^0(g) + h\nu \rightarrow Hg^0(ads,II)$	Lab study, Xe-lamp	Air/snow	258 - 313	6.9×10^{-5} s^{-1}, 5.0×10^{-5} s^{-1}	Dommergue et al., 2007
$HgS^{2-} + h\nu \rightarrow Hg^0(aq)$	Field	Water/air	ambient	Not significant	Zhang, 2006
$Hg^0/Hg(CH_3)_2 \rightarrow Hg^0(g)/Hg(CH_3)_2(g)$	Field study	Soil/air	ambient	< 1 nmol m^{-2} h^{-1}	Schuter et al., 2000
$Hg^0(ads, II) \rightarrow Hg^0(g)$	Field study	Temperate, boreal	ambient	55 pmol m^{-2} h^{-1}	Grigal, 2002
		Contaminated	ambient	~6500 pmol m^{-2} h^{-1}	
$Hg^0(g) \rightarrow Hg^0(ads, II)$	Field study	Open, temperate Air/soil	ambient	12 pmol m^{-2} h^{-1}	
		Temperate Forest Air/ soil	ambient	47 pmol m^{-2} h^{-1}	
		Throughfall/ litterfall	ambient	190 nmol m^{-2} a^{-1}	
$Hg^0(ads, II) \rightarrow Hg^0(ads, II)$	Lit. Cit. Field study	Soil sequestration	ambient	25 nmol m^{-2} a^{-1}	
$Hg^0(aq) \rightarrow Hg^0(g)$		Water/air		25 pmol m^{-2} h^{-1}	Zhang, 2006
	Model/Field study	Forest soild/air	ambient	250 pmol m^{-2} h^{-1}	Lindberg et al., 1992
$Hg^{(II)}[Cl^-/NO_3^-]$ + surface sediment $\rightarrow Hg^0(aq)$	Field	Lake water/ sediment	~298	~10^{-5} – 10^{-6} min^{-1}	Peretyazhko et al., 2007

(continued)

Table 15.3 (continued)

Surfaces	Type of expt	Interface	Temp (K)	Rates/rate constants/results; No general units	Reference
$Hg^0(aq) \rightarrow Hg^0(g)$	Model fit, empirical data	Water/air	298	5.7 pmol m^{-2} h^{-1} loss	Hines and Brezonik, 2004
	Field study	Lake water/air	298?	16 pmol m^{-2} h^{-1} loss (0.6% h^{-1})	Amyot et al., 1994
$Hg^0(aq) + h\nu \rightarrow Hg(g)$	Lab study	River water/air	ambient	0.2×10^{-4} s^{-1}	Amyot et al., 1997
		Sea water/air		0.3×10^{-4} s^{-1}	
$Hg^0(ads) + Br \rightarrow HgBr$	Field study	In snow (20-60 cm deep)	~263	2×10^{-11} cm^3 $molec^{-1}$ s^{-1}	Fain et al., 2006[97]
$Hg(g) + Cl_2(g) \rightarrow (HgCl)_n(s)$	Lab	Air/surface	525	50 second reaction	Medhekar et al., 1979
	Lab	Air/water		Some enhancement	Skare and Johansson, 1992
	Lab	Air/water +sulfite	298	6.1×10^9 M^{-1} s^{-1}	Roy and Rochelle, 2004
$Hg^0(g) + Cl_2(aq) \rightarrow Hg^{2+} + 2Cl^-$	Abs	Water + NaOCl	298	1.7×10^{15} M^{-1} s^{-1} (too fast?)	Zhao and Rochelle, 1999
	Abs	Water + NaOCl	328	1.4×10^{17} M^{-1} s^{-1}	
$Hg^0(g,ads) + O_2(g,ads) \rightarrow$ Product	Langmuir-Hinshelwood mech.	Air/Fly ash surface	373-573	$1.5 - 6.5 \times10^{-12}$ $(cm^{-3}\ molec^{-1})^{0.5}$ s^{-1}	Hall et al., 1991
$Hg^0(g) + SO_2 + O_2 \rightarrow HgSO_4(s)$	Lab	Air/Pt	348-673	3.5 mg Hg/hr (348-600K)	Schofield, 2004
$Hg^0(g) \rightarrow$ Product	Field test	Air/Pd	450	93% oxidation	EPRI, 2005
		Air/SCR catalyst		62% oxidation	
		Air/TMT-15 catalyst*		Inconclusive; intended to prevent re-emission of Hg	Blythe, 2006
$Hg^0(g) + SO_2 + NO_2 + HCl \rightarrow Hg(s)^{2+}$	Lab	Air/Fly Ash	453	~30 % Hg(g) oxidation	Norton et al., 2003
$Hg^0(g) \rightarrow Hg(ads)$	Lab	Air/Fly Ash + carbon	293, 313	More mercury adsorption at 20°C than 40°C	Hwang et al., 2002
$Hg^0(g) + h\nu \rightarrow$ prod.	Lab, Xenon lamp	Air/quartz surface	293	1.2×10^{-5} s^{-1}	Sheu and Mason, 2004
$Hg^0(g) + h\nu + H_2O(l) \rightarrow$ prod				4.0×10^{-5} s^{-1}	
$Hg^0(g) + h\nu + NaCl \rightarrow$ prod				1.6×10^{-3} s^{-1}	
$Hg^0(g) + h\nu + NaCl + H_2O(l) \rightarrow$ prod				1.7×10^{-3} s^{-1}	
$HgCl_2(g) + H_2 \rightarrow Hg_2Cl_2(s) + 2HCl$	Lab, laser	N_2/Stainless steel surface	473	Unkonwn mech	Wang et al., 1983

$Hg^0(g) + (HCl) \rightarrow Hg(ads)$	Lab	N_2/Stainless steel surface or PTFE teflon	423	HCl enhances Hg^0 removal SS: ($0 \rightarrow 44$ ng) SS: ($66 \rightarrow 128$ ng)	Turchi, 2000
$Hg(g) \rightarrow Hg(ads)$	Lab work	N_2 and trace gas/Gold	411	25 nm thick Au sheet absorbs for 33 min vs. 5 min for 2.5 nm sheet. Hg penetrates Au.	Turchi, 2000
$HgCl_2(g) \rightarrow HgCl_2(ads)$	Simulated flue gas	N_2/gold Trace gas	422	Acid gases (HCl or NO_2) + SO_2 reduce ads. cap. $HgCl_2$ adsorbs; no rxn. Similar to carbon surf. ads.	
$Hg^{2+}(aq)$ + N719-TiO_2	Lab expt	Water/N719-TiO2	~298	Hg^{2+} binding constant: $3 \times 10^5\ M^{-1}$ 65% scavenging eff.	Li et al., 2007
$Hg(NO_3)_2(aq) \rightarrow Hg_{ads}(II)$	Lab	Water/Fly ash	303	Freundlich parameters: k = 1.230, 1/n = 0.361 @ pH = 4.2, > 90% adsorption	Sen and De, 1987
$Hg^0(II)_{aq} \rightarrow Hg(II)_{ads}$	Lab	Water/Activated carbon		Freundlich parameters:	Mohan et al., 2000
			300	k = 0.1427, 1/n =0.71	
			318	k = 0.0663, 1/n =0.75	
			338	k = 0.01073, 1/n =1.38	
$Hg^0(g) + Cl_2 \rightarrow HgCl_2$	Lab	Air/Au surface	448-498	40-60 % oxidation k ~ 10^{-8} cm^3/(molec/s) (473 K) ? Langmuir-Hinshelwood mechanism proposed	Zhao et al., 2006
$Hg(NO_3)_2(aq) + 2e^- \rightarrow HgAu(s)$	Lab	Au-coated microparticle surface	298	0.35 V causes AuHg amalgam formation	Barrosse-Antle, 2007
$Hg^0(g) + (H_2S) \rightarrow HgS(s) + S(s)$	Simul. flue gas	Flue gas/Fe_2O_3(N)	353	H_2S initiates Hg removal rxn up to 65% Hg loss in stream. -no effect from H_2 or CO. H_2O reduces Hg adsorbance.	Wu et al., 2006
$Hg^0(g) + HCl \rightarrow HgO(s)$	Lab, simul. flue comb.	Fe_2O_3, Fe_2O_3-$Ca(OH)_2$, FeS_2, Fe_2O_3(1% wt)/TiO_2 surfaces	353	Fe_2O_3: 50% removal Fe_2O_3/TiO_2: 80% Fe_2O_3-$Ca(OH)_2$: 70% FeS_2: 60% HCl(g) suppressed Fe_2O_3 activity only	Wu et al., 2008
$Hg^0(g) + UV \rightarrow HgO(s)$	Lab, simul. flue comb.	SiO_2-TiO_2 surface	408	99% removal	Li et al., in press

(continued)

Table 15.3 (continued)

Surfaces	Type of expt	Interface	Temp (K)	Rates/rate constants/results; No general units	Reference
$Hg^0(g) + \Delta \rightarrow HgO(s)$	Lab, simul. flue comb.	Extensive list of metal oxide surface mixtures, Mars-Maessen mech.	410	Cr_2O_3/Al_2O_3, MnO_2/Al_2O_3, and MoS_2 show high Hg adsorption capacities.	Granite et al., 2000
$Hg^0(g) + \Delta \rightarrow HgO(s)$	Lab, simul. flue comb.	Various metal surface catalysts. Het. rate constants measured.	411	Rank: Ir > Ir/HCl > Darco > Thief/HCl (in terms of oxidation efficiency)	Presto et al., 2006
$HgBr_2(s) + Ag_2WO_4(s)$ $HgWO_4(s) + AgBr(s)$	Lab, glass tube diffusion	Solid-state reaction	413 – 463	$K = 1.10 \times 10^{-4}$ cm h^{-1} @ 120°C $(\text{thickness})^2 = Kt$	Jain and Beg, 1995
$HgCl_2(s) + Ag_2WO_4(s)$ $HgWO_4(s) + AgCl(s)$			438 – 481	$K = 2.25 \times 10^{-4}$ cm h^{-1} @ 145°C	
$Hg^0(g) \rightarrow Hg^0(ads)$	Lab, over activated C	Ptolemais lignite + S	323	300 – 900 ng Hg/mgAC addn of S incrss. ads	Skodras et al., 2007
$HgCl_2(g) \rightarrow Hg^0(ads)$	Lab	Air/Cysteine over silica	298 – 398	Capture eff CE: 12mg/g < CE < 33 mgHg/g	Abu-Daabes and Pinto, 2005

*TMT-15 is a 15 % aqueous solution of Trimercapto-s-triazine, a trisodium salt ($C_3N_3S_3Na_3$)

Additional information:

1. For comparative discussion of $Hg^0(g)$ oxidation surface mechanisms (Langmuir-Hinshelwood, Eley-Rideal, Mars-Maessen), see Presto and Granite, 2006.
2. For large-scale fluxes to/from ocean, land, and anthropogenic sources, see Lindberg et al, 2007.
3. For detailed photochemical information of mercury reactions in solid, liquid, and gas phases, see Zhang, 2000.
4. Concentrations of $[Hg^0(aq)]$ in various northern lakes: Peretyazhko et al, 2006.
5. Large compilation of Henry's Law values for atmospherically relevant species: Sander, 1999.
6. Compilation of aqueous rates of $Hg^0(aq)$ oxidation by several combinations of semiquinones, Cl^-, fulvic acid, saline water, UV radiation, DOC, fresh water, lake water, sea water, NO_2 NO_3^-,
7. Other possible catalysis surfaces: Au, Ag, Cu, Ir, Pd, C (fly ash, thief carbon) (Presto and Granite, 2006), Fe_2O_3, iron oxide—Ca(OH), FeS_2 (Wu et al., 2008), TiO_2-SiO_2 (Li et al., in press). Expansive catalysts can often be deposited over a bed of Al_2O_3, TiO_2, or other comparatively cheaper metal oxides.
8. $Hg^{2+}(aq)$ complexes with: CH_3^-, Cl^-, H_2O, HCO_3^-, CO_3^-, SO_4^-, OH^-, fulvic acid, humic acid, oxalate, citrate,…
9. Detailed discussion of Hg emissions and deposition to/from soil, see Schluter [95], Gigal [104], and Gabriel [195] (includes some details of aquatic Hg distribution).
10. Estimated cost of removing mercury from coal fire plant (using SCR catalyst): 29,000 USD/lb (Turchi, 2000). Aerosol pH = 2-5; Surface water (lake, river, ocean)

The disadvantage of the relative rate is that the calculated reaction rate constant is only as good as the original value of the reaction rate constant for the reference molecule used, and this why most detailed relative rate studies include several reference molecules to overcome this challenge. Another disadvantage is the complexity of the reactants and enhanced potential for side reactions. This challenge can be overcome with careful experimental setups and additional targeted experiments to minimize and characterize the extent of undesired reactions. An advantage of a detailed relative study is that one can readily perform the experiments under simulated tropospheric conditions, and also the reaction chambers can be coupled several state-of-the-art instruments for simultaneous analysis which allows detailed product analysis as well as kinetic determinations.

15.2.2 $Hg^0 + O_3$ and $Hg^0 + HO$

The advantage of the absolute method is clearly the fact that there is no need for incorporation of errors due to the reference molecules. However, in many absolute studies, one can follow merely one or two reactants, and considering the complexity of mercury reactions, and the extent of secondary reactions, the calculated values may be affected. Another challenge is some absolute studies are performed at lower pressure than tropospheric boundary layer pressure (~ 740 Torr) and concentrations orders of magnitude higher than tropospheric levels. Hence the data obtained under such conditions must be properly corrected for the ambient tropospheric situation, particularly in the case of complex mercury adduct reaction, and given the lack of detailed product analysis, and different carrier gases, this is not trivial. However, as it shown in Pal and Ariya (2004), both relative and absolute studies of the same reaction can yield the same values of rate constants within the experimental uncertainties and thus increase the confidence in the overall result. Ozone is an important atmospheric constituent and due to its atmospheric abundance, $O_3 + Hg^0$ has been a target of several laboratory studies. We have previously studied ozone-addition with elemental mercury under dry conditions (Pal and Ariya, 2004). The net reaction is written:

$$Hg_{(g)} + O_{3(g)} \overrightarrow{k_{net}} HgO_{(s)} + O_{2(g)} \quad (1)$$

Calvert and Lindberg (2005) suggest reaction (1) could be proceed by an addition of ozone, followed by a re-arrangement into the linear species OHgOO:

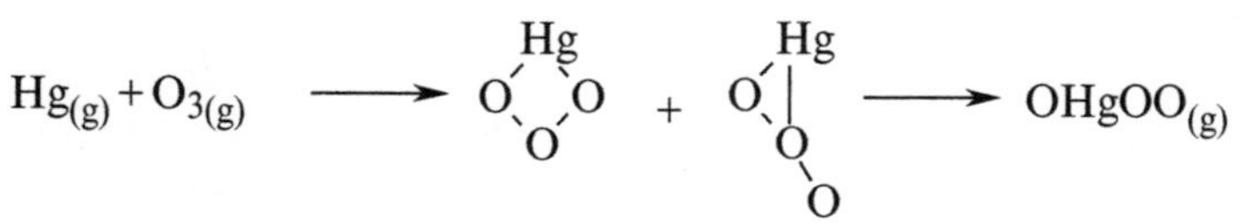

(2)

The reaction may be followed by dissociation into O_2 and $HgO_{(g)}$, the latter precipitating immediately to $HgO_{(s)}$.

$$OHgOO_{(g)} \rightarrow HgO_{(g)}(\rightarrow HgO_{(s)}) + O_{2(g)} \quad (3)$$

The dissociation and precipitation are essentially irreversible steps. As shown in Table 15.1, the apparent rate constant, k_{net}, reaction (1), was previously found by our group (Pal and Ariya, 2004) to be $(7.5 \pm 0.9) \times 10^{-19}$ cm^3 molecule^{-1} s^{-1}, in good agreement with Sumner et al. (2005): $(6.4 \pm 2.3) \times 10^{-19}$ cm^3 molecule^{-1} s^{-1} (performed in a much larger 17 m^3 chamber where the heterogeneous reactions were significantly reduced by direct increase in surface-to-volume ratios). Our rate constant was found to be larger than an earlier study by Hall (1995) $(0.3 \pm 0.2) \times 10^{-19}$ cm^3 molecule^{-1} s^{-1}, and smaller than both Schroeder et al.'s (1991) value 49×10^{-19} cm^3 molecule^{-1} s^{-1} (no error reported), and Iverfeldt and Lindqvist's (1986) value, 20×10^{-19} cm^3 molecule^{-1} s^{-1} (no error reported). In a new study by our group (Snider et al., 2008), for the first time, we have examined the effects of two atmospherically relevant polar compounds, $H_2O_{(g)}$ and $CO_{(g)}$, on the absolute observed rate coefficients of the O_3-initiated oxidation rate of $Hg^0_{(g)}$, at 296 ± 2 K using gas chromatography coupled to mass spectrometry (GC-MS). In CO-added experiments, we observed a significant increase in the reaction rate that could be explained by pure gas-phase chemistry. In contrast, we found the apparent rate constant, k_{net}, varied with the surface-to-volume ratio (0.6 to 5.5 L flasks) in water-added experiments. We have observed small increases in k_{net} for nonzero relative humidity, RH < 100%, but substantial increase at RH < 100%. Product studies were performed using mass spectrometry and high resolution transmission electron microscopy coupled to an electron dispersive spectrometer (HRTEM-EDS). A water/surface/ozone independent ozone oxidation rate was estimated to be $(6.2 \pm (1.1; t\sigma/\sqrt{n}) \times 10^{-19}$ cm^3 molecule^{-1} s^{-1}. There is furthermore ± 20% accumulated uncertainties associated with the ensemble of the experimental setup used in this study.

In our previous study by Pal and Ariya (2004), we have observed HgO from gas-phase aerosols and as deposits using mass spectrometry techniques. Please note that our methodology could not allow us to evaluate the phase of the HgO at that stage. Most products were obtained as condensed matter deposited on the reaction walls, and with some aerosols identified on 2 micron filters. We had also obtained a very minor amount of mercury containing compounds from the gas phase (including suspended matter, i.e. aerosols that were not collected on 2 micron filter which included finer particles). We could not then identify the phase of these identified mercury compounds assumed to be HgO. Please also note that in chemical kinetics for gas phase oxidation reaction, we use terminologies such as "products observed in the course of gas-phase reaction of …", it does not refer to the product as a gas, it just refers to the fact shat the initial "reactants" were in the gas phase. In our recent study, as depicted in Figure 15.2, using high-resolution microscopy technique, we confirmed that HgO product is indeed solid ($HgO_{(s)}$). Our results gave evidence for enhanced chain growth of $HgO_{(s)}$ on a carbon grid at RH = 50%. Clearly, due to importance of this reaction, further laboratory kinetic and mechanistic studies are desired.

Hydroxyl radical (HO) is considered to be the dominant daytime cleanser of the atmosphere. The major formation pathway in the troposphere is considered to involve photolysis of ozone followed by the reaction with water vapour:

$$O_3 + h\upsilon\,(\text{predominantly } \lambda \leq 320\text{nm}) \rightarrow O(^1D) + O_2 \tag{4}$$

$$O(^1D) + H_2O \rightarrow 2\,HO \tag{5}$$

In addition to the photolysis pathway, there are some dark reactions (ozonolysis) that have been proposed to be of significance at night or during the winter (Ariya et al., 2000). Typical background concentrations of ozone range from 20-30 ppbv (1 ppbv = 2.45×10^{10} cm^{-3}), and can peak to a few hundred ppb during smog situations (Finlayson-Pitts and Pitts, 1999).

To date, there is very limited kinetic data on HO + $Hg^0_{(g)}$ as shown in Table 15.1. The results of (Sommar et al., 2001) are in excellent agreement with the recent results (Ariya et al., 2004). These reported values are both lower than Bauer et al. upper limit evaluations for HO initiated oxidation reaction. Again HgO was observed as product, and we can now confirm that it is $HgO_{(s)}$ (Figure 15.2). Further kinetic and mechanistic studies of this reaction are desirable.

15.2.3 $NO_3 + Hg^0_{(g)}$

The nitrate radical, NO_3, is an important intermediate in the night time chemistry of the atmosphere. Upon sunrise, nitrate ions undergo photolysis to NO_2 or NO (Finlayson-Pitts and Pitts, 1999). Temperature dependence kinetics of elemental mercury, as well as dimethyl mercury, with NO_3 have been studied (Table 15.1). Sommar et al. (1997) employed a fast flow-discharge technique to study these reactions and obtained a second order rate constant value of 4 × 10 $^{-15}$ cm^3 molec^{-1} s^{-1}.

15.2.4 X_2/X/XO (X = Cl, Br, and I) + $Hg^0_{(g)}$

Mercury and halogen interaction has been experimentally studied under atmospheric conditions as summarized below.

Methyl iodide was shown to be non-reactive toward Hg^0 under atmospheric conditions ($k < 1 \times 10^{-21}$ cm^3 molecules^{-1} s^{-1}) (Tokos et al., 1998).

The first study of Hg^0 and chorine atoms was published in 1968 (Horne et al., 1968). HgCl was measured by time resolved absorption spectroscopy in the temperature range 383 – 443 K

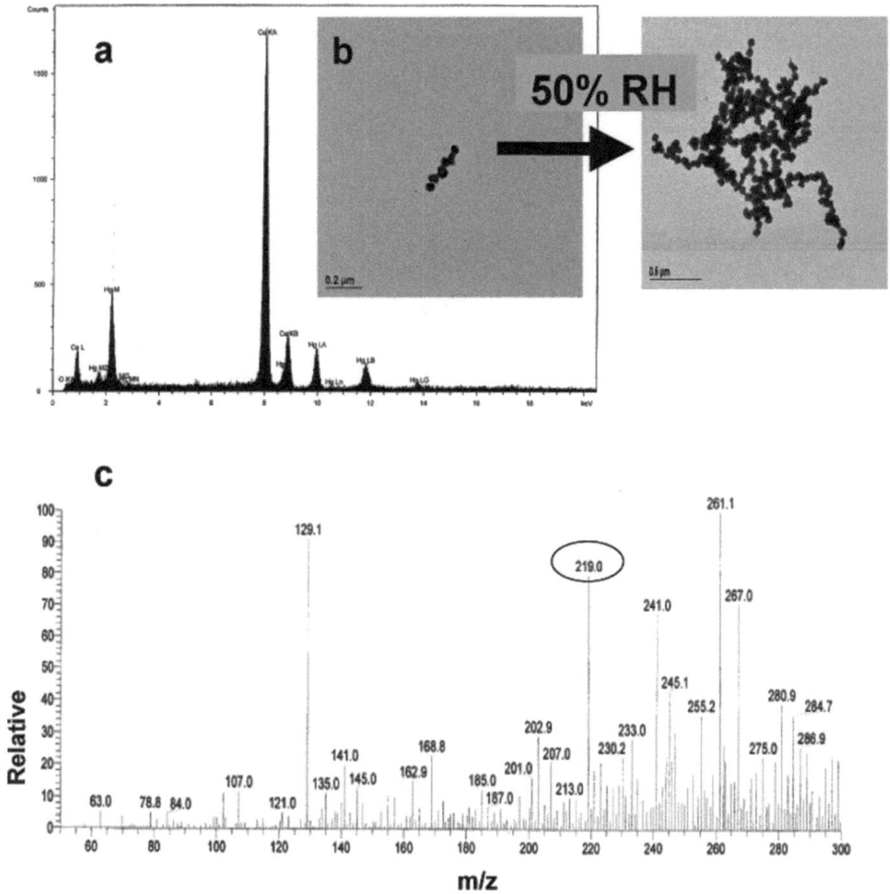

Figure 15.2 **a)** Energy dispersive spectroscopy (EDS) image of HgO **b)** Comparative HRTEM image of HgO deposit at RH = 0% and 50%, and **c)** CI of HgO product at RH = 0% and 50%. From Snider et al., 2008

$$\text{Hg+CL} \rightarrow \text{HgCL} \quad (6)$$

The rate constant, k_1 for the reaction of mercury with chlorine atoms was then derived to be 5.0×10^{-11} cm^3 $molecules^{-1}$ s^{-1} in 720 Torr CF_3Cl and 1.5×10^{-11} cm^3 $molecules^{-1}$ s^{-1} in 10 Torr CF_3Cl + 710 Torr Ar. The authors (Horne et al., 1968) mentioned that k_1 has an uncertainty of a factor of three because of the accumulation of experimental errors in evaluating the separate terms and the rate constant can be considered to be more accurate than the order of magnitude when the results is transferred to atmospheric conditions.

Molecular chlorine was suggested to have a relatively modest reaction rate, 4×10^{-16} cm^3 $molecules^{-1}$ s^{-1} (Schroeder et al, 1991; Menke and Wallis, 1980; Medhekar et al., 1979; Skare and Johansson, 1992; Seigneur et al, 1994) though the reaction was found to be strongly surface catalysed, (Medhekar et al., 1979; Skare and Johansson, 1992) and the experimental value should be considered as an upper limit.

In 2002, extensive kinetic and product studies on the reactions of gaseous Hg^0 with molecular and atomic halogens (X/X_2 where X = Cl, Br) have been performed at atmospheric pressure (750 ± 1 Torr) and room temperature (298 ± 1 K) in air and N_2 and published (Ariya et al., 2002). Kinetics of the reactions with X/X_2 were studied using both relative and absolute techniques. Cold vapour atomic absorption spectroscopy (CVAAS) and gas chromatography with mass spectroscopic detection (GC-MS) were the analytical methods applied. The measured rate constants for the reactions of Hg^0 with Cl_2, Cl, Br_2, and Br were $(2.6 \pm 0.2) \times 10^{-18}$, $(1.0 \pm 0.2) \times 10^{-11}$, $< (0.9 \pm 0.2) \times 10^{-16}$, and $(3.2 \pm 0.3) \times 10^{-12}$ cm^3 $molecule^{-1}$ s^{-1}, respectively. Thus Cl_2 and Br_2 are not important reactants in the troposphere for the Cl_2 and Br_2 concentrations reported in literature. Please note that in the case Br reactions, in our laboratory, due to the existing experimental conditions, we only could deploy one reference molecules for the relative rate studies. As explained above, in relative studies the evaluation of the rate constants is very much dependant on the accuracy of the values of that the reference reactions. Hence, any challenge with the values of the reference reaction significantly affects the value of the reaction of interest, in this case, Br + Hg^0. At this stage, in our laboratories, we have acquired additional the state-of-the-art facilities to revisit the Br-atom reactions using both absolute and relative techniques.

Chlorine and bromine atoms were generated using UV and visible photolysis of molecular chlorine and bromine, respectively, in addition to UV ($300 \leq \lambda \leq 400$ nm) photolyis of chloroacetyl chloride and dibromomethane. The reaction products were analyzed in the gas-phase, in the suspended aerosols and on the wall of the reactor using MS, GC-MS and inductively coupled plasma mass spectrometry (ICP-MS). The major products identified were $HgCl_2$ and $HgBr_2$ adsorbed on the wall. Suspended aerosols, collected on the micron filters, contributed to less than 0.5% of the reaction products under the experimental conditions. Studies by Sumner et al. (2005) revisited both reactions using a 17.3 m^3 environmental chambers equipped with fluorescent lamps and sun lamps to mimic environmental reactions, and evaluated the rate constants to be in the order of 10^{-12} cm^3 $molecule^{-1}$ s^{-1} and 10^{-11} cm^3 $molecule^{-1}$ s^{-1} for reactions of Br and Cl, respectively. Another research group (Donahue et al. 2005, 2006)) has reported two other kinetic data sets for Cl and Br reactions using a pulsed laser photolysis-pulsed laser induced fluorescence spectroscopy (Table 15.1). These data sets are obtained using pseudo-first order conditions with respect to halogens or mercury. The authors of these studies indicate an uncertainty estimation of ±50% in the rate coefficients due to the determination of absolute concentrations of chlorine and Br atoms. These reactions are reported slower (factors of 9-15) that the other laboratory studies. Hence, further studies of these reactions are strongly desired.

Reactions of mercury with halogen oxide radicals drew major attention in the light of satellite BrO column measurements as well as simultaneous mercury and ozone depletion in the planetary boundary layer (Richter et al., 1998; 2002, Muller et al., 2002; Van Roozendael et al., 2002; Goddsite et al., 2004). Experimental studies of XO reactions are very scarce. To our knowledge there is only one published laboratory kinetic study on the reaction of BrO with elemental mercury (Raofie and Ariya, 2003) during which, using the relative rate methods, the room temperature bimolecular rate constant for BrO + $Hg^0_{(g)}$ was estimated to lie within the range $10^{-15} < k < 10^{-13}$ cm^3 $molecule^{-1}$ s^{-1}. The faster end of this range makes BrO a significant potential contributor to mercury depletion events in the Arctic. This is however in contradiction with theoretical calculations, see next section. A report was published on the first experimental product study of BrO-initiated oxidation of elemental mercury at atmospheric pressure of ~ 740 Torr and T = 296 ± 2 K (Raofie and Ariya, 2004). The authors used chemical ionization and electron impact mass spectrometry, gas chromatography coupled to a mass spectrometer, a MALDI-TOF mass spectrometer, a cold vapour atomic fluorescence spectrometer, and high-resolution transmission electron microscopy coupled to energy dispersive spectrometry. BrO radicals were formed using visible and UV photolysis of Br_2 and CH_2Br_2 in the presence of ozone. They analyzed the products in the gas phase, on suspended aerosols and on wall deposits, and identified HgBr, HgOBr or HgBrO, and HgO as reaction products. Experimentally, they were unable to distinguish between HgBrO and HgOBr. The existence of stable $Hg^{(I)}$ in form of HgBr, along with $Hg^{(II)}$ upon BrO-initiated oxidation of Hg^0, suggests that in field studies it is fundamental to selectively quantify various mercury species in mercury aerosols and deposits. The majority of mercury containing products were identified as deposits, however, aerosols accounted for a substantial portion of products. It is noteworthy that we anticipate the possibility of transformation of $Hg^{(I)}$ to $Hg^{(II)}$ at high humidity levels though care must be taken as previously mentioned to extrapolate the results to ambient concentration levels of the reactants. No definite conclusions on the potential primary or secondary reactions of BrO, can be made at this stage. Even considering one order of magnitude uncertainties in the existing kinetic data, Br reactions make it the likely radical to explain elemental mercury depletion in the Arctic. Two independent studies (Ariya et al., 2004; Goodsite et al., 2004) confirm this conclusion. The existing kinetic results indicate that the direct BrO impact is less important than Br, but further studies are required to examine this conclusion. For example, A Saiz-Lopez et al. (2007) have discovered via long path DOAS measurements, significant amounts of iodine oxide (IO) above the Antarctic ice, and that bromine persists there for several months throughout the summer, thus giving rise to a greater oxidizing effect than formerly though possible, given observations in the Arctic. However, there is a recent product studies on iodine compounds with elemental mercury, which are very much similar to those reported by Br counterparts (Snider et al., 2008). The reactions of molecular iodine are shown to be very slow to be significant in the atmosphere. Further kinetic studies on I and IO are desired.

15.3 Theoretical Evaluation of Kinetic Data

The possibility of theoretically predicting the thermochemistry of mercury-containing species of atmospheric interest is of strong importance due to the paucity of accurate experimental information. They also serve fundamentally to further comprehend the complex reaction mechanisms. Accurate ab initio studies for measurements such as heats of formation, reaction enthalpies, and activation energies are particularly challenging, particularly in light of the large nuclear charge (80) and large number of electrons intrinsic to mercury. There is a detailed review on ab-initio thermochemical and kinetic studies on mercury reactions (Ariya and Peterson, 2005) and hence we discuss previous studies only in relation to experimental results. The existing theoretical kinetic data are also shown in Table 15.1. Ab initio calculations have to include e.g., careful choice of electron correlation method, treatment of relativistic effects, basis set truncation errors, etc., in order to obtain accurate kinetic data. The latter depends intimately on the underlying potential energy surface. A rigorous calculation of the rate coefficient for a given reaction generally involves either quantum scattering or classical trajectory calculations, which in turn require a global or semi-global potential energy surface (PES) calculated by ab initio methods. While these treatments are feasible for relatively small systems depending on the required accuracy of the underlying PES, most studies employ more approximate treatments of the reaction dynamics, e.g., transition state theory (TST) or RRKM theory (Rice-Ramsberger-Kassel-Marcus theory).

To the knowledge of the authors there are three studies by Goodsite et al, Khalizov et al, and Tossell (Goodsite et al., 2004, Khalizov et al., 2003, Tossell, 2003). The three groups looked on the reaction system of reactions:

$$Hg + X \rightarrow HgX \quad (7)$$

$$HgX \rightarrow Hg + X \quad (-7)$$

$$HgX + X \rightarrow HgXr_2 \quad (8)$$

$$Hg + XO \rightarrow HgO + X \quad (9)$$

where X is either Cl, Br or I.

Goodsite et al. (2004) and Khalizov et al. (2003) used the Gaussian 98 suite of programmes and the RRKM method. Goodsite et al. (2004) used the Steven Basch Kraus triple spilt CEP-121G basis set that account for some relativistic effects in the inner electrons of Hg. Khalizov et al. (2003) used two different basis sets. The first basis set was LanL2DZ which inner electrons substituted by effective core potentials and double- quality valence function for the heavier elements. The second basis set employs the ECP60MWB pseudo potential. Tossel used GAMESS, Gaussian 94 and Gaussian 98 software for the calculations also including relativistic effects.

Reaction 9 was investigated for X=Br to be endothermic and most probably without any importance in the atmosphere whereas reaction 2 is exothermic (Goodsite et al., 2004, Khalizov et al., 2003, Tossell, 2003). At room temperature Reaction 9 in all the studies was calculated to be exothermic. Khalizov et al. (2003) concluded that Hg + Br might be the dominant processes for atmospheric mercury depletion episodes (AMDE) occurring during Arctic Spring. This conclusion is further supported by Goodsite et al., (2004) that studied the temperature dependence of the reaction and showed that the HgBr intermediate is stabilised towards uni-molecular degradation at low temperatures which permits the addition of another Br (reaction 3) and thus $HgBr_2$ is a possible candidate for the formation of the otherwise unknown RGM.

From the standpoint of theoretical quantum chemistry, accurate calculations on molecular species involving mercury are particularly challenging in comparison to light, main group elements. In part this is due to its large number of electrons like any late transition metal atom, but the main difficulties lie in the treatment of its strong relativistic effects that to a large degree dictate its chemistry. Fortunately one can account for these effects very conveniently and accurately by use of modern relativistic pseudopotentials (PPs), which are also referred to as effective core potentials (ECPs). Nearly all ab initio calculations involving mercury employ the PP approximation to recover both scalar and vector relativistic effects. The former includes the mass-velocity and Darwin terms of the relativistic Hamiltonian while the latter is dominated by the spin-orbit interaction. By using relativistic PPs, much of the machinery of state-of-the-art quantum chemistry that has been so successful for lighter elements can be utilized with only few modifications for mercury-containing species. In regards to prediction of accurate molecular structures (better than 0.01 Å in bond lengths) and thermochemistry (accuracies at or below 4 kJ/mol), the strategy can be generalized as outlined below.

(i) The most accurate PP parameters available for mercury and perhaps other heavy atoms in the system of interest should be used. Those recently developed by the Stuttgart group (Figgen et al., 2005, Peterson, 2003) have been adjusted to multiconfigurational Dirac-Hartree-Fock calculations and appear to be the best choice at the present time for mercury and heavy main group elements, e.g., Br, I, Pb, etc.

(ii) Gaussian basis sets that have been matched to the PP(s) being used in the calculation should be chosen carefully. For the newer Stuttgart PPs mentioned above, full series of correlation consistent basis sets, e.g., cc-pV*n*Z-PP (*n*=D, T, Q, 5), are now available (Ariya and Peterson, 2005, Balabanov and Peterson, 2003) and should be used if at all possible. These have the unique property of systematically converging computed quantities to the complete basis set limit as successive members of the series are used. This effectively removes this source of error in the calculation and is essential for accurate error estimates and eliminating fortuitous error cancellations that can lead to inaccurate predictions.

(iii) In terms of the choice of electron correlation method, for thermochemistry and equilibrium structures the coupled cluster method, CCSD(T), has been shown to provide very accurate results for mercury species. For large scale potential energy surfaces or excited electronic states, multireference configuration interaction (MRCI) approaches must generally be used. Recent examples involving Hg include the low-lying electronic states of HgO and HgS, (Cressiot et al., 2007; Peterson et al., 2007; Shepler and Peterson, 2003) a quasiclassical trajectory study of the Hg+Br recombination reaction (Shepler et al., 2007), and a global potential energy surface for $HgBr_2$ (Shepler et al., 2005). While density functional theory (DFT) is a very popular approach in quantum chemistry due to its low scaling in terms of computational cost, it has not been shown to yield particularly accurate results for mercury-containing species. For example, previous large basis set DFT results (Khalizov et al., 2003) for the reaction enthalpies of $Hg+Br_2$ and Hg+Br differed by nearly a factor of two from the analogous (presumably accurate) CCSD(T) values.

(iv) While the relativistic PP will automatically account for scalar relativistic effects, some additional calculations incorporating spin-orbit coupling are generally warranted for mercury-containing systems. There are several avenues available for these calculations, but this remains one of the greater challenges for the accurate treatment of heavy-atom molecules and is not as amenable to the non-expert user. The reader is referred to Shepler and Peterson (2003) and Shepler et al., (2005) for some representative applications to mercury-containing systems.

Accurate quantum chemistry calculations using the methods briefly outlined above have played an important role in our current understanding of the oxidation of gas phase mercury in the atmosphere. High level ab initio calculations (Peterson et al., 2007; Shepler and Peterson, 2003) have conclusively demonstrated that $HgO_{(g)}$ was not sufficiently stable to facilitate the reaction of Hg with the BrO radical. Before these calculations, oxidation of mercury by BrO was thought to be strongly exothermic since the experimental bond dissociation energy of Hg^0 was estimated to be near 50 kcal/mol instead of the value of about 4 kcal/mol conclusively obtained by theory. State-of-the-art quantum chemical calculations on a variety of mercury-halogen species (Goodsite et al., 2004; Khalizov et al., 2003; Tossell, 2003; Balabanov and Peterson, 2003; Shepler et al., 2005, 2007) led to the mechanism whereby the formation of the HgBr radical by atom-atom recombination is the rate determining step in the gas phase oxidation of mercury in the troposphere. Recently the quasiclassical trajectory approach has been used on an accurate HgBr-Ar potential energy surface (Shepler et al., 2007) to determine thermal rate coefficients for the recombination reaction over a wide range of temperatures. The results were in excellent agreement with recent laboratory measurements. Analogous calculations have also been carried out recently for the $Hg+Br_2$ and HgBr + Br reactions on an accurate global HgBr2 potential energy surface (Shepler et al., 2005). The latter surface exhibited a large barrier to insertion of Hg into the Br_2 bond, which provided a rationale for the very slow rate measured experimentally for that reaction. Reliable rate coefficient calculations for Hg + Br and HgBr + Br have also been carried out using RRKM methods (Goodsite et al., 2004).

In addition to the determination of thermal rate coefficients, one of the main contributions of quantum chemistry is the prediction and subsequent characterization of new products and intermediates from proposed mercury oxidation mechanisms. Since high accuracy methods can predict heats of formation and bond dissociation energies to better than 4 kJ/mol accuracy even for mercury compounds, this is more than sufficient to determine the stability of novel mercury species. One such study, which is ongoing in one of our research groups, is the characterization of the mercury hypohalite species, e.g., HgBrOBr, which have been proposed (Calvert and Lindberg, 2003) to be formed by the reaction of HgBr with halogen monoxides but have not yet been observed by experiment. Our initial investigations find that they are thermodynamically very stable species.

15.4 Reactions at Interfaces: Heterogeneous Reactions

One of the first steps for mercury to undergo in a surface reaction is adsorption. There are two principal modes of adsorption of mercury molecules on any surface. The basis of distinction is the nature of the bonding between the molecule and the surface. In physical adsorption (physi-sorption), the bonding is by weak Van der Waals - type forces. There is no significant redistribution of electron density in either the molecule or at the substrate surface. In a chemisorption process a chemical bond, involving substantial rearrangement of electron density, is formed between the adsorbate and substrate. The nature of this bond may lie anywhere between the extremes of virtually complete ionic or complete covalent character, and hence it is significantly stronger than physical adsorption (40-800 kJ/mol in comparison to 5-40 kJ/mol) (Atkins and de Paula, 2002). There are a few ways to distinguish physisorption and chemisorption. The temperature over which chemisorption occurs can be only over a small surface, but is almost unlimited. However, for physisorption the temperature range is around condensation point of a gas such as Hg^0. Physisorption is generally reversible, non-dissociative, potentially multilayer and fast, whereas chemisorption is dissociative and often include an activated process with wide range kinetic desorption and limited to monolayers. To distinguish the type of adsorption, one can evaluate the vibrational frequency of substrate-adsorbate bond, or shift in energy or intensity of the valence orbitals in the substrate and adsorbate surface. For most mercury environmental surface studies shown in Table 15.1, the fundamental difference between chemi- and physi-sorption are not yet evaluated, and should be studied in future.

Environmental interfaces are very dynamic with respect to Hg cycling. The main surfaces interacting with the air compartment are soil, vegetation, snow, ocean and lake surfaces. These interfaces are sites of redox reactions and Hg exchange with the atmosphere. We here present an overview of Hg behavior at these interfaces, with respect to its reactivity and evasional flux.

15.4.1 Lake Surface

Lake surfaces represent about 1% of landmass surfaces, and are therefore not major players in controlling global fluxes (Mason and Sheu, 2002). However, evasion of Hg^0 from surfaces can significantly alter the Hg budget in these systems, with a potential impact on the contamination of fish. Results from a whole-ecosystem loading experiment (METAALICUS, Mercury Experiment to Assess Atmospheric Loading in Canada and the United States) have established that 45% of newly deposited Hg could be transformed near the water/air interface of a small boreal lake and returned to the atmosphere (Amyot et al., 2004; Southworth et al., 2007). For one of the greatest freshwater systems, Lake Superior, Rolfhus et al., (2003) estimated that Hg evasion from the lake surface completely counterbalanced atmospheric Hg deposition.

Mercury transformations at the air/lake interface are usually dominated by the photoreduction of $Hg^{(II)}$ to Hg^0 (Amyot et al., 1994). This production of Hg^0 typically displays both diel and seasonal patterns with maxima under sunlit and warm conditions (Zhang, 2006). Photoreduction of $Hg^{(II)}$ can be induced by UV and, to a lesser extent, by visible radiation. The fact that visible radiation can induce this reduction suggests that DOC chromophores may be involved (Fitzgerald et al., 2007). Filtration experiments have shown that this photoreduction can be homogeneous. It can be mediated by iron(III) (Zhang and Lindberg, 2001) and humic acids (Allard and Arsenie, 1991). However, it can also be biologically-mediated (Siciliano et al., 2002). The relative importance of these mechanisms will differ with pH, light attenuation and DOC levels at the surface. An in-depth review on this topic is presented in Zhang and Wong, 2007.

Of lesser importance in lakes, (photo)oxidation of Hg^0 to $Hg^{(II)}$ has been observed and also follows a diel cycle (Garcia et al., 2005). This oxidation is mainly promoted by the UV-A waveband and can be driven by the formation of strong Hg oxidizing agents (e.g. OH radicals) or be indirectly caused by the photoproduction of hydrogen peroxide which, in turn, regulates microbial oxidation processes (Siciliano et al., 2002).

Current models do not predict well the formation of Hg^0 and its evasion from lake surfaces. Processes occurring in the surface microlayer need to be better assess in order to establish the actual Hg^0 gradient at all water/air interfaces (for lakes, oceans and estuaries).

15.4.2 Surface of Oceans

Oceanic surfaces are a major site of Hg exchange at the global scale, with evasional fluxes accounting for about 39% of global Hg emissions (Mason and Sheu, 2002). Hg at this interface undergoes similar transformations to those described for lakes. Rolfhus and Fitzgerald (2004) estimated that about 70% of volatile Hg formed in coastal seawaters was of photochemical origin, 20% came from bacterial processes, and 10% from uncharacterized dark reduction.

In addition, two major differences can be highlighted between freshwater and saltwater interfaces. First, in addition to Hg^0, ocean waters also contain significant concentrations of another highly volatile species, dimethylHg. This species formed at depth can be brought up to the air/seawater interface by upwelling currents in coastal areas. Second, oxidation of Hg^0 to $Hg^{(II)}$ is far more prevalent in saltwaters and will hamper the evasional fluxes of Hg^0. This oxidation is photo-induced and promoted by halogen chemistry both above and below the water/air interface (Sheu and Mason; 2004, Lalonde et al., 2001). In Sheu and Mason (2004), aqueous NaCl/NaBr salts were photolyzed in the presence of $Hg^0_{(g)}$. It was discovered that Hg^0(g), in the presence of water, salt, and under a Xe-lamp, the oxidation rate constant increased 100-fold compared with irradiated salt-free water. Work was done at ambient temperatures in a quartz container. Mechanisms proposed involved volatilization of halogen species, which then react with mercury. Many secondary reactions of mercury were also considered (i.e. those with OH, BrO, ClO, and O_3), generally initiated by the presence of salt and light energy. Sheu and Mason (2004) also note reactions of Hg + Br were 25 faster than with Cl radicals.

Accidental Hg reduction by marine microorganisms has been proposed as a significant source of Hg^0 in the mixing layer for a long time (Mason et al., 1995). There is evidence the bacterial mercuric reductase enzyme (MerA) will reduce MeHg and inorganic $Hg^{(II)}$ species to $Hg^0(g)$ in Arctic coastal and marine environments (Poulain et al., 2007). It is noted this reduction is an apparently deliberate self-preservation of certain biota against methylmercury contamination in water. In addition to photoreduction recycling mercury, bacteria are capable of re-volatilizing the metal at comparable levels even with only 1% of cells active.

As mentioned for lakes, very few studies have focus on the sea surface microlayer, even though microscale processes in this layer may have an important impact on evasional fluxes.

Using *ab-initio* chemistry, Shepler et al., (2007b) noted water microsolvation (using 1-3 water molecules) favored the oxidation of mercury in the presence of bromine.

$$Hg + Br \rightarrow HgBr \tag{10}$$

$$HgBr + Br \rightarrow HgBr_2 \tag{11}$$

$$HgBr + Br \rightarrow Hg + Br_2 \tag{12}$$

Reactions (10) and (11) were found more favorable in the presence of water, whereas reaction (12) was less favorable when solvated. They conclude it is probable the effects of ice, snow, and water surfaces enhance the scavenging of mercury by halogens.

15.4.3 Snow Surface

The role of the snow surface on the reactivity of Hg and its release to the atmosphere has been discussed for polar regions in Chapter 9 of Steffen et al. (2007). It has been demonstrated that in suburban and remote temperate areas, about 50% of newly deposited Hg is returned back to the atmosphere within 24 to 48 hours (Lalonde et al., 2002). This release results from the photoreduction of $Hg^{(II)}$ to Hg^0 in the snowpack, mostly induced by UV-B radiation (Lalonde et al., 2003). The processes leading to this reduction have not been elucidated.

In forested areas, the canopy has a significant impact on the behavior of Hg in the underlying snowpack (Poulain et al., 2007b). Snow under canopy has typically higher Hg levels than snow from open areas (e.g. frozen lake surfaces); photoreduction of $Hg^{(II)}$ followed by evasion is less efficient in forested areas because of light attenuation by the canopy. Poulain et al. (2007c) calculated net winter gain of Hg in snow under canopies dominated by conifers whereas, under a deciduous canopy, the pool of Hg stored at the end of the winter was comparable to that of wet deposition. Coniferous trees were both a source of Hg to the forest floor (via throughfall) and an obstacle to Hg photoreduction in underlying snow. Snow over lakes acted as a winter source of Hg to the atmosphere. Whereas most Hg deposited by snow on lakes is lost before snowmelt, Hg deposited on the forested watershed is largely retained in snowpacks.

Snow can house a number of different $Hg^{(II)}$ species, i.e. HgC_2O_4, $Hg(OH)_2$, HgOHCl, HgO (Ferrari et al., 2002), and possibly others. It is clear from experiments that $Hg^{(II)}$ on snowpacks is photo-reduced by natural sunlight (Dommergue et al., 2007). Mercury over snow originates from atmospheric $Hg^0_{(g)}$ through dry deposition (Schroeder et al., 1998) and oxidation mainly via O_3, BrO, and Br (Ariya et al., 2004). Concentrations of arctic mercury on snowpacks are guided by incoming and outgoing fluxes, which depend on light intensity and oxidant concentration, respectively. Oxidation of mercury over snowpacks is part of a dynamic system of ice, snow, ozone, UV-Vis light, Cl, and Br radicals (see Lindberg et al., 2002). It is believed that aerosol ice surfaces catalyze oxidation of $Hg^0_{(g)}$ to HgO or $HgBr_2/Cl_2$ during the Arctic spring (Lindberg et al., 2002). Oxidation is aided by the destruction ozone by Br over ice and formation of Hg^0 oxidants (Ariya et al., 2004; Oltmans et al., 1989).

Methylmercury ($MeHg^{(I)}$) has been observed in Arctic snowpacks (St. Louis, et al., 2007). The origins of $MeHg^{(I)}$ are aqueous (oceans, lakes), however its volatility is low: $K_p = [HgMe_{(g)}]/[HgMe_{(aq)}] = 2 \times 10^{-5}$, at 298 K (Schluter, 2000). By contrast, volatility of dimethylmercury (Me_2Hg) is much higher: $K_p = 0.31$ (Schluter, 2000). St Louis et al., (2007) hypothesize MeHg originates from nearby ocean sources as $MeHg_2$, then converts to MeHgCl in the salty snow. They note there is a positive correlation between total mercury and Cl concentrations over snowpacks. The salinity of Arctic snow can range anywhere between 20 – 2000 mg l^{-1} (Ariya et al., 2004).

Snow spiked with hydrogen peroxide was observed to enhance $Hg^{(II)}$ deposition five-fold under natural Arctic springtime sunlight (Lahoutifard et al., 2006). The mechanism of oxidation is not known, though it is suggested equilibrium can be formed with chlorine in acidic conditions:

$$H_2O_2 + 2H^+ + Hg^0 + 2Cl^- \leftrightarrow 2H_2O + HgCl_2 \quad (13)$$

Samples were spiked with 50 μM H_2O_2, similar to natural concentrations (30 μM). Hence H_2O_2 was suggested to play a significant role in Hg^0 oxidation under UV light hat should be further confirmed. Bromine, as well as Cl, can oxidize Hg^0

in snow. Fain et al. (2006) calculated from field samples a mercury + bromine oxidation rate constant very similar to lab studies; 2×10^{-11} cm^3 molecule $^{-1}$ s^{-1} at -10 °C.

Poulain et al., (2007) observe 100-fold higher concentrations of total mercury in snow found near Arctic sea/ice boundaries than inland. They note melting of the snow/ice during springtime further enhances mercury deposition (until enhanced light intensity of the spring re-volatilizes condensed mercury). Douglas et al., (2008) observe various crystalline morphologies of snow will exhibit varying degrees of $Hg^0_{(g)}$ scavenging, with up to an order of magnitude difference in deposited concentrations. Heterogeneous mercury reactions evidently depend on surface morphology in addition to surface species present. Aspmo et al., (2006) observed $Hg^0_{(g)}$ concentrations over sea ice, noting some increase in concentration (1.82 ng m^{-3}) compared with background north Atlantic ocean levels (1.53 ng m^{-3}). Analyses were done in the summer and spring, leading to the possibility of re-emission of $Hg^{(II)}$ over ice and snow. Depth profiles of mercury concentration over snow show generally higher levels than atmospheric background levels. Despite observed rapid mercury depletion events in the polar regions (Schroeder et al., 1998; Ebinghaus et al., 2002), but the over all fate is an subject of debate (Steffen et al., 2002). The total volume of mercury entering the Arctic circle is calculated to be about 300 Mg per year, via global model simulation (Ariya et al., 2004). This influx is largely scavenged over ice, snow, and water via bromine explosions (Tackett et al., 2007). However, this deposition in part is rapidly reduced to $Hg^0_{(g)}$ later during Arctic springtime (Steffen et al., 2002).

Frost flowers have been a known source of halogen from sea ice for several years (Kaleschke et al., 2004). Now it is possible these ice crystals provide via high surface areas a scavenging of mercury (Douglas et al., 2008) It was found crystals formed in the vapour phase have higher mercury concentrations (2-10 times as much) than snow deposits. The only difference between frost flowers and snow deposits would appear to be their morphology; hence the surface design (diamond dust, surface hoar, blowing snow, glass trays) may also affect kinetics of Hg(II/I). A previous study already supposed Br radicals are released from the sea ice crystals (Gauchard et al., 2005), implying heterogeneous reactions are responsible in part for mercury oxidation. However, we point out that the Br/BrO + Hg^0 oxidation itself is gaseous. Trace species affecting Hg reduction/oxidation in snow include, but are not limited to, Br^-, Cl^-, microbes, and ice/snow morphology.

15.4.4 Soil Surface

Mercury air-soil exchange is an important component of the Hg cycle at regional and global scales (Grigal, 2002). Hg^0 volatilization from soils has been correlated to soil Hg concentration (Gustin et al., 1999), soil moisture (Gustin and Lindberg, 2005), atmospheric oxidants (Gustin et al., 2005), meteorological conditions (barometric pressure, temperature, wind speed and turbulence, and solar radiation).

Under low atmospheric Hg concentrations, barren soils can act as Hg sources to the atmosphere during the day or sinks of Hg at night (Xin and Gustin, 2007). Photochemical processes are likely the main driver of Hg^0 formation and evasion when the substrate is moist or after rain events (Edwards et al., 2001), whereas solar-induced thermodesorption of Hg^0 is probably more important under dry conditions (Poissant et al., 2004). The sorption properties of soils will be dictated by the mineralogical composition. For instance, the presence of kaolinite, montmorillonite, and goethite in soils has been shown to enhance the sorptive capacity of soils (Edwards et al., 2001).

Since soils can significantly differ in their sorption capacity and their reactivity, and current evasional estimates are site-specific, the overall global fluxes associated with soils are still poorly constrained and need further assessment.

15.4.5 Vegetation Surface

Vegetated areas are key players in global Hg cycling. According to Mason and Sheu (2002), net Hg evasion from land is 8 Mmol yr^{-1}; emissions from vegetated areas (forest, prairies and farmland) are estimated at 9 Mmol yr^{-1} and uptake of Hg^0 by plant drives a depositional flux of -7 Mmol yr^{-1}. Because of the magnitude of these vegetation fluxes, a far better understanding of these surfaces is needed to constrain flux estimates.

The plant/air interface is a site of both passive and active exchange of $Hg^{(II)}$ and Hg^0. For example, atmospheric particulate Hg and reactive gaseous Hg can be absorbed on leaf surfaces after dry deposition (Hanson et al., 1995). Stomata can actively take up atmospheric Hg^0 (Lee et al., 2000). This assimilated Hg can come either from passing air masses or from soil Hg emissions below the canopy; in the latter case, this uptake results in a fast cycling of Hg within the forest. In contaminated sites, plants can translocate Hg from soils to leaves, with some Hg being released through stomata or through litterfall. The leaf surface has also been shown to be a site of photochemical transformations of deposited $Hg^{(II)}$ to Hg^0, followed by its evasion to the atmosphere. The UV band was shown to be the most efficient radiation in this reduction.

15.4.6 Carbon (Fly Ash, Charcoal)

Knowledge that coal combustion is a source for mercury dates back over 35 years (Joensuu, 1971). Residues of coal combustion in industrial power plants generate fly ash, composing mostly of SiO_2 and Al_2O_3. Fly ash composition and morphology make it suitable for zeolite synthesis (Querol et al., 1997). Flue gas may be comprised of CO_2, O_2, CO, NO, NO_2, SO_2, H_2S, HCl, NH_3, N_2O, and Hg (Hall et al., 1995). Incomplete combustion ($T < 400$ °C) leads to carbon in fly ash, usually enhancing mercury adsorption (Pavlish et al., 2003). Presence of carbon also leads to high "Loss on ignition" (LOI), which is defined in the context of coal combustion as the fly ash weight loss at a given (elevated) temperature. Thus the carbon content may be expelled when sufficiently heated, possibly taking absorbed mercury with it. Presence of carbon was found to increase the BET surface area of fly ash, enhancing adsorption (Hower et al., 2000). Surface area per unit mass, and per unit area in the flue, is significant in describing adsorption. Carbon content ranges from 6 to 850 $m^2 g^{-1}$, with ~70 $m^2 g^{-1}$ in charcoal (Chen, 2007). The BET surface area of iron oxide is 62 $m^2 g^{-1}$ (Wu et al., 2006). Soot/fly ash may disperse globally; fly ash has been shown to be a component of Arctic aerosols (Daisey et al., 1981). It is possible carbon in aerosols will affect mercury oxidation rates under environmental conditions.

Presto and Granite (2006) have efficiently summarized the significant contributions carbon, metal, metal oxide, and other surfaces in simulated and experimental coal-combustion conditions. We attempt to avoid duplication of their review material by further updating this subject, though some reference material of theirs is necessarily highlighted. We also refer the reader to Pavlish et al., (2003) for an earlier review of mercury capture in power plants.

Activated carbon can absorb mercury in aqueous solutions as well gaseous systems (Namasivayam and Kadirvelu, 1999; Ranganathan, 2002; Yardim et al., 2003). Sen and De (1987) found that aqueous $Hg(NO_3)_2$ was readily adsorbed by fly ash at a pH = 3.5 – 4.5. At a pH of 5, $Hg^{(II)}$ was hypothesized to transform into $Hg(OH)_2$ over the carbon (Namasivayam and Kadirvelu, 1999). Chen (2007) noted that $H_2O_{(g)}$ did not affect mercury oxidation. Some experiments have noted humidity negatively affecting oxidation (Menke and Wallis, 1980; Seigneur et al., 1994).

It is clear that many factors affect the adsorption - hence redox - reactions of mercury over fly ash. The most important trace elements affecting the oxidation rate are HCl, ClO, and Cl_2. Mechanistically, we suspect that the majority of oxidation in the presence of fly ash or carbon is heterogeneous based on the evidence of Presto and Granite (2006). Temperature is a significant factor in oxidation rate; optimal values must be achieved to balance reaction efficiency and total adsorption. Hall et al., (1995) discovered a mixture of oxygen and mercury at 100-300 °C would react in the presence of fly ash or carbon. There was a measurable oxidation rate constant of $\sim 10^{-4}$ s^{-1}. Surface kinetics have been postulated to obey a Langmuir - Hinshelwood mechanism, where both mercury and oxygen adsorb onto the carbon surface before reacting. A temperature of about 200 °C was found optimal. Xu et al., (2008) (see also Change and Offen, 1995) compared mercury oxidation by different pathways using a combination of kinetic modeling and *ab initio* chemistry over a carbon surface. They conclude that Hg^0 + ClO reactions may be more significant at T > 130°C than mercury reactions with either Cl_2 or HCl.

We conclude this section by stating the use of fly ash or charcoal in removing mercury is not cost-effective (Change and Offen, 1995), varying between 14,000 – 38,000 USD/lb Hg. The useful temperature range is not wide for carbon; the peak efficiency temperature is ~200 °C (Hall et al., 1995). Carbon is not effective at high temperatures (> 400 °C) due to its LOI. Some studies find temperature to be inversely proportional to Hg^0 removal (Dunham et al.,

2003); it is found that carbon at 20 °C absorbs Hg^0 better than at 40 °C (Hwang et al., 2002). Fly ash is stable at high temperatures, however efficiency also decreases with increasing temperature. Fly ash also does not efficiently oxidize mercury unless other additives (HCl,H_2S) are present. Although carbon/fly ash injection is a very natural method to removing $Hg^0_{(g)}$, it remains the engineers' and physical chemists' goal to achieve improved mercury absorbency by more robust and cheaper adsorbents.

15.5 Open Questions and Future Directions

The lack of knowledge in the gas phase and liquid phase looks insignificant in comparison to the lack of knowledge at the interfaces, and thus heterogeneous reactions. The knowledge of mercury chemical, physical and biological interactions at environmental surfaces is scarce at best. It is now evident that the existence of the surfaces, different types of surfaces, and different environmental conditions can alter the transformation of mercury in pure gas phase or aqueous phase. However, the quantification of the impact of surfaces is yet to be understood. The challenges facing surface chemistry includes:

1. As far as the theoretical calculations are concerned, one of the major challenges is the accurate inclusion of spin-orbit coupling effects, particularly for large molecules and clusters. Advances are currently being made in the area of two-component DFT theory and this may very well be a promising avenue for incorporating these effects. Of course the methods outlined above are also mostly limited to gas phase calculations. Accurate theoretical treatment of condensed phase system continues to be a great challenge. Both cluster models and ab initio molecular dynamics methods will certainly play a large role in future studies of the heterogeneous reactivity of mercury.
2. Despite the novel positive acquisitions of knowledge from experimental and theoretical studies of gas-phase elemental mercury chemistry there are still large gaps before a complete understanding of the fate of mercury in the atmosphere is obtained. It is essential to provide kinetic data, information about formed products.
3. There are some limited studies on the kinetics of gas-phase elemental mercury oxidation on surfaces (e.g., Lee et al, 2004; Flora et al., 1998; Vidic et al., 1998). However, experimental studies on uptake or kinetics of heterogeneous reactions of mercury on various environmentally relevant surfaces such as ice, snow, and aerosols and biomaterials, are needed.
4. Lack of knowledge of detailed mercury chemical speciation in the field studies. Currently, the existing techniques are quite poor in providing detailed chemical structure of mercury compounds at the environmental interfaces as at the matter of fact even in atmosphere, water and snow. The operational definitions are used to discern amongst different functional groups, however, as they are not based

on fundamental understanding of physical and chemical structures of molecules, it is very difficult to use them adequately for proper understanding of surface chemistry and physics of mercury. Further development of targeted techniques for detailed mercury analysis is essential.

5. Currently, the knowledge of chemical reactions involving mercury compounds in aerosols and clouds is limited, and sometimes contradictory, liquid phase chemistry kinetic data. However, there is an urgency of for research on heterogeneous mercury reactions at fundamental theoretical, kinetic and dynamic studies, as well as proper incorporation in atmospheric modeling.
6. Fundamental surface sciences during the last several decades have achieved break through understanding of interfaces at molecular and cluster levels. It is wise for mercury scientists to take advantage of this existing body of knowledge including techniques such as various types of electron microscopy (e.g. transmission to electron force) to further understand the physical property of the surfaces, and the nature of the bonds between substrate and surface, as well as substrate-substrate configuration changes upon interactions with surfaces. This case is particularly valid for surfaces such as snow, as well as aerosols and cloud droplets. It is of outmost interest to understand the mechanism(s) on or wihin these surface reactions.
7. The nature of diffusion of mercury species in surfaces and interfaces (e.g., snow/ice) should be characterized.
8. The importance of so-called "micro-layer" within the interface in relation to the entire surface should be studied.
9. There is an amazing range of biological surfaces available for mercury transformation. Reactions are shown to occur on the surfaces or be altered within the biological bodies. The detailed chemical transformation of such reactions implicating biological transformation of mercury and its impact on physical and chemical characteristics of mercury compounds in environment is a fascinating field of studies that should be attempted from nano to macro scales.
10. We know presently full well that to grasp the mercury transformation on this planet, the knowledge of pure gas, or condensed-phase physics and chemistry will not suffice. The feedbacks of gas phase on surfaces or liquid/solid/heterogeneous phase on environmental surfaces are ought to be characterized. The impact of heterogeneity on surfaces in local, regional and global scales is ought to be understood.
11. Anthropogenic activities in the domains of new materials and nanotechnology, has produces novel surfaces as product of by-product of such activities. These molecules are in addition to oxidized transition metals (Fe, Mn, V, Cu, Ti), noble metals (Au, Pd, Ag, Cu) and metal oxides, glass type structures that are known to be involved in mercury transformations or its removal. There is not much known on the interactions of human-made novel surfaces with mercury compounds. As anthropogenic activities currently represent the major mercury emission in the atmosphere, the importance of these surfaces on Hg transformation should be understood.

Acknowledgements We acknowledge Natural Science Foundation of Canada (NSERC) and Canadian Foundation for innovation (CFI), McGill University and Italian Ministry of Environment for financial support

References

Abu-Daabes, M.A. and N.G. Pinto, Synthesis and characterization of a nano-structured sorbent for the direct removal of mercury vapor from flue gases by chelation. Chemical Engineering Science, 2005. 60(7): p. 1901–1910.

Agarwal, H., C.E. Romero, and H.G. Stenger, Comparing and interpreting laboratory results of Hg oxidation by a chlorine species. Fuel Processing Technology, 2007. 88(7): p. 723–730.

Allard, B. and I. Arsenie, Abiotic Reduction of Mercury by Humic Substances in Aquatic System - an Important Process for the Mercury Cycle. Water Air and Soil Pollution, 1991. 56: p. 457–464.

Amyot, M., et al., Sunlight-Induced Formation of Dissolved Gaseous Mercury in Lake Waters. Environmental Science & Technology, 1994. 28(13): p. 2366–2371.

Amyot, M., G.A. Gill, and F.M.M. Morel, Production and loss of dissolved gaseous mercury in coastal seawater. Environmental Science & Technology, 1997. 31(12): p. 3606–3611.

Amyot, M., D. Lean, and G. Mierle, Photochemical formation of volatile mercury in high Arctic lakes. Environmental Toxicology and Chemistry, 1997. 16(10): p. 2054–2063.

Amyot, M., et al., Formation and evasion of dissolved gaseous mercury in large enclosures amended with (HgCl2)-Hg-200. Atmospheric Environment, 2004. 38(26): p. 4279–4289.

Ariya, P. A.; Jobson, B. T.; Sander, R.; Niki, H.; Harris, G. W.; Hopper, J. F.; Anlauf, K. G., Measurements of C-2-C-7 hydrocarbons during the Polar Sunrise Experiment 1994: Further evidence for halogen chemistry in the troposphere. Journal of Geophysical Research-Atmospheres, 1998. 103(D11): p. 13169–13180.

Ariya, P.A., J.F. Hopper, and G.W. Harris, C-2-C-7 hydrocarbon concentrations in arctic snowpack interstitial air: Potential presence of active Br within the snowpack. Journal of Atmospheric Chemistry, 1999. 34(1): p. 55–64.

Ariya, P.A., R. Sander, and P.J. Crutzen, Significance of HOx and peroxides production due to alkene ozonolysis during fall and winter: A modeling study. Journal of Geophysical Research-Atmospheres, 2000. 105(D14): p. 17721–17738.

Ariya, P.A., A. Khalizov, and A. Gidas, Reactions of gaseous mercury with atomic and molecular halogens: Kinetics, product studies, and atmospheric implications. Journal of Physical Chemistry A, 2002. 106(32): p. 7310–7320.

Ariya, P.A., A.P. Dastoor, M. Amyot, W.H. Schroeder, L. Barrie, K. Anlanf, F. Raofic, A. Ryzhkov, D. Davignon, J. Lalongde and A. Steffen, The Arctic: a sink for mercury. Tellus Series B-Chemical and Physical Meteorology, 2004. 56(5): p. 397–403.

Ariya, P. and K. Peterson, Chemical Transformation of Gaseous Elemental Hg in the Atmosphere. In: Dynamics of Mercury Pollution on Regional and Global Scales (N. Pirrone and K. Mahaffey eds.) 2005: p. 261–294.

Ariya, Parisa A. Henrik Skov and Michael Evan Goodsite, Review of Applications of Theoretical Methods and Experimental Studies to Evaluate the Oxidation of Gaseous Elemental Mercury in the Atmosphere, Accepted under minor revisions, Advances in Quantum Chemistry, 55, ACADEMIC PRESS, 2008, ISBN-13: 978-0-12-374335-0

Aspmo, K., et al., Mercury in the atmosphere, snow and melt water ponds in the North Atlantic Ocean during Arctic summer. Environmental Science & Technology, 2006. 40(13): p. 4083–4089.

Atkins, P. and J. de Paula, Atkins' Physical Chemistry. 7th Edition ed. 2002, New York: Oxford University Press.

Balabanov, N.B. and K.A. Peterson, Mercury and Reactive Halogens: The Thermochemistry of Hg + {Cl2, Br2, BrCl, ClO, and BrO}. J. Phys. Chem. A, 2003. 107(38): p. 7465–7470.

Barkay, T., S.M. Miller, and A.O. Summers, Bacterial mercury resistance from atoms to ecosystems. FEMS Microbiology Reviews, 2003. 27(2-3): p. 355–384.

Barrie, L. A.; Bottenheim, J. W.; Schnell, R. C.; Crutzen, P. J.; Rasmussen, R. A. et al., Ozone Destruction and Photochemical-Reactions at Polar Sunrise in the Lower Arctic Atmosphere. Nature, 1988. 334(6178): p. 138–141.

Barrosse-Antle, L.E., et al., The expansion/contraction of gold microparticles during voltammetrically induced amalgamation leads to mechanical instability. New Journal of Chemistry, 2007. 31(12): p. 2071–2075.

Bauer, D., et al., Gas phase elemental mercury: a comparison of LIF detection techniques and study of the kinetics of reaction with the hydroxyl radical. Journal of Photochemistry and Photobiology a-Chemistry, 2003. 157(2-3): p. 247–256.

Blythe, G.M., Field Testing of a Wet FGD Additive for Enhanced Mercury Control – Pilot-scale Test Results. 2006, URS Corporation: Austin, Texas.

Brooks, S.B., et al., The mass balance of mercury in the springtime arctic environment. Geophysical Research Letters, 2006. 33(13): p.-.

Buxton, G.V., et al., Critical-Review of Rate Constants for Reactions of Hydrated Electrons, Hydrogen-Atoms and Hydroxyl Radicals (.Oh/.O-) in Aqueous-Solution. Journal of Physical and Chemical Reference Data, 1988. 17(2): p. 513–886.

Calvert, J.G. and S.E. Lindberg, A modeling study of the mechanism of the halogen-ozone-mercury homogeneous reactions in the troposphere during the polar spring. Atmospheric Environment, 2003. 37(32): p. 4467–4481.

Calvert, J.G. and S.E. Lindberg, Mechanisms of mercury removal by O-3 and OH in the atmosphere. Atmospheric Environment, 2005. 39(18): p. 3355–3367.

Change, R. and G.R. Offen, Mercury emission control technologies: An EPRI synopsis. Journal: Power Engineering (Barrington); Journal Volume: 99; Journal Issue: 11; Other Information: PBD: Nov 1995, 1995: Size: pp. 51–56; Other: PL:.

Chen, X., Impacts of Fly Ash Composition and Flue Gas Components on Mercury Speciation in Civil and Environmental Engineering. 2007, University of Pittsburgh: Pittsburgh. p. 87.

Cressiot, C., et al., Stability of the HgS molecule and spectroscopy of its low lying electronic states. Molecular Physics, 2007. 105(9): p. 1207–1216.

Daisey, J.M., R.J. Mccaffrey, and R.A. Gallagher, Polycyclic Aromatic-Hydrocarbons and Total Extractable Particulate Organic-Matter in the Arctic Aerosol. Atmospheric Environment, 1981. 15(8): p. 1353–1363.

Dommergue, A., E. Bahlmann, R. Ebinghaus, C. Ferrari and C. Boutron, Laboratory simulation of Hg-0 emissions from a snowpack. Analytical and Bioanalytical Chemistry, 2007. 388(2): p. 319–327.

Donohoue, D.L., D. Bauer, and A.J. Hynes, Temperature and pressure dependent rate coefficients for the reaction of Hg with Cl and the reaction of Cl with Cl: A pulsed laser photolysis-pulsed laser induced fluorescence study. Journal of Physical Chemistry A, 2005. 109(34): p. 7732–7741.

Donohoue, D.L., et al., Temperature and pressure dependent rate coefficients for the reaction of Hg with Br and the reaction of Br with Br: A pulsed laser photolysis-pulsed laser induced fluorescence study. Journal of Physical Chemistry A, 2006. 110(21): p. 6623–6632.

Douglas, T.A., et al., Influence of Snow and Ice Crystal Formation and Accumulation on Mercury Deposition to the Arctic. Environ. Sci. Technol., 2008.

Dunham, G.E., R.A. DeWall, and C.L. Senior, Fixed-bed studies of the interactions between mercury and coal combustion fly ash. Fuel Processing Technology, 2003. 82(2-3): p. 197–213.

Ebinghaus, R., H.H. Kock, C. Temme, J.W. Einax, A.G. Löwe, A. Richter, J.P. Burrows and W.H. Schroeder Antarctic Springtime Depletion of Atmospheric Mercury. Environ. Sci. Technol., 2002. 36(6): p. 1238–1244.

Edwards, J.R., R.K. Srivastava, and J.D. Kilgroe, A study of gas-phase mercury speciation using detailed chemical kinetics. Journal of the Air and Waste Management Association ; ISSUE: 51; PBD: Jun 2001, 2001: 869–877.

Evaluate the Oxidation of Gaseous Elemental Mercury in the Atmosphere Advances in Quantum Chemistry, 2008. in press.

F Fain, X., et al., Fast depletion of gaseous elemental mercury in the Kongsvegen Glacier snowpack in Svalbard. Geophysical Research Letters, 2006. 33(6).

ain, X., et al., Diurnal production of gaseous mercury in the alpine snowpack before snowmelt. Journal of Geophysical Research-Atmospheres, 2007. 112(D21): p. 311.

Ferrari, C.P., et al., Mercury speciation in the French seasonal snow cover. Science of the Total Environment, 2002. 287(1-2): p. 61–69.

Ferrari, C.P., A. Dommergue, and C.F. Boutron, Gaseous mercury distribution in interstitial air of snow pack in Station Nord, Greenland. Evidence of permanent mercury depletion event in the air of snow during polar sunrise. Journal De Physique Iv, 2003. 107: p. 459–462.

Ferrari, C.P., et al., Profiles of Mercury in the snow pack at Station Nord, Greenland shortly after polar sunrise. Geophysical Research Letters, 2004. 31(3): P.-.

Figgen, D., et al., Energy-consistent pseudopotentials for group 11 and 12 atoms: adjustment to multi-configuration Dirac-Hartree-Fock data. Chemical Physics, 2005. 311(1-2): p. 227–244.

Finlayson-Pitts, B.J. and J.N. Pitts, Chemistry of the Upper and Lower Atmosphere: Theory, Experiments, and Applications. 1999: Elselvier. 969.

Fitzgerald, W.F., C.H. Lamborg, and C.R. Hammerschmidt, Marine biogeochemical cycling of mercury. Chemical Reviews, 2007. 107(2): p. 641–662.

Flora, J.R.V., et al., Modeling powdered activated carbon injection for the uptake of elemental mercury vapors. Journal of the Air & Waste Management Association, 1998. 48(11): p. 1051–1059.

Gabriel, M.C. and D.G. Williamson, Principal Biogeochemical Factors Affecting the Speciation And Transport of Mercury through the terrestrial environment. Environmental Geochemistry and Health, 2004. 26(4): p. 421–434.

Garcia, E., M. Amyot, and P.A. Ariya, Relationship between DOC photochemistry and mercury redox transformations in temperate lakes and wetlands. Geochimica Et Cosmochimica Acta, 2005. 69(8): p. 1917–1924.

Gårdfeldt K., J. Sommar, D. Strömberg and X. Feng,, Oxidation of atomic mercury by hydroxyl radicals and photoinduced decomposition of methylmercury in the aqueous phase. Atmospheric Environment, 2001. 35(17): p. 3039–3047.

Gardfeldt, K. and M. Jonsson, Is bimolecular reduction of Hg(II) complexes possible in aqueous systems of environmental importance. Journal of Physical Chemistry A, 2003. 107(22): p. 4478–4482.

Gauchard, P.-A., et al., Study of the origin of atmospheric mercury depletion events recorded in Ny-Alesund, Svalbard, spring 2003. Atmospheric Environment, 2005. 39(39): p. 7620–7632.

Goodsite, M.E., J.M.C. Plane, and H. Skov, A theoretical study of the oxidation of Hg-0 to HgBr2 in the troposphere. Environmental Science & Technology, 2004. 38(6): p. 1772–1776.

Granite, E.J., H.W. Pennline, and R.A. Hargis, Novel sorbents for mercury removal from flue gas. Industrial & Engineering Chemistry Research, 2000. 39(4): p. 1020–1029.

Greig, G., H.E. Gunning, and O.P. Strausz, Reactions of Metal Atoms. II. The Combination of Mercury and Bromine Atoms and the Dimerization of HgBr. The Journal of Chemical Physics, 1970. 52(7): p. 3684–3690.

Grigal, D.F., Inputs and outputs of mercury from terrestrial watersheds: a review. Environmental Reviews, 2002. 10(1): p. 1.

Gustin, M. and S. Lindberg, Terrestial Hg Fluxes: Is the Next Exchange Up, Down, or Neither?, in Dynamics of Mercury Pollution on Regional and Global Scales:, N. Pirrone and K.R. Mahaffey, Editors. 2005. p. 241–259.

Gustin, M.S., et al., New insights into mercury exchange between air and substrate. Geochimica Et Cosmochimica Acta, 2005. 69(10): p. A700–A700.

Gustin, M.S., S.E. Lindberg, and M.A. Allan, Special Section: Constraining mercury emissions from naturally enriched surfaces: Assessment of methods and controlling parameters (Mercury Flux) - Preface. Journal of Geophysical Research-Atmospheres, 1999. 104(D17): p. 21829–21830.

Habibi, M.H., G. Habibian, and M.A. Haghighipor, Photocatalytic reduction and recovery of inorganic mercury compounds as environmental pollutants in aquatic system using TiO2 suspension. Fresenius Environmental Bulletin, 2003. 12(7): p. 808–812.

Hall, B., P. Schager, and E. Ljungström, An experimental study on the rate of reaction between mercury vapour and gaseous nitrogen dioxide. Water, Air, & Soil Pollution, 1995. 81(1): p. 121–134.

Hall, B., P. Schager, and J. Weesmaa, The homogeneous gas phase reaction of mercury with oxygen, and the corresponding heterogeneous reactions in the presence of activated carbon and fly ash. Chemosphere, 1995. 30(4): p. 611–627.

Hall, B., P. Schager, and O. Lindqvist, Chemical-Reactions of Mercury in Combustion Flue-Gases. Water Air and Soil Pollution, 1991. 56: p. 3–14.

Hall, B., The gas phase oxidation of elemental mercury by ozone. Water, Air, & Soil Pollution, 1995. 80(1): p. 301–315.

Hanson, P.J., et al., Foliar exchange of mercury vapor: Evidence for a compensation point. Water, Air, & Soil Pollution, 1995. 80(1): p. 373–382.

Hines, N.A. and P.L. Brezonik, Mercury dynamics in a small Northern Minnesota lake: water to air exchange and photoreactions of mercury. Marine Chemistry, 2004. 90(1-4): p. 137–149.

Horne, D.G., R. Gosavi, and O.P. Strausz, Reactions of Metal Atoms .I. Combination of Mercury and Chlorine Atoms and Dimerization of Hgcl. Journal of Chemical Physics, 1968. 48(10): p. 4758.

Horvath, O., E. Bodnar, and J. Hegyi, Photoassisted oxidative degradation of surfactants and simultaneous reduction of metals in titanium dioxide dispersions. Colloids and Surfaces a-Physicochemical and Engineering Aspects, 2005. 265(1-3): p. 135–140.

Hower, J.C., et al., Mercury Capture by Distinct Fly Ash Carbon Forms. Energy Fuels, 2000. 14(1): p. 224–226.

Hwang, J.Y., X. Sun, and Z. Li, Unburned Carbon from Fly Ash for Mercury Adsorption: Separation and Characterization of Unburned Carbon. The Journal of Minerals and Materials Characterization and Engineering, 2002. 1(1): p. 39–60.

Iverfeldt, A. and O. Lindqvist, Atmospheric Oxidation of Elemental Mercury by Ozone in the Aqueous Phase. Atmospheric Environment, 1986. 20(8): p. 1567–1573.

Jain, A. and M.A. Beg, Kinetics and mechanism of solid state reactions of silver tungstate with mercuric bromide and mercuric chloride. Polyhedron, 1995. 14(15-16): p. 2293–2299.

Joensuu, O.I., Fossil Fuels as a Source of Mercury Pollution. Science, 1971. 172(3987): p. 1027–1028.

Kaleschke, L., et al., Frost flowers on sea ice as a source of sea salt and their influence on tropospheric halogen chemistry. Geophysical Research Letters, 2004. 31(16): -.

Khalizov, A.F., et al., A theoretical study on the reactions of Hg with halogens: Atmospheric implications. Journal of Physical Chemistry A, 2003. 107(33): p. 6360–6365.

Lahoutifard, N., L. Poissant, and S.L. Scott, Scavenging of gaseous mercury by acidic snow at Kuujjuarapik, Northern Quebec. Science of The Total Environment, 2006. 355(1-3): p. 118–126.

Lalonde, J.D., A.J. Poulain, and M. Amyot, The role of mercury redox reactions in snow on snow-to-air mercury transfer. Environmental Science & Technology, 2002. 36(2): p. 174–178.

Lalonde, J.D., et al., Photo-induced Hg(II) reduction in snow from the remote and temperate Experimental Lakes Area (Ontario, Canada). Journal of Geophysical Research-Atmospheres, 2003. 108(D6): -.

Lalonde, J.D., et al., Photooxidation of Hg0 in artificial and natural waters. Environmental Science & Technology, 2001. 35(7): p. 1367–1372.

Lalonde, J.D., et al., Photoinduced oxidation of Hg-0 (aq) in the waters from the St. Lawrence estuary. Environmental Science & Technology, 2004. 38(2): p. 508–514.

Lee, X., G. Benoit, and X.Z. Hu, Total gaseous mercury concentration and flux over a coastal saltmarsh vegetation in Connecticut, USA. Atmospheric Environment, 2000. 34(24): p. 4205–4213.

Lee, T.G., P. Biswas, and E. Hedrick, Overall Kinetics of Heterogeneous Elemental Mercury Reactions on TiO2 Sorbent Particles with UV Irradiation. Ind. Eng. Chem. Res., 2004. 43(6): p. 1411–1417.

Lin, C.J. and S.O. Pehkonen, Aqueous free radical chemistry of mercury in the presence of iron oxides and ambient aerosol. Atmospheric Environment, 1997. 31(24): p. 4125–4137.

Lin, C.J. and S.O. Pehkonen, Oxidation of elemental mercury by aqueous chlorine (HOCl/OCl-): Implications for tropospheric mercury chemistry. Journal of Geophysical Research-Atmospheres, 1998. 103(D21): p. 28093–28102.

Lin, C.-J. and S.O. Pehkonen, The chemistry of atmospheric mercury: a review. Atmospheric Environment, 1999. 33(13): p. 2067–2079.

Li, X., et al., Functionalized titania nanoparticles for mercury scavenging. Journal of Materials Chemistry, 2007. 17(19): p. 2028–2032.

Li, Y. and C.-Y. Wu, Kinetic Study for Photocatalytic Oxidation of Elemental Mercury on a SiO2–TiO2 Nanocomposite. Environmental Engineering Science, 2007. 24(1): p. 3–12.

Li, Y., P. Murphy, and C.-Y. Wu, Removal of elemental mercury from simulated coal-combustion flue gas using a SiO2-TiO2 nanocomposite. Fuel Processing Technology. In Press, Corrected Proof.

Lindberg, S.E., et al., Atmosphere-Surface Exchange of Mercury in a Forest - Results of Modeling and Gradient Approaches. Journal of Geophysical Research-Atmospheres, 1992. 97(D2): p. 2519–2528.

Lindberg, S.E., S. Brooks, C-J Lin, K. J. Scott, M. S. Landis, R.K. Stevens, M.E. Goodsite and A. Richter, Dynamic Oxidation of Gaseous Mercury in the Arctic Troposphere at Polar Sunrise. Environ. Sci. Technol., 2002. 36(6): p. 1245–1256.

Lindberg, S., R. Bullock, R. Ebinghaus, D. Engstrom, X. Feng, W. Fitzgerald, N. Pirrone, E. Prestbo, and C. Seigneur, A Synthesis of Progress and Uncertainties in Attributing the Sources of Mercury in Deposition. AMBIO: A Journal of the Human Environment, 2007. 36(1): p. 19–33.

Lindqvist, O. and H. Rodhe, Atmospheric Mercury - a Review. Tellus Series B-Chemical and Physical Meteorology, 1985. 37(3): p. 136–159.

Macdonald, R.W., T. Harner, and J. Fyfe, Recent climate change in the Arctic and its impact on contaminant pathways and interpretation of temporal trend data. Science of The Total Environment, 2005. 342(1-3): p. 5–86.

Mason, R.P., F.M.M. Morel, and W.F. Fitzgerald, The biogeochemical cycling of elemental mercury: Anthropogenic influences. Geochimica et Cosmochimica Acta ; Vol/Issue: 58:15, 1994: Pages: 3191–3198.

Mason, R.P., F.M.M. Morel, and H.F. Hemond, The Role of Microorganisms in Elemental Mercury Formation in Natural-Waters. Water Air and Soil Pollution, 1995. 80(1-4): p. 775–787.

Mason, R.P. and G.R. Sheu, Role of the ocean in the global mercury cycle. Global Biogeochemical Cycles, 2002. 16(4).

Mason, R.P., N.M. Lawson, and G.R. Sheu, Mercury in the Atlantic Ocean: factors controlling air-sea exchange of mercury and its distribution in the upper waters. Deep-Sea Research Part Ii-Topical Studies in Oceanography, 2001. 48(13): p. 2829–2853.

Medhekar, A.K., et al., Surface catalyzed reaction of Hg + Cl2. Chemical Physics Letters, 1979. 65(3): p. 600–604.

Menke, R. and G. Wallis, Detection of mercury in air in the presence of chlorine and water vapor. American Industrial Hygiene Association Journal, 1980. 41(2): p. 120 – 124.

Mohan, D., et al., Kinetics of mercury adsorption from wastewater using activated carbon derived from fertilizer waste. Colloids and Surfaces A: Physicochemical and Engineering Aspects, 2000. 177(2-3): p. 169–181.

Morel, F.M.M., A.M.L. Kraepiel, and M. Amyot, The chemical cycle and bioaccumulation of mercury. Annual Review of Ecology and Systematics, 1998. 29: p. 543–566.

Muller, M.D., et al., Neural network scheme for the retrieval of total ozone from Global Ozone Monitoring Experiment data. Applied Optics, 2002. 41(24): p. 5051–5058.

Munthe, J., Z.F. Xiao, and O. Lindqvist, The Aqueous Reduction of Divalent Mercury by Sulfite. Water Air and Soil Pollution, 1991. 56: p. 621–630.

Munthe, J. and W.J. Mcelroy, Some Aqueous Reactions of Potential Importance in the Atmospheric Chemistry of Mercury. Atmospheric Environment Part a-General Topics, 1992. 26(4): p. 553–557.

Munthe, J., The Aqueous Oxidation of Elemental Mercury by Ozone. Atmospheric Environment Part a-General Topics, 1992. 26(8): p. 1461–1468.

Namasivayam, C. and K. Kadirvelu, Uptake of mercury (II) from wastewater by activated carbon from an unwanted agricultural solid by-product: coirpith. Carbon, 1999. 37(1): p. 79–84.

Niki, H., et al., A Long-Path Fourier-Transform Infrared Study of the Kinetics and Mechanism for the Ho-Radical Initiated Oxidation of Dimethylmercury. Journal of Physical Chemistry, 1983. 87(24): p. 4978–4981.

Norton, G.A., et al., Heterogeneous oxidation of mercury in simulated post combustion conditions. Fuel, 2003. 82(2): p. 107–116.

Nriagu, J.O., Mercury Pollution from the Past Mining of Gold and Silver in the America. Science of the Total Environment, 1994. 149(3): p. 167–181.

O'Driscoll, N.J., et al., Gross photoreduction kinetics of mercury in temperate freshwater lakes and rivers: Application to a general model of DGM dynamics. Environmental Science & Technology, 2006. 40(3): p. 837–843.

Oltmans, S.J., et al., Seasonal Surface Ozone and Filterable Bromine Relationship in the High Arctic. Atmospheric Environment, 1989. 23(11): p. 2431–2441.

P'yankov, V.A., Kinetics of the Reaction of Mercury Vapors with Ozone. Journal of General Chemistry of USSR. , 1949. 19: p. 187–192.

Pal, B. and P.A. Ariya, Gas-phase HO center dot-Initiated reactions of elemental mercury: Kinetics, product studies, and atmospheric implications. Environmental Science and Technology, 2004. 38(21): p. 5555–5566.

Pal, B. and P.A. Ariya, Studies of ozone initiated reactions of gaseous mercury: kinetics, product studies, and atmospheric implications. Physical Chemistry Chemical Physics, 2004. 6(3): p. 572–579.

Pavlish, J.H., et al., Status review of mercury control options for coal-fired power plants. Fuel Processing Technology, 2003. 82(2-3): p. 89–165.

Pehkonen, S.O. and C.J. Lin, Aqueous photochemistry of mercury with organic acids. Journal of the Air & Waste Management Association, 1998. 48(2): p. 144–150.

Peretyazhko, T., et al., Formation of dissolved gaseous mercury in a tropical lake (Petit-Saut reservoir, French Guiana). Science of The Total Environment, 2006. 364(1-3): p. 260–271.

Peterson, K.A., B.C. Shepler, and J.M. Singleton, The group 12 metal chalcogenides: an accurate multireference configuration interaction and coupled cluster study. Molecular Physics, 2007. 105(9): p. 1139–1155.

Peterson, K.A., Systematically convergent basis sets with relativistic pseudopotentials. I. Correlation consistent basis sets for the post-d group 13-15 elements. Journal of Chemical Physics, 2003. 119(21): p. 11099–11112.

Philippidis, G.P., et al., Effect of Gene Amplification on Mercuric Ion Reduction Activity of Escherichia-Coli. Applied and Environmental Microbiology, 1991. 57(12): p. 3558–3564.

Poissant, L., et al., Atmospheric mercury speciation and deposition in the Bay St. Francois wetlands. Journal of Geophysical Research-Atmospheres, 2004. 109(D11): -.

Poulain, A.J., et al., Biological and Chemical Redox Transformations of Mercury in Fresh and Salt Waters of the High Arctic during Spring and Summer. Environ. Sci. Technol., 2007. 41(6): p. 1883–1888.

Poulain, A.J., et al., Mercury distribution, partitioning and speciation in coastal vs. inland High Arctic snow. Geochimica Et Cosmochimica Acta, 2007c. 71(14): p. 3419–3431.

Poulain, A.J., et al., Potential for mercury reduction by microbes in the high arctic. Applied and Environmental Microbiology, 2007. 73(7): p. 2230–2238.

Poulain, A.J., V. Roy, and M. Amyot, Influence of temperate mixed and deciduous tree covers on Hg concentrations and photoredox transformations in snow. Geochimica Et Cosmochimica Acta, 2007b. 71(10): p. 2448–2462.

Prairie, M.R., et al., An Investigation of Tio2 Photocatalysis for the Treatment of Water Contaminated with Metals and Organic-Chemicals. Environmental Science & Technology, 1993. 27(9): p. 1776–1782.

Presto, A.A. and E.J. Granite, Survey of Catalysts for Oxidation of Mercury in Flue Gas. Environ. Sci. Technol., 2006. 40(18): p. 5601–5609.

Presto, A.A., et al., A kinetic approach to the catalytic oxidation of mercury in flue gas. Energy & Fuels, 2006. 20(5): p. 1941–1945.

Querol, X., et al., Synthesis of Na-zeolites from fly ash. Fuel, 1997. 76(8): p. 793–799.

Ranganathan, K., Adsorption of Hg(II) ions from aqueous chloride solutions using powdered activated carbons. Carbon, 2003. 41(5): p. 1087–1092.

Raofie, F. and P.A. Ariya, Kinetics and product study of the reaction of BrO radicals with gaseous mercury. Journal De Physique IV, 2003. 107: p. 1119–1121.

Raofie, F. and P.A. Ariya, Product study of the gas-phase BrO-initiated oxidation of Hg-0: evidence for stable Hg1+ compounds. Environmental Science and Technology, 2004. 38(16): p. 4319–4326.

Raofie, F., G. Snider, and P.A. Ariya, The Reaction of Gaseous Mercury with Molecular Iodine, Atomic Iodine and Iodine Oxide Radicals: Kinetics, Product Studies, and the Atmospheric Implication. Canadian Journal of Chemistry, 2008. (accepted).

Richter, A., et al., GOME measurements of stratospheric and tropospheric BrO. Remote Sensing of Trace Constituents in the Lower Stratosphere, Troposphere and the Earth's Surface: Global Observations, Air Pollution and the Atmospheric Correction, 2002. 29(11): p. 1667–1672.

Richter, A., et al., GOME observations of tropospheric BrO in northern hemispheric spring and summer 1997. Geophysical Research Letters, 1998. 25(14): p. 2683–2686.

Rodríguez, S., et al., A mechanistic model for mercury capture with in situ-generated titania particles: role of water vapor. J Air Waste Manag Assoc. , 2004. 54(2): p. 149–156.

Rolfhus, K.R., et al., Distribution and Fluxes of Total and Methylmercury in Lake Superior. Environ. Sci. Technol., 2003. 37(5): p. 865–872.

Rolfhus, K.R. and W.F. Fitzgerald, Mechanisms and temporal variability of dissolved gaseous mercury production in coastal seawater. Marine Chemistry, 2004. 90(1-4): p. 125–136.

Roy, S. and G.T. Rochelle, Simultaneous absorption of mercury and chlorine in sulfite solutions. Chemical Engineering Science, 2004. 59(6): p. 1309–1323.

Saiz-Lopez, A., et al., Boundary layer halogens in coastal Antarctica. Science, 2007. 317(5836): p. 348–351.

Sander, R., Compilation of Henry's Law Constants for Inorganic and Organic Species of Potential Importance in Environmental Chemistry (Version 3). 1999.

Schlüter, K., Review: evaporation of mercury from soils. An integration and synthesis of current knowledge. Environmental Geology, 2000. 39(3): p. 249–271.

Schofield, K., Let them eat fish: hold the mercury. Chemical Physics Letters, 2004. 386(1-3): p. 65–69.

Schroeder, W., G. Yarwood, and H. Niki, Transformation processes involving mercury species in the atmosphere — results from a literature survey. Water, Air, & Soil Pollution, 1991. 56(1): p. 653–666.

Schroeder, W.H., G. Yarwood, and H. Niki, Transformation Processes Involving Mercury Species in the Atmosphere - Results from a Literature Survey. Water Air and Soil Pollution, 1991. 56: p. 653–666.

Schroeder, W.H. and J. Munthe, Atmospheric mercury-An overview. Atmospheric Environment, 1998. 32(5): p. 809–822.

Schroeder, W. H., Anlauf, K. G., Barrie, L. A., Lu, J. Y., Steffen, A., Schneeberger, D. R., and Berg, T.:, Arctic springtime depletion of mercury. Nature, 1998. 394(6691): p. 331–332.

Seig, K., Adsorption of Hg(II) ions from aqueous chloride solutions using powdered activated carbons. Carbon, 2003. 41(5): p. 1087–1092.

Seigneur, C., W. Jacek, and C. Elpida, A Chemical Kinetic Mechanism for Atmospheric Inorganic Mercury. Environ. Sci. Technol, 1994. 28(9): p. 1589–1597.

Seiler, W., C. Eberling, and F. Slemr, Global Distribution of Gaseous Mercury in the Troposphere. Pure and Applied Geophysics, 1980. 118(4): p. 964–974.

Sen, A.K. and A.K. De, Adsorption of Hg(II) by coal fly ash. Water Research, 1987. 21(8): p. 885–888.

Senior, C.L., et al., Gas-phase transformations of mercury in coal-fired power plants. Fuel Processing Technology, 2000. 63(2-3): p. 197–213.

Shepler, B.C. and K.A. Peterson, Mercury Monoxide: A Systematic Investigation of Its Ground Electronic State. J. Phys. Chem. A, 2003. 107(11): p. 1783–1787.

Shepler, B.C., et al., Aqueous Microsolvation of Mercury Halide Species. J. Phys. Chem. A, 2007b. 111(44): p. 11342–11349.

Shepler, B.C., N.B. Balabanov, and K.A. Peterson, *Ab Initio Thermochemistry Involving Heavy Atoms: An Investigation of the Reactions Hg + IX (X = I, Br, Cl, O)*. J. Phys. Chem. A, 2005. 109(45): p. 10363–10372.

Shepler, B.C., N.B. Balabanov, and K.A. Peterson, *Hg plus Br à HgBr recombination and collision-induced dissociation dynamics*. Journal of Chemical Physics, 2007. 127(16).

Sheu, G.-R. and R.P. Mason, An Examination of the Oxidation of Elemental Mercury in the Presence of Halide Surfaces. Journal of Atmospheric Chemistry, 2004. 48(2): p. 107–130.

Siciliano, S.D., N.J. O'Driscoll, and D.R.S. Lean, Microbial reduction and oxidation of mercury in freshwater lakes. Environmental Science & Technology, 2002. 36(14): p. 3064–3068.

Skare, I. and R. Johansson, Reactions between mercury vapor and chlorine gas at occupational exposure levels. Chemosphere, 1992. 24(11): p. 1633–1644.

Skodras, G., I. Diamantopoujou, and G.P. Sakellaropoujos, Role of activated carbon structural properties and surface chemistry in mercury adsorption. Desalination, 2007. 210(1-3): p. 281–286.

Skubal, L.R. and N.K. Meshkov, Reduction and removal of mercury from water using arginine-modified TiO2. Journal of Photochemistry and Photobiology a-Chemistry, 2002. 148(1-3): p. 211–214.

Slemr, F., G. Schuster, and W. Seiler, Distribution, speciation, and budget of atmospheric mercury. Journal of Atmospheric Chemistry, 1985. 3(4): p. 407–434.

Sliger, R.N., J.C. Kramlich, and N.M. Marinov, Towards the development of a chemical kinetic model for the homogeneous oxidation of mercury by chlorine species. Fuel Processing Technology, 2000. 65-66: p. 423–438.

Snider, G., F. Raofie, and P.A. Ariya, Effects of Relative Humidity and CO(g) on the O3-initiated Oxidation Reaction of Hg0(g): Kinetic & product studies. Physical Chemistry Chemical Physics, 2008. (accepted, with revisions).

Sommar, J., et al., On the Gas Phase Reactions Between Volatile Biogenic Mercury Species and the Nitrate Radical. Journal of Atmospheric Chemistry, 1997. 27(3): p. 233–247.

Sommar, J., et al., A kinetic study of the gas-phase reaction between the hydroxyl radical and atomic mercury. Atmospheric Environment, 2001. 35(17): p. 3049–3054.

Southworth, G., et al., EVASION OF ADDED ISOTOPIC MERCURY FROM A NORTHERN TEMPERATE LAKE. Environmental Toxicology and Chemistry, 2007. 26(1): p. 53–60.

St.Louis, V.L., et al., Methylated Mercury Species in Canadian High Arctic Marine Surface Waters and Snowpacks. Environ. Sci. Technol., 2007. 41(18): p. 6433–6441.

Steffen, A., et al., Atmospheric mercury concentrations: measurements and profiles near snow and ice surfaces in the Canadian Arctic during Alert 2000. Atmospheric Environment, 2002. 36(15-16): p. 2653–2661.

Steffen, A., et al., A synthesis of atmospheric mercury depletion event chemistry linking atmosphere, snow and water. Atmos. Chem. Phys. Discuss., 2007. 7(4): p. 10837–10931.

Sumner, A., et al., Environmental Chamber Studies of Mercury Reactions in the Atmosphere, in Dynamics of Mercury Pollution on Regional and Global Scales:. 2005. p. 193–212.

Tackett, P.J., et al., A study of the vertical scale of halogen chemistry in the Arctic troposphere during Polar Sunrise at Barrow, Alaska. Journal of Geophysical Research-Atmospheres, 2007. 112(D7): P.-.

Tokos, J.J.S., et al., Homogeneous gas-phase reaction of Hg[degree sign] with H2O2, 03, CH3I, AND (CH3)2S: Implications for atmospheric Hg cycling. Atmospheric Environment, 1998. 32(5): p. 823–827.

Tossell, J.A., Calculation of the Energetics for Oxidation of Gas-Phase Elemental Hg by Br and BrO. J. Phys. Chem. A, 2003. 107(39): p. 7804–7808.

Turchi, C.S., Novel Process for Removal and Recovery of Vapor-Phase Mercury, in Other Information: PBD: 29 Sep 2000. 2000, ADA Technologies, Inc. : Littleton, CO. p. Size: 57 pages.

Van Roozendael, M., et al., Intercomparison of BrO measurements from ERS-2 GOME, ground-based and balloon platforms. Remote Sensing of Trace Constituents in the Lower Stratosphere, Troposphere and the Earth's Surface: Global Observations, Air Pollution and the Atmospheric Correction, 2002. 29(11): p. 1661–1666.

Vidic, R.D., M.T. Chang, and R.C. Thurnau, Kinetics of vapor-phase mercury uptake by virgin and sulfur-impregnated activated carbons. Journal of the Air & Waste Management Association, 1998. 48(3): p. 247–255.

Wang, R.G., M.A. Dillon, and D. Spence, A phenomenological study of heterogeneous chemical reactions of mercuric chloride on heated stainless steel surfaces[sup a)]. The Journal of Chemical Physics, 1983. 79(2): p. 1100–1101.

Wang, Z. and S.O. Pehkonen, Oxidation of elemental mercury by aqueous bromine: atmospheric implications. Atmospheric Environment, 2004. 38(22): p. 3675–3688.

Whalin, L.M. and R.P. Mason, A new method for the investigation of mercury redox chemistry in natural waters utilizing deflatable Teflon(R) bags and additions of isotopically labeled mercury. Analytica Chimica Acta, 2006. 558(1-2): p. 211–221.

Widmer, N.C., et al., Practical Limitation of Mercury Speciation in Simulated Municipal Waste Incinerator Flue Gas. Combustion Science and Technology, 1998. 134(1): p. 315 – 326.

Wilcox, J., et al., Theoretically Predicted Rate Constants for Mercury Oxidation by Hydrogen Chloride in Coal Combustion Flue Gases. Environ. Sci. Technol., 2003. 37(18): p. 4199–4204.

Wu, S., M. Azhar Uddin, and E. Sasaoka, Characteristics of the removal of mercury vapor in coal derived fuel gas over iron oxide sorbents. Fuel, 2006. 85(2): p. 213–218.

Wu, S., et al., Development of iron-based sorbents for Hg0 removal from coal derived fuel gas: Effect of hydrogen chloride. Fuel, 2008. 87(4-5): p. 467–474.

Xiao, Z.F., D. Stromberg, and O. Lindqvist, Influence of Humic Substances on Photolysis of Divalent Mercury in Aqueous-Solution. Water Air and Soil Pollution, 1995. 80(1-4): p. 789–798.

Xin, M. and M.S. Gustin, Gaseous elemental mercury exchange with low mercury containing soils: Investigation of controlling factors. Applied Geochemistry, 2007. 22(7): p. 1451–1466.

Xu, M., et al., Kinetic calculation and modeling of trace element reactions during combustion. Powder Technology, 2008. 180(1-2): p. 157–163.

Yamamoto, M., Stimulation of elemental mercury oxidation in the presence of chloride ion in aquatic environments. Chemosphere, 1996. 32(6): p. 1217–1224.

Yardim, M.F., et al., Removal of mercury (II) from aqueous solution by activated carbon obtained from furfural. Chemosphere, 2003. 52(5): p. 835–841.

Zepp, R.G., J. Hoigne, and H. Bader, Nitrate-Induced Photooxidation of Trace Organic-Chemicals in Water. Environmental Science & Technology, 1987. 21(5): p. 443–450.

Zhang, H. and S.E. Lindberg, Sunlight and iron(III)-induced photochemical production of dissolved gaseous mercury in freshwater. Environmental Science & Technology, 2001. 35(5): p. 928–935.

Zhang, H., Photochemical Redox Reactions of Mercury, in Recent Developments in Mercury Science. 2006. p. 37–79.

Zhang, L. and M.H. Wong, Environmental mercury contamination in China: Sources and impacts. Environment International, 2007. 33(1): p. 108–121.

Zhao, L.L. and G.T. Rochelle, Mercury absorption in aqueous hypochlorite. Chemical Engineering Science, 1999. 54(5): p. 655–662.

Zhao, Y., et al., Application of Gold Catalyst for Mercury Oxidation by Chlorine. Environ. Sci. Technol., 2006. 40(5): p. 1603–1608.

Zheng, C., et al., Kinetic mechanism studies on reactions of mercury and oxidizing species in coal combustion. Fuel, 2005. 84(10): p. 1215–1220.

Chapter 16
Importance of a Global Scale Approach to using Regional Models in the Assessment of Source-Receptor Relationships for Mercury

O. Russell Bullock Jr. and Lyatt Jaeglé

Summary Regional atmospheric models simulate their pertinent processes over a limited portion of the global atmosphere. This portion of the atmosphere can be a large fraction, as in the case of continental-scale modeling, or a small fraction, as in the case of urban-scale modeling. Regional modeling of any air pollutant requires that the meteorological and chemical conditions at the boundaries of the model domain be taken into account, especially if the pollutants are long-lived. It was once a common practice for the boundary concentrations of mercury and its reactants to be specified using time-constant values based on limited historical observation. These values were often invariant in the horizontal and vertical dimensions too. This relatively simple procedure for establishing boundary concentrations could be justified based on the previous notion that mercury was a rather inert and long-lasting air pollutant. However, with the subsequent discovery of rapid physical and chemical transformations of atmospheric mercury and significant concentrations of oxidized mercury far removed from known emission sources, the presumption of mercury as an inert substance has generally disappeared. The effect of intercontinental transport is now treated with greater concern in regional atmospheric mercury modeling. Global mercury models are now commonly used to define boundary values for regional mercury modeling. However, the global and regional models must use consistent information for emissions, surface physiology and meteorology to achieve consistent simulation results and associated source-receptor relationships. There is certainly a need for international cooperation on field research and numerical model development to supply the tools needed for confident assessment of source-receptor relationships for mercury on both global and regional scales.

16.1 Introduction

Regional models are typically used for atmospheric modeling when fine spatial detail is desired over a specific area of interest and there is insufficient computing power available to calculate the model over the entire global extent of the atmosphere.

N. Pirrone and R. Mason (eds.), *Mercury Fate and Transport in the Global Atmosphere*,
DOI: 10.1007/978-0-387-93958-2_16, © Springer Science+Business Media, LLC 2009

Regional models are also used when there is a lack of necessary information outside of the region of interest. Regardless of motivation, regional models have been used to identify and assess the sources responsible for a variety of atmospheric contaminants including mercury. However, regional models typically require a specification of conditions at the lateral boundaries of their model domains. When these lateral boundary conditions are found to be important to the simulation of conditions within the model domain, a compromise must be made between model fidelity and model practicality. In order to make the simulations more realistic and credible, the model domain must be expanded to the largest extent possible while still allowing the necessary model calculations to be perfomed.

Some examples of regional models that have been used to simulate atmospheric Hg are the Acid Deposition and Oxidant Model (ADOM) (Petersen et al., 2001), the Community Multiscale Air Quality model (CMAQ) (Bullock and Brehme, 2002), the Danish Eulerian Hemispheric Model (DEHM) (Christensen, 1997), the Eulerian Model for Air Pollution (EMAP) (Syrakov, 1995; Ilyin et al., 2002), the Regional Modeling System for Aerosols and Deposition (REMSAD) (ICF, 2005), the Trace Element Analysis Model (TEAM) (Pai et al., 1997) and the Hybrid Single Particle Lagrangian Integrated Trajectory model (HYSPLIT) (Cohen et al., 2004). However, the HYSPLIT model uses a Lagrangian modeling framework to track pollutant emissions from within its modeling domain and does not treat boundary fluxes of mercury. The following discussion deals only with models using Eulerian frameworks where boundary fluxes are considered in the treatment of global transport.

Mercury (Hg) is known to be a global pollutant due to the long atmospheric lifetime of gaseous elemental Hg (GEM). As a global pollutant, trans-boundary flows of Hg into regional modeling domains are an obvious concern. The mean atmospheric lifetime of GEM is believed to be on the order of months or longer. This long lifetime implies slow deposition where only a small fraction of the GEM transported into a regional model domain is likely to deposit before it is eventually transported out. For this reason, some early modeling assessments of source attribution and trans-national-boundary flux were conducted using rather simple estimates of the boundary concentrations of GEM based on the assumption that they have little effect on the simulated deposition and source-receptor relationships of Hg (US EPA, 1997; Bullock, 2000). However, two additional Hg species are measurable in ambient air, namely reactive gaseous Hg (RGM) and particulate Hg Hg(p), and both are known to deposit from the atmosphere at a significant rate. Boundary values for RGM and Hg(p) concentrations have been a concern since the early days of atmospheric mercury modeling, but they were often assumed to be negligible at regional boundaries far removed from industrial sources. Since then, rapid atmospheric conversion of GEM to oxidized forms was first reported by (Schroeder et al., 1998) and has been reported subsequently at a number of other polar and circumpolar locations. Evidence of significant concentrations of Hg(p) at high altitude has also been reported (Murphy et al., 2006). Because of these recently discovered opportunities for the creation and long-range transport of oxidized forms of Hg, boundary concentrations for all three measurable Hg species have become a greater concern for regional modeling of atmospheric Hg.

To address these concerns, global modeling with relatively coarse spatial resolution has been used as a preliminary step to define lateral boundary conditions for fine-resolution regional Hg modeling (Seigneur et al., 2001; Seigneur et al., 2004; US EPA, 2005). This technique can provide more realistic, time-resolved boundary information to regional models. Of course, global models are subject to the same uncertainties regarding their simulation of emission, transformation and deposition of Hg as are regional models.

Basic scientific uncertainties about the true behavior of atmospheric Hg are discussed in previous chapters of this volume. As discussed below, boundary concentrations derived from global models can differ significantly, as can the responses of the regional model when exposed to these differing boundary conditions.

16.2 Previous Testing and Application

Results from a study of atmospheric Hg models conducted by the Meteorological Synthesizing Centre - East (MSC-E) from 2000 to 2005 provide a good example of how regional modeling can be used to assess source-receptor relationships (Ryaboshapko et al., 2007). Three individual countries roughly equal in size but located in different parts of Europe were selected by MSC-E for source-receptor calculations: the UK, Poland and Italy.

Figure 16.1 (from Ryaboshapko et al., 2007) shows the range of model simulation results for total Hg deposition to these three countries during February 1999, August 1999 and for the entire year of 1999. The regional model simulations

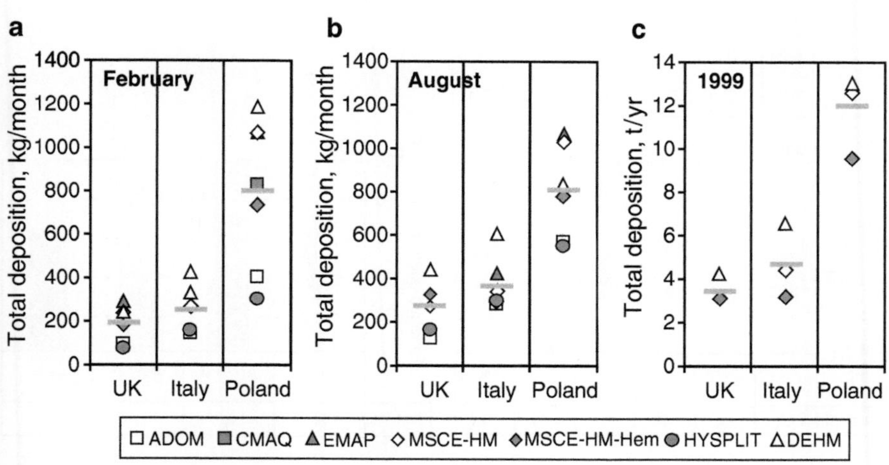

Figure 16.1 Total Hg deposition to the UK, Italy and Poland: (**a**) in February 1999; (**b**) in August 1999; and (**c**) in the whole year 1999. Symbols depict modelling results, light grey dashes show the ensemble average

were also used to calculate four parameters: (NAS) deposition caused by National Anthropogenic emission Sources of a given country; (EAS) deposition caused by all European Anthropogenic emission Sources except the anthropogenic sources of a given country; (GNR) deposition caused by Global anthropogenic sources (excluding European sources), Natural sources and Re-emission; and (ROF) Relative Out-Flow determined as a fraction of national anthropogenic emissions transported outside a given country. Two models (ADOM and CMAQ) stored their output data in such a way that EAS and GNR could not be calculated separately and had to be combined into a single parameter. These source-receptor parameters as calculated for Poland are shown in Figure 16.2 (from Ryaboshapko et al., 2007).

The range of results shown here for Poland is typical of those found for the U.K. and Italy. The differences in the simulated monthly and annual deposition amounts to each of the three countries are certainly significant, as are the differences in source attribution and relative outflow parameters. Because the regional models applied in this first Hg model intercomparison study were free to use different data for meteorology, boundary concentrations, and concentrations of Hg reactants within their individual modeling domains, it was not possible to determine if the simulation differences were primarily due to variations in basic process modeling or if they were simply due to the use of different boundary concentrations or other types of input data.

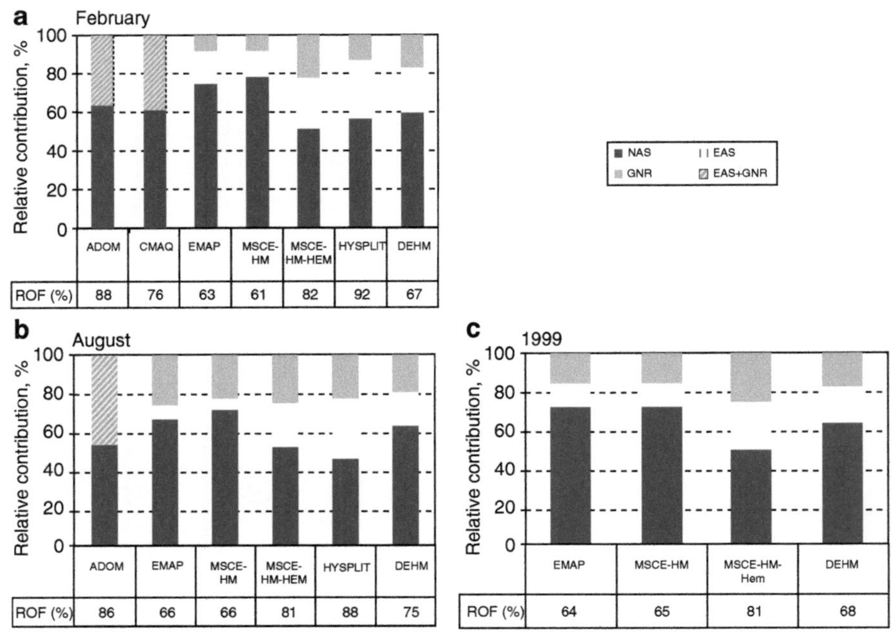

Figure 16.2 Contribution of national anthropogenic (NAS), European anthropogenic (EAS), and global, natural and re-emission sources (GNR) to Hg deposition over Poland: (**a**) in February 1999; (**b**) in August 1999; and (**c**) in the whole year 1999. Relative output flow (ROF) is also presented for each country

16.2.1 Introduction of Dynamic Global Modeling for Boundary Conditions

Concurrently with the MSC-E study, researchers at Atmospheric and Environmental Research, Inc., (AER) developed a model nesting approach for Hg using a global chemical transport model (CTM-Hg) and an imbedded regional model covering North America (Seigneur et al., 2001). The global CTM-Hg described in Shia et al. (1999) provided a horizontal resolution of 8° latitude and 10° longitude and a vertical resolution of 7 layers between the Earth's surface and ~12 km altitude and 2 layers between ~12 km and 30 km altitude. The regional model employed was the Trace Element Analysis Model (TEAM) described in Pai et al. (1997). It provided a horizontal resolution of 100 km and a vertical resolution of 6 layers between the Earth's surface and 6 km altitude. Both models were used to simulate the year of 1998. Figure 16.3 (from Seigneur et al., 2001) shows a comparison of annual average total gaseous Hg (TGM) as simulated by the TEAM along with available measurements for 6 U.S. states.

Because TGM includes both Hg^0 and RGM, knowledge about its concentration in air provides only limited guidance for estimating Hg deposition flux. Atmospheric concentrations of RGM are usually a small but highly variable fraction (0.1-10%) of the TGM concentration and RGM is much more readily deposited than Hg^0. Figure 16.4 (from Seigneur et al., 2001) shows a comparison of simulated and observed annual wet deposition fluxes of Hg for 12 U.S. states. The observations were obtained from the Mercury Deposition Network (MDN) described by (Vermette et al., 1995).

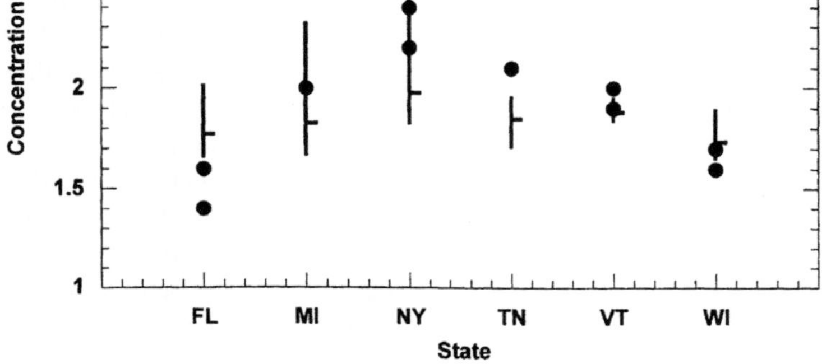

Figure 16.3 Comparison of simulated annual-average gas-phase mercury concentrations (*ng m^{-3}*) with measured data for 1998

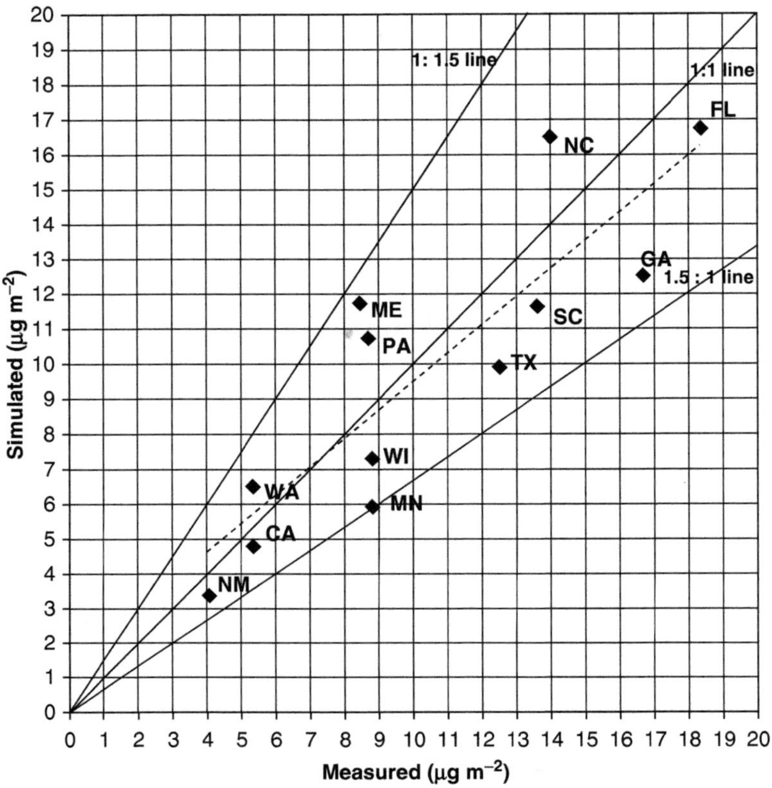

Figure 16.4 Comparison of simulated and measured 1998 wet deposition fluxes ($\mu g\ m^{-2}$) of total mercury over the United States (CA, California; FL, Florida; GA, Georgia; ME, Maine; MN, Minnesota; NC, North Carolina; NM, New Mexico; PA, Pennsylvania; SC, South Carolina; TX, Texas; WA, Washington; WI, Wisconsin)

The observed values shown are the average of all MDN observations for each state where 8 of the 12 states had 2 or more MDN observations. These results certainly demonstrated some modeling skill for Hg wet deposition. However, as discussed in Seigneur et al. (2001), a number of uncertainties affected these modeling results. Some of those uncertainties had to do with global "background" emissions and atmospheric transformations of Hg during intercontinental transport. The lack of information about dry deposition of the Hg species simulated was also identified as a source of uncertainty. Nonetheless, this first integration of regional and global modeling was hailed as an advance towards a better modeling approach for atmospheric Hg. However, the models themselves were recognized to be deficient with regard to scientific process information for Hg and this deficiency remains an issue in atmospheric Hg modeling today.

Another example of using global models to provide boundary data for regional modeling of atmospheric mercury is modeling performed by the U.S. EPA to investigate proposed reductions in mercury emissions from coal-fired electric generating units (EGUs) within the U.S. For this analysis, the CMAQ model was used to simulated the year 2001 with and without future Hg emission reductions. Input data for initial and boundary concentrations of Hg and other pertinent air pollutants were derived from the global GEOS-Chem model (Selin et al., 2007). This approach has now been adopted by the U.S. EPA and it represents an advance from earlier CMAQ modeling of Hg where time-invariant boundary values were used (Bullock and Brehme, 2002).

To evaluate the accuracy of the CMAQ 2001 simulation, results for annual wet deposition of total Hg were compared to observations from the MDN in much the same way as Seigneur et al. (2001) had done for the TEAM model. However, many more MDN observations were available for the 2001 time frame of the US EPA study. Also, each individual observation was compared to its corresponding model result without any state-by-state averaging. The results of this comparison are shown in Figure 16.5. Discounting one comparison for the MDN station in British Columbia (BC) where the model input data for precipitation were found to be grossly inaccurate, the CMAQ base-case simulation resolved about 60% of the observed variance in annual Hg wet deposition. Figure 16.6 shows the simulated reduction in Hg deposition from all Hg emission controls expected to be in force by 2020. Significant reductions in Hg deposition are confined to a rather small fraction of the continental U.S. This suggests that much of the Hg deposition in the relatively unaffected areas might be attributable to Hg transported into the modeling domain. The following section

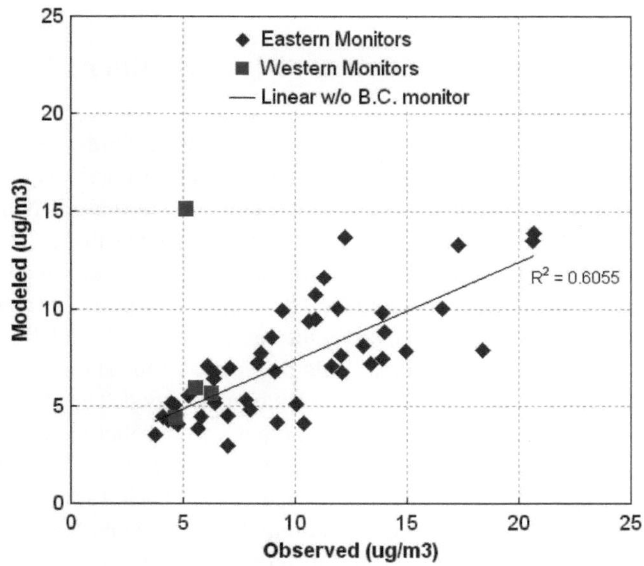

Figure 16.5 Comparison of CMAQ-simulated total Hg wet deposition for 2001 to observations from the Mercury Deposition Network (MDN). The result for the MDN monitor in British Columbia (B.C.) is discounted in the calculation of the least-squares linear fit and the R^2 statistic because of errors in the precipitation input data for that location

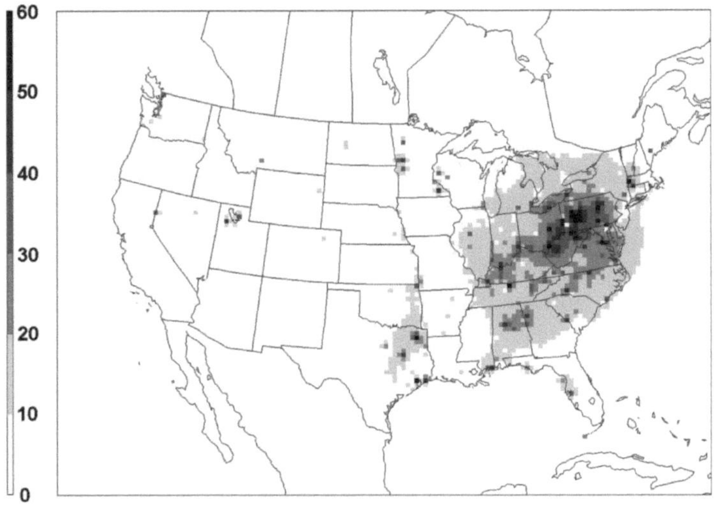

Figure 16.6 CMAQ-simulated results for the percent reduction in mercury deposition by 2020 when proposed regulations are fully implemented

discusses an investigation of the importance of boundary fluxes of Hg to simulations with the CMAQ model and two other regional models.

16.3 Testing Model Sensitivities to Intercontinental Transport

The North American Mercury Model Intercomparison Study (NAMMIS) was organized by the US EPA to apply regional-scale atmospheric Hg models in a tightly constrained testing environment with a modeling domain in North America where standardized measurements of Hg wet deposition are available (Bullock et al., 2008a; 2008b). The intent was to have all regional models in the study use exactly the same input data for initial and boundary conditions, meteorology and emissions, and to have all models applied to exactly the same horizontal modeling domain so that the effects of differing input data could be reduced, thus allowing the effects of differing scientific process treatments to be better understood. The ultimate goal was to determine which scientific process uncertainties are contributing most to observed discrepancies in model simulations of Hg deposition. Three regional atmospheric Hg models were applied: CMAQ, REMSAD, and TEAM. Each was applied to simulate the entire year of 2001 using three different initial condition and boundary condition (IC/BC) data sets, each developed from a different global model. The global models used were the CTM-Hg and the GEOS-Chem models (both described above) and the Global/Regional Atmospheric Heavy Metals (GRAHM) model described in Dastoor and Larocque (2004). All regional modeling results were compared to observed Hg wet deposition data from the MDN and special event-based monitoring at the Proctor Maple Research Center (PMRC) near Underhill, VT (Keeler et al., 2005).

The global models were not the primary subjects of the study, but they did exhibit significant differences in their simulated air concentrations of Hg^0, RGM and particulate Hg at the lateral boundaries of the regional modeling domain for the NAMMIS. Thus, the three IC/BC data sets provided an opportunity to investigate the effect of uncertainty regarding intercontinental transport of Hg species. Figure 16.7 shows the annual average air concentrations of Hg^0, RGM and particulate Hg across the western boundary of the regional domain as determined from the CTM, GEOS-Chem and GRAHM global simulations. While the patterns of Hg^0 air concentration are somewhat different, it is unlikely that these differences would affect the resulting deposition pattern for total Hg across the regional model domain since Hg^0 is deposited quite slowly under most conditions. However, the differences in RGM and particulate Hg concentration patterns may very well affect the total-Hg deposition pattern and the modeling accuracy as compared to observations. Figure 16.8 shows the resulting R^2 correlation statistic for simulated annual Hg wet deposition compared to observed values for each of the regional model simulations conducted in the NAMMIS. Figure 16.9 shows the mean fractional bias for each simulation while Figure 16.10 shows the mean fractional error. In addition to model-to-model differences in these statistics, these figures also show a considerable sensitivity of all three models to the IC/BC data sets used.

The US EPA has performed additional qualitative analysis of the CMAQ modeling results for Hg wet deposition obtained from the three IC/BC data sets from the NAMMIS (Bullock, 2008). This analysis looked at the concentration patterns of Hg^0, RGM and particulate Hg along all four lateral boundaries for each IC/BC set. It was shown that Hg^0 concentrations at the boundaries had a slight and somewhat negative correlation to the magnitude of the wet deposition flux, both at locations near the boundaries and within the interior of the modeling domain.

This is because the lower Hg^0 concentrations simulated by the global models were associated with higher RGM and particulate Hg concentrations as a product of simulated oxidation of Hg^0 to gaseous and aerosol $Hg^{(II)}$ forms. Indeed, the analysis of CMAQ sensitivity to RGM at the boundaries showed obvious positive correlations with simulated Hg wet deposition, especially near the boundaries. The CMAQ showed less sensitivity to particulate Hg concentrations at the boundaries as compared to RGM, but positive correlations with total-Hg wet deposition were noted.

In general, the NAMMIS and follow-on studies have shown that RGM concentrations specified at the lateral boundaries of regional modeling domains can have a significant effect on the intensity of Hg wet deposition simulated, not only near the boundary but also in interior regions of the domain. Lateral boundary concentrations specified for other mercury species may also impact simulations of wet deposition. It is difficult to draw firm conclusions about the effect of Hg^0 boundary concentrations brought about by simulated or actual intercontinental transport without also considering the effect of boundary fluxes of important oxidants of mercury (e.g., ozone, hydroxyl radical, halogens).

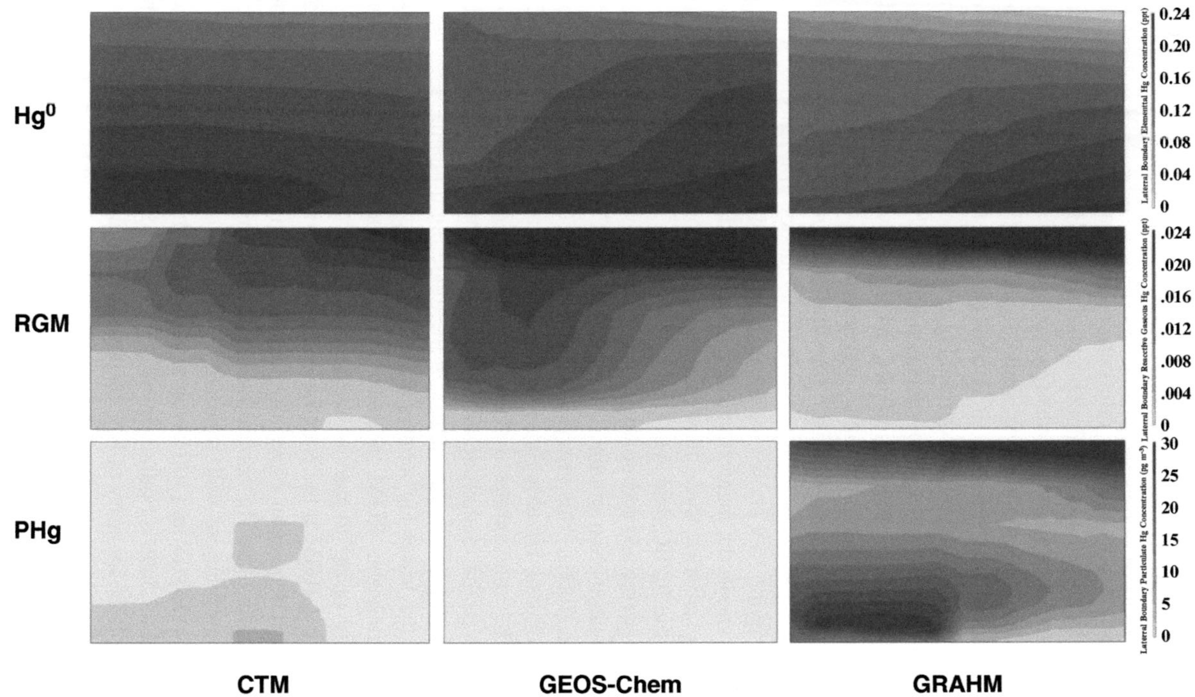

Figure 16.7 Annual average air concentrations of Hg^0, RGM and particulate Hg across the western boundary of the NAMMIS regional domain as determined from the CTM, GEOS-Chem and GRAHM global simulations

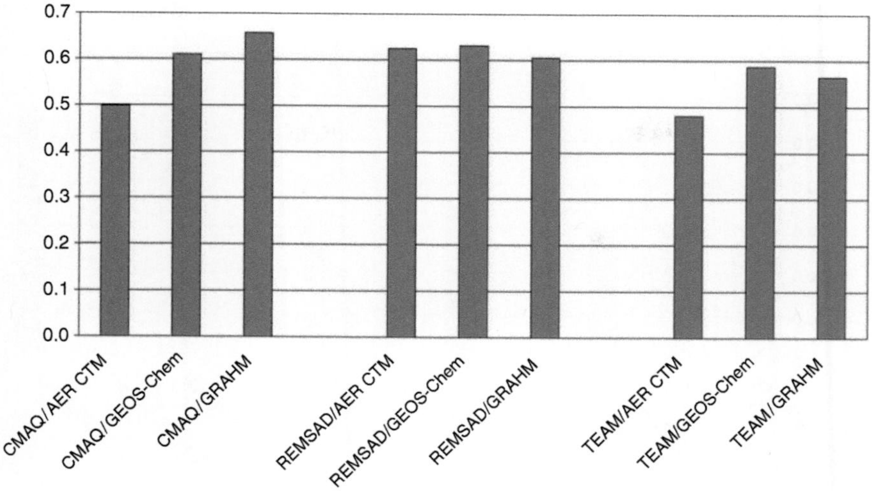

Figure 16.8 R^2 correlation statistics for simulated annual Hg wet deposition compared to observed values for each of the regional model simulations conducted in the North American Mercury Model Intercomparison Study

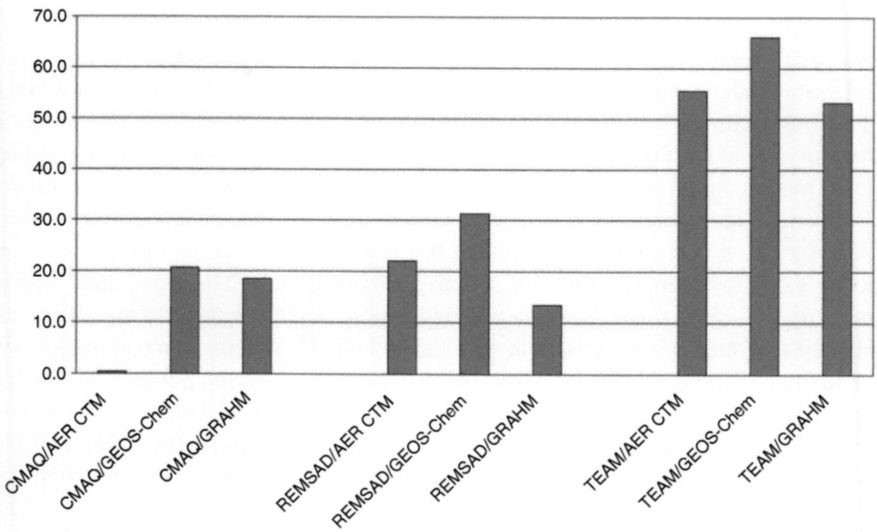

Figure 16.9 Mean fractional bias (percent) for simulated annual Hg wet deposition compared to observed values for each of the regional model simulations conducted in the North American Mercury Model Intercomparison Study

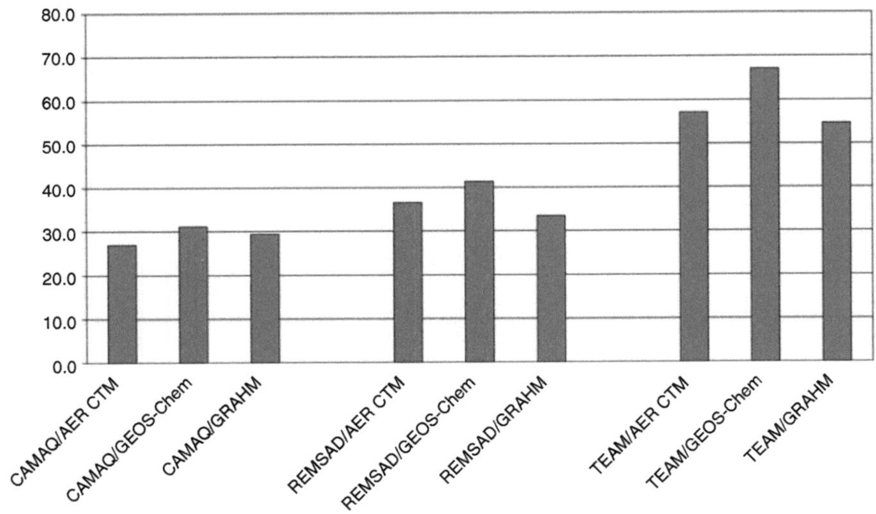

Figure 16.10 Mean fractional error (percent) for simulated annual Hg wet deposition compared to observed values for each of the regional model simulations conducted in the North American Mercury Model Intercomparison Study

16.4 Future Research and Policy Implications

Numerical simulation modeling has played a critical role in the advancement of our scientific understanding of many environmental and ecological issues. As we make new observations and gain knowledge about specific aspects of atmospheric mercury behavior, that knowledge is incorporated into numerical models by adding to or modifying its formulations. These new formulations are tested when model simulations are compared to new independent observations. This iterative process of observation and model development has provided us with the models we have today. Certainly, there remain some serious gaps in our understanding of atmospheric mercury emission, transformation and deposition. Scientific uncertainty affects all atmospheric mercury models, both regional and global. Whether current models are good enough to be used as a basis for environmental protection policy is as much a political question as a scientific one. It is interesting to note that, for the regional models applied in the NAMMIS, the lowest mean fractional error for annual Hg wet deposition was around 30%, which is comparable to the inter-annual variability seen in Mercury Deposition Network observations over the past few years (see http://nadp.sws.uiuc.edu/mdn/maps/). This suggests that the modeling error from the best models currently available introduces a similar level of uncertainty as does the choice of the particular annual period to be simulated.

Gaps in our understanding of chemical and physical processes are discussed in the previous chapter. Here we have shown that regional model simulations of atmospheric mercury are sensitive to the boundary concentrations of Hg specified as input data. Global models do not require spatial boundary information by nature of their definition. During the last few years, it has become common practice for regional models to be supplied their necessary boundary information based on global model simulations. However, we have shown here that current global models do not agree on the Hg concentrations that the regional models are sensitive to. There is an obvious need for a globally common framework for observation and modeling and a unified assessment of global-scale mercury transport.

In the past 10 years, new and improved information on sources and behaviour of atmospheric mercury has certainly improved our basic scientific understanding. For example, the discovery of gaseous mercury depletion events in circum-polar locations and the subsequent determination that halogens from sea water were chemically oxidizing elemental mercury to forms that rapidly deposited to the local snowpack. High altitude measurements of discrete mercury species have found that elemental mercury air concentrations tend to decrease with height, and that the concentrations of oxidized forms of Hg tend to increase with height, suggesting that significant oxidation of Hg is taking place in the middle and/or upper levels of the troposphere. Laboratory experiments have been used to estimate kinetic rate constants that all models need for their simulations of chemical transformations. Increases in computing speeds and memory capacity have allowed the application of fine-scale modeling over larger spatial and temporal domains. However, we still cannot simulate global scale atmospheric mercury at the fine spatial resolution required for environmental assessment of specific water bodies where Hg contamination is a problem.

One aspect of using global models to provide boundary information for regional models deserves special attention for future development. Global and regional models need to use consistent meteorological data in their respective simulations. For example, all models simulate higher concentrations of RGM in air masses that have not been subject to wet deposition over a long period of time. If a global model simulates no precipitation and high RGM concentrations at a regional model boundary where the regional model simulates some precipitation, the regional model will show excessively large wet deposition fluxes near that boundary. In reality, truly dry air masses do not develop precipitation instantaneously. Instead, cloud water droplets are gradually formed where significant quantities of RGM could be dissolved, chemically reduced to Hg^0 and released back to the air before precipitation and wet deposition could occur. Wind flow patterns also need to be consistent between the global and regional models. The global model might simulate wind flow from an area of concentrated emissions within the regional model domain towards the boundary while the regional model instead simulates wind flow in the opposite direction. This results in the regional model simulating in-flow and deposition of Hg from internal sources. In the opposite case, where the regional model simulates out-flow, the boundary concentrations from the global model are of no consequence at all. So the average effect of inconsistent wind flow between the global and regional models is an artificial recirculation of emissions from within the regional domain. This problem of inconsistent meteorological data can be avoided through the use of nested-grid meteorological modeling where the global model and the regional model are essentially the same model being applied with the same meteorological information on two different domains with different spatial resolutions within those domains.

Obviously, source-receptor relationships simulated by regional models are influenced by the boundary air concentrations specified for all pertinent pollutants. Regional models may be able to provide useful information about the relative importance of the Hg emission sources within their modeling domains without detailed and accurate boundary flux information. However, the importance of Hg emission sources located outside these domains cannot be assessed with any confidence without accurate boundary concentration data for all Hg species and their reactants. The fact that global models do not agree in their simulated concentrations for pertinent species is certainly cause for concern. We must develop standardized ambient sampling techniques for atmospheric Hg that can be deployed all around the globe and at all levels of the atmosphere, not just at the surface. Then we must actually deploy them with a consistent and continuous coverage of Earth's atmosphere in order to know if any global model is capable of providing realistic boundary data to regional models.

References

Bullock, O.R., 2000. Modeling assessment of transport and deposition patterns of anthropogenic mercury air emissions in the United States and Canada. *Science of the Total Environment* 259, 145–157.

Bullock, O.R., Brehme, K.A., 2002. Atmospheric mercury simulation using the CMAQ model: formulation description and analysis of wet deposition results. *Atmospheric Environment* 36, 2135–2146.

Bullock, O. R., 2008b. The effect of lateral boundary values on atmospheric mercury simulations with the CMAQ model. In: Carlos Borrego and Ana Isabel Miranda, editors. *Air Pollution Modeling and Its Applications XIX.* Springer, Dordrecht, The Netherlands, in press.

Bullock, O.R., Atkinson, D., Braverman, T., Civerolo, K., Dastoor, A., Davignon, D., Ku, J.-Y., Lohman, K., Myers, T., Park, R., Seigneur, C., Selin, N.E., Sistla, G., Vijayaraghavan, K., 2008a. The North American Mercury Model Intercomparison Study (NAMMIS). Part 1: Study description and model-to-model comparisons. *Submitted to Journal of Geophysical Research*

Christensen, J.H., 1997. The Danish Eulerian hemispheric model - A three-dimensional air pollution model used for the Arctic. *Atmospheric Environment* 31, 4169–4191.

Cohen, M., Artz, R., Draxler, R., Miller, P., Poissant, L., Niemi, D., Ratte, D., Deslauriers, M., Duval, R., Laurin, R., Slotnick, J., Nettesheim, T., McDonald, J., 2004. Modeling the atmospheric transport and deposition of mercury to the Great Lakes. *Environmental Research* 95, 247–265.

Dastoor, A.P., Larocque, Y., 2004. Global circulation of atmospheric mercury: A modeling study. *Atmospheric Environment* 38, 147–161.

ICF, 2005. User's Guide to the Regional Modeling System for Aerosols and Deposition (REMSAD), Version 8. ICF Consulting/SAI, San Francisco, CA.

Ilyin, I., Ryaboshapko, A., Afinogenova, O., Berg, T., Hjellbrekke, A.-G., Lee, D.S., 2002. Lead, cadmium and mercury transboundary pollution in 2000. EMEP/MSC-E Report 5/2002, Meteorological Synthesizing Centre - East, Moscow, Russia, 131 pp. http://www. msceast. org. html.

Keeler, G. J., Gratz, L., Al-Wali, K., 2005. Influences on the long-term atmospheric mercury wet deposition at Underhill, Vermont. *Ecotoxicology* 14, 71–83.

Murphy, D.M., Hudson, P.K., Thomson, D.S., Sheridan, P.J., Wilson, J.C., 2006. Observations of mercury-containing aerosols. *Environmental Science & Technology* 40, 3163–3167.

Pai, P., Karamchandani, P., Seigneur, C., 1997. Simulation of the regional atmospheric transport and fate of mercury using a comprehensive Eulerian model. *Atmospheric Environment* 31, 2717–2732.

Petersen, G., Bloxam, R., Wong, S., Munthe, J., Kruger, O., Schmolke, S.R., Kumar, A.V., 2001. A comprehensive Eulerian modelling framework for airborne mercury species: model development and applications in Europe. *Atmospheric Environment* 35, 3063–3074.

Ryaboshapko, A., Bullock, O.R., Christensen, J., Cohen, M., Dastoor, A., Ilyin, I., Petersen, G., Syrakov, D., Travnikov, O., Artz, R.S., Davignon, D., Draxler, R.R., Munthe, J., Pacyna, J., 2007. Intercomparison study of atmospheric mercury models: 2. Modelling results vs. long-term observations and comparison of country atmospheric balances. *Science of the Total Environment* 377, 319–333.

Schroeder, W.H., Anlauf, K.G., Barrie, L.A., Lu, J.Y., Steffon, A., Schneeberger, D.R., Berg, T., 1998. Arctic springtime depletion of mercury. *Nature* 394, 331–332.

Seigneur, C., Karamchandani, P., Lohman, K., Vijayaraghavan, K., Shia, R.-L., 2001. Multiscale modeling of the atmospheric fate and transport of mercury. *Journal of Geophysical Research* 106(D21), 27,795-27,809.

Seigneur, C., Vijayaraghavan, K., Lohman, K., Karamchandani, P., Scott, C., 2004. Global source attribution for mercury deposition in the United States, *Environmental Science & Technology* 38, 555–569.

Selin, N.E., Jacob, D.J., Park, R.J., Yantosca, R.M., Strode, S., Jaegle, L., Jaffe, D., 2007. Chemical cycling and deposition of atmospheric mercury: Global constraints from observations. *Journal of Geophysical Research* 112, D02308, doi:10.1029/2006JD007450.

Shia, R.L., Seigneur, C., Pai, P., Ko, M., Sze, N.D., 1999. Global simulation of atmospheric mercury concentrations and deposition fluxes. *Journal of Geophysical Research* 104(D19), 23,747-23,760.

Syrakov, Dimiter, Prodanova, Maria and Slavov Kiril (2005)Bulgarian Emergency Response System: Description and Ensemble Performance NATO Science Series Series IV: Earth and Environmental Series Advances in Air Pollution Modeling for Environmental Security Proceedings of the NATO Advanced Research Workshop on Advances in Air Pollution Modeling for Environmental Security Borovetz, Bulgaria 8–12 May 2004-10.1007/1-4020-3351-6_33

US EPA, 1997. Mercury Study Report to Congress. Fate and Transport of Mercury in the Environment, Vol. III. EPA-452/R-97-005, US Environmental Protection Agency, US Government Printing Office, Washington, DC.

US EPA, 2005. Technical Support Document for the Final Clean Air Mercury Rule: Air Quality Modeling, EPA Office of Air Quality Planning and Standards, Research Triangle Park, NC 27711, March 2005. (http://www.epa.gov/ttn/atw/utility/aqm_oar-2002-0056-6130.pdf)

Vermette, S., Lindberg, S., and Bloom, N., 1995. Field tests for a regional Mercury Deposition Network - Sampling design and preliminary test results. *Atmospheric Environment* 29, 1247–1251.

Chapter 17
Global Mercury Modelling at Environment Canada

Ashu P. Dastoor and Didier Davignon

Summary We describe the recent developments of Environment Canada's atmospheric mercury model (GRAHM) and its application to the intercontinental source-receptor relationships of mercury. The model includes 2188 Mg yr^{-1} global anthropogenic emissions, 1600 Mg yr^{-1} terrestrial emissions and 2600 Mg yr^{-1} oceanic emissions). Transport, chemical transformation and deposition of Hg^0, $Hg^{(II)}$ and Hg(p) are simulated in GRAHM within a meteorological assimilation and forecasting system. Current version of the GRAHM includes GEM oxidation by ozone in the troposphere and halogen oxidation in the Polar and the marine boundary layers. It also includes dynamic exchange of mercury fluxes at air-snow/ice interface. The model simulates springtime atmospheric mercury depletion events (AMDEs) and the net accumulation of mercury in snow in the Polar Regions. We performed one reference simulation with emissions as above and four perturbation simulations with 20% reduced anthropogenic emissions over East Asia, South Asia, Europe and North Africa and North America. 20% reduction in anthropogenic emissions of mercury over East Asia, South Asia, Europe and North Africa and North America represent 7.7%, 1.6%, 2.5% and 1.3% reduction in global anthropogenic emissions respectively. The deposition over East Asia, South Asia, Europe and North America are reduced by 13.5%, 7.9%, 8.3% and 4.3% due to the emission reductions within the same regions. Deposition in North America is found to be most affected by the emission reductions in other regions and the deposition in East Asia is least affected by outside reductions. The deposition in the Arctic is nearly equally sensitive to the unit emission reductions in Europe and East Asia and is most sensitive in springtime due to the high deposition related to AMDEs.

17.1 Introduction

Hg is on the priority list of a large number of international agreements, conventions and national advisories aimed at the protection of the environment including all compartments, human health and wildlife (e.g. CLRTAP, AMAP, UN-ECE, HELCOM, OSPAR and many more). The mercury methylation process is shown to be strongly correlated with loadings of inorganic mercury to the aquatic system

N. Pirrone and R. Mason (eds.), *Mercury Fate and Transport in the Global Atmosphere*,
DOI: 10.1007/978-0-387-93958-2_17,

from the atmosphere (Hammerschmidt et al., 2004). Ambient measurements of mercury reveal a vast global pool of mercury in the atmosphere up to the tropopause revealing its global nature and yet at the same time, atmospheric measurements of mercury in polar regions, in marine boundary layer and in the upper troposphere have shown that elemental mercury can be rapidly oxidized to more hygroscopic forms and deposited at much shorter time scales. In the atmosphere, mercury is mainly present as gaseous elemental mercury (GEM), reactive gaseous mercury (RGM) and particulate mercury (Hg(p)). GEM is the most dominant form of mercury in the atmosphere with the longest life time (0.5 – 2 yr) and RGM and Hg(p) have a significantly shorter lifetime (few days - week) and are deposited rapidly via dry and wet deposition (Lin et al., 2006). Emissions of mercury from natural as well as anthropogenic sources, multiple time scales of mercury in the atmosphere and photochemically and biologically driven revolatilization of mercury at the surface render understanding of global cycling of mercury a challenging subject. The impact of confounding factors such as climate change and changes in other pollutants in the atmosphere add to the complexity of determining the impact of reducing emissions of mercury on deposition.

In this chapter, we briefly describe the Environment Canada's global mercury model GRAHM and provide results on model simulations to assess the impact of reduced anthropogenic emissions from four main source regions of mercury in Northern Hemisphere.

17.2 Model Description

The Global/Regional Atmospheric Heavy Metals Model (*GRAHM*) is an Eulerian, multi-scale meterological and mercury simulation model which is developed by including atmospheric mercury dynamical, physical and chemical processes on-line into Canadian operational weather forecasting and data assimilation model *GEM* (Global Environmental Multiscale Model). Details of the model GRAHM are described in Dastoor and Larocque (2004), Ariya et al. (2004) and Dastoor et al. (2008). Transport, transformation and surface exchange of three mercury species, namely, gaseous elemental mercury (GEM), gaseous divalent mercury (RGM) and particulate mercury (Hg(p)) are simulated in the model. For this study, the anthropogenic mercury emissions were updated to year 2000 (Pacyna et al., 2006). Terrestrial and oceanic emissions of GEM from direct natural sources and from previously deposited mercury were introduced based on the global mercury budget study by Mason and Sheu (2002). Land based natural emissions were spatially distributed according to the natural enrichment of mercury and re-emissions were distributed according to the distribution of total deposition of mercury for historic years. Oceanic emissions in the model are spatially distributed according to the deposition and primary production distribution. Seasonal and diurnal dependence is added to both land and oceanic emissions as a function of solar irradiance.

Following the discussion in Calvert et al. (2005) and the observational constraints of mercury measurements on global scale, we have included slow oxidation of GEM by O_3 reported in Hall, (1995) as the primary oxidation of GEM in the gas phase. Monthly averaged and diurnally varying concentrations of O_3 are specified from a detailed tropospheric chemistry model MOZART (Horowitz et al., 2003). The oxidation product is equally partitioned between RGM and Hg(p). Oxidation of GEM by O_3, OH and reactive chlorine and reduction of RGM by S(IV) and via photolysis of $Hg(OH)_2$ are part of the aqueous phase redox chemistry in the model. Sorption of aqueous mercury species to elemental carbon aerosols is also included in the model. GEM is dry deposited only over forest regions due to its low solubility and it is modelled using a seasonally dependent constant dry deposition velocity in the range of 0.001-0.03 cm s^{-1}. Dry deposition of RGM is parameterized utilizing the multiple resistance analogy. Dry deposition velocity for Hg(p) is parameterized as a function of particle size and density. Both gaseous and particle dry deposition velocities are functions of micro-meteorological conditions, land-use types, surface wetness, snow/ice and surface roughness characteristics. GRAHM, being an online model, has the advantage of using detailed cloud micro-physical parameters which are simulated at each time step by the model cloud parameterization schemes which are used for the formation of wet deposition in the model. Mercury removal through the conversion of hydrometeors to precipitation as well as through below cloud scavenging are included in the model.

At polar sunrise, gaseous elemental mercury (GEM) undergoes an exceptional dynamic exchange in the air and at the snow surface, during which GEM can be rapidly removed from the atmosphere (the so called atmospheric mercury depletion events (AMDEs), as well as re-emitted from the snow within a few hours to days in the Polar Regions. During AMDEs, gaseous elemental mercury is converted to hygroscopic mercury species (reactive gaseous mercury and total particulate mercury). Experimental and theoretical studies demonstrate that reactions of GEM with BrO and Br are the most likely mechanism of AMDEs in the Polar Regions (Steffen et al., 2008). Similar to the Polar Regions, diurnally varying elevated concentrations of RGM are also observed in remote marine boundary layer (MBL) suggesting RGM production photochemically by reactions with halogens liberated from sea-salt particles (Laurier et al., 2003). Halogen oxidation of mercury over oceans and the Polar Regions can potentially alter the life time of GEM in the atmosphere and its intercontinental long-range transport. GRAHM includes GEM oxidation by halogen species in the Polar Regions and the MBL. We have included GEM reactions with BrO, Br_2, Br, Cl_2 and Cl using reaction rate coefficients from Ariya et al. (2004). Reported Hg^0-Br reaction rate coefficients range from 1.1×10^{-12} to 3.6×10^{-13} cm^3 $molecules^{-1}$ s^{-1} (Donohoue et al., 2006). Our choice of Hg^0-Br reaction rate coefficient (Ariya et al., 2004) is based on the best correlation estimates between model simulated AMDEs and observed AMDEs at Alert, Canada for multiple years. We have assumed part of the oxidation product of mercury reactions by halogens as RGM and partly as TPM in the absence of full understanding of the heterogeneous chemistry of mercury. Since the dry deposition velocities of RGM and TPM are much larger compared to GEM, and both, RGM and TPM, are readily deposited to snow, lack of accuracy in partitioning between RGM and TPM is expected to have small impact on the deposition estimates of mercury to the snow.

Sources and processes responsible for bromine liberation in the Arctic boundary layer are not fully understood. Sea-salt aerosols, sea-salt deposits on snow, new sea-ice surfaces and frost flowers have been suggested as sources. In the absence of a complete understanding of these processes, the model includes GOME (Global Ozone Monitoring Experiment) satellite derived monthly average Br/BrO concentrations into the boundary layer of the model. A strong photo-reduction and evasion of mercury deposited to the snowpacks following AMDEs has been observed to reduce significantly the accumulation of mercury of in snowpacks (Poulain et al. 2004; Kirk et al. 2006). Poulain et al. (2004) also found evidence for re-oxidation of reduced GEM in the high Arctic snowpacks due to the presence of oxidative molecules (organic and halides) in the marine/coastal polar surface snow during springtime. Present knowledge on the Hg (photo) redox chemistry in the snow is very limited for model parameterization. We have included evasion of mercury from the snowpacks as a function of solar radiation reaching the surface including the influence of clouds and surface temperature with an assumption that it occurs near the surface from the newly deposited mercury.

17.3 Results and Discussion

The model resolution for this study was $2^{\circ}\times2^{\circ}$ in the horizontal and 28 levels up to 10 hPa in the vertical with higher resolution in the boundary layer. Model simulations were performed for the years 2001 with an initial spin-up for three years with the above configuration. The model was evaluated against the globally observed data. The mercury budgets in various compartments in the model were found to be well balanced. The model simulated yearly average of surface air GEM concentration is 1.8 ng m^{-3} in Northern Hemisphere and 1.3 ng m^{-3} in Southern Hemisphere which are within the range of measured concentrations. North to South gradient of ambient GEM concentrations over the Atlantic and Pacific oceans simulated by the model and its comparison with the observations is shown in Figure 17.1. It was found to be improved in this version of the model compared to the previous version as a result of the addition of halogen mercury chemistry in the MBL.

Figure 17.2 shows seasonal average surface air GEM concentrations and observed and model simulated time series of GEM at Alert, Canada Highest GEM concentrations above 2.4 ng m^{-3} are found over East Asia. The GEM concentrations are highest in wintertime over Europe and North America. Enhanced Northward transport of mercury from Europe in winter and from East Asia in spring is evident in the seasonal average concentrations. Most dramatic seasonal cycle is simulated over the Arctic. In the springtime when AMDEs are active the GEM concentrations are lowest and recover to the background values in fall.

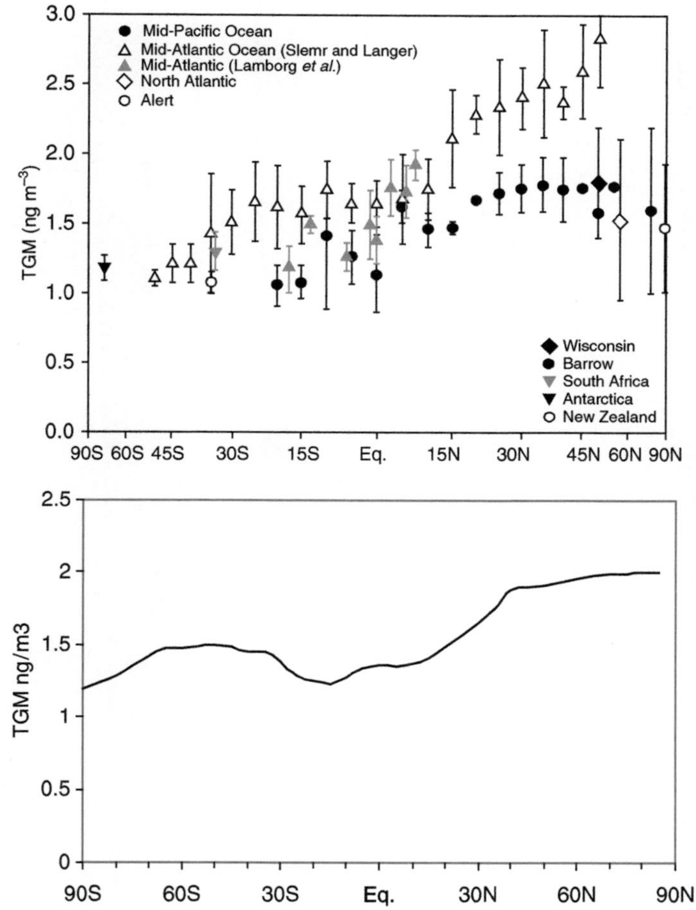

Figure 17.1 Inter-hemispheric gradient of TGM from observations (top) (from Lamborg et al., 2002) and GRAHM model simulation (bottom)

Atmospheric mercury measurements have been conducted in the Arctic at Alert, Barrow, Amderma, Ny-Alesund, and Station Nord (Steffen et al., 2008). Longest record of GEM measurements in the Arctic is available at Alert, Nunavut, which has been continuous since 1995. Rapid exchange of mercury related to AMDEs in springtime is seen in the model simulation and in the observation (Figure 17.1). Strong re-emission events are also observed and simulated in the late spring related to the higher surface temperatures. Br and BrO concentrations used in the model are climatological averages with dependence on the solar influx, boundary layer height and surface temperature. However, despite these limitations, the model captures 1-2 weeks AMDE cycles and the interannual variability in GEM concentrations. These results suggest significant role of meteorological processes such as

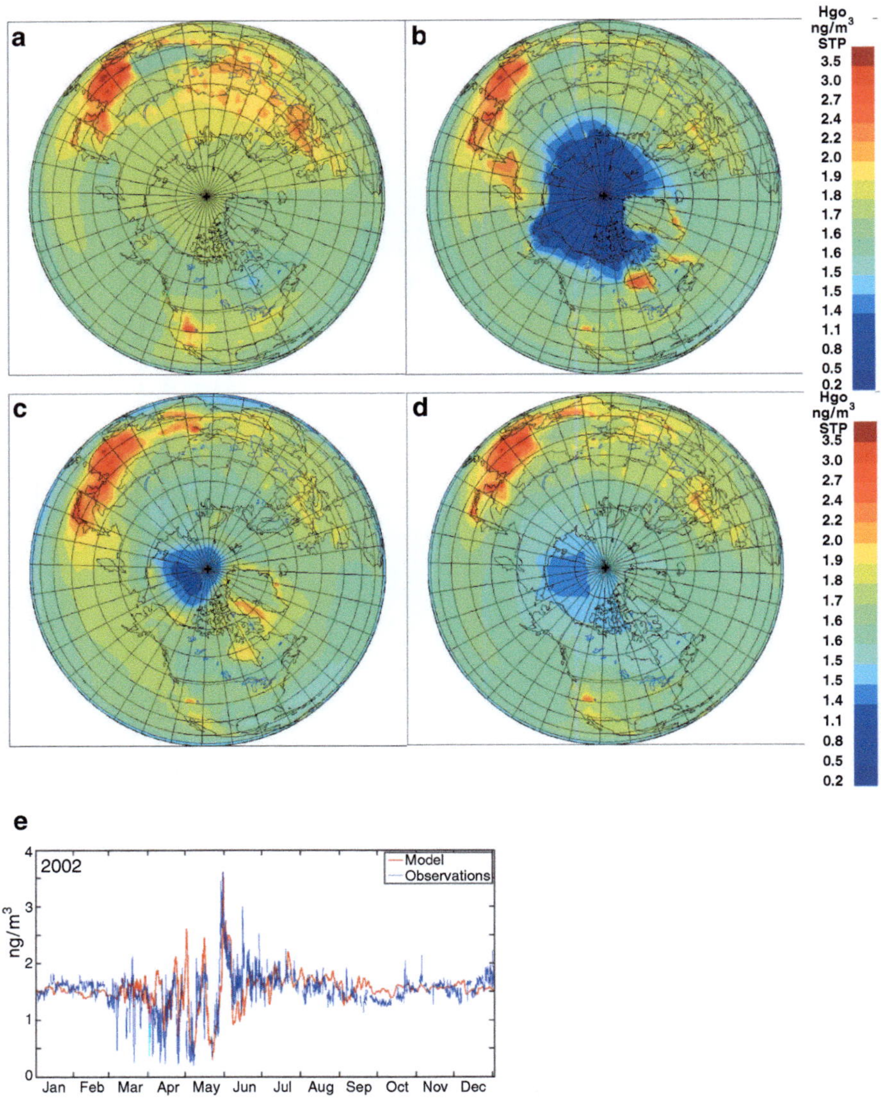

Figure 17.2 GRAHM simulated average surface air Hg^0 concentrations (*ng standard m*$^{-3}$) for (**a**) winter, (**b**) spring, (**c**) summer and (**d**) fall and (**e**) observed and model simulated time series of surface air Hg^0 concentrations at Alert, Canada for 2002

transport, boundary layer height, solar radiation reaching ground, clouds, temperature inversion and surface temperature in establishing AMDEs. Measured median concentrations of GEM at Alert are 1.61 ng m^{-3} in winter, 1.34 ng m^{-3} in spring, 1.78 ng m^{-3} in summer and 1.53 ng m^{-3} in fall. Model simulated average median surface air concentrations of GEM at Alert are 1.51 ng m^{-3} in winter, 1.35 ng m^{-3} in spring, 1.75 ng m^{-3} in summer and 1.52 ng m^{-3} in fall. Model simulated GEM and RGM air concentrations at other locations are found to be in the same range as reported in the limited measurements reported in the literature (Steffen et al., 2008).

In addition to the above reference simulation, we conducted four perturbation experiments with 20% reduction in anthropogenic emissions in four major source regions as defined by The Task Force on Hemispheric Transport of Air Pollution. These regions are: East Asia (EA: 15-50N; 95-160E), Europe and North Africa (EU: 25-65N; 10W-50E), South Asia (SA: 5-35N; 50-95E), and North America (NA: 15-55N; 125-60W). Each of these regions represent 841 Mg, 276 Mg, 179 Mg and 143 Mg of total anthropogenic mercury emissions annually respectively. The annual total deposition of mercury from all mercury emissions and the reduction in total deposition of mercury for the four perturbation experiments are illustrated in Figure 17.3.

The wet and dry depositions account for 45% and 55% of total deposition over land and 64% and 36% of total deposition over ocean respectively. Deposition is seen to be high over the mercury source regions and the Polar Regions. The annual total deposition in the four source regions of the perturbation experiments are 593 Mg for East Asia, 232 Mg for Europe and North Africa, 191 Mg for South Asia and 281 Mg for North America.

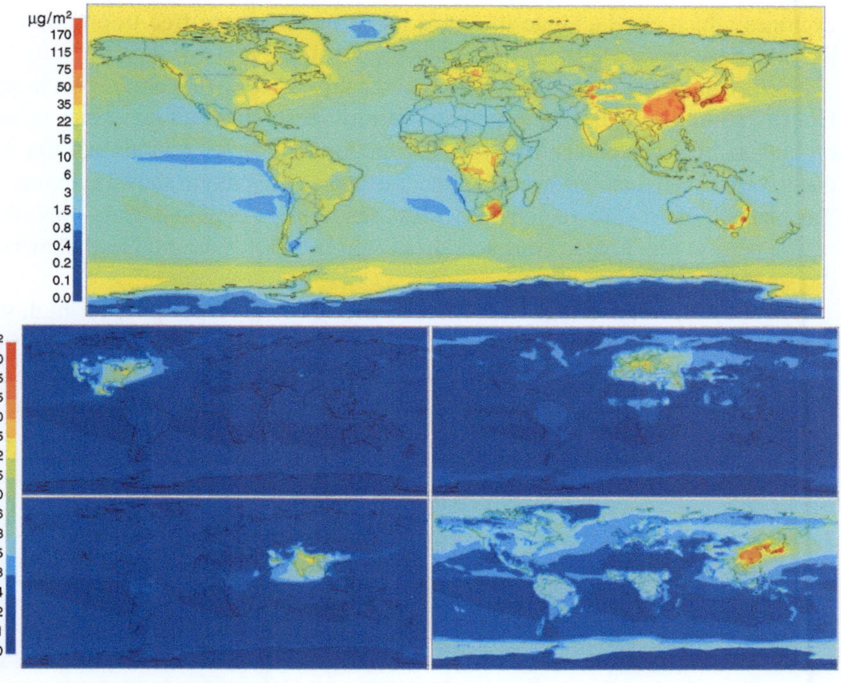

Figure 17.3 Total mercury deposition for year 2001 from reference simulation (top) and reduction in deposition due to 20% reduction in anthropogenic emissions (scaled to 100% in the figure) from North America (middle left), Europe and North Africa (middle right), South Asia (bottom left) and East Asia (bottom right)

The deposition over the Polar Regions is enhanced by the springtime AMDEs. The model estimates deposition of 243 Mg yr^{-1}, and evasion of 145 Mg yr^{-1} resulting in net atmospheric flux of 98 Mg yr^{-1} to the Arctic Ocean (Outridge et al. 2008). The wet deposition simulated by the model is in the range of measured deposition data for North America and Europe. The deposition reduction due to 20% reduction of anthropogenic emissions is seen maximum over the source regions and the regions of high precipitation.

The deposition reductions in the four regions due to reductions in anthropogenic emissions in each of the four regions are summarized in Table 17.1. Wet and dry deposition of mercury mainly occurs as RGM and Hg(p). RGM and Hg(p) emitted in each of the four regions (East Asia: 43%, Europe: 41%, South Asia: 47% and North America: 41%) is mostly deposited regionally due to its short life time of the order of days. Long range transport of elemental mercury to the receptor regions and its oxidation leads to the deposition contribution from intercontinental sources.

Figure 17.4 presents the sensitivities of the surface air GEM concentrations and column GEM burden at receptor regions including the Arctic to unit emissions in each of the four source regions averaged for the four seasons. Surface air is seen to be most sensitive to regional emissions whereas column mercury is most sensitive to the long range transport from East Asia for most regions. The East Asian emissions are located on the eastern seaboard of Asia in the region of mid-latitude cyclones which can facilitate lifting of the pollution to the free troposphere from the boundary layer where it can be transported efficiently.

Surface air in the Arctic region is most sensitive to long range transport in winter from all source regions. The deposition sensitivities to the unit emissions in source regions and the percentage deposition reduction at receptor regions and the Arctic due to emission reductions in each of the four source regions are shown is Figure 17.5. The depositions in all regions are most sensitive to regional emissions. Arctic region is most sensitive to emissions from Europe, East Asia and North America in the springtime.

The AMDEs related high deposition in the Arctic is responsible for high sensitivity during spring. The impact of emission reduction in remote regions to local deposition is maximum for North America where deposition is reduced by 2.8% due to 20% emission reduction in East Asia. Outside impact is minimum for East

Table 17.1 Reduction in mercury deposition (*Mg* yr^{-1}) over four receptor regions and the Arctic due to 20% reduction in emissions from four source regions. First column shows the reduction in emissions for the four source regions

Receptors Sources	Emission reduction	South Africa	East Asia	Europe	North America	Arctic
South Asia	35.7	15.1	2.3	0.7	1.0	0.9
East Asia	168.2	3.6	80.0	4.2	7.1	8.0
Europe	55.1	1.8	1.8	19.2	1.7	2.7
N. America	28.6	0.4	0.8	0.8	12.2	1.2

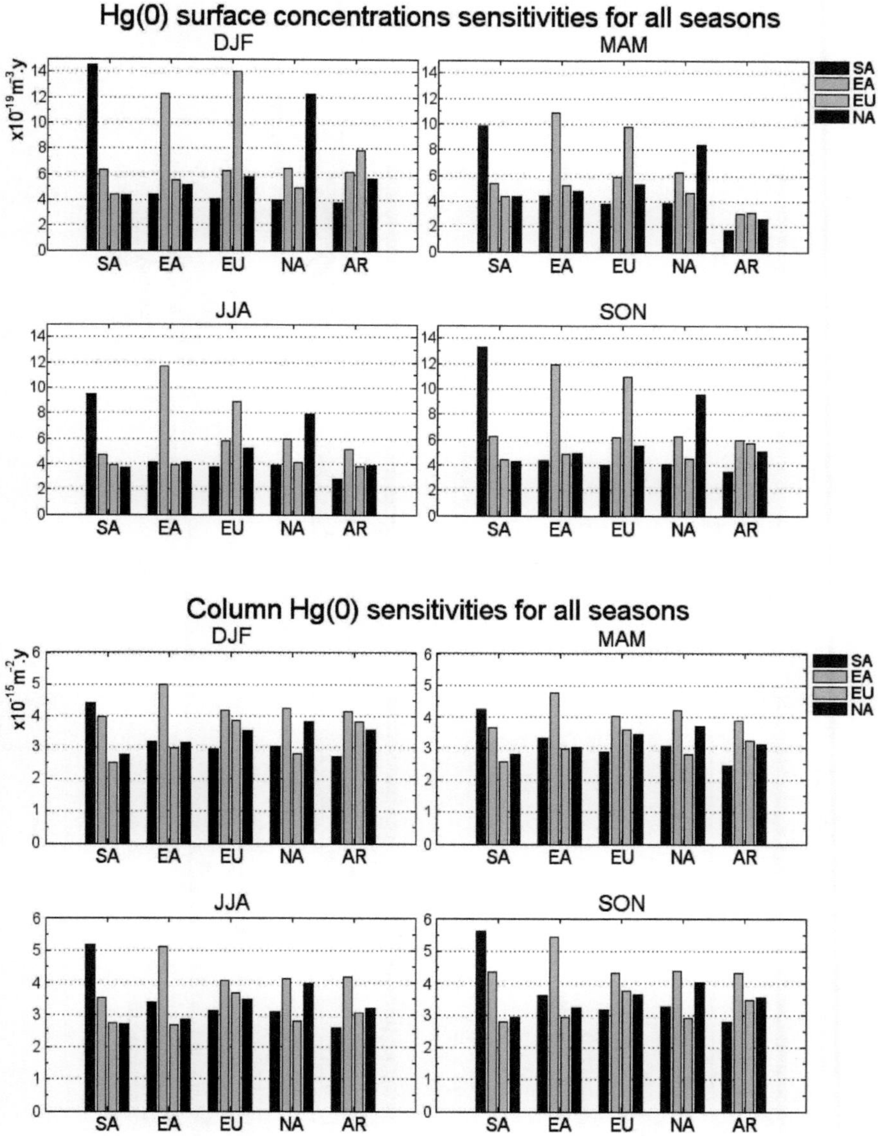

Figure 17.4 The surface air GEM concentration (top) and column GEM burden (bottom) sensitivities at receptor regions (x axis) and the Arctic to unit emission reductions from the source regions defined as South Asia (SA), East Asia (EA), Europe and North Africa (EU) and North America (NA)

Asia where 71% of deposition comes from local sources. The maximum deposition reduction to the Arctic comes from the reductions in East Asian emissions because of its highest emissions.

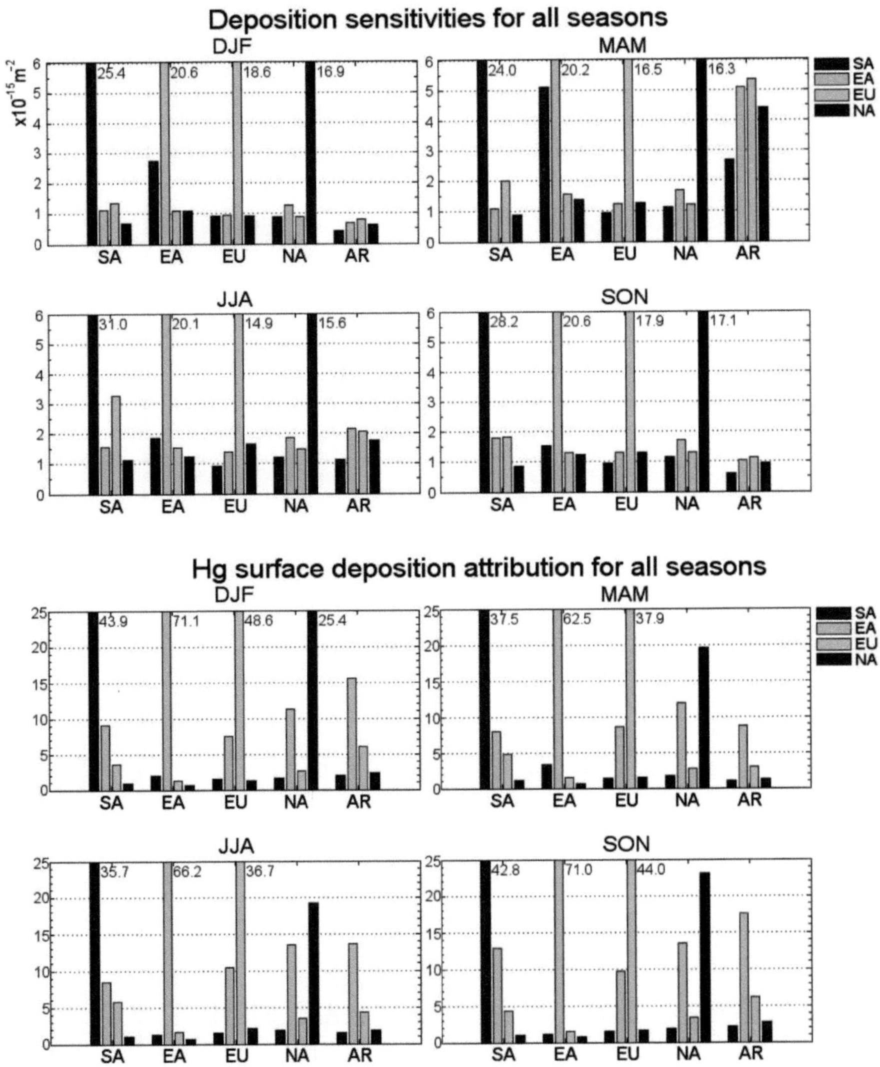

Figure 17.5 Mercury deposition sensitivities at receptor regions and the Arctic to unit emission reductions from the source regions defined as South Asia (SA), East Asia (EA), Europe and North Africa (EU) and North America (NA) (top). Percentage change in mercury deposition at receptor regions and the Arctic from 20% reduction in emissions at the source regions defined as above (the values are scaled to 100% reduction in the figure)

Numerous experimental and modelling studies have demonstrated the transpacific transport of air pollutant from East Asia to North America. The main mechanism for the rapid transport of Asian pollution is considered to be the lifting of boundary layer pollution into the free troposphere by mid-latitude cyclones. Once in the free troposphere over the western Pacific, Asian pollution can be transported undiluted to the northeastern Pacific in westerlies within 5 - 10 days. The subsiding air in anticyclones can bring the pollution to the lower levels affecting western North America and the Arctic. Convection and orographic lifting are also important mechanism in transporting Asian pollution. The transport across the Pacific is complex and usually involves several mid-latitude cyclones and associated warm conveyor belts. Springtime is found to be the most active period for episodic transport of Asian pollution to North America. East Asian mercury emissions are roughly half of the anthropogenic emissions globally.

The deposition contribution from East Asia to North America is estimated to be ~14% in this study. Jaffe et al., (2005) identified several Asian outflow of mercury at Hedo Station, Okinawa, Japan and the Mt. Bachelor Observatory in central Oregon, USA during spring 2004. They observed mean GEM concentrations of 2.04 ng m^{-3} at Hedo Station which is higher than the Northern Hemispheric background value of 1.8 ng m^{-3} due to the impact of Asian ouflow. They identified several long-range transport episodes at Mt. Bachelor. One large episode was observed around 25 April where the peak total Hg concentrations reached ~2.5 ng m^{-3} which is 0.7 ng m^{-3} above the background value. They found GEM/CO ratio at Mt. Bachelor to be very similar to the GEM/CO ratio found at Hedo Station. Using GEM/CO ratio and CO emissions, they estimated mercury emissions from Asia two times higher than the Asian mercury emissions estimates by Pacyna et al. (2006). In order to investigate the origin of high concentrations of mercury at Mt. Bachelor and the transport mechanism during the episode of April 25, 2004, we performed a series of GRAHM simulations at various resolutions. We found that the model simulated the observed episode best at resolution $0.25° \times 0.25°$ lat-long (Figure 17.6).

We performed simulations with all global emissions, only Asian anthropogenic emissions and other anthropogenic emissions. The top panel in the figure shows GEM mixing ratios at 500 hPa from all emissions on April 25, 2004. The bottom panel shows GEM mixing ratios at 500 hPa from only Asian emissions. Transpacific transport of mercury and significantly high concentrations of GEM along western coast of North America on April 25 are well simulated by the model. The model simulation with only Asian emissions reproduces the episodic transpacific transport of mercury from East Asia to North America and high concentrations of mercury around Oregon on April 25 suggesting East Asia to be the origin of this episode. Although, the depth of the peak in mercury concentrations at Mt. Bachelor is reproduced in all emissions simulation, it is underpredicted in only Asian emissions. Comparison of the two simulations suggests significant contribution of re-emissions and/or natural emissions in addition to the contribution from direct anthropogenic emissions from Asia.

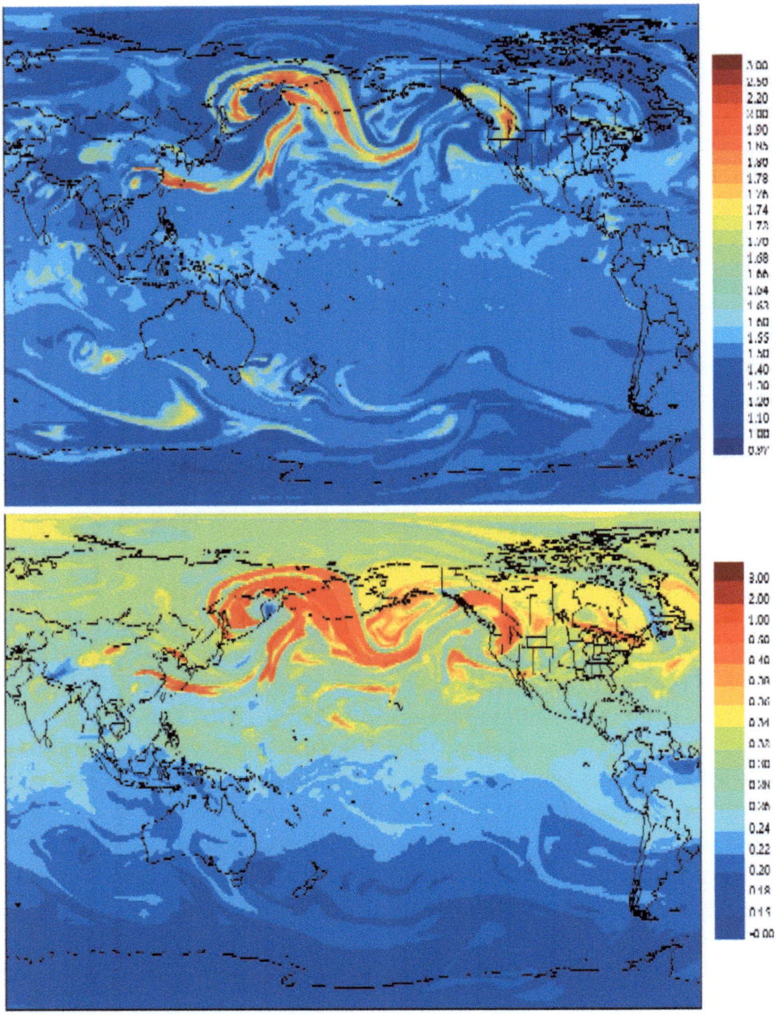

Figure 17.6 GRAHM air concentrations of mercury (*ng standard* m^{-3}) on 18 April 25, 2004 at 500 mb showing episode of Asian outflow of mercury reaching N. America which was observed at Mt. Bachelor in central Oregon. The top anthropogenic panel shows simulation from all emissions and the bottom panel shows simulation from anthropogenic Asian emissions

17.4 Uncertainties and Future Research

Deposition of mercury to a region from remote sources involves long range transport of GEM and its conversion to hygroscopic mercury species (RGM and Hg(p)). Current understanding of atmospheric chemistry and kinetics are based on limited laboratory and theoretical studies. Some of the reactions are not clearly defined and reaction products are not well understood. Although, several gas phase oxidation pathways including O_3, OH and Br have been suggested for removal of GEM in the troposphere by the laboratory studies, their relative importance in establishing the observed tropospheric life time of GEM is not clear. Fast oxidation rates measured for some of these reactions suggest presence of significant reduction processes of mercury in the atmosphere which are yet to be identified. Comprehensive understanding of the mercury chemistry and adequate measurements of mercury species in free troposphere are required for improving the mercury chemical mechanism in the model. We have included the halogen mercury chemistry in the MBL and the Polar boundary layer which relies on halogen concentration estimates from observations. Model development will be continued to include representation of halogen liberation mechanism and halogen chemistry in the model. Accurate understanding and modelling of mercury exchange at the land/ocean/air interface is required to improve the characterization of long range transport of mercury for better attribution of sources of mercury in deposition. Presently, GRAHM includes dynamic exchange of mercury between snow and air, whereas land and ocean are assumed to have longer response time compared to the simulation period. We are investigating coupling of GRAHM with terrestrial and ocean models to improve the surface interactions in the model. The dry deposition of mercury to different surfaces and wet deposition over the oceans are not constrained by observations in the model due to the lack of such measurement data. Poor speciation of anthropogenic emissions is another source of uncertainty in attributing the emission sources to the deposition in the model.

References

Ariya, P., A. Dastoor, M. Amyot, W. Schroeder, L. Barrie, K. Anlauf, F. Raofie, A. Ryzhkov, D. Davignon, J. Lalonde, A. Steffen, 2004. Arctic: A sink for mercury. *Tellus*, 56B, 397–403.

Calvert, J.G., S.E. Lindberg, 2005. Mechanisms of mercury removal by O3 and OH in the atmosphere. *Atmospheric Environment*, 39, 3355–3367.

Dastoor, A.P., Y. Larocque, 2004. Global circulation of atmospheric mercury: A modeling study. *Atmospheric Environment*, 38, 147–161.

Dastoor, A.P., D. Davignon, N. Theys, M. Roozendael, A. Steffen and P.A. Ariya, 2008. Modeling Dynamic Exchange of Gaseous Elemental Mercury at Polar Sunrise. Environmental Science and Technology (in press).

Donohoue, D.L., D. Bauer, B. Cossairt, and A. J. Hynes, 2006. Temperature and pressure dependent rate coefficients for the reaction of Hg with Br and the reaction of Br with Br: A pulsed laser photolysis-pulsed laser induced fluorescence study. *Journal Physical Chemistry A*, 110, 6623–6632.

Hall, B., 1995. The gas phase oxidation of elemental mercury by ozone. *Water, Air, and Soil Pollution*, 80, 301–315.

Hammerschmidt CR, W.F. Fitzgerald, 2004. Geochemical controls on the production and distribution of methylmercury in near-shore marine sediments. *Environmental Science and Technology*, 38, 1487–1495.

Horowitz, L. W., S. Walters, D. L. Mauzerall, L. K. Emmons, P.J. Rasch, C. Granier, X. Tie, J.-F. Lamarque, M.G. Schultz, G.P. Brasseur, 2003. A global simulation of tropospheric ozone and related tracers : Description and evaluation of MOZART, version 2, *Journal of Geophysical Research* , 108(D24), 4784, doic 10.1029/2002JD002853.

Jaffe, D., E. Prestbo, P. Swartzendruber, P. Weiss-Penzias, S. Kato, A. Takami, S. Hatakeyama and Y. Kajii, 2005, Export of atmospheric mercury from Asia. *Atmospheric Environment*, 39, 3029–3038.

Kirk, J.L., V.L. St. Louis, M.J. Sharp, 2006. Rapid reduction and reemission of mercury deposited into snowpacks during atmospheric mercury depletion events at Churchill, Manitoba, Canada. *Environmental Science and Technology*, 40, 7590–7596.

Laurier, F. J. G., R. P. Mason, L. Whalin, S. Kato, 2003. Reactive gaseous mercury formation in the North Pacific Ocean's marine boundary layer: A potential role of halogen chemistry. *Journal of Geophysical Research*, 108, D17, art # 4529.

Lin, C.-J., P. Pongprueksa, S.E. Lindberg, S.O. Pehkonen, D. Byun, C. Jang, 2006. Scientific uncertainties in atmospheric mercury models I: Model science evaluation. *Atmospheric Environment*, 40, 2911–2928.

Mason, R. P., G.-R Sheu, 2002. The role of the ocean in the global mercury cycle. *Global Biogeo. Cycles*, 16, art. # 1093.

Outridge, P.M., R.W. Macdonald, F. Wang, G.A. Stern and A.P. Dastoor, 2008. A mass balance inventory of mercury in the Arctic Ocean. *Environmental Chemistry*, 5, 89–111. doi:10.1071/EN08002.

Pacyna, E. G., Pacyna, J. M., Steenhuisen, F., Wilson, S., 2006. Global anthropogenic mercury emission inventory for 2000. *Atmospheric Environment*, 40, 4048–4063.

Poulain, A.J., J.D. Lalonde, M. Amyot, J.A. Shead, F. Raofie, P.A. Ariya, 2004. Redox transformations of mercury in an Arctic snowpack at springtime. *Atmospheric Environment*, 38, 6763–6774.

Steffen, A., Douglas, T., Amyot, M., Ariya, P., Aspmo, K., Berg, T., Bottenheim, J., Brooks, S., Cobbett, F., Dastoor, A., Dommergue, A., Ebinghaus, R., Ferrari, C., Gardfeldt, K., Goodsite, M.E., Lean, D., Poulain, A., Scherz, C., Skov, H.,Sommar, J., Temme, C., 2008. A synthesis of atmospheric mercury depletion event chemistry in the atmosphere and snow. *Atmospheric Chemistry and Physics*, 8, 1445–1482.

Chapter 18
The Geos-Chem Model

Lyatt Jaeglé, Sarah A. Strode, Noelle E. Selin, and Daniel J. Jacob

Summary We examine the response of deposition to decreases in anthropogenic emissions using the GEOS-Chem global atmosphere-ocean-land mercury simulation. Total global mercury sources in the model are 9230 Mg yr^{-1} (3400 Mg yr^{-1} anthropogenic, 650 Mg yr^{-1} biomass burning, 2180 Mg yr^{-1} land emissions, 3000 Mg yr^{-1} ocean emissions). Our atmospheric simulation describes the cycling of mercury through the surface ocean and land reservoirs. The model includes atmospheric oxidation of Hg^0 by OH and O_3, and in-cloud reduction of $Hg^{(II)}$. Wet and dry deposition account for 32% and 68% of the global sink, respectively. The lifetime of mercury against deposition is 0.6 years. We conduct four sensitivity simulations where anthropogenic emissions are reduced by 20% over East Asia, Europe, South Asia, and North America, leading to decreases in global deposition of -3.5, -0.9, -0.8, and -0.5% respectively. One third of the deposition decrease occurs in the source regions, and the rest is distributed globally due to decreased long-range transport of Hg^0 and subsequent oxidation to $Hg^{(II)}$. Regional decreases in deposition within the source regions range from -12% for East Asia (60% of depositions due to local emissions) to -3% for North America (where only 15% of deposition is due to local emissions). When normalized by total emissions, we find that Hg deposition in the Arctic is more sensitive to decreases in European emissions compared to decreases in East Asian, North American, or South Asian emissions. Our estimates of the distribution of deposition and its response to decreases in anthropogenic emissions are limited by uncertainties in the speciation of anthropogenic emissions, redox chemistry of atmospheric mercury, the role of dry deposition, and the cycling efficiency of mercury in the ocean and land reservoirs.

18.1 Introduction

The dual character of mercury as a local and global pollutant makes it difficult to assess the relative contributions of regional and global sources to deposition. Anthropogenic emissions of mercury as short-lived reactive gaseous ($Hg^{(II)}$) and particulate (Hg(p)) mercury forms lead to rapid deposition near source regions.

N. Pirrone and R. Mason (eds.), *Mercury Fate and Transport in the Global Atmosphere*, DOI: 10.1007/978-0-387-93958-2_18,

If mercury is emitted in its long-lived elemental form (Hg^0), it can be transported on global scales, oxidized to $Hg^{(II)}$, and then deposited far from source regions. In addition, dry deposition of Hg^0 itself can affect receptor regions. Cycling of anthropogenic mercury through land and ocean surfaces further complicates source attribution of mercury.

In this chapter, we use the GEOS-Chem global mercury simulation to examine the response of deposition to decreased anthropogenic emissions for four regions in the Northern Hemisphere. We examine source-receptor relationships from regional to global scales, and assess the role of cycling of anthropogenic mercury in the surface ocean and land reservoirs.

18.2 Model Description

The GEOS-Chem chemical transport model (Bey et al., 2001) simulates mercury in the atmosphere-ocean-land system. We use here model version 7.04 (http://www.harvard.as.edu/chemistry/trop/geos). The atmospheric mercury simulation is described in Selin et al. (2007) and has a horizontal resolution of 4 degrees latitude by 5 degrees longitude and 30 vertical levels. It is driven by assimilated meteorological observations from the NASA Goddard Earth Observing System (GEOS-4). Three atmospheric mercury species are simulated: elemental mercury (Hg^0), oxidized mercury ($Hg^{(II)}$), and refractory particulate mercury (Hg(p)).

Emissions are shown in Table 18.1 and Figure 18.1 and described in Selin et al. (2007). Anthropogenic emissions are based on the Pacyna et al. (2006) inventory for year 2000, with updates described in Selin et al. (2008): 50% increase in Asia, consistent with estimates of Jaffe et al. (2005) and Strode et al. (2008), and a 30%

Table 18.1 Budget of mercury in the GEOS-Chem model for the globe and the four HTAP regions

	Global	East Asia	Europe	South Asia	N. America
Total Sources (*Mg yr*$^{-1}$)	9230	1980	720	680	780
Anthropogenic	3400	1340	380	300	190
Biomass Burning	650	70	10	70	30
Land[1]	1700	290	230	110	320
Prompt Land recycling	480	80	60	60	50
Ocean[2]	3000	200	40	140	190
Total Sinks (*Mg yr*$^{-1}$)	9230	1080	560	630	900
Dry Deposition[3]	6300	750	470	410	590
Wet Deposition	2930	340	90	220	310
Mercury burden (*Mg*)	5600				
Mercury burden (*yr*)	0.6				

[1] The land source includes geogenic, evapotranspiration, and soil emissions.

[2] The ocean source listed in the table is the net ocean evasion via gas exchange.

[3] Dry deposition does not include the dry deposition of Hg^0 over the ocean, which is treated as part of the ocean term.

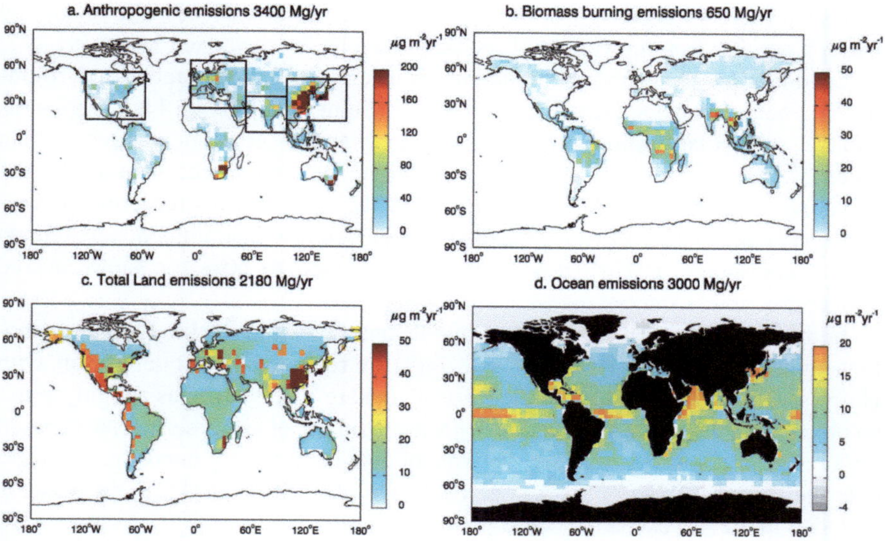

Figure 18.1 Global distribution of annual emissions ($\mu g\, m^{-2} yr^{-1}$) in the GEOS-Chem model: A) Anthropogenic, B) biomass burning, C) land, and D) ocean emissions. The domains for the four HTAP regions are indicated on panel a)

increase in the rest of the world in order to obtain an unbiased simulation. These modifications add 760 Mg yr^{-1} to the 2190 Mg yr^{-1} Pacyna et al., (2006) inventory. Our inventory also includes a 450 Mg yr^{-1} source of Hg^0 from artisanal mining, resulting in global anthropogenic emissions of 3400 Mg yr^{-1} (2215 Mg yr^{-1} Hg^0, 935 Mg yr^{-1} $Hg^{(II)}$, and 250 Mg yr^{-1} Hg(p)). Biomass burning emissions account for 650 Mg yr^{-1}, and are based on a monthly climatological biomass burning CO inventory (Duncan et al., 2003) scaled with a 2.1×10^{-7} Hg^0/CO emission ratio.

The atmospheric model is fully coupled with a mixed-layer slab ocean model described in Strode et al. (2007). The ocean model includes interactions of the mixed layer with the atmosphere and deep ocean, as well as conversion of among elemental, divalent, and nonreactive aqueous mercury species. The model is constrained to reproduce mean observed aqueous concentrations. We find a net Hg^0 global net ocean evasion flux of 3000 Mg yr^{-1} Hg^0. Re-emission of previously deposited mercury accounts for 89% of our ocean emissions, with the remaining fraction coming from upwelling of mercury from below the mixed layer.

The atmospheric model is fully coupled with a mixed-layer slab ocean model described in Strode et al. (2007). The ocean model includes interactions of the mixed layer with the atmosphere and deep ocean, as well as conversion of among elemental, divalent, and nonreactive aqueous mercury species. The model is constrained to reproduce mean observed aqueous concentrations. We find a net Hg^0 global net ocean evasion flux of 3000 Mg yr^{-1} Hg^0. Re-emission of previously deposited mercury accounts for 89% of our ocean emissions, with the remaining fraction coming from upwelling of mercury from below the mixed layer.

Land-atmosphere exchange of mercury is mechanistically parameterised, resulting in a net Hg^0 land source of 2180 Mg yr^{-1} (Selin et al., 2008). Land emissions include soil volatilisation, evapotranspiration, and a geogenic source, which together account for 1700 Mg yr^{-1}. In addition, the land source includes partial recycling of recently deposited mercury (480 Mg yr^{-1}). For this recycling component, we assume that 20% of $Hg^{(II)}$ and Hg(p) deposited to land immediately returns to the atmosphere as Hg^0 (this number is increased to 60% for snow-covered surfaces).

In the atmosphere Hg^0 is oxidized by OH and O_3, while in clouds $Hg^{(II)}$ is photochemically reduced to Hg^0. $Hg^{(II)}$ and Hg(p) are removed by wet deposition, with Hg(p) scavenged as a water soluble aerosol, and $Hg^{(II)}$ scavenged as a highly water soluble gas (Selin et al., 2007). Dry deposition to land of all three species is based on a resistance in series scheme. The model also includes dry deposition of $Hg^{(II)}$ over the oceans by loss on sea-salt aerosols. Dry deposition of Hg^0 to the ocean is included in the gas exchange ocean parameterisation.

In the simulations presented here, we use meteorological observations for the year 2001, with a 5-year spin-up period (repeating the year 2001) for the reference simulation. In addition, we conduct four perturbation experiments where primary anthropogenic emissions are reduced by 20% for each of the following regions: East Asia (EA: 15-50° N; 95-160° E), Europe and North Africa (EU: 25-65° N; 10° W-50° E), South Asia (SA: 5-35° N; 50-95° E), and North America (NA: 15-55° N; 125-60° W). The definition of these regions follows the guidelines of the Task Force on Hemispheric Transport of Air Pollution (TF HTAP) perturbation experiments (http://aqm.jrc.it/HTAP/). For each 20% perturbation experiment, we run the model for 5 additional years and present the results for the last year.

18.3 Results/Discussion

In this section, we briefly present results from our reference simulation (section 18.3.1), and then discuss the results of the perturbation experiments, focusing on the response of Hg deposition rates to emission reductions (section 18.3.2) and on the effects on land and ocean cycling (section 18.3.3).

18.3.1 Reference Simulation

Annual mean distributions of surface concentrations of Hg^0 and $Hg^{(II)}$ + Hg(p) are presented in Figure 18.2 (top panels). In the model, globally averaged surface concentrations of total mercury are 1.5 ng m^{-3}, with $Hg^{(II)}$ + Hg(p) accounting for 0.03 ng m^{-3}. Enhanced concentrations of Hg^0 are found over industrial regions, reaching

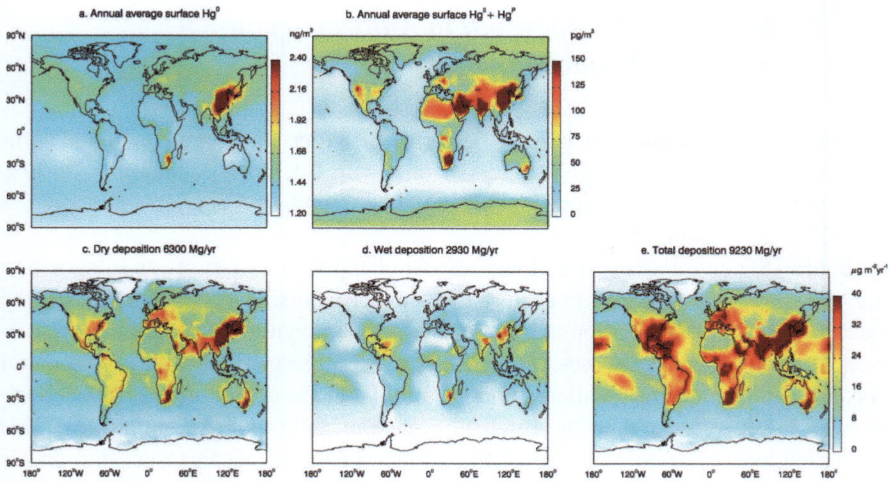

Figure 18.2 Annual mean distribution of surface concentrations of a) Hg^0($ng\ m^{-3}$) and b) $Hg^{(II)}$ + Hg(p) ($pg\ m^{-3}$). The bottom three figures show the annual mean distribution of deposition fluxes ($\mu g\ m^{-2} yr^{-1}$): c) Dry deposition, d) Wet deposition, and e) Total deposition

>2.4 ng m^{-3} over East Asia (Figure 18.2), which accounts for 39% of global anthropogenic emissions (see Table 18.1). $Hg^{(II)}$ and Hg(p) are emitted by anthropogenic sources and elevated concentrations are generally located over industrial regions. Elevated $Hg^{(II)}$ levels also occur over mountains and deserts, where enhanced mixing of free tropospheric air brings down high $Hg^{(II)}$. Over deserts, weak wet deposition of $Hg^{(II)}$ further contributes to the enhancement. Concentrations of $Hg^{(II)}$ increase with altitude in the model because of the source from oxidation of the global Hg^0 pool and longer lifetime of $Hg^{(II)}$, consistent with observations (Swartzendruber et al., 2006). Low $Hg^{(II)}$ concentrations occur over oceans, where loss on sea-salt aerosols dominates dry deposition.

Selin et al. (2007) have conducted detailed comparisons of the model to surface observations of Hg^0, $Hg^{(II)}$, and Hg(p) and find reasonable agreement. The model reproduces mean annual concentrations total gaseous mercury (TGM = Hg^0 + $Hg^{(II)}$) concentrations at land sites, but underestimates cruise observations in the North Atlantic and North Pacific by 25%. The model interhemispheric gradient for TGM is 1.2, consistent with land-based observations, but lower than the 1.5 gradient inferred from cruise observations. The reason for the discrepancy is unclear. The model captures the observed seasonal cycle of TGM at surface sites in the northern hemisphere.

Globally, wet deposition results in a 2930 Mg yr^{-1} sink of mercury (93% as $Hg^{(II)}$ and the rest as Hg(p)), while dry deposition yields a 6300 Mg yr^{-1} sink (26% Hg^0, 73% $Hg^{(II)}$, 1% Hg(p)). The overall lifetime of mercury in the model against deposition is 0.6 years (Table 18.1). The spatial distribution of deposition is shown on the bottom panels of Figure 18.2. The model reproduces the observed spatial and temporal distribution of wet deposition over the U.S. (Selin and Jacob, 2008).

18.3.2 Response of Deposition to Anthropogenic Emission Reductions

Together, the four regions where the perturbation experiments are conducted comprise 65% of global anthropogenic emissions, with EA alone accounting for 39% (1340 Mg yr^{-1}) of global anthropogenic emissions, while NA accounts for only 6% (190 Mg yr^{-1}) (Table 18.1).

Table 18.2 summarizes the global annual changes in emissions and deposition for the 20% anthropogenic perturbation experiments. The magnitude of the decrease in global deposition is proportional to the magnitude of the initial reduction in anthropogenic sources. Thus EA has the largest effect both globally and locally, and NA has the smallest effect. We find that the decrease in global deposition is 25-30% larger than the initial decrease in anthropogenic emissions. This amplification is due to response of the surface land and ocean reservoirs, and is discussed in more detail in section 3.3.

Table 18.3 summarizes the absolute changes in deposition over land for each pair of source-receptors and for the globe. For a given region, Hg deposition is due

Table 18.2 Global annual change in sources and sinks for the four perturbation simulations

	ΔEast Asia	ΔEurope	ΔSouth Asia	ΔNorth America
Change in emissions (*Mg* yr^{-1})	-326	-85.4	-71.8	-46.6
Anthropogenic (-20%)	-262	-69	-56.5	-37.3
Land[1]	-20	-6.9	-5.2	-3.7
Ocean[1]	-44	-9.5	-10.1	-5.6
Change in deposition (*Mg* yr^{-1})	-323	-84.7	-71.3	-46.2
Dry Deposition	-220	-59.5	-45.8	-30.5
Wet Deposition	-103	-25.2	-25.5	-15.7
Amplification factor $\frac{\Delta Deposition}{\Delta Anthropogenic}$	1.23	1.23	1.26	1.24

[1]The change in ocean and land emissions is calculated interactively in response to the 20% reduction in anthropogenic emissions for each region.

[2]The last row shows the amplification factor, which we define as the change in deposition divided by the change in anthropogenic emissions.

Table 18.3 Absolute change (in *Mg* yr^{-1}) in deposition over land for each pair of source-receptor regions and for the globe.

Receptors Sources	EastAsia	Europe	South Asia	N. America	Global (land only)	Global (all)
East Asia	-76	-13	-9.9	-15	-167	-323
Europe	-2.7	-23	-3.3	-3.6	-48	-85
South Asia	-2.4	-3.4	-18	-2.6	-36	-71
North America	-1.1	-2.1	-1.2	-14	-25	-46

Each line corresponds to the results of a perturbation experiment, while the columns show the response in the receptor regions. The last 2 columns show the global decrease in deposition over land only and everywhere.

to a combination of deposition of local anthropogenic emissions of $Hg^{(II)}$ and HgP and of oxidation of the global Hg^0 pool to $Hg^{(II)}$. Because of their short lifetimes, most of the anthropogenic $Hg^{(II)}$ and Hg(p) emissions are deposited near sources, while the anthropogenic Hg^0 can be transported globally and contribute to the global Hg^0 pool.

Overall, we find that 24-31% of the total change in deposition occurs over the land in the source regions (Table 18.3), another 5-10% occurs over the oceans in source regions. The remaining change in deposition is distributed globally as a result of the decrease in the global Hg^0 pool.

Figure 18.3 show the global distribution in the deposition decrease (in percent) for each perturbation experiment relative to the reference simulation, while Figure 18.4 zooms in on the deposition decrease over emission regions.

The percentage decreases in deposition over land are also summarized in Table 18.4. This percentage decrease in deposition over source regions reflects the relative contribution from local anthropogenic Hg emissions and from the global Hg^0 pool. Thus, if local deposition were solely controlled by anthropogenic emissions within the region, we would expect a 20% decrease in anthropogenic emissions to result in a 20% decrease in deposition in that region. Instead, we find decreases ranging from -12% to -3%. The EA region displays the largest reduction in local deposition

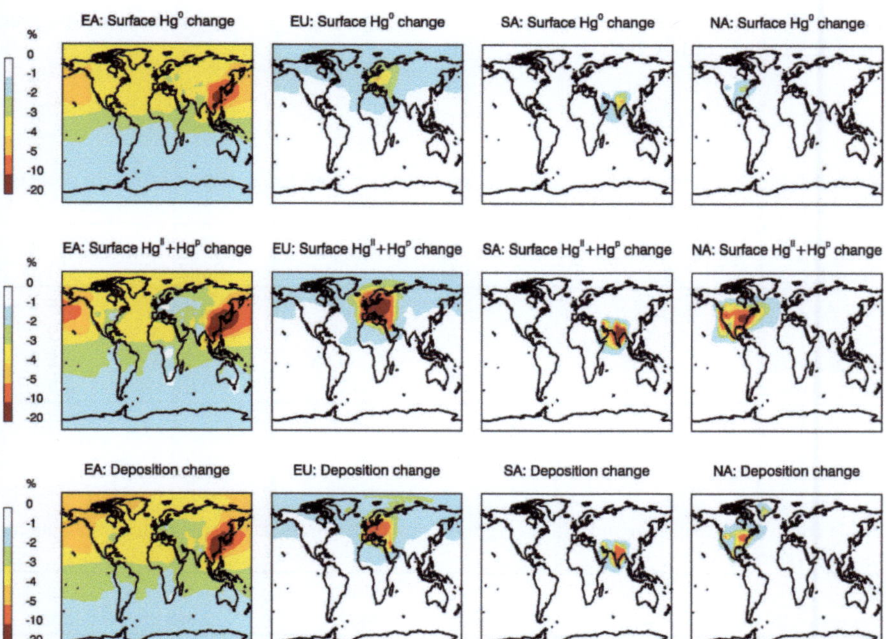

Figure 18.3 Percentage change in annual surface concentrations of Hg^0 and $Hg^{(II)}$ + Hg(p) (top 2 rows) and in deposition (bottom row) for each perturbation simulation. The perturbation simulations are shown from left to right: East Asia (EA), Europe and North Africa (EU), South Asia (SA), and North America (NA)

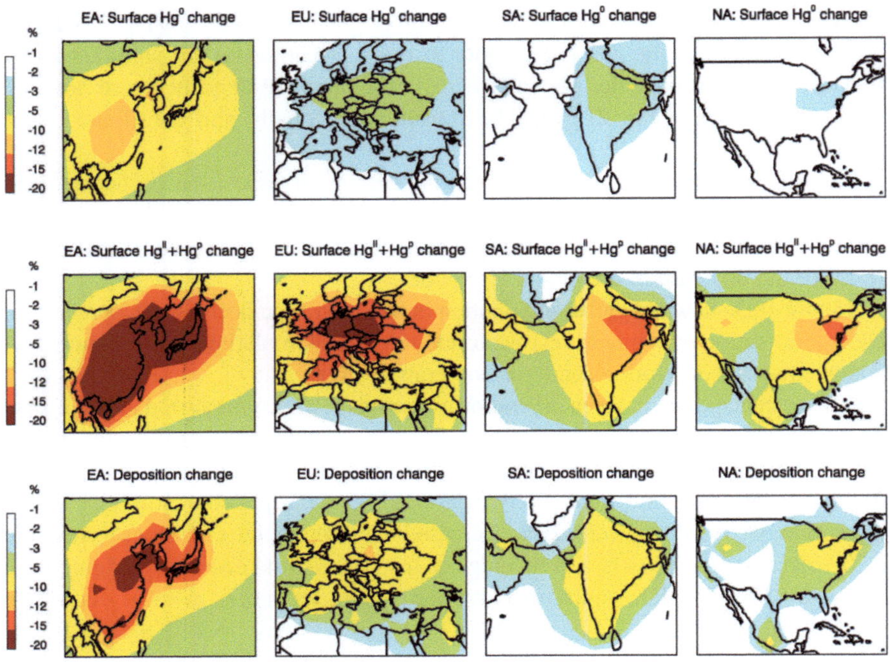

Figure 18.4 Same as Figure 18.3, but focusing on the source regions where anthropogenic emissions are reduced by 20%

Table 18.4 Mean relative change (%) in deposition over land for each pair of source-receptor regions and for the globe

Receptor: Source:	East Asia	Europe	South Asia	N.America	Global (land only)	Global (all)
East Asia	-12.3	-2.8	-2.7	-3.2	-4.0	-3.5
Europe	-0.43	-4.7	-0.88	-0.76	-1.2	-0.9
South Asia	-0.39	-0.7	-4.8	-0.55	-0.9	-0.8
North America	-0.18	-0.44	-0.33	-3.0	-0.6	-0.5

Table 18.5 Relative change in Hg^0, and $Hg^{(II)}$ + Hg(p) concentrations (%) for each pair of source-receptor region

Receptor: Source:	East Asia	Europe	South Asia	N. America	Global
East Asia	-7.6, -14	-3.2, -2.2	-3.1, -2.4	-3.5, -2.8	-3.1, -3.6
Europe	-0.8, -0.4	-2.5, -7.3	-0.9, -1	-0.9, -0.7	-1, -1.4
South Asia	-0.5, -0.3	-0.6, -0.9	-2.1, -6.5	-0.5, -0.5	-0.6, -0.9
North America	-0.3, -0.2	-0.5, -0.3	-0.4, -0.3	-1.1, -5.8	-0.4, -0.6

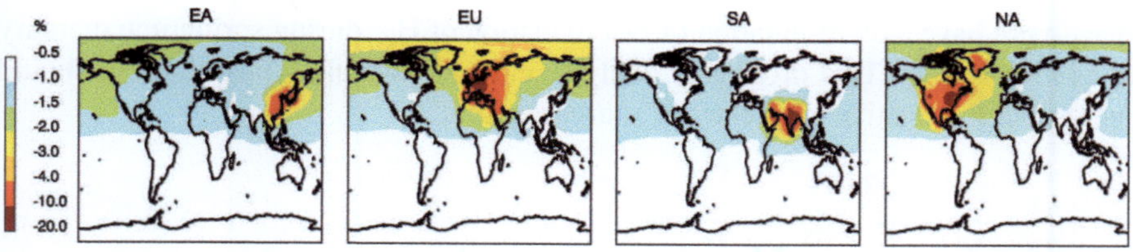

Figure 18.5 Change in annual deposition. Same as bottom panel in Figure 18.3, but scaled to a 100 Mg yr^{-1} anthropogenic emission decrease for each region

(-12%), this implies that about 60% (obtained by dividing 12 by 20) of the deposition in EA is the result of local emissions. In contrast, for NA we only find a 3% decrease in deposition, thus about 15% of the deposition in that region is due to NA anthropogenic emissions. The rest comes from oxidation of the global Hg^0 pool. This is consistent with Selin and Jacob (2008), who found that a significant fraction of wet deposition over NA results from scavenging of $Hg^{(II)}$ in the free troposphere above 850 hPa. For the EU and SA regions, we find local decreases in deposition of 5%, implying that regional emissions account for ~25% of local deposition.

Globally averaged, decreases in deposition range from 3.5% for a reduction in EA anthropogenic emissions, to less than 1% for reductions in the NA source (Table 18.4). Regions with low anthropogenic sources are the most affected by outside influence. For example, a 20% reduction in EA emissions leads to a larger decrease in NA (-3.2%) than a 20% reduction in local anthropogenic emissions in NA (-3%).

What is the relative efficiency of different regions in affecting deposition over remote regions? We assess this by normalizing the decreases in deposition to a 100 Mg yr^{-1} decrease in anthropogenic emissions. We thus multiply the deposition decreases from Figure 18.3 by global scaling factors of 0.4, 1.4, 1.8, 2.7 for EA, EU, SA, and NA, respectively. This assumes a linear response of deposition to emission changes (we verified that it is a reasonable assumption by conducting perturbation experiments varying the decrease in emissions). The results are displayed in Figure 18.5.

We see that a region of particular sensitivity to emission reductions is the Arctic (defined here as latitudes poleward of 65 N). Reductions in anthropogenic emissions from EA, EU, SA, EA contributed to deposition decreases in the Arctic of -3.8%, -1.55%, -0.53%, -0.54%, respectively. When we normalize these numbers to 100 Mg/yr decrease in anthropogenic emissions, we obtain -1.4%, -2.2%, -0.9%, -1.4%. We thus find that European emissions lead to a deposition decrease that is 60% larger than either EA or NA. Compared to SA, European emissions are a factor of 2.4 more effective. This is consistent with our general understanding of transport pathways to the Arctic (Stohl et al., 2002; Travnikov, 2005). Thus is would seem that decreases in European mercury emissions will be the most effective on a per Mg basis at reducing Hg contamination in the Arctic. However, anthropogenic European emissions have already been strongly reduced over the past decades, and at present they are three times smaller than emissions from EA. Thus, EA emissions have the strongest influence on current Arctic deposition. Note that we do not have a representation of fast oxidation of Hg^0 during springtime mercury depletion events. This process is likely to enhance overall deposition, but should not affect the relative efficiency of source regions.

18.3.3 Response of Land and Ocean Emissions

Our coupled land-atmosphere-ocean mercury simulation allows us to examine how reduced anthropogenic emissions affect the rapid cycling of mercury in the land and ocean reservoirs. Atmospheric deposition is the main source of mercury to the surface ocean. Thus, a decrease in deposition of anthropogenic mercury leads to a decrease in the concentrations of mercury in the surface ocean, which in turn results in a decrease in the net ocean-atmosphere flux. Because of the rapid cycling of Hg between the atmosphere and surface ocean, a new steady-state is reached within 2-3 years in our perturbation simulations. Similarly, prompt recycling of land emissions responds rapidly through the recycling of deposited mercury, as we assume steady-state between local deposition and land recycling. Over longer timescales (decades to centuries) exchange of Hg with the thermocline, and land cycling through evapotranspiration and volatilization will also respond, but we do not take these changes into account here.

Table 18.2 summarizes the reduction in land and ocean emissions for each anthropogenic perturbation simulation. We find that the response of these reservoirs leads to a significant amplification of the initial reduction in primary anthropogenic emissions. For example, decreasing East Asian anthropogenic emissions by 20% (-262 Mg yr^{-1}) leads to a 44 Mg yr^{-1} reduction in ocean emissions and a 20 Mg/yr reduction in land emissions. The resulting change in Hg cycling leads to an overall decrease in emissions (and thus deposition) that is 1.25 times larger than the initial change in anthropogenic emissions. Similar amplification factors (1.23-1.26) are found for the other sources regions.

The spatial distribution of the reduction in land and ocean emissions is shown in Figure 18.6. We find that half of the reduction in land emissions occurs within the source region, in response to local decreases in deposition. The remaining 50% occurs outside these source regions, as a result of decreased long-range transport of Hg^0 anthropogenic emissions, conversion to $Hg^{(II)}$, and deposition.

For ocean emissions, we see significant decreases in the coastal regions near sources (N. American East coast, Northwest Pacific, Mediterranean, Indian Sea), but the decreases are distributed more globally, with 75% of the reduced ocean emissions occurring outside the source regions (Figure 18.6). The areas with large decreases are concentrated in the Tropics, where high deposition rates and oceanic biological productivity lead to rapid reduction of deposited aqueous $Hg^{(II)}$ to aqueous Hg^0, which is then released to the atmosphere. At high latitudes, the cold water is undersaturated in Hg^0 and the ocean is a net sink for Hg^0 (Figure 18.1). Thus, as atmospheric Hg^0 decreases in our perturbation experiments, the ocean becomes less undersaturated, leading to a net increase in air-sea exchange in the high-latitude North Atlantic Ocean and Southern Ocean.

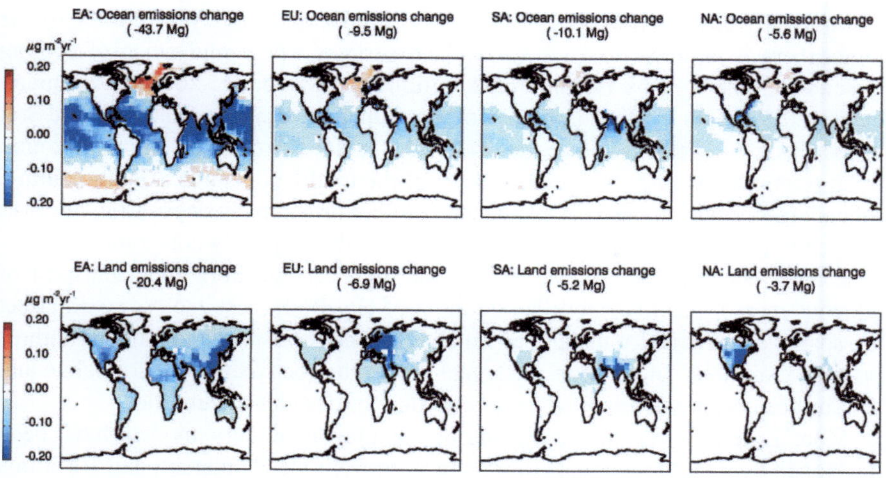

Figure 18.6 Distribution of absolute change ($\mu g\ m^{-2}\ yr^{-1}$) in annual mean ocean (top), and land (bottom) emissions for each perturbation simulation

18.4 Uncertainties in Model Results and Future Research

The distribution of deposition and its response to decreases in anthropogenic emissions is affected by a number of key uncertainties: speciation of anthropogenic emissions, reduction-oxidation chemistry of mercury in the atmosphere and clouds, precipitation, dry deposition, and cycling in land and ocean reservoirs. We briefly discuss each of these uncertainties below and emphasize areas of future research.

The speciation of anthropogenic emissions as $Hg^0/Hg^{(II)}/Hg(p)$ is poorly understood, as few source test measurements exist. In our emission inventory, $Hg^{(II)}$ + Hg(p) account for 35% of global anthropogenic emissions. This number increases to 45% for regions dominated by coal combustion. If a higher fraction of Hg were to be emitted as reactive mercury, this would lead to an increase in local deposition of $Hg^{(II)}$ + Hg(p) and reduced long-range transport of Hg^0. Assumptions about the redox chemistry of Hg in the atmosphere will also affect our results. Our chemistry scheme is very simple compared to other more complex model implementations (Ryaboshapko et al., 2002). Given the poor knowledge of rate constants, our approach has been to only implement two oxidation reactions (O_3 and OH) and adjust the rate of in-cloud photochemical reduction to best reproduce observed global Hg concentrations and seasonal variation at Northern mid-latitudes. Currently, we do not take into account halogen oxidation of Hg^0, which can be significant in polar regions, marine boundary layer, and upper troposphere (Holmes et al., 2006), but we will implement this in the future.

Dry deposition plays a dominant role in our simulated global deposition (Table 18.1). Few measurements of Hg dry deposition velocities exist. Our simulation of a large Hg^0 dry deposition over land and $Hg^{(II)}$ deposition over the oceans by sea-salt uptake, while consistent with the existing few measurements, is thus poorly constrained.

The fraction of newly deposited Hg that is available for cycling back to the atmosphere is assumed to be 20% for land surface, in the mid-range of cycling efficiencies inferred by isotopic observations (Amyot et al., 2004; Hintelmann et al., 2002). This cycling is likely to be a function of vegetation type and soil properties, which would induce variability that is not currently taken into account. Such prompt recycling is a small fraction of the total land source. The remaining, much larger, land sources will also respond to perturbations in deposition, but over a much longer time-scale, introducing a lag-time that is poorly constrained.

For the ocean emissions, we do not take into account horizontal transport of aqueous mercury and neglect riverine input of Hg. Both these factors could affect the spatial distribution of the air-sea flux of Hg^0 and its response to reduced anthropogenic emissions. In addition, exchange of mercury between the thermocline and the surface ocean will result in a response time of decades to centuries.

More generally, an increasing number of global mercury models have been developed over the past 5 years (as can be seen from the chapters in part III of this report), but the observational constraints on these models remain limited. In particular very few measurements of speciated mercury in the free troposphere exist. Such measurements, conducted on aircraft over continental scales would help quantify the distribution and chemical transformation of mercury in the free troposphere and provide crucial validation for global and regional models.

References

Amyot, M., G., Southworth, S.E. Lindberg, H. Hintelmann, J.D. Lalonde, N. Ogrinc, A.J. Poulain and K.A. Sandilands, 2004. Formation and evasion of dissolved gaseous mercury in large enclosures amended with $^{200}HgCl_2$, *Atmospheric Environment*, 38, 4279-4289.

Bey, I., et al., 2001. Global modeling of tropospheric chemistry with assimilated meteorology: Model description and evaluation, *Journal of Geophysical Research-Atmospheres*, 106, 23,073-23,096.

Duncan, B. N., R.V. Martin, A. Staudt, R. Yevich, and J.A. Logan, 2003. Interannual and seasonal variability of biomass burning emissions constrained by satellite observations, *Journal of Geophysical Research-Atmospheres*, 108, 4040, doi:10.129/2002JD002378.

Hintelmann, H., R. Harris, A. Heyes, J. Hurley, C. Kelly, D. Krabbenhoft, S. Lindberg, J.W.M. Rudd, K. Scott and V. St. Louis, 2002. Reactivity and mobility of new and old mercury deposition in a boreal forest ecosystem during the first year of the METAALICUS study, *Environmental Science and Technology*, 36, 5034-5040.

Holmes, C. D., D. J. Jacob, and X. Yang, 2006. Global lifetime of elemental mercury against oxidation by atomic bromine in the free troposphere. *Geophysical Research Letters*, 33, L20808, doi:10.1029/2006GL027176.

Jaffe, D., E. Prestbo, P. Swartzendruber, P. Weiss-Penzias, S. Kato, A. Takami, S. Hatakeyama, and Y. Kajii, 2005. Export of atmospheric mercury from Asia, *Atmospheric Environment*, 39, 3029-3038.

Pacyna, E.G., J.M. Pacyna, F. Streenhuisen, and S. Wilson, 2006. Global anthropogenic mercury emission inventory for 2000. *Atmospheric Environment*, 40, 4038-4063.

Ryaboshapko, A., et al., 2002. Comparison of mercury chemistry models. *Atmospheric Environment*, 36, 3881-3898.

Selin, N.E., D.J. Jacob, R.J. Park, R.M. Yantosca, S. Strode, L. Jaeglé and D. Jaffe, 2007. Chemical cycling and deposition of atmospheric mercury: Global constraints from observations. *Journal of Geophysical Research-Atmospheres*, 112, D02308, doi:10.1029/2006JD007450.

Selin, D.J. Jacob, R.M. Yantosca, S. Strode, L. Jaeglé, and E.M. Sunderland, 2008. Global 3-D land-ocean-atmosphere model for mercury: Present-day versus preindustrial cycles and anthropogenic enrichment factors for deposition. Global Biogeochemical Cycles, doi:10.1029/2007GB003040.

Selin, N.E., and D.J. Jacob, 2008. Seasonal and spatial patterns of mercury wet deposition in the United States: Constraints on the contribution from North American anthropogenic sources. Atmospheric Environment, doi:10.1016/j.atmosenv.2008.02.069

Stohl, A., et al., 2002. On the pathways and timescales of intercontinental air pollution transport. *Journal of Geophysical Research-Atmospheres*, 107(D23), 4684, doi:10.1029/ 2001JD001396.

Strode, S., L. Jaeglé, N.E. Selin, D.J. Jacob, R.J. Park, R.M. Yantosca, R.P. Mason, and F. Slemr, 2007. Air-Sea Exchange in the Global Mercury Cycle. *Global Biogeochemical Cycles*, 21, GB1017, doi:10.1029/2006GB002766.

Strode, S., L. Jaeglé, D.A. Jaffe, P.C. Swartzendruber, N.E. Selin, C. Holmes, and R.M. Yantosca, 2008. Trans-Pacific transport of mercury. Journal of Geophysical Research-Atmospheres, doi:10.1029/2007JD009428.

Swartzendruber, P., D.A. Jaffe, E.M. Presbo, P. Weiss-Penzias, N.E. Selin, R. Park, D.J. Jacob, S. Strode, and L. Jaeglé, 2006. Observations of reactive gaseous mercury in the free troposphere at the Mount Bachelor Observatory. *Journal of Geophysical Research-Atmospheres*, 111, D24301, doi:10.1029/2006JD007415.

Travnikov, O., 2005. Contribution of the intercontinental atmospheric transport to mercury pollution in the Northern Hemisphere. *Atmospheric Environment*, 39, 7541-7548.

Chapter 19
The ECHMERIT Model

Gerlinde Jung, Ian M. Hedgecock, and Nicola Pirrone

Summary A global atmospheric circulation model, ECHMERIT, with coupled meteorology and atmospheric chemistry including mercury has been developed. The model is designed to study in detail the atmospheric physical and chemical processes which influence the atmospheric lifetime of mercury, and therefore its global deposition patterns. ECHMERIT, based on the Global Circulation Model (GCM) ECHAM5 differs from most global mercury models in that the emissions, chemistry, transport and deposition are coupled on-line to the GCM. Many atmospheric chemistry models treat meteorological and chemical processes separately, in giving the meteorological conditions to the CTM in relatively coarse temporal resolutions. Their coupling in ECHMERIT avoids temporal and spatial interpolation, which permits more accurate representation of the interaction between chemical and meteorological phenomena, providing more highly resolved (temporal and spatial) meteorological fields. The coupling of the modules has been achieved in a modular way keeping the model as flexible as possible. This flexibility allows the model to be run with different levels of complexity. An example of a process for which a more detailed impact study could improve understanding of the atmospheric Hg cycle is biomass burning, which involves emission, reaction, gas-particle partitioning and deposition of Hg. At present ECHMERIT is being used within the HTAP Task Force to contrast and compare the performance of different models in a series of passive tracer transport experiments. Modelling experiments with extensive Hg chemistry, with both wet and dry deposition included, to assess the consequences of emission scenario changes for different Hg source areas are being evaluated.

19.1 Introduction

The global extent of elemental Hg in the atmosphere is clear evidence for its long-range transport and the relative stability of its concentration, particularly in background sites suggests that its atmospheric lifetime and/or the balance between deposition and re-emission processes are such that the atmospheric concentration is

N. Pirrone and R. Mason (eds.), *Mercury Fate and Transport in the Global Atmosphere*, DOI: 10.1007/978-0-387-93958-2_19, © Springer Science+Business Media, LLC 2009 maintained almost at a uniform level both spatially and temporally (chapter 9). Local variations in anthropogenic emissions, biomass burning and emissions from contaminated sites as well as variations in deposition rates, for example the extreme case of polar atmospheric mercury depletion events (AMDEs) produce spatial and temporal perturbations of average concentrations. However with the exception of major sources, this variability appears not to have a major effect on local average atmospheric Hg concentrations. Some meteorological influences, which affect atmospheric transport patterns, produce seasonal variations in Hg concentration in areas with very few or no local sources, but even these variations tend to be similar from year to year, bound to typical seasonal large-scale circulation patterns. With the exception of AMDEs the concentration of Hg remains within a narrow range of values in each hemisphere, the northern hemisphere having a higher average value (1.5-1.7 ng m^{-3}) than the southern (1.0-1.3 ng m^{-3}) (chapter 9, chapter 10), as a result of the majority of anthropogenic emission sources being located in the northern hemisphere and slow interhemispheric exchange. Additionally, possibly the atmospheric lifetime of Hg in the southern hemisphere is smaller due to a higher concentration of bromine following a higher sea-salt aerosol production in the southern hemisphere caused by higher wind speeds (Yang et al., 2005). Exactly how this consistancy arises is of interest because it would appear to suggest that Hg is relatively unreactive in the atmosphere, whilst at the same time recent kinetics measurements and also field measurements indicate that Hg can be oxidized really quite rapidly under certain conditions and oxidized Hg compounds are deposited far more rapidly than elemental Hg. As is described in other chapters of this report (chapter 16) when attempting to model Hg concentrations and deposition fluxes using regional transport models the model boundary conditions exert a profound influence over the results obtained. Therefore global Hg models also serve to provide boundary conditions for more highly resolved regional models. However, global models until now have had some difficulty in reproducing observed Hg concentrations (and therefore an estimate of the atmospheric Hg budget) for a number of reasons. The major uncertainties in global models are emission (and re-emission) inventories and atmospheric chemistry, both of which have been called into questioned in modelling studies (e.g. Lohman et al., 2008; Seigneur et al., 2006). Certainly changes in anthropogenic emissions, some of which have been rapid, especially in Asia make the problem more difficult. The presence of quite different rate constants for the same reaction in the peer reviewed literature compound the problem for modellers (chapter 14). ECHMERIT, being modular in design is easily configured to test individual processes, and different parameterizations of individual processes with varying levels of approximation so that not only the importance of a given process but also the way it is represented can be assessed globally, whether it is the gas phase chemistry, aqueous phase chemistry, deposition, interaction of Hg with biomass burning plumes etc. In this way ECHMERIT will be able to provide feedback to experimentalists on the processes which are key to the atmospheric Hg cycle and require further study and also to regional modellers: providing insight of which processes may need particular attention (biomass burning, halogen production from sea salt aerosol, for example) in the specific region being modelled, in order to obtain accurate concentration fields and deposition flux estimates for ecosystems to determine the impact of Hg pollution on a given area and its inhabitants.

19.2 Model Description

The new global chemistry and transport model *ECHMERIT* is a fully-coupled model, based on the Atmospheric General Circulation Model (AGCM) *ECH*AM5, and a *MER*cury chemistry module, developed at the Institute for Atmospheric Pollution of the National Research Council (CNR-IIA) of *It*aly.

The atmospheric physics part of the global-scale model ECHMERIT is based on the fifth generation global climate model ECHAM5, developed and maintained at the MPI (Max-Planck Institute) in Hamburg. The performance of ECHAM5 has been tested extensively (Roeckner et al., 2003). It is a pseudo-spectral GCM, which has been widely used to investigate atmospheric responses to various greenhouse-gas emission scenarios. The nudging routine implemented in ECHAM5 is used to relax ECHAM5 meteorology fields towards ECMWF reanalysis data, which is especially necessary if running longer-term model simulations for the reproduction of real meteorological conditions. The basic prognostic variables of the meteorological core of the model are vorticity, divergence, temperature, logarithm of surface pressure and the mixing ratios of the various water species. Radiation includes an annual, as well as a diurnal cycle of solar forcing and is calculated with a 2 hour time step. Absorption due to water vapour, CO_2, and O_3 are taken into account, as well as scattering by cloud water or ice in the cloudy part of the column and scattering and absorption due to aerosols according to Mie-theory.

Gridscale cloud water content and cloud ice content are calculated from their respective budget equations, including transport of cloud water and a simplified representation of microphysical processes such as condensation, evaporation, formation of cloud droplets through coalescence, and sedimentation of ice crystals. Sub-gridscale cloud formation is also parameterized. A bulk mass flux scheme is used to represent convective precipitation processes. Deep, mid level and shallow convection are considered. Organized entrainment is calculated from buoyancy, organized detrainment is computed from a spectrum of clouds detraining at different heights. The soil and land surface model comprises the budgets of soil heat and water, snow cover and the heat budget of land ice. It takes into consideration the stomatal control of surface evapotranspiration through vegetation, interception, and the dependence of the sensible heat flux on snow coverage in a highly parameterized form.

In the current version of ECHMERIT a gas phase chemistry scheme based on the CBM-Z mechanism (Zaveri and Peters, 1999) and including gas phase Hg chemistry is implemented, making use of the KPP (Kinetic Pre-Processor) (Damina-Iordache et al., 2002) to provide the source code necessary to solve the system of Ordinary Differential Equations (ODEs) representing the mechanism.

KPP is a useful for two important reasons. It is very simple to change the chemical scheme, re-run KPP and obtain the appropriate code to be linked to ECHMERIT, it also has a sparse Jacobian option which when used makes the code obtained faster to run. The chemical mechanism also includes aqueous phase chemistry (and gas-aqueous phase exchange) to simulate the partitioning of atmospheric constituents between phases and their reactions in cloud droplets. Mass transfer to and from the aqueous phase is modelled as two distinct unidirectional reactions as described in Pirrone et al. (2000). Precipitation is in many regions the major deposition pathway for Hg, as not only is Hg oxidised within droplets, oxidised Hg compounds present in the atmosphere are also very readily scavenged by clouds and rain. The Hg chemistry in the model includes the gas phase oxidation of Hg by O_3, OH and Cl, and aqueous phase oxidation by O_3, OH and HOCl, as well as aqueous phase complexation and the reduction of $HgSO_3$. The entire mechanism now includes 121 chemical species and represents gas, as well as aqueous-phase chemistry, with a total number of 288 chemical reactions. To solve the chemical ODE system the computationally efficient second order Rosenbrock solver was chosen. In order to save computational time and to avoid model instabilities for grid cells with a low water content ECHMERIT distinguishes between wet and dry chemistry. The wet chemistry module is only called in case of cloud water content exceeding a specific pre-defined threshold value.

Transport is calculated, using a flux-form semi-Lagrangian advection scheme, that is already implemented in ECHAM5 for passive tracer variables, as well as water vapour. It is a flux-form scheme and therefore mass conserving and it preserves linear tracer correlations. The tracers are transported through advection, convection and vertical diffusion. The different transport processes are simulated separately using an operator splitting method (Aghedo et al., 2008). In ECHMERIT 27 species are transported and deposited.

The dry deposition scheme, applied to the lowermost model layer, is based on the commonly applied multiple-resistance approach of Wesely and Hicks (1977). The deposition velocity is calculated as a function of aerodynamic resistance, boundary layer resistance and surface resistance. Within ECHMERIT an approach similar to that of Kerkweg et al. (2006) is used, which takes into consideration bulk properties of the respective surfaces, without accounting for removal processes occurring in different layers of the canopy. Deposition velocities are calculated from turbulent transfer, vegetation activity and uptake rates on soil, water and snow/ice. Then a scaling of all species resistances to those of SO_2 and O_3 as in Wesely (1989) is applied. This makes the scheme easily extendible to other chemical species. Basic resistances values for SO_2 and O_3 are taken from Kerkweg et al. (2006). The partitioning of land grid cells into fractions of snow/ice, bare soil, water/wet skin and vegetation, that is already implemented in ECHAM5 is taken into consideration to account for subgridscale land-surface characteristics and their impacts on friction velocity, roughness length etc. Therefore grid-average deposition velocities are calculated as the area-weighted average of the deposition velocities for each subgrid fraction. Landuse dependent values such as LAI, roughness length, etc. are directly transferred from ECHAM5 to the deposition module. Aerosol dry deposition velocities are calculated, considering both, dry deposition and gravitational settling as sedimentation, according to the approach of Slinn (1982).

A simple approach was chosen to calculate wet deposition of the transported chemical species. Below-cloud, as well as in-cloud scavenging are considered. The parameterization calculates the loss through wet deposition as proportional to the concentration of air pollutants. The factor of proportionality, that is the scavenging rate, depends on an assumed scavenging efficiency, the total rainfall intensity (grid-scale and subgridscale), cloud water content and species solubility according to Henry's law, a mean cloud or rain droplet radius and rain droplet fall velocity, according to the approach of Seinfeld and Pandis (1998). No wet deposition is calculated for species with low solubilities (expressed through a Henry's law constant of less than 100 M atm^{-1}).

Emissions from emission inventories e.g. the POET emissions (Granier et al., 2005) are included within a preprocessing step interpolated to the model grid, using the mass conserving remapping tool of the Climate Data Operators (CDO) (Schulzweida et al., 2007) and are subsequently added to the chemical species concentration with each advection time step. Anthropogenic emissions of mercury species for the year 2000 from Pacyna et al. (2006) are used, which are speciated as elementary, divalent and particulate mercury. Mercury emissions from soils depend on surface temperature. The approach for a simulation of mercury emissions from the vegetation canopy depends on canopy evapotranspiration, which is calculated within the meteorology part of the model, and the mercury concentration in the surface soil waters, which is currently set to a constant value of 5×10^5 ng m^{-3}. Emissions from forest fires are included, in mapping an annual mean value of 650 Mg yr^{-1} Hg (chapter 8) to the spatial and temporal distribution of CO biomass burning emissions from the POET emissions inventory (Granier et al., 2005). Ocean emissions are also mapped according to biogenic CO emissions from the ocean, assuming a global annual sum of 3000 Mg yr^{-1} of Hg to be emitted from the ocean. A certain fraction (20%) of wet and dry deposition of RGM, $Hg^{(II)}$ and Hg(p) are assumed to be directly reemitted to the atmosphere over land areas. Re-emissions over the oceans are included within the global annual oceanic emissions.

As in the spectral transform model ECHAM5, the horizontal resolution can range from T21 (5.06 degrees) to T159 (0.75 degrees). In the vertical the model is discretized with a hybrid-sigma-pressure system with from 19 to 31 non-equidistant layers. The upper layer is at 10 hPa. Initial and lower boundary sea surface temperatures (SST) are reanalysed monthly averaged observational SST data from the AMIP (Atmospheric Model Intercomparison Project). Land use characteristics, such as roughness length, leaf area index (LAI) etc. are directly derived from ECHAM5

For the present purpose of the HTAP model intercomparison, the meteorology-chemistry model allows an almost direct comparison with other global models as the Hg chemistry included is similar. The model intercomparison will permit differences arising from model formulation to be diagnosed. In the future however it will be necessary to review the Hg chemistry mechanism, especially in light of the conclusions from Chapter 14 of this report.

19.3 Results/Discussion

Mercury concentrations are in the following given as ppq (parts per quadrillion), which is 1×10^{-6} ppb and under standard surface conditions 1 ng m^{-3} equals 112 ppq.

19.3.1 Sensitivity Analysis of Chemical Mechanisms

When using global models the first priority is to ensure that the model gives results, such as atmospheric concentrations of important chemical species and deposition fluxes which are in line with observations. This may sound rather obvious but in reality it is not necessarily easy to achieve. While for some atmospheric constituents there are well known constraints in terms of emissions, reaction rate constants and a well organised observational network with high temporal and spatial resolution with which a model may be compared, ozone would be a good example, this is definitely not the case for mercury. The anthropogenic emissions of mercury are the cause of some debate and however much those who compile emission databases do their best to be up to date, the actual emissions are changing. Natural emissions and the re-emission of previously deposited Hg are even less well defined. There are a number of rate constant determinations for important atmospheric reactions (both in the gas and aqueous phases) which are not in agreement, and unfortunately the spatial and temporal range of measurements of Hg and its compounds in the atmosphere is seriously limited. Regional models, although now tending toward the use of output from global models for their boundary conditions, were previously constrained by unvarying boundary conditions which had a large and linear influence on the Hg concentrations within their modelling domain (Chapter 16). Global models cannot be constrained in the same way, and can, if the balance between emission, re-emission, atmospheric redox reactions and deposition is not appropriate give atmospheric Hg concentration distributions which bear no resemblance to reality. During the development of ECHMERIT a number of variations of the Hg chemical mechanism were investigated and it was found that the use the different reaction rate constants published in the literature for the same reaction profoundly influenced the results of the simulations.

The reaction between ozone and elemental Hg is a key factor in determining the rate at which elemental Hg is removed from the atmosphere. Should it be shown, as has been suggested (Calvert & Lindberg, 2005), that this reaction is not of particular relevance in the atmosphere then all global (and regional) mercury models presently in use would need to be rewritten! Assuming that it is important the reaction rate used in models profoundly influences the distribution of Hg in model simulations.

The gas phase reaction between elemental Hg and OH which is used in atmospheric Hg models has also been questioned in the literature (Calvert & Lindberg, 2005), with some authors suggesting that it does not occur under atmospheric conditions (Chapter 14). Using a chemical mechanism which does not include the

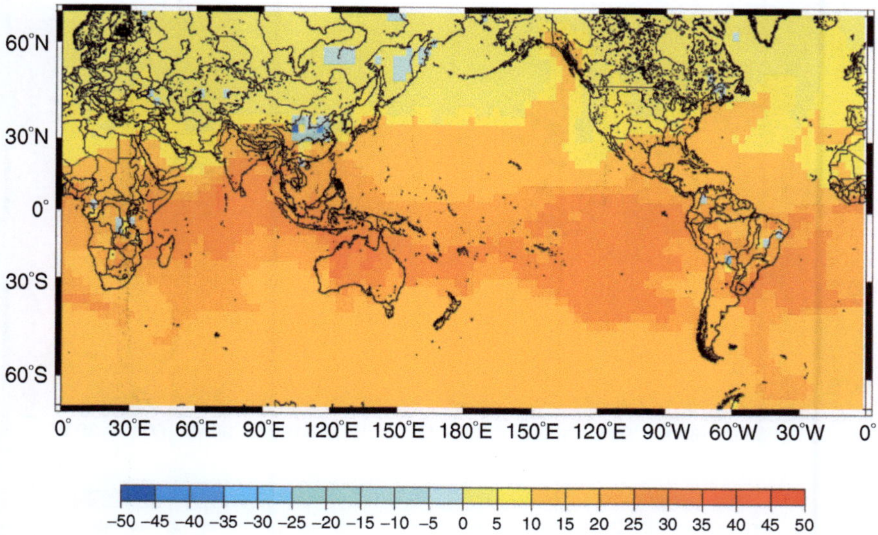

Figure 19.1 Difference (ppq) in Hg^0 concentrations between a run excluding and one including the reaction of OH with Hg^0, both using the Hg + O_3 reaction rate constant of Hall (1995)

reaction between OH and elemental Hg and uses the reaction rate constant for the Hg + O_3 reaction published by Hall (1995), it is possible to obtain the homogeneous atmospheric elemental Hg distribution which the available measurements support. Using the same chemical mechanism, but including the OH reaction with Hg^0 leads to a pattern of concentration differences in elemental Hg, particularly noticeable in equatorial regions as illustrated in Figure 19.1. Major decreases of Hg^0 are to be found in a zone around and south of the equator, corresponding to (for the simulated month of January) the zone of elevated OH levels during the day. Replacing the Hg + O_3 reaction rate constant with that of Pal and Ariya (2004) and including the OH reaction as well gives a distribution of differences with respect to the run excluding OH and with the O_3 reaction rate constant of Hall (1995), which is illustrated in Figure 19.2.

It is clear that there is a much lower concentration of elemental Hg in tropical regions and there also appears to be little evidence of intercontinental transport, which is to be expected if Hg is being rapidly oxidised and therefore more rapidly deposited.

The higher ozone levels over of the northern hemispheric warm current regions of the oceans (Figure 19.3) and in the tropics, as well as the higher temperatures in the tropics, given the temperature dependence of the Hg + O_3 rate constant (Pal and Ariya, 2004) combine to produce a pronounced latitudinal gradient in the capacity of the troposphere to oxidise Hg which in the simulations corresponds to a latitudinal gradient in the elemental Hg concentration in the lowest model layer; something which has never been observed (Figure 19.4). The possible effect of the gas phase reduction of RGM by CO (Pongprueksa et al., 2008) is currently being investigated.

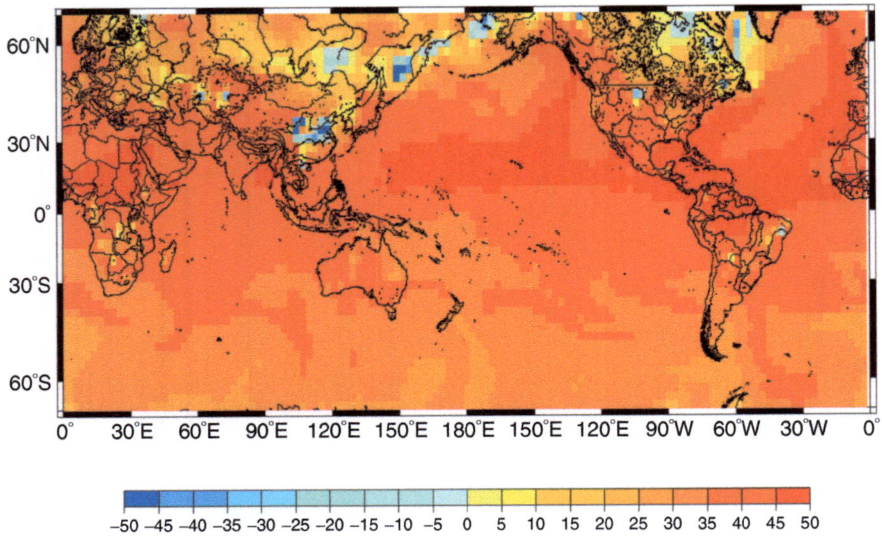

Figure 19.2 Difference (ppq) in Hg^0 concentrations between a run including the $Hg + O_3$ reaction rate constant of Hall (1995) and excluding the OH reaction and one run including OH reaction and using the $Hg + O_3$ reaction rate constant of Pal and Ariya (2004)

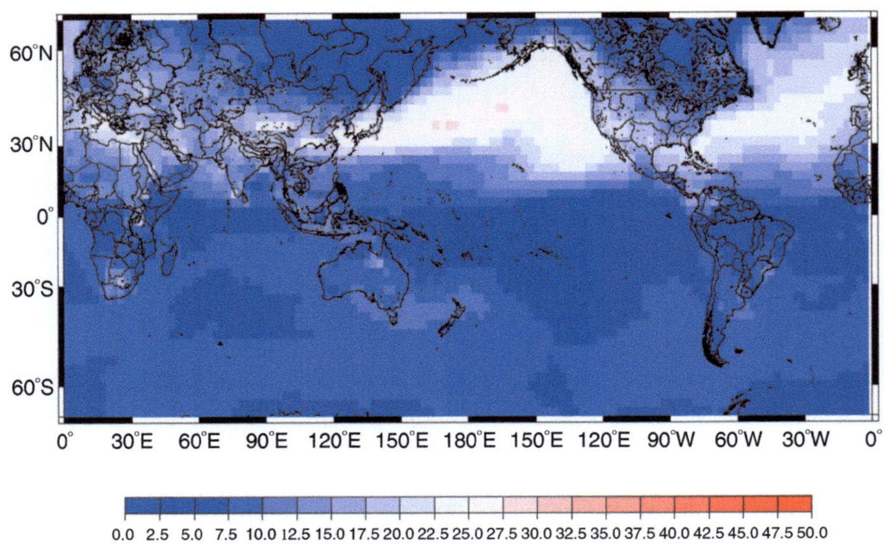

Figure 19.3 Monthly mean ozone concentration (ppb), January 2001

The other effect that the removal of the OH reaction from the mechanism has is to give RGM concentrations in the lowest model layer which show very little diurnal variation, even in the MBL where measurements show that in the absence of

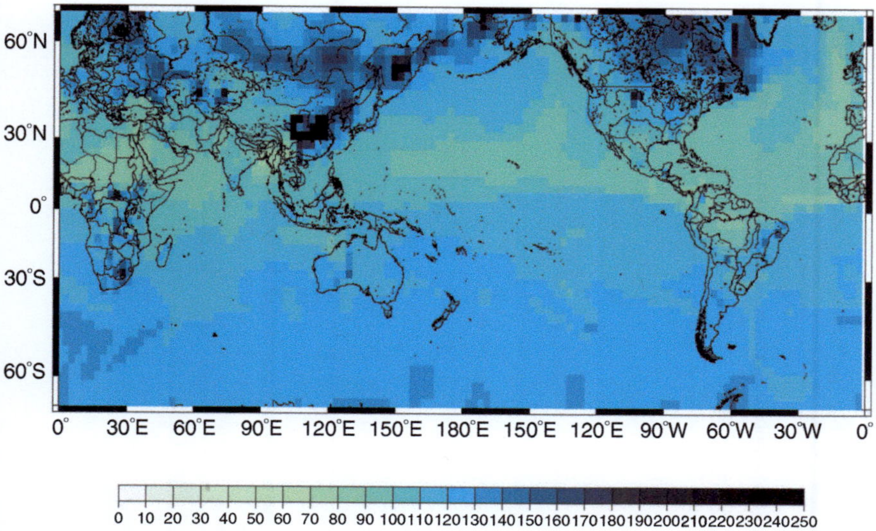

Figure 19.4 Hg^0 concentration (ppq) using the $Hg^0 + O_3$ reaction rate of Pal and Ariya (2004) and including the Hg^0 + OH reaction

precipitation RGM shows distinct variations with maxima towards midday and minima at midnight. It has been suggested that this is due to oxidation by halogens (particularly Br) which are released from sea salt aerosols.

It is now well established that halogen containing compounds oxidise Hg, during AMDEs the concentrations of O_3 and OH reach extremely low levels and yet elemental Hg continues to be oxidised often leading to concentrations below the detection limit of the automatic instruments commonly used for atmospheric Hg sampling. The inclusion of Br in the chemical mechanism employed in global Hg models represents a major obstacle, because of the complexity of the chemical mechanism which would need to be used and the uncertainty in the magnitude of the sea salt source of bromine.

19.3.2 Model Evaluation

For a meaningful simulation of the oxidation of Hg to RGM in current atmospheric Hg chemistry schemes, one of the most important species is ozone, (in its own right and due to the the formation of OH) which therefore needs to be simulated with accuracy. Six hour simulated ozone concentrations for winter (February) and summer (June) of the year 2001 for the lowest model layer are compared to observational surface stations of the EMEP measurement network Figures 19.5 and 19.6). It is demonstrated, that the ozone levels are generally represented well. In particular, the influence of synoptic disturbances are represented within the model run, as

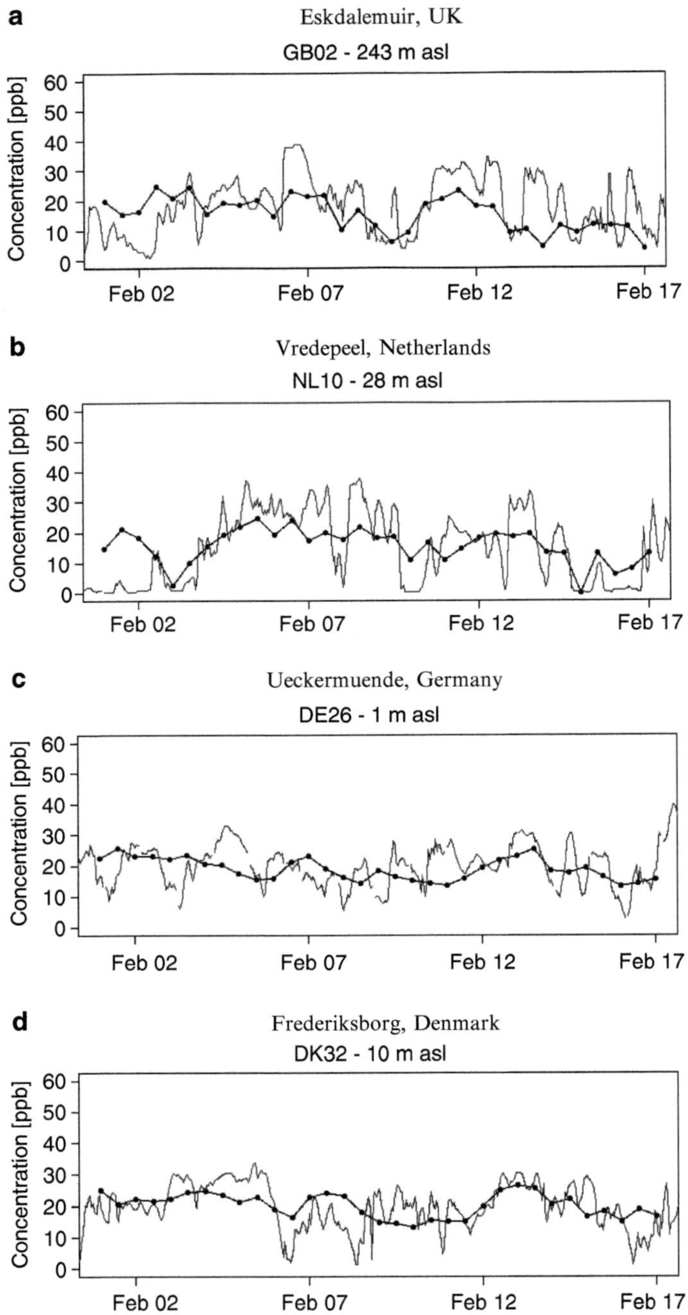

Figure 19.5 Simulated O_3 concentrations (ppb) versus observations, February 2001

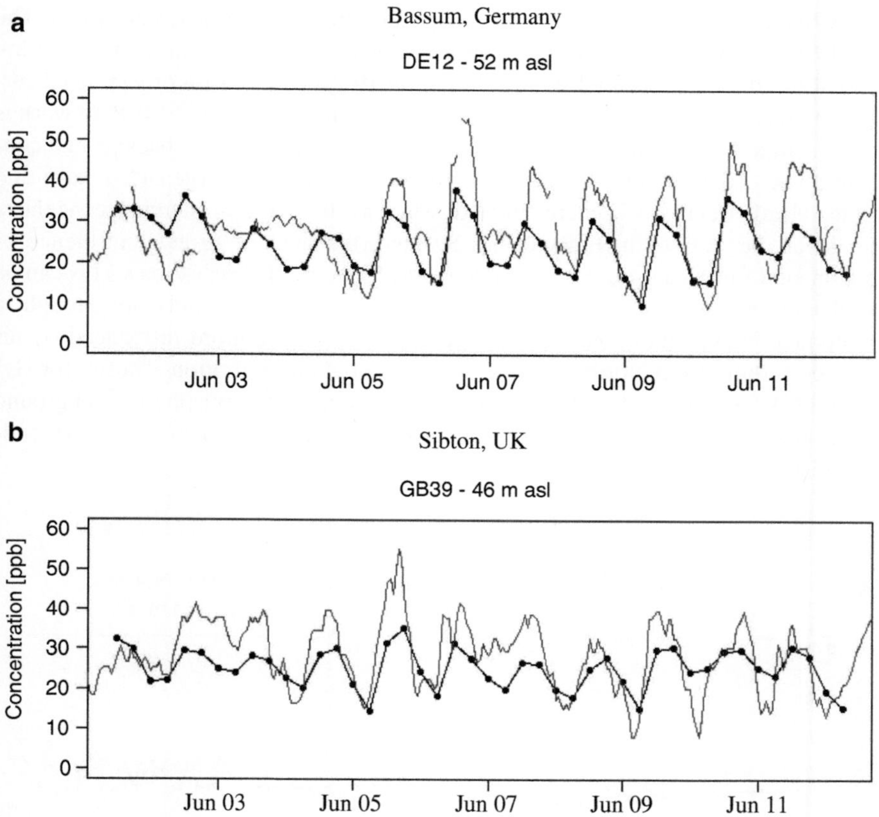

Figure 19.6 Simulated O_3 concentrations (ppb) versus observations, June 2001

can be seen by the good agreement of synoptic scale (days-week) variability of ozone concentrations. High frequency variability in the model is underestimated, especially in winter. In summer (Figure 19.6) a pronounced, photochemically dependent, daily cycle is simulated, with a good agreement of duration and frequency of high ozone levels, but a slight underestimation in magnitude.

Partly this can be explained by the coarse resolution of a global model, where the grid box value represents the volume mean species concentration, which is then compared to an observed point value. Therefore observed daily minimum and maximum values will often not be reproduced, as they strongly depend on subgrid-scale effects (e.g. induced through topography, land-sea contrasts). Similar behaviour was observed as well e.g. by Duncan et al. (2008) for the GMI CTM. Hence an additional validation on basis of ozone columns, daily, monthly and annual means is being performed as well.

A first evaluation of simulated mercury concentrations was also performed. The model was therefore initialized with an assumed initial (background) concentration and allowed to spin-up for 1 month. In Figure 19.7 a comparison of simulated surface mercury concentrations with observed concentrations of the EMEP network is demonstrated. Generally a quite good agreement is observed. The background concentration of TGM for the two background stations Spitsbergen (Norway) and Mace Head (Ireland) is underestimated, whereas the range of mercury concentrations for the stations in Finland and Sweden, that have a stronger influence of continental European sources, is in better agreement with observations. This can be explained by the fact that a spin-up period of 1 month is possibly not enough to reproduce background concentrations at stations that are more influenced by air masses coming from North America, and the initial concentrations chosen for Hg^0 are with 1.5 ng m^{-3} slightly lower than observed northern hemispheric background concentrations. Therefore a longer spin-up period, as is used in the HTAP full-chemistry simulations is necessary.

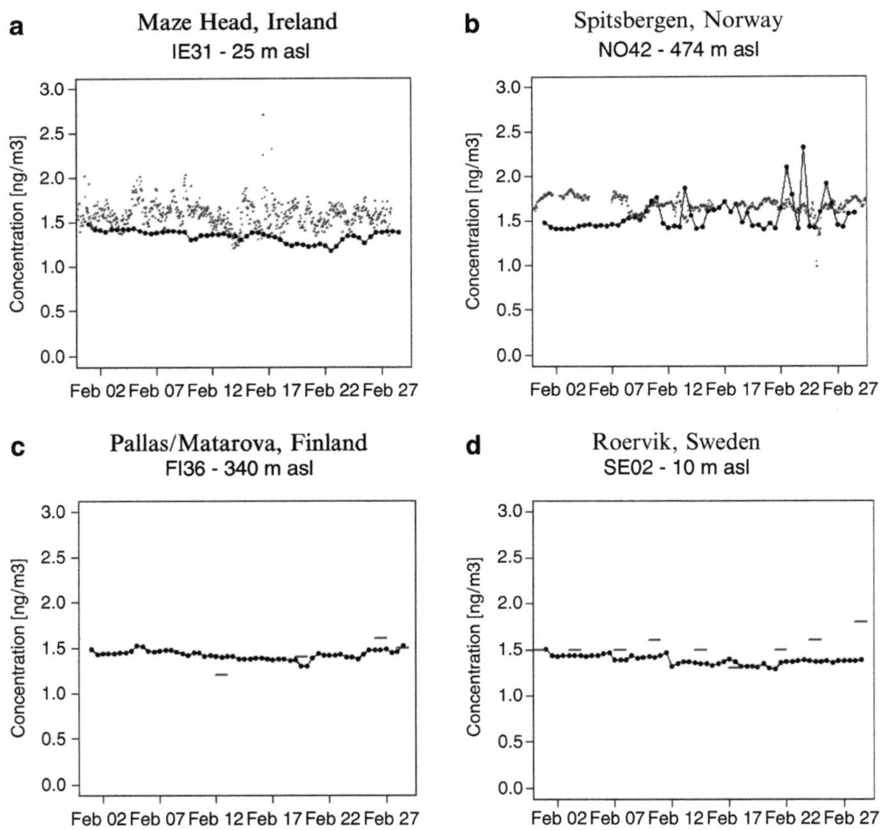

Figure 19.7 Simulated TGM concentrations (*ng m^{-3}*) versus observations, February 2001

19.3.3 Tracer Transport Studies

Within the framework of the HTAP (Hemispheric Transport of Air Pollutants) model intercomparison study, passive tracer experiments were performed to identify source-receptor relationships for the northern hemisphere. Therefore four different source and receptor regions were defined, as demonstrated in Figure 19.8. The regions are the following, EU for Europe, NA for North America, EA for East Asia and SA for South Asia (more or less restricted to the Indian subcontinent). As passive tracers different artificial tracers were defined, in order to represent certain chemical species. Amongst the predefined tracers a CO-like tracer with a lifetime of 50 days and 3 different Hg-like tracers were defined with lifetimes of 180, 360 and 540 days. The destruction of the species was represented by a simple exponential decay with the respective lifetimes. No complex chemistry and neither wet nor dry deposition are included in this experiment. The simulations were performed with a spin-up period of 1 year for the CO-like tracers and a spin-up of 2 years for the Hg-like tracers. The simulation year (2001) was run, making use of the nudging routine of ECHAM5, for a relaxation of the simulated meteorology to ECMWF ERA40 reanalysis data. Monthly mean emission data are used for CO, whereas for the Hg-like tracers only annual means of anthropogenic emissions were available (Tables 19.1 and 19.2). Clearly the South and East Asian source regions show a

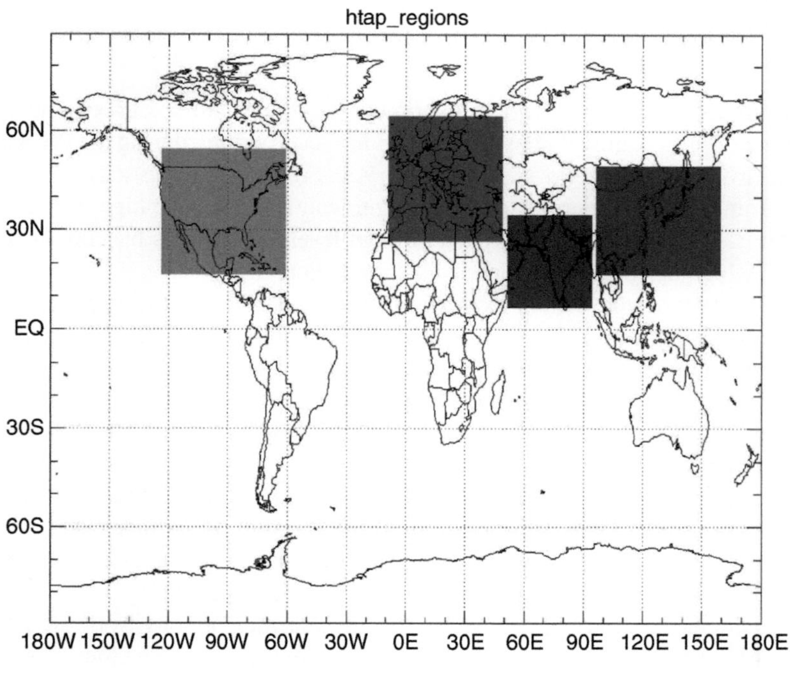

Figure 19.8 Source and receptor regions for HTAP experiments

Table 19.1 Monthly mean emission of CO for source regions in total and as percentage of the global value

	EA *(Tg)*	EA *(%)*	SA *(Tg)*	SA *(%)*	EU *(Tg)*	EU *(%)*	NA *(Tg)*	NA *(%)*	Gobal *(Tg)*
Jan	14.99	14	11.95	11	6.48	6	6.37	6	109.20
Feb	13.26	16	10.21	12	5.80	7	5.97	7	85.10
Mar	14.01	18	19.12	25	6.66	9	7.21	9	77.24
Apr	13.41	16	23.73	28	6.94	8	8.96	11	83.97
May	9.68	14	6.48	9	6.74	10	9.29	13	69.19
Jun	7.72	11	4.63	7	6.17	9	8.23	12	67.70
Jul	6.80	8	3.81	5	6.85	8	8.05	10	83.03
Aug	8.09	8	5.04	5	7.65	7	8.64	8	105.34
Sep	9.58	11	6.52	7	7.14	8	8.48	9	90.23
Oct	11.98	16	8.35	11	7.08	9	8.67	12	75.14
Nov	12.94	17	9.88	13	6.82	9	7.66	10	77.29
Dec	14.60	15	11.62	12	6.73	7	6.63	7	96.30

Table 19.2 Annual mean anthropogenic emissions of Hg^0 for source regions in total and as percentage of the global value

	EA *(Mg)*	EA *(%)*	SA *(Mg)*	SA *(%)*	EU *(Mg)*	EU *(%)*	NA *(Mg)*	NA *(%)*	Global *(Mg)*
Annual	474.38	37	94.70	7	160.86	13	78.36	6	1276.72

much higher intra-annual variability of CO emissions, than Europe and North America. This is mainly due to pronounced biomass burning during the dry season, with a strong maximum in April for South Asia. In the monsoon season CO emissions from these two source regions are significantly lower. The high overall emissions of CO in August are partly due to high levels of biomass burning in central Africa, which lies outside the defined source regions.

The anthropogenic mercury emissions used in the tracer experiment have no annual cycle. Table 19.2 shows that more than one third (37%) of global anthropogenic emissions of Hg occur within the East Asian source region. The second largest source region is Europe which accounts for 13% of global emissions. The different lifetimes of the Hg-like tracers lead to distinctly different distributions and rates of mixing of the tracers in the atmosphere (compare Figure 19.9). The tracer concentrations of the Hg-like tracer with a 360 day lifetime, from the different source regions in the mid-troposphere (~500 hPa) and in the surface layer are illustrated in Figures 19.10 and 19.11, respectively. Clearly the dominating contribution of Hg deriving from EA and EU on the globe can be evaluated.

The simulated contributions to the specific receptor regions from the defined source regions for CO as well as Hg-like tracers, are summarized in Table 19.3. The values were calculated for the entire tropospheric column, excluding the lowermost

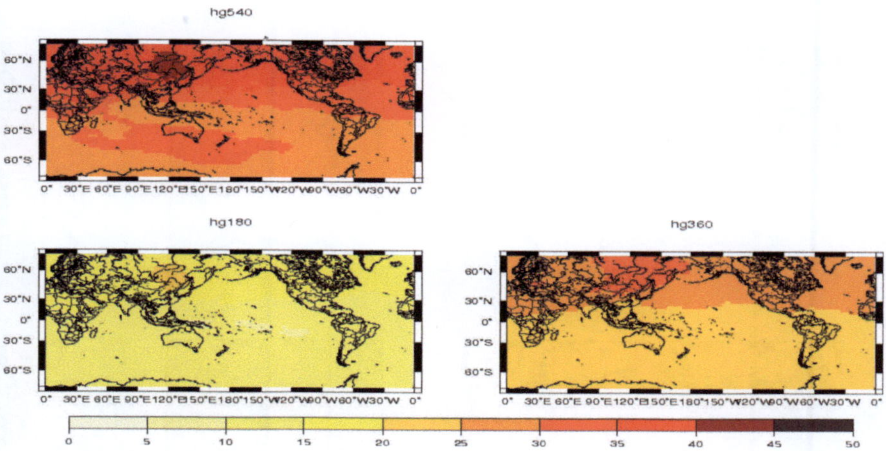

Figure 19.9 Hg-like tracers with different lifetimes (days), concentration (ppq) in 500 hPa

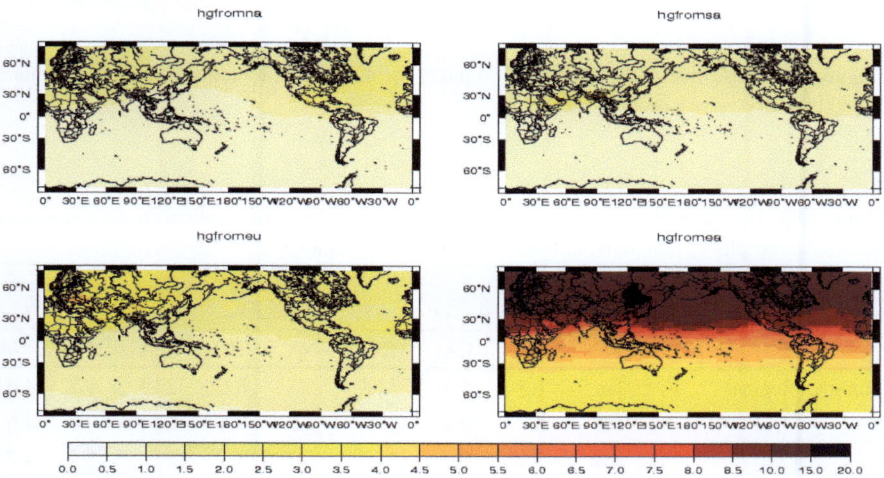

Figure 19.10 Hg-like tracer (lifetime 360 days) for 4 different source regions, concentration (ppq) in 500 hPa, July 2001

model level. The most pronounced differences for the different receptor regions between summer and winter, that can be found for the CO-like tracer are a higher contribution of the not specifically considered source regions (mostly in the southern hemisphere). This can be partly traced back to the changed atmospheric circulation patterns and a stronger inflow from the South to the receptor regions in the northern hemisphere. Furthermore in July a minor influence of east, as well as

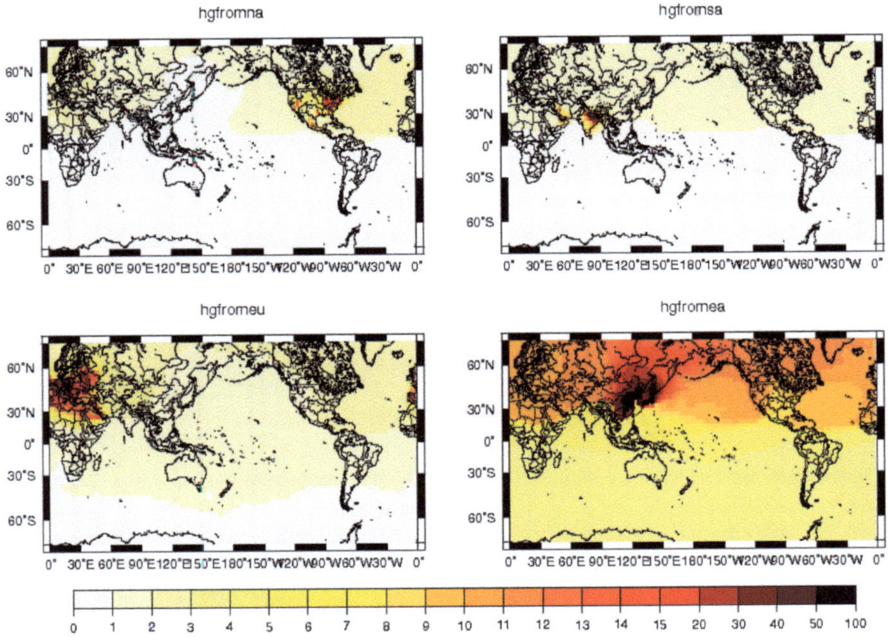

Figure 19.11 Hg-like tracer (lifetime 360 days) for 4 different source regions, concentration (ppq) in surface layer, July 2001

Table 19.3 Hg tracer experiment monthly percentages of tracer concentrations in the atmosphere above the receptor region deriving from a specific source region

			CO			Source Regions		*Hg*		
			EU	EA	SA	NA	EU	EA	SA	NA
Receptor Regions	SA	jan	6.3	15.5	32.1	5.5	8.6	27.5	9.7	4.6
		feb	5.1	9.8	23.3	4.5	8.4	27.6	9.2	4.6
		mar	5.2	9.9	21.2	5.4	8.3	27.6	8.1	4.7
		apr	5.2	10.0	25.4	5.5	8.0	27.0	7.9	4.5
		may	5.7	9.2	28.5	7.1	8.1	26.4	8.1	4.6
		jun	10.5	9.8	24.8	8.7	9.4	26.3	9.2	4.4
		jul	12.2	8.7	19.1	8.0	9.5	27.1	9.5	4.2
		aug	10.1	6.9	19.0	6.7	8.5	26.6	9.6	4.1
		sep	8.7	5.3	20.7	6.5	8.5	26.1	9.6	4.4
		oct	5.6	4.2	25.4	5.3	7.7	24.6	9.6	4.3
		nov	4.2	5.4	29.3	5.1	7.0	24.8	9.8	4.3
		dec	3.6	6.3	25.7	4.5	6.9	25.1	8.7	4.3

(continued)

Table 19.3 (continued)

		Source Regions							
		CO					*Hg*		
		EU	EA	SA	NA	EU	EA	SA	NA
EA	jan	8.2	35.1	16.7	8.0	8.5	40.7	5.7	4.6
	feb	6.5	31.4	14.4	6.1	8.4	40.0	5.8	4.4
	mar	6.5	28.5	18.6	5.5	8.6	39.1	5.6	4.4
	apr	4.7	25.9	33.8	4.3	7.9	38.5	6.0	4.2
	may	5.0	26.3	26.8	5.4	7.5	37.6	6.4	4.0
	jun	5.4	30.4	17.4	6.0	6.9	39.8	5.8	3.7
	jul	5.5	28.4	10.8	5.9	6.7	40.1	5.2	3.6
	aug	4.4	28.5	7.7	5.5	6.2	40.9	4.8	3.7
	sep	4.9	25.9	9.2	6.6	6.6	38.7	5.3	4.1
	oct	5.7	30.3	13.2	7.1	7.0	39.2	5.6	4.2
	nov	6.4	31.1	11.5	7.9	7.4	39.1	5.0	4.3
	dec	6.8	31.7	12.4	7.4	7.7	39.0	5.0	4.4
EU	jan	24.1	21.4	14.0	17.8	15.3	37.5	5.3	6.4
	feb	20.8	23.1	13.1	15.4	15.0	36.9	5.3	6.6
	mar	19.3	22.3	13.3	15.2	14.4	36.9	5.3	6.5
	apr	20.5	19.7	17.1	16.0	15.0	36.3	5.1	6.5
	may	19.9	19.3	20.0	17.8	14.7	36.9	5.0	6.4
	jun	21.1	17.1	13.3	21.9	13.9	36.2	5.0	6.4
	jul	24.9	13.8	7.9	24.3	14.4	35.5	4.8	6.5
	aug	25.8	11.5	4.4	23.8	14.3	35.6	4.5	6.6
	sep	27.5	11.9	2.8	22.7	15.1	35.5	3.9	6.5
	oct	24.3	16.1	4.0	23.3	13.8	36.2	4.0	6.6
	nov	23.6	20.2	6.6	21.1	13.7	37.1	4.3	6.4
	dec	24.9	21.4	8.8	17.3	14.7	35.7	4.5	6.2
NA	jan	7.2	26.1	16.0	19.8	8.3	38.6	5.6	7.2
	feb	5.6	24.9	14.7	16.5	8.4	37.4	5.9	6.9
	mar	4.6	24.0	17.1	16.6	8.2	37.6	5.8	6.9
	apr	4.0	21.5	24.9	18.7	8.1	37.1	5.8	6.7
	may	4.1	20.8	25.6	21.5	7.9	37.2	5.7	6.7
	jun	4.5	20.3	14.9	27.3	7.6	36.9	5.3	6.9
	jul	4.5	16.1	7.8	33.3	7.6	36.4	4.7	7.8
	aug	5.0	13.5	4.1	32.1	7.9	35.9	4.3	7.7
	sep	4.6	14.7	3.4	30.8	7.4	35.9	4.1	7.5
	oct	5.4	19.2	5.8	27.9	7.5	37.2	4.3	7.1
	nov	5.1	23.0	9.7	24.8	7.4	37.1	4.8	7.0
	dec	4.7	25.2	11.8	19.6	7.2	36.6	5.0	6.6

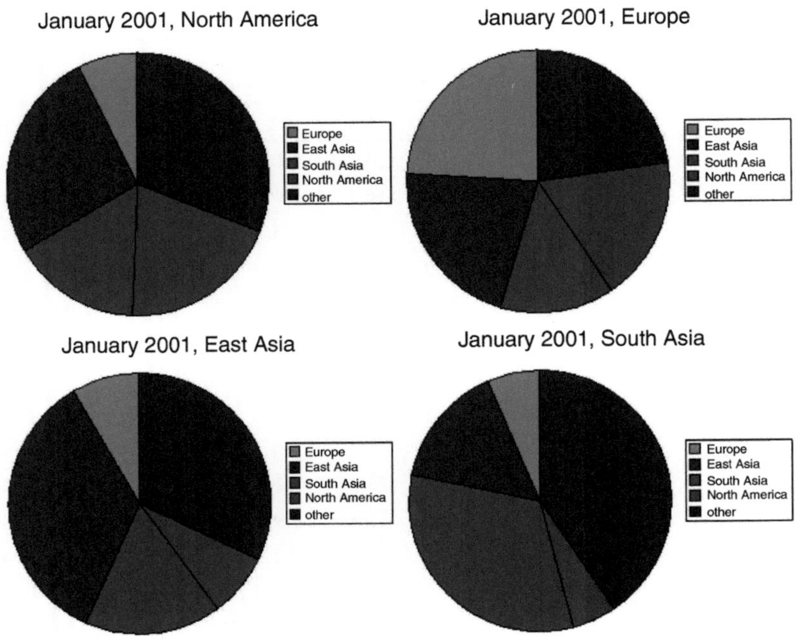

Figure 19.12 CO-like tracer: contribution of source regions to receptor regions, January 2001

south Asian sources to all receptor regions can be delineated (compare Figures 19.12 and 19.13). The North American and European contribution in contrast does not show a comparable large intra-annual variability, as do the emissions. One very clear signal of a breakdown of South Asian outflow can be observed from the month of July to December. This is to one part due to the strong decrease in CO emissions from biomass burning with the beginning of the monsoon season in June/ July. On the other hand, a change in the atmospheric circulation with the monsoon causes a change in transport pathways. Due to dominating westerly winds and also determined by the lifetime of CO, the fraction of CO arriving in Europe originating from North America is almost as large as that from East Asia.

Looking at the results for the mercury-like tracer (lifetime 360 days) in Figure 19.14 draws a different picture. In almost all receptor regions, the largest contributions are due to only two source regions: Europe and East Asia. These are the regions with the largest annual emissions and additionally the longer lifetime of the Hg-like tracer with respect to the CO-like tracer leads to a better mixing and a longer range transport of this tracer. From Table 19.3 it becomes obvious, that also the intra-annual differences are smaller. This effect can, to a large extent, be contributed to the missing annual cycle in the emission data, but is also due to the longer lifetimes and therefore better mixing of Hg in the atmosphere.

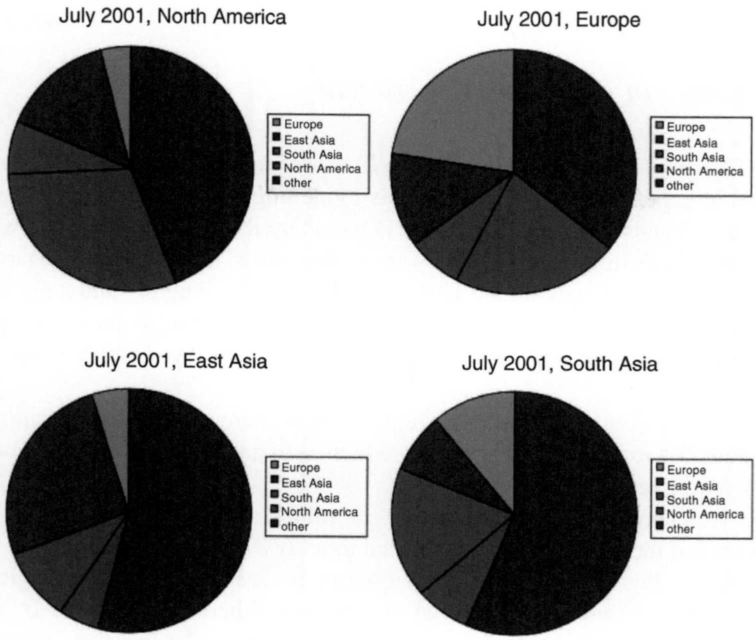

Figure 19.13 CO-like tracer: contribution of source regions to receptor regions, July 2001

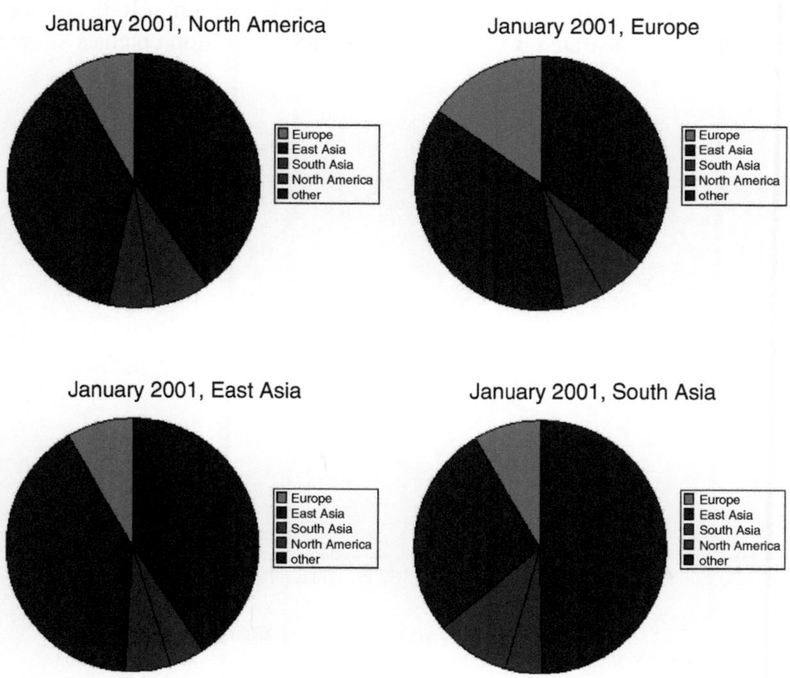

Figure 19.14 Hg-like tracer: contribution of source regions to receptor regions, January 2001

19.3.4 Emission Reduction Experiment

ECHMERIT will also contribute to the HTAP model intercomparison with emission reduction experiments. To illustrate this modelling approach, short, preliminary runs of 3 months (January-March 2001) including the full chemistry and wet, as well as dry deposition mechanisms were performed. These runs were started with a globally uniform background Hg^0 concentration of 1.5 ng m^{-3} and a pre-run of one month. Therefore these experiments have to be seen as preliminary, example runs that serve to illustrate the capability of the model to be used for this kind of experiment.

A control simulation with unchanged emissions of mercury species was run, as well as a reduction experiment with a 20% reduction of anthropogenic emissions in all defined source regions (see Figure 19.8).

Figure 19.15 shows the change (control run minus reduced emissions) in Hg^0 concentrations in the surface model layer due to a 20% emission reductions in all source regions. Clearly the largest decrease can be seen in the source regions themselves, but a signal is found over the entire northern hemisphere. In the southern hemisphere hardly any change is simulated, due to the fact that southern hemispheric emission were not changed and additionally because inter-hemispheric (global) mixing occurs on a time scale of about one year.

The impact of 20% emission reductions on the deposition of mercury species is demonstrated in an example, focusing on dry and wet deposition of RGM (compare Figure 19.16). It is illustrated, that both wet and dry deposition changes show the

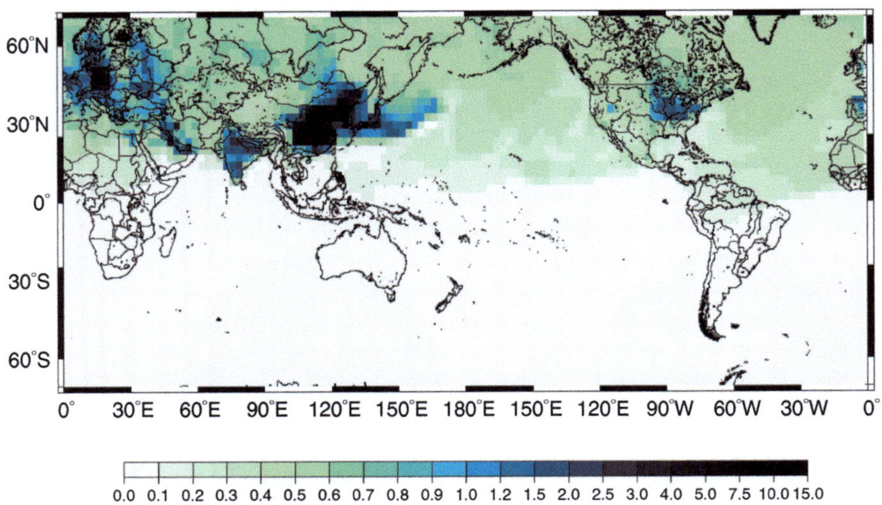

Figure 19.15 Changes (control run minus reduction experiment) in Hg^0 concentration (ppq) with a reduction of 20% in all 4 source regions, Jan-Mar 2001

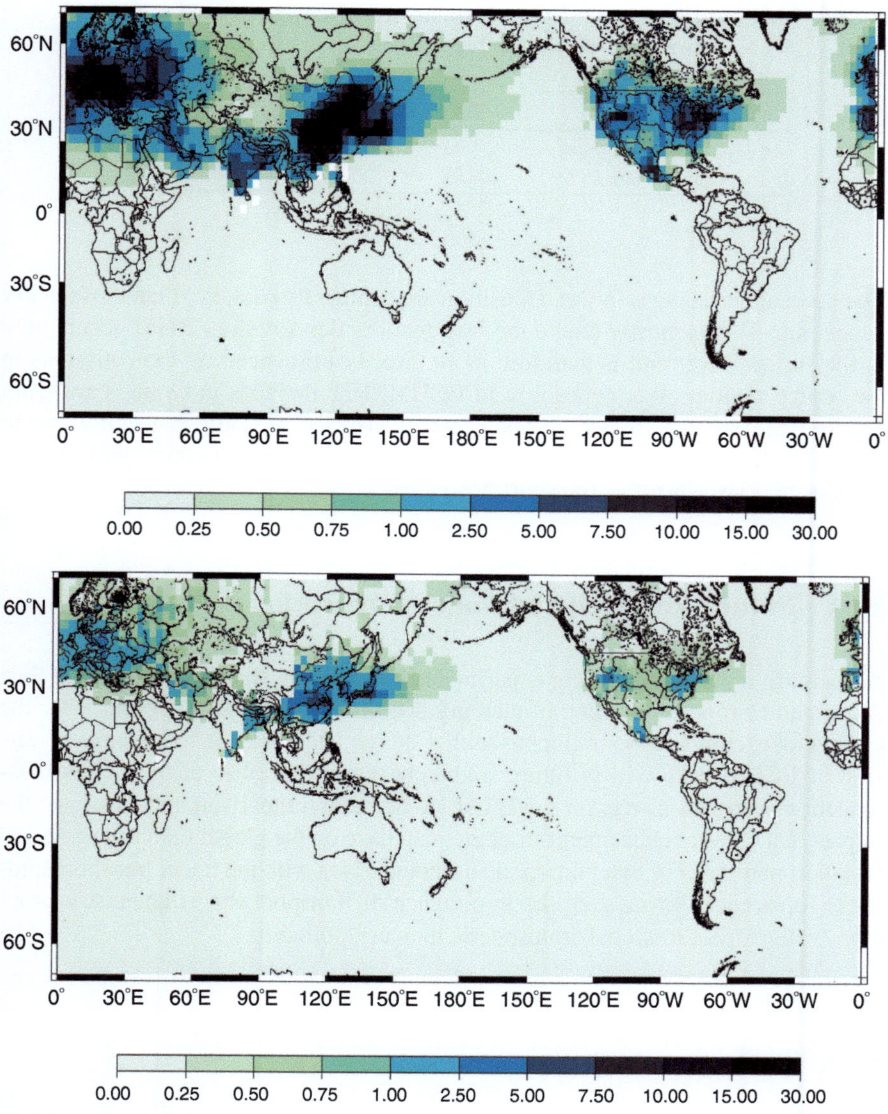

Figure 19.16 Changes (control run minus reduction experiment) in wet (below) and dry (up) deposition due to a 20% emission reduction in all 4 source regions, Jan-Mar 2001

strongest change near emission sources. Clearly wet deposition change (as well as wet deposition itself) is negligible in desert regions like the Sahara. For the three month period, a reduction in mercury emissions for all four source regions leads to a strong signal in dry deposition change over the regions of East Asia (-8.98%) and Europe (-4.86%) (compare Table 19.4). This reflects once more the importance of these two regions, and the fact that the largest emissions derive from there.

Table 19.4 Changes (control run minus reduction experiment) in wet and dry deposition (%) in the receptor regions for a 20% emission reduction in all source regions, Jan-Mar 2001

	EA	SA	EU	NA
wet deposition change (%)	1.91	0.56	0.93	0.43
dry deposition change (%)	8.98	2.87	4.86	1.92

The percentage changes in wet deposition are comparably smaller than those in dry deposition. This is mostly due to the fact, that dry deposition of RGM occurs only in the surface layer and is therefore more directly influenced by the emissions in the source regions. Wet deposition in ECHMERIT includes in-cloud scavenging and is therefore determined to some extent also by mercury concentrations in higher model levels, up to 10-15 km height, which also reflect emission processes in other regions as well as transport processes.

19.4 Future Research and Policy Implications

Already from the passive tracer experiments it is obvious, that the influence of East Asian and European emission of mercury species are of major importance for the entire globe, compared to pollutants with a shorter lifetime. The full chemistry version of ECHMERIT will in future studies be a useful tool to analyse emissions-reduction scenarios as e.g. for the HTAP model intercomparison, to investigate the impact that emission abatement strategies can have on the global mercury cycle and transport patterns. For that purpose also a comparison with the tracer transport studies is foreseen, to investigate the importance of transport versus chemistry processes with respect to global atmospheric mercury pollution.

References

Aghedo A., Rast S. & Schultz M. G. (2008): Sensitivity of tracer transport to model resolution, forcing data and tracer lifetime in the general circulation model ECHAM5; *Atmospheric Chemistry and Physics Discussions*; Vol. 8: p. 137–160.

AMIP: http://www.pcmdi.llnl.gov/projects/amip/AMIP2EXPDSN/BCS/amip2bcs.php.

Calvert J.G. & Lindberg S.E. (2005): Mechanisms of mercury removal by O3 and OH in the atmosphere; *Atmospheric Environment*; Vol. 39: p. 3355–3367.

Damina-Iordache V., Sandu A., Damian-Iordache M., Carmichael G. R. & Potra F. A. (2002): The kinetic preprocessor KPP: a software environment for solving chemical kinetics; *Computers Chemical Engineering*; Vol. 26: p. 1567–1579.

Duncan B. N., West J. J., Yoshida Y., Fiore A. M., & Ziemke J. R. (2008): The influence of European pollution on ozone in the Near East and northern Africa; *Atmospheric Chemistry and Physics Discussions*; Vol. 8: p. 1913–1950.

Granier C., Guenther A., Lamarque J., Mieville A., Muller J., Olivier J., Orlando J., Peters J., Petron G., Tyndall G. & Wallens S. (2005): POET, a database of surface emissions of ozone precursors; techreport; available on the internet at: http://www.aero.jussieu.fr/projet/ ACCENT/ POET.php.

Hall B. (1995): The phase oxidation of elemental mercury by ozone. *Water, Air and Soil Pollution*; Vol. 80: p. 301–315.

Kerkweg A., Buchholz J., Ganzeveld L., Pozzer A., Tost H. & Joeckel P. (2006): Technical Note: an implementation of the dry removal processes DRY DEPosition and SEDImentation in the Modular Earth Sumodel System (MESSy); *Atmospheric Chemistry and Physics Discussions*; Vol. 6: p. 6853–6901.

Lohman K., Seigneur C., Gustin M. & Lindeberg S. (2008): Sensitivity of the global atmospheric cycle of mercury to emissions; *Applied Geochemistry;* accepted for publication.

Pacyna E.G., Pacyna J.M., Steenhuisen F. & Wilson S. (2006): Global anthropogenic mercury emission inventory for 2000. *Atmospheric Environment*, Vol. 40 (22): p. 4048–4063.

Pal B. & Ariya P. (2004): Studies of ozone initiated reactions of gaseous mercury: kinetics, product studies, and atmospheric implications. *Physical Chemistry Chemical Physics*; Vol. 6: p. 572–579.

Pirrone N., Forlano L. & Hedgecock I.M. (2000): Role of the ambient aerosol in the atmospheric processing of semivolatile contaminants: a parameterized numerical model (Gas-Particle Partitioning (GASPAR)). *Journal of Geophysical Research*; Vol. 105(D8): p. 9773–9790.

Pongprueksa P., Lin C.-J., Lindberg S.E., Jang C., Braverman T., Bullock O.R., Ho T.C. & Chu H.-W. (2008): Scientific uncertainties in atmospheric mercury models III: Boundary and initial conditions, model grid resolution, and $Hg^{(II)}$ reduction mechanism; *Atmospheric Environment*; Vol. 42: p. 1828–1845.

Roeckner E., Baeuml G., Bonaventura L., Brokopf R., Esch M., Giorgetta M., Hagemann S., Kirchner I., Kornblueh L., Manzini E., Rhodin A., Schlese U., Schulzweida U., Tompkins A., (2003): The atmospheric general circulation model ECHAM5: Model description. Max Planck Institute for Meteorology, Technical Report, 349.

Schulzweida U., Kornblueh L. & Quast R. (2007): CDO User's Guide, Climate Data Operators, Version 1.0.7; techreport; www.mpimet.mpg.de/fileadmin/software/cdo/.

Seigneur C., Vijayaraghavan K. & Lohmann K. (2006): Atmospheric mercury chemistry: Sensitivity of global model simulations to chemical reactions; *Journal of Geophysical Research – Atmospheres*; Vol. 111, D22306, doi:10.1029/2005JD006

Seinfeld C. P. & Pandis S. N. (1998): Atmospheric Chemistry and Physics, From Air Pollution to Climate Change.

Slinn, W. G. N., (1982) Prediction for particle deposition to vegetative canopies. *Atmospheric Environment*; Vol.16: p. 1785–1794.

Wesely M. L. (1989): Parameterization of the surface resistances to gaseous dry deposition in regional-scale numerical models; *Atmospheric Environment*; Vol. 23: p. 1293–1304.

Wesely M. L. & Hicks B. B. (1977): Some factors that affect the deposition rates of sulfur dioxide and similar gases on vegetation; *Journal of Air Pollution Control Assessment*; Vol. 27: p. 1110–1116.

Yang X., Cox R.A., Warwick N.J., Pyle J.A., Carver G.D., O'Connor F.M. & Savage N.H. (2005): Tropospheric bromine chemistry and its impacts on ozone: A model study; *Journal of Geophysical Research*; Vol. 110, D23311, doi:10.1029/2005JD006244.

Zaveri R. A. & Peters L. K. (1999): A new lumped structure photochemical mechanism for large-scale applications; *Journal of Geophysical Research*; Vol. 104(D23): p. 30.387–30.415.

Chapter 20
The EMEP/MSC-E Mercury Modeling System

Oleg Travnikov and Ilia Ilyin

Summary The EMEP/MSC-E hemispheric chemical transport model (MSCE-HM-Hem) and its regional version (MSCE-HM) are applied for operational calculations of mercury transboundary pollution within the European region and in the Northern Hemisphere. This chapter contains examples of the models application for assessment of mercury atmospheric dispersion and deposition both on hemispheric and regional scales. Model simulations of mercury atmospheric dispersion in the Northern Hemisphere have been performed for the period 1990-2004. Long-term changes of mercury deposition during this period have been evaluated for different continents and regions of the Northern Hemisphere. Obtained modelling results have been compared with long-term monitoring data from various national and international networks. Besides, intercontinental transport of mercury as well as sensitivity of mercury deposition in the Northern Hemisphere to emission reduction in different continents have been estimated.

20.1 Introduction

Mercury is among the priority pollutants considered under the Convention on Long-Range Transboundary Air Pollution (CLRTAP). Along with other heavy metals it was included into the Protocol on Heavy Metals – an international binding instrument under the Convention regulating atmospheric emissions of these pollutants and aimed at reduction of their adverse effects on human health and the environment. Scientific support of development and implementation of the Protocol is provided by the Cooperative Programme for Monitoring and Evaluation of the Long-range Transmission of Air Pollutants in Europe (EMEP). To meet requirements of the Protocol the Meteorological Synthesizing Centre - East of EMEP (EMEP/MSC-E) develops and applies appropriate chemical transport models to provide Parties to the Convention with information on transboundary fluxes and deposition of heavy metals in Europe including mercury.

This chapter contains examples of application of the EMEP/MSC-E chemical transport models for assessment of mercury atmospheric dispersion and deposition

N. Pirrone and R. Mason (eds.), *Mercury Fate and Transport in the Global Atmosphere*,
DOI: 10.1007/978-0-387-93958-2_20, © Springer Science+Business Media, LLC 2009

both on hemispheric and regional scales. Particular attention is paid to evaluation of long-term changes of mercury deposition in different parts of the Northern Hemisphere and to estimates of relative importance of the intercontinental atmospheric transport for mercury deposition and their sensitivity to emission reduction in various continents.

20.2 Model Description

The EMEP/MSC-E hemispheric chemical transport model (MSCE-HM-Hem) and its regional version (MSCE-HM) are applied for operational calculations of mercury and other heavy metals transboundary pollution within the European region and in the Northern Hemisphere in connection with the EMEP programme and other activities relating to the CLRTAP. Detailed description of the models is available in (Travnikov and Ryaboshapko, 2002; Travnikov and Iliyn, 2005; Travnikov, 2005). A brief overview of the model formulation is presented below.

The hemispheric MSCE-HM-Hem model is a three-dimensional Eulerian type chemical transport model driven by off-line meteorological data. The model domain covers the whole Northern Hemisphere with resolution 2.5° × 2.5° (Figure 20.1(a)). The terrain-following pressure-based vertical coordinate consists of eight irregular layers up to the lower stratosphere. The model considers mercury emissions from anthropogenic and natural sources, transport in the atmosphere, chemical transformations both in gaseous and aqueous phases, and deposition to the ground. The regional version of the model (MSCE-HM) is formulated on the polar stereographic projection and covers the European region and surrounding areas by a regular grid with 50×50 km spatial resolution at 60°N (Figure 20.1b). Both models have the same formulation of major physical and chemical processes and are coupled by the one-way nesting. Daily mean concentrations of three atmospheric mercury forms (Hg^0, $Hg^{(II)}_{gas}$, Hg_{part}) pre-calculated with the hemispheric model are used for definition of initial and boundary conditions for the regional model.

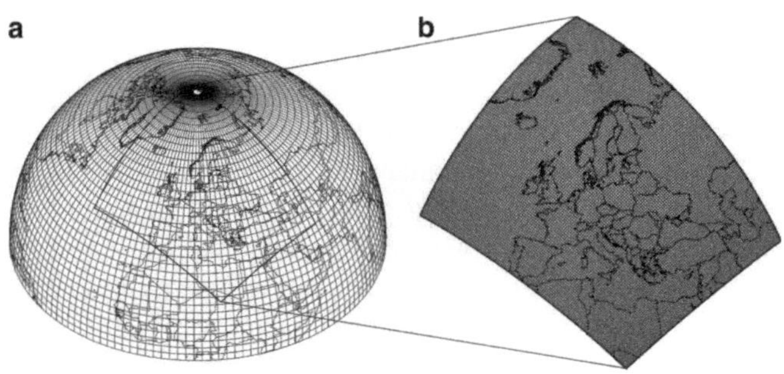

Figure 20.1 Hemispheric (**a**) and the EMEP regional (**b**) grids of the MSCE-HM-Hem and MSCE-HM models

The chemical scheme of mercury transformation in the atmosphere is based on the kinetic mechanism developed by Petersen et al. (1998). The original scheme was optimized in order to accelerate the model performance for operational calculations. For this purpose very slow reactions were neglected, whereas very fast ones were replaced by appropriate equilibriums. The model description of physical and chemical transformations of mercury in the atmosphere includes dissolution in cloud droplets, gas-phase and aqueous-phase oxidation by ozone, chlorine and hydroxyl radical, aqueous-phase reduction via decomposition of sulphite complexes, formation of chloride complexes, and adsorption by soot particles in cloud water. A summary of all chemical transformations included to the model is presented in Table 20.1.

The monthly mean data on air concentration of chemical reactants involved in reactions with mercury are taken from Wang et al. (1998) for ozone, Chin et al. (1996) for sulphur dioxide, and Spivakovsky et al. (2000) for hydroxyl radical. The original data were interpolated to the model grid. In order to take into account the diurnal cycle of OH radical we assume zero concentration at night and concentrations proportional to the cosine of the solar zenith angle during daytime. Besides, air concentrations of OH were decreased by a factor of 2 in the cloud environment and below clouds to account for reduction of photochemical activity (Seigneur et al., 2001). Besides, following Seigneur et al. (2001) we adopt air concentration of molecular chlorine in the lowest model layer over the ocean to be 100 ppt at night and 10 ppt during daytime and zero concentration over land. Model description of removal processes includes dry deposition and wet scavenging. The dry deposition scheme is based on the resistance analogy approach (Wesely and Hicks, 2000) and allows taking into account deposition to different land cover types (forests, grassland, water surface etc.). Dry deposition of particles to vegetation is described using the theoretical formulation by Slinn (1982) and fitted to experimental data (Ruijgrok et al., 1997; Wesely et al., 1985).

The parameterization of dry deposition to water surfaces is based on the approach suggested by Williams (1982) taking into account the effects of wave breaking and aerosol washout by seawater spray. Parameters of dry deposition of gaseous oxidised mercury are chosen as for those of nitric acid (zero surface resistance) because of similar solubility of these substances. Besides, dry deposition of gaseous elemental mercury to vegetation is considered. The deposition velocity of this form varies from 0 to 0.03 cm s^{-1} depending on surface temperature, solar radiation and vegetation type. The model distinguishes in-cloud and sub-cloud wet scavenging of particulate species and highly soluble reactive gaseous mercury based on empirical data. Besides, the precipitation rate is scaled for convective precipitation according to Walton et al. (1988) to take into account fractional coverage of a grid cell with precipitating clouds.

The hemispheric MSCE-HM-Hem model is driven by off-line meteorological parameters supplied by the low atmosphere diagnostics system (SDA) developed in co-operation with Hydro-meteorological Centre of Russia (Rubinstein and Kiktev, 2000). The system provides 6-hour weather prediction data along with estimates of the atmospheric boundary layer parameters and covers the Northern Hemisphere with the model spatial resolution. NCAR/NCEP re-analysis data is utilized as the

Table 20.1 Summary of mercury transformations included into the model

Reactions and equilibriums	k or H	Units	Reference
$Hg^0_{(gas)} + O_3(gas) \rightarrow Hg(II)_{(part)} + products$	2.1×10^{-18} exp(-1247/T)	cm^3 molecule^{-1} s^{-1}	Hall, 1995
$Hg^0_{(gas)} + {}^{\cdot}OH_{(gas)} \rightarrow Hg(II)_{(part)} + products$	8.7×10^{-14}	cm^3 molecule^{-1} s^{-1}	Sommar et al., 2001
$Hg^0_{(gas)} + Cl_2(gas) \rightarrow Hg(II)_{(gas)}$	2.6×10^{-18}	cm^3 molecule^{-1} s^{-1}	Ariya et al., 2002
$Hg^0_{(aq)} + O_3(aq) \rightarrow Hg^{2+}_{(aq)} + products$	4.7×10^7	M^{-1}s^{-1}	Munthe, 1992
$Hg^0_{(aq)} + {}^{\cdot}OH_{(aq)} \rightarrow Hg^{2+}_{(aq)} + products$	2.4×10^9	M^{-1}s^{-1}	Gårdfeldt et al., 2001
$Hg^0_{(aq)} + Cl(l)_{(aq)} \rightarrow Hg^{2+}_{(aq)} + products$	2×10^6	M^{-1}s^{-1}	Lin and Pehkonen, 1999
$Hg^{2+}_{(aq)} + 2SO^{2-}_{3(aq)} \rightarrow Hg(SO_3)^{2+}_{(aq)}$	$1.1\times10^{-21}[SO_{2(gas)}]^{2}\times10^{4pH}$ *	s^{-1}	Petersen et al., 1998
$Hg(SO_3)^{2-}_{2(aq)} \rightarrow {}_qHg^0_{(aq)} + products$	4.4×10^{-4}	s^{-1}	Petersen et al., 1998
$Hg_nCl_{m(dis)} \leftrightarrow Hg^{2+}_{(aq)}$	f([Cl$^-$]) **	1	Lurie, 1971
$Hg_nCl_{m(dis)} \leftrightarrow Hg_nCl_{m(soot)}$	0.2	1	Petersen et al., 1998
$Hg(SO_3)^{2-}_{2(dis)} \leftrightarrow Hg(SO_3)^{2-}_{2\ (soot)}$	0.2	1	Petersen et al., 1998
$Hg^0_{(gas)} \leftrightarrow Hg^0_{(aq)}$	1.76×10^{-23}T exp(9.08(T$_0$/T-1))	M cm^3 molecule^{-1}	Andersson et al., 2004
$HgCl_{2(gas)} \leftrightarrow HgCl_{2(aq)}$	1.75×10^{-16}T exp(18.75(T$_0$/T-1))	M cm^3 molecule^{-1}	Ryaboshapko et al., 2001
$O_{3(gas)} \leftrightarrow O_{3(aq)}$	1.58×10^{-24}T exp(7.8(T$_0$/T-1))	M cm^3 molecule^{-1}	Sander, 1997
${}^{\cdot}OH_{(gas)} \leftrightarrow {}^{\cdot}OH_{(aq)}$	3.41×10^{-21}T exp(17.72(T$_0$/T-1))	M cm^3 molecule^{-1}	Jacobson, 1999
$Cl_{2(gas)} \leftrightarrow Cl(l)_{(aq)}$	4.48×10^{-16}	M cm^3 molecule^{-1}	Lin and Pehkonen, 1999

* $[SO_{2(gas)}]$ is in ppbv

$$** \frac{[Hg_nCl_{m(dis)}]}{[Hg^{2+}_{(aq)}]} = \frac{[Cl^-_{(aq)}]}{1.82\cdot10^{-7}} + \frac{[Cl^-_{(aq)}]^2}{6.03\cdot10^{-14}} + \frac{[Cl^-_{(aq)}]^3}{8.51\cdot10^{-15}} + \frac{[Cl^-_{(aq)}]^4}{8.51\cdot10^{-16}}$$

input information for the system. The regional MSCE-HM model is driven by meteorological data pre-processed by the MM5 - Fifth Generation Penn State/NCAR Mesoscale Model (Grell et al., 1995). The pre-processor utilizes the NCAR/NCEP re-analysis or ECMWF data as the input information and provides 6-hour weather prediction data with the same spatial resolution as that of the transport model.

Anthropogenic emissions data for long-term calculations during the period 1990-2005 were prepared utilizing global mercury emission datasets available for the years 1990, 1995, and 2000 (CGEIC website; Pacyna et al., 2003; Pacyna et al., 2006). For other years of the period the linear interpolation was applied, except for emissions for years 2001-2004 which were taken equal to those in 2000 because of absence of more recent data. However, this assumption evidently does not allow taking into account significant growth of mercury emissions in China since 2000 (see Chapters 2 and 3). Estimated long-term changes of total mercury anthropogenic emissions in the Northern Hemisphere and their spatial distribution in 2000 are presented in Figure 20.2.

As seen, relative contribution of European and North American sources to total hemispheric emission significantly decreased in the first half of the period while the contribution of Asian sources increased. Thus, the overall emission changed slightly between 1990 and 2004.

Natural emission and re-emission of mercury from soil and seawater was taken into account using global estimates by Lamborg et al. (2002). Spatially resolved emission flux was obtained by distribution of the global emission values over Earth's surface depending on the soil temperature for emissions from land and proportional to the primary production of organic carbon for emissions from the oceans (Travnikov and Ryaboshapko, 2002). It was expected that mercury is emitted from natural surfaces in elemental gaseous form. The temperature dependence of soil emission was described by an Arrhenius type equation with empirically derived activation energy about 20 kcal mol^{-1} (Kim et al., 1995; Carpi and Lindberg, 1998;

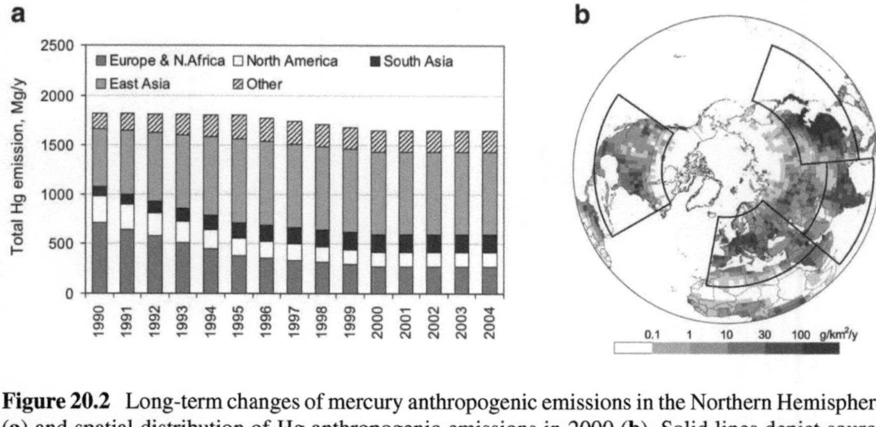

Figure 20.2 Long-term changes of mercury anthropogenic emissions in the Northern Hemisphere (**a**) and spatial distribution of Hg anthropogenic emissions in 2000 (**b**). Solid lines depict source regions selected for the analysis

Poissant and Casimir, 1998; Zhang et al., 2001). Evasion of Hg from geochemical mercuriferrous belts (Gustin et al., 1999) was assumed to be 10 times higher then that from background soils. Monthly mean satellite based data on the ocean primary production of carbon (Behrenfeld and Falkowski, 1997) were utilized to distribute the natural mercury emission flux over the ocean.

Two-years model spin-up was performed to fill up the atmosphere with mercury. To take into account the inter-hemispheric transport of mercury a fixed gradient of elemental mercury concentration of 0.05 ng m^{-3} degree was set at the equator. This value was obtained from approximation of measurement data from the ocean cruises (Slemr, 1996).

20.3 Results and Discussion

Model simulations of mercury atmospheric dispersion in the Northern Hemisphere were performed for the period 1990-2004. Long-term changes of mercury deposition during this period were evaluated for different continents and regions of the Northern Hemisphere. Obtained modelling results were compared with long-term monitoring data from various national and international networks (EMEP, NADP/MDN, CAMnet etc.) Besides, intercontinental transport of mercury as well as sensitivity of mercury deposition in the Northern Hemisphere to emission reduction in different continents were estimated. Major modelling results are discussed below.

20.3.1 Spatial distribution and long-term trends

Spatial patterns of total mercury deposition in the Northern Hemisphere and, with higher resolution, in Europe are illustrated in Figure 20.3 for the year 2001.

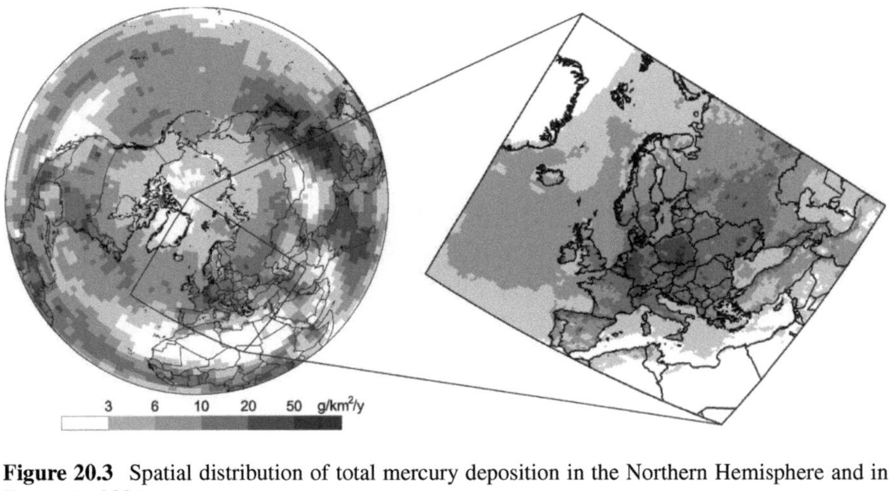

Figure 20.3 Spatial distribution of total mercury deposition in the Northern Hemisphere and in Europe in 2001

In general, spatial distribution of deposition well corresponds to the emission field (Figure 20.2b): Elevated deposition fluxes (more than 20 g km^{-2} yr^{-1}) were in Europe, North America, Eastern and Southern Asia. Mercury deposition in these regions are defined to a greater extent by primary short-lived mercury forms (oxidized gaseous and particulate mercury) directly emitted to the atmosphere from local anthropogenic sources. On the other hand, deposition over remote regions depend on *in situ* oxidation of elemental mercury transported from various continents.

Therefore, considerable deposition fluxes over some parts of the Atlantic and Pacific oceans (up to 12 g km^{-2} yr^{-1}) are result of combination of two factors: relatively high concentrations of the oxidants (tropospheric ozone first of all) and elevated precipitation amount in these regions. Here we consider only long-term deposition of mercury to the surface and do not take into account its fast circulation between air and seawater in the marine boundary layer (e.g. Laurier et al., 2003; Hedgecock and Pirrone, 2004). The lowest deposition fluxes were predicted over desert regions in Africa and in some areas of the Arctic mostly because of small precipitation amount. However, it should be noted that chemical mechanism of mercury depletion events (MDEs) was not taken into account in this study. Thus, one can expect that mercury deposition in the Arctic is considerably higher, particularly, in the coastal areas.

Estimates of long-term changes of mercury deposition to different continents and regions of the Northern Hemisphere are presented in Figure 20.4. The most significant decrease of deposition during the considered period took place in

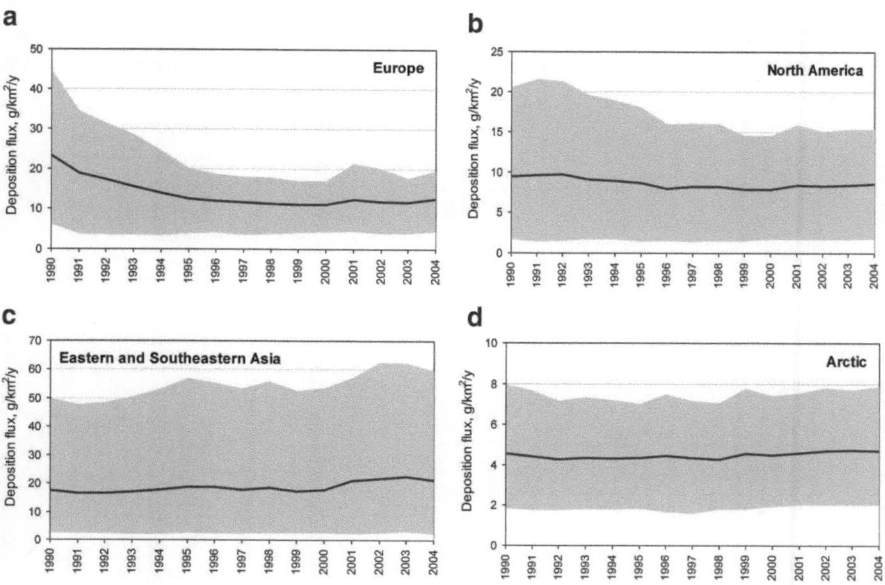

Figure 20.4 Long-term changes of mercury deposition flux in Europe (**a**), North America (**b**), Eastern and Southeastern Asia (**c**), and in the Arctic (**d**). Solid line presents average flux over the region; shaded area shows 90%-confidence interval of the flux variation over the region

Europe: the average deposition flux decreased by a factor of two, whereas the highest deposition fluxes decreased by a factor of three (Figure 20.4(a)). The reason for that is considerable emission reduction in Europe in this period (see Figure 20.2a). Changes of mercury deposition in North America (Figure 20.4(b)) are less pronounced because of smaller emission reduction and more essential contribution of mercury transport from other continents (in particular, from East Asia). In agreement with changes of anthropogenic emissions, deposition of mercury increased in Eastern and Southeastern Asia (Figure 20.4(c)). However, the increase was smoothed by decrease of emissions in other parts of the Northern Hemisphere. In the beginning of the period deposition levels in Eastern and Southeastern Asia was comparable with those in Europe and almost twice higher than in North America, whereas by the end of the period deposition in Asia became the highest in the Northern Hemisphere. Estimated mercury deposition in the Arctic did not change significantly during the 15-years period (Figure 20.4(d)) and they are several times lower than those in major industrialized continents. However, as it was mentioned above, these estimates do not take into account effect of the MDEs and, therefore, underestimate actual deposition in the Arctic. It should be also noted that some increase of mercury deposition estimated for last years of the period (2001-2004) was caused solely by changes of meteorological conditions since emissions were taken unchanged for these years.

Estimates performed by the regional version of the model with more detailed emissions for Europe (*WebDab*, 2006) demonstrate continuous decrease of deposition in Europe during the whole period including the last years (Figure 20.5(a)). In particular, Figure 20.5 shows trends of mercury deposition changes to different land use categories in Europe. As seen the largest deposition values and the most significant decrease are characteristic of urban areas. It is evident that they are located in the

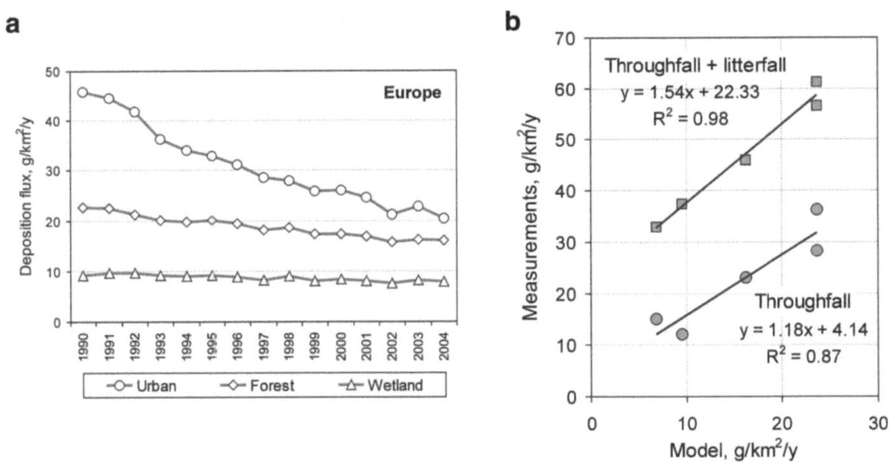

Figure 20.5 Long-term changes of mercury deposition flux to different landuse categories in Europe (**a**) and comparison of calculated mercury deposition to forests with throughfall measurements at forest sites in Europe (**b**)

immediate vicinity of main anthropogenic sources and reflect emission dynamics. On the other hand, changes of deposition to other sensitive land use categories (forest, wetland) are less pronounced because they are located more remotely from direct emission sources and more affected by transport form other continents and natural/re-emission sources. Another aspect of these estimates is connected with current understanding of mercury exchange with vegetation. Analysis of contemporary models parameterization of mercury deposition to forest reveal that simulated total (wet and dry) deposition cannot explain overall input of mercury to the forested ecosystems (Munthe, 2005). Particularly, Figure 20.5(b) presents comparison of modelled deposition fluxes with measurements at several forested ecosystems in Europe (Schwesig et al., 1999; Schwesig and Matzner, 2000; Porvari and Verta, 2003; Lee et al., 2000; Munthe et al., 1995). As seen from the figure, the model successfully reproduces throughfall deposition flux of mercury, however, it fails taking into account litterfall component of deposition, which is defined by mercury accumulated in foliage. Thus, mercury deposition simulated by the model likely significantly underestimate actual total input of mercury to forested ecosystems.

20.3.2 Evaluation of Modelling Results

In order to evaluate the modelling results calculated mercury concentrations in air and precipitation were compared with available measurements. For this purpose we used long-term observations from the EMEP monitoring network in Europe (Aas and Breivik, 2006) as well as the NADP /MDN (http://nadp.sws. uiuc.edu/mdn/) and CAMNet (Kellerhals et al., 2003) networks in North America. Besides, measurements at the Arctic stations (Barrow, Alert, Amderma), data from the MAMCS project (Pirrone et al., 2003) and some measurements from Asia (Tan et al., 2000; Kim et al., 2002) were involved in the comparison. Location of monitoring sites used in the model evaluation is shown in Figure 20.6(a). Results of the comparison of annual mean values for the period 1990-2004 are presented in Figures 20.6(b) and 20.6(c).

As seen, modelled concentrations of total gaseous mercury well agree with observed ones. Discrepancy between calculated and measured values does not exceed 20% with the exception of high values at Chinese and Korean sites which are somewhat underestimated by the model. On the other hand, the model tends to overestimate observed mercury concentration in precipitation. Besides, scattering of modelled and observed concentration in precipitation is more significant than in the case of air concentration. Nevertheless, difference between the calculated and measured values does not exceed a factor of two.

An example of comparison of measured and modelled long-term variation of mercury concentration in air and precipitation during the period 1990-2004 at some monitoring sites is shown in Figure 20.7. Both model and observations demonstrate relatively low variation of total gaseous mercury concentration in air (Figures 20.7(a) and 20.7(b)).

The model somewhat overestimate relatively low concentrations (below 1.5 ng m^{-3}) at station Pallas in Finland. Besides, it does not catch deep depressions of the concentration

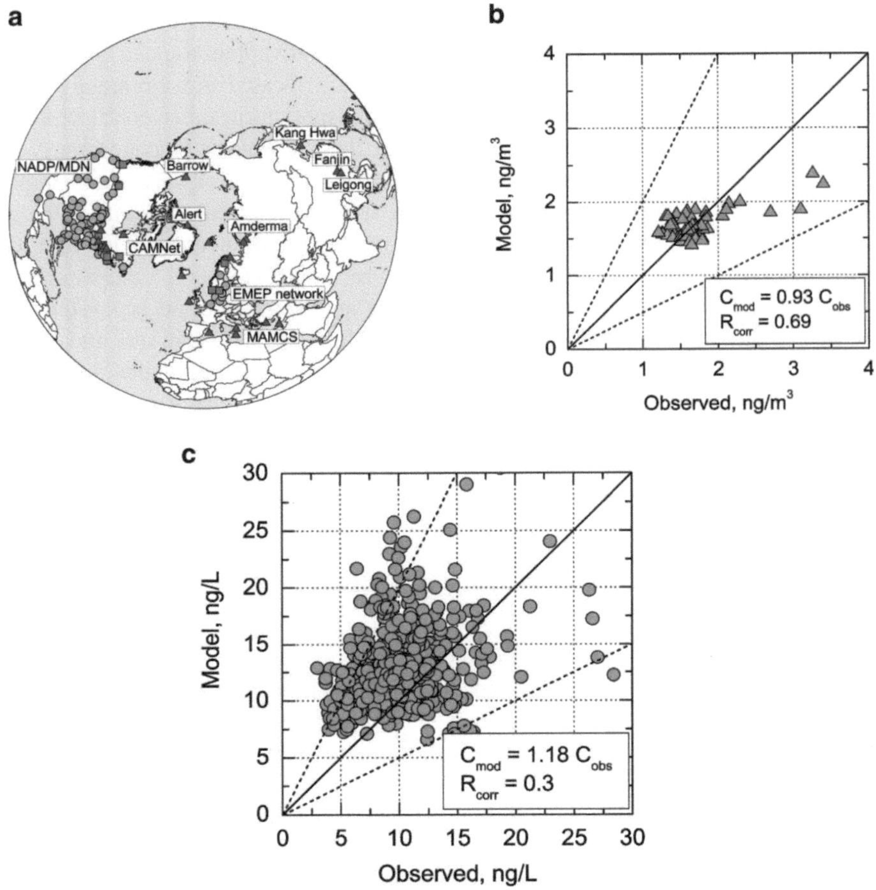

Figure 20.6 Location of monitoring sites used in the model evaluation (**a**), calculated vs. measured values of mean annual concentration of total gaseous mercury (**b**) and mercury concentration in precipitation (**c**). Dashed lines depict two-fold difference interval

during springtime because of MDEs at the Arctic station Zeppelin located at Spitsbergen. Both measurements and the model do not show significant changes of air concentrations between 1990 and 2004. The model successfully reproduces variation of mercury concentration in precipitation (Figures 20.7(c) and 20.7(d)). The variation exhibits pronounced seasonal cycle with maximum in summer and minimum in winter, which is defined by higher concentrations of photo oxidants during summertime.

20.3.3 Intercontinental Transport

As it was demonstrated above mercury deposition in many cases is significantly affected by transport from other continents (except areas located in the immediate vicinity of direct emissions). To evaluate the influence of intercontinental transport

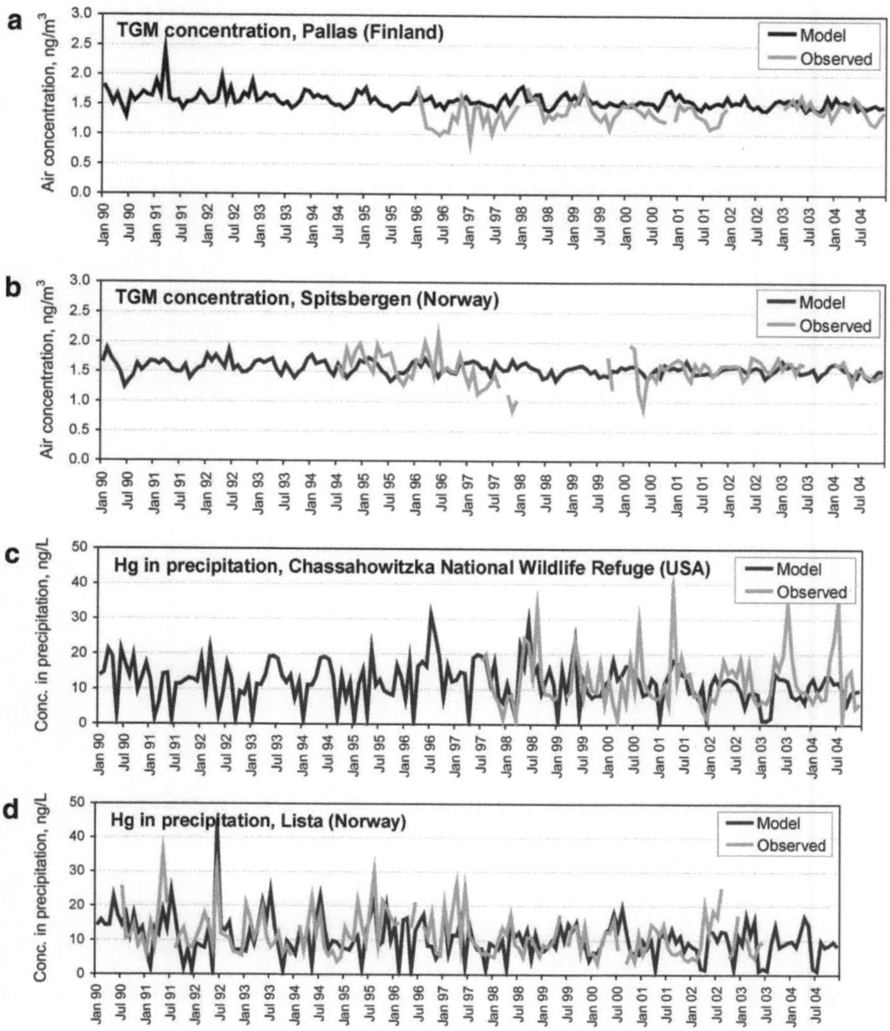

Figure 20.7 Modelled vs. measured long-term variation of monthly mean total gaseous mercury concentration in air (**a, b**) and total mercury concentration in precipitation (**c, d**) at some monitoring sites in Europe (Aas and Breivik, 2006) and North America (http://nadp.sws.uiuc.edu/mdn/)

on mercury deposition in the Northern Hemisphere and its sensitivity to emission reduction in different continents we performed model runs with anthropogenic emissions decreased by 20% in four major source regions – Europe, North America, East Asia and South Asia (see Figure 20.2 (b)) – with respect to the base case for the year 2001.

Figure 20.8 illustrates spatial patterns of simulated decrease mercury deposition in the Northern Hemisphere due to 20% emission reduction in the mentioned above regions. Mercury deposition in the Northern Hemisphere are the most sensitive to emission reduction in East Asia (Figure 20.8(a)) because of large relative contribution

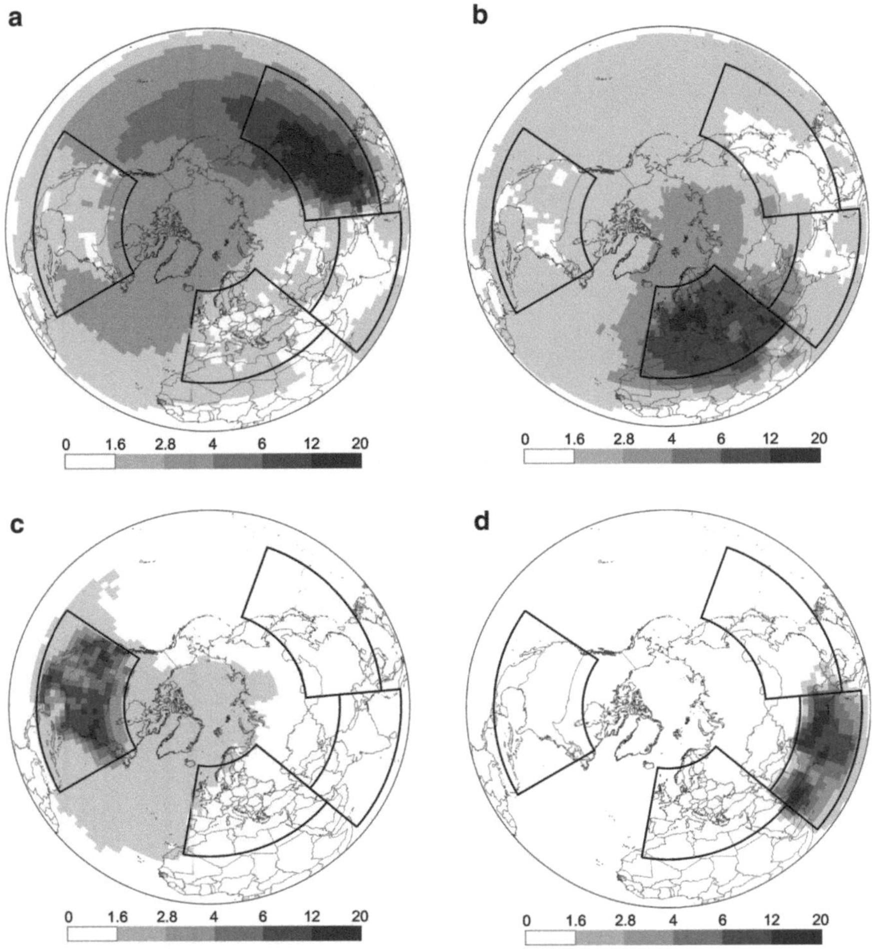

Figure 20.8 Spatial distribution of relative decrease of mercury deposition due to 20% emission reduction in East Asia (**a**), Europe (**b**), North America (**c**), and South Asia (**d**)

of sources located in this region to the global mercury emissions. Deposition of mercury decreases considerably not only in Eastern Asia but also over the North Pacific, North America and the Arctic. Emission reduction in Europe also results in considerable deposition decrease (apart from in Europe itself) in the North Atlantic, the Arctic and Northern Africa (Figure 20.8(b)). Decrease of mercury emissions in North America partly affects deposition in the North Atlantic and the Arctic (Figure 20.8(c)). Emission reduction in Southern Asia has the lowest effect on decrease of mercury deposition in the Northern Hemisphere (Figure 20.8(d)). The probability distribution of the deposition decrease values over the Northern Hemisphere with respect to 20% emission reduction in four selected regions is characterized quantitatively in Figure 20.9(a). As seen from the figure the most probable values of the

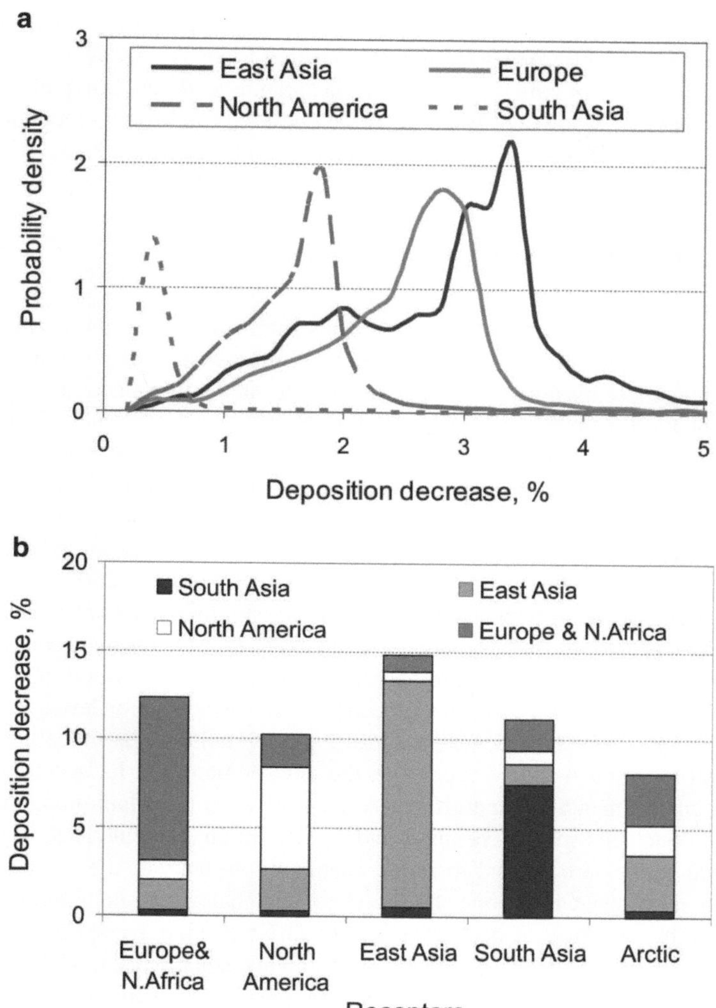

Figure 20.9 Probability distribution of mercury deposition decrease over the Northern Hemisphere due to 20% emission reduction in the selected source regions (**a**) and contribution of the source regions to the deposition decrease in different receptor regions (**b**)

deposition decrease are 3.2%, 2.8%, 1.8%, and 0.4% for East Asia, Europe, North America and South Asia, respectively.

The relative contribution of different source regions to the deposition reduction of mercury over various receptor regions is illustrated in Figure 20.9(b). As seen deposition in Europe is most sensitive to reduction of anthropogenic emissions from its own sources but also in Eastern Asia and North America. Deposition decrease in East and South Asia is also mostly defined by reduction of emissions from own sources. On the other hand, mercury deposition in North America only partly depends on emissions from own sources but also to a great extent on those from East Asian and European sources. Deposition to the Arctic are almost equally sensitive to changes of emissions in East Asia and in Europe and also partly to those in North America. Total deposition decrease due to 20% emission reduction in all four regions varies from 8% in the Arctic to 15% in East Asia.

20.4 Uncertainty and Future Research

The analysis presented above analysis contains significant uncertainties. First of all, one of the most essential input data for mercury atmospheric dispersion modelling – anthropogenic emissions – remain incomplete and rather uncertain. The most resent global estimates of anthropogenic emissions relate to the year 2000 (Pacyna et al., 2006) and do not include some important parameters such as temporal variation of emissions. Therefore, further development of mercury atmospheric modelling requires update and improvement of global emissions data. Another important information affecting quality of mercury modelling is data on natural emission and re-emission of mercury. According to available estimates (Mason and Sheu, 2002; Strode et al., 2007) mercury evasion from land and the ocean significantly contributes to total mercury emission to the atmosphere and even likely exceeds the anthropogenic one. Therefore, reliable spatially and temporally resolved data on natural emission and re-emission are critical for estimates of realistic mercury deposition levels. A possible solution of this problem is development of a global multi-media modelling system taking into account mercury cycling in the environment and its accumulation in different environmental compartments. Among the most uncertain model processes one could select atmospheric chemistry and dry deposition (particularly, to vegetation). Additional studies are needed to refine chemical constants of the most important oxidation reactions with ozone, hydroxyl radical, halogens and improvement of model parameterisation of dry deposition of different mercury species. Available long-term measurement data do not allow complete evaluation of modelled mercury concentration and deposition levels because they are mostly located in North America and Europe. There are large territories in Africa and Asia not covered by observations at all and only episodic measurements are available from the oceans. Besides, very scarce measurement data is available on mercury dry deposition and speciated air concentrations. Development of a global monitoring network of long-term observation including measurements of different mercury species could significantly improve evaluation of modelling results.

References

Andersson, M., Wängberg, I., Gårdfeldt, K., Munthe, J., 2004. Investigation of the Henry's low coefficient for elemental mercury. Proceedings of the 7th Conference "Mercury as a global pollutant". RMZ – Materials and Geoenvironment, Ljubljana, June 2004

Aas W. and Breivik K., 2006. Heavy metals and POP measurements, 2004. EMEP/CCC-Report 7/2006, Kjeller, Norwegian Institute for Air Research, Oslo. (www.nilu.no/ projects/ccc/ reports.html)

Ariya P.A., Khalizov A., Gidas A. (2002) Reactions of gaseous mercury with atomic and molecular halogens: kinetics, product studies, and atmospheric implications. *J. Phys. Chem.* 106, 7310–7320

Behrenfeld M.J. and Falkowski P.G. (1997) Photosynthetic derived from satellite-based chlorophyll concentration. *Limnol. Oceanogr.*, 42(1), 1–20

Carpi A. and Lindberg S. E. (1998) Application of a teflonTM dynamic flux chamber for quantifying soil mercury flux: tests and results over background soil. *Atmos. Environ.* 32(5), 873–882

CGEIC website, http://www.ortech.ca/cgeic/

Chin M., Jacob D.J., Gardner G.M., Forman-Fowler M.S., Spiro P.A., Savoie D.L. (1996) A global three-dimensional model of tropospheric sulfate, *J. Geophys. Res.* 101, 18667–18690

Gårdfeldt K., Sommar J., Strömberg D., Feng X. (2001) Oxidation of atomic mercury by hydroxyl radicals and photoinduced decomposition of methylmercury in the aqueous phase. *Atmos. Environ.* 35, 3039–3047

Grell GA, Dudhia J, and Stauffer DR. (1995) A description of the Fifth-Generation Penn State / NCAR Mesoscale Model (MM5). NCAR Technical Note NCAR/TN-398+STR. Mesoscale and Microscale Meteorology Division, National Center for Atmospheric Research, Boulder, Colorado, 122 pp.

Gustin M.S., Lindberg S., Marsik F., Casimir A., Ebinghaus R., Edwards G., Hubble-Fitzgerald C., Kemp R., Kock H., Leonard T., London J., Majewski M., Montecinos C., Owens J., Pilote M., Poissant L., Rasmussen P., Schaedlich F., Schneeberger D., Schroeder W., Sommar J., Turner R., Vette A., Wallschlaeger D., Xiao Z., and Zhang H. (1999) Nevada STORMS project: Measurement of mercury emissions from naturally enriched surfaces. *J. Geophys. Res.* 104(D17), 21831–21844

Hall B. (1995) The gas phase oxidation of mercury by ozone. *WASP* 80, 301–315

Jacobson, M. Z., 1999. Fundamentals of atmospheric modeling. Cambridge University Press. 656 p.

Hedgecock I.M., Pirrone N. (2004) Chasing Quicksilver: Modelling the Atmospheric Lifetime of $Hg^0_{(g)}$ in the marine boundary layer at various latitudes. *Environ. Sci. Technol.* 38, 69–76

Kellerhals M., Beauchamp S., Belzer W., Blanchard P., Froude F., Harvey B., McDonald K., Pilote M., Poissant L., Puckett K., Schroeder B., Steffen A., Tordon R. (2003) Temporal and spatial variability of total gaseous mercury in Canada: results from the Canadian Atmospheric Mercury Measurement Network (CAMNet). *Atmos. Environ.* 37(7), 1003–1011

Kim K.-H., Lindberg S. E. and Meyers T. P. (1995) Micrometeorological measurements of mercury vapor fluxes over background forest souls in eastern Tennessee. *Atmos. Environ.* 29(2), 267–282

Kim, K.-H., Kim, M.-Y., Kim, J., Lee, G. (2002) The concentrations and fluxes of total gaseous mercury in a western coastal area of Korea during late Mart 2001. *Atmos. Environ.* 36, 3413–3427

Lamborg C.H., Fitzgerald W.F., O'Donnell J., Torgersen T. (2002) A non-steady state compartmental model of global-scale mercury biogeochemistry with interhemispheric atmospheric gradients. *Geochim. Cosmochim. Acta* 66, 1105–1118

Laurier F.J.G., Mason R.P., Whalin L., Kato S. (2003) Reactive gaseous mercury formation in the North Pacific Ocean's marine boundary layer: A potential role of halogen chemistry. *J. Geophys. Res.* 108(D17), 4529

Lee Y.H., Bishop K.H., Munthe J. (2000) Do concepts about catchment cycling of methylmercury and mercury in boreal catchments stand the test of time? Six years of atmospheric inputs and runoff export at Svartberget, northern Sweden. *Sci. Tot. Environ.* 249, 11–20

Lin C.-J., Pehkonen S. O. (1999) The chemistry of atmospheric mercury: a review. *Atmos. Environ.* 33, 2067–2079

Lurie Yu. Yu. [1971] Handbook for Analytical Chemistry. Khimiya, Moscow, 454 p.

Mason R.P. and Sheu G.-R. (2002) Role of the ocean in the global mercury cycle. *Glob. Biogeochem. Cycles* 16(4), 1093, doi:10.1029/2001GB001440

Munthe J. (1992) The aqueous oxidation of elemental mercury by ozone. *Atmos. Environ.* 26A, 1461–1468

Munthe J., Hultberg H., Iverfeldt Å. (1995) Mechanisms of deposition of methylmercury and mercury to coniferous forests. *WASP* 80, 363–371

Munthe (2005) personal communication

Pacyna J, Pacyna E, Steenhuisen F, Wilson S. (2003) Mapping 1995 global anthropogenic emissions of mercury. *Atmos. Environ.* 37(S1), S109–S117

Pacyna E. G., Pacyna J. M., Steenhuisen F. and Wilson S. (2006) Global anthropogenic mercury emission inventory for 2000. *Atmos. Environ.* 40(22), 4048–4063

Pirrone N., Ferrara R., Hedgecock I. M., Kallos G., Mamane Y., Munthe J., Pacyna J. M., Pytharoulis I., Sprovieri F., Voudouri A., Wangberg I. (2003) Dynamic Processes of Mercury Over the Mediterranean Region: results from the Mediterranean Atmospheric Mercury Cycle System (MAMCS) project. *Atmos. Environ.* 37(S1), 21–39

Petersen G., Munthe J., Pleijel K., Bloxam R., Vinod Kumar A. (1998) A comprehensive Eulerian modeling framework for airborne mercury species: development and testing of the tropospheric chemistry module (TCM). *Atmos. Environ.* 32, 829–843

Poissant L. and Casimir A. (1998) Water–air and soil–air exchange rate of total gaseous mercury measured at background sites. *Atmos. Environ.* 32(5), 883–893

Porvari, P., Verta, M. (2003) Total and methyl mercury concentrations and fluxes from small boreal forest catchments in Finland. *Environ. Pollut.* 123(2), 181–191

Rubinstein K., Kiktev D. (2000) Comparison of the atmospheric low-layer diagnostic system (SDA) for pollution transfer modelling at MSC-East (Moscow) and MSC-West (Oslo). *Environmental Modelling & Software* 15, 589–596

Ruijgrok W., Tieben H., Eisinga P. (1997) The dry deposition of particles to a forest canopy: a comparison of model and experimental results. *Atmos. Environ.* 31, 399 – 415

Ryaboshapko A., Ilyin I., Bullock R., Ebinghaus R., Lohman K., Munthe J., Petersen G., Segneur C., Wangberg I. (2001) Intercomparison study of numerical models for long-range atmospheric transport of mercury. Stage I: Comparison of chemical modules for mercury transformations in a cloud/fog environment. EMEP/MSC-E Technical report 2/2001, Meteorological Synthesizing Centre – East, Moscow, Russia (www.msceast.org/publications.html)

Sander R. (1997) Henry's law constants available on the Web. EUROTRAC Newsletter 18, 24-25 (www.mpch-mainz.mpg.de/~sander/res/henry)

Schwesig D., Ilgen G., Matzner E. (1999) Mercury and methylmercury in upland and wetland acid forest soils of a watershed in NE-Bavaria, Germany. *WASP* 113, 141–154.

Schwesig D., and Matzner E. (2000) Pools and fluxes of mercury and methylmercury in two forested catchments in Germany. *Sci. Total Environ.* 260, 213–223

Seigneur C., Karamchandani P., Lohman K., Vijayaraghavan K., Shia R.-L., 2001. Multiscale modeling of the atmospheric fate and transport of mercury. *J. Geophys. Res.* 106, 27795–27809

Slemr F. (1996) Trends in atmospheric mercury concentrations over the Atlantic ocean and the Wank Summit, and the resulting concentrations on the budget of atmospheric mercury. In: Baeyens W., Ebinghaus R. Vasiliev O. (Eds.), Global and Regional Mercury Cycles: Sources, Flaxes and Mass Balances, pp. 33–84. NATO-ASI-Series, Kluwer Academic Publishers, Dordrecht, The Netherlands.

Slinn W.G.N. (1982) Predictions for particle deposition to vegetative canopies. *Atmos. Environ.* 16, 1785–1794

Spivakovsky C.M., Logan J.A., Montzka S.A., Balkanski Y.J., Foreman-Fowler M., Jones D.B.A., Horowitz L.W., Fusco A.C., Brenninkmeijer C.A.M., Prather M.J., Wofsy S.C., McElroy M.B. (2000) Three-dimensional climatological distribution of tropospheric OH: Update and evaluation. *J. Geophys. Res.* 105, 8931–8980

Sommar J., Gårdfeldt K., Strömberg D., Feng X. (2001) A kinetic study of the gas-phase reaction between the hydroxyl radical and atomic mercury. *Atmos. Environ.* 35, 3049–3054

Strode S.A., Jaegle L., Selin N., Jacob D.J., Park R.J., Yantosca R.M., Mason R.P., Slemr F. (2007) Air-sea exchange in the global mercury cycle. *Glob. Biogeochem. Cycles* 21(4), GB1017

Tan H., He J.L., Liang L., Lazoff S., Sommer J., Xiao Z.F., Lindqvist O. (2000) Atmospheric mercury deposition in Guizhou, China. *Sci. Total Environ.* 259, 223–230

Travnikov O, Ryaboshapko A. (2002) Modelling of mercury hemispheric transport and depositions. EMEP/MSC-E Technical Report 6/2002, Meteorological Synthesizing Centre - East, Moscow, Russia. (www.msceast.org/publications.html)

Travnikov O, Ilyin I. (2005) Regional model MSCE-HM of heavy metal transboundary air pollution in Europe. EMEP/MSC-E Technical Report 6/2005, Meteorological Synthesizing Centre - East, Moscow, Russia. (www.msceast.org/publications.html)

Travnikov O. (2005) Contribution of the intercontinental atmospheric transport to mercury pollution in the Northern Hemisphere. *Atmos. Environ.* 39, 7541–7548

Walton J.J., MacCracken M.C. and Ghan S.J. (1988) A global-scale Lagrangian trace species model of transport, transformation, and removal processes. *J. Geophys. Res.* 93(D7), 8339–8354

Wang N., Logan J.A., Jacob D.J. (1998) Global simulation o ftropospheric O3-Nox-hydrocarbon chemistry, 2., Model evaluation and global ozone budget. *J. Geophys. Res.* 103, 10727–10755

WebDab (2006), http://webdab.emep.int/

Wesely M.L., Cook D.R., Hart R.L. (1985) Measurements and parameterization of particulate sulfur dry deposition over grass. *J. Geophys. Res.* 90, 2131–214

Wesely M.L. and Hicks B.B. (2000) A review of the current status of knowledge on dry deposition. *Atmos. Environ.* 34, 2261–22

Williams R.M. (1982) A model for the dry deposition of particles to natural water surfaces. *Atmos. Environ.* 16, 1933–1938

Zhang H., Lindberg S. E., Marsik F. J., and Keeler G. J. (2001) Mercury air/surface exchange kinetics of background soils of the Tahquamenon River watershed in the Michigan upper peninsula. *WASP* 126, 151–169

Chapter 21
The AER/EPRI Global Chemical Transport Model for Mercury (CTM-HG)

Christian Seigneur, Krish Vijayaraghavan, Kristen Lohman, and Leonard Levin

Summary Mercury (Hg) has an atmospheric residence time on the order of 1 year (Schroeder and Munthe, 1998). Therefore, it can be transported over long distances and the development of source-receptor relationships requires modeling tools that are compatible with fate and transport processes at global scales. Furthermore, the assessment of the potential impact of mercury emission sources at regional scales requires knowledge of the upwind concentrations of mercury species because those upwind "background" concentrations are quite influential for modeling the atmospheric fate and transport of mercury at continental and regional scales. Since there is a paucity of data to specify such boundary conditions, particularly aloft, it is more reliable to obtain such boundary conditions from a global simulation, contingent upon satisfactory performance of the global model. To that end, the AER/EPRI global chemical transport model for mercury (CTM-Hg) was developed to simulate the global cycling of atmospheric mercury. We present first a general description of the model, followed by more detailed discussions of the chemical mechanism and emission inventory. Then, results of a performance evaluation with some available data are presented. Finally, we present results for the contribution of four source regions to atmospheric mercury deposition in those regions.

21.1 Description of the CTM-Hg

The global Hg model (Shia et al., 1999; Seigneur et al., 2001, 2004) is based on the three-dimensional (3-D) CTM developed at the Goddard Institute for Space Studies (GISS), Harvard University, and the University of California at Irvine. The 3-D model provides a horizontal resolution of 8° latitude and 10° longitude and a vertical resolution of nine layers ranging from the Earth's surface to the lower stratosphere. Seven layers represent the troposphere (between the surface and ~12 km altitude), and two layers the stratosphere (between ~12 km and 30 km altitude). Transport processes are driven by the wind fields and convection statistics calculated every 4 hours (for 1 year) by the GISS general circulation model (Hansen et al., 1983). This 1-year data set is used repeatedly for multi-year simulations until steady state is achieved. Steady state is typically achieved after 17 years.

N. Pirrone and R. Mason (eds.), *Mercury Fate and Transport in the Global Atmosphere*,
DOI: 10.1007/978-0-387-93958-2_21,

The global mercury emission inventory used to drive the model is for a datum of 2000; it is described in detail below. The chemical kinetic mechanism represents the current state of the science (Ryaboshapko et al., 2002; Seigneur et al., 2006; Seigneur et al., 2007); it is also described in detail below. Dry and wet deposition fluxes are simulated as follows (Seigneur et al., 2004). The dry deposition velocity of $Hg^{(II)}$ is selected to be 0.25 cm s^{-1}. The dry deposition velocity of Hg^0 is 0.01 cm s^{-1} over land and, because of its low solubility, 0 over the oceans. The Hg(p) deposition velocity is 0.1 cm s^{-1} over land and 0.01 cm s^{-1} over water. Wet deposition is calculated using the cloud droplet chemical concentrations and the precipitation patterns. For below-cloud scavenging, no scavenging of Hg^0, 100% scavenging of $Hg^{(II)}$ and 50% scavenging of Hg(p) are assumed.

21.2 Emission Inventory

The base emission inventory is presented in Table 21.1 (Lohman et al., 2008). Anthropogenic emissions account for about one-third of the total emission rate of about 6,600 Mg yr^{-1} and are dominated by Asia (about half of the world anthropogenic emissions). Natural emissions account for about one-sixth of the total emissions, which corresponds to a ratio of current to pre-industrial mercury emissions of about three, which is consistent with current estimates (Lindberg et al, 2007). Emissions of previously deposited mercury accounts for half the total emission inventory, because it is assumed that, on average, 50% of deposited mercury can be emitted back to the atmosphere.

The impact of uncertainty in the global Hg emission inventory was investigated with the global CTM-Hg (Lohman et al., 2008). A base simulation was developed to reflect "best estimates". Three sensitivity simulations reflecting uncertainty over different categories of emissions were conducted. In each sensitivity scenario, a change was made to the emission inventory, but all other parameters remained the same as in the base scenario. The first sensitivity scenario increased global natural emissions with a net increase in overall background emissions of 16%. The second

Table 21.1 Global Mercury Emissions (*Mg yr*$^{-1}$) for 2000 (Lohman et al., 2008)

Sources	Emission rate
North America	138
South & Central America	92
Europe	239
Asia	1204
Africa	407
Oceania	125
Total direct anthropogenic	2205
Natural emissions (land and water)	1067
Emissions of previously deposited Hg	3272
Total	6544

doubled anthropogenic emissions from China. The third included the increased background emissions from the first sensitivity simulation and also doubled all anthropogenic emissions. The base simulation and the first two sensitivity simulations all produce results that are compatible with observations in terms of ambient Hg^0 concentrations at remote sites and the North-South total gaseous mercury (TGM) gradient in the Atlantic Ocean. On the other hand, the third sensitivity scenario demonstrates that increasing background emissions by 16% and doubling anthropogenic emissions produces unrealistic results: the North-South TGM gradient is not reproduced correctly and, when the modeled results are compared to measurements at remote sites, the modeled results show a consistent over prediction by at least 0.4 ng m^{-3} in all regions. However, such a scenario cannot be ruled out because current uncertainties in the redox cycle of Hg (see discussion of mercury chemistry below) and in the potential for soils and vegetation to act as significant sinks can lead to a plausible scenario.

In summary, the uncertainty in Hg emissions at the global scale is estimated to be about a factor of two. Reducing those uncertainties will require more accurate emission inventories of anthropogenic sources, particularly for countries that have been identified as large Hg emitters (e.g., China) and for source categories that are not currently included in emission inventories but could be significant. In addition, better characterization of background emissions (i.e., both natural emissions and the emissions of previously deposited Hg) is needed.

21.3 Atmospheric Chemistry of Mercury

Table 21.2 presents the atmospheric transformations among inorganic mercury species that are simulated in the model. They include the gas-phase oxidation of Hg^0 to $Hg^{(II)}$, the aqueous-phase oxidation of Hg^0 to $Hg^{(II)}$, the aqueous-phase reduction of $Hg^{(II)}$ to Hg^0, various aqueous-phase equilibria of $Hg^{(II)}$ species and the adsorption of $Hg^{(II)}$ to PM. This atmospheric chemistry of mercury shows that aqueous-phase reactions (those that occur in clouds and fogs) can lead to either oxidation of $Hg^{(0)}$ to $Hg^{(II)}$ or reduction of $Hg^{(II)}$ to Hg^0. Such reduction-oxidation cycles affect the overall atmospheric lifetime of mercury. The chemical atmospheric lifetime of Hg^0 is currently believed to be three to four months (Radke et al., 2007). However, in non-precipitating clouds, $Hg^{(II)}$ may be reduced back to Hg^0, thereby extending the lifetime of mercury in the atmosphere. It is, therefore, important to differentiate between the chemical lifetime of a mercury species, which may range from several hours to several days for $Hg^{(II)}$ and Hg(p) and is a few months for Hg^0, and the overall atmospheric lifetime of mercury (that can cycle among the various species), presently estimated to be on the order of one year (1.2 years in the CTM-Hg).

The chemical species reacting with Hg are input to the model with a spatial resolution consistent with the model's grid spacing and a monthly temporal resolution. Some species concentrations were obtained from previous model simulations whereas others were obtained from measurements reported in the literature. The concentrations of O_3, SO_2, and OH were obtained from Wang et al. (1998), Chin et al. (1996) and

Table 21.2 Equilibria and reactions of atmospheric mercury

Equilibrium Process or Chemical Reaction	Equilibrium or Rate Constant	Reference
Hg^0 (g) ↔ Hg^0 (aq)	0.11 M atm^{-1}	Sanemasa, 1975; Clever et al., 1985
$HgCl_2$ (g) ↔ $HgCl_2$ (aq)	1.4×10^6 M atm^{-1}	Lindqvist and Rodhe, 1985
$Hg(OH)_2$ (g) ↔ $Hg(OH)_2$ (aq)	1.2×10^4 M atm^{-1}	Lindqvist and Rodhe, 1985
$HgCl_2$ (aq) ↔ Hg^{2+} + 2 Cl^-	10^{-14} M^2	Sillen and Martell, 1964
$Hg(OH)_2$ (aq) ↔ Hg^{2+} + 2 OH^-	10^{-22} M^2	Sillen and Martell, 1964
Hg^{2+} + SO^{2-}_3 ↔ $HgSO_3$	2.1×10^{13} M^{-1}	van Loon et al., 2001
$HgSO_3$ + SO^{2-}_3 ↔ $Hg(SO^{2-}_3)$	1.0×10^{10} M^{-1}	van Loon et al., 2001
$Hg^{(II)}$ (aq) ↔ $Hg^{(II)}$ (p)	34 L g^{-1}	Seigneur et al., 1998
Hg^0 (g) + O_3 (g) ↔ $Hg^{(II)}$ (g)	3×10^{-20} cm^3 $molec^{-1}s^{-1}$	Hall, 1995
Hg^0 (g) + HCl(g) → $HgCl_2$(g)	10^{-19} cm^3 $molec^{-1}$ s^{-1}	Hall and Bloom, 1993
Hg^0 (g) + H_2O_2 (g) → $Hg(OH)_2$ (g)	8.5×10^{-19} cm^3 $molec^{-1}s^{-1}$	Tokos et al., 1998
Hg^0 (g) + Cl_2(g) → $HgCl_2$(g)	2.6×10^{-18} cm^3 $molec^{-1}s^{-1}$	Ariya et al., 2002
Hg^0 (g) + OH(g) → $Hg(OH)_2$(g)	8.7×10^{-14} cm^3 $molec^{-1}s^{-1}$	Sommar et al., 2001
Hg^0 + BrO → HgO + Br	1.5×10^{-14} cm^3 $molec^{-1}s^{-1}$	Raofie and Ariya, 2003
Hg^0 + Br → HgBr	3.6×10^{-13} cm^3 $molec^{-1}s^{-1}$	Donohoue et al., 2006
HgBr → Hg^0 + Br	7.9×10^{-3} s^{-1}	Goodsite et al., 2004
HgBr + Br → $HgBr_2$	2.5×10^{-10} cm^3 $molec^{-1}s^{-1}$	Goodsite et al., 2004
HgBr + OH → HgBrOH	2.5×10^{-10} cm^3 $molec^{-1}s^{-1}$	Goodsite et al., 2004
Hg^0 (aq) + O_3 (aq) → Hg^{2+}	4.7×10^7 M^{-1} s^{-1}	Munthe, 1992
Hg^0 (aq) + OH (aq) → Hg^{2+}	2.0×10^9 M^{-1} s^{-1}	Lin and Pehkonen, 1997
$HgSO_3$ (aq) → Hg^0 (aq)	0.0106 s^{-1}	van Loon et al., 2000
$Hg^{(II)}$ (aq) + HO_2 (aq) → Hg^0 (aq)	1.7×10^4 M^{-1} s^{-1}	Pehkonen and Lin, 1998
Hg^0 (aq) + HOCl (aq) → Hg^{2+}	2.09×10^6 M^{-1} s^{-1}	Lin and Pehkonen, 1998
Hg^0 (aq) + OCl^- → Hg^{2+}	1.99×10^6 M^{-1} s^{-1}	Lin and Pehkonen, 1998

$Hg^{(II)}$ refers to divalent Hg species; see references for temperature and pressure dependence.

Spivakovsky et al. (1990), respectively. The 3-D concentrations of HO_2 and H_2O_2 were obtained from C. Spivakovsky (private communication, 2001). These concentration fields vary spatially and temporally with monthly mean values applied to the model's 3-D grid. The aqueous concentrations of OH and HO_2 are calculated from the gas-phase concentrations using their temperature-dependent Henry's law constants (Jacobson, 1999). The gas-phase concentrations of OH and HO_2 are reduced by factors of 2 and 10, respectively, in grid cells containing clouds to account for reduced photochemical activity and heterogeneous chemistry within clouds (Jacob, 2000; Jaeglé et al., 2001). The OH and HO_2 concentrations are assumed to be zero at night because they are produced primarily via photochemical reactions. The HCl concentration is assumed to be 1.2×10^{10} molecules cm^{-3} at the surface decreasing to 10^8 molecules cm^{-3} at 10 km altitude (Graedel and Keene, 1995). The Cl_2 concentration is assumed to be zero over land, 100 ppt at the surface and 50 ppt aloft at night, and 10 ppt during the day over the oceans (Spicer et al., 1998). The BrO concentrations are obtained from satellite data (Burrows et al., 1999) with vertical profiles from a modeling study (Yang et al., 2005). The Br concentrations are assumed to be 1% of the BrO concentrations. The particulate matter concentration in cloud droplets is assumed to be 0.02 g L^{-1} (Seigneur et al., 1998).

The global CTM-Hg was used to investigate the effect on the global mercury cycle of several alternative reaction kinetics or reaction pathways (Seigneur et al., 2006). These reactions included the gas-phase oxidation of Hg^0 by O_3, the gas-phase oxidation of Hg^0 by OH, the aqueous-phase reduction of $Hg^{(II)}$ by HO_2 radicals, a pseudo-first-order gas-phase reduction of $Hg^{(II)}$ and the gas-phase reduction of $Hg^{(II)}$ by SO_2. The global simulation results obtained with the current assumptions made regarding the emissions and removal rates of Hg species suggest the following.

The kinetics of the oxidation of Hg^0 by O_3 (Pal and Ariya, 2004) is fast and would require a commensurate but yet unidentified reduction reaction to lead to realistic Hg concentrations. An increase in Hg emissions by a factor of two or three (i.e., within a plausible range of uncertainty) to compensate for the faster kinetics does not lead to realistic Hg^0 concentrations because the North/South Hg^0 concentration gradient that has been observed over the Atlantic ocean is not reproduced. This kinetics may include both homogeneous and heterogeneous processes and should, therefore, be considered as an upper limit under atmospheric conditions.

The reduction of $Hg^{(II)}$ by HO_2 in aqueous solution has been challenged by Gårdfeldt and Johnson (2003). However, a reduction reaction with an overall rate similar to that of the reduction of $Hg^{(II)}$ by HO_2 is needed to balance the oxidation of Hg^0 by OH and O_3 currently used in models. Alternatively, if the gas-phase oxidation of Hg^0 by O_3 and OH is lower than currently assumed in models, as suggested by some theoretical considerations (Calvert and Lindberg, 2005), then, the kinetics of the reduction of $Hg^{(II)}$ should be decreased accordingly. For example, the reduction of $Hg^{(II)}$ by HO_2 is not needed if the gas-phase oxidation of Hg^0 by OH is eliminated. However, eliminating the Hg^0 gas-phase oxidation by both O_3 and OH (or assuming that they both have a negligible contribution to Hg^0 oxidation overall) does not lead to realistic results. One can obtain Hg^0 concentrations that appear to have realistic levels overall by jointly eliminating the $Hg^{(II)} + HO_2$ reaction and increasing the Hg^0 dry deposition velocity (within its plausible range of uncertainty); however, the simulation results do not reproduce the North/South $Hg^{(0)}$ gradient observed over the Atlantic ocean. Therefore, some gas-phase oxidation of Hg^0 by oxidants such as O_3 and OH is needed to obtain realistic Hg^0 concentrations.

The proposed reduction of $Hg^{(II)}$ in power plant plumes can be represented by a reaction of $Hg^{(II)}$ with SO_2 (Lohman et al., 2006); this reaction does not affect the global cycling of Hg significantly and, consequently, it is not included in the base chemical kinetic mechanism of the CTM-Hg.

The kinetics of the OH and O_3 gas-phase reactions should continue to be investigated as they are potentially key reactions for Hg^0 oxidation. We also recommend that reactions that could reduce $Hg^{(II)}$ continue to be investigated.

21.4 Model Performance Evaluation

Figure 21.1 presents surface concentrations of Hg^0 over the globe for our base simulation (Lohman et al., 2008). The surface Hg^0 concentrations display a strong latitudinal gradient with background concentrations mostly in the range of 1.2 to

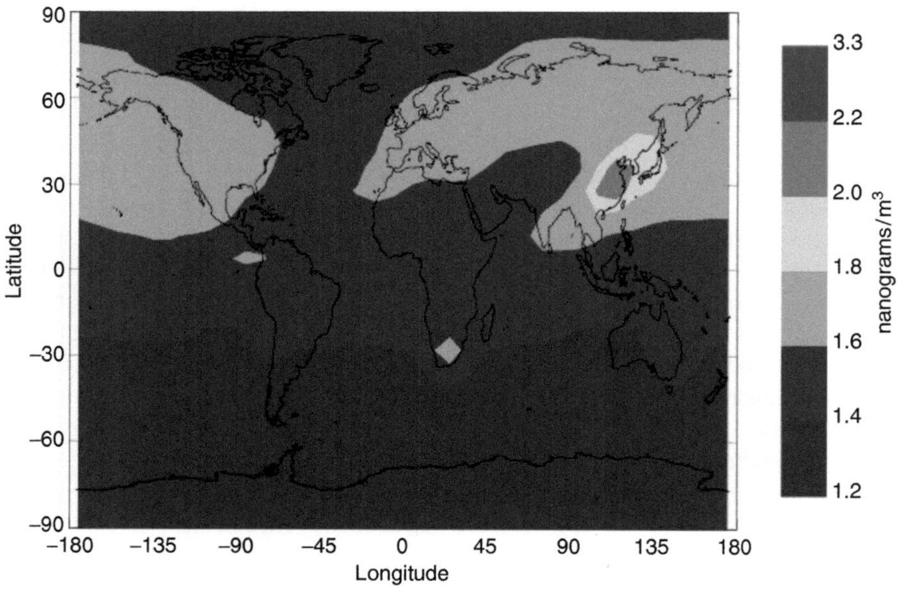

Figure 21.1 Modeled global Hg^0 concentrations (*ng m*$^{-3}$)

1.6 ng m^{-3} in the southern hemisphere and mostly in the range of 1.6 to 1.8 ng m^{-3} in the northern hemisphere. This latitudinal gradient is consistent with data from Temme et al. (2003). Concentrations above 1.9 ng m^{-3} are simulated over the large source areas of East Asia. In the southern hemisphere, South Africa shows up as a large source area with Hg^0 concentrations up to 1.7 ng m^{-3}.

The $Hg^{(II)}$ concentrations (not shown) display stronger spatial variations than the Hg^0 concentrations due to their stronger correlations with source areas such as South Africa, North America, Europe and Asia. The highest $Hg^{(II)}$ concentrations (in the range of 200 to 300 pg m^{-3}) are simulated over eastern China, due to high anthropogenic emissions in Asia. The Hg(p) concentrations (not shown) are solely of anthropogenic origin in the model and, therefore, they provide footprints of the major source areas. Concentrations of Hg(p) in East Asia are in the range of 100 to 200 pg m^{-3}.

Table 21.3 presents a comparison of simulated mercury concentrations with background concentrations measured at remote sites.

The model is within 0.1 ng m^{-3} (i.e., about 6%) of the measurements for the eastern North American sites. The model is also in good agreement at the western North American sites. The model underestimates at the Japanese site by 8%. At the northern European sites, the model is within 10% of the measurements at three sites but overestimates by 20% at the Finnish site. The model reproduces the measurements well at the Arctic sites but overestimates by 20% at the Antarctic site.

Figure 21.2 presents mercury deposition in the base simulation over the globe in terms of dry deposition of Hg^0, dry deposition of $Hg^{(II)}$ and wet deposition of $Hg^{(II)}$

Table 21.3 Comparison of measured and modeled concentrations of Hg^0 (*ng* adapted from Lohman et al., 2008; see the source for the measurement references)

Site	Measurements	Base	Scenario 1	Scenario 2	Scenario 3
Eastern NA.	1.50 – 1.83	1.60 – 1.65	1.75 – 1.82	1.73 – 1.78	2.25 – 2.42
Western NA.	1.43 – 1.77	1.69 – 1.70	1.94 – 2.00	1.83 – 1.85	2.44 – 2.52
Western Europe	1.72 – 1.75	1.62	1.80	1.75	2.32
Northern Europe	1.35 – 1.76	1.61 – 1.64	1.78 – 1.95	1.74 – 1.76	2.33 – 2.38
Japan	2.04	1.87	2.09	2.14	2.92
Arctic	1.32 – 1.80	1.53 – 1.64	1.66 – 1.79	1.64 – 1.76	2.11 – 2.29
Antarctic	0.99	1.29	1.41	1.35	1.71

NA.: North America

(wet deposition of Hg^0 is negligible and is not reported here). Dry deposition of Hg^0 occurs over land as dry deposition of Hg^0 over water is assumed to be negligible due to its very low solubility in water. Dry deposition of $Hg^{(II)}$ occurs also predominantly over land because $Hg^{(II)}$ emission sources are land-based; the largest $Hg^{(II)}$ dry deposition fluxes are simulated over East Asia. Wet deposition of Hg occurs downwind of large anthropogenic source areas where there is significant precipitation; as a result, the largest Hg wet deposition fluxes are simulated over Southeast Asia.

21.5 Source/Receptor Relationships

The global CTM-Hg has been applied in combination with a regional mercury model, the Trace Element Analysis Model (TEAM), to estimate for the first time the contribution of various source categories to atmospheric mercury deposition in the United States (Seigneur et al., 2004). The source categories included anthropogenic emissions from each continent (Africa, Asia, Europe, North America, Oceania, and South and Central America) and natural emissions from oceans and land as a single source category.

These results are summarized in Table 21.4 for average mercury deposition over the contiguous United States. North American anthropogenic sources were estimated to contribute only 30%. The largest contribution to mercury deposition in the United States was calculated to be natural sources with 33% (which is consistent with the assumption made in the model that current emissions are three times natural emissions).

Asian anthropogenic emissions were calculated to contribute 21% on average to U.S. mercury deposition; at specific receptors located throughout the United States, Asian emissions contributed from 5% (in the Northeast) to 36% (in the South). Since this modeling study was conducted, other modeling studies (Travnikov, 2005; Selin et al., 2007) and experimental studies (Jaffe et al., 2005) have confirmed the important contribution of Asian emissions to mercury concentrations and deposition in the United States.

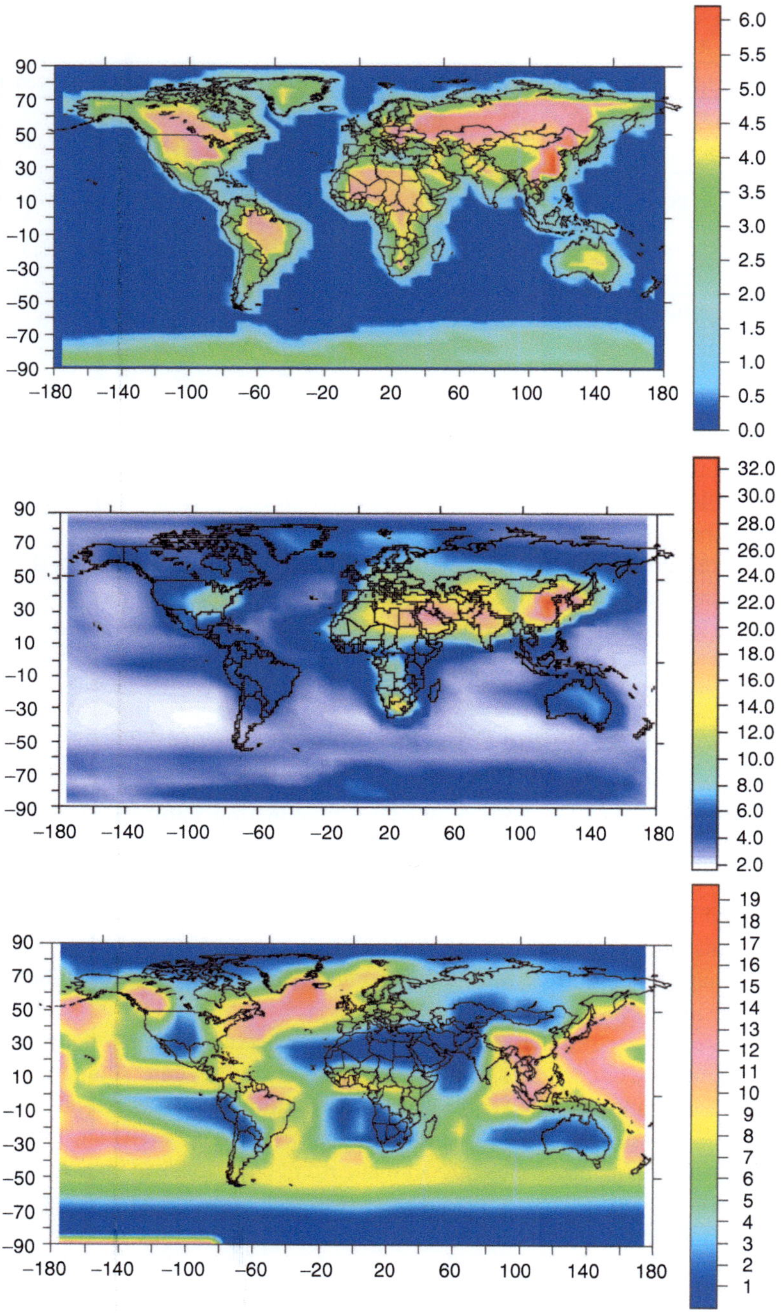

Figure 21.2 Atmospheric deposition of mercury (*μg m*$^{-2}$*yr*$^{-1}$) in the base simulation: (top) Hg^0 dry deposition; (middle) $Hg^{(II)}$ dry deposition; (bottom) $Hg^{(II)}$ wet deposition

In this study, this global CTM-Hg was applied to simulate the contributions of anthropogenic emissions from four regions (South Asia, East Asia, Europe and North Africa, and North America; see Figure 21.3) to mercury deposition in those regions and over the entire globe. Four global simulations were conducted where the anthropogenic emissions of each region were reduced by 20%. In addition, a fifth simulation was conducted where the anthropogenic emissions of all four regions were reduced by 20% each.

The results are presented in Figure 21.4 for each of these five simulations (the effect of each emission reduction scenario is presented as the difference between the base simulation and the emission scenario simulation; therefore, the mercury

Table 21.4 Relative contributions of anthropogenic source areas and natural emissions to atmospheric Hg deposition in the contiguous United States (adapted from Seigneur et al., 2004). (Totals may not add to 100% due to round-off.)

Source area	Hg Wet deposition	Hg Dry deposition	Hg Total deposition
North America	24%	43%	30%
South America	4%	2%	4%
Europe	8%	4%	7%
Asia	26%	11%	21%
Oceania	1%	0%	1%
Africa	5%	2%	4%
Natural	32%	36%	33%

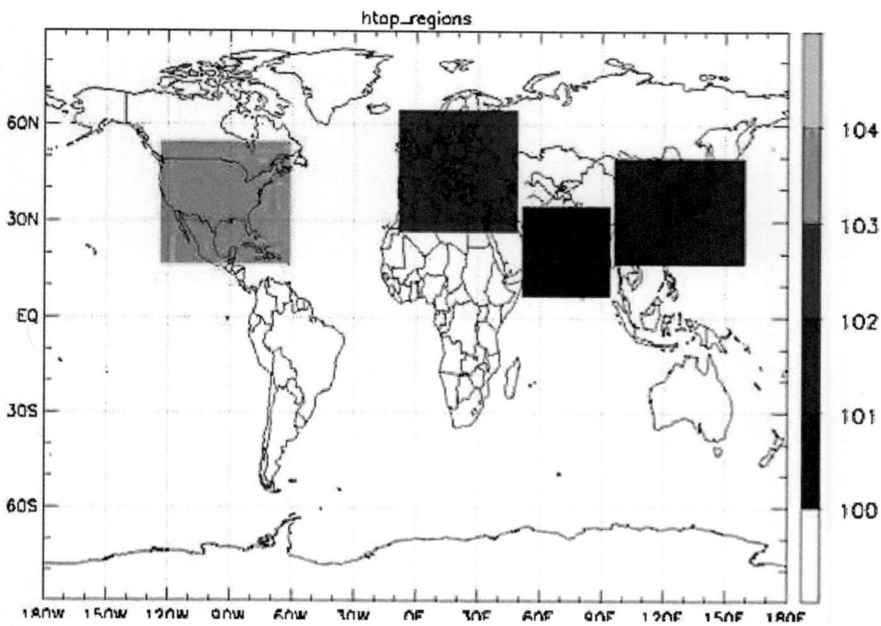

Figure 21.3 UNEP source areas used in the global emission scenarios

total deposition fluxes correspond to the contribution of 20% of the anthropogenic emissions of the source region of interest). Table 21.5 summarizes the results for annual mercury deposition in each of those four regions, over all four regions together and over the entire globe. The effect of a 20% reduction in mercury anthropogenic emissions in all four regions is equal to the sum of the four individual regional contributions because the mercury atmospheric system is linear.

Reduction of anthropogenic emissions in one region has the largest impact on mercury deposition in that region. Reduction of mercury emissions in East Asia has the largest effect because anthropogenic emissions in that region are the largest: a reduction of anthropogenic emissions by 20% in East Asia leads to a reduction in mercury deposition of 7% in that region and more than 2% reduction in mercury deposition in the other three regions. A reduction of 20% in anthropogenic emissions from Europe and North Africa leads to nearly 4% reduction on average in

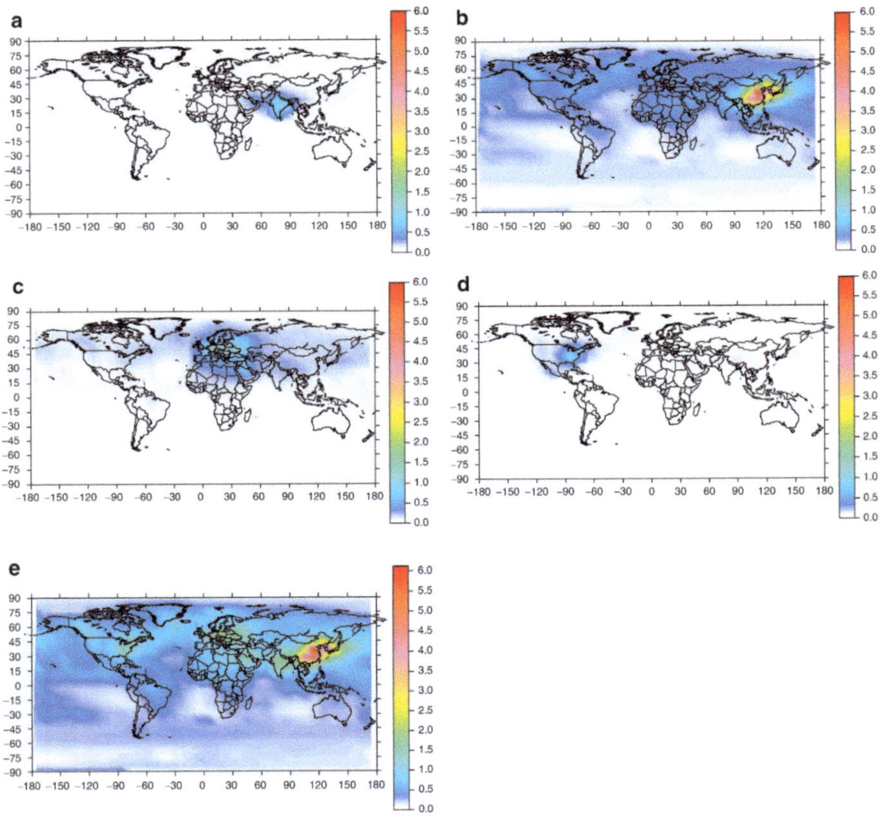

Figure 21.4 Contribution of *20%* of anthropogenic emissions from a region to total atmospheric mercury deposition (*μg m*$^{-2}$ *yr*$^{-1}$): (**a**) South Asia, (**b**) East Asia, (**c**) Europe and North Africa, (**d**) North America, (**e**) all four regions

Table 21.5 Hg deposition fluxes from the base simulation and reductions in deposition due to 20% reductions in anthropogenic Hg emissions in the four UNEP regions

		Reduction (%)				
Region	Base (%)	S. Asia	E. Asia	Europe+ N. Africa	N. America	All areas
S. Asia	18.9	2.67	2.16	0.87	0.33	6.04
E. Asia	24.1	0.43	7.27	0.70	0.28	8.68
Europe + N. Africa	18.1	0.35	2.10	3.67	0.39	6.52
N. America	14.3	0.35	2.54	0.81	1.55	5.25
All 4 regions	18.5	0.65	4.02	1.57	0.64	6.89
World	11.1	0.40	2.40	0.86	0.38	4.04

mercury deposition in that region. A reduction of 20% in anthropogenic emissions from South Asia leads to more than 2% reduction on average in mercury deposition in that region. A reduction of 20% in anthropogenic emissions from North America leads to nearly less than 2% reduction on average in mercury deposition in North America. A 20% reduction in anthropogenic emissions from South Asia, Europe and North Africa, and North America leads to less than 1% reduction in mercury deposition in the other regions (i.e., outside the region where the emissions were reduced). These results highlight the importance of anthropogenic emissions from East Asia on mercury deposition in the northern hemisphere.

21.6 Conclusion

The AER/EPRI global chemical transport model for mercury (CTM-Hg) was developed to simulate the global cycling of atmospheric mercury. Because the atmospheric lifetime of mercury is so long (on the order of 1 yr) mercury is considered a global pollutant and thus emissions of mercury in one part of the globe may impact deposition at another. The uncertainties of the emission inventories and the chemical mechanisms will impact the reliability of the global modeling results, which will in turn impact the uncertainty of the ensuing regional modeling. The current uncertainty in total Hg emissions at the global scale is about a factor of two. Reducing those uncertainties will require refining the emission inventories of anthropogenic sources, in particular, emissions for countries that have been identified as large Hg emitters (e.g., China) and emissions from source categories that are not currently included in emission inventories but could be significant. Also, better characterization of both natural emissions and the emissions of previously deposited Hg is needed. In addition to refining the emission inventory, improvements in the understanding of the Hg atmospheric chemical cycle are key to improving the results of global Hg modeling. The OH, O_3 and bromine gas-phase Hg^0 oxidation reactions should continue to be investigated as well as the reactions that could reduce $Hg^{(II)}$.

Global modeling provides insights into the sources that contribute to mercury deposited at a particular location. In particular, the modeling results presented here highlight the importance of anthropogenic emissions from East Asia to mercury deposition in the northern hemisphere.

References

Ariya, P.A., A. Khalizov and A. Gidas, 2002. Reactions of gaseous mercury with atomic and molecular halogens: kinetics, product studies, and atmospheric implications, *J. Phys. Chem.*, 106, 7310-7320.

Burrows, J.P. et al., 1999. The Global Ozone Monitoring Experiment (GOME): Mission concept and first scientific results, *J. Atmos. Sci.*, 56, 151-175.

Calvert, J.G. and Lindberg, S.E., 2005. Mechanisms of mercury removal by O_3 and OH in the atmosphere. *Atmos. Environ.*, 39, 3355–3367.

Chin, M., D.J. Jacob, G.M. Gardner, M.S. Forman-Fowler, P.A. Spiro and D.L. Savoie, 1996. A global three-dimensional model of tropospheric sulfate, *J. Geophys. Res.*, 101, 18,667-18,690.

Clever, H., S.A. Johnson and E.M. Derrick, 1985: The solubility of mercury and some sparingly soluble mercury salts in water and aqueous solutions, *J. Phys. Chem. Ref. Data*, 14, 631-680.

Donohoue, D.L.; Vauer, D.; Cossairt, B.; Hynes, A., 2006. Temperature and pressure dependent rate coefficients for the reaction of Hg with Br and the reaction of Br with Br: A pulsed laser photolysis - pulsed laser induced fluorescence study, *J. Phys, Chem. A.*, 110, 6623-6632.

Gårdfeldt, K. and M. Johnson, 2003. Is bimolecular reduction of Hg(II)-complexes possible in aqueous systems of environmental importance? *J. Phys. Chem.*, 107, 4478-4482.

Goodsite, M.E., J.M.C. Plane and H. Skov, 2004. A theoretical study of the oxidation of Hg^0 to $HgBr_2$ in the troposphere, *Environ. Sci. Technol.*, 38, 1772-1776.

Graedel, T.E. and W.C. Keene, 1995. Tropospheric budget of ractive chlorine, *Global Biogeochem. Cycles*, 9, 47-77.

Hall, B. and N. Bloom, 1993: Report to EPRI, Palo Alto, California, USA.

Hall, B., 1995: The gas-phase oxidation of elemental mercury by ozone, *Water Air Soil Pollut.*, 80, 301-315.

Hansen, J.; Russel, G.; Rind, D.; Stone, P.; Lacis, A.; Lebedeff, S.; Ruedy, R.; Travis, I., 1983. Efficient three-dimensional global models for climate studies: models I and II, *Mon Weather Rev.* 111, 609-662.

Jacob, D.J., 2000. Heterogeneous chemistry and tropospheric ozone, *Atmos. Environ.*, 34, 2131-2159.

Jacobson, M.Z., 1999. Fundamentals of Atmospheric Modeling, Cambridge University Press, New York.

Jaeglé, L., D.J. Jacob, W.H. Brune and P.O. Wennberg, 2001. Chemistry of HOx radicals in the upper troposphere, *Atmos. Environ.*, 35, 469-489.

Jaffe, D., E. Prestbo, P. Swartzendruber, P. Weiss-Penzias, S. Kato, A. Takami, S. Hatakeyama and Y. Kajii, 2005: Export of atmospheric mercury in Asia, *Atmos. Environ.*, 39, 3029-3038.

Lin, C.J. and S.O. Pehkonen, 1997: Aqueous-free radical chemistry of mercury in the presence of iron oxides and ambient aerosol, *Atmos. Environ.*, 31, 4125-4137.

Lin, C.J. and S.O. Pehkonen, 1998: Oxidation of elemental mercury by aqueous chlorine (HOCl/OCl-): Implications for tropospheric mercury chemistry, *J. Geophys. Res.*, 103, 28093-28102.

Lindberg, S., O.R. Bullock, R. Ebinghaus, D. Engstrom, X. Feng, W. Fitzgerald, N. Pirrone, E. Prestbo and C. Seigneur, 2007. A synthesis of progress and uncertainties in attributing the sources of mercury in deposition, *Ambio*, 36, 19-33.

Lindqvist, O. and H. Rodhe, 1985: Atmospheric mercury – a review, *Tellus*, 37B, 136-159.

Lohman, K., C. Seigneur, E. Edgerton and J. Jansen, 2006. Modeling mercury in power plant plumes, *Environ. Sci. Technol.*, 40, 3848-3854.

Lohman, K., C. Seigneur, M. Gustin and S. Lindberg, 2008. Sensitivity of the global atmospheric cycling of mercury to emissions, *Appl. Geochem.*, 23, 454-466.

Munthe, J., 1992: The aqueous oxidation of elemental mercury by ozone, Atmos. *Environ., Part A*, 26, 1461-1468.

Pal, B. and P.A. Ariya, 2004. Studies of ozone initiated reactions of gaseous mercury: kinetics product studies, and atmospheric implications, *Phys. Chem. Chem. Phys.*, 6, 572-579

Pehkonen, S.O. and C.J. Lin, 1998: Aqueous photochemistry of divalent mercury with organic acids, *J. Air Waste Manage. Assoc.*, 48, 144-150.

Radke, L.F., H.R. Friedli and B.G. Heikes, 2007. Atmospheric mercury over the NE Pacific during spring 2002: Gradients, residence time, upper troposphere lower stratosphere loss, and long-range transport, J. Geophys. Res., 112, D19305, doi:10.1029/2005JD005828.

Raofie, F. and P.A. Ariya, 2003. Reactions of BrO with mercury: kinetic studies, *J. Phys. IV France*, 107, 1119-1121.

Ryaboshapko, A., R. Bullock, R. Ebinghaus, I. Ilyin, K. Lohman, J. Munthe, G. Petersen, C. Seigneur and I. Wängberg, 2002: Comparison of mercury chemistry models, *Atmos. Environ.*, 36, 3881-3898.

Sanemasa, I., 1975: The solubility of elemental mercury vapor in water, *Bull. Chem. Soc. Jpn.*, 48, 1795-1798.

Schroeder, W.H. and J. Munthe, 1998. Atmospheric mercury – An overview, *Atmos. Environ.*, 32, 809-822.

Seigneur, C., H. Abeck, G. Chia, M. Reinhard, N.S. Bloom, E. Prestbo and P. Saxena, 1998: Mercury adsorption to elemental carbon (soot) particles and atmospheric particulate matter, *Atmos. Environ.*, 32, 2649-2657.

Seigneur, C., P. Karamchandani, K. Lohman, K. Vijayaraghavan and R.-L. Shia, 2001: Multiscale modeling of the atmospheric fate and transport of mercury, *J. Geophys. Res.*, 106, 27795-27809.

Seigneur, C., K. Vijayaraghavan, K. Lohman, P. Karamchandani and C. Scott, 2004: Global source attribution for mercury deposition in the United States, *Environ. Sci. Technol.*, 38, 555-569.

Seigneur, C., K. Vijayaraghavan and K. Lohman, 2006. Atmospheric mercury chemistry: sensitivity of global model simulations to chemical reactions, J. Geophys. Res., 111, D22306, doi:10.1029/2005JD006.

Seigneur, C., K. Lohman, K. Vijayaraghavan, R. Balmori and L. Levin, 2007. Effect of bromine chemistry on mercury deposition in the United States, Air Quality VI: International conference on Mercury, Trace Elements, SO3 and Particulate Matter, and Greenhouse Gases, 24-27 September 2007, Arlington, Virginia.

Selin, N.E., D.J. Jacob, R.J. Park, M. Yantosca, S. Strode and L. Jaeglé, 2007. Chemical cycling and deposition of atmospheric mercury: Global constraints from observations, J. Geophys., Res., 112, D02308, doi:10.1029/2006JD007450.

Shia, R.L., C. Seigneur, P. Pai, M. Ko, and N.D. Sze, 1999. Global simulation of atmospheric mercury concentrations and deposition fluxes, *J. Geophys. Res.*, 104, 23747-23760.

Sillen, G.L. and A.E. Martell, (Eds.), 1964: Stability constants of metal ion complexes, Spec. Publ. Chem. Soc., 17, 754.

Sommar, J., K. Gårdfeldt, D. Strömberg and X. Feng, 2001: A kinetic study of the gas-phase reaction between the hydroxyl radical and atomic mercury, *Atmos. Environ.*, 35, 3049-3054.

Spicer, C.W., E.G. Chapman, B.J. Finlayson-Pitts, R.A. Plastridge, J.M. Hubbe, J.D. Fast and C.M. Berkowitz, 1998. Unexpectedly high concentrations of molecular chlorine in coastal air, *Nature*, 394, 355-356.

Spivakovsky, C.R., R. Yevich, J. Logan, S. Wofsy, M. McElroy and M. Prather, 1990. Tropospheric OH in the three-dimensional chemical tracer model: An assessment based on the observations of CH3Cl3, *J. Geophys. Res.*, 95, 18,442-18,439.

Temme, C., F. Slemr, R. Ebinghaus and J.W. Einax, 2003. Distribution of mercury over the Atlantic Ocean in 1996 and 1999-2001, *Atmos. Environ.*, 37, 1889-1897.

Tokos, J.J.S., B. Hall, J.A. Calhoun and E.M. Prestbo, 1998: Homogeneous gas-phase reaction of Hg0 with H2O2, O3, CH3I, and (CH3)2S: Implications for atmospheric Hg cycling, *Atmos. Environ.*, 32, 823-827.

Travnikov, O., 2005. Contribution of the intercontinental atmospheric transport to mercury pollution in the Northern Hemisphere, *Atmos. Environ.*, 39, 7541-7548.

van Loon, L., E. Mader and S.L. Scott, 2000: Reduction of the aqueous mercuric ion by sulfite: UV spectrum of $HgSO_3$ and its intramolecular redox reactions, *J. Phys. Chem.*, 104, 1621-1626.

van Loon, L.L., E.A. Mader and S.L. Scott, 2001: Sulfite stabilization and reduction of the aqueous mercuric ion: kinetic determination of sequential formation constants, *J. Phys. Chem.*, 105, 3190-3195

Wang, Y., J.A. Logan and D.J. Jacob, 1998. Global simulation of tropospheric O3-NOx-hydrocarbon chemistry- 2. Model evaluation and global ozone budget, *J. Geophys. Res.*, 103, 10,727-10,755.

Yang, X.; Cox, R.A.; Warwick, N.J.; Pyle, J.A.; Carver, G.D., O'Connor, F.M., Savage, N.H., 2005. Troposhperic bromine chemistry and its impacts on ozone: A model study, J. Geophys. Res., 110, D23311, doi:10.1029/2005JD006244.

List of Figures

List of Tables

Acronyms

AAGR	Average Annual Growth Rate
ABC	Atmospheric Brown Clouds project
ACAP	Arctic Council Action Plan to Eliminate Pollution of the Arctic
ACE	Aerosol Characterization Experiment
ACI	Activated Carbon Injection
ADOM	Acid Deposition and Oxidant Model
AGCM	Atmospheric General Circulation Model
ALRT	Asian Long-Range Transport
AMA	Alkali Manufacturers Association
AMAP	Arctic Monitoring and Assessment Program
AMDEs	Atmospheric Mercury Depletion Events
AMIP	Atmospheric Model Intercomparison Project
AMNET	Atmospheric Mercury Network
APCD	Air Pollution Control Device
APLA	Amalgamated Prospectors and Leaseholders Association of Western Australia
ASGM	Artisanal Small scale Gold Mining
ATSR	Along Track Scanning Radiometer
AVHRR	Advanced Very High Resolution Radiometer
AZSP	Artisanal zinc Smelting Process
BET	Brunauer-Emmett-Teller theory
BHEL	Bharat Heavy Electricals
BOF	Basic Oxygen Furnace
CAMNet	Canadian Atmospheric Mercury Network
CAMR	Clean Air Mercury Rule
CARIBIC	Civil Aircraft for Regular Investigation of the Atmosphere Based on an Instrumented Container
CASA	Carnegie-Ames-Stanford-Approach biogeochemical model
CASM	Communities and Small Scale Mining Secretariate of the World Bank

CBL	Chesapeake Biological Laboratory
CEC	Commission for Environmental Cooperation of Mexico
CETEM	Centro de Tecnologia Mineral
CGEIC	Canadian Global Emissions Interpretation Centre
CGMMN	Coordinated Global Mercury Monitoring Network
CIDA	Canadian International Development Agency
CIFFC	Canadian Interagency Forest Fire Centre
CKD	Cement Kiln Dust
CLTRAP	Convention on Long-Range Transport of Air Pollutants
CMAQ	Community Multiscale Air Quality Modeling System
CNCI	Cement and Concrete Institute
CoMSA	Chamber of Mines of South Africa
COMTRADE	United Nations Commodity Trade Statistics Database
CPCB	Central Pollution Control Board
CSE	Centre for Science and Environment
CTM	Chemical Transport Model
CVAAS	Cold Vapour Atomic Absorption Spectroscopy
CVAFS	Cold-Vapor Atomic Fluorescence Spectrometry
DEAT	Department of Environmental Affairs and Tourism
DEHM	Danish Eulerian Hemispheric Model
DF	Digital Filtration technique
DFT	Density Functional Theory
DGM	Dissolved Gaseous Hg
DGVM	Dynamic Global Vegetation Model
DLI	Defense Language Institute
DME	Department of Minerals and Energy of South Africa
DME	Department of Minerals and Energy, South Africa
DOC	Dissolved Organic Carbon
DTI	Department of Trade and Industry
DWAF	Department of Water Affairs and Forestry
EAF	Electric Arc Furnace
EC	European Commission
ECCCEY	Editorial Committee of China Mechanical Industry Yearbook
ECCNMI	Editorial Committee of China Nonferrous Metals Industry
ECE	Economic Commission for Europe
ECHAM	European Centre Hamburg Model
ECHMERIT	ECHAM5-Mercury chemistry module
ECMWF	European Centre for Medium-Range Weather Forecasts
EEA	European Environment Agency
EF	Emission Factor
EIA	Energy Information Administration
EMAP	Eulerian Model for Air Pollution
EMEP	Co-operative programme for monitoring and evaluation of long-range transmissions of air pollutants in Europe

EMEP	European Monitoring and Evaluation Programme
EP	Electrolytic Process
EPTRI	Environment Protection, Training & Research Institute
ESP	Electrostatic Precipitator
EZF	Electric Zinc Furnace
FA	Fly ash emission control
FAO	Food and Agriculture Organization
fAPAR	Photosynthetically Active Radiation absorbed by the green plant canopy
FF	Fabric Filter
FGD	Flue Gas Desulphurisation
FRA	Forest Resource Assessment
GAW	Global Atmospheric Watch station
GBA	Global Burnt Areas
GDP	Gross Domestic Product
GEF	Global Environment Facility
GEM	Gaseous Elementary Mercury
GEOS-CHEM	Goddard Earth Observing System - Chemical model
GFED	Global Fire Emission Database
GLOBSCAR	Global Burn Scars
GMA	Global Mercury Assessment
GMP	Global Mercury Project
GOI	Ministry of Coal Controllers Organization
GOME	Global Ozone Monitoring Experiment
GRAHM	Global/Regional Atmospheric Heavy Metals model
HELCOM	Helsinki Commission for protecting the Baltic Sea from pollution
HRTEM-EDS	High Resolution Transmission Electron Microscopy - Electron Dispersive Spectrometer
HTAP	Hemispheric Transport of Air Pollutants
HYSPLIT	Hybrid Single Particle Lagrangian Integrated Trajectory model
IAWG	International Ash Working Group
ICES	International Council for the Exploration of the Sea
ICMGP	International Congresses on Mercury as a Global Pollutant
ICP-MS	Inductively Coupled Plasma - Mass Spectrometer
IEA	International Energy Agency
IFFN	International Forest Fire News
IGAC	International Global Atmospheric Chemistry project
IGBP	International Geosphere-Biosphere Programme
ISAM	Integrated Science Assessment Model
ISP	Imperial Smelting Process
ITC	International Institute for Geoinformation Science and Earth Observation

JST	Japanese Standard Time
LAI	Leaf Area Index
LIDAR	Laser Imaging Detection and Ranging
LPJ	Lund-Postdam-Jena
LRTAP	Long-Range Transboundary Air Pollution
LST	Local Standard Time
MALDI-TOF	Matrix-Assisted Laser Desorption/Ionization - Time of Flight
MAMCS	Mediterranean Atmospheric Mercury Cycle System
MBL	Marine Boundary Layer
MDEs	Mercury Depletion Events
MDN	Mercury Deposition Network
MECAWEx	Integrated mercury cycling, transport, and air-water exchange model
MERCYMS	An Integrated Approach to Assess the Mercury Cycle into the Mediterranean Basin
METAALICUS	Mercury Experiment To Assess Atmospheric Loading in Canada and the United States
MMSD	Mining, Minerals and Sustainable Development
MODIS	Moderate Resolution Imaging Spectroradiometer
MOE	Mercury Over Europe
MOZART	Model of Ozone and Related Tracers
MRA	Mining Review Africa
MSC-E	Meteorological Synthesizing Centre – East
MSW	Municipal Solid Waste
MWI	Municipal waste incinerators
NADP	National Atmospheric Deposition Program
NBS	National Bureau of Statistics of China
NCAR	National Center for Atmospheric Research
NCEP	National Centers for Environmental Prediction
NEI	National Emission Inventory of USA
NEMA	National Electrical Manufacturers Association
NEWMOA	North-east Waste Management Officials' Association
NGO	Non-Governmental Organization
NIFC	National Interagency Fire Center of USA
NIOSH	National Institute for Occupational Safety and Health
NMOC	Non-Methane Organic Compounds
NOAA	National Oceanic and Atmospheric Administration
NPP	Net Primary Production
OPC	Ordinary Portland Cement
OSPAR	Oslo and Paris Convention Convention for the Protection of the Marine Environment of the North-East Atlantic
PALMS	Particle Analysis by Laser Mass Spectroscopy

PAR	Photosynthetically Active Radiation
PC	Pulverized Coal
PM	Particulate Matter
PMP	Pyro-Metallurgic Process
POPs	Persistent Organic Pollutants
PVC	Polyvinyl Chloride
REMSAD	Regional Modeling System for Aerosols and Deposition
RGM	Reactive Gaseous Mercury
RRKM	Rice-Ramsberger-Kassel-Marcus theory
RZSP	Retort Zinc Smelting Process
SAISI	South African Iron and Steel Institute
SAMA	South African Mercury Assessment Programme
SDA	Swiss Development Agency
SeaWIFS	Sea-Viewing Wide Field-of-View Sensor
SoFAMMS	South Florida Atmospheric Mercury Monitoring Study
SPOT	Système Pour l'Observation de la Terre (Earth Observation System)
STORMS	Study and Tests of the Release of Mercury From Soils
TAM	Total Airborne Mercury
TEAM	Trace Element Analysis Model
TED	Trade and Environment Database
TF-HTAP	Task Force on HTAP
TGM	Total Gaseous Mercury
TPM	Total Particulate Mercury
TRACE-P	TRAnsport & Chemical Evolution over the Pacific
TSP	Total suspended particulate
TST	Transition State Theory
UBC	University of British Columbia
UNDP	United Nations Development Programme
UNECE-LRTAP	UNECE Convention on Long-Range Transboundary Air Pollution
UNECE	United Nations Economic Commission for Europe
UNEP	United Nations Environmental Programme
UNESCAP	United Nations Economic and Social Commission for Asia and the Pacific
UNIDO	United Nations Industrial Development Organization
US-DOE	US Department of Energy
USEPA	United State Environmental Protection Agency
USGS	United State Geological Survey
UTC	Coordinate Universal Time
VCM	Vinyl Chloride Monomer
VIRS	Visible and Infrared Scanner
VOC	Volatile Organic Compound

WCI	World Coal Institute
WHO	World Health Organization
WRF	Weather Research & Forecasting model
WWF	World Wildlife Fund
ZMWG	Zero Mercury Working Group

Index

MIX
Papier aus verantwortungsvollen Quellen
Paper from responsible sources
FSC® C105338
FSC
www.fsc.org

If you have any concerns about our products,
you can contact us on
ProductSafety@springernature.com

In case Publisher is established outside the EU,
the EU authorized representative is:
Springer Nature Customer Service Center GmbH
Europaplatz 3, 69115 Heidelberg, Germany

Printed by Libri Plureos GmbH
in Hamburg, Germany